X-RAY MICROSCOPY

Related Titles from AIP Conference Proceedings

521   Synchrotron Radiation Instrumentation: Eleventh US National Conference
Edited by Piero Pianetta, John Arthur, and Sean Brennan, May 2000, 1-56396-941-6

506   X-Ray and Inner Shell Processes: 18th International Conference
Edited by R. W. Dunford, D. S. Gemmell, E. P. Kanter, B. Krässig, S. H. Southworth,
and L. Young, February 2000, 1-56396-713-8

503   Two Dimensional Correlation Spectroscopy
Edited by Y. Ozaki and I. Noda, March 2000, 1-56396-916-5

500   The Physics of Electronic and Atomic Collisions: XXI International
Conference
Edited by Yukikazu Itikawa, Kazuhiko Okuno, Hiroshi Tanaka, Akira Yagishita, and
Michio Matsuzawa, February 2000, 1-56396-777-4

417   Synchrotron Radiation Instrumentation: Tenth US National Conference
Edited by E. Fontes, December 1997, 1-56396-742-1

To learn more about these titles, or the AIP Conference Proceedings Series, please
visit the webpage **http://www.aip.org/catalog/aboutconf.html**

X-RAY
MICROSCOPY

Proceedings of the Sixth International Conference

Berkeley, CA   2–6 August 1999

EDITORS
Werner Meyer-Ilse
Tony Warwick
David Attwood
Lawrence Berkeley National Laboratory

Melville, New York
AIP CONFERENCE PROCEEDINGS ■ 507

Editors:

Werner Meyer-Ilse[†]

Tony Warwick
David Attwood
Lawrence Berkeley National Laboratory
One Cyclotron Road
Berkeley, CA 94720
U. S. A.

E-mail: warwick@lbl.gov
      DTAttwood@lbl.gov

[†] deceased

L.C. Catalog Card No. 00-101916
ISBN 1-56396-926-2
ISSN 0094-243X
Printed in the United States of America

CONTENTS

ADVANCES IN SOFT X-RAY MICROSCOPY

MULTI-KEV X-RAY MICROSCOPY

BIOLOGICAL SCIENCES

MATERIALS AND SURFACE SCIENCES

ENVIRONMENTAL AND SOIL SCIENCES

INSTRUMENTATION DEVELOPMENT

NOVEL APPROACHES TO COHERENT IMAGING

NANOMETER X-RAY OPTICS

COMPACT SOURCES

PREFACE

This book is based on presentations at the Sixth International Conference on X-Ray Microscopy, XRM99, which took place in Berkeley, California, August 2-6, 1999. It contains reviews and status reports from the international community of practitioners of x-ray microscopy.

In addition to providing the latest information on the development of x-ray microscopy techniques, the program of the meeting was designed to emphasize scientific accomplishments from the various fields in which x-ray microscopy and spectromicroscopy are yielding new information.

The applications of these techniques have broadened significantly since the preceding meeting in this series, in Würzburg, Germany during August 1996. This present meeting included sessions on applications to biological science, materials science, and molecular environmental science. In each of these areas x-ray microscopy offers special capabilities to image beyond what can be seen with visible light, to measure the spatial distribution of physical states and chemical species, and to explore heterogeneous processes in realistic environments. We saw, very clearly at this meeting, areas of productive scientific research with enormous potential for growth.

The advent of new soft x-ray synchrotron light sources at many locations around the world has provided further capacity to develop these potentials for growth. The incubation of programs at the various centers takes time. We saw at this meeting that the original programs are mature and that large new programs are becoming productive at these new synchrotron sources.

The development of labeling techniques and the use of cryogenic mitigation against sample damage have significantly advanced the applications of x-ray microscopy in bioscience.

Numerous spatially resolved studies of physically and chemically heterogeneous materials were reported, finding real application to materials problems.

Notable progress was reported in x-ray imaging experiments using multi-keV x-rays, phase contrast imaging, dark field techniques, and tomography. Hard x-ray zone plates are available and in use.

It seems clear that the productivity of x-ray microscopy and spectromicroscopy will increase substantially between now and the next meeting in this series, which will be held in Grenoble, France, in 2002.

The page allocation for contributions to this volume was tight, so many of the papers represent just a glimpse of the work that has been done. But all contributions are published in this volume.

We were saddened by the death or our colleague Werner Meyer-Ilse, in whose name an award was established that will be presented at each occasion of the International Conference on X-Ray Microscopy to a young scientist whose work over the preceding three years represents an outstanding contribution to the field.

We would like to thank all the participants and authors for an enjoyable meeting, and for their excellent cooperation in producing this record of the proceedings.

Werner Meyer-Ilse

Tony Warwick

David Attwood

Berkeley, California, October 1999

Attendees at the Sixth International Conference on X-Ray Microscopy at Berkeley, CA, August 1999

Werner Meyer-Ilse

To Werner Meyer-Ilse
August 18, 1955—July 14, 1999

It is with sadness that we dedicate the proceedings of this conference to our lost colleague and friend Werner Meyer-Ilse. Werner was Chairman of the International Program Committee and was to be one of the co-chairmen of this conference, the subject of which was his life's research interest and work. He died shortly before the conference as the result of a car crash during a trip to his native Germany to attend a conference and to visit his family.

He was 43. He is survived by his wife, Andrea, and his daughters Eva and Julia, who live in Lafayette, California.

Werner, a microscopist, was one of the brightest lights at the LBNL Advanced Light Source, a staff scientist for the Center for X-Ray Optics, and the man who directed the design and construction of ALS Beamline 6.1.2, also known as XM-1, a direct-imaging transmission x-ray microscope acknowledged as one of the finest instruments of its kind in the world. Once it was built, Werner oversaw all the research conducted with it. He once said "My personal interest is to develop x-ray microscopy and I am only able to do this in a collaborative manner...therefore all of the users of this beamline, from all their different scientific fields, become my collaborators in developing the technology."

At the time of the accident that led to his death, Werner was in Germany to deliver a talk on x-ray microscopy in Hamburg, and on the train ride back to the airport in Frankfurt, he stopped in his home town of Göttingen to visit family members. He was born there on August 18, 1955, grew up on a nearby farm, and received both of his degrees in physics from the Georg-August Universität at Göttingen. His wife Andrea is also from Göttingen and they were high school sweethearts.

Werner first came to LBNL in 1989 as a consultant to work on the design of an x-ray microscope for the Advanced Light Source, then under construction. He returned on a more permanent basis in February, 1992 to work at the Center for X-Ray Optics on the development of the XM-1 and future high-resolution x-ray microscope beamlines.

Werner thoroughly enjoyed his colleagues and collaborations at the CXRO and the ALS at LBNL, and was enthusiastic about his involvement in every aspect of the design and construction of the x-ray microscope and his involvement in the scientific investigations that it made possible.

ADVANCES IN SOFT X-RAY MICROSCOPY

Visualization of 30 nm Structures in Frozen–Hydrated Biological Samples by Cryo Transmission X–Ray Microscopy

G. Schneider, B. Niemann, P. Guttmann, D. Weiß,
J.-G. Scharf*, D. Rudolph, G. Schmahl

*Institut für Röntgenphysik, Georg–August Universität Göttingen,
Geiststraße 11, 37073 Göttingen, Germany*
**Medizinische Klinik und Poliklinik, Abteilung Gastroenterologie und Endokrinologie,
Georg–August Universität Göttingen, Robert–Koch–Straße 40, 37075 Göttingen, Germany*

Abstract. A new object stage with extremely low thermal drift at -170°C was developed for the cryo transmission X–ray microscope (cryo–TXM) at the electron storage ring BESSYI (Berlin). The new set–up enables high resolution studies of frozen–hydrated cells and was applied in investigations of cryogenic Kupffer cells from a rat liver. The ultrastructure and numerous X–ray dense vacuoles are resolved allowing a more comprehensive interpretation of data obtained by TEM studies. Furthermore, the cryo–TXM has been recently used for non–destructive computed tomography of intact frozen–hydrated objects. The resolution obtainable in TXM micrographs is limited significantly by the photon density applied to illuminate an object. The contrast transfer of the TXM was evaluated including the real X–ray optical elements with the help of a so–called multiple plane–wave model which is based on Fourier optics. It allowed to optimize the X–ray optical set–up for best contrast transfer and to minimize the photon density required to detect ice–embedded protein structures. However, the results show that details in biological objects smaller than 30 nm in size, e.g. single chromatin fibers in cell nuclei, can only be visualized if a drastically increased photon flux of the X–ray source is available from undulator insertion devices of electron storage rings. Furthermore, for this purpose new condenser concepts like a rotating condenser and highly efficient X–ray objectives with smallest zone structures of 20 nm have to be employed. This progress in the instrumentation will enable new applications ultimately resulting in artifact–free high–resolution images of radiation sensitive biological samples.

INTRODUCTION

X–ray microscopy is already a well established technique to study biological structures as small as 40 – 50 nm in size. Below this resolution level, many interesting questions arise in biology. For example the higher order structure of the eucaryotic chromosome is based on a long term filamentous nucleoprotein unit called the 30 nm fiber. How this fiber is arranged in the native chromosome in cell nuclei is of

CP507, *X-Ray Microscopy: Proceedings of the Sixth International Conference,*
edited by W. Meyer-Ilse, T. Warwick, and D. Attwood
© 2000 American Institute of Physics 1-56396-926-2/00/$17.00

great biological interest [1]. Because of its possibility to image intact cells with natural absorption contrast between organic structures and water, X–ray microscopy is well suited for 3D structural analysis of chromosomes and nuclei.

X–ray microscopy has made important progress during the last years. The resolution of X–ray objectives as well as their diffraction efficiency has been improved significantly. Currently, zone plates (ZPs) with smallest zones of about 20 nm are available [2–4]. The problem of radiation damage was solved by introducing the cryo technique to stabilize hydrated biological structures [5]. In addition, it was demonstrated that 3D reconstructions can be obtained from X–ray micrographs taken at different viewing angles [6,7]. These earlier 3D experiments were performed with dried samples at room temperature. With the new instrumentation presented in this work, computed tomography of frozen–hydrated biological objects became feasible [8]. For the further development of TXMs a theoretical understanding of the image formation process under the conditions of the real X–ray optical arrangement is of vital interest. Therefore, a model describing the image contrast and the photon density required to detect small ice–embedded protein structures is presented.

SET–UP OF THE CRYO–TXM AT BESSY I FOR 2D AND 3D STUDIES OF FROZEN–HYDRATED SAMPLES

Experiments with frozen–hydrated biological samples demonstrated a very high structural stability allowing to perform multiple imaging [5]. This offers the possibility to perform tomography with intact cryogenic biological cells. For 3D studies of frozen–hydrated objects it was necessary to develop a new cryo stage for the TXM. Furthermore, the new cryo set–up was designed for minimal thermal drift by an improved temperature stabilization of the cryogenic nitrogen gas surrounding the samples.

For data acquisition in computed tomography the specimens have to be rotated perpendicular to the optical axis. Due to the short focal length of high–resolution ZPs very compact rotatable object holders are required for large tilt–angles. Two types of object holder were developed to fulfil this requirement:

- small strips with a maximum width of 200 μm and holes of about 50 μm in diameter drilled in the center of the strip

- thin glass capillaries with 10 μm in diameter and a glass wall thickness of less than 0.4 μm

The small strip has the advantage that it can be used for comparatively extended flat cells with the drawback that the ice–layer thickness to be transmitted during imaging scales with $1/\cos\theta$ where θ denotes the tilt–angle. Consequently, reconstructions can only be performed with data sets from a limited sector of dial of about $\pm75°$. However, to preserve the 3D spatial frequency spectrum and to avoid

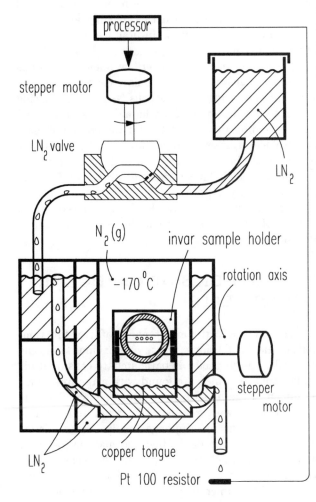

FIGURE 1. Frontal view of the cryogenic object stage. The cryo–box is cooled by liquid nitrogen (LN$_2$). To ensure a constant temperature inside the box, the hollow side–walls (made of copper) and the bottom of the box are filled to constant height levels with the help of overflow openings. The LN$_2$ flow to the box is regulated with a cryo–valve. As soon as LN$_2$ drops on the resistor, it cools down and the flow through the cryo–valve is reduced by the processor. By heating the X–ray window in front of the X–ray objective which is close to the cryogenic object, it is possible to regulate the object temperature continuously from room temperature to LN$_2$ temperature. The cryo object holder is mounted magnetically inside the box and can be adjusted perpendicular to the optical axis with an accuracy of 100 nm. Computed tomography requires that the tilt–angle of the sample has to be adjusted with an accuracy of about 0.1°. To permit such an accuracy a stepper motor including a reduction gear was installed outside the cryo–box to rotate the cryogenic objects.

artifacts in the reconstructed volume, images from the full angular range of viewing angles (±90°) have to be taken. Using rotationally symmetric glass capillaries as object holders overcomes the angular limitation but has the main disadvantage to restrict the size of the objects as they have to fit into the capillaries.

A scheme of the new cryogenic object stage adapted to the TXM is shown in figure 1. A dewar supplies liquid nitrogen (LN$_2$) through a self–developed cryo–valve to a copper cryo–box which encloses and thermally insulates the object. First the side–chambers of the cryo–box are filled up to the height of an internal overflow tube which is connected to the bottom of the cryo–box. The maximum LN$_2$ level is controlled by another overflow opening. Underneath this opening an overflow detector consisting of a simple Pt100 resistor is located. In practice, the LN$_2$ flow into the box is only slightly larger than required to compensate the loss of LN$_2$ by evaporation. The surplus of LN$_2$ – having passed the overflow openings – drops onto and cools down the Pt100 resistor which is used to regulate electronically the cryo–valve with a stepper motor. The constant flow of gaseous N$_2$ generated inside the box by evaporation ensures that no water vapor from humid air is introduced through slits at the cover of the cryo–box to the inside and thereby suppresses a contamination of the object with ice–crystals. Furthermore, this N$_2$ gas flow and the LN$_2$ cooled side–walls guarantee that the object can be kept at cryogenic temperature with an accuracy of about 0.1°C. Thus significant drifts (≥ 0.5 nm/s) of the sample are avoided during the X–ray exposure. The new cryo set–up was used for tomographic imaging of frozen–hydrated algae [8], for 2D investigations of labeled cell nuclei [9], and studies of frozen–hydrated Kupffer cells of a rat liver which are presented in the following section.

X–RAY MICROSCOPIC STUDIES OF CRYOGENIC LIVER CELLS

Hepatic sinusoidal cells play a critical role in the maintenance of liver function, under both physiological and pathological conditions. Four types of sinusoidal cells have been identified: endothelial cells, Kupffer cells, hepatic stellate cells, and pit cells. Kupffer cells are hepatic macrophages which are found within the liver sinusoid. Their location within the lumen of the hepatic sinusoid makes Kupffer cells the first macrophages to come into contact with gut–derived foreign and potentially noxious materials. In common with other macrophages, main characteristics of Kupffer cells are endocytosis with destruction of the ingested material, antigen presentation, and the secretion of biologically active products. Kupffer cells possess the typical morphology and behaviour of macrophages. The endocytotic capacity of these cells is very high and is enabled by numerous pinocytotic and phagocytotic vesicles. These ultrastructural features, together with abundance of a variety of lysosomal enzymes present in Kupffer cells, reflect the prominent role of these cells in the degradation of particles taken up from the bloodstream.

X–ray microscopy as well as electron microscopy were used to characterize the ultrastructure of the Kupffer cells [10]. These recent X–ray microscopic studies were performed with chemically fixed cells at room temperature. Now cryo–TXM was used to compare chemically fixed cells with initially living cells which were only rapidly cooled in liquid ethane. In figure 2 a Kupffer cell is depicted which was

6

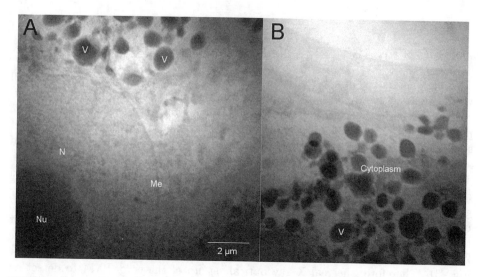

FIGURE 2. X–ray micrograph taken at 2.4 nm wavelength of a Kupffer cell which was chemically fixed with 1.5% glutaraldehyde in buffer and washed with destilled water before rapid cooling in liquid ethane. A: The nuclear membrane (Me), the cell nucleus (N) with the nucleolus (Nu) and several X–ray dense vesicles (V) which are close to the cell nucleous. B: Cytoplasm with vesicles and filament structures.

fixed with 1.5% glutaraldehyde in buffer and washed in destilled water before shock–freezing in liquid ethane. As demonstrated in this micrograph, Kupffer cells have a well–developed vacuolar apparatus with numerous vacuoles. Some of them are X–ray transparent in a crescent–shaped area within the vacuole; others were uniformly dense as already demonstrated by conventional TXM at room temperature. Note that the nuclear membrane is smooth (Fig. 2A) and in general the transparency seems to be homogeneous outside the cell (Fig. 2B). First experiments were done to image rapidly cooled Kupffer cells in culture medium. However, the X–ray micrographs (not shown here) show artifacts depending on the sample thickness. Due to an imperfect vitrification, ice–crystal segregation causes an inhomogeneous X–ray transparency through the culture medium. To avoid such freezing artifacts, it is planned to use high–pressure vitrification.

To study the sequence of phagocytotic processes in Kupffer cells, it is necessary to use living cells, which are still capable of endocytosing X–ray dense particles inside their vesicles. This has been demonstrated first by X–ray microscopy studies of chemically fixed Kupffer cells at room temperature [10]. However, the kinetics of phagocytosis in Kupffer cells cannot be studied with chemical fixation, because the cell activity cannot be stopped fast enough. Shock–freezing occurs within milliseconds in cells, therefore, cryo–TXM studies will enable studies on sequences of the process of phagocytosis in Kupffer cells.

CONTRAST TRANSFER OF THE CRYO–TXM

Two parameters are of substantial importance in order to detect an object detail with a given signal–to–noise ratio: image contrast mode and photon density required to illuminate the object. Earlier model calculations on these parameters were performed for amplitude contrast imaging [11] and phase contrast imaging of wet samples [12]. However, all models described in the literature neglect the limited spatial frequency transfer of ZP objectives which is due to their finite apertures. Therefore, the obtained data are based on ideal imaging conditions assuming a contrast transfer of one for all spatial frequencies. In this work, results of an imaging theory called *multiple plane–wave* (MPW) model are presented which makes use of methods developed in Fourier optics. The MPW model takes into account the object illumination by the condenser, the limited numerical aperture of the ZP objective and the annular phase–plate in the back–focal plane of the ZP (see Fig. 3). Therefore, the real X–ray optical set–up of the cryo–TXM to detect an ice–embedded protein structure can be simulated for different contrast modes like amplitude contrast (AC), phase contrast (PC) and dark field imaging. In figure 4 the image contrast of an ice–embbeded protein grating is plotted for relevant X–ray optical set–ups.

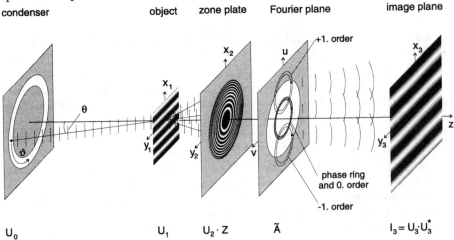

FIGURE 3. Model illustrating the imaging process in the different planes of a TXM. As it assumes an incoherently illuminated condenser, the object illumination can be represented by oblique plane–waves emerging from the condenser. The ZP objective collects the diffracted light from the object, e.g. an ice–embbeded protein grating, and produces an enlarged image of the object in the image plane. Note that the spectrum of the object transmission filtered by the ZP is obtained in the Fourier plane. The dots in the Fourier plane correspond to the spectral components obtained for the propagation direction of the shown plane–wave after having passed the grating object and the ZP objective [14].

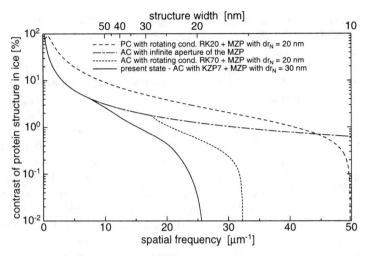

FIGURE 4. Image contrast as a function of the spatial frequency for a model protein grating with the composition $C_{94}H_{139}N_{24}O_{31}S$ and density $\rho_p=1.35\ g/cm^3$ in vitreous ice. Parameters: X–ray objectives with outermost zone widths of $dr_N = 20$ and 30 nm, diffraction efficiency of the micro zone plates $\eta^{MZP} = 1$ and wavelength $\lambda=2.4$ nm. The image contrast and the obtainable spatial resolution depend critically on the object illumination by the condenser (KZP7, RK70 and RK20) and the numerical aperture of the micro zone plate (MZP). Note that phase contrast is necessary to detect very small, weakly absorbing features below about 30 nm in size.

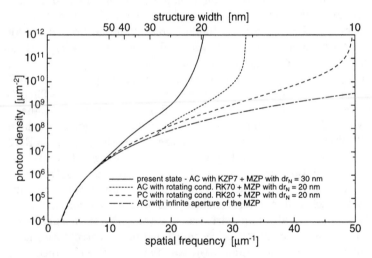

FIGURE 5. Photon densities required to visualize an protein grating embedded in an ice layer of 10 μm thickness with a signal–to–noise ratio SNR = 3. The calculations are performed for the same optics as in figure 4. The plots show that the X–ray optical set–up has to be optimized and an X–ray source which provides a high photon flux has to be used to obtain images of low photon noise in several seconds exposure time.

In all calculations it was assumed that the grating object has a thickness of half the grating period. In the present state the X–ray optical set–up of the TXM at BESSY I consist of the condenser ZP KZP7 which has an outermost zone width of 54 nm and a ZP objective which has an outermost zone width of 30 nm. Protein structures smaller than 50 nm show a significantly smaller amplitude contrast in the image if the calculations are performed for real optics – in contrast to the results obtained earlier with models which assumed ideal optics (see also figures 4 and 5 which include the earlier results for comparison). If we choose a ZP objective with an outermost zone width of 20 nm and use a condenser called RK70 which illuminates only zone structures of 68 – 72 nm width of the ZP objective, the amplitude contrast is close to the natural absorption contrast down to 30 nm protein structures size, which is essential for applications with quantitative measurements. For example, in nano–tomography the linear absorption coefficients of the object structures are evaluated by a reconstruction algorithm. If the recorded image contrast is lower than the real natural element contrast then the reconstructed linear absorption coefficients do not reflect the correct object material parameters, e.g., the object density. From this point of view the contrast transfer obtained with the condenser RK70 is well suited for computed tomography.

In the near future the cryo–TXM will be installed at an undulator beam–line of the new electron storage ring BESSY II. Up to now at BESSY I radiation from a bending magnet fully illuminates the large condenser KZP7 which has a comparatively large inner central stop to generate the required hollow cone illumination of the object [13]. This condenser is no longer suited, because the undulator radiation is much more collimated. To solve this problem the use of an off–axis ZP linear monochromator and synthesis of the hollow cone illumination of the object by a rotating mirror system was proposed, which also improves the monochromaticity and the homogeneity of the object illumination [15].

The aperture of this rotating condenser system in the new cryo–TXM at BESSY II can be adjusted to any aperture of the objective [16]. Therefore, it is possible to realize condensers like RK70 as well as condenser with higher aperture like RK20 which illuminates only the 20 nm zone width region of a high spatial resolution ZP objective. It can be seen in figure 4 that the condenser RK20 which matches the aperture of the objective is well suited for highest spatial resolution. Note that this calculation was performed for optimized phase contrast with an annular nickel phase–plate of 460 nm thickness in the Fourier plane to enhance the image contrast which is important especially for weakly absorbing structures.

The MPW model was also applied to calculate the photon density in the object plane which is necessary to detect an ice–embedded protein structure with a signal–to–noise ratio SNR = 3. Figure 5 shows the results which were obtained for the same X–ray optical arrangements as discussed above. The plot shows that the required photon density for imaging structures below 50 nm in size depends critically on the apertures of the X–ray optical elements. To minimize the required photon density and consequently the related dose in the sample, it is especially important to use objectives with high apertures and phase contrast imaging (see Fig. 5). In comparison with results obtained by the earlier models which assumed an infinite aperture of the objective, the plots of figure 5 show that significantly higher photon densities are required especially if structures below 30 nm feature size have to be imaged. Therefore, the installation of TXMs at third generation electron storage rings, e.g. BESSY II, which have undulator insertion devices as X–ray sources of drastically increased brilliance and photon flux is even more important.

ACKNOWLEDGEMENTS

The technical assistance of P. Nieschalk, J. Herbst, T. Lehmann, H. Düben, S. Hupe, and S. Zachmann is gratefully acknowledged. This work was supported by BMBF under contract number 05 SL8MG11 and the Deutsche Forschungsgemeinschaft under contracts Schm1118/5-1 and Schm1118/5-2.

REFERENCES

1. E. Lattman, "X–Ray Microscopy in the Study of Biological Structure: A Prospective View" in *X–ray Microscopy III* edited by D. Sayre, M. Howells, J. Kirz, H. Rarback, Springer Series in Optical Sciences **56**, 1988, pp. 384-390.
2. G. Schneider, T. Schliebe, H. Aschoff, *J. Vac. Sci. Technol. B* **13**, 2809-2812 (1995).
3. S. Spector, C. Jacobsen, D.M. Tennant, *J. Vac. Sci. Technol. B* **15**, 2872-2875 (1997).
4. M. Peuker, "Nickel Zone Plates for Soft X–ray Microscopy", this volume
5. G. Schneider, B. Niemann, P. Guttmann, D. Rudolph, and G. Schmahl, *Synchrotron Radiation News* **8**, 19-28 (1995).
6. W. S. Haddad, J. E. Trebes, A. H. Anderson, R. A. Levesque & L. Yang, *Science* **266**, 1213-1215 (1994).
7. J. Lehr, *Optik* **104**, 166-170 (1997).
8. D. Weiss, G. Schneider, B. Niemann, P. Guttmann, D. Rudolph, and G. Schmahl, "Tomographic Imaging of Cryogenic Biological Specimens with the X-ray Microscope at BESSY I", this volume
9. S. Vogt, G. Schneider, A. Steuernagel, J. Lucchesi, E. Schulze, D. Rudolph, and G. Schmahl, "X-ray microscopic Imaging of Labeled Nuclear Cell Structures", this volume
10. J.-G. Scharf and G. Schneider, *Journal of Microscopy* **193**, 250-256 (1999).
11. D. Sayre, J. Kirz, R. Feder, D. M. Kim, E. Spiller, *Ultramicroscopy* **2**, 337-349 (1977).
12. D. Rudolph, G. Schmahl and B. Niemann, "Amplitude and Phase Contrast in X-Ray Microscopy" in *Modern Microscopies, Techniques and Applications* edited by A. Michette and P. Duke, Plenum Press, London, 1990, pp. 59-67.
13. G. Schmahl, D. Rudolph, B. Niemann, P. Guttmann, J. Thieme, G. Schneider, C. David, M. Diehl, T. Wilhein, *Optik* **93**, 95-102 (1993).
14. G. Schneider, *Ultramicroscopy* **75**, 85-104 (1998).
15. B. Niemann, "High Numerical–Aperture X-Ray Condensers for Transmission X-Ray Microscopes" in *X-ray Microscopy and Spectromicroscopy* edited by J. Thieme, G. Schmahl, D. Rudolph, E. Umbach, Springer Verlag, 1998, pp. IV45-55.
16. B. Niemann, P. Guttmann, D. Hambach, G. Schneider, D. Weiß, G. Schmahl, "The Condenser–Monochromator with Dynamical Aperture Synthesis for the TXM at an Undulator Beamline at BESSY II", this volume

Recent Developments In Scanning Microscopy at Stony Brook

C. Jacobsen, S. Abend, T. Beetz, M. Carlucci-Dayton, M. Feser, K. Kaznacheyev, J. Kirz, J. Maser[1], U. Neuhäusler[2], A. Osanna, A. Stein, C. Vaa, Y. Wang[1], B. Winn, S. Wirick

Department of Physics & Astronomy, State University of New York at Stony Brook, USA
[1] Present address: Advanced Photon Source, Argonne National Lab, USA
[2] Also: Institute for X-ray Physics, Universität Göttingen, Germany

Abstract. Recent activities in scanning transmission x-ray microscopy at Stony Brook are outlined.

INTRODUCTION

Scanning transmission x-ray microscopes have been developed at Stony Brook since 1980 [1]. We summarize here our recent activities in scanning microscopy using undulator radiation. The Stony Brook STXMs make use of zone plates fabricated in a collaboration with Lucent Technologies Bell Labs [2, 3], and are used by a number of researchers as described in these proceedings.

THE NSLS X-1A MICROSCOPY BEAMLINES

The X-1 undulator at the National Synchrotron Light Source is a high-brightness source of soft x rays [4]. Upgrades to the storage ring, and improvements in its operating mode, have led to a twentyfold increase in source brightness over the past decade. Even so, the horizontal source output is into about 100 spatially coherent "modes," while a diffraction-limited scanning microscope can only use one spatially coherent mode [5]. As a result, we are able to operate two scanning microscopes at the same time that a spectroscopy beamline (X-1B) is taking beam. The increased horizontal divergence of the source also means that the undulator peaks are spread out to a width of ~20 eV, so that XANES spectra can be acquired without a need to adjust the undulator tuning parameter K (see e.g., [6]).

In 1996, our microscopy beamline underwent a substantial upgrade [7] that included important contributions from H. Ade, C. Buckley, M. Howells, S. Hulbert, I. McNulty, and T. Oversluizen. The upgraded beamline (see Fig. 1) has a separate monochromator for each microscope end station, with a spectral resolving power of between 1000 and 5000, depending on entrance slit width and photon energy. A

CP507, *X-Ray Microscopy: Proceedings of the Sixth International Conference,*
edited by W. Meyer-Ilse, T. Warwick, and D. Attwood
© 2000 American Institute of Physics 1-56396-926-2/00/$17.00

stigmatic source is provided with adjustable size, allowing each scanning microscope endstation control over the tradeoff between flux and spatial resolution.

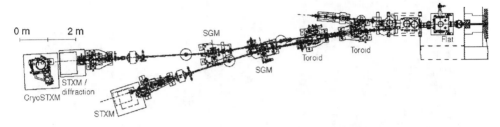

Figure 1. Schematic of the X-1A undulator beamlines at the National Synchrotron Light Source, Brookhaven National Laboratory. The undulator (with 37 periods of 8 cm length) is located about 12 m to the right of the "Flat" mirror at the right end of this diagram.

THE CRYO SCANNING TRANSMISSION X-RAY MICROSCOPE

Radiation damage is a limitation in studies of some specimens, especially wet biological specimens. The use of cryo methods for radiation damage protection in electron microscopy is well established [8-10], and cryo has begun to be employed in x-ray microscopy with great success [11-14].

At Stony Brook, we have developed a cryo scanning transmission x-ray microscope that makes use of a transmission electron microscope-type specimen holder (see Fig. 2) [13, 15]. With this system, we are able to take specimens that have been prepared on standard 3 mm electron microscope grids and load them into the microscope at temperatures below -150°C, for imaging at temperatures of about -165°C (see Fig. 3).

Figure 2. The cryo STXM. The x-ray beam enters through the beam pipe at left. At right is shown a closeup of a cryo holder rotated for tomography experiments; the holder tip can be seen extending across a square cutout in the piezo scanning stage inside the vacuum chamber.

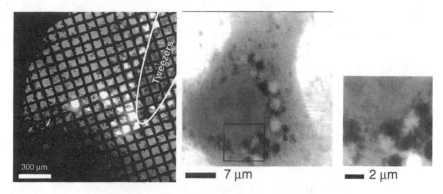

Figure 3. Grids of frozen hydrated cells can be loaded into the cryo STXM, and large area scans can be taken to identify regions with appropriate ice thickness (left). Higher resolution images can then be taken of smaller areas (center, right). Some ice crystal formation can be seen in the medium surrounding the specimen. From Maser *et al.* [15].

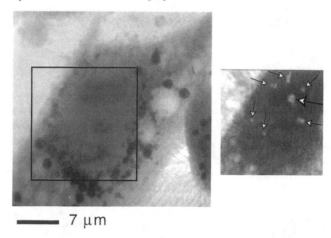

Figure 4. Several submicron regions of a frozen hydrated fibroblast were exposed to a dose of ~10^{10} Gray. The image at left was taken *afterwards*, and shows no obvious sign of radiation damage. Upon slow warming of the sample, the free radicals created by exposure were able to diffuse and react, creating holes in the specimen (at right). From Maser *et al.* [15].

We have carried out studies of mass loss in the cryo STXM. In these studies, we find that radiation doses of up to about 10^{10} Gray can be tolerated by frozen hydrated fibroblasts before noticeable changes are observed (see Fig. 4) [15]. We have also used the cryo STXM to obtain the first 3D reconstruction of a frozen hydrated eucaryotic cell (see Fig. 5) [16].

14

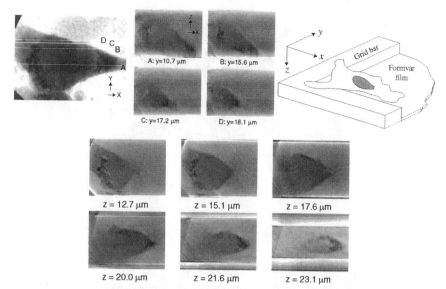

Figure 5. Tomography of a frozen hydrated fibroblast using cryo STXM. At top left is the 0° tilt image of a frozen hydrated fibroblast. The tomographic dataset, consisting of images from -55° to +60° in 5° tilt increments acquired over a 36 hour period, was reconstructed using ART. Y slices from the indicated planes into the paper are shown at top middle, and sections from different depth planes along the beam direction are shown at bottom left. The orientation of these axes, and of the cell, is shown at top right. In several cases, lipid-rich vesicles can be resolved in the reconstruction which lie on top of each other in the 0° tilt image. From Wang et al., [16].

ROOM TEMPERATURE SCANNING MICROSCOPY

While the cryo STXM offers important new capabilities for imaging, tomography, and spectroscopy of biological specimens, the majority of the studies carried out at X-1A make use of our older STXM [17, 18] to examine room temperature specimens. First tests of a new replacement for this microscope, plus improved x-ray detectors, are described in these proceedings [19].

Both the room temperature STXM and the cryo STXM have available a new mode of operation where a sequence of images is acquired at various photon energies [20]. These image "stacks" are then aligned to remove any position shifts from image to image, and absorption spectra can then be obtained from single image pixels or from user-defined regions of pixels (see Fig. 6).

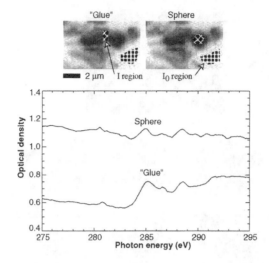

Figure 6. Images (top) and absorption spectra (bottom) of a region of the interplanetary dust particle L2009J4. A spectromicroscopy "stack" of images was taken over an energy range of 270 to 310 eV; the particular image shown twice above was taken at 296.25 eV., with a pixel size of 24 nm. From such an image sequence, one can define incident flux I_0 regions, and calculate the absorption spectrum for all other pixels or for groups of pixels as a transmitted flux I region. The larger grains of material (including the spherical grain at right) in this interplanetary dust particle thin section show relatively little carbon content, while the material surrounding the spheres (labeled "glue" in the image) shows more pronounced carbon content, as indicated by the increase in absorption near the carbon edge in the "glue" spectrum. Specimen prepared by L. Keller and imaged by G. Flynn *et al.*; for details see [20].

APPLICATIONS OF THE X-1A MICROSCOPES

Besides those applications briefly outlined here, the scanning transmission x-ray microscopes at X-1A are used for studies of polymers (see *e.g.*, [21] and papers in these proceedings), colloid chemistry and environmental science (see e.g., [22] and papers in these proceedings), and other research areas described in these proceedings.

ACKNOWLEDGEMENTS

We gratefully acknowledge support from the Office of Biological and Environmental Research, U.S. DoE under contract DE-FG02-89ER60858, the National Science Foundation under grants DBI-9605045 and ECS-9510499, and the Alexander von Humboldt Foundation (Feodor-Lynen Fellowship, JM). This work was carried out at the National Synchrotron Light Source at Brookhaven National Laboratory, which is supported by the U.S. Department of Energy.

REFERENCES

1. Rarback, H., et al., Scanning X-ray Microscopy -- First Tests with Synchrotron Radiation, in Scanned Image Microscopy, E.A. Ash, Editor. 1980: London. p. 449-456.

2. Spector, S., C. Jacobsen, and D. Tennant, Process optimization for production of sub-20 nm soft x-ray zone plates. Journal of Vacuum Science and Technology, 1997. B 15(6): p. 2872-2876.

3. Tennant, D., et al. Electron beam lithography of Fresnel zone plates using a rectilinear machine and trilayer resists. These proceedings.

4. Rarback, H., et al., Coherent radiation for x-ray imaging---the soft X-ray undulator and the X1A beamline at the NSLS. Journal of X-ray Science and Technology, 1990. 2: p. 274-296.

5. Jacobsen, C., J. Kirz, and S. Williams, Resolution in soft x-ray microscopes. Ultramicroscopy, 1992. 47: p. 55-79.

6. Jacobsen, C., et al., The X-1A Scanning Transmission X-ray Microscope: Optics and Instrumentation, in X-ray Microscopy IV, V.V. Aristov and A.I. Erko, Editors. 1994, Bogorodskii Pechatnik: Chernogolovka, Russia. p. 304-321.

7. Winn, B., et al., X1A: second generation undulator beamlines serving soft x-ray spectromicroscopy experiments at the NSLS. Reviews of Scientific Instruments, 1996. 67(9): p. 1-4.

8. Taylor, K. and R. Glaeser, Electron diffraction of frozen, hydrated protein crystals. Science, 1974. 106: p. 1036-1037.

9. Taylor, K.A. and R.M. Glaeser, Electron microscopy of frozen hydrated biological specimens. Journal of Ultrastructure Research, 1976. 55: p. 448-456.

10. Echlin, P., Low-Temperature Microscopy and Analysis. 1992, New York: Plenum Publishing.

11. Schneider, G., et al., Cryo x-ray microscopy. Synchrotron Radiation News, 1995. 8(3): p. 19-28.

12. Schneider, G. and B. Niemann, Cryo x-ray microscopy experiments with the x-ray microscope at BESSY, in X-ray Microscopy and Spectromicroscopy, J. Thieme, et al., Editors. 1998, Springer-Verlag: Berlin. p. I-25-34.

13. Maser, J., et al., Development of a cryo scanning x-ray microscope at the NSLS, in X-ray Microscopy and Spectromicroscopy, J. Thieme, et al., Editors. 1998, Springer-Verlag: Berlin. p. I-35-44.

14. Schneider, G., Cryo x-ray microscopy with high spatial resolution in amplitude and phase contrast. Ultramicroscopy, 1998. 75: p. 85-104.

15. Maser, J., et al., Soft x-ray microscopy with a cryo STXM: I. Instrumentation, imaging, and spectroscopy. Journal of Microscopy (in press)

16. Wang, Y., et al., Soft x-ray microscopy with a cryo STXM: II. Tomography. Journal of Microscopy (in press).

17. Jacobsen, C., et al., Diffraction-limited imaging in a scanning transmission x-ray microscope. Optics Communications, 1991. 86: p. 351-364.

18. Zhang, X., et al., Micro-XANES: chemical contrast in the scanning transmission x-ray microscope. Nuclear Instruments and Methods in Physics Research, 1994. A 347: p. 431-435.

19. Feser, M., et al. Instrumentation advances and detector development with the Stony Brook scanning transmission x-ray microscope. These proceedings.

20. Jacobsen, C., Wirick, S., Flynn, G., Zimba, C. Soft x-ray spectroscopy from sub-100 nm regions. Journal of Microscopy (in press).

21. Ade, H., et al., X-ray spectroscopy of polymers and tribological surfaces at beamline X-1A at the NSLS. Journal of Electron Spectroscopy and Related Phenomena, 1997. 84: p. 53-72.

22. Neuhäusler, U., et al., Soft x-ray spectromicroscopy on solid stabilized emulsions. Colloid & Polymer Science, 1999. 277: p. 719-726.

The X-Ray Microscopy Facility At The ESRF : A Status Report

J. Susini, R. Barrett, B. Kaulich, S. Oestreich, M. Salomé

European Synchrotron Radiation Facility, BP-220, F38043 Grenoble Cedex, France

Abstract. ID21 is a beamline dedicated to X-ray imaging and spectro-microscopy in the 0.2-7 keV energy range. Initiated four years ago, the beamline construction is almost completed and the beamline is now entering into the first operational phase. The beamline is installed on a low beta straight section which is equipped with three undulators and serves two independent end-stations on two separate branch-lines. The scanning X-ray Microscope, served by the "direct" branch-line, equipped with two fixed-exit monochromators and the full-field imaging transmission X-ray microscope, served by the side-branch and optimised for imaging techniques in the 3-7 keV range. Both microscopes use zone-plates as focussing lenses. This paper describes the beamline architecture and provides some figures on the current performance of the beamline.

INTRODUCTION

Traditionally sub-micron X-ray microscopy was developed using rather soft X-ray energies, in particular in the so-called water window region [1]. The parallel advances in the developments both of X-ray sources and micro-focussing optics for higher energies open now the opportunity for new applications: the low source emittance of third generation synchrotron machines, coupled with the versatility of new types of insertion devices, enable the control of the brightness, the emitted spectrum, polarization and degree of coherence of the photon beam. In parallel the technique of micro-fabrication (Fresnel and Bragg-Fresnel lenses) and development of alternative focussing methods (refractive lenses, mirrors) allows access to much higher energies [2]. Although the multi-keV range has been so far rather unexplored, these energies offer several scientific opportunities: i) Access to K-absorption edges (and emission lines) of medium-light elements (e.g. Al, Si, P, S, Cl, K, Ca, Sc, Ti, V, Cr, Mn, Fe) and L and M edges (emission lines) of heavier elements (e.g. Au, Ag, ...) for micro-spectroscopies in absorption or fluorescence modes. ii) At shorter wavelengths (8-2 Å) the radiation is more penetrating and gives better conditions for differential phase contrast and cryo-microscopy on thick samples ($10 < t$ (μm) < 100). Furthermore the zone-plate based microscope benefits from longer focal lengths and larger depth of focus. iii) Finally the control of polarisation associated with the access to the K-edges of transition metals and L-edges of rare earth metals open the way for of micro-focussed X-ray Magnetic Circular Dichroic measurements. Based on the above considerations and to exploit the ESRF source characteristics, the project of building an X-ray microscopy beamline was approved in 1993 and started at the end of 1994.

CP507, *X-Ray Microscopy: Proceedings of the Sixth International Conference*,
edited by W. Meyer-Ilse, T. Warwick, and D. Attwood
© 2000 American Institute of Physics 1-56396-926-2/00/$17.00

The beamline welcomed its first users at the end of 1997 and is now fully operational. After a brief overview of the source parameters, the design of the beamline will be described with emphasis on the source performance and optical design. A detailed description of the two end-stations, the scanning X-ray microscope (SXM) and the full-field transmission X-ray microscope (TXM), is the subject of other papers of this conference.

THE SOURCE

The ESRF machine is routinely operated at 6 GeV electron energy with a maximum current of 200 mA. The horizontal source emittance is $3.8 \ 10^{-9}$ m.rad and the coupling is usually 1%. The X-ray microscopy beamline is installed on a 4.8 m long low beta straight section the parameters of which are given in table 1. The choice of a low beta section (small source but larger divergence, see table 2) was driven by the following considerations. Firstly, the small horizontal source allows the zone plate to be used in a diffraction limited regime even for long focal lengths. Secondly, the scanning X-ray microscope benefits from higher coherent brilliance produced by such a source (at the ESRF low beta sections have a better phase-space matching than the high beta sections). Furthermore the larger horizontal divergence offers the possibility to split the beam for parallel operation of the two branch-lines and limits the heat load on the first optical components by spreading the power over a larger surface. However, although the low emittance of the beam is best suited for the operation of the SXM, it complicates the numerical aperture matching between the illumination and objective zone-plate for the full-field microscope [3].

TABLE 1. Electron Source parameters.

	Horizontal	Vertical
Beta function [m]	0.5	2.73
RMS source size [μm]	57	10.3
RMS source divergence [μm]	88.3	3.8

TABLE 2. FWHM Photon source parameters for 3 energies.

	7.0 keV	2.0 keV	0.4 keV
Horizontal source size [μm]	47	47	47
Vertical source size [μm]	12	12	12
Horizontal source divergence [μrad]	82	85	92
Vertical source divergence [μrad]	12	20	44

The straight section is equipped with 3 different insertion devices (see table 3). As shown in figure 2 the use of two linear undulators, a U42 and a W80, allows the full energy range to be covered while maintaining the total power to manageable levels (always below 700 W through an aperture of 5x5 mm^2 placed at 28 m from the source) without compromising the available flux. At low energy (large wavelength) the lower brilliance is largely compensated by an increase of the lateral coherence. The helical undulator produces 100% circularly polarised light between 3.4 and 6.6 keV [4]. However, the degree of circular polarisation remains higher than 80% for energies below 3.4 keV. Furthermore most of the outgoing power produced by a helical undulator is concentrated in the fundamental peak and the absence of harmonics is an appreciable attribute for specific experiments.

TABLE 3. Insertion Device parameters.

Parameters	U42	W80	HU52
Type	linear undulator	linear wiggler	helical undulator
Period [mm]	42	80	52
Number of poles	38	20	59
Length [m]	1.53	1.60	1.60
B_o [T]	0.54	0.80	0.88 (B_{ox}) and 0.99 (B_{oy})
K_{max}	2.1	5.9	1.21 (K_x) and 1.83 (K_y)
P_{max} [kW]	2.0	4.6	0.8

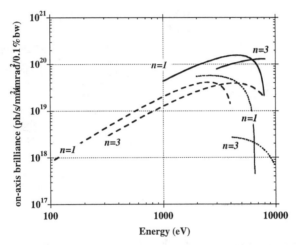

FIGURE 1. On–axis brilliance of the fundamental (n=1) and third harmonic (n=3) of the three insertions devices used on ID21. The continuous, long-dashed and point-dashed lines correspond to the U42, W80 and HU52 respectively. The parameters of these insertion devices are given in table 3.

OPTICAL LAYOUT

The general layout is shown in figure 3. The beamline is windowless (i.e. no beryllium window separates it from the storage ring) and therefore necessitates a UHV environment. The white beam is first conditioned by a 2-bounce fixed-exit mirror device consisting of two parallel silicon mirrors. The glancing angle is tunable from about 7 to 20 mrad using either Pt, Si, or Ni reflective coatings, thus allowing harmonic rejection better than three orders of magnitude for energies between 1-7 keV while ensuring overall transmission greater than 75 %. Furthermore, the mirrors are horizontally deflecting in order to preserve the vertical beam coherence. The two mirrors are water cooled from the sides and the maximum power absorbed in the first mirror is about 700 W. This system, by an efficient damping of the high-order harmonics and separation of the bremsstrahlung from the synchrotron beam, allows a significant reduction of the shielding required for the downstream beam transport. The next optical component is a multilayer acting as a beam-steerer (or beam-splitter) to direct the beam on the side branch (or giving the possibility to operate the two branch-lines simultaneously).

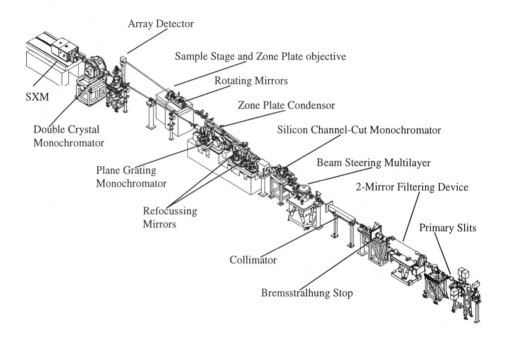

FIGURE 2. General layout of the beamline with the main optical components of the two branch-lines and end-stations.

The scanning X-ray microscope branch-line

Access to a very broad energy range between 0.2-7 keV, with high flux and high spectral resolution, requires the use of two separate monochromators. The above energy range is thus split into two ranges:

i) The "high" energy region (E > 2 keV), covered by use of a fixed-exit double crystal monochromator. For experiments requiring high spectral resolution Silicon (111) crystals are used (see figure 3a) while for flux demanding experiments Ni/B4C multilayers can be used ($\Delta E/E = 0.01$)(see figure 3b). Although these multilayers are now optimized for higher energies, they provide an easy way to access energies between 2 and 1 keV (see figure 3c). In order to avoid an amplification of the residual beam instability, the monochromator is positioned as close as possible to the SXM (1.5 m). A focussing geometry is possible and the source is thus imaged onto the SXM pinholes via a sagittally focussing cylindrical mirror and a vertically focussing spherical mirror.

ii) The "low" energy range (0.25 < E (keV) < 1.5), which will be efficiently covered by a plane grating monochromator device with an anticipated resolution of about 5000 (see figure 4a). This system associates a holographic grating, tuned by a single rotation, with a vertically focussing spherical mirror [5]. In this configuration, the pinhole aperture used as source for the SXM acts here as exit "slits" of the grating. The first grating is currently under commissioning. Preliminary results are presented in figure 4b.

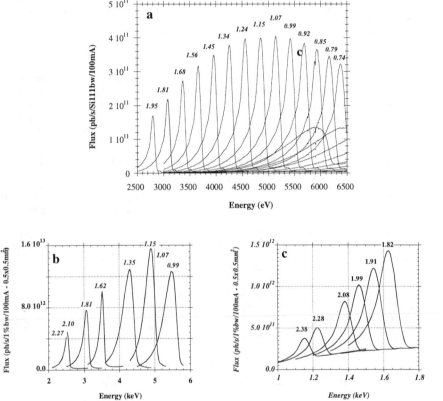

FIGURE 3. Measured fluxes through an aperture of 0.5x0.5 mm2 located at 28 m from the source and normalized for a machine current of 100 mA. For each undulator peak (n=1) the K parameter value is given. a) U42 with the DCM (Si[111]), b)U42 with DCM(Ni/B₄C multilayers) and c) W80 with DCM(Ni/B₄C multilayers).

Special care has been taken to allow easy swapping between the two configurations involving different refocussing mirrors. A detailed description of the optical layouts is given in references [6] and [7].

A separate paper (this conference) is devoted to the presentation of the SXM end-station and the first experimental results [8].

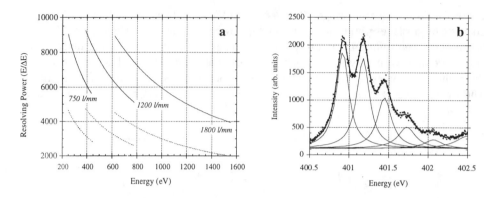

FIGURE 4. a) Anticipated spectral resolving power of the 3 holographic gratings for two figure slope errors of the optical components of the plane grating monochromator: the 1μrad rms./component (plain line) and the 2μrad rms./component (dashed line). b) first recorded spectrum using the vibrational states of the N_2 1s→π* transitions. A 750mm/l grating was used with a resolving power of about 4000.

The transmission full-field X-ray microscope branch-line

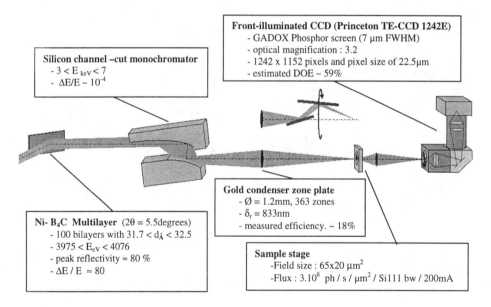

FIGURE 5. Optical layout of the side-branch which serves the TXM end-station. The main parameters of each component are given. The high resolution imaging configuration requires rotating mirrors to increase the numeral aperture.

The side branch serves a full-field imaging or transmission X-ray microscope (TXM) designed for absorption as well as Zernike phase contrast modes in the photon energy range of 3-7 keV [9]. The optical scheme is similar to the principle adopted for the TXM at BESSY and at the ALS but differs by the use of a Silicon [111] monochromator providing an energy resolution of about $\Delta E/E \sim 5 \ 10^{-4}$ (see figure 5). This enables combination of imaging and spectro-microscopy techniques. As already mentioned in the introduction the main complication in the design comes from the very low emittance of the source. The use of a single zone plate as condenser gives a very low numerical aperture which significantly affects the resolution of the microscope. The adopted strategy is to insert between the zone-plate condenser and the sample a rotating 2-mirror device which generates quasi-incoherent illumination [10].

ZONE-PLATE R&D PROGRAM

As discussed in the previous paragraphs, the X-ray microscopy beamline covers several disciplines over a rather broad energy range. Therefore various types of zone plates are currently requested for each kind of application. i) Fluorescence or diffraction applications in a multi-keV energy range require large diameter and medium-resolution (thus long focal length) zone-plates, offering high flux and a better access to the sample. ii) Full-field X-ray microscopy requests high resolution zone-plate lens objectives and large diameter zone-plate lens condensers. In all cases the efficiency of the lens must be as high as possible, and for practical reasons, the use of apodized zone-plates (with a central stop) is desirable. Several collaborations were therefore initiated with different laboratories with particular emphasis on high efficiency zone-plates for the 2-7 keV range on one hand and high-resolution zone plates for the Ca K-edge region (~ 4 keV) on the other. The most representative results are summarised in table 4. For each type of zone-plate, the diameter D, the outmost zone width d_m, the number of zones N and the measured efficiency ε are given.

TABLE 4. examples of zone-plate parameters used on ID21.

Contact	material	D [mm]	d_m [nm]	N	ε [%]
E. Di Fabrizio (Trieste, Italy) [9]	Au	1200	826	363	18 [6keV]
E. Di Fabrizio (Trieste, Italy) [11]	Ni	150	500	70	57 [7keV]
H. Kihara (Kansai University, Japan) [12]	Ta	1040	180	1722	17 [5keV]
P. Charalambous (KCL, UK) [13]	W	68	128	60	8 [3.5keV]
M. Panitz (Göttingen, Germany) [14]	Au	65	70	300	11 [4keV]

CONCLUSION AND OUTLOOK

The X-ray microscopy facility, which associates the two types of microscopes (a transmission XM and a scanning XM), on the same beamline, is a unique facility which is now moving into full operation. The SXM has been demonstrated to perform well in both fluorescence and transmission modes in the 2-7 keV range with a Silicon monochromator. First spectra at 400eV were also obtained with the plane grating monochromator but further commissioning is necessary to improve the beam stability and outgoing flux. The TXM meets the expected performance in term of resolution, flux and ease of use. Finally the development of both microscopes benefits greatly from active collaborations aiming to produce well suited zone-plates. Further development and in-house research programs will be initiated in the up coming months, in particular concerning cryo-microscopy, micro-tomography and micro-XMCD, in parallel with a full user program.

ACKNOWLEDGEMENTS

The authors would like to thank L. Andre, R. Baker, J.L. Berclaz, G. Berruyer, S. Blanchard, F. Demarcq, G. Rostaing, M. Soulier and F. Thurel for their invaluable technical assistance. The conception of the beamline has benefited greatly from discussions with many colleagues including C. Buckley, P. Charalambous, E. Di Fabrizio and B. Niemann.

REFERENCES

1. Kirz, J., Jacobsen, C., and Howells, M.; *Quarterly Reviews of Biophysics*, **28** (1), (1995).

2. See papers in recent XRM conferences : e.g. X-ray Microscopy and Spectromicroscopy, edited by J. Thieme et al., Berlin Heidelberg : Springer Verlag (1996) and these proceedings (1999).

3. Oestreich, S., Niemann, B., in : X-ray Microscopy and Spectromicroscopy, edited by J. Thieme et al., Berlin Heidelberg : Springer Verlag, 1998, pp.VI.77 – VI.81.

4. Elleaume, P., Chavanne, J., Nucl. Instr. and Methods, **A304**, (1991) pp. 719-724 .

5. Delcamp, E., Polack, F., Lagarde, B., and Susini, J. , *Proceedings SPIE conference*, **2856**, Denver (1996).

6. Susini, J., and Barrett, R., in : X-ray Microscopy and Spectromicroscopy, edited by J. Thieme et al., Berlin Heidelberg : Springer Verlag, 1998, p I.45 – I.54.

7. Barrett, R. , J., Susini, J., in "X-ray Microfocussing : Applications and Techniques, edited by I. McNulty, Proc. SPIE **3449**, pp. 80-90 (1998).

8. Barrett, R., Kaulich, B., Salome, M., Susini, J., *this conference*.

9. Kaulich, B., Niemann, B., Rostaing, G., Oestreich, S., Salome, M., Barrett, R., Susini, J., *this conference*.

10. Oestreich, S., Niemann, B., Rostaing, G., Kaulich, B., Barrett, R., and Susini, J., *this conference*.

11. Fabrizio, E., Romanato, F., Gentili, M., Cabrini, S., Kaulich, B., Susini, J., Barrett, R., *Nature to be published* (1999).

12. Takemoto, K., Kihara, H., this conference.

13. 6. Charalambous, P., *this conference*.

14. M. Panitz, G. Schneider, M. Pauker, D. Hambach, B. Kaulich, S. Oestreich, J. Susini, G. Schmahl; *this conference*.

Growth Of Thin Metal Films Studied By Spectromicroscopy

Th. Schmidt[1,2], B. Ressel[2], S.Heun[2], K.C. Prince[2] and E. Bauer[3]

[1]Experimentelle Physik II, Universität Würzburg, Germany,
[2]ELETTRA, Sincrotrone Trieste, Italy
[3]Department of Physics and Astronomy, ASU, Tempe, USA,

Abstract. The many possibilities of SPELEEM are particularly interesting for the study of thin film growth processes. This is demonstrated for the heteroepitaxial growth of thin metal films in the presence of co-adsorbates. Two model systems are chosen:. Fe on W(100) and Pb on Si(111). The Stranski-Krastanov growth mode of both systems is changed to a quasi-Frank-van der Merwe growth mode by a surfactant (Pb) and an interfactant (Au), respectively..

INTRODUCTION

Spectromicroscopy has become one of the most promising techniques which require the high brilliance of the present third-generation synchrotron radiation sources. The combination of tunable synchrotron radiation with a parallel imaging low energy electron microscope (LEEM) into the Spectroscopic Photo Emission and Low Energy Electron Microscope (SPELEEM) allows to investigate both, structural and electronic properties on the ten nanometer length scale. In contrast to photo emission microscopes which use soft x-ray-generated secondary electrons for imaging contrast, the SPELEEM is equipped with two additional features: a 180° imaging electron energy analyzer and an electron gun. The first enables energy selection of the photoemitted electrons and the second imaging with reflected electrons. Examples for the wide field of applications of this instrument are on the one hand fundamental studies of thin film growth, surface reactions and other surface phenomena and on the other hand the characterization of real industrial samples such as quantum dots or field effect transistors [1]. In this paper we demonstrate the capabilities of this multi-method instrument by a study of thin metal growth in the presence of co-adsorbates. Thin metal overlayers are important both in technology (e.g. contacts, diodes, interconnect, magnetic memory devices) and science (e.g. new quantum effects due to reduced dimensionality). Unfortunately ultra-thin film grow rarely in the desired monolayer-by-monolayer growth mode (Frank-van der Merwe, FM mode, Fig.1a). In general, three-dimensional islands grow from the very beginning (Volmer-Weber, VW mode, Fig. 1c) or after an initial mono- or multilayer has grown (Stranski-Krastanov, SK mode, Fig. 1b) [2]. [Fig.1(a)-(c)]. The three-dimensional growth can be switched to a

CP507, X-Ray Microscopy: Proceedings of the Sixth International Conference,
edited by W. Meyer-Ilse, T. Warwick, and D. Attwood
© 2000 American Institute of Physics 1-56396-926-2/00/$17.00

FM-like growth by using either a surfactant [3] (Fig.1d) or an interfactant [4] (Fig. 1e) which both are deposited before the film growth. By definition the surfactant modifies the surface of the growing film and remains on it during the growth. Therefore in the ideal case complete segregation of the surfactant species is necessary. The interfactant on the other hand affects the interface between the substrate and the growing film. In this case, segregation is undesirable.

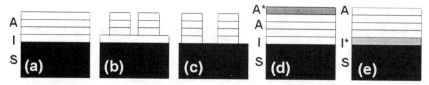

FIGURE 1. Epitaxial growth modes: (a) Frank-van der Merwe, (b) Stranski-Krastanov, (c) Volmer-Weber and possible influence on the growth by (d) a surfactant and (e) an interfactant. A=adsorbate, I=Interface, S=substrate, A*=surfactant and I*=interfactant.

EXPERIMENTAL SETUP AND METHODS

The SPELEEM instrument is a precursor instrument of a commercial version [5] and was installed for a period of nearly three years at the undulator beamline 6.2LL of ELETTRA. The instrument, the electron optics and the method are described in detail in Ref. [6]. The sample can be heated in the ultra high vacuum main chamber to 2000 K and cooled to about 260 K. Six ports point at a grazing incidence angle of 15° at the sample surface for in-situ deposition and illumination by an UV-lamp (hν ≈ 5 eV) and by the soft x-ray synchrotron light. The optics of the beamline produces at the sample a spot of about 50 μm x 30 μm at a maximum photon flux of $5x10^{12}$/s with the photon energy resolution set to 30 meV [7].

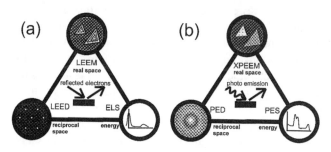

FIGURE 2. The basic surface analysis combination in the SPELEEM: (a) reflected and (b) photo emitted electrons (From [1]).

The superior merits of this instrument are the high lateral resolution of 22 nm in spectromicroscopy and 8 nm in LEEM combined with the good energy resolution of the imaging analyzer (< 0.5 eV), fast image acquisition in the video mode and slow image acquisition at 1 to 60 sec./image with a Peltier cooled 16-bit slow scan CCD camera. In contrast to other Photo Emission Electron Microscopes (PEEM), the special electron optics of the SPELEEM enables a multi-method characterization of the specimen surface. The design allows fast switching between three fundamental modes: imaging of (a) the surface (microscopy), (b) the angular distribution, and (c) the dispersive plane of the analyzer (spectroscopy). Using reflected electrons this corresponds to LEEM for structural sensitive imaging of mono-atomic steps, superstructure domains etc., LEED determining the surface structure and ELS (Electron Loss Spectroscopy) [Fig. 2(a)]. For photo-emitted electrons the methods are XPEEM (soft X-ray excited PEEM) for element specific and/or electronic state sensitive imaging, PED (Photo Electron Diffraction) for determination of the local structure of the emitter and PES (Photo Electron Spectroscopy) for the analysis of the elemental composition and/or the electronic state the surface [Fig.2(b)]. Additional methods are available depending on the kinetic energy of the electrons, the photon energy, spin-polarization of the electron gun, light polarization, the type of emission (Auger, thermo-emission) etc. Apertures in the intermediate image plane and in the angular distribution plane and the energy slits improve on the one hand the imaging quality by aberration reduction but also produce contrast or lateral sensitivity. In dark field LEEM a superstructure LEED spot is selected by an aperture so that only the surface area with this superstructure appears bright. In XPEEM a narrow energy slit is necessary to select the desired electrons, for example valence electrons for imaging. With the aperture in the intermediate image plane a surface area of ≤ 1 μm diameter can be selected from which the PED pattern can be obtained. Spectra can be achieved either directly by imaging the dispersive plane or by analysis of image stacks in XPEEM or PED, at fixed photon energy and scanned kinetic energy of the electrons (fixed electron and scanned photon energy is possible, too).

RESULTS

We present the results of two experiments: in the first part we investigate the segregation of the surfactant Pb during the growth of Fe on W(100) and in the second part the behavior of Au as an interfactant in the growth of Pb on Si(111).

Growth of Fe on W(100) pre-deposited with 2 ML Pb

Fe grows on W(100) in the SK mode: two pseudomorphic monolayers are followed by three-dimensional islands. The behavior of the Fe 3d signal during the growth shows that this can be changed to quasi-FW growth if Pb is deposited first. In Fig.3 2 ML Pb were deposited on W(100) forming three-dimensional islands on a complete Pb monolayer. Their diameter is about 100 nm with a height of 30 nm, determined from the shadow of the 15° incident synchrotron light [Fig. 3 left side]. Their lateral spreading can nicely be recorded with XPEEM images taken with Pb 5d electrons during the growth [Fig. 3]

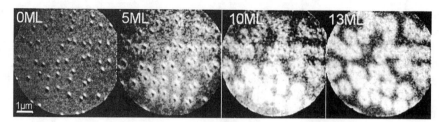

FIGURE 3. Spreading of Pb during the growth of Fe on a Pb-covered W(100) surface at 278 K. The XPEEM images are taken with Pb 5d-electrons during the growth. hv=70.5eV, E_{kin}=49eV (From [1]).

After the growth the resulting surface was analyzed by LEEM and spectroscopic XPEEM. At least three different structures can be identified according the three different intensity level in LEEM [Fig.4(a)]. The Pb 5d and Fe 3d XPEEM images [Fig.4(b), (c)] showing a contrast inversion are two images of a stack of 50 with scanned kinetic energy. The spectra to the right are from the regions marked 'a' and 'b'.

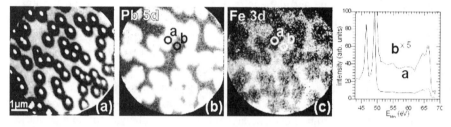

FIGURE 4. 13 ML thick Fe film grown at 278 K on a Pb-covered W(100) surface. (a) LEEM image taken at 8.3eV (top), (b,c) XPEEM images taken with Pb 5d and Fe 3d electrons, respectively. Right side: spectra from areas (a,b), hv=70.5eV. The surface region in the LEEM image is shifted (From [1]).

The Pb and Fe peak ratios of the different areas together with the observation in Fig. 3 show that the segregation of the Pb monolayer is hindered: less than 50 percent of it is able to swim on the surface during the growth of 13 ML Fe. On the other hand the three-dimensional Pb crystals act as a Pb sources during the Fe growth, which keeps the Pb concentration constant around the crystals.

Growth of Pb on Si(111) pre-deposited with Au

Whereas on the clean Si(111)-7x7 surface Pb grows in SK mode, the growth changes drastically if a submonolayer of Au is deposited at 850 K beforehand, producing a Au-√3 reconstruction. The resulting layer-by-layer growth shows in LEEM images [Fig.5 top] the nucleation of new terraces (white dots) and the generation of regions with two-dimensional periodic lattice modulation (dark areas in the 2-4 ML images). The substrate step structure is reproduced due to the perfectness of the growth. The Au 4f and the Si 2p photo electron signal decrease as expected during the Pb growth but the Pb 5d signal decreases also after the initial increase up to one monolayer in spite of the monolayer by monolayer growth [Fig.5 bottom].

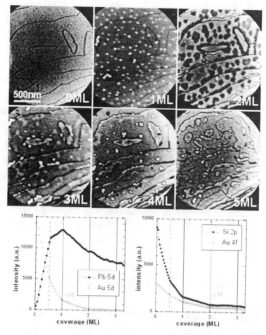

FIGURE 5. Au-interfactant-mediated growth of Pb on a Si(111) surface. LEEM images taken with 8eV electrons (top), and the photoeletron signal of the Pb 5d and Au 5d electrons (bottom left, hν = 72eV) and of the Au 4f and Si 2p electrons (bottom right, hν = 133eV) as a function of Pb coverage [1]

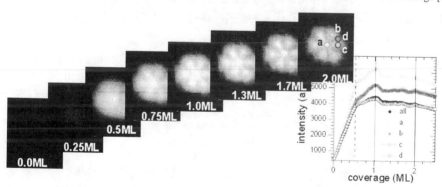

FIGURE 6. PED pattern taken during the growth of Pb on the Si(111)-Au√3 surface, hν = 63 eV, using the Pb $5d_{3/2}$ peak. Full angular space.

The photo electron diffraction pattern recorded during the Pb growth shows an overall decrease of intensity after 1 ML [Fig.6], not only in the forward direction so that the decrease of the Pb 5d signal cannot be attributed to diffraction changes with increasing thickness. It is likely that the intensity from one monolayer is higher than that from thicker layers because of a higher (electron) reflectivity of the underlying Au√3 structure compared to the Pb-bulk. As a consequence photo electron emitted in the direction of the substrate are reflected and can contribute to the signal. The results, however, clearly show that Pb grows perfectly in the quasi-FW mode and that Au remains at the interface.

CONCLUDING REMARKS AND OUTLOOK

The combination of high spatial and energy resolution together with multi-method characterization at a high image acquisition rate compared to other methods allows a comprehensive study of fundamental surface science processes and the analysis of technical devices in the sub-micron range. Several projects of this type have been started recently, the most ambitious one being the SMART microscope [8], an instrument of the same principle, which is expected to reach even 1nm spatial resolution at 0.1eV energy resolution with a more sophisticated electron optics and a high brilliance undulator beamline.

ACKNOWLEDGEMENTS

The development of the basic instrument was financially supported by the Volkswagen Foundation. The experimental project at ELETTRA was partly funded by the BMBF (within the SMART collaboration) under contracts 05 644 WWA and 05 SL8 WW1, partly supported in-house. T.S. was supported financially by the Training and Mobility of Researcher (TMR) program of the European Community (contract no. ERB FMBI-CT 96-1749) and E.B. acknowledges a NATO travel grant.

REFERENCES

1. Bauer, E., and Schmidt, Th., "Multi-Method High Resolution Surface Analysis with Slow Electrons" in *High-Resolution Imaging and Spectrometry of Materials*, edited by M. Rühl and F. Ernst, Springer Series in Material Science, to be published

2. Bauer, E., *Z. Kristallogr.* **110**, 372 (1958)

3. Copel, M., Reuter, M.C., Kaxiras E., and Tromp, R., *Phys. Rev. Lett.* **63**, 632 (1989)

4. Jalochowski, M., and Bauer, E., *J. Appl. Phys.* **63**, 3501 (1988)

5. ELMITEC Elektronenmikroskopie GmbH, Am Kaiser Wilhelm Schacht 1, D-38678 Clausthal-Zellerfeld, Germany, FAX:++49-5323-78931

6. Bauer, E., Rep. Prog. Phys. **57**, 895 (1994); Schmidt, Th., Heun, S., Slezak, J., Diaz, J., Prince, K.C., Lilienkamp, G., Bauer, E., *Surf. Rev. Lett.* **5**, 1287 (1998)

7. Schmidt, Th., et al. *J. Synch.Rad.* **6**, 957 (1999)

8. Fink, R. et al., *J. Electron Spectr. Rel. Phen.* **84**, 231 (1999)

Composition Mapping considerations in scanning soft x-ray microscopy

C.J. Buckley

Department of Physics, King's College London, Strand, London WC2R 2LS, U.K.

Abstract. The effects of distortion to absorption measurements made in the soft x-ray region are considered in this paper. The effects of background radiation, spectral contamination, resolving power and detector non-linearity is discussed within the context of what is required for quantitative compositional mapping.

INTRODUCTION

Quantitative chemical state mapping (compositional mapping) can be achieved via scanning x-ray transmission microscopy (STXM). The technique, introduced by Zhang[1] to map and measure DNA and protein in sperm heads, requires that transmission measurements are made at a series of energies where the values of the mass absorption coefficients of the principal components are known. Good signal to noise ratios are obtained in the maps when the absorption coefficients at the chosen energies are significantly different from each other and the transmission measurements are made accurately and with sufficiently good photon statistics to cleanly resolve these differences.

Mapping four components has been achieved recently and was reported by this author at the conference which gave rise to these proceedings. Where larger numbers of specimen components are involved, the accurate measurement of x-ray absorption by the specimen and reference materials is of utmost importance. The technique of quantitative compositional mapping is non trivial due to the pit falls introduced by phenomena which can give rise to inaccurate measurements of absorption. These distortions are non-linear by nature and can result in considerable amounts of artifact being introduced into compositional maps. The phenomena behind these distortions are discussed in this paper and recommendations are made on the levels of these phenomena which are acceptable in order to produce accurate compositional maps.

CP507, *X-Ray Microscopy: Proceedings of the Sixth International Conference*,
edited by W. Meyer-Ilse, T. Warwick, and D. Attwood
© 2000 American Institute of Physics 1-56396-926-2/00/$17.00

Compositional mapping via absorption differences

For mono-energetic soft x-rays the absorption A by a specimen containing n components in the region sampled by the x-ray beam will be given by

$$A = \sum_{i=1}^{n} \mu_i \rho_i t \qquad (1)$$

Where μ_i, and ρ_i are the mass absorption coefficient and density of the ith component respectively and t is the sample thickness at the beam position. In order to map the mass thickness' (the mass of a component per sampled specimen area) of the individual components it is necessary to measure the absorption accurately and also to obtain accurate values of the mass absorption coefficients. If this is done for a number of x-ray energies equal to or greater than the number of principal components, then the mass thickness' of the different components may be solved for by a variety of numeric methods – e.g. see Hitchcock $et.\ al.$[2], S. Urquhart $et.\ al.$[3] and Osanna $et.\ al.$[4] in these proceedings. However, even small errors in the measured absorption coefficients or the measured absorption at a pixel will propagate to considerable inaccuracies in the compositional maps. Accurate measurement of the x-ray absorption by specimens is essential, but is prone to error due to experimental considerations. A working rule of thumb is that accurate compositional maps will be obtained when net distortions to absorption values are kept below a few percent. Artifacts will appear in maps when absorption coefficient values and absorption measurements at pixels have errors which are greater than this. The sources of error and their effects are discussed below.

Sources of inaccuracy in absorption measurements

When a well calibrated monochromator is used with sufficient numbers of photons, there are four main phenomena which will lead to artifacts in compositional maps due to inaccurate absorption measurements. These occur in absorption measurements in either the reference spectra, the images or both. They are: background radiation, higher spectral order contamination, spectral band pass distortions and detector non-linearity. These are considered in turn below. Distortions introduced by x-ray beam damage are specimen specific and are not discussed here.

Distortions due to background radiation

Background radiation will distort absorption measurements. Assuming that the detector and associated electronics are configured such that they do not inject a background signal, a major source of background can be leakage of radiation through the zone plate's central stop and order selection aperture (OSA) arrangement. This will bathe the specimen in a radiation disc the size of the OSA and the measured transmission will be a combination of that for the first order focus at the chosen pixel and that for the specimen integrated over the disc foot-print. For a zone plate used in positive first order, an OSA which has an open diameter of one third of the zone plate is often used. Thus, the background will consist of the number of photons incident on this diameter times the transmission of the central stop material to this radiation. The specimen may attenuate this swath of radiation, but in many cases most of it will be transmitted. This radiation will be additive to the focused radiation transmitted by the specimen, in which case the transmission, τ, will be

$$\tau = \frac{I_{total}}{I_{0\,total}} = \frac{I_0 e^{-\mu\rho t} + B}{I_0 + B} \tag{2}$$

where I_0 is the incident focussed radiation, I is the transmitted focussed radiation, and B is the unfocussed radiation leaked through the central stop. If this background radiation is considered as a fraction, k, of the incident focused radiation ($B = kI_0$), then (2) becomes

$$\tau = \frac{e^{-\mu\rho t} + k}{(1 + k)} \tag{3}$$

The true absorption of the specimen at the focused probe position will be $\mu\rho t$ while the interpreted absorption will be that of the natural log of equation 3. The ratio of the interpreted to true absorption is plotted in figure 1 for a range of absorption lengths and percentages of background radiation.

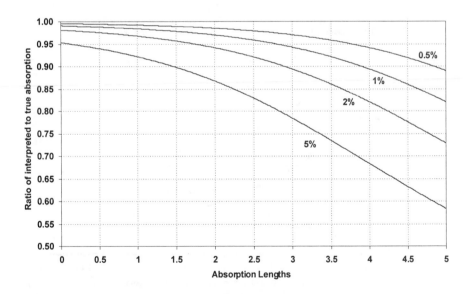

FIGURE 1. The ratio of what would be measured as absorption to true absorption in the presence of different amounts of background radiation.

The transmission of the central stop plays an important roll in providing distortion free measurements. The fractional area of a one-third zone plate diameter OSA is about 10% of the zone plate's area. As the first order efficiency of many soft x-ray zone plates is about 10%, this gives an approximate equal contribution of radiation from the first and zero orders to the specimen in the absence of a central stop. Thus if the zone plate illumination is truly monochromatic, the background percentage can be calculated directly from the stop transmission. E.g. a 0.2 micron thick gold stop would provide a background of about 0.5% at the carbon K absorption edge and about 2% at the oxygen K edge. The effect of this radiation can be seen in figure 1. In practice, the background is likely to have larger values than these due to the presence of higher order spectral contamination. This is discussed below.

Distortions due to the presence of higher spectral order radiation

Most monochromator systems will pass some x-rays at multiples of the selected fundamental energy. The percentage higher order contamination reaching the specimen will vary depending on the design of the beamline. Pre and post grating mirror schemes are chosen to minimise the percentage of higher order radiation passed but a compromise must be reached in order to maintain efficient throughput and flexibility of energy tunability leading to some amount of spectral contamination from higher orders. The zone stop can effectively reduce the much of unfocussed background due to both higher and fundamental spectral order radiation if made thick enough (see figure 2).

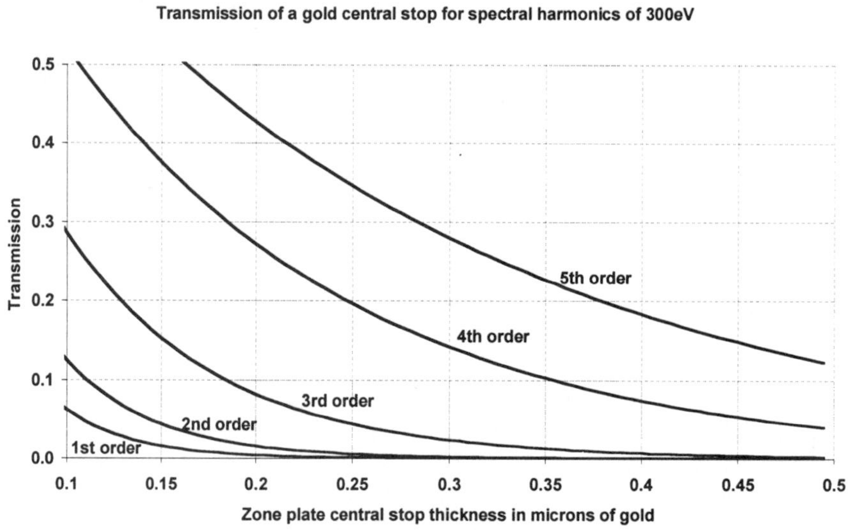

Transmission of a gold central stop for spectral harmonics of 300eV

FIGURE 2. The transmission of a gold central stop to spectral harmonics of 300eV radiation. A gold stop of thickness 0.5 microns will be sufficient to prevent distortions to absorption measurements at the carbon K absorption edge, though a greater thickness will be required at the oxygen K edge.

A sufficiently thick central stop will prevent unfocussed radiation from reaching the specimen. However, a percentage of the higher spectral order radiation will be incident on the specimen due to the diffraction properties of the zone plate. A carefully positioned OSA can help reduce the contamination by screening spectral orders which are focussed at a fraction of the zone plate's first order focal length. However, the OSA can not block any higher order spectral radiation where the spectral order and the zone plate diffraction order are the same I.e., the nth order spectral radiation will be focussed by the zone plate's nth diffraction order such that their focus cones coincide. This means that the first order zone plate focus cone will always contain N_T photons such that

$$N_T = \sum_{i=1}^{\infty} N_i \xi_i \qquad (4)$$

Where N_i is the number of photons incident on the zone plate from the ith spectral order and ξ_i is the zone plate's i^{th} diffraction order efficiency to the ith harmonic radiation. It is fortunate that this summation quickly tails to zero after the third order as the positive order efficiency of Fresnell zone plates is diminished by an inverse power law according to order. As an example, if a beam line passes 5% of second order spectral contamination to a zone plate made of 0.15 micron thick nickel which has a first order efficiency of 10% and a second order efficiency of 2% (due to uneven mark to space ratio of the zones) , then the net contribution of the higher spectral order will at focus will be 1%. A similar calculation can be done for third spectral order radiation which gives contribution of about 0.15%. These higher energies will typically be poorly attenuated by the specimen and will give rise to an effective background thereby distorting absorption measurements as indicated in figure 1.

Distortions due to resolving power effects

Distortions introduced by monochromator resolving power (band pass) effects can be extremely detrimental to compositional mapping. This effect has been considered in detail in a previous publication[5]. The effect is so detrimental because the measured absorption in the presence of band pass effects is a function of both the energy width of the absorption feature and the optical thickness. This means that the effect will primarily show up at near edge absorption fine structure (NEXAFS) peaks where the degree of distortion is greatest for narrow peaks. Figure 3 illustrates this effect. Figure 4 shows the distortion to the measurement of absorption at NEXAFS peaks for peaks as a fraction of absorption peak energy width to the band pass width of the monochromator. Clearly the band pass should be less than one third of the peak width. This translates to a spectral purity of 0.03% or better for most NEXAFS peaks in the soft x-ray region.

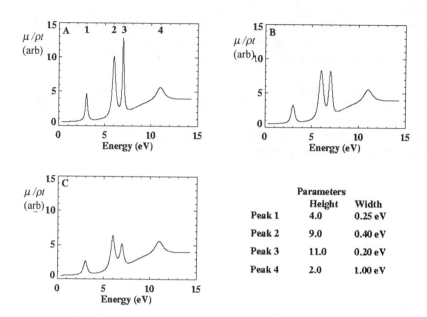

FIGURE 3. Illustration of distortion to absorption values due to monochromator resolving power effects. A: The undistorted spectrum. B: The spectrum returned by a monochromator with a band pass (FWHM) of 0.3eV with specimen of a maximum absorption equivalent to 1 absorption length. C: As for B but with a maximum absorption of 10 absorption lengths.

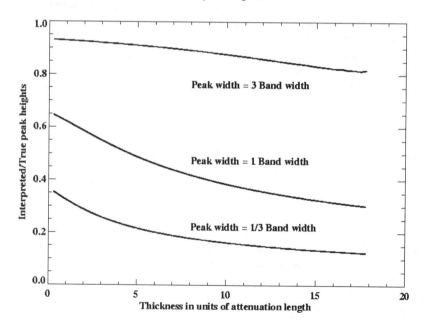

FIGURE 4. Distortion to measured absorption peak height due to resolving power effects.

All detectors will become non linear at some upper level of input radiation flux. In the case of single wire gas counter detectors, the upper limit on counting rate imposed by space charge effects[6] is a few MHz. The degree of non linearity can be assessed experimentally by exposing the detector to increasing amounts of x-ray flux (e.g. by increasing the opening of a beam aperture) while alternately introducing and removing an x-ray beam attenuation. The non-linearity can then be inspected by plotting the count rate with no attenuation against that with the attenuator. Figure 5 shows results taken in this way for the single wire gas counter detector on the STXM[7] at the X1a beam line of the NSLS.

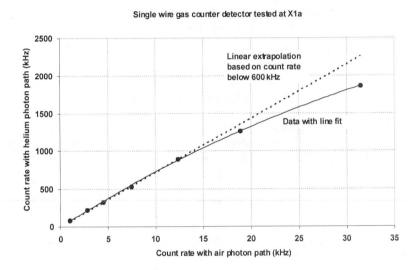

FIGURE 5. Assessment of the non linearity of a gas counter x-ray detector at the STXM[7] at the NSLS.

The distortion to count rate will provide a diminished signal for low absorptions. In particular, this can adversely affect all readings of the incident flux, upon which the absorption measurements are based. However if the detector's performance is calibrated, corrections may be applied via extrapolation (e.g. dotted line).

Summary and Conclusions

The degree of tolerance to the absorption measurement distortions will largely depend on the differences in absorption values of the principal components. The more principal components there are, the smaller the differences are likely to be and the tolerance will be less. However, in order to apply some working guidlenes for successful mapping, the following criteria are suggested as needing to be met if reliable quantitative compositional maps are to be formed. These are:

1) The resolving power of the soft x-ray monochromator needs to be 0.03% or better.

2) The beam line should deliver no more than 2% second order spectral contamination to the zone plate and no more than 5% of third spectral order.

3) The central stop should transmit no more than 0.1% of any radiation. A practical thickness to achieve this in the soft x-ray range is one micron of gold.

4) Detector linearity should be fully characterised at the time of the experiment and non-linearities compensated for by extrapolation.

Acknowledgements

The author would like to thank Sue Wirick, Chris Jacobsen and Janos Kirz for assistance with and use of their x-ray microscope at the NSLS and also for the positive and productive atmosphere they create around their beam line.

References

1. X. Zhang, R. Balhorn, J. Mazrimas and J. Kirz. 1996. *Mapping and measuring DNA to protein ratios in mammalian sperm head by XANES imaging*. J. Struct. Biol. 116 335-344.

2. A.P. Hitchcock, T. Tyliszczak, Y.M. Heng, R. Cornelius, J.L. Brash and H. Ade. X-ray *Spectromicroscopy Studies of Protein-Polymer Interactions* - These proceedings.

3. S.G. Urquhart, H.W. Ade, E.G. Rightor, G.E. Mitchell, A.P. Hitchcock, M. Rafailovich, J. Sokolov and A.J. Dias. *Spectroscopic Aspects of the Analytical X-ray Spectromicroscopy of Polymers* -These proceedings.

4. A. Osanna et al. *Spectromicroscopy with the Stony Brook cryo STXM* - These proceedings.

5. Buckley C.J. and Zhang X. *The Characterisation and Compensation of the Thickness Effect in Quantitative NEXAFS Measurements*. SPIE publishing, Volume 3449, 1998, pp 208-219.

6. Buckley, "X-Ray Detectors", in *X-Ray Science and Technology*, edited by A.G. Michette and C.J. Buckley, IOP publishing, 1993, pp 207-253.

7. C. Jacobsen, S. Williams, E. Anderson, M.T. Brown, C.J. Buckley, D. Kern, J.Kirz, M.Rivers and X. Zhang. 1991, *Diffraction-limited imaging in a scanning-transmission x-ray microscope*, Optics Communications, 86, 3:0351-36.

Real-Time Phase-Contrast X-Ray Imaging Using Two-Dimensionally Expanded Synchrotron Radiation X-Rays at the BL24XU (Hyogo-BL) of the SPring-8

Y. Kagoshima, Y. Tsusaka, J. Matsui, K. Yokoyama, K. Takai, S. Takeda,
K. Kobayashi[*], H. Kimura[*], S. Kimura[*] and K. Izumi[*]

Faculty of Science, Himeji Institute of Technology
3-2-1 Kouto, Kamigori, Ako, Hyogo 678-1297, Japan
*Fundamental Research Laboratories, NEC Corporation
34 Miyukigaoka, Tsukuba, Ibaraki 305-8501, Japan

Abstract. Phase-contrast x-ray imaging has been studied for materials, biological and medical sciences at the Hyogo-prefectural beamline (BL24XU) of the SPring-8. Its optical system consists of a successive arrangement of horizontal and vertical $(+,-)$ silicon double crystals taking asymmetric Bragg reflection. A living insect and a frog were observed in real time at the photon energy of 15 keV. Boundary structures in samples were clearly observed with much higher contrast than those obtained in absorption-contrast imaging. The beamline BL24XU, optical system and experimental results are described.

INTRODUCTION

Image contrast in "absorption-contrast" imaging results from differences in absorption of a sample. It holds an essential problem; namely, a higher radiation dose is unavoidable in order to obtain higher image contrast. Furthermore, the absorption contrast in a hard x-ray region is not necessarily sufficient for investigating samples composed of light elements. On the other hand, based on the fact that the phase shift cross section for light elements is almost a thousand times larger than the absorption one in a hard x-ray region, "phase-contrast" imaging has recently been studied actively. The real-time phase-contrast x-ray imaging using sophisticated crystal optics combined with the third generation synchrotron radiation source [1] is presented.

THE BL24XU (HYOGO-BL)

The Hyogo-prefectural beamline, BL24XU, has been built and commissioned in 1998 as the first contract beamline of the SPring-8.[2] Its x-ray source is the Figure-8 undulator, which produces horizontally or vertically polarized x-rays with integer or half-odd-integer harmonics.[3] A schematic drawing of the beamline is shown in Fig. 1.

CP507, *X-Ray Microscopy: Proceedings of the Sixth International Conference,*
edited by W. Meyer-Ilse, T. Warwick, and D. Attwood
© 2000 American Institute of Physics 1-56396-926-2/00/$17.00

The beamline has three monochromators and three experimental hutches for different purposes. It adopts so-called "Troika" conception to provide simultaneous and independent operation of three experiments. Two upstream monochromators (A and B) are double crystal monochromators using diamond crystals. Their first crystal works as a beam splitter, with which the troika conception can be performed. The monochromator C is a silicon double crystal monochromator. By removing the first crystal from the optical axis, the white undulator radiation can be also introduced into the hutch C. The hutch A is for protein crystal structure analysis, the hutch B is for grazing incidence x-ray diffraction, fluorescence x-ray analysis, *in-situ* observation of surface crystal structure during MOCVD epitaxial growth, and the hutch C is for x-ray imaging applications. The present experiment was performed in the hutch C.

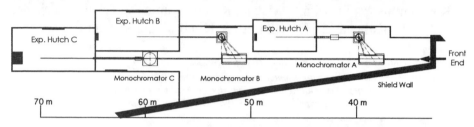

FIGURE 1. Schematic Drawing of BL24XU.

THE OPTICAL SYSTEM AND REAL-TIME OBSERVATION

The optical system is shown in Fig. 2. The photon energy of the fundamental harmonic peak of the undulator was tuned to be 15 keV. Its polarization was a horizontal linear one. The x-ray beam was first collimated by a four-quadrant slit with an aperture size of 1 mm $\times$ 1 mm installed in the front end at a distance of 30 m from the source point. The beam was next monochromatized by a horizontal dispersion silicon double-crystal monochromator (monochromator C) with 111 symmetric reflection. The beam was then horizontally expanded two times by (+ , −) arrangement of the asymmetric reflection with a case of a low incidence beam angle at a distance of about 70 m from the source. Finally, the beam was vertically expanded in the same way as in the horizontal case. Since the crystal surface and the diffraction plane chosen were 100 and 511, respectively, the asymmetry factor b was 0.207. A living pillbug *(armadillidium vugare)* was observed in real time using an x-ray camera and recorded on a videotape. Figure 3(a) shows an example of the phase contrast image. Boundary structures are clearly observed with good contrast. A living frog was also observed in real time. Examples of the captured images are shown in Figs. 3(b) and 3(c). It can be seen, as indicated by white arrows in Fig. 3(a) and by lung cells in Fig. 3(c), that the high contrast is located in regions enclosing air. This is due to the fact that the density difference between the organs and air is much larger than that between different organs. According to the analysis of the image blurring at the edge structure, the spatial resolution was estimated to be about 15 μm, which is almost equal to the theoretical resolution of 18 μm.[1]

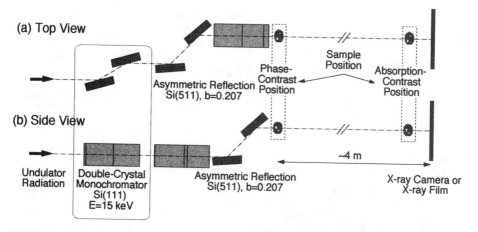

FIGURE 2. Optical System for Phase-Contrast X-Ray Imaging. (a) Top View and (b) Side View. The original beam size was expanded by using asymmetric Bragg reflection in both vertical and horizontal directions. Absorption-contrast imaging can also be performed simply by placing the sample just in front of the image sensor.

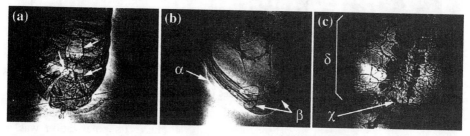

FIGURE 3. Phase-Contrast X-Ray Images of Living Samples. (a) Pillbug, (b) and (c) Frog. White arrows in (a) indicate the existence of air in the sample. In (b) and (c), eyeballs (α), nostrils (β), spine (χ) and cell-like structures inside the lungs (δ) are clearly observed with good contrast.

DIRECT MAGNIFICATION MOTHOD

Another approach as microscopy has been done as a joint work with NEC (K.K, H.K, S. K and K.I). Being put before the crystals, the sample is directly magnified by the asymmetric reflections as shown in Fig. 4. The merit of this method is that each crystal can play a role of an analyzer. Therefore, the phase information of the sample can be extracted by tuning the analyzer crystal appropriately.[4] As a demonstration of this method, a magnified image of Cu #2000 mesh was taken and is shown in Fig. 5. According to the intensity profile the resolution seems to be around 10 μm. The blurring of the image may be explained by the extinction distance of the crystals. The blurring is estimated be ~20 μm for π-polarization and ~13 μm for σ-polarization in the present optical system. The asymmetric reflections with total reflection arrangement can avoid that effect. In the case of the total reflection, the crystal surface, of course, must be perfectly flat.

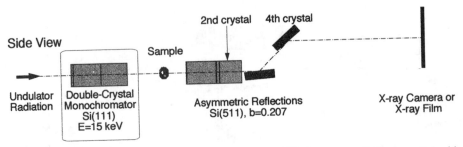

FIGURE 4. Optical System for Direct Magnification Using Asymmetric Bragg Reflections. Only side view is shown. The 2nd and the 4th crystals can work as an analyzer crystal.

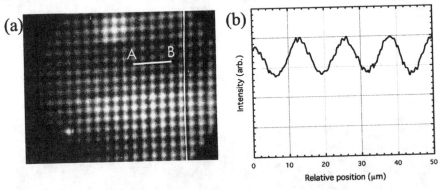

FIGURE 5. (a) Magnified Image of Cu #2000 Mesh and (b) Intensity Profile along A-B in (a).

ACKNOWLEDGEMENTS

The authors would like to express their gratitude to Hyogo prefecture for supporting the projects. They also thank to all SPring-8 staffs for the operation of the storage ring. This work has been carried out according to the proposal number of C99A24XU-001N.

REFERENCES

1. Kagoshima, Y., Tsusaka, Y., Yokoyama, K., Takai, K., Takeda, S. and Matsui, J., *Jpn. J. Appl. Phys.* **38**, L470-L472 (1999).

2. Matsui, J., Kagoshima, Y., Tsusaka, Y., Katsuya, Y., Motoyama, M., Watanabe, Y., Yokoyama, K., Takai, K., Takeda. S., and Chikawa, J., "Hyogo BL(BL24XU)" in *SPring-8 Annual Report 1997*: Japan Synchrotron Radiation Research Institute, 1998, pp. 125-130.

3. Tanaka, T., and Kitamura, H., *Nucl. Instrum. Method.* **A364**, 368-373 (1995).

4. Davis, T. J., Gureyev, T.E., Gao, D., Stevenson, A. W. and Wilkins, S. W., *Phys. Rev. Lett.* **74**, 3173-3176 (1995).

The transmission X-ray microscope end-station at the ESRF

B.Kaulich*, B.Niemann†, G.Rostaing*, S.Oestreich*, M.Salome*, R.Barrett*, and J.Susini*

*ESRF, ID21 - X-ray microscopy beamline, BP220, F-38043 Grenoble Cedex, France
†IRP, University of Goettingen, Geiststrasse11, D-37073 Goettingen, Germany

Abstract. A full-field imaging or transmission X-ray microscope (TXM) working in the photon energy of 3-7 keV was built at the ID21 beamline of the ESRF and is operational since the beginning of this year. The TXM is designed to work in absorption as well as in Zernike phase contrast modes and also offers the possibility for micro-spectroscopic investigations (XAS, element mapping). In this contribution, we give a technical description and characterization of the TXM end-station and discuss its imaging performance.

INTRODUCTION

The optical scheme of the microscope follows the principle of the TXM at BESSY [1] and the XM-1 microscope at the ALS [2]. Zone plates are used to condense the beam onto the sample and for the objective lens. The major difference is that a channelcut monochromator with an energy resolution of 0.7 eV is included in the setup for spectro-microscopy applications [3]. The other difference is a consequence of the small emittance of third generation synchrotron radiation sources which makes the use of a condenser as unique beam condensing element difficult due to the resulting low numerical aperture illumination. Therefore, a rotating two mirror assembly [4] can be introduced in the condenser system to match the numerical apertures between condenser and imaging objective and to generate quasi-incoherent illumination. A detailed description of the optical layout of the TXM branch is included in a contribution to this conference [5]. Therefore, we focus on the technical description of the TXM.

THE TXM END-STATION

The strategy adopted to the ID21 TXM was to privilege mechanical stability for high spatial resolution imaging and high modularity for the different intended

CP507, *X-Ray Microscopy: Proceedings of the Sixth International Conference,*
edited by W. Meyer-Ilse, T. Warwick, and D. Attwood
© 2000 American Institute of Physics 1-56396-926-2/00/$17.00

imaging modes. A compact design was essential in order to maintain sufficient space for a sample environment in air for ease of exchange and later implementation of a cryo-, tomography- or user specific stages with high accuracy and reproducibility of (re-)positioning. A visible light microscope for pre-alignment of the sample and finding regions of interest prior to X-ray radiation is included in the design. All mechanical stages are outside the vacuum to allow easy access and to avoid vacuum impacts in regions of fragile optical components. The microscope consists of three basic modules: A sample stage in air which is movable in xyz (x defines the X-ray beam axis), a yz-movable vacuum entrance window stage which can also be used as a support for additional optical elements, and a xyz-movable stage housing the micro zone plate objective lens which is in vacuum. A three-dimensional cross-section of the design is shown in Fig.1 with the X-ray beam direction from the lower left. In order to satisfy space constraints and nevertheless maintain

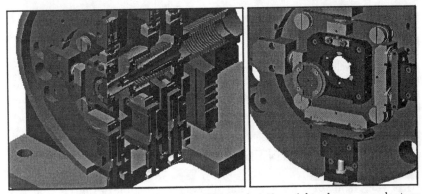

FIGURE 1. left scheme: cross section of the TXM end-station; right scheme: sample stage with kinematic mount sample support

a relatively compact and stable system with the necessary accuracies, specially adapted translation stages were built in-house. Special care was taken in the guiding quality of each stage and stable coupling of the sample and the zone plate stage in order to avoid parasitic movements between them during imaging. The same type of stage was chosen for the sample , the entrance window and the zone plate stage. As shown in figure 1 (right side), the mobile chariot moves perpendicular to the beam on the precisionly machined translation base via four thrust bearings at each edge of the chariot. In order to minimize pitch and roll, the chariot is constrained by spring washers. The zone plate stage is additionally constrained by a vacuum force introduced by unmatched bellow diameters (see Fig.1). The translation is driven by Newfocus Picomotors and transferred to the chariot by intermediate thrust plates which are guided by polymer cages. At the opposite side of each motor drive, LVDT encoders are mounted to allow measurement of the position of the chariot. For movements along the optical axis, the whole yz-translation body is moved on ramps driven by stepper motors. The quality of this translation is ensured by three parallel ball-bush guides.

The working TXM is shown in Fig.2. The left image shows the rotating mirrors (front) for the matching of the numerical aperture of the condenser system to the numerical aperture of the micro zone plate objective lens, and the TXM end-station at the back. The rotating mirrors and TXM end-station are mounted on independent granite support tables. The rotating mirror assembly can be moved on the granite to align a visible light microscope in the beam position.for sample pre-alignment. dose to it. The right image of Fig.2 shows the TXM end-station with the exit tube of the condenser system in the center and behind the sample support and the entrance window tube.

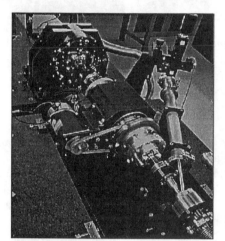

FIGURE 2. Left image: the TXM with rotating mirrors, visible light microscope for pre-adjustments and the end-station; right image: the TXM end-station

To determine the spatial resolution and stability of the TXM setup, a well characterized test structure was imaged. This 400 nm thick gold object consists of a variable line-spacing grating with line widths from about a micron down to 85 nm. The grating was imaged by a Au micro zone plate with an outermost zone width of 70 nm and a focal length of 15 mm at 4 keV. Both, test grating and micro zone plate were generated at the Institute for X-ray Physics at the University of Goettingen, Germany [6]. The beam was condensed onto a sample by a 1.2 mm diameter gold ZP generated by E.diFabrizio [7]. The measured photon flux in the sample illuminating spot is 3×10^8 ph/s/Si$\langle 111 \rangle$BW/200mA at 4 keV. The measured spot size (FWHM) is 65 μm in the horizontal and 20 μm in the vertical direction. The rotating mirror system was not used in these initial tests.. Fig.3 shows an X-ray image of the test grating acquired with an exposure time of 1 s. The field of view is 21.8 μm $\times$ 13.5μm. The smallest lines of 85 nm can clearly be resolved with good contrast. In order to test the stability of the setup, the same structure was imaged with an exposure time of 30 min without seeing any changes in the spatial resolution or contrast.

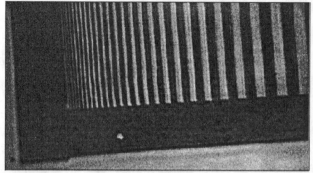

FIGURE 3. X-ray image of a Au test grating with a smallest line width of 85 nm, the exposure time was 1 s, the field of view is 21.8 μm×13.5μm. The 85nm structures can clearly be resolved with good contrast

ACKNOWLEDGEMENTS

The authors would like to thank L.Andre, R.Baker, G.Berruyer, A.Koch, F.Demarq, D.Fernandez, P.Noe, S.Ohlsson, D.Rolhion, M.Soulier, F.Thurel, and H.Witsch for their invaluable technical assistance and contributions during the construction and installation of the TXM at the ID21 beamline. The performance of this microscope is strongly correlated to recent developments of phase zone plates for keV X-rays by E.diFabrizio, M.Panitz and P.Charamboulos. The conception of the microscope has benefitted greatly from many hours of discussions with many collaborators, especially G.Schmahl, G.Schneider, T.Wilhein, and W.Meyer-Ilse.

REFERENCES

1. Niemann,B.; *in X-ray microscopy IV* eds.: V.V.Aristov and A.I.Erko, Bogorodski Pechatnik, Moscow (1994)
2. Meyer-Ilse,W.; *in X-ray microscopy and Spectromicroscopy* eds.: J.Thieme et.al, Springer Verlag (1998)
3. Oestreich,S., Kaulich,B., and Susini,J.; *Rev.Sci.Instr.*, Vol.70, **4** (April 1999)
4. Oestreich,S., Niemann,B., Rostaing,G., Kaulich,B., Barrett,R., and Susini,J.; *see contribution to this conference*
5. Susini,J., Barrett,R., Kaulich,B., Oestreich,S., and Salome,M., *see contribution to this conference*
6. Panitz,M., Schneider,G., Peuker,M., Hambach,D., Kaulich,B., Oestreich,S., Susini,J., Schmahl,G.; *see contribution to this conference*
7. di Fabrizio,E., Grella,L., Gentili,M., Peschiaroli,L., Mastrogiacomo,L., and Maggiora,R.; *J.Vac.Sci.Technol. B*, **14**(1998)
8. Kaulich,B., Wilhein,T., Oestreich,S., Salome,M., Barrett,R., and Susini,J.; *submitted to Appl.Phys.Letters*

Magnetization Imaging Using Scanning Transmission X-Ray Microscopy

J. B. Kortright,[1] S.-K. Kim,[1] H. Ohldag,[2] G. Meigs,[3] and A. Warwick[3]

[1]*Materials Sciences Division, Lawrence Berkeley National Laboratory, Berkeley, CA, USA*
[2]*Heinrich-Heine University, Duesseldorf, Germany*
[3]*Advanced Light Source, Lawrence Berkeley National Laboratory, Berkeley, CA, USA*

Abstract. A Faraday magneto-optical effect filter has been developed to convert linearly to elliptically polarized undulator radiation to provide magnetic circular dichroism (intensity) contrast from magnetic samples in an existing scanning transmission x-ray microscope. Strong magnetic contrast is observed across a 180° domain wall in a 33 nm thick demagnetized Fe film with in-plane magnetization, and clear structure is associated with the domain wall. By rotating the sample through large angles, magnetization components normal to the film are clearly observed and are associated with the domain wall structure, confirming that it is a cross-tie wall. Strengths and weaknesses of this approach compared to others are discussed.

INTRODUCTION

Large resonant element-specific magneto-optical (MO) effects, coupled with spatial resolution on the order of tens of nanometers make soft x-ray microscopy a potentially attractive tool for spatially resolved studies of magnetic materials.[1] Approaches to magnetic microscopy using soft x-rays fall into two categories, electron-based and photon-based. Electron-based approaches utilize incident photons to produce absorption contrast through magnetic circular dichroism (MCD), but images are obtained by focussing photo-emitted electrons.[2-4] Electron-based techniques are thus near-surface sensitive and are not generally compatible with strong and varying applied magnetic fields. The first photon-based magnetic imaging used an imaging transmission zone-plate microscope to demonstrate the ability to image magnetization reversal in applied fields with sensitivity to the entire thickness of a film having perpendicular magnetic anisotropy.[5,6] Here we report early results of magnetic imaging using a scanning transmission zone-plate microscope, and demonstrate additional strengths of photon-based techniques for soft x-ray magnetic studies.

EXPERIMENT

An existing scanning transmission x-ray microscope (STXM) on a linearly polarized undulator[7] was used to obtain images reported here. Two distinct magnetic contrast mechanisms were considered, and to date one has been implemented.

CP507, *X-Ray Microscopy: Proceedings of the Sixth International Conference,*
edited by W. Meyer-Ilse, T. Warwick, and D. Attwood
© 2000 American Institute of Physics 1-56396-926-2/00/$17.00

FIGURE 1. Incident linearly polarized light becomes elliptically polarized on transmission through the polarizing filter. Reversing the sign of $\mathbf{k} \cdot \mathbf{M}$ reverses the transmitted helicity. The zone plate lens focusses on the sample, that is raster scanned to produce an image using the transmitted intensity. The 7 mm working distance allows for sample rotations from normal to 45° incidence angle.

The large Faraday rotation of linear polarization near $L_{2,3}$ resonances of Fe, Co, and Ni present one clear approach to obtaining phase (or polarization) contrast, since tunable linear polarizers that sense Faraday rotation exit.[8-10] It was simpler, however, to use an existing polarimeter[11] to develop a circular polarizing transmission filter to convert linear to elliptical polarization.[12]

This polarizing filter consists of a soft magnetic film deposited onto a Si_3N_4 membrane with magnetization M saturated along the direction that maximizes $\mathbf{k} \cdot \mathbf{M}$, where $\mathbf{k}$ is the wavevector of the transmitted radiation (see Fig. 1). Incident linearly polarized radiation is a coherent superposition of equal amounts of + and − helicity circular polarization components. Thus by tuning to L_3 or L_2 peaks of magnetic circular dichroism (MCD) spectra, the differential absorption of these circular components yields an elliptically polarized transmitted beam. Reversing the direction of M reverses the helicity of the transmitted elliptical polarization. Results presented here were obtained at the Fe L_3 edge by mechanically rotating the filter, although in the future M will be reversed electromagnetically. The transmitted elliptical polarization is not pure circularly polarized, but has both circular and linear components; this may have consequences in interpreting some contrast features as seen below. The polarizing filter was upstream of the sample in a polarimeter capable of measuring transmitted polarization. However it could equally well be positioned downstream of the sample. This circular polarizing filter is the first optical *application* of resonant MO effects in the x-ray range. Transmission MO effects are generally referred to as Faraday effects, and this filter is thus a Faraday modulator in the x-ray range. It converts linear to elliptical polarization when tuned to the MCD peaks as used here. It also rotates linear polarization when tuned just off of the MCD peaks, according to Kramers-Kronig transformation relating the real and imaginary parts of the Faraday MO response.[8]

Initial magnetization images using the STXM and polarizing filter studied demagnetized Fe films, both because of expected strong contrast and because relatively much is known about magnetization structure in soft films such as these. Fe films with different thickness were sputtered through 0.4 mm by 0.4 mm stencil masks onto Si_3N_4 membranes yielding large patterned samples that were demagnetized in an ac magnetic field before introduction into the microscope. Figure 1 shows the experimental geometry. The spatial resolution of the zone plate focus was roughly 200 nm at the time that images below were obtained. This will improve to 30 nm or less with improvements in microscope stability and zone plates. The polarizing filter used for these measurements was 12 nm thick. Since maximum MCD contrast coincides with maximum absorption at the L_3 peak, roughly an order of magnitude is lost to absorption in both the polarizer and sample using this scheme.

RESULTS AND DISCUSSION

Normalization of transmitted intensity images to a quantitative absorption scale is a prerequisite to obtaining spatially resolved quantitative magnetization information. Four measured values are used in this normalization. The intensity transmitted for each helicity, $I_{+/-}$, is first normalized by an upstream intensity signal from the order sorting aperture, $I^0_{+/-}$, to account for incident intensity variations with time to yield the intensity transmitted trough the membrane and magnetic sample, $I_{+/-} / I^0_{+/-}$. The same quantity is measured in a region of the sample consisting of just the SiNx membrane with no Fe sample. Dividing the sample ratio by the membrane only ratio yields the transmission through just the magnetic film sample, $T_{+/-} = \exp[-\mu_{+/-} t]$ where t is its thickness and $\mu_{+/-}$ its absorption coefficient for opposite helicity elliptical polarization. Assuming that the absorption of opposite helicity circular components is proportional to $k \cdot M$, then the absorption model developed in ref. [12] for the polarizing filter can be extended to describe the absorption of opposite helicity circular components in both the filter and sample. The absorption of each helicity circular component in each film must be explicitly included, and the magnetization in the sample is allowed to vary in direction while that in the polarizer is assumed to lie along the saturation direction.

Strong magnetic contrast in a 33 nm thick demagnetized Fe film is seen in several images in Fig. 2. These images were taken at 45° incident angle in a plane horizontal to the image (see inset), and reveal large 180° domains, with average magnetization direction as noted by arrows, separated by a domain wall. All images are of the same 15 μ x 15 μ field. In this figure (a) and (b) show absorption for opposite helicity obtained as described above and then normalized by the average absorption to give $\mu_+ t / \langle \mu t \rangle$ and $\mu_- t / \langle \mu t \rangle$, respectively. Here $\langle \mu t \rangle$ is a single value giving the spatially-averaged and helicity-averaged absorption. (c) is the spatially-resolved average absorption normalized by this same spatially-averaged absorption, or $(\mu_+ t + \mu_- t) / \langle \mu t \rangle$. (d) is the difference in absorption between each helicity, plotted as $(\mu_+ t - \mu_- t) / \langle \mu t \rangle + 1$. These normalizations yield contrast about unity, as shown in the adjacent gray scales. Aside from these normalizations, boxcar averaging of each pixel with its four neighbors was applied to reduce noise.

Large contrast from 180° domains that reverses with helicity is observed in these images. The image contrast is consistent with that predicted using the absorption

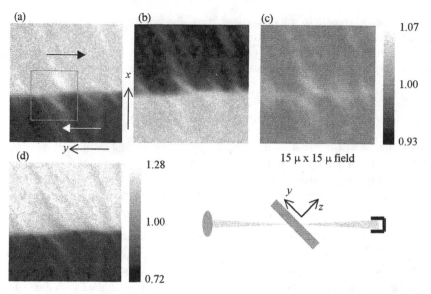

Figure 2. Images of magnetization across a 180° domain wall in a demagnetized 33 nm Fe film. Radiation was incident on the sample at 45° as noted. (a) and (b) are absorption images for opposite helicity elliptical polarization, (c) their sum, and (d) their difference. See text for normalization.

model mentioned above for two films. Additional features in the images include an offset in the domain wall by roughly 0.5 μ centered in the box in image 2 (a), and symmetric variations in absorption contrast associated with this offset that extend outward into the domains. Similar but weaker features are spaced roughly equally along the domain wall and extend into the adjacent domains. It is well known that soft magnetic films in this thickness range exhibit hybrid structure containing both Néel wall segments (in which M rotates mostly in-plane across the wall) and Bloch wall segments or lines (in which M rotates out-of-plane across the wall).[13] The features in Fig. 2 are thus possibly associated with hybrid domain wall structures, in particular cross-tie walls consisting of periodic Bloch lines resulting in a lowering of magnetostatic energy compared to a simple 180° Néel wall.

To further investigate the magnetization structure associated with the complex wall the sample was rotated to normal incidence, and images were collected of the same region as in Fig. 2. Resulting normal incidence images, normalized as above, are in Figure 3. Although contrast is much reduced, magnetic contrast is clearly visible especially at the wall region, with some weak features extending into the domains. The contrast in normal incidence images must come from regions having a net component of magnetization perpendicular to the film. A clear periodicity of about 4 μ is observed in the spacing of these regions of normal M, as is a somewhat regular shape that changes systematically at each feature with helicity. The absorption difference with helicity in Fig. 3(d) reveals an asymmetry in absorption above and below the wall, and thus a corresponding assymetry in the z component of magnetization component of magnetization across the wall. The sets of images in Figs. 2 and 3 are somewhat offset, and we believe that the feature highlighted in the square in each image is the same feature based on their high degree of spatial correlation.

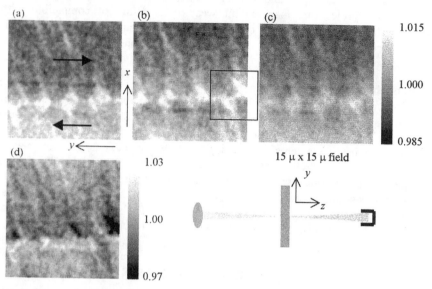

Figure 3. Images of magnetization across a 180° domain wall in a demagnetized 33 nm Fe film. Radiation was incident on the sample at 45° as noted. (a) and (b) are absorption images for opposite helicity elliptical polarization, (c) their sum, and (d) their difference. See text for normalization.

Weak contrast features are observed in the spatially resolved average absorption images in Figs. 2c and 3c. In each case, brighter features indicating higher absorption are associated with the structure at the domain wall. In a model were MCD absorption contrast is strictly proportional to $k \cdot M$, we would expect no contrast in polarization averaged absorption images. Possibilities that may explain this contrast include experimental artifacts such as asymmetry in the degree of circular polarization for opposite helicities. Another possibility is higher order MO effects such as linear dichroism.

General features associated with hybrid domain wall structures such as that observed here have been observed by several other techniques, including optical Kerr effect and Lorentz imaging. Likewise general models have been put forth to describe such structures.[13] Nonetheless there appears to be a gap between the ability of existing imaging techniques to provide quantitative, 3-dimensional vector magnetization information with high spatial resolution.

CONCLUSIONS AND FUTURE OUTLOOK

A Faraday circular polarizer provides a simple approach to obtaining strong magnetic contrast. While this simplicity is an advantage, there are also limitations to this approach. One limitation is that the spectral range over which magnetic imaging can be performed is limited to selected absorption lines of a few ferromagnetic elements. This both precludes full spectroscopy to gain chemical sensitivity and apply sum rule analysis for spin and orbital moments, and also imaging studies of magnetism in species that may exhibit weak magnetism not compatible with a polarizing filter. Other limitations come from possible complications resulting from impure polarization states incident on the sample. Photon-based microscopy on an elliptically polarizing undulator beamline will avoid these limitations in the future.

Results shown here clearly demonstrate the ability of photon-based soft x-ray magnetic imaging techniques to resolve 3-d magnetic structure over length scales from 200 (and less) nm and up. Other attributes of photon-based soft x-ray magnetic microscopy include imaging in applied fields, imaging deeply buried layers and interfaces, and using elemental specificity to study interactions between magnetic layers. While other techniques can image magnetic materials well, no other techniques share the same attributes as photon-based magnetic soft x-ray microscopy. These techinques thus appear to have great promise as a tool to quantify the spatial dependence of the complete magnetization vector in the study of many important problems involving small features, buried layers, and interacting systems.

ACKNOWLEDGEMENTS

This work was supported by the Director, Office of Science, Office of Basic Energy Sciences, Materials Sciences Division of the U.S. Department of Energy under Contract No. DE-AC03-76SF00098.

REFERENCES

1. Kortright, J. B., Awschalom, D. D., Stöhr, J., Bader, S. D., Idzerda, Y. U., Parkin, S. S., P., Schuller, Ivan K., and Siegmann, H.-C., "Research Frontiers in Magnetic Materials at Soft X-Ray Synchrotron Radiation Facilities," *J. Magn. Magn. Mater.*, in press.
2. Stöhr, J., Wu, Y., Hermsmeier, B. D., Samant, M. G., Harp, G. R., Koranda, S., Dunham, D. and Tonner, B. P., Science **259**, 658 (1993).
3. Schneider, C. M., *J. Magn. Magn. Mater.*, **156**, 94-98 (1996).
4. Hillebrecht, F. U., Spanke, D., Dresselhaus, J., and Solinus, V., *J. Electron Spectrosc.* **88**, 189-200 (1997).
5. Fischer, P., Shütz, G., Schmahl, G., Guttmann, P., and Raasch, D., *Z. Physik B* **101**, 313-316 (1996).
6. Fischer, P., Eimüller, T., Shütz, G., Guttmann, P., Schmahl, G., Pruegl, K., and Bayreuther, G., *J. Phys. D: Appl. Phys.* **31**, 649-655 (1998).
7. Warwick, T., et al., *Rev. Sci. Instrum.* **69**, 2964-2973 (1998).
8. Kortright, J. B., Rice, M., Carr, R., *Phys. Rev. B* **51**, 10240-10243 (1995).
9. Kortright, J. B., Rice, M., Kim, S.-K., Walton, C. C., and Warwick, T. *J. Magn. Magn. Mater.* **191**, 79-89 (1999).
10. Kortright, J. B., Rice, M. and Franck, K. D., *Rev. Sci. Instrum.* **66**, 1567-1569 (1995).
11. This polarimeter is conceptually similar to that described in ref. 10, but is designed to be an in-line device that radiation can pass through to downstream experiments.
12. Kortright, J. B., Kim, S.-K., Warwick, T. and Smith, N. V., *Appl. Phys. Letters* **71**, 1446-1448 (1997).
13. Hubert, A., and Schäfer, R., *Magnetic Domains: The Analysis of Magnetic Microstructures*, Berlin, Springer-Verlag, 1998, pp. 238-271.

Development of a Soft X-ray Dark-Field Imaging Microscope

Hidekazu Takano and Sadao Aoki

Institute of Applied Physics, University of Tsukuba, 1-1-1 Tennoudai, Tsukuba, Ibaraki 305-8573, Japan.

Abstract. A soft x-ray dark-field imaging microscope with Wolter-type mirrors was constructed in a laboratory scale. An annular aperture stop was used to cut off the undeviated beam, which satisfied the dark-field condition. The image was formed with only scattered x-rays from the object. It provides not only absorption but also phase shift information. Especially, the x-rays that have particular diffraction angle form the dark-field images in this system because Wolter mirror has a thin annular aperture. And the structures of 110~460nm in the object was emphasized for 5.2nm x-rays.

1. INTRODUCTION

Soft x-ray microscopy has been expected to be a promising analytical tool to study organic materials including biological specimens. Most of the soft x-ray microscopes use amplitude contrast to obtain an x-ray image within the "water-window" region. One of the other approaches to image an object of very low amplitude contrast is a dark-field (DF) x-ray microscope. It can detect very small objects or a slight difference of thickness by using scattering x-rays and it is very sensitive to high spatial frequency like an edge and a particle. Some DF microscopes were built with scanning mode (1,2). However, a DF signal is so small that it requires long exposure time. In order to overcome this disadvantage we constructed a soft x-ray DF imaging microscope by using a laser-produced-plasma x-ray source and Wolter-type mirrors. And typical soft x-ray DF images were obtained for the first time (3). Those images showed that contrast of the DF image was not so similar to that of the bright-field (BF) image, and some contrast enhancement was seen in the DF image.

In this paper we present a qualitative explanation of the DF image of a test pattern and an application to a biological specimen.

2. OPTICAL SYSTEM

A schematic diagram of the optical system is shown in Fig. 1. It consists of a laser-produced-plasma x-ray source (Nd-YAG laser, 1064nm, 8ns), a Wolter-type condenser and an objective mirror (4,5) and a back-illuminated CCD camera. An aluminum solid target was selected for the laser-produced-plasma that yielded several

CP507, *X-Ray Microscopy: Proceedings of the Sixth International Conference,*
edited by W. Meyer-Ilse, T. Warwick, and D. Attwood
© 2000 American Institute of Physics 1-56396-926-2/00/$17.00

line spectra near 5.2nm. In order to produce a DF image, two aperture stops, one of which was at the exit pupil of the condenser and the other was at the entrance pupil of the objective, were used (Fig. 2). Thus the undeviated beam could not enter the objective mirror. The effective field of view was about 280µm in diameter, which was determined by the distance between the specimen and the objective aperture. The BF image of the same field of view could be obtained by taking off the objective aperture stop.

In our system, the width of the annular aperture of the Wolter mirrors is restricted so small that the number of the diffraction orders of the deviated x-rays which enter the objective mirror is limited. As a result, contrast of some structure a with specific period is emphasized in the image. The angular apertures of the condenser and the objective are 137.3~164.3mrad, and 175.7~184.8mrad, respectively. So the x-rays whose diffraction angles are 11.4~47.5mrad, and 313.0~349.1mrad contribute to forming the DF image. In the case of 5.2nm x-rays (strongest line for Al plasma), the emphasized structure periods are about 16nm and 110~460nm.

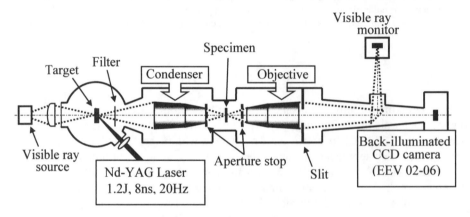

FIGURE 1. Schematic diagram of the soft x-ray microscope system.

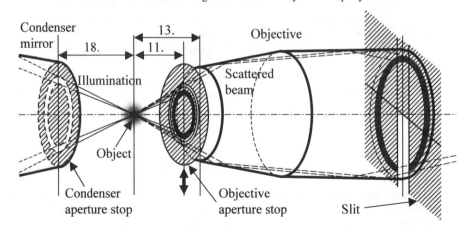

FIGURE 2. Arrangement of two apertures for the DF condition.

3. EXPERIMENT

The emphasis of the specified structure was investigated. A slit put at the exit of the objective mirror restricts the aperture in sagittal direction. Then only the beam deviated by the sample in meridional direction can enter the objective mirror. A test

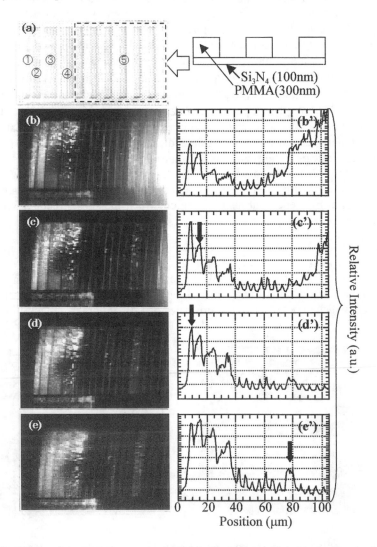

FIGURE 3. Images of the PMMA test sample and the intensity profiles: (a). An image taken by optical microscope (differential interference). The regions indicated by ①~⑤ include line and space patterns of 0.4, 0.6, 1.0, 2.0, 10.0μm pitch, respectively. (b)~(e). Soft x-ray DF images obtained by changing the collecting angle of the diffraction x-rays from the sample. Then the maximum spatial periods of (b), (c), (d), (e) are 610nm, 430nm, 340nm, 230nm, respectively. The images were taken with the laser power of 0.21J and the exposure was 20 shots. (b')~(e'). The intensity profiles that correspond to each image.

pattern shown in Fig. 3(a) was used. The regions indicated as ①~⑤ in the figure include line and space patterns of 0.4, 0.6, 1.0, 2.0, 10.0μm pitch. The effective diffraction orders of the scattered x-rays, which enter the objective mirror, were changed by moving the objective aperture stop (Fig. 2). Figures 3(b)~(e) show the DF images of the test sample when the minimum angles between the illumination and the deviated x-rays that passed through the objective aperture stop were 8.6, 12.0, 15.4, 22.2mrad, respectively. Figures 3(b')~(e') show the intensity profiles corresponding each image. Then the maximum spatial periods which contribute the images of Fig. 3(b)~(e) were 610nm, 430nm, 340nm, 230nm, respectively for 5.2nm x-ray. In Fig. 3(c), the intensity ratio of the 0.6μm period pattern, indicated by an arrow in Fig. 3(c'), to the 0.4μm period pattern decreases against Fig. 3(b). It is because that the diffracted beam at the 0.6μm period pattern can not enter the objective mirror in the DF condition of Fig. 3(c). The decrement in intensity for the 0.4μm period pattern in Fig. 3(d) is similarly explained. The region indicated by an arrow in Fig. 3(e') is also interesting. While only the edge brightens in Fig. 3(c), the inside brightens in Fig. 3(e). From the active diffraction angle in Fig. 3(e), it is conjectured that the region consists of 90~230nm structures. In the DF image, not only the edges or particles that consist of high spatial frequency but also the region which includes structures with specific periods is emphasized.

As an application of the DF microscope to the biological sample, epidermis of broad beans was observed. Figures 4(a) and 4(b) shows the BF and DF images. The DF image could be obtained with only 1 shot exposure. Reversed contrast can be seen between the BF and the DF images. And there are some emphasized structures in the DF image, which may be due to the structures of 110~460nm.

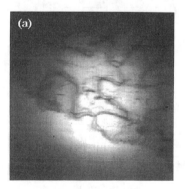 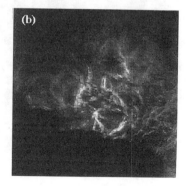

FIGURE 4. Soft x-ray images of epidermis of broad beans: (a). A BF image (Laser power: 0.17J, Exposure: 1shot). (b). A DF image (Laser power: 1.2J, Exposure: 1shot).

4. CONCLUSION

Contrast emphasis of the specific structure could be explained with the DF images of the periodic test pattern. The DF image of the biological specimen was obtained with 1 shot exposure. The spatial resolution in this system was restricted to 700nm by the pixel size of the CCD camera and the objective magnification. Better resolution will be possible by using a high spatial resolution detector (6). The DF image make it possible to observe organisms in a solution even outside the "water window" region, where the ratio of δ (the real part of the refractive index) to β (the imaginary part) becomes large. Generally, the shorter the x-ray wavelength is, the greater the contribution of δ becomes. That's why the DF method in the hard x-ray region will have much more advantages than in the soft x-ray region.

ACKNOWLEDGEMENTS

We are grateful to Nikon Inc. for the developments of the Wolter mirrors under collaboration. This work was partially supported by the Grants-in-Aid for Scientific Research Fund from Ministry of Education, Science, and Culture of Japan (No. 08405010).

REFERENCES

1. Morrison, G. R., and Browne, M. T., *Rev. Sci. Instrum* **63(1)**, 611- 614 (1992).

2. Chapman, H. N., Jacobsen, C., and Williams, S., *Ultramicroscopy* **62**, 191-213 (1996).

3. Takano, H., Yokota, K., and Aoki, S., submitted to *Jpn. J. Appl. Phys.*

4. Wolter, V. H., *Ann. Phys.* **10**, 94-112 (1952).

5. Onuki, T., Sugisaki, K., and Aoki, S., *SPIE* **1720**, 258-263 (1992).

6. Aoki, S., Ogata, T., Sudo, S., and Onuki, T., *Jpn. J. Appl. Phys.* **31**, 3477-3480 (1992).

X-ray Microscopy of Rubber Modified Poly(methyl methacrylate) Blends Produced by Cryogenic Mechanical Alloying

A. P. Smith, R. J. Spontak, C. C. Koch and H. Ade*

*Departments of Materials Science and Engineering and *Physics*
North Carolina State University, Raleigh NC 27695

Abstract. X-ray microscopy has been utilized to characterize the morphology of polymer blends produced by cryogenic mechanical alloying. It is found that mechanical alloying yields intimately mixed blends with the degree of mixing increasing with increasing milling time. In the absence of chemical reaction or physical compatibilization, intimate mixing is compromised when the blends are post-annealed or processed at high temperatures for extended times.

INTRODUCTION

As the need for and requirements placed on multifunctional polymeric materials continually rises, polymer blends and alloys have become increasingly utilized in polymer engineering applications.[1] One of the intrinsic problems with most polymer blends is their immiscibility, which complicates the production and retention of fine-scale mixing.[2] Although numerous strategies have been developed to increase miscibility of blend components, these often require addition of expensive modifiers or special processing protocols that may degrade the polymer or produce unwanted side effects.[3,4] A method of increasing the degree of mixing in polymer blends without the above complications is through solid-state blending, wherein the polymers are unable to flow and, therefore, phase-separate. To investigate the efficacy of this technique in preparing novel polymeric blends, we have utilized cryogenic mechanical alloying (high-energy ball milling of two or more components) to mix poly(methyl methacrylate) (PMMA) with poly(ethylene-*alt*-propylene) (PEP) in order to modify the properties of brittle PMMA.[5] Scanning transmission x-ray microscopy (STXM) has been utilized to study the blend phase morphology and its evolution during annealing. Results from this study are summarized here.

EXPERIMENTAL

Commercial PMMA ($\overline{M}_n$=255 kg/mol, $\overline{M}_w/\overline{M}_n$=4.07) was purchased from Aldrich, whereas the PEP ($\overline{M}_n$=63.4 kg/mol, $\overline{M}_w/\overline{M}_n$=1.06) was custom-synthesized by living anionic polymerization. Three grams of these polymers were mixed in a 25/75 w/w

CP507, *X-Ray Microscopy: Proceedings of the Sixth International Conference,*
edited by W. Meyer-Ilse, T. Warwick, and D. Attwood
© 2000 American Institute of Physics 1-56396-926-2/00/$17.00

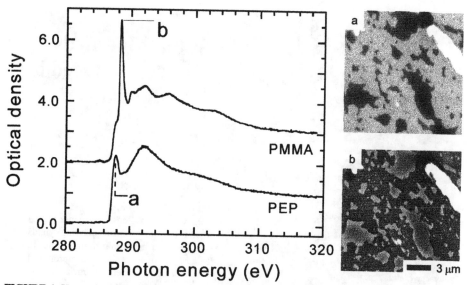

FIGURE 1. X-ray absorption spectra of PMMA and PEP. By acquiring images at 288.0 (a) and 288.4 eV (b), the spatial distribution of the blend components can be unambiguously identified as shown.

PEP/PMMA ratio within a hardened steel vial, and 30 g of ball bearings were added. The vial was sealed in high-purity Ar and subsequently placed in a SPEX 8000 mixer/mill and violently shaken at −180°C (so that both components are solids) for up to 10 hrs. Bearing-bearing and bearing-wall impacts serve to reduce the polymers to very fine, intimately mixed powders. These powders were consolidated into microscopy samples at temperatures up to 200°C for times up to 60 min. The solid samples were cryoultramicrotomed at −100°C to produce thin sections measuring approximately 100 nm thick. These thin sections were then examined in the Stony Brook STXM. Details of the microscope[6] and its use for polymer applications[7] are provided elsewhere in this volume.

RESULTS AND DISCUSSION

Figure 1 shows C K-edge x-ray absorption spectra obtained from PMMA and PEP showing distinct differences that can be exploited to obtain the morphology of these blends. The dominant feature in the PMMA spectra is the sharp peak at 288.4 eV due to excitation of the C=O carbon atom, while the PEP spectra has only a relatively broad peak at 288.0 eV. By imaging at these energies, micrographs that unambiguously show the phase morphology of the blends can be obtained, as demonstrated by the images in Fig. 1. These micrographs have been acquired at photon energies of 288.0 eV (a, PEP dark) and 288.4 eV (b, PMMA dark) from a sample milled for 5 hrs and consolidated at 125°C. These images clearly reveal the location of the two blend components, as well as a hole in the sample (white area in

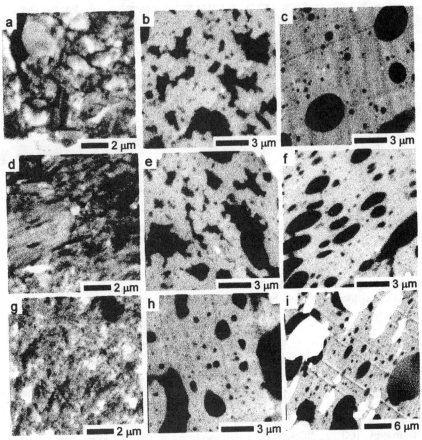

FIGURE 2. X-ray micrographs obtained at 288.0 eV from 25/75 PEP/PMMA blends milled for 1 (a-c), 5 (d-f) and 8 (g-i) hrs and consolidated at 25°C (a, d, g), 125°C (b, e, h) and 200°C (c, f, i) showing the initial morphologies of the blends and the effect of temperature on each morphology.

both images). Images subsequently presented here are acquired at 288.0 eV so that the PEP always appears dark.

One of the most basic questions regarding the use of mechanical alloying to generate novel polymeric blends is the initial morphology of the material and the effect of post-annealing on this morphology. Figure 2 shows x-ray micrographs obtained from blends alloyed for 1 (a-c), 5 (d-f) and 8 (g-i) hrs and consolidated at 25°C (a, d, g), 125°C (b, e, h) and 200°C (c, f, i) for 5 min. The samples compacted at 25°C have been pressed under high pressure and the PEP is presumed to have flowed around, and outlined, the immobile PMMA. This is evident in Fig. 2a where the dark PEP encapsulates PMMA particles that range in size from 300 nm to 3 μm. As the milling time increases, the PMMA domains become smaller until they consistently reach 200-300 nm for the sample milled for 8 hrs (Fig. 2g). This series of images indicates that the PMMA/PEP blends become more intimately mixed as the milling time is increased.

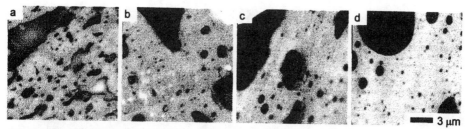

FIGURE 3. X-ray micrographs of a 25/75 PEP/PMMA blend milled for 8 hrs and annealed at 150°C for (a) 1, (b) 5, (c) 15 and (d) 60 min showing phase coarsening of the blends. Note the retention of the small PEP domains (200 nm) even after extended annealing.

Once the temperature of the blends becomes sufficiently high to allow mobility of the PMMA phase, the morphology dramatically changes, as also seen in Fig. 2. The PEP forms domains within a continuous PMMA matrix for the blends annealed at 125 and 200°C. For the samples consolidated at 125°C and milled for 1 or 5 hrs (Figs. 2b, 2e), the PEP domains range in size from 200 nm to 5 μm across and retain a jagged appearance. This feature suggests that the annealing temperature is not sufficiently high to permit significant reorganization of the polymers in such a short time. The same sample milled for 8 hrs (Fig. 2h), however, consists of domains that exhibit much smoother interfaces. Improved molecular mobility is consistent with the observation that the PMMA glass transition temperature decreases substantially with increasing milling time.[8] These micrographs demonstrate that the characteristics of the initial morphology is, for the most part, retained after annealing at 125°C. As the temperature is increased to 200°C, however, the additional chain mobility results in substantial changes in blend morphology.

The jaggedness of the PEP domains in the samples milled for 1 and 5 hrs (Figs. 2c, 2f) is completely lost, as the domains become more spherical in appearance. The sizes of the PEP dispersions do not change very much, ranging from 200 nm to 3 μm across in these blends. In the blend milled for 8 hrs (Fig. 2i), the size of the PEP domains increases dramatically up to 10 μm across, and the PEP flows from the thin sections. Recall that holes in the sections appear white in x-ray images. An increase in the size of the PEP domains indicates that sufficient molecular mobility exists for the inherent blend miscibility to drive continued phase separation in this blend. These images likewise demonstrate that, while the intimate morphology initially produced by mechanical alloying is retained at low temperatures and short milling times, annealing or further processing at elevated temperatures may promote significant phase separation within the blends.

To investigate the degree to which the blends continue to phase-separate, we have performed additional studies by annealing the blends at elevated temperatures for extended times. Figure 3 shows x-ray micrographs obtained from the sample milled for 8 hrs and annealed (a) 0 (quenched upon reaching the target temperature), (b) 5, (c) 15 and (d) 60 min at 150°C. The sample immediately quenched consists of PEP domains possessing a jagged interface, again suggesting that insufficient time has been allowed for significant reorganization of the blend. As the annealing time is increased,

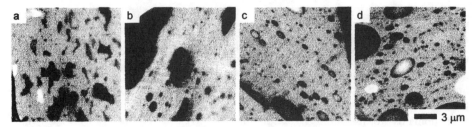

FIGURE 4. X-ray micrographs of a 25/75 PEP/PMMA blend cryomilled 8 hrs and annealed 15 min at (a) 125, (b) 150, (c) 175 and (d) 200°C. While these images show the fine-scale blend morphology, more representative micrographs of the large-scale morphology are included in Fig. 5.

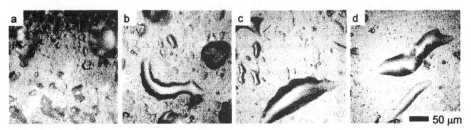

FIGURE 5. Optical micrographs acquired in reflection from the faced surface of the blends shown in Fig. 4 and illustrating the large-scale PEP domains existing within these blends.

the PEP domains become smoother and increase in size up to 10 μm across, revealing that phase coarsening can occur even at this relatively low temperature when long annealing times are employed. Results of annealing the blend milled for 8 hrs and annealed for 15 min at several different temperatures are displayed in Fig. 4. Here, x-ray micrographs collected at 288.0 eV are presented for samples annealed at (a) 125, (b) 150, (c) 175 and (d) 200°C. Again, at the lowest annealing temperature, the PEP domains exhibit a jagged appearance. As the temperature is increased, the domains become smoother and increase in size to some extent.

Of particular interest in these micrographs is the retention of the smallest PEP domains, even at the highest temperatures. The presence of these domains may indicate that the classic mechanism of Ostwald ripening, in which large particles grow at the expense of small particles, is not solely responsible for the observed phase evolution. It is important to recognize that these micrographs show the fine scale morphology. They are, however, somewhat misleading in that the large-scale morphology is ignored due to the fragility of the thin sections. A more representative picture of the blend morphology is provided by optical micrographs obtained in reflection of the faced bulk films after microtoming. Figure 5 shows representative optical micrographs of the blend milled for 8 hrs and annealed according to the same conditions listed in Fig. 4. At 125°C, the PEP domains measure less than about 25 μm across. As the temperature is increased, however, the PEP domains grow significantly and, in some cases, exceed 200 μm across. These images demonstrate that considerable phase coarsening occurs within these blends when they are annealed at elevated temperatures for extended times.

CONCLUSIONS

The chemical sensitivity of x-ray microscopy has been utilized in the present study to ascertain the phase morphology of PEP/PMMA blends prepared by cryogenic mechanical alloying. Results obtained here indicate that mechanical alloying produces intimately mixed blends with the degree of mixing increasing with increasing milling time. For blends post-annealed or processed at temperatures above the PMMA T_g, PEP domains form within a continuous PMMA matrix and possess a jagged appearance at low temperatures. These domains become smoother with increasing temperature. An increase in either temperature under isochronal conditions or milling time under isothermal conditions induces a marked increase in the size of the PEP domains. In the limit of high annealing temperature and long milling time, PEP domains are observed to exceed 200 μm in size. While mechanical alloying constitutes a viable means of producing intimately mixed polymer blends, such intimate mixing may be sorely compromised when, in the absence of chemical reaction or physical compatibilization, blends are post-annealed or processed at elevated temperatures for extended times.

ACKNOWLEDGEMENTS

We are indebted to Dr. S.D. Smith for providing the PEP used in this study. X-ray microscopy data were collected using the Stony Brook STXM instrument developed by the group of J. Kirz and C. Jacobsen at SUNY Stony Brook, with financial support from the Office of Biological and Environmental Research, DOE, under contract DE-FG02-89ER60858, and NSF under grant DBI-960-5045. S. Spector and C. Jacobsen of SUNY Stony Brook and D. Tennant of Lucent Technologies developed the zone plates with support from NSF under grant ECS-951-0499. NSF Young Investigator Award (DMR-945-8060) supports H.A. and A.P.S. We thank Mr. D.A. Winesett and Dr. S.G. Urquhart for technical assistance.

REFERENCES

1. Utracki, L. A., *Polymer Alloys and Blends* Hanser, Berlin, 1990.

2. Sperling, L. H., *Polymeric Multicomponent Materials* John Wiley & Sons, New York, 1997.

3. Macosko, C. W., Guegan, P., Khandpur, A. K., *et. al., Macromolecules* **29**, 5590 (1996).

4. Bates, F. S., *Science* **251**, 898 (1991).

5. Smith, A. P., Spontak, R. J., Ade, H. *et al., Adv. Mater.* **11**, 1277 (1999).

6. Jacobsen, C., these proceedings.

7. Ade, H., these proceedings.

8. Smith, A. P., Shay, J. S., Spontak, R. J., *et al., Polymer*, submitted.

MULTI-KEV X-RAY MICROSCOPY

X-Ray Microtomography (μCT) Using Interferometric Phase Contrast

Ulrich Bonse[a], Felix Beckmann[b], and Theodor Biermann[a]

[a]Institute of Physics, University of Dortmund, Germany
[b]Hamburger Synchrotronstrahlunglabor (HASYLAB) at Deutsches Elektronen-Synchrotron (DESY), Hamburg, Germany

Abstract. Phase-contrast tomography is capable of characterizing organic tissues, e.g. rat trigenimal nerve or brain matter by accurate measurement of mass density ρ_m. The presently attained accuracy is of the order of 10^{-3} relative. Aspects of importance for reaching this accuracy are discussed. We found an influence on the phase-measuring interferometer through heating caused by absorption of x-ray photons from the synchrotron radiation beam. Means to avoid this source of error are described.

STATE OF THE ART OF μCT

X-ray computerized tomography (CT) is three-dimensional (3D) imaging using x-rays. Image contrast is brought about through local variations of the x-ray absorption coefficient μ within the specimen. In medicine this method is widely used employing hard x-rays generated by standard sealed-off x-ray tubes. The in-vivo spatial resolution achieved is of the order of 150 to 200 μm. In materials research, where the radiation dose delivered to the specimen is usually not limiting the exposure time, two to three times better spatial resolution is state of the art.

When synchrotron radiation (SR) with five to ten orders of magnitude higher brilliance than radiation obtainable from x-ray tubes became available the extension of CT to microtomography (μCT) occurred quite fast - at least in the materials research sector. Spatial resolutions of 1 to 5 μm [1] were soon achieved. With hard x-ray synchrotron-radiation μCT is normally performed in the parallel-projection mode, i.e. the specimen is parallel-projected to an x-ray area detector with no *x-ray optical* magnification. Therefore, spatial resolution depends mainly on the area detector's spatial resolution which for a while stalled at 1 to 2 μm. However, quite promising detectors with sub-micrometer resolution are presently under development [2]. At the same time, with increasing resolution, the sample size shifted from the meter regime down to the mm regime, since, with a given size of the processing computer, the total number of voxels containing independent information remains the same. Decreasing sample and voxel size, however, entails less absorption contrast. This is especially the case with specimens from Biology and Medicine which usually consist to their greater part or even completely of light elements.

CP507, *X-Ray Microscopy: Proceedings of the Sixth International Conference,*
edited by W. Meyer-Ilse, T. Warwick, and D. Attwood
© 2000 American Institute of Physics 1-56396-926-2/00/$17.00

It did not take long for physicists to realize that the *x-ray phase shift* inflicted by the specimen is much better suited for imaging structures consisting of light elements than the structures' *absorption*. The essential reason for this is that phase shift varies proportional ρ_e, the electron charge density. ρ_e is itself proportional to the order number Z and the number density N of atoms. Contrary to this, absorption (except near absorption edges) varies roughly proportional $(\rho_e)^p$ with $3 \leq p \leq 4$. Hence, in the regime of small ρ_e values, i.e. light elements, the $\sim \rho_e$ relationship maintains good detectability of small ρ_e differences whereas $\sim (\rho_e)^p$ does not. Another strength of phase contrast is that in the case of a specimen with a large span of different ρ_e values (e.g. a tungsten wire covered by a layer of strontium oxide) the contrast of either component stays within the dynamic range of the detector and thus remains manageable whereas with absorption contrast this is not so.

But how to achieve x-ray phase contrast? One of two methods that have been developed is employing x-ray interferometry [3]. The other is making direct use of the coherence properties of SR [4]. Here the interferometric technique [5-8] will briefly be described. Besides yielding detailed images of the specimen it provides the essential ability of normal CT, i.e. to map as function of position x,y,z quantitatively the *absorption properties* $\mu(x,y,z)$, now also for mapping $\rho_e(x,y,z)$.

Quantitative Phase Measurement With SR

A typical beamline well suited to phase-contrast µCT (p-µCT) is BW2 of HASYLAB at DESY in Hamburg. SR generated by the high intensity wiggler of BW2 with 56 magnetic poles is monochromatized by a silicon (111) double crystal monochromator. The special feature of BW2 is *high brightness* (5×10^{10} ph s^{-1} mm^{-2} in the beam behind the monochromator at 20 keV) *combined with a relatively large beam cross section* (e.g. 10 mm and more in the horizontal by 3 mm in the vertical direction) to permit tomographic parallel projections of *samples of a reasonable size*.

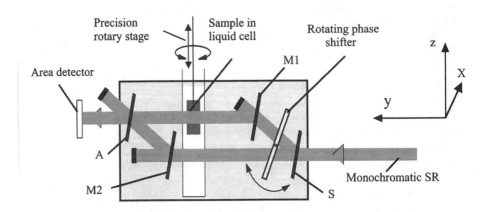

FIGURE 1. Interferometer part of the beamline when set for p-µCT. The rotary stage is for taking projections at e.g. 0.5 degree angular increments and for moving the specimen in and out of the beam.

The outlay and function of the x-ray interferometer is illustrated in Fig. 1. Four crystal wafers, S (splitter), M1, M2 (mirrors), and A (analyzer), all parts of a monolithic single crystal of silicon, split the incoming monochromatiç x-ray beam into two coherent beams and recombine them again by perfect-crystal diffraction. One beam is passing the specimen and the other is serving as reference beam The resulting interference pattern is converted to visible light by means of a fluorescent screen and then imaged to a CCD using standard light optics. It may be noted that, in order to make best use of the available photon flux, the interferometer is oriented like the monochromator to diffract in the vertical plane. Interferometric phase-shift determination is based on recording for every projection angle two sets of each at least three interference patterns with overall phase differences of 0, $2\pi/3$, $4\pi/3$, respectively, between the interfering beams of the interferometer [7]. One set is recorded with and the other without the specimen in the beam, the latter set serving as reference. P-μCT is measuring the *electronic charge* density ρ_e of a specimen. If performed carefully changes of the order of 10^{-3} ρ_e or even below can be detected. Assuming the average specimen to have about an equal number of neutrons and protons or a known ratio between them, one can relate ρ_e to the *mass* density ρ_m in an unique way. The capability of p-μCT to characterize specimens precisely by their density variations has already been used in studies of cancerous tissues [8,9].

Density Measurement Of Nerve Tissue With P-μCT

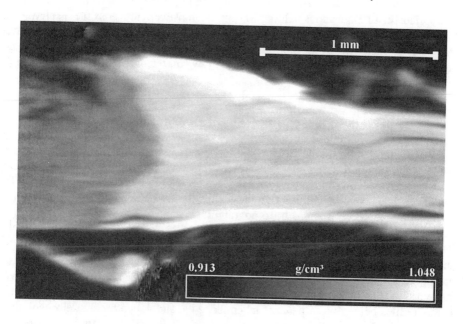

FIGURE 2. Tomographic section of rat trigeminal nerve. Section is 5.4 μm thick. 12 keV.

71

Fig. 2 gives an example for quantitative measurement of mass density ρ_m by p-μCT. Shown is a tomographic section along the axis of the intracranial part of a rat trigeminal nerve studied for revealing the early stages of chemically induced tumor growth in the peripheral nerve system [9]. Values of ρ_m ranging from 0.913 to 1.048 g/cm³ have been ascribed to gray values as indicated by the gray-shade bar at the bottom. The less dense part to the left is brain and the denser to the right is nerve tissue. Because of their slightly larger ρ_m, several bundles of nerve fibers following the horizontal nerve axis all across the nerve/brain interface are discernable. Cancerous tissue has also larger density. In principal, by comparing interface near tissues from rats that were exposed to the carcinogen with tissues of not exposed rats, locations of early tumor growth can be identified. The ρ_m differences detected were about 5 mg/cm³ (14 mg/cm³) for brain tissue (adjacent nerve tissue), respectively, indicating that the tumor growth initiates in the nerve tissue close to the nerve/brain interface [9]. Since the ρ_m changes are so small it is essential for the p-μCT density measurement to be very accurate.

To achieve the claimed sensitivity and accuracy, excellent stability and reproducibility of interference patterns is mandatory. Typical causes of instabilities are mechanical vibrations or drifts and thermal fluctuations of .the four wafers of the interferometer with respect to each other. However, mechanical shifts or temperature changes *equal* and *common to all wafers* have - to first order - no influence on interference patterns. Therefore, thermal and mechanical disturbances acting on the interferometer from the *surrounding* have found to be excludable by providing appropriate insulation and shielding. There is, however, the possible experimental error caused by internal heating of interferometer wafers through photons absorbed from the diffracted x-ray beams during exposure. We have studied this effect and describe in the following its magnitude and how the error resulting from it can be minimized.

Influence Of Heating Caused By Photon Absorption

The first indication that heating of the interferometer wafers really does occur was found when taking a series of exposures of the same projections at different 'delay times' after the x-ray shutter was opened. A typical example is shown in Fig. 3. The left image was taken 3 s and the right 45 s after shutter opening at 20 keV. A carbon-fiber reinforced epoxy specimen (tomographic axis vertical) covers four fifths of the right part of each frame. The remaining fifth on the left displays the empty beam intensity profile. Inside the specimen, at the lower left some differences between left and right interference pattern can immediately be seen. For a quantitative comparison profile scans were performed in locations indicated by black lines. The scan with contour 6 in Fig. 3 runs from the empty beam region up the left edge of the specimen. Three profiles belonging to patterns taken 3 s, 24 s, and 45 s after shutter opening are shown in Fig. 4. By including also profiles taken at intermediate times we find that contrast (Fig. 5) and spacing of fringes (Fig. 6) increase systematically with delay time. The largest changes occur within the first 5 s. After this we see a kind of exponential decay of profile differences and contrast improvement until about 45 s to 60 s when the patterns are found to be stable again. As a first practical consequence we gave up closing the x-ray shutter between exposures, although it would reduce the radiation dose applied to the specimen.

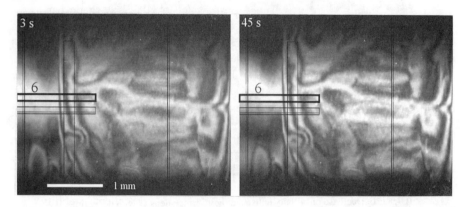

FIGURE 3. Same projection, however taken 3s and 45s after shutter opening, respectively. 20 keV.

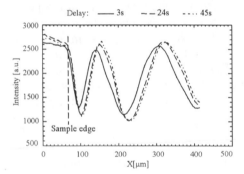

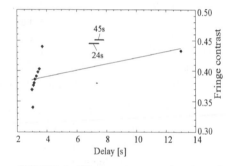

FIGURE 4. Measured fringes of projections shown in Fig. 3. Note relative shift of profiles as function of delay time.

FIGURE 5. Fringe contrast as function of delay. Contrast increases when shutter was longer open before exposure is taken.

The observations can be interpreted in the following way: Before the shutter opens the temperatures of wafers S, M1, M2, and A are at equilibrium (and probably equal to within a few 10^{-2} K or even less). On shutter opening, commencing photon absorption causes wafer temperatures to rise. However, since different wafers see different x-ray intensities and spectral windows the rate of temperature changes cannot be the same for all wafers, and it takes some time until a new equilibrium is reached. Wafers S and M2 absorb from the full spectral band which is offered by the beamline monochromator, whereas wafers M1 and A see only the much narrower band which has already been reflected once by S and M2, respectively. Furthermore, wafers sitting in second, third or fourth position must absorb less power than wafers in positions in front of them. Changing temperature differences mean a changing overall built-in moiré pattern in the interferometer, and that is what we observe.

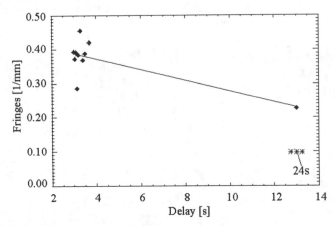

FIGURE 6. Decrease of initially enhanced moiré fringe density with delay time.

Assuming a steady flux of photons a new equilibrium distribution of wafer temperatures should establish itself again when the time elapsed after the shutter was opened has become long enough. However, a stable moiré cancels from the phase measurement because projections are corrected with the empty-beam reference projection before entering the evaluation procedure. For accurate phase measurement it is therefore necessary (and sufficient!) to insure thermal *equilibrium* of the interferometer over the duration of the pair of exposures with and without the sample in the beam, i.e. typically for 30 s to 60 s. It is *not* necessary that all wafers work at the *same* temperature nor that the built-in moiré is the same for different exposure pairs.

The amount of temperature changes caused by photon absorption can be estimated as follows: Denoting by Λ the *spacing* of moiré fringes we find from Fig. 6 the measured temperature induced increase of *spatial frequency* Λ^{-1} *of fringes* to be $\Lambda^{-1} \approx 0.3$ mm^{-1}. For changes of Bragg plane spacing $\Delta d \ll d$ we may use the simple relationship $d^2/\Delta d = \Lambda = 1/0.3$ mm from which $\Delta d/d = 0.3$ mm$^{-1} \times d$ [mm] is calculated. For the Silicon 220 reflection we have $d \approx 1.9 \times 10^{-7}$ mm so that $\Delta d/d \approx 5.7 \times 10^{-8}$ is the temperature induced relative change of the Bragg plane spacing. Using the thermal expansion coefficient of Silicon $\alpha \approx 3 \times 10^{-6}$ K^{-1} we can calculate from $\Delta d/d$ the temperature difference of wafers caused by photon absorption. We find it to be of the order of $\Delta T \approx 5.7 \times 10^{-8} / 3 \times 10^{-6} \approx 0.02$ K. This number can be checked from the other side, i.e. from the absorbed photon power in combination with the known geometry of the interferometer. According to the brightness of SR of $\approx 5 \times 10^{10}$ ph s^{-1} mm^{-2} at 20 keV after the monochromator, a typical beam 8 mm wide and 3 mm high carries $\approx 1.2 \times 10^{12}$ ph/s. Using the normal absorption coefficient for Silicon at 20 keV, $\mu = 1.014$ mm^{-1}, we calculate a wafer 0.5 mm thick to absorb $1.2 \times 10^{12} \times \{1 - \exp(-0.507)\} \approx 5 \times 10^{11}$ ph/s of 20 keV energy each. This corresponds to an absorbed power $P \approx 5 \times 10^{11} \times 20 \times 10^3 \times 1.602 \times 10^{-19}$ W ≈ 1.6 mW. We assume the wafer to conduct P across its cross section $A \approx 8$ mm$\times$0.5 mm$^- = 4$ mm^2 to the massive silicon basis of the interferometer. If the wafer is hit at a typical height above the basis of h ≈ 4 mm, the path length L over which the heat is to be conducted $L \approx h = 4$ mm. Applying standard formalism of heat conduction we calculate the resulting equilibrium temperature raise ΔT of the wafer as $\Delta T = P\, h\, A^{-1}\, \lambda^{-1}$, where $\lambda \approx 157$ W m^{-1} K^{-1} is the heat conductivity of Silicon. With these numbers we calculate $\Delta T \approx 0.01$ K which is reasonably close to the estimate made above on the basis of the heat-induced lattice change measurable shortly after shutter opening when equilibrium of the 'new' wafer temperatures has not yet been reached again.

We expect that by modifying the exact shape of wafers and base of the interferometer monolith it will be possible to avoid transient temperature differences resulting from photon heating altogether. This will help to improve the accuracy and sensitivity of x-ray interferometric densitometry and its applications.

ACKNOWLEDGEMENTS

We like to thank Prof. M.F. Rajewsky, Institute of Cell Biology (Cancer Research) for providing the specimen shown in Fig. 2. We also thank W. Drube and H. Schulte-Schrepping, HASYLAB at DESY, Hamburg, for help when using beamline BW2. Financial support by the Ministry of Science and Research of the Land NRW, Düsseldorf, is also gratefully acknowledged.

REFERENCES

1. Bonse, U., Nußhardt, R., Busch, F., Pahl, R., Kinney, J.H., Johnson, Q.C., Saroyan, R.A. , and Nichols, M.C., *J. Materials Science* **26**, 4076 - 4085 (1991).

2. Koch, A., Raven, C., Spanne, P., and Snigirev, A., *J. Opt. Soc. A.* Vol. **15**, 1940-1951(1998).

3. Bonse, U., and Hart, M., *Appl. Phys. Let.* **6**, 155-156 (1965).

4. Cloetens, P., Ludwig, W., Van Dyck, D., Guigay, J.P., Schlenker, M., and Baruchel, J., *Proc. SPIE* **3772**, in press (1999).

5. Beckmann, F., Bonse, U., Busch, F., Günnewig, O., and Biermann, T., *HASYLAB Annual Report, Part 2*, 691-692 (1995).

6. Momose, A., *Rev. Sci. Instrum.*, **66**, 622-628 (1995).

7. Beckmann, F., Bonse, U., Busch, F., and Günnewig, O., *J. Comput. Assist. Tomogr.* **21**, 539-553 (1997).

8. Momose, A., Takeda, T., Itai, Y., and Hirano, K., *Nature Medicine* **2**, 473-475 (1996).

9. Beckmann, F., Heise, K., Kölsch, B., Bonse, U., Rajewsky, M.F., Bartscher, M., and Biermann, T., *Biophysical Journal*, **76**, 98-102 (1999).

Coherent High Energy X-ray Optics for Imaging, Diffraction and Spectroscopy

Irina Snigireva and Anatoly Snigirev

European Synchrotron Radiation Facility (ESRF), B. P. 220, 38043 Grenoble, France

Abstract. The present status of X-ray optics in high-energy domain, an area in which much progress has been made in recent years, is reviewed with respect to lateral resolution, flux, imaging capability. The latest results on development of new focusing devices such as compound refractive lenses are shown. Hard X-ray microscopy techniques realized at the ESRF undulator beamline ID22 are presented. Some recent application in imaging, spectroscopy and microdiffraction are briefly described.

INTRODUCTION

Third-generation synchrotron radiation sources such as the European Synchrotron Radiation Facility (ESRF) offer very intense ($\sim 10^{13}$ ph/sec/Si-111 BW) and highly collimated (divergence ~ 10 μrad) X-ray beams in the energy range of 6-100 keV. The really new feature of the 3rd generation SR sources is high spatial coherence, which results from a very small source size (S $\sim$ 25 microns) and from large source-to-object distance (R $\sim$ 50 m) [1]. With such a beam, high-resolution optics is spatially coherently illuminated and forms a diffraction limited X-ray spot. These advances give rise to tremendous progress in the development of microfocusing optics, and more fine optics like Bragg-Fresnel (BFO) [2-6], Fresnel zone plates (FZP) [7,8] and compound refractive lenses (CRL) [9-15] is now widely applied. Phase contrast imaging and holography are other remarkable achievements offered by coherent hard X-rays [16-23]. Passing through the sample the laser-like X-ray beam with sub-microradian angular spread allows the recording of both the amplitude and phase of the scattered waves via the interference between coherent background (direct beam) and diffracted beam. In parallel with the improvement of synchrotron X-ray sources and optics, X-ray detectors with submicrometer spatial resolution have been developed. Taken together these components have made possible the convergence of imaging, diffraction and spectroscopy within X-ray microscopy as in the domain of high-energy.

COHERENT HIGH ENERGY X-RAY OPTICS

We have collected hard X-ray microfocusing devices in the Table 1. Practically all mentioned optical systems have overcome the micrometer barrier and reached a sub-

CP507, *X-Ray Microscopy: Proceedings of the Sixth International Conference,*
edited by W. Meyer-Ilse, T. Warwick, and D. Attwood
© 2000 American Institute of Physics 1-56396-926-2/00/$17.00

micrometer resolution. Taking into account coherent properties of the X-ray beam, the coherence preservation is precisely, an essential feature, which is required of the focusing optics. This is because the coherent optics like Fresnel zone plates (FZP), Bragg Fresnel optics (BFO) and compound refractive lenses (CRL) are the most

TABLE 1. Focusing optics for hard X-rays (6 < E < 100 keV).

	Bent mirror or crystal			Capillaries	Waveguides	Fresnel optics	Bragg-Fresnel Optics		Refractive optics
				Kreger 1948	Feng et al 1993	Baez 1952	Aristov et al, 1986 [2]		Snigirev et al, 1996 [9]
	mirror	multilayer	crystal				ML	crystal	
	Kirkpatrick Baez, 1948	Underwood Barbee, 1986	Johann, Johansson, 1931-1933						
accept	~1 mm	~1 mm	1 mm	0.1 mm	~10 µm	~1 mm	~1 mm		~1 mm
gain	10^4	10^4	10^4	10^2	<10	10^4	10^4		10^2-10^3
E	<20keV	<80keV	<100keV	<20keV	<20keV	<20keV	<100keV		<100keV
ΔE/E	w. b.	10^{-2}	10^{-4}-10^{-6}	10^{-3}	10^{-3}	10^{-3}	10^{-2}	10^{-4}-10^{-6}	10^{-2}-10^{-3}
flux(ph/s)	10^8 (9keV)	10^{10} (12keV)	10^9 (90keV)	10^9 (8keV)	10^9 (10keV)	10^9 (9keV)	10^9 (10keV)	10^9 (10keV)	10^9 (20keV)
resol.	0.7 µm [24]	0.7 µm [25]	1.2 µm [26]	0.05 µm [27]	0.15 µm [28]	0.15 µm [7]	0.9 µm [29]	0.5 µm [4]	0.3 µm [12]

interesting. It is evident from creating principles, that BFO is an ideal coherent optics. A linear Bragg Fresnel lens (Fig. 1) is acting as a focusing monochromator producing

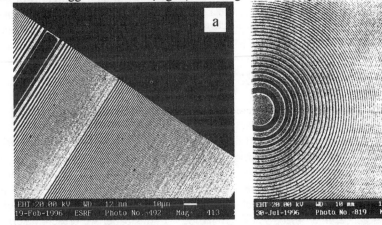

FIGURE 1. SEM images of linear BFL composed of 1st and 3rd order of diffraction (a) and FZP widely used at ID22 provided by IESS, CNR, Rome, Italy, Dr. M. Gentilli and Dr. E. Di Fabrizio) (b).

a cylindrical wave front therefore, BFO is widely used for high-resolution diffraction studies [5,6]. The FZP and CRL are used in transmission geometry that considerably simplifies the alignment of both optics and sample, and location an area of interest in the sample. Significant progress was achieved in development of FZP for the last years, nanometer focusing was demonstrated [7,8]. Apart of microprobe applications FZP is very promising device for Fourier transform holography [30].

Compound refractive optics made from low-Z materials have been proposed and already applied for focusing high energy X-rays [9-11]. Cylindrical CRLs are installed for beam conditioning in the front-ends of some beamlines at the ESRF and other facilities [31]. However it is difficult to jump with cylindrical CRLs to sub-μm focusing while they suffer from spherical aberration. We produced parabolic CRLs in polycrystalline aluminium (Fig. 2) by pressing technique [12-15]. They are genuine imaging devices, similar to glass lenses for visible light. The lenses focus in two directions. They are virtually free of spherical aberration. They are very robust, easy to align and they can be used in an energy range from 6 keV to 100 keV. They do not deteriorate in the white beam of an ESRF undulator. A spot size of 0.3 μm at 20keV has been measured. the ID22 undulator beamline. The gain of the Al lenses varies between 50 and 100 [12-15].

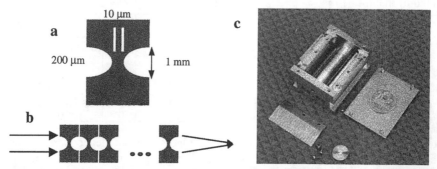

FIGURE 2. Schematic sketch of a parabolic compound refractive lens. The individual lenses (a) are stacked behind each other to form a compound refractive lens (b) in the specially design holder (c).

HIGH-ENERGY X-RAY MICROSCOPY.

Hard X-ray microscopy can not compete with electron and soft X-ray microscopy in terms of resolution, but rather offers unique advantages including the ability to image thick samples in natural environment. It demand little or no specimen preparation, and can be used to observed local composition, chemical state and structure High-energy microscopy is realized at undulator ID22 Micro-Fluorescence, Imaging and Diffraction (μ-FID) beamline at the ESRF [32]. Applications of high-energy microscopy are in several fields including biology, medicine, material science, environmental science, microelectronics and geology. In this paper we present a series of examples to indicate the nature and flavor of the type of experiments performed at the beamline

Phase Contrast Imaging and Tomography with Coherent Illumination

Using the coherent properties of the ESRF beams it is possible to realize a phase contrast imaging directly from the sample in the transmission geometry [16-21]. The technique is closely similar to in-line holography, where detector records the interference between roughly spherical waves scattered by the sample with plane reference waves coming directly from the illuminating source. Development of high-resolution on-line detectors [19] has made possible to combine the phase contrast imaging with computed microtomography (CMT) and reconstruction strategies. A drastically improved image contrast, as compared to absorption CMT (Figure 3), was obtained when imaging a wet human coronary artery specimen. In the tomograms previously invisible detail could be visualized with absorbed doses below the level where radiation damage impedes the imaging. The results indicate a considerable potential of the in-line holographic CMT method in biomedical microscopy.

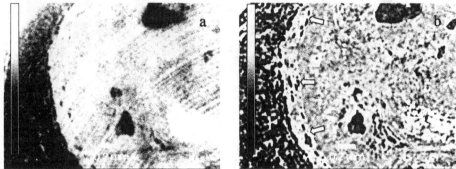

FIGURE 3. Absorption (a) and phase contrast (b) tomographic sections through a coronary artery. The phase contrast image demonstrates that the inner surface of the blood vessel wall is delineated as a thin line. The imaging was performed with 28 keV X-rays and a sample-to-detector distance of 66.5 cm. The image is 1.9 mm wide.

The resolution of the phase contrast imaging is limited by the detector (0.7 μm [19]). Future trends to improve resolution will be determined by magnification with X-ray optics [12,13]. To increase the flux in intensity limited techniques, an undulator beam reflected on a mirror ("pink" beam) can be used instead of monochromatic beam. Despite of slight beam coherence degradation, phase contrast imaging is still possible. Intensity gain by factor 100 opens applications of "pink" beam for hard X-ray microscopy and time resolved imaging [33]. First experiments for in situ studies of dendritic grows in PbSn alloys was successfully performed [34] with the fast readout FreLoN CCD camera [35]. Quantitative phase contrast imaging in holographic mode and non-interferometric phase retrieval technique is under development [21].

Phase contrast imaging can be developed from transmission to reflection geometry including diffraction topography [22]. Diffraction topography (Bragg geometry) with coherent illumination was successfully applied for phase mapping of the domain structure of a ferroelectric material like lithium niobate [18]. It was shown that sensitivity of phase contrast diffraction topography could be enhanced by varying the object-to-detector distance. In addition, the exit wave information lost in the vicinity of the sample position can be retrieved from number of images at various distances and it should be possible to solve the inverse problem of internal structure determination [23].

Full Field Microscopy with Coherent Illumination

To achieve submicrometer resolution in imaging mode one has to magnify the image prior registration using BFO [4], FZP or parabolic CRLs. The sketch of the CRL based X-ray microscope is shown in Fig. 4. At present, it is possible to image an area of

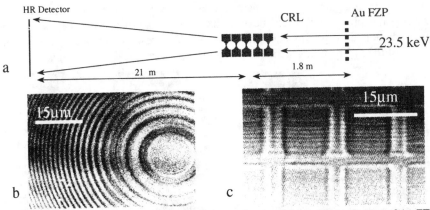

FIGURE 4. Experimental setup for CRL based X-ray microscope (a) and X-ray Images of Au-FZP (b) and Au-grid (c). Images were recorded at 23.5 keV. Fine structures 0.3 μm size is clearly visible.

about 300 μm in diameter with resolving power of 0.3μm [12]. The microscope operates in the energy range up to 50 keV, allowing for magnifications up to 50. Images of biological and soil samples were taken [14]. The X-ray microscope is obviously compatible with tomographic imaging.

Microspectroscopy

X-ray microspectroscopy can provide information on the spatial distribution, oxidation state chemical environment of trace elements. Hard X-rays open access to the K-edges of heavy elemnts (Z>14), the fluorescence yield is significantly greater, and hard X-rays suffer less attenuation along the optical pass to and from the sample [36]. Moreover, the sample preparation is far simpler and working in the air makes possible to study the samples in their natural hydration state. The straightforward applications for fluorescence mapping in fluid inclusions, micrometeorites, and cancer cell clusters, were realized. Micro-XANES analysis was used to determine the oxidation state of Mn at various locations inside a sample of corroded Roman glass. The brown color of the corroded glass is related to the increased Mn or Fe content. It was shown that Mn

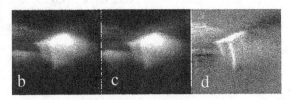

FIGURE 4. Optical image of the glass (a). X-ray fluorescence maps of a 0.3x0.3 μm area inside the corrosion body were recorded at 6.55 keV - Mn^{2+} excitation (b) and 6.564 keV - crest energy for Mn^{4+} (c). By subtracting both images from each other after scaling an image of the distribution of Mn^{4+} was obtained (d) (courtesy of K. Janssens).

is only present as MnO_2-presipitates in between the thin layers forming the corrosion body (Fig. 4). Fluorescence micro tomography in pencil-beam geometry combined with simplified algebraic or filtered backprojection reconstruction methods provides high quality and precision elemental images [37].

Microdiffraction

Hard X-rays play the leading role in studies of crystalline materials. X-ray microdiffraction has two orders of magnitude greater sensitivity to crystallographic strain than electron microscopy and can determine crystal structure, crystalline texture and crystalline strain distribution at the micron scale. Coherent microoptics removes the restrictions on structural perfection and allows to carry out high-resolution microdiffraction analysis on both perfect [5] or nearly perfect single crystals [6] and single grains. The first experiment on single grain analysis applying CRL was performed on an Interstitial Free Titanium steel after different thermomechanical treatments (recrystallization, cold-rolling, annealing) with a 30 μm grain size [38]. Local dislocation density, residual stresses, intragranular misorientation, can be obtained from diffraction line profile analysis. One of the attractive features of the optical setups using BFO, FZP and CRL is the well-defined shape of the focus beam and that the intensity distribution in the focus spot has a high degree of azimutal symmetry. This is very desirable for SAXS measurements. With rather long focal length focusing elements, there is enough space around the sample to introduce special environmental apparatus like cryostats, furnaces or high-pressure cells.

CONCLUSION

The combination of third-generation synchrotron radiation sources, high performance optics and detectors allows to transform limited microprobe techniques to the full-scale hard X-ray microscopy with sub-micrometer resolution involving imaging, spectroscopy and diffraction. X-ray microscopy permits comprehensive studies in biomedicine, material science, geology, astrophysics and environmental science. The field of X-ray microscopy is rapidly growing with many challenges and opportunities for those involved.

REFERENCES

1. Snigirev, A., *SPIE Conference Proceedings*, **2856**, 26-33, (1996).

2. Aristov, V. V., Snigirev, A. A, Basov, Yu., A, Nikulin, A. Yu, "X-ray Bragg optics", in *"Short Wavelength Coherent Radiation: Generation and Applications"*, edited by D.T. Attwood and J. Bokor, AIP Conference Proceedings 147, New York: American Institute of Physics, 1986, pp. 253-259.

3. Snigireva, I, Souvorov, A, Snigirev, A, Bragg Fresnel Optics for high energy X-ray microscopy techniques at the ESRF, in *"X-ray Microscopy and Spectroscopy"* edited by J. Thieme, G. Schmahl, D. Rudolph, E. Umbach, Berlin Heidelberg, Springer-Verlag, 1998, pp. IV37 – IV44.

4. Snigirev A., Snigireva I., Bosecke P., Lequien S., Schelokov I., *Optics Comm.*, **135**, 378-384, (1997).

5. Iberl, A., Schuster, M., Gobel, H., Meyer, A., Baur, B., Matz, R., Snigirev, A., Snigireva, I., Freund, A., Lengeler, B., Heinicke, H., *J. Phys. D: Appl. Phys.*, **28**, A200-A205, (1995).

6. Biermann, H., Grossmann, B. V., Mechsner, S., Mughrabi, H., Ungar, T., Snigirev, A., Snigireva, I., Souvorov, A., Kocsis, M., Raven, C., *Scripta Materiala*, **37**, 1309-

7. Yun, W, Lai, B, Cai, Z, Masser, J, Legnini, D, Gluskin, E, Chen, Z, Krasnoperova, A, Vladimirsky, Y, Cerrina, F, Di Fabrizio, E, Gentili, M, *Rev. Sci. Instrum.* **70**, 2238-2241, (1999).

8. Yun, W., Pratt, S. T., Miller, R. M., Cai, Z., Hunter, D. B., Jarstfer A. G., Kemner, K. M., Lai, B., Lee, H. R., Legnini, D. G. Rodrigues W., Smith, Ch., *J. Synchrotron. Radiat.*, **5**, 1390-1395, (1999).

9. Snigirev, A., Kohn, V., Snigireva, I., Lengeler, B., *Nature*, **384**, 49-51, (1996).

10. Snigirev, A., Kohn, V., Snigireva, I., Souvorov, A., Lengeler, B., *Applied Optics*, **37**, 653-662, (1998).

11. Lengeler, B., Tummler, J., Snigirev, A., Snigireva, I., Raven, C., *Journal of Applied Physics*, **84**, 5855-5861, (1998).

12. Lengeler, B., Schroer, C. G., Richwin, M., Tummler, J., Drakopoulos, M., Snigirev, A., Snigireva, I., *Applied Physics Letters*, **74**, 3924-3926, (1999).

13. Lengeler, B., Schroer, C. G., Tummler, J., Benner, B., Richwin, M, Snigirev, A., Snigireva, I., Drakopoulos, M., *J. Synchrotron Rad.*, **6**, 1153-1165, (1999).

14. Schroer, C. G., Lengeler, B., Benner, B., Tummler, J., Gunzler, F., Drakopoulos, M., Weitkamp, T., Snigirev, A., Snigireva, I. this issue.

15. Schroer, C. G., Lengeler, B., Benner, B., Tummler, J, Gunzler, F., Drakopoulos, M., Simionovici, A., Snigirev, A., Snigireva, I., this issue.

16. Snigirev, A., Snigireva, I., Kohn, V., Kuznetsov, S., Schelokov, I., *Rev. Sci. Instrum.*, **66**, 5486-5492, (1995).

17. Raven, C., Snigirev, A., Snigireva, I., Spanne, P., Souvorov, A., Kohn, V., *Appl. Phys. Letters*, **69**, 1826-1828, (1996).

18. Hu, Z. H., Thomas, P. A., Snigirev, A., Snigireva, I. Souvorov, A., Smith, P. G. R., Ross, G. W., Teat, S., *Nature*, **392**, 690-693, (1998).

19. Koch, A., Raven, C., Snigirev, A., Spanne P., *J. Opt. Soc. Am.*, **A 15**, 1940-1951, (1998).

20. Spanne, P., Raven, C., Snigireva, I., Snigirev, A., *Physics in Medicine and Biology*, **44**, 741-749, (1999).

21. Gureyev, T. E., Raven, C., Snigirev, A., Snigireva, I., Wilkins, S. W., *J. Phys. D: Appl. Phys.* **32**, 563-567, (1999).

22. Kuznetsov, S., Snigireva, I., Souvorov, A., Snigirev, A., *Phys, Stat. Sol.(a)*, **172**, 3-13, (1999).

23. Drakopoulos, M., Hu, Z. H., Kuznetsov, S., Snigirev, A., Snigireva, I., Thomas, P. A., *J. Phys. D: Appl. Phys.*, **32**, A160-165, (1999).

24. Iida A., Hirano K., *Nucl. Instrum. & Methods B*, **114**, 149-153, (1996).

25. Underwood J. H., Thomson A. C., Kortright J. B., Chapman K. C., Lunt D., *Rev. Sci. Instrum.* Abstract **67**, 3359, (1996).

26. Lienert U., Schulze C., Honkimaki V., Tschentscher Th., Garbe S., Hignette O., Horsewell A., Lingham M., Poulsen H. F., Thomsen N. B., Ziegler E., *J. Synchrotron Rad.*, **5**, 226-231, (1998).

27. Bilderback D., Hoffman S. A., Thiel D. J., *Science*, **263**, 201-203, (1994).

28. Cedola A., Di Fonzo S., Jark W., and Lagomarsino S., *J. of Phys. D, Appl. Phys*, **32**, 1 (1999)

29. Chevallier P., Dhez P., Legrand F., Erko A., Agafonov Yu., Panchenko L. A., Yakshin A., *J. of Trace and Micriprobe Techniques*, **14**, 517-539, (1996).

30. Leitenberger, W., Snigirev, A., this issue.

31. Elleaume, P., *Nucl. Instrum. & Methods*, **A412**, 483-506, (1998). 1314, (1997).

32. ESRF Website: http://www.esrf.fr/exp_facilities/ID22/

33. Weitkamp, T., Raven, C., Snigirev, A. *SPIE Conference Proceedings*, **3772**, 311-317, (1999).

34. Mathiesen R. H., Arnberg L., Mo F., Weitkamp T., Snigirev A., to be published in *Phys. Rev.Lett.*.

35. Labiche J. C., Segura-Puchdes J., van Brusel D., Moy J. P., *ESRF Newslett*, **25**, 41-43, (1996).

36. Adams, F., Janssens, K., Snigirev, A., *J. of Anal. At. Spectr.*, **13**, 319-331, (1998).

37. Simionovici A., Chukalina M., Drakopoulos M., Snigireva I., Snigirev A., Schroer Ch., Lengeler B., Janssens K., Adams F., this issue.

38. Castelnau O., Chauveau T., Drakopoulos M., Snigirev A., Snigireva I., Schroer C., Ungar T., Proceedings of ECRS V conf. Delft (Holland) 28-30 September, (1999).

Phase-Contrast Hard X-ray Imaging Microscope with Wolter Mirror Optics

N. Watanabe, S. Aoki, H. Takano, K. Yamamoto, A. Takeuchi, H. Tsubaki, and T. Aota

Institute of Applied Physics, University of Tsukuba, 1-1-1 Tennoudai, Tsukuba, Ibaraki, 305-8573, Japan

Abstract. A Zernike-type one-dimensional phase-contrast x-ray microscope with a Wolter mirror was developed and tested. A Cu wire or an Al phase plate was placed at the back focal plane of the Wolter mirror objective. A dark-field image of a Cu mesh could be obtained at 9.00 keV by using the Cu wire as a direct beam stop, and a phase-contrast image could be obtained at 8.97 keV by using the same optics. A polyethylene telephthalate (PET) fiber of 9 µm in diameter could be obtained at 8.97 keV by using the Al phase plate, which could seldom be observed without the phase plate. These results agreed fairly well with simulated images.

INTRODUCTION

In visible light optics, a Zernike-type phase-contrast microscope is widely used to observe unstained biological specimens which are almost transparent in bright-field imaging. In soft x-ray region below 1 keV, the same type of a phase-contrast microscope was developed (1). In hard x-ray region above several keV, several types of phase-contrast imaging methods have been developed, such as projection methods with coherent x-ray source (2-4) or collimator and analyzer crystals (5), a Bragg-Fresnel optics with coherent illumination (6), and phase-contrast tomography (7) based on an x-ray interferometer (8). Using these methods, it has been demonstrated that phase-contrast is much higher than absorption contrast especially for a specimen that consists of light elements. However, the lack of lens systems has prevented the development of a Zernike-type phase-contrast microscope for hard x-rays. We developed and tested this type of a phase-contrast hard x-ray microscope with a Wolter mirror optics.

HARD X-RAY MICROSCOPE WITH A WOLTER MIRROR

A Wolter mirror is a grazing-incidence optics which consists of a combination of axisymmetric hyperboloidal and ellipsoidal surfaces with a common focus. Figure 1 shows the parameters of the Wolter mirror objective we designed and fabricated by using the glass replica technique (9). The mirror had a narrow annular aperture 200 mm apart from the object point. The diameter and the width of the aperture were 10

CP507, *X-Ray Microscopy: Proceedings of the Sixth International Conference,*
edited by W. Meyer-Ilse, T. Warwick, and D. Attwood

mm and 0.16 mm, respectively. The grazing angle was 7 mrad and the magnification ratio was 10. X-rays up to approximately 11 keV could be reflected by the Pt-coated surface. The spatial resolution was estimated to be about 10 μm at 6.4 keV (Fe Kα) from the experiment of the full-field x-ray fluorescence microscope with this mirror (10).

The Wolter mirror has fixed object and image points determined by its ellipsoidal and hyperboloidal foci as shown in Fig. 1. However, when a parallel x-ray beam is incident onto a part of the annular aperture of the mirror, the beam is focused onto a small spot of several microns in diameter upstream the image plane like a refractive lens. We referred the plane included this focused point to be the back focal plane of the Wolter mirror. If a phase modulator is placed at this focused point, only the undeviated direct beam transmitted through a specimen can be modulated without the influence of the deviated beam by diffraction, reflection, and refraction. This is the basic idea of a Zernike-type phase-contrast microscope and a phase modulation by a specimen can be converted to an intensity distribution.

A phase contrast x-ray microscope based on this concept was constructed at BL3C2, the Photon Factory, Japan. Figure 2 shows a schematic of the optical system. The storage ring energy was 2.5 GeV and the ring current was about 300 mA. Parallel monochromatic x-rays at around 9 keV were incident on a specimen. The x-ray beam size was restricted to 130 μm × 400 μm by a slit in front of a specimen. The beam transmitted through a specimen was incident on a part of the annular aperture of the mirror. The image of a specimen was focused on a CCD camera (TI, TC-215) cooled by a Peltier element down to -30 centigrade. The pixel size is 12 μm × 12 μm. The x-ray path between the slit and the detector was stayed in air. The back focal plane of the mirror was located 180 mm downstream of the mirror, where a phase modulator was placed.

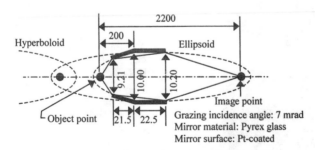

FIGURE 1. Schematic of the Wolter mirror objective. Unit: mm.

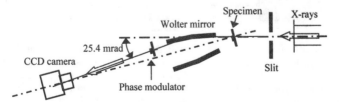

FIGURE 2. Optical system of a phase contrast x-ray microscope installed at the Photon Factory.

Figure 3 shows a calculated spot diagram and an experimentally obtained focused x-ray image at the back focal plane at 9 keV. The calculated spot size was smaller than 1 μm. The synchrotron radiation source was located 35 m upstream of the optical system and its positron beam size was σ_x=0.74 mm and σ_y=0.26 mm. Suppose that x-ray source size is 2.35 σ_x and 2.35 σ_y, the reduced image size at the back focal plane is 8.9 μm horizontally and 3.1 μm vertically. Even if the influence of the source size was taken into account, the focused image size in the horizontal direction in Fig. 3(b) was much worse compared with the calculation. This seems to be due to the figure error of the mirror. Figure 4(b) shows an image of a copper mesh at 9.00 keV. To extend the dynamic range of the CCD camera, a 2 s-exposure was accumulated 10 times. The horizontal wires could be observed. However, the vertical wires could not

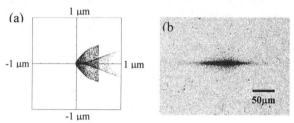

FIGURE 3. (a) Calculated spot diagram at the back focal plane of the Wolter mirror in the case of a parallel beam (130×400 μm). (b) Focused x-ray image corresponding to the calculation.

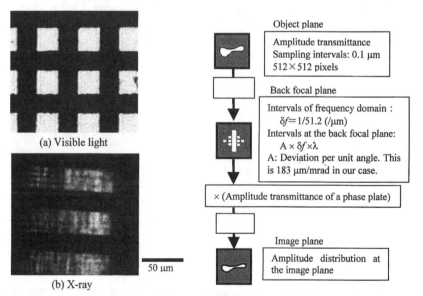

FIGURE 4. Images of a Cu mesh with 51 μm pitch. (a) Optical microscope image. (b) Bright-field x-ray image at 9.00 keV. The image was taken by 10 times accumulation of 3 s exposure.

FIGURE 5. Simulation method for phase contrast imaging. It is supposed that a specimen is illuminated by parallel rays and the objective can focus the parallel rays onto a point at the back focal plane of the objective.

be observed. The horizontal and vertical line-shaped artifacts in Fig. 4(b) seem to be due to the figure error and surface roughness of the mirror. The resolution along the vertical direction was estimated to be 4.6 μm from the edge profile, which was better than the resolution when the whole aperture was used (10). From these results, we concluded that a small part of the annular aperture of the mirror could focus an x-ray image only in the direction perpendicular to the meridional plane of the mirror. Then, a one-dimensional phase modulator that consisted of a copper wire or an aluminum foil was placed at the back focal plane to cover the focused x-ray spot shown in Fig. 3(b). This was one-dimensional phase-contrast imaging that could detect a phase modulation in the vertical direction.

Experimentally obtained images were compared with theoretically calculated images. The simulation method followed the Abbe theory of image formation (11) as shown in Fig. 5. It was supposed that an incident x-ray beam was parallel and coherent, an objective can focus the incident beam onto a point at the back focal plane, and the entrance pupil was not limited. The displacement of a focused point at the back focal plane per unit angle of the incident beam was determined to be 183 μm/mrad from ray-tracing calculation.

PHASE-CONTRAST IMAGING WITH A DIRECT BEAM MODULATOR OF A COPPER WIRE

The transmittance of a copper of 25 μm in thickness is 0.18 % at 9.00 keV, and 44 % at 8.97 keV. Then, a copper wire of 25 μm in diameter works as a direct beam stop at 9.00 keV or a phase modulator at 8.97 keV. To evaluate the effect of the direct beam modulator, a copper wire of 25 μm in diameter was placed at the back focal plane of the Wolter mirror. And a copper mesh with 51 μm pitch and the width of each wire of 21 μm was imaged at 9.00 keV and 8.97 keV.

Figure 6(a) shows the image at 9.00 keV. This was a dark-field image because the undeviated direct beam was excluded by the stop. Sharp edge contrast could be obtained. The line profile of the image was compared with that of a simulated image in Fig. 6(b) and (c). The edge contrast of the experimentally obtained profile is not so sharp compared with the simulation result. This seems to be due to the low resolution of the objective mirror.

Figure 7(a) shows the image at 8.97 keV. This image is very different from Fig. 6(a) because the direct beam was not excluded but their phases were advanced. The calculated maximum phase shift is 2.4 wavelength. Then, the phase-contrast image could be obtained. The line profile of the image was compared with that of a simulated image in Fig. 7(b) and (c). The high frequency terms in the simulation cannot be seen in the experimentally obtained profile. However, the dark edge contrast in the image can be successfully explained from the simulated profile.

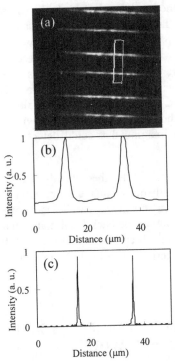

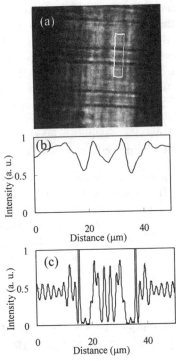

FIGURE 6. (a) Dark-field x-ray image of a Cu mesh with 51 μm pitch at 9.00 keV. A direct beam stop of a Cu wire was used. Exposure time: 50 s × 10 times accumulation. (b) The line profile of the white rectangle area of the image. (c) A simulated intensity profile of a Cu wire of 21 μm in diameter, which was the same as the mesh wires.

FIGURE 7. (a) Phase-contrast x-ray image of a Cu mesh with 51 μm pitch at 8.97 keV. A phase modulator of a Cu wire was used. Exposure time: 8 s × 10 times accumulation. (b) The line profile of the white rectangle area of the image. (c) A simulated intensity profile of a Cu wire of 21 μm in diameter, which was the same as the mesh wires.

PHASE-CONTRAST IMAGING WITH AN ALUMINUM PHASE PLATE

If a specimen has a slightly lower index of refraction than that of the surround, the wave after transmission through the specimen is advanced. The difference between the advanced wave and its surroundings is a sine curve, which is advanced in phase by almost 1/4 wavelength with respect to the surrounding wave. It represents the light deviated by diffraction at the specimen. If this deviated diffraction wave is artificially advanced by 3/4 wavelength, this wave interferes with the surrounding wave constructively and the specimen appears brighter than its surround (12).

To observe a weekly absorbing specimen, a slit-like phase plate that consisted of two aluminum foils of 15 μm in thickness was made. The spacing between two foils was 9 μm. The phase plate was placed at the back focal plane of the mirror to allow an undeviated direct beam pass through this spacing. The phase shift and transmittance of the foil is calculated to be 0.74 wavelength and 87 % at 8.97 keV, respectively. Then, the phase plate works as a 3/4 wavelength phase plate.

A polyethylene telephthalate (PET) fiber of 9 μm in diameter was observed at 8.97 keV. The calculated maximum absorption and phase shift are 0.55 % and 0.24 wavelength, respectively. Figure 8 shows the bright-field image without the phase plate and the phase-contrast image with the phase plate. In the bright-field image in Fig. 8(b), the fiber can seldom be observed. However, the phase-contrast image in Fig.

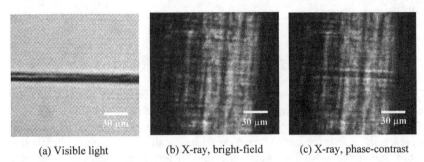

(a) Visible light   (b) X-ray, bright-field   (c) X-ray, phase-contrast

FIGURE 8. Images of a polyethylene telephthalate fiber of 9 μm in diameter. (a) Optical microscope image. (b) Bright-field x-ray image at 8.97 keV. (c) Phase-contrast x-ray image at 8.97 keV. An aluminum slit-like phase plate of 15 μm in thickness was used. Exposure time of x-ray images: 2 s × 10 times accumulation.

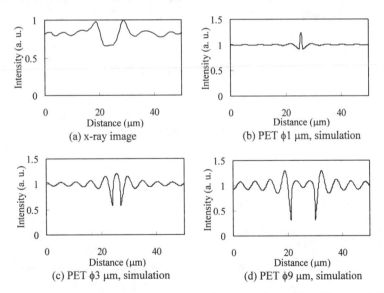

FIGURE 9. Experimentally obtained and calculated intensity profiles of images of PET fibers. (a) Intensity profile of the x-ray image in Fig. 8(c). (b) Profile of a simulated image of a PET fiber of 1 μm in diameter. (c) Profile of a simulated image of a PET fiber of 3 μm in diameter. (d) Profile of a simulated image of a PET fiber of 9 μm in diameter.

8(c) has fairy good contrast. The fiber can be seen as a dark line in Fig. 8(c), that is different from the above mentioned theory. Thus, the intensity profile of the phase contrast image was compared with theoretically simulated ones of PET fibers of several different diameters as shown in Fig. 9. The simulated profile of a 1 μm PET fiber in Fig. 9(b) shows bright contrast, that is expected from the theory. Increasing the diameter, the profile becomes more complicated. When the diameter is 9 μm, the simulated profile in Fig. 9(d) shows dark edge contrast. On the other hand, the profile of the x-ray image in Fig. 9(a) shows dark contrast over the whole wire. It seems that the dark area of the edges of the fiber was spread because of the low resolution of the optical system, and the relatively bright area at the center of the fiber in Fig. 9(d) was diminished.

FUTURE DEVELOPMENT

The phase contrast x-ray microscope with the Wolter mirror optics and the phase plate could image an almost transparent specimen, such as the PET fiber at 8.97 keV. The phase contrast images agreed fairly well with the theoretical calculations. However, only one-dimensional image could be obtained. It is due to figure errors of the objective mirror. To distinguish the undeviated direct beam and the deviated diffraction beam along the horizontal direction in Fig. 3(b) at the back focal plane of the mirror, it is necessary to make the horizontally focused spot size to be as small as the vertical one. If the focused spot size can be made so small, a point-like phase plate or a ring-shaped phase plate can be used for two-dimensional phase-contrast imaging. These are currently under development.

ACKNOWLEDGEMENTS

The authors thank Prof M. Ando, X. Zhang, and H. Sugiyama for their support during the experiments at the Photon Factory. This work was supported by the Grant-in-Aid for Scientific Research (A) No. 08405010 from the Ministry of Education, Science, Sports and Culture of Japan.

REFERENCES

1. Schmahl, G., Rudolph, D., Guttmann, P., Schneider, G., Thieme, J., and Niemann, B., *Rev. Sci. Instrum.* **66**, 1282-1286 (1995).

2. Snigirev, A., Snigireva, I., Kohn, V., Kuznetsov, S., and Schelokov, I., *Rev. Sci. Instrum.* **66**, 5486-5492 (1995).

3. Wilkins, S. W., Gureyev, T. E., Gao, D., Pogany, A., and Stevenson, A. W., *Nature* **384**, 335-338 (1996).

4. Lagomarsino, S., Cedola, A., Cloetens, P., Fonzo, S. Di, Jark, W., Soullié, G., and Riekel, C., *Appl. Phys. Lett.* **71**, 2557-2559 (1997).

5. Davis, T. J., Gao, D., Gureyev, T. E., Stevenson, A. W., and Wilkins, S. W., *Nature* **373**, 595-598 (1995).

6. Snigirev, A., Snigireva, I., Bösecke, P., Lequien, S., and Schelokov, I., *Opt. Commun.* **135**, 378-384 (1997).

7. Momose, A., Takeda, T., Itai, Y., and Hirano, K., *Nature Medicine* **2,** 473-475 (1996).

8. Bonse, U., and Hart, M., *Appl. Phys. Lett.* 6, 155-156 (1965)

9. Onuki, T., Sugisaki, K., and Aoki, S., *Proc. SPIE* 1720, 258-263 (1992).

10. Aoki, S., Takeuchi, A., and Ando, M., *J. Synchrotron Rad.* 5, 1117-1118 (1998).

11. Goodman, J. W., *Introduction to Fourier Optics 2nd edition*, New York: The McGraw-Hill Companies, Inc., 1996, pp.128-130.

12. Bennett, A. H., Osterberg, H., Jupnik, H., and Richards, O. W., *Phase Microscopy Principles and Applications* , New York: John Wiley & Sons, Inc., 1951, pp.13-36.

Spectromicroscopy Using An X-Ray Microprobe At SPring-8 BL39XU

Shinjiro Hayakawa and Takeshi Hirokawa

Applied Physics and Chemistry, Faculty of Engineering, Hiroshima University
Higashi-Hiroshima, Hiroshima 739-8527, Japan

Yohichi Gohshi

National Institute for Environmental Studies, Onogawa, Tsukuba, Ibaraki 305-0053, Japan

Motohiro Suzuki and Shunji Goto

SPring-8, Mikazuki, Hyogo 679-5198, Japan

Abstract. Brilliant undulator radiation from a super photon ring 8 GeV (SPring-8) is utilized for spectromicroscopy in hard x-ray region. Design parameters of the x-ray microprobe dedicated for BL39XU are described. X-ray beam less than 1 µm is expected by using a Kirkpatrick and Baez mirror system, and both conventional and high resolution x-ray fluorescence spectrometers are available by using a Si(Li) detector and a wavelength dispersive spectrometer equipped with a flat analyzer crystal and a PSPC. Preliminary experimental results of the x-ray microprobe system show possibilities of trace characterization with spatial resolution.

INTRODUCTION

Characterization of trace elements with spatial resolution is becoming more important in many fields of science and technologies because functions or mechanism of advanced materials, integrated devices, biological sections and cells are closely related to localized trace component in many cases. To handle with these problems, synchrotron radiation (SR) excited x-ray fluorescence (XRF) analysis can provide local elemental composition of bulk materials without giving significant damage to the sample. Energy tunability of the SR enables selective excitation of an element of interest, and the minimum detection limit (MDL) of the XRF analysis using an energy dispersive spectrometer (EDS) has reached to sub-ppm or sub-pg in relative or absolute amount [1]. Moreover, the energy dependence of the XRF yield can provide x-ray absorption fine structure (XAFS) of a trace element [2], and chemical state or local structure of the trace element can be discussed with this method.

Energy tunable x-ray microprobes (or scanning x-ray microscope) composed of total reflection mirrors were developed by using the second generation SR sources [3-7], and there have been a lot of research activities in materials science and biological applications with the spatial resolution ranging from 1 to 10 µm.

Recently, high energy storage rings of extremely small emittance has been constructed to produce undulator radiation in the hard x-ray region. Gain of several orders of magnitude in brilliance has enabled us to achieve improvements both in sensitivity and spatial resolution. This paper describes x-ray microprobe system dedicated at SPring-8 BL39XU.

CP507, *X-Ray Microscopy: Proceedings of the Sixth International Conference,*
edited by W. Meyer-Ilse, T. Warwick, and D. Attwood
© 2000 American Institute of Physics 1-56396-926-2/00/$17.00

X-RAY MICROPROBE SYSTEM AT BL39XU

BL39XU is equipped with an in-vacuum-type linear undulator, and the tunable range of x-ray energy is 6 to 25 keV by utilizing the first and the third harmonics of undulator radiation. A brilliance of more than 1×10^{19} photons/s/mrad^2/mm^2 (0.1 % bandwidth) is expected with a 100 mA storage-ring current. The transport channel of the beamline is equipped with a rotated-inclined double-crystal monochromator and a Pt-coated horizontal beam deflection mirror, and the details of the beamline is described previously[8].

An x-ray microprobe system is placed in a experimental hutch which is located between 46 m and 52.5 m from the source. Figure 1 shows the schematic view of the x-ray microprobe system. The present system uses slits or a pinhole to define beam size on the sample, and the beam size ranges from 1 mm to 10 μm. The beam spot of less than 1 μm will be expected by using a Kirkpatrick and Baez (crossed elliptical) mirror [9]. To collect XRF signals, both a conventional EDS using a Si(Li) detector and a wavelength-dispersive spectrometer (WDS) are attached. Considering that the XRF is from a small area, fairly good energy resolution can be realized by dispersing an XRF spectrum with a flat analyzer crystal. The dispersed spectrum can be detected by a position-sensitive proportional counter (PSPC). The central Bragg angle is fixed to be 21 degree with the standard setup, and the range of x-ray energy can be changed by using an appropriate analyzer crystal mounted on the crystal revolver. To realize better energy resolution, WDS system is composed of a θ-2θ stage with a longer sample-PSPC distance. These spectrometers can be switched by rotating the sample surface.

Figure 2 shows Cr Kα XRF image of a Cr coated resolution test pattern measured with the beam defined with a 10 μm pinhole. Periods of the patters are 200 μm, 80 μm, 40 μm, 20 μm and 10 μm from the bottom to the top on the image, and the patters of 20 μm period is clearly resolved.

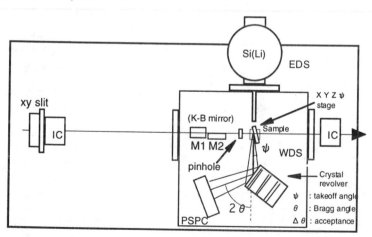

Figure 1. X-ray microprobe system equipped with an EDS and a WDS XRF spectrometers.

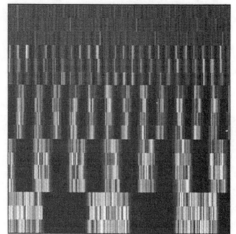

Figure 2 Cr XRF image of a resolution test pattern.

Kirkpatrick and Baez mirror

Figure 3 shows parameters of the Kirkpatrick and Baez (K-B) mirror under fabrication. Glancing angle of these two mirrors is 4 mrad, and the mirror surface will be coated with Rh. X-rays up to 20 keV will be focused onto the sub micron beam spot. When x-rays are focused with the aspherical total reflection mirrors, the focused beam size is strongly affected with the slope error of the mirror in many cases. Therefore, a smaller mirror-to-focus distance is employed at the expense of the acceptance of the mirror. The effective aperture is only 120 μm in horizontal and vertical direction. However, the expected photon flux is more than 10^{11} photons/s within this aperture.

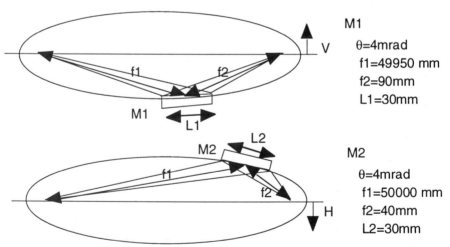

M1
θ=4mrad
f1=49950 mm
f2=90mm
L1=30mm

M2
θ=4mrad
f1=50000 mm
f2=40mm
L2=30mm

Figure 3. Schematic view of the Kirkpatrick and Baez mirror.

Wavelength dispersive spectrometer

The energy resolution of the spectrometer ($\Delta E/E$) can be described by the following expressions [10].

$$\Delta E/E = \cot \theta_B \Delta \theta \tag{1}$$

$$\Delta \theta^2 = \Delta \theta_1^2 + \Delta \theta_2^2 + \Delta \theta_3^2 + \omega^2 \tag{2}$$

$$\Delta \theta_1 = d/L \tag{3}$$

$$\Delta \theta_2 = s/L \tag{4}$$

$$\Delta \theta_3 = \arcsin \frac{\sin \theta_B}{\sqrt{1+\phi_v^2}} - \theta_B \tag{5}$$

where θ_B is the Bragg angle, d the effective horizontal beam size of the x-ray source, L the optical path length of the central beam between the sample and the PSPC, s the spatial resolution of the PSPC and ω the intrinsic width of diffraction. $\Delta \theta_1$ is divergence in the plane of diffraction caused by the finite source size, $\Delta \theta_2$ is the angular resolution of the PSPC and $\Delta \theta_3$ is the angular deviation from the Bragg angle caused by the vertical divergence of x-rays. When the longer path length is employed, the effects of $\Delta \theta_1$ and $\Delta \theta_2$ can be negligible, and the energy resolution will be determined by the intrinsic width of the diffraction and the vertical divergence ϕ_v.

In the standard setup θ_B is 21 deg and L=200 mm, and the energy resolution is mainly determined by the spatial resolution of the PSPC. Figure 4 shows XRF spectra of stainless steel foil (SUS304) measured with several analyzer crystals. In this measurement spatial resolution of the PSPC was slightly worse than the designed value (200 μm), and the FWHM of Cr Kα peak with a Si(111) analyzer crystal was 35 eV.

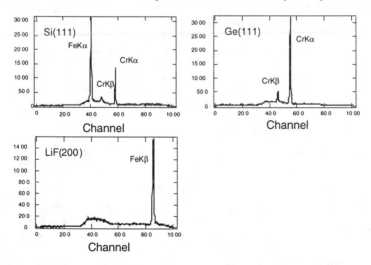

Figure 4   WD XRF spectra of stainless steel foil (SUS304) of 8 μm in thickness.

Figure 5 shows XRF spectra of the same sample measured with a Ge (333) analyzer crystal employing longer path geometry. FWHM of the Fe Kα1 is about 3.8 eV. The energy resolution of the spectrometer was sufficient to observe satellite line which appears as a shoulder in the lower energy side of the Kβ main line. Hämäläien et al. [11] have reported high resolution Kβ x-ray fluorescence spectra of MnO and MnF$_2$ by using a Johan-type Si(440) crystal spectrometer [12]. A high resolution x-ray fluorescence spectrometer enables selective detection of a satellite line eliminated from the main line, and the spin-selective x-ray absorption spectroscopy was reported by selectively monitoring different regions of the Kβ emission as a function of the incident x-ray energy[11-14]. Spin selective XAFS of small area or spin selective x-ray imaging will be realized with the combination of the WD spectrometer and the x-ray microprobe.

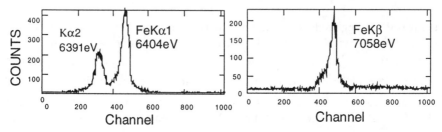

Figure 5. High resolution Fe Kα and Kβ XRF spectra measured from a SUS304 foil. A Ge(333) analyzer crystal was used with the path length of 800 mm .

XRF IMAGING OF TRACE ELEMENTS

Figure 6. shows line scan images of a single crystal of synthetic diamond grown under high pressure with metallic solvent of Fe$_{55}$Ni$_{29}$Co$_{16}$. The beam size was 150 μm in this measurement. As reported previously trace impurities are incorporated from the solvent, and Ni and Co are selectively dissolved into {111} growth sector[15]. Their concentrations in {111} growth sector are around 30 ppm and 3 ppm for Ni and Co, and the concentrations in {100} growth sector is less than the detection limit (0.1 ppm). Ni and Co XRF images show clear contrast between two {111} growth sectors and one {100} growth sector, and it is clear that the trace element imaging of sub ppm level has been realized using the brilliant undulator radiation from the SPring-8.

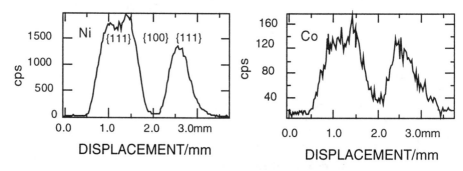

Figure 6. Line scan XRF images of a synthetic diamond.

ACKNOWLEDGMENTS

We are grateful for the help of members of spectrochemical group for design and construction of the beamline. This work was partially supported by a Grand-in-Aid for Scientific Research No. 08555210 from the Ministry of Education, Science, Sports and Culture of Japan. Experiments were carried out under the approval of the SPring-8 advisory committee (Proposal number 1997B0131, 0132, 1998A0303).

REFERENCES

1. A. Iida and Y. Gohshi, in Handbook on Synchrotron Radiation Vol.4, Elsevier, Ch.9 (1991).
2. S. Hayakawa, Y. Gohshi, A. Iida, S. Aoki and K. Sato, Rev. Sci. Instrum. 62, 2545-2549 (1991).
3. S. Hayakawa, A. Iida, S. Aoki and Y. Gohshi, Rev. Sci. Instrum. 60, 2452-2455 (1989).
4. S. Hayakawa, Y. Gohshi, A. Iida, S. Aoki and M. Ishikawa, Nucl. Instrum. Methods Phys. Res., B49, 555-560 (1990).
5. Y. Suzuki and F. Uchida, Jpn. J. Appl. Phys., 30, 1127-1130 (1991).
6 A. Iida and T. Noma, Nucl. Instrum. Methods Phys. Res., B82, 129-138 (1993).
7. S. Hayakawa and Y. Gohshi, in Application of Synchrotron Radiation to Materials Analysis, Elsevier, Ch. 3 (1996).
8. S. Hayakawa, S. Goto, T. Shoji, E. Yamada and Y. Gohshi, J. Synchrotron Rad., 5, 1114-1116 (1998).
9. P. Kirkpatrick and A. V. Baez, J. Opt. Soc. Am. 9, 766- 770(1948).
10. S. Hayakawa, A. Yamaguchi, Y. Gohshi, T. Yamamoto, K. Hayashi, J. Kawai and S. Goto, Spectrochim. Acta. B, 54B, 171-178(1999).
11. K. Hämäläien, C.-C. Kao, J. B. hastings, D. P. Siddons, L. E. Berman, V. Stojanoff and S. P. Cramer, Phys. Rev., B 46 14274-14277 (1992) .
12. V. Stojanoff, K. Hämäläien, D. P. Siddons, J. B. Hastings, L. E. Berman, S. Cramer and G. Smith, Rev. Sci. Instrum., 63 1125-1127 (1992).
13. M. M. Grush, G. Christou, K. Hämäläien and S. P. Cramer, J. Am. Chem. Soc., 117, 5895-5896 (1995) .
14. G. Peng, X. Wang, C. R. Randall, J. A. Moore and S. P. Cramer, Appl. Phys. Lett., 65 2527-2529 (1994).
15. S. Hayakawa et al., Trans. Mat. Res. Soc. Jpn., 14B, 1559-1562(1994).

Design of X-Ray Interferometer for Phase-Contrast X-Ray Microtomography

Atsushi Momose and *Keiichi Hirano

Advanced Research Laboratory, Hitachi, Ltd., Hatoyama, Saitama 350-0395, Japan
Institute of Structure Materials Science, Tsukuba, Ibaraki 305-0801, Japan

Abstract. We tested an X-ray interferometer fabricated to improve the spatial resolution of phase-contrast X-ray computed tomography (CT). The main factor limiting the spatial resolution is the diffraction phenomenon in the crystal wafer of the interferometer. Improvement of the spatial resolution was attempted by thinning the wafer. A preliminary test was performed by checking interference patterns. Thinning the wafer from 1 to 0.24 mm showed clear improvement in the image quality.

INTRODUCTION

Since the invention of an X-ray interferometer by Bonse and Hart,[1] phase-contrast X-ray imaging has been attempted in the hard X-ray energy region.[2] Because materials consisting mainly of low-Z elements can be observed with extremely high sensitivity, this technique is attractive for investigating internal structures in organic matter without the need for staining. In the early stage, however, experiments were performed by recording interference patterns on films. Then, contrast was occasionally too complicated to extract structural information. Furthermore, a perfect interferometer had to be used, otherwise built-in contrast due to the deformation of an interferometer disturbed the image.

Recently, the technique of phase-shifting interferometry has been introduced in the hard X-ray region.[3] X-ray phase maps can be measured and they enable quantitative image recognition. Because the built-in contrast derived from the deformation of an interferometer can be measured as a background phase map, a real phase map related to a sample can be obtained by subtraction. Moreover, phase-shifting X-ray interferometry led us to phase-contrast X-ray CT[3] that reveals three-dimensional structures in organic matter. With recent progress of digital X-ray area detectors and synchrotron X-ray sources, phase-contrast X-ray CT is fully practical and has been used to observe animal[5,6] and human[7] tissues.

In this paper, we concentrate on the issue of the spatial resolution of phase-contrast X-ray CT. With our apparatus,[8] the typical spatial resolution is around 30 μm. In our previous paper,[9] the main factor limiting the spatial resolution was pointed out, and a design of an interferometer was presented to reduce the influence of this factor. This paper presents a preliminary test result of an interferometer fabricated according to this design.

CP507, *X-Ray Microscopy: Proceedings of the Sixth International Conference*,
edited by W. Meyer-Ilse, T. Warwick, and D. Attwood
© 2000 American Institute of Physics 1-56396-926-2/00/$17.00

SPATIAL RESOLUTION

Figure 1 shows a triple Laue-case X-ray interferometer used in our apparatus. The entire body is monolithically cut out from an ingot of a perfect silicon crystal. There are three crystal wafers formed with the equal gaps between them. When the incident X-rays satisfy the Bragg diffraction condition for the first wafer (S), the beam is split into two coherent beams by means of Laue diffraction. Each beam is split by the center wafer (M), and, the two beams traveling inside are split by the third wafer (A) and overlap. When a sample is placed in one of the beam paths, an interference pattern appears in the beams outgoing from the third wafer (A).

A sample is put in a cell filled with water and rotated for a CT scan. The cell is placed in one of the beam paths and a phase shifter is placed in the other (reference) beam path. The plate phase shifter can be rotated to change the phase of the reference beam for phase-shifting interferometry.

The factors limiting the spatial resolution in the case of phase-contrast X-ray CT using an X-ray interferometer are the resolution of an X-ray area detector, X-ray source size, and wafer thickness of the interferometer. Recently, X-ray area detectors with a spatial resolution of a few microns have been available and the X-ray source size is also fully small at 3rd-generation synchrotron radiation facilities. The last factor, wafer thickness, is currently responsible for the spatial resolution in phase-contrast X-ray CT, and in this paper discussion is concentrated on reducing blurring caused by this factor.

We observe X-ray interference patterns that appear because the X-ray wavefront is distorted due to the phase shift caused by a sample. Because X-rays travel in the direction perpendicular to the wavefront, generation of interference fringes means that X-rays are refracted. In many cases, this refraction is negligible because the deflection angle is normally below 10 μrad. However, because of the diffraction at the crystal wafer (A) located between a sample and an X-ray area detector in our setup, we can no longer neglect the refraction when attempting to achieve phase-contrast microtomography.

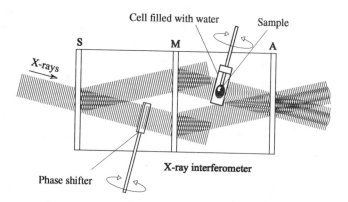

FIGURE 1. X-ray interferometer and setup for phase-contrast X-ray CT.

99

Here, we will consider the path of X-rays propagating inside the crystal wafer. When a plane wave impinges on a wafer satisfying the Bragg diffraction condition exactly, the path is parallel to the corresponding lattice plane. However, the path changes sensitively when the X-ray incident angle deviates slightly from the exact Bragg angle, according to the theory of dynamical diffraction of X-rays.[11] The deviation of the incident angle is amplified roughly by 10^4 times in the crystal. Therefore, the X-ray refraction by a sample causes the extraordinary shift of the exit point from the back surface of the wafer (A), as illustrated in Fig. 2(a). Consequently, fringes are distorted and visibility decreases.

The shift S of the exit point can be a representative variable indicating the spatial resolution and is approximately proportional to the thickness T of the wafer (A). Figure 2(b) shows S as a function of the deviation of the incident angle $\Delta\theta$ calculated for Si(220) diffraction of 17.7-keV X-rays. In Fig. 2(b), the spacing l of interference fringes produced by interfering with the undeflected reference beam is also shown. For example, when $\Delta\theta = 2$ μrad and $T = 1$ mm, S exceeds 30 microns and the fringe spacing produced in the condition is also about 30 microns. As a matter of fact, we can no longer observe interference fringes in such a case. Therefore, $\Delta\theta$ should be small enough to satisfy $l > |S|$ in a whole image. The range of $\Delta\theta$ that satisfies the requirement expands by decreasing T.

This burring effect is anisotropic because the beam deflection due to Bragg diffraction occurs only in the direction parallel to the scattering plane. Therefore, in phase-contrast X-ray CT, a sample is rotated around the axis parallel to the scattering plane. Then, sectional images are reconstructed on planes perpendicular to the scattering plane, minimizing the blurring effect.[3] However, since real samples have complex three-dimensional structures, the blurring effect remained to some extent. This is the main factor limiting the spatial resolution of phase-contrast X-ray CT.

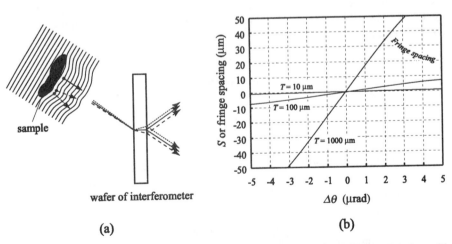

(a) (b)

FIGURE 2. (a) Schematic drawing showing that beam deflection caused by refraction is enhanced by a crystal wafer, and (b) the shift S of beam exit point from the back surface of the wafer and corresponding fringe spacing as functions of the deviation of the incident angle $\Delta\theta$.

100

FIGURE 3. X-ray interferometer fabricated to improve the spatial resolution. The central part (15 mm in diameter) of wafer (A) is thinned from the outside by chemical etching. The etched area is 0.24-mm thick.

DESIGN OF INTERFEROMETER

According to the estimation described above, we designed and fabricated an X-ray interferometer with a thin wafer.[9, 10] The center part of the wafer (A) was only etched to avoid degrading mechanical stability (Fig. 3). The remaining marginal part of the wafer supported the thinned area. Only the outside surface was etched, and therefore the spacing between the wafers of the interferometer was unchanged, which is important for ensuring the coherency of the system. The diameter of the etched area was about 15 mm. More details about fabrication process have been described elsewhere.[10]

RESULTS AND DISCUSSION

The fabricated X-ray interferometer was preliminarily tested by observing interference patterns with an X-ray zooming tube.[12] The test was carried out at the SPring-8, Japan. Figures 4(a) to (d) are the first results obtained for a copper mesh which had holes 85-µm in diameter with a 125-µm pitch. In this case, the generated contrast is mainly due to absorption. However, the blurring effect and its reduction by thinning a wafer are clearly seen. Although images of holes blurred in the direction parallel to the scattering plane (indicated by arrows) when T was 1 and 2 mm, each hole could be resolved with an interferometer whose wafer was thinned down to 0.24 mm.

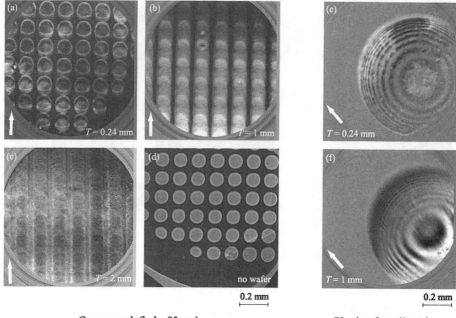

Cupper mesh (hole: 85 μm) Plastic sphere (1 mm)

FIGURE 4. A copper mesh (a - d) and a plastic sphere (e, f) observed with X-ray interferometers. Holds in the mesh become resolved in the direction parallel to the scattering plane (indicated by arrows) by thinning the wafer down to 0.24 mm. Interference fringes produced by a plastic sphere also became visible in the direction parallel to the scattering plane by thinning the wafer.

To investigate the improvement when phase contrast is dominant, we observed a plastic sphere 1-mm in diameter as shown in Figs. 4(e) and (f). When $T = 1$ mm, interference fringes almost smeared out in the direction parallel to the scattering plane. Furthermore, fringes are not concentric circles, i. e. distorted. However, when $T = 0.24$ mm, all of the interference fringes became visible and were less distorted.

Thus, the improvement in image quality was fully demonstrated by thinning a wafer. As the next step, we will check the improvement of the spatial resolution in phase maps and phase-contrast X-ray CT images. Some stains seen in Figs. 4 are due to the nonuniformity in the surface finish of the etched area. Improvement of the fabrication process is also needed.

In biological observation, the spacing of generated interference fringes is normally larger than 0.1 mm. In such a case, micrometer resolution can be achieved by thinning the wafer below 0.1 mm.

CONCLUSIONS

To improve the spatial resolution attainable with phase-contrast X-ray CT, an X-ray interferometer with a partially etched wafer was developed. The performance was preliminarily checked by observing interference patterns. Remarkable improvements in visibility and image distortion were demonstrated by comparing interference patterns obtained when the thicknesses of the wafer were 1 and 0.24 mm. The improvement in the image quality is promising for phase-contrast microtomography although some improvements in the fabrication technique are still needed. We expect phase-contrast microtomography will be a tool for investigating three-dimensional structures in biological soft tissues with the spatial resolution nearly a few microns.

ACKNOWLEDGMENTS

This study is being conducted using Special Coordination Funds for Promoting Science and Technology from the Science and Technology Agency of the Japanese Government. The experiments were performed under the approval of Spring-8 committee 1999A0079-NOM-np.

REFERENCES

1. Bonse, U., and Hart, M., *Appl. Phys. Lett.* **6**, 155-156 (1965)

2. Ando, M., and Hosoya, S., in *Proc. 6th International Conference of X-ray Optics and Microanalysis*, edited by G. Shinoda, K. Kohra, and T. Ichinokawa, Tokyo, Univ. Tokyo Press, 1972, pp. 63-68.

3. Momose, A., *Nucl. Instrum. Meth.* A**352**, 622-628 (1995)

4. Momose, A., and Fukuda, J., *Med. Phys.* **22**, 375-380 (1995)

5. Momose, A., Takeda, T., Itai, Y., and Hirano, K., *Nature Medicine* **2**, 473-475 (1996)

6. Beckmann, F., Bonse, U., Busch, F., and Günnewig, O., *J. Comput. Assist Tomogr.* **21**, 539-553 (1997)

7. Momose, A., Takeda, T., Itai, Y., and Hirano, K., in: *X-ray Microscopy and Spectromicroscopy*, edited by J. Thieme et al., Berlin, Springer-Verlag, 1998; II-207 - 212.

8. Momose, A., Takeda, T., Itai, Y., Yoneyama, A., and Hirano, K., *J. Synchrotron Rad.* **5**, 309-314 (1998)

9. Momose, A., and Hirano, K., *Jpn. J. Appl. Phys.* **38**, Suppl. 38-1, pp. 625-629 (1999)

10. Hirano, K., and Momose, A., *Jpn. J. Appl. Phys.*, submitted.

11. Batterman, B.W., and Cole, H., *Mod. Phys.* **36,** 681-717 (1964)

12. Kinoshita, K., Matsumura, T., Inagaki, Y., Hirano, N., Sugiyama, M., and Kihara, H., *SPIE Proc.* **1741**, 287-293 (1992)

BIOLOGICAL SCIENCES

Localization Of Proteins And Nucleic Acids Using Soft X-ray Microscopy

Carolyn A. Larabell[*], Deborah Yager[*], and Werner Meyer-Ilse[#]

[*]Life Sciences Division and [#]Center for X-ray Optics, Lawrence Berkeley
National Laboratory, Berkeley, CA 94720

Abstract. The high-resolution soft x-ray microscope (XM-1) at the Advanced Light Source was used to examine whole, hydrated mammalian cells, both chemically fixed and rapidly frozen and viewed in a cryostage. Using x-ray microscopy, high contrast information about the organization of the cytoplasm and nucleus of these cells was revealed at unsurpassed resolution. It is important to note that cryo-fixed cells have been examined in a state that most closely resembles their natural environment in that the cells were not exposed to chemical fixatives or chemical contrast enhancement reagents. We also used the power of soft x-ray microscopy to examine the localization of proteins and nucleic acids in whole, hydrated cells using silver-enhanced, immunogold labeling techniques. With this approach, we have obtained information about the distribution of such molecules with respect to cellular ultrastructure at five times better resolution than light microscopy. The power of soft x-ray microscopy to provide superb resolution information about the subcellular localization of proteins and nucleic acids places it in a commanding position to contribute to our understanding of the numerous molecules being identified through modern molecular biology techniques.

INTRODUCTION

Imaging cells using a variety of microscopy techniques has provided information about the organization of cells and subcellular structures that is critical for our understanding of cellular function. The challenge has been to obtain the best resolution morphological information about cells that are examined in a state most closely resembling their natural environment. Soft x-ray microscopy is proving to be a very powerful method in that one can examine whole, hydrated cells, avoiding potential artifacts introduced by the dehydration, embedding and sectioning that is required for electron microscopy. By using a cryostage, we can examine cells that have been rapidly frozen and viewed in a state more closely approximating that seen in living cells. These cells are free of chemical fixation artifacts and, since they are viewed by utilizing the contrast provided by examining them in the water window, do not require chemical enhancement agents. Furthermore, immunolabeling techniques can be used to obtain information about the distribution of proteins and nucleic acids at five times better resolution than light microscopy. We describe in this paper the use of soft x-ray microscopy to visualize cellular structures in whole, hydrated and cryo-fixed cells as well as the power of immunocytochemistry to localize specific molecules.

CP507, *X-Ray Microscopy: Proceedings of the Sixth International Conference,*
edited by W. Meyer-Ilse, T. Warwick, and D. Attwood
© 2000 American Institute of Physics 1-56396-926-2/00/$17.00

IMAGING MAMMALIAN CELLS USING X-RAY MICROSCOPY

We used the soft x-ray microscope (XM-1) at the Advanced Light Source to examine whole, hydrated mammalian cells. Cells were grown on silicon nitride windows, fixed in 2% glutaraldehyde, then examined in the fully hydrated state. These cells reveal excellent ultrastructural details of the cell nucleus and cytoplasm. The nuclear membrane in the cell shown in Figure 1 is not present since this cell is in mitosis. Instead, we see two clusters of chromosomes in the center of the cytoplasm as they begin moving to opposite regions of the cytoplasm.

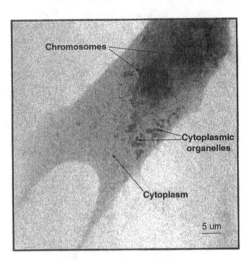

FIGURE 1. Mouse 3T3 fibroblast that was fixed in 2% glutaraldehyde and viewed in the hydrated state. The cell is undergoing division and the separating chromosomes are apparent as they begin moving to opposite poles. Numerous organelles are seen in the cytoplasm.

CRYO X-RAY MICROSCOPY OF MAMMALIAN CELLS

The ultimate goal of cell biologists/microscopists is to obtain information about the structure of cells at the best possible resolution with minimal perturbation of the cellular ultrastructure during processing. Rapid freezing techniques have long been recognized as an excellent way to achieve this goal, and several rapid freezing devices are commercially available and routinely utilized by a small number of laboratories. Unfortunately, techniques for examining these well-frozen cells in ways that do not require subsequent, potentially damaging processing have been more difficult to develop. Viewing rapidly frozen whole cells using high voltage electron microscopes, which provide superb resolution, has not been possible since these microscopes are restricted to imaging very thin (< 1 μm) objects. As a consequence, the well-frozen cells must then be processed using techniques such as freeze-substitution and low-temperature embedding followed by preparation of thin sections. The use of soft x-ray cryo-microscopy, however, provides an excellent solution to this problem since whole cells up to 10 μm thick can be examined following cryo-fixation. We used a cryo-fixation apparatus and cryo-stage built by the Center for X-ray Optics (CXRO) and in operation at XM-1 at the ALS (see Meyer-Ilse et al., this issue) to examine initially live mammalian cells, such as the 3T3 fibroblast shown in Figure 2. The nuclear membrane surrounding the nucleus is well preserved, as are numerous cytoplasmic organelles such as the long, tubular mitochondria. These cells were remarkably stable during viewing and demonstrated no apparent radiation damage, even after repeated imaging with the x-ray microscope (data not shown here; see Meyer-Ilse et al., this issue). Imaging rapidly frozen whole cells using cryo-tomography will enable us to obtain unique three-dimensional information about cells and interactions of intracellular organelles.

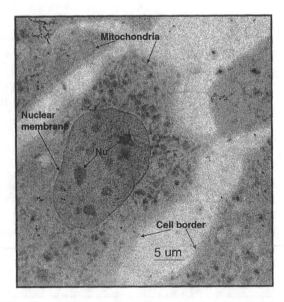

FIGURE 2. Mouse 3T3 fibroblast cells that were rapidly frozen and examined in the cryostage. The nucleus contains several nucleoli (Nu) and is surrounded by a distinct nuclear membrane. Numerous vesicles and organelles, such as mitochondria are seen in the cytoplasm.

IMMUNOLOCALIZATION OF PROTEINS IN MAMMALIAN CELLS

A powerful method for identifying structure-function relationships of cells and proteins in cells is to determine the subcellular location of these proteins using immunocytochemistry. This requires chemical fixation of the cells followed by incubation in antibodies that recognize a specific protein followed by a tagged probe detectable by the microscope. We used a technique routinely used for electron microscopy and recently used for x-ray microscopy (1,2) that involves silver-enhancement of gold-tagged antibodies to examine the distribution of proteins in whole hydrated cells. The cells were fixed in 2% paraformaldehyde, 0.1% glutaraldehyde, and 0.1% Triton X-100 in a cytoskeletal buffer. They were then treated with SuperBlock to prevent non-specific binding of antibodies to other proteins, incubated in anti-tubulin primary antibodies, rinsed, then incubated in secondary antibodies tagged with 1.4 nm gold particles. After antibody labeling the cells were fixed in 2% glutaraldehyde to stabilize them, then incubated in silver to enhance the size of the gold particles for viewing in the x-ray microscope. The labeled microtubule network can be seen coursing through the cytoplasm of a mammary epithelial cell (Figure 3). This network, which is critical for transport of organelles and molecules throughout the cell, is composed of 25 nm diameter tubules that are visualized as a result of their decoration with heavy dense precipitates of silver.

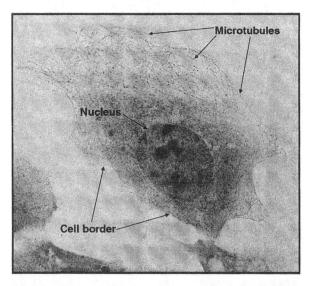

FIGURE 3. Microtubule network in a mouse mammary epithelial cell. Cells were fixed, labeled with primary antibodies against tubulin, followed by secondary antibodies tagged with 1.4 nm gold particles that were subsequently enhanced with silver.

The nucleus of the cell is extremely thick and has been difficult to study using electron microscopy. Therefore, attempts to examine nuclear structure have relied on the use of extensive extraction procedures. We examined the distribution of a splicing factor in cells that had been prepared by mild permeabilization prior to fixation. Using this approach, we can detect numerous clusters of splicing factor in the nucleus in addition to several nucleoli in each nucleus (Figure 4).

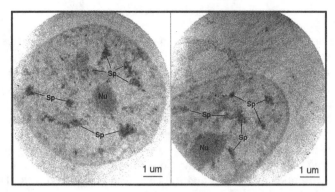

FIGURE 4. Localization of splicing factor in the nucleus of whole, hydrated cells. Permeabilized cells were fixed, incubated in primary antibodies against splicing factor then secondary, gold-labeled antibodies that were subsequently silver-enhanced for visualization in the x-ray microscope.

IMMUNOLOCALIZATION OF NUCLEIC ACIDS IN MAMMALIAN CELLS

It is also important to know the subcellular distribution of nucleic acids. RNAs, for example, are often positioned in distinct cellular locations where large numbers of proteins are needed upon short notice. We are using in situ hybridization to examine the distribution of RNAs in whole cells (Figure 5). Cells were fixed in 2% paraformaldehyde, 0.1% glutaraldehyde and 0.1% Triton X-100 in a cytoskeleton buffer, incubated in digoxigenin-labeled actin mRNA, then incubated in gold-labeled anti-digoxigenin antibodies that were subsequently enhanced with silver. The cells were then examined in the x-ray microscope and demonstrated a uniform distribution of actin mRNA throughout the cytoplasm and some mRNA particles in the nucleus. The control cell, which was not exposed to mRNA, contains no label.

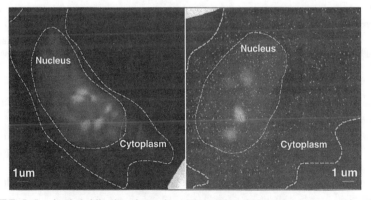

FIGURE 5. In situ hybridization for actin mRNA. Blue dots in the image on the right represent particles of digoxigenin-tagged actin mRNA that have been localized using antibodies against digoxigenin followed by silver-enhanced, gold labeled secondary antibodies. The image on the left is the control, which was not exposed to the digoxigenin-tagged actin mRNA, and is free of label.

SUMMARY

We have used soft x-ray microscopy to examine whole, hydrated cells that were either chemically fixed or rapidly frozen and examined using a cryo-stage. These images reveal ultrastructural details of whole, hydrated cells at 36 nm resolution - five times better resolution than possible with visible light microscopy. The use of the cryo-stage enabled us to examine the ultrastructure of cells that were initially living and most closely resemble the state of cells within tissues. We are now in a position to use cryo-tomography to obtain unique three-dimensional images of the subcellular structural organization of intact cells in a way not possible using any other form of microscopy. We have also utilized the power of x-ray microscopy to examine the subcellular distribution of proteins and RNAs in whole, hydrated cells. This technique will be of critical importance in the post-genomic era as we face the daunting task of determining the function of the vast number of genes and gene products identified as a result of modern molecular biology techniques.

ACKNOWLEDGEMENTS

We gratefully acknowledge the staff of Berkeley Lab's Center for X-ray Optics for their assistance in operation of the microscope. We also thank Dr. Mina Bissell, Director of the Life Sciences Division at Berkeley Lab, for her enthusiastic support. This research is supported by the United States Department of Energy Office of Biological and Environmental Research and the Laboratory Directed Research and Development Program of the E. O. Lawrence Berkeley National Laboratory under the Department of Energy contract No. DE-AC03-76SF00098.

REFERENCES

[1] G. Schmahl, P. Guttmann, G. Schneider *et al.*, *Phase Contrast Studies of Hydrated Specimens with the X-ray Microscope at BESSY* (Bogorodskii Pechatnik Publishing Company, Chernogolovka, 1994).

[2] H. N. Chapman, C. Jacobsen, and S. Williams, "A Characterisation of Dark-Field Imaging of Colloidal Gold Labels in a Scanning Transmission X-Ray Microscope," Ultramicroscopy **62** (3), 191-213 (1996).

[3] Meyer-Ilse, W., Denbeaux, G., Johnson, L. E., Bates, W., Lucero, A., and Anderson, E. H. The High Resolution X-ray Microscope XM-1.

Morphological Studies of Human Sperm Using the Aarhus X-ray Microscope

Joanna V. Abraham-Peskir[1], Eric Chantler[2], Peter Guttmann[3], Tage Hjort[4], Robin Medenwaldt[1], Christine McCann[2], Erik Uggerhøj[1], Thomas Vorup-Jensen[4]

1 ISA, Institute for Storage Ring Facilities, University of Aarhus, DK-8000, Denmark
2 Academic Unit of Obstetrics and Gynaecology and Reproductive Health Care, St Mary's Hospital, University of Manchester, Manchester M13 0JH, UK
3 Forschungseinrichtung Röntenphysik, Georg-August-Univerität Göttingen and Berliner Elektronenspeicherring-Gesellschaft für Synchrotronstrahlung m.b.H (BESSY), Germany
4 Department of Medical Microbiology and Immunology, University of Aarhus, DK-8000, Denmark
E-mail: jabraham@ifa.au.dk

Abstract. Using the Aarhus transmission X-ray microscope we have shown that the mitochondria of human spermatozoa can exist in two morphologically distinct states. We have also discovered new structures on the human spermatozoon surface. These structures manifest as clear vesicular bodies associated with specific membrane domains. They can occur around the acrosomal segment, the mid-piece region or at the basal region. Prior to our findings they were not described in the literature, even though they were clearly visible by light microscopy and ubiquitous among populations of sperm from fertile donors. We report on our findings and subsequent endeavours to elucidate the function of these fascinating structures.

In the early 1990s it was demonstrated by various groups that sperm could be imaged by XM (1, 2, 3), and that X-ray absorption near edge spectroscopy (XANES) in combination with scanning XM could also be applied (4). More recently, XM has been applied directly to the study of sperm (5, 6, 7, 8). The human sperm head is an ideal size for XM studies: the head (approx. 5 x 3 µm) and mid-piece fit into the imaging field of view (a fixed value of 10-15 µm). Therefore, the need for constructing composite images is eliminated. DNA is densely packed into the head resulting in high photoelectric absorption compared to the surrounding medium, and consequently high contrast images are obtained. Sperm are relatively radiation resistant, visible structural damage is not seen on multiple exposures. Therefore, sperm can be investigated without chemical fixation or the use of cryo-techniques.

Currently there are two ongoing research programmes in which the Aarhus XM plays a major role. With Eric Chantler's group from Manchester University, UK, we are investigating sperm membranes and with Tage Hjort's group from Aarhus University, Denmark, we are investigating sperm mitochondria.

CP507, *X-Ray Microscopy: Proceedings of the Sixth International Conference,*
edited by W. Meyer-Ilse, T. Warwick, and D. Attwood
© 2000 American Institute of Physics 1-56396-926-2/00/$17.00

Vesicular Bodies Associated With Specific Membrane Domains

With Eric Chantler's group we described for the first time vesicular bodies associated with specific membrane domains of human sperm (7). Sperm have five specific membrane domains, the acrosomal segment, equatorial segment, basal, mid-piece, and tail. Each domain has specific properties, in terms of fluidity, lipid and cholesterol content.

Vesicular bodies had low X-ray absorbency (Figure 1) unlike residual bodies known as cytoplasmic droplets which are usually re-absorbed before ejaculation, but retained on immature sperm. The vesicular bodies we described had not been previously reported in the literature. Initially they were thought to be radiation-induced artefacts. However, the size, shape, and location did not change after multiple exposures. More significantly vesicular bodies were seen clearly by light microscopy (LM).

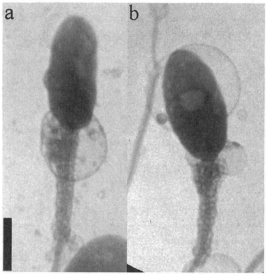

FIGURE 1. Aarhus XM images of fully-hydrated human sperm, showing vesicular bodies associated with **a** the mid-piece region, and **b** the acrosomal segment. Scale bar =2 μm.

We investigated 24 human semen samples from 18 donors of proven fertility supplied by CRYOS International Sperm Bank Ltd., Aarhus. All samples had sperm with vesicular bodies associated with the mid-piece. Figure 2 shows the number of mid-piece vesicles for each sample investigated. The mean number of sperm with mid-piece vesicles was 30 out of 100, counted within one hour post-ejaculation. Donor 12 had a high incidence of vesicles on three separate occasions. This could indicate that the occurrence of vesicular bodies may be of clinical significance. So far, we have only looked at donors of proven fertility. However, we plan to investigate samples from infertile donors.

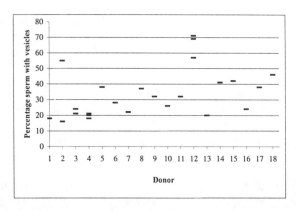

FIGURE 2. Percentage of human spermatozoa with vesicular bodies associated with the mid-piece region, counted by DIC light microscopy, 24 samples from 18 donors of proven fertility.

Vesicular bodies associated with the acrosomal segment were commonly seen on sperm in seminal plasma by XM, but not by LM. However, we were able to chemically induce vesicular bodies associated with the acrosomal region and then visualise them by LM. Intracellular calcium increases as the sperm matures (9). Alkalisation within the sperm head, combined with a transient influx of calcium act to initiate calcium efflux from an internal calcium store, leading to sustained elevation of internal calcium and to the acrosome reaction (10). Exposing the sperm to calcium ionophore A23187 chemically induces the processes leading to acrosome reaction. Vesicular bodies associated with the acrosomal segment were seen by LM after addition of the ionophore to a sample that had been removed from seminal plasma. Using a chlortetracycline (CTC) fluorescence assay (11), we were able to monitor sperm maturation and distinguish acrosome-reacted sperm. By this method we saw that as the sperm matured, head vesicles appeared, and after acrosome reaction they disappeared.

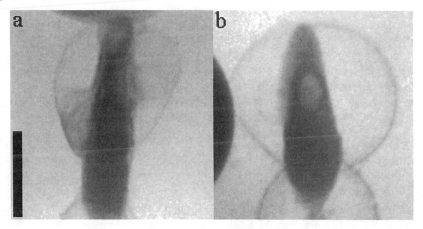

FIGURE 3. Aarhus XM images of fully-hydrated human sperm heads. **a** The acrosomal cap (arrow) has become detached and material exudes within the vesicular body. **b** Vesicular body sticking to a Dynosphere spacer. Scale bar =2 µm.

Figure 3a shows a sperm head undergoing acrosome reaction. The acrosomal cap is perforated and material is leaking out. Note that this is occurring inside the vesicular body associated with the acrosomal region. The head vesicles also appeared to be sticky. Figure 3b shows a head vesicle adhering to a Dynosphere.

If vesicular bodies play a specific role in the sperm maturation process, then one could hypothesise that similar structures exist on sperm of other mammalian species. Vesicular bodies were found to be associated with the mid-piece region of bovine sperm were seen by LM (Figure 4a) and XM (Figure 4b). Vesicular bodies associated with the acrosomal segment were also seen by XM (Figure 4c).

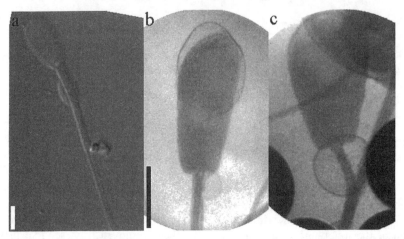

FIGURE 4. Bull sperm. **a** DIC LM image showing vesicular body associated with the mid-piece region. **b, c** Aarhus XM images showing vesicular bodies associated with (b) the acrosomal segment and (c) the mid-piece region. Scale bars = 3 μm.

Mitochondria

The processes involved in the stages that lead up to induction of the acrosome reaction are poorly understood. As the sperm matures it acquires the capacity to fertilise, thus reaching a state termed capacitation. Hyperactivated swimming patterns are a characteristic of capacitation. However, little is known about the capacitation process. Many studies have understandably concentrated on the head during this process, with less emphasis on mid-piece changes.

Preliminary studies of human sperm, before and after exposure to capacitating conditions, were made using the XM at BESSY I, Germany. Morphological changes to the mitochondria of the mid-piece region were seen. This work was extended with Tage Hjort's group from Aarhus University. Using a larger cohort of donors, a distinct morphological change in the mid-piece region of human sperm after exposure to capacitating conditions was verified from Aarhus XM images (8). In the fresh ejaculate the sperm mitochondria were tightly wrapped around the mid-piece (Figure 5a). The mitochondria became loosely wrapped or distended after exposure to capacitating conditions (Figure 5b). Figure 5c shows a model of the axoneme and mitochondria, morphologically transformed to the loosely-wrapped state. In seminal plasma, hardly any of the sperm had mitochondria that were loosely wrapped. However, after exposure to capacitating conditions almost all the sperm had mitochondria that were loosely wrapped or distended (8).

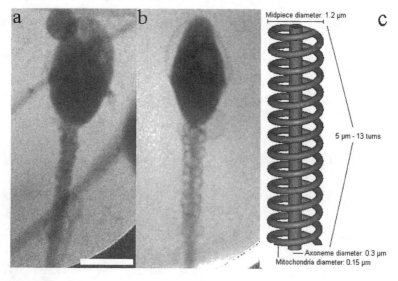

FIGURE 5. Aarhus XM images of human sperm taken from Human Reproduction 14 (1999) 880-884. **a** In seminal plasma showing mitochondria tightly wrapped around the axoneme. **b** After exposure to capacitating conditions, mitochondria are loosely wrapped. **c** Model of the mid-piece region showing loosely-wrapped mitochondria. Scale bar = 2 µm.

That mitochondria change morphology is in line with the general concept that mitochondria are mobile and plastic, changing shape in a matter of minutes depending of the prevailing physiological conditions and during spermiogenesis (12, 13). However, this has not been previously observed for the mitochondria of post-ejaculatory sperm, and could be linked to the observation that after capacitation sperm change their motility pattern to a hyperactivated state. Studies are continuing to elucidate the biochemical processes involved during this morphological change.

ACKNOWLEDGEMENTS

We thank CRYOS International Sperm Bank Ltd. for there continuing support, and Stuart Lunt for providing valuable assistance.

REFERENCES

1. Da Silva, L. B., Trebes, J. E., Balhorn, R., Mrowka, S. J., Anderson, E. H., Attwood, D. T., et al., *Science* **258**, 269-271 (1992).

2. Loo, B. W., Williams, S., Meizel, S., and Rothman, S. S., *J. Microscopy* **166**, RP5-6 (1992).

3. Tomie, T., Shimizu, H., Majima, T., Yamada, M., Kanayama, T. Kondo, H. et al., *Science* **252**, 691-693 (1991).

4. Ade, H., Zhang, X., Cameron, S., Costello, C., Kirz, J., and Williams, S., *Science* **258**, 972-5 (1992).

5. Balhorn, R., Corzett, M., Allen, M. J., Lee, C., Barbee, T. W., and Koch, J.A., et al., *Soft X-ray Microscopy* **SPIE 1741**, 374-385 (1992).

6. Zhang, X., Balthorn, R., Mazrimas, J., and Kirz, J., *J. Struct. Biol.* **116**, 335-344 (1996).

7. Abraham-Peskir, J., Chantler, E., McCann, C., Medenwaldt, R., and Ernst, E., *Med. Sci. Res.* **26**, 663-667 (1998).

8. Vorup-Jensen, T., Hjort, T., Abraham-Peskir, J. V., Guttmann, P., Jensenius, J. C., Uggerhøj, E., and Medenwaldt, R., *Hum. Reprod.* **14**, 880-884 (1999).

9. Baldi, E., Casano, R., Falsetti, C., Krausz, C., Maggi, M., and Forti, G., *J. Androl.* **12**, 323-330 (1991).

10. Florman, H. M., Corron, M. E., Kim, T. D., and Babcock, D. F., *Dev. Biol.* **152**, 304-314 (1992).

11. Lee, M. A., Trucco, G. S., Bechtol, K. B., Wummer, N., Kopf, G. S., Blasco, L., and Storey, B. T., *Fertil. Steril.* **48**, 649-658 (1987).

12. Woolley, D. M., *J. Cell Sci.* **6**, 865-879 (1970).

13. Otani, H., Tanaka, O., Kasai, K., and Yoshioka, T., *Anat. Rec.* **222**, 26-33 (1988).

Development And Evaluation Of Cryo-imaging Of Unicellular Algae Using Soft X-ray Transmission Microscopy: Ultrastructure And Elemental Analysis.

T.W.Ford[1], W.Meyer-Ilse[2] & A.D.Stead[1].

[1]Division of Biology, School of Biological Sciences, Royal Holloway, University of London, Egham, Surrey TW20 0EX, UK. [2]Center for X-Ray Optics, Lawrence Berkeley National Laboratory, One Cyclotron Road, Berkeley, CA94720 USA.

Abstract. Living, untreated cells are very radiation-sensitive when exposed to soft X-rays, which creates problems for microscopy. Using increased soft X-ray fluence (and hence reduced exposure times) or fixation of cells with glutaraldehyde did not provide realistic solutions. However, rapid freezing of samples preserved structural integrity, increased radiation resistance (up to 2×10^7 Gy or more) and enhanced the resolution and contrast of the image. Oxygen mapping was also achieved by sequential imaging of a frozen cell either side of the oxygen edge.

1. INTRODUCTION

The ultrastructure of the unicellular green alga *Chlamydomonas reinhardtii* has been extensively studied by transmission electron microscopy (TEM) and living cells of this organism have also been examined by soft X-ray contact microscopy (SXCM) and soft X-ray transmission microscopy (SXTM). Images from both these techniques generally support the appearance under TEM though differences have been reported, in particular the presence of relatively large, X-ray dense spheres within the cell. The preferred technique for examining living cells is SXTM since this provides an immediate image for evaluation and the potential for data collection and elemental analysis. However, it appears to have a rather narrow window of exposure which will produce images of sufficient resolution but without obvious radiation damage to the cell ultrastructure. Intracellular structures, in particular the spheres, are very sensitive to soft X-rays and dosages in excess of 10^6 Gy (commonly exposures in excess of 2 seconds) have produced clear structural damage (1). Possible solutions would be to reduce the exposure time or improve the radiation stability of the sample.

2. IMAGING LIVING AND FIXED *CHLAMYDOMONAS* CELLS

Recently zone plates with a much greater efficiency have been installed on XM-1 allowing

CP507, *X-Ray Microscopy: Proceedings of the Sixth International Conference,*
edited by W. Meyer-Ilse, T. Warwick, and D. Attwood

a reduction in the imaging times for living *Chlamydomonas* cells. The shortest exposure which produced a visible image was 0.005A.s.with a calculated dose of $4x10^5$Gy though the resolution was poor with only the cell outline and spheres visible.

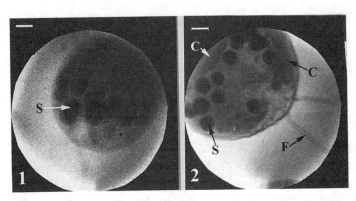

FIGURES 1&2. (1) Image of living untreated cell following 0.05A.s exposure ($4x10^5$Gy) and (2) frozen cell following 0.2A.s. exposure ($0.8-2x10^7$Gy). C-chloroplast; F-flagellum; S-sphere. Bar=1µm.

A second exposure of 0.005A.s. two minutes later showed no evidence of damage to spheres. Single exposures of 0.05A.s. (equivalent to $4x10^5$Gy) produced slightly better resolution (Fig. 1) and exposures up to 0.2A.s. (equivalent to $1.6x10^6$Gy) were possible without damage to the spheres, though sequential imaging of the same cell was not possible without visible radiation damage.

Cells which are chemically fixed should be more resistant to radiation damage though the fixation process itself is likely to introduce structural artefacts into the resulting image (2). Cells of *Chlamydomonas* were fixed in 1.25% glutaraldehyde for approximately 5 minutes before transfer to fresh culture medium. Images of these cells following 0.1A.s. exposure (equivalent to $1x10^6$Gy) showed clear damage to the spheres which was not obvious in fresh cells following the same exposure.

3. IMAGING FROZEN *CHLAMYDOMONAS* CELLS

Calculations suggest that frozen biological material can withstand up to four orders of magnitude increases in radiation without detectable structural damage (3). This would allow not only longer exposure, but also repeat exposures of the same specimen. However, freezing biological specimens can itself introduce ultrastructural damage. Several methods are currently employed in electron microscopy techniques (4) and the cryo-preservation of algae is routinely used in culture collections (5).

Water can account for 40-90% of a biological sample though micro-organisms such as unicellular algae usually have a much lower water content. When such samples are cooled below zero, ice crystals form which can damage the ultrastructure of cells. Ultrastructure is preserved best in the vitreous state though this requires rapid freezing at rates of $>2 \times 10^5$°C.sec^{-1}. In practice, the accepted criterion for freezing speed is that it should reduce ice crystal size to a point where there is no visible structural distortion (which will depend on the resolution of the microscopy).Ultrastructural damage can be reduced by including a cryoprotectant such as glycerol though this has the potential for introducing artefacts into the image. Cell organelles vary in their sensitivity to ice crystal damage with mitochondria, endoplasmic reticulum and the Golgi apparatus showing best preservation whilst the nucleus is the most sensitive.

For imaging *Chlamydomonas* using XM-1 at the ALS, a monolayer of cells sandwiched between two silicon nitride windows was frozen by jets of helium, cooled by passage through liquid nitrogen, onto each window. This achieved freezing rates of 10^3°C.s^{-1} and a final sample temperature of -150°C which was maintained during imaging.

Frozen cells of *Chlamydomonas* could be exposed to dosages of $0.8-2 \times 10^7$Gy without any visible radiation damage (Fig 2) This image shows clearly the insertion of the flagella into the anterior X-ray-lucent end of the cell, thylakoids in the chloroplast and a number of intact, X-ray-dense spheres in the anterior cytoplasm. This exposure is in excess of that which caused visible destruction of spheres in living cells.

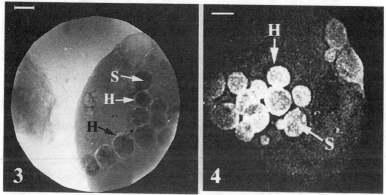

FIGURES 3&4. Images of slow-frozen cells (3) following 0.5A.s. exposure (1×10^7Gy) and (4) showing oxygen distribution by sequential imaging at 2.4nm and 2.3nm. H-halo; S-spheres. Bar=1μm.

However, one problem with holding cells at these very low temperatures is that ice formation occurs on the outside of the silicon nitride windows of the sample holder. In an attempt to avoid this, slow freezing of samples was carried out at between 10 and 10^2°C.s^{-1} with the cells being held at only -50°C during imaging. Spheres in such slow-frozen cells appear intact after a dosage of 1×10^7Gy (Fig. 3) showing similar radiation resistance as fast frozen cells (Fig. 2) and greater resistance than living cells (Fig. 1). However, even after the first exposure, the spheres had a clear, soft X-ray-transparent halo around them which was not seen in images of fast-frozen cells. This could be due to shrinkage of the spheres during the slower freezing process or radiation damage due to the sample being held at only -50°C rather than -150°C.

An important attribute of SXTM is the ability to perform elemental mapping on hydrated, as opposed to dried or fixed cells, and so avoid concerns about the loss or movement of mobile elements during the preparation procedures. By using frozen specimens with their higher radiation resistance, it should be possible to collect sufficient data without radiation-induced ultrastructural damage. Although the data have not been fully analysed, images taken of slow-frozen cells either side of the oxygen absorption edge show that the spherical inclusions are prominent features of the cells when the wavelength used is 2.4nm but these features show increased X-ray density when the wavelength used is 2.3nm (the absorption edge of oxygen), suggesting that these spheres are rich in oxygen. An oxygen map produced from these two images shows the spheres to be particularly oxygen-rich (Fig. 4).

4. CONCLUSIONS

Imaging of living, untreated *Chlamydomonas* cells is possible up to a radiation dose of 1.6×10^6Gy without visible radiation damage. However, sequential imaging of the same cell is only possible at much lower dosage (typically 5×10^4Gy). Fixation of cells with glutaraldehyde is not a viable option since even a short period of fixation caused damage to spheres. Freezing of the cells increased their resistance to radiation damage substantially and doses of up to 2×10^7Gy produced no visible damage. The resolution and contrast was much improved in these cells with several ultrastructural features clearly visible. Reducing the speed of freezing and holding the sample at a higher temperature reduced the ice formation on the silicon nitride windows but appeared to cause some shrinkage of cell contents though the spheres remained intact. By imaging the same cell either side of the oxygen edge it was possible to produce a map of oxygen concentration within the cell which suggested that the spheres were particularly oxygen-rich.

5. ACKNOWLEDGEMENTS

The advice and assistance of Bill Bates, Greg Denbeaux, Lewis Johnson and Angelic Lucero is gratefully acknowledged

6. REFERENCES

1.Ford,T.W., Page,A.M., Meyer-Ilse,W., Brown,J.T., Heck,J. and Stead,A.D. *X-ray Microscopy and Spectromicroscopy,* Berlin, Springer. 1998 pp. II,185-II,190.
2.Galway,M.E., Heckman,J.W., Hyde,G.J. and Fowke,L.C. *Methods in Cell Biology Vol. 49*, Academic Press, 1995, ch. 1, pp. 3-19.
3. Scheider,G. and Niemann,B. *X-ray Microscopy and Spectromicroscopy,* Berlin, Springer. 1998 pp. I,25-I,34 .
4. Severs,N.J. and Shotton,D.M. *Cell Biology; a Laboratory Handbook 2nd Edition Vol. 3* Academic Press, 1998, pp. 299-309.
5. Crutchfield,A.L.M., Diller,K.R. and Brand,J.J. *European Journal of Phycology* **34**, 43-52 (1999).

Tomographic Imaging Of Cryogenic Biological Specimens With The X-ray Microscope At BESSY I

D. Weiß, G. Schneider, B. Niemann,
P. Guttmann, D. Rudolph, G. Schmahl

Georg-August-Universität, Institut für Röntgenphysik (IRP), Geiststr. 11, 37073 Göttingen, Germany

Abstract. Soft X-ray microscopy employs the natural absorption contrast between water and protein in the 2.34 - 4.38 nm wavelength region with a resolution down to 30 nm. The large depth of focus of the Fresnel zone plates used as X-ray objectives permits tomographic reconstruction based on the microscopic images. High-resolution images require a high specimen radiation dose, and a large number of images taken at different viewing angles is needed for tomographic reconstruction. Therefore, cryo microscopy is necessary to preserve the structural integrity of hydrated biological specimens during image acquisition. The cryo transmission X-ray microscope at the electron storage ring BESSY I (Berlin) was used to obtain a tilt series of images of the frozen-hydrated green alga *Chlamydomonas reinhardtii*. The living specimens were inserted into borosilicate capillaries, then rapidly cooled by plunging into liquid nitrogen. The capillary specimen holders allow image acquisition over the full angular range of 180°. The reconstruction shows details inside the alga down to 60 nm size and conveys a detailed impression of the specimen structure. This technique is expected to be applicable to a wide range of biological specimens, such as the cell nucleus. It offers the possibility of imaging the three-dimensional structure of hydrated biological specimens close to their natural living state.

INTRODUCTION

Classical computed tomography (CT) calculates the three-dimensional (3D) distribution of the linear absorption coefficient (LAC) of an object based on a series of geometric projections of the object absorption taken at different tilt angles. X-rays are well suited to acquire such projections because the refraction due to changes in the refractive index of the object is negligible. However, the resolution of the projections is limited to a few microns [1]. This restriction can be overcome by using an objective to acquire magnified images of the object.

Soft X-ray microscopy has been used to visualize 30 nm structures in frozen-hydrated biological specimens [2]. Because of the large depth of focus of several microns of the Fresnel zone plates used as X-ray objectives, a magnified image of an object acquired with an X-ray microscope can be used as an approximation of a projection of the object at this high resolution. Computed tomography based on X-ray

CP507, *X-Ray Microscopy: Proceedings of the Sixth International Conference,*
edited by W. Meyer-Ilse, T. Warwick, and D. Attwood
© 2000 American Institute of Physics 1-56396-926-2/00/$17.00

microscopic images has already been demonstrated on micro-fabricated gold patterns using a scanning transmission X-ray microscope [3], and on the mineralized sheaths of bacteria *Leptothrix ochracea* using the transmission X-ray microscope at the electron storage ring BESSY I [4].

At 2.4 nm wavelength, the linear absorption coefficient of protein is about an order of magnitude larger than that of water, providing a natural contrast mechanism for hydrated biological specimens. In order to apply CT to these radiation-sensitive biological specimens, the LAC distribution must be preserved during the acquisition of the tilt series. This can be achieved by imaging at cryogenic temperatures, where the proteins are embedded in a matrix of vitreous ice which prevents the diffusion of the products of radiation damage and effectively preserves the LAC distribution at the microscopic resolution.

EXPERIMENTAL SETUP

The cryo transmission X-ray microscope (CTXM) at BESSY I images hydrated biological specimens at cryogenic temperatures and thus eliminates radiation-induced object damage during image acquisition [5]. The cryo stage has now been modified to permit rotation of a cryo object around an axis perpendicular to the optical axis [6]. An important prerequisite for artifact-free reconstruction is a sufficiently large angular range of the tilt series, ideally spanning 180 degrees. Suitable object holders are manufactured using a micropipette puller. A borosilicate glass tube is locally heated, then pulled until it breaks at its thinnest point and produces a tapering glass capillary with an inner diameter of ca. 10 μm and a wall thickness of 0.3 - 0.4 μm. Borosilicate glass was chosen for its low X-ray absorption at 2.4 nm wavelength. The rotationally symmetric object holder serves as a container for a liquid medium in which many biological specimens can be suspended. The capillary force propels medium and specimens to the very tip of the capillary. The X-ray transmission of the water-filled capillary tip is approx. 15 - 20% at 2.4 nm wavelength.

VITRIFICATION AND DATA ACQUISITION

Once filled, the capillary tip is rapidly cooled by plunging into liquid nitrogen or liquid ethane. Liquid nitrogen (LN_2) is easily handled but it exhibits the Leidenfrost phenomenon, and average local cooling rates do not exceed 500 K/s, whereas cooling in liquid ethane permits up to 15.000 K/s [7]. Vitrification must take place with a cooling rate of at least 10.000 K/s, otherwise microscopic ice crystals form and may destroy the specimen ultrastructure [7].

A suspension of living green algae *Chlamydomonas reinhardtii* was injected into the capillary holder and, in this first experiment, cooled by plunging into LN_2. The object holder was transferred to the CTXM immersed in LN_2, and images of an alga were acquired for 42 tilt angles spanning 185 degrees. To improve the signal-to-noise ratio of the intensity data, three images were acquired and added for each tilt angle. After image acquisition, the images were aligned to a common axis of rotation. For this purpose colloidal gold spheres of 60 nm diameter were added to the alga suspension. The gold spheres adhere to the inner glass wall and provide a set of high-contrast fiducial markers. Image alignment was performed using an algorithm proposed by Penczek et al. for aligning electron micrographs for electron tomography, where a 3D marker model is iteratively constructed by minimizing the sum of the squared differences between the measured marker positions and the corresponding positions reprojected from the 3D marker model [8].

RESULTS

The linear absorption coefficient of the alga was reconstructed using a multiplicative algebraic technique [4]. The reconstructed volume conveys a clear impression of the inner structure of the specimen (Fig. 1). The prominent features of the cell (cell walls, fibrous chloroplast, pyrenoid with starch hull imbedded in the chloroplast) are known from electron microscopic studies of dried thin sections [9]. Another striking feature are several X-ray dense spherical vesicles. There is evidence of flagellar roots (cf. Fig. 1, bottom of the lower left panel), but not of the flagellae themselves. In the lumen of the chloroplast, where there should be the nucleus and nucleolus, the reconstruction shows a honeycomb-like structure, probably a consequence of the imperfect vitrification.

The edge sharpness of the glass wall and the full width at half maximum (FWHM) of the colloidal gold spheres give an indication of the resolution currently obtained in the reconstruction. The FWHM of the gold spheres ranges from 70 to 100 nm, for a nominal diameter of 60 nm; the glass wall edge sharpness corresponds to a FWHM of 60 to 70 nm.

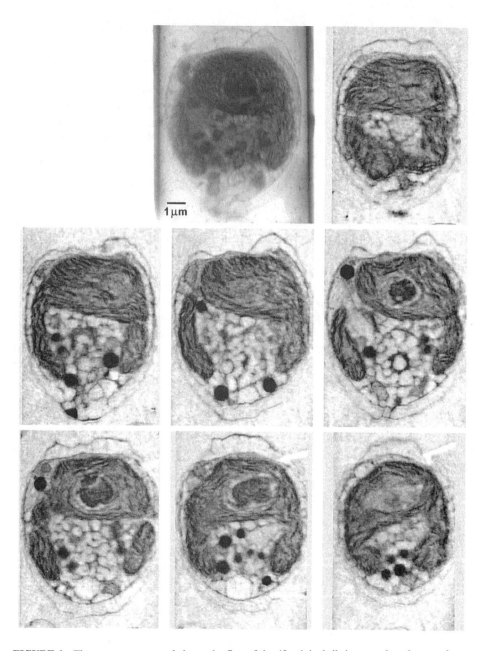

FIGURE 1. The upper center panel shows the first of the 42 original tilt images, the other panels are parallel slices through the reconstructed volume, viewed at the same viewing angle as the original image, and spaced at 0.6 µm. Low absorption has been mapped to light gray, and high absorption to dark gray.

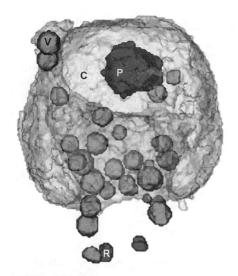

FIGURE 2. Watershed segmentation of the reconstructed specimen yields the identifiable organelles: chloroplast (C), pyrenoid (P), spherical vesicles (V), and flagellar roots (R) (visualization using the Amira system developed at ZIB.)

TABLE 1. Quantitative analysis of some specimen organelles, based on segmented volume data (cf. Fig. 2), showing the absolute and relative volume as well as the average linear absorption coefficient (LAC). To exclude the borders of the segmented regions from the LAC average, the regions were eroded with a 3 pixel radius before averaging.

	vol. [μm^3]	vol. [%]	LAC [μm^{-1}]
chloroplast	129.13	47.54	0.385(91)
pyrenoid	3.42	1.26	0.441(66)
spherical vesicles	4.00	1.47	0.711(86)
flagellar roots	0.14	0.05	0.206(39)
other	134.91	49.67	0.187(80)
total	271.60	100.00	0.306(137)

To perform quantitative analysis on the organelles of the specimen, the reconstructed volume was segmented using 'watershed segmentation' [10]. The resulting regions are compact and disjunct and can be visualized as surfaces (Fig. 2). Based on this segmentation, the organelles can be characterized in regard to their volume and average LAC (Table 1). The X-ray dense spherical vesicles show the highest LAC value; however the expected value of approx. 1 μm^{-1} of protein is not reached. The protein is hydrated at a level below the reconstruction resolution; this lowers the apparent LAC.

Evaluating the volume outside the alga gives a LAC value for ice of 0.098(14) μm^{-1}. Based on the atomic scattering factors tabulated by Henke [11], theoretical LAC values of 0.109(2) μm^{-1} for ice and 1.35(14) μm^{-1} for a typical protein $C_{94}H_{139}N_{24}O_{31}S$ with $\rho_{protein} = 1.35(14)$ g/cm^3 were calculated. With the complete-cell average LAC of 0.306(137) μm^{-1}, this yields a (dry) protein content of the alga of 17(11)%.

CONCLUSIONS

The cryo transmission X-ray microscope at BESSY I was modified to permit the acquisition of a tilt series of images of a cryogenic biological specimen. Borosilicate glass capillaries were used as specimen holders with unlimited angular range. A series of 42 tilt images spanning 185° was acquired of a specimen of the green alga *Chlamydomonas reinhardtii*, which, in this first experiment, had been cooled using liquid nitrogen. The images were aligned to a common axis of rotation using colloidal gold spheres as fiducial markers.

Based on the X-ray microscopic images, the specimen was reconstructed using a multiplicative algebraic technique. The reconstruction shows the 3D specimen structure with a resolution approaching that of the microscopic images. The internal organelles of the specimen present themselves as known from light and electron microscopic studies, with the exception of the nucleus which was probably destroyed during the cooling process. The spherical vesicles are highly absorbing; the usual electron-microscopic staining techniques do not show this. The reconstructed volume permits quantitative analysis of the volume and average linear absorption coefficient of the organelles.

ACKNOWLEDGEMENTS

The authors would like to thank the staff of the Sammlung von Algenkulturen at Göttingen for providing the specimens, P. Nieschalk for excellent workshop support, and the staff at BESSY I for good working conditions. This work was supported by the Deutsche Forschungsgemeinschaft under contracts Schm1118/5-1 and Schm1118/5-2.

REFERENCES

1. Bonse, U., and Busch, F., *Prog. Biophys. molec. Biol.* **65**, 133-169 (1996).

2. Schneider, G., *Ultramicroscopy* **75**, 85-104 (1998).

3. Haddad, W. S., McNulty, I., Trebes, J. E., Anderson, E. H., Levesque, R. A., and Yang, L., *Science* **266**, 1213-1215 (1994).

4. Lehr, J., *Optik* **104**, 166-170 (1997).

5. Schneider, G., Niemann, B., Guttmann, P., Rudolph, D., and Schmahl, G., *Synchrotron Radiation News* **8**, 19 (1995).

6. Schneider, G., Niemann, B., Guttmann, P., Weiß, D., Scharf, J.-G., Rudolph, D., and Schmahl, G., "Visualization of 30 nm Structures in Frozen-Hydrated Biological Samples by Cryo Transmission X-Ray Microscopy", this volume.

7. Echlin, P., *Low-Temperature Microscopy and Analysis*, New York: Plenum Press, 1992, pp. 32-71.

8. Penczek, P., Marko, M., Buttle, K., and Frank, J., *Ultramicroscopy* **60**, 393-410 (1995).

9. Harris, E. H., *The* Chlamydomonas *Sourcebook*, San Diego: Academic Press, 1988, pp. 65-99.

10. Beucher, S., and Meyer, F., "The Morphological Approach to Segmentation: The Watershed Transformation" in *Mathematical Morphology in Image Processing*, edited by E. R. Dougherty, New York: Marcel Dekker, 1992, pp. 433–481.

11. Henke, B. L., Gullikson, E. M., and Davis, J. C., *Atomic Data and Nuclear Data Tables* **54**, 342 (1993).

The High Resolution X-ray Microscope, XM-1

W. Meyer-Ilse[1], G. Denbeaux[1,2], L.E. Johnson[1], W. Bates[1],
A. Lucero[1], E. H. Anderson[1]

[1]Center for X-ray Optics, Lawrence Berkeley National Laboratory, Berkeley California 94720, USA
[2]Physics Department, Duke University, Durham NC 27708, USA

Abstract. We give an overview of the activities at the high-resolution x-ray microscope XM-1 at the Advanced Light Source, including both scientific programs and instrumental enhancements. The instrument is being actively used in many fields including biology, environmental and material sciences. A new high efficiency condenser zone plate and precision computer control of the microscope allow users to obtain many hundreds of images in a day. Further developments at XM-1 include a cryogenic sample stage for sample preservation and plans for the implementation of a cryo-tilt stage to capture stereoscopic information.

THE MICROSCOPE

The XM-1 x-ray microscope was built in 1994 by the Center for X-ray Optics (CXRO) at Lawrence Berkeley National Laboratory to provide a high throughput of high spatial resolution transmission images, from a wide variety of thick (< 10 micron) samples (1). Much of the heritage of XM-1 is derived from the microscope pioneered by the University of Göttingen as they share similar optical configurations (2). An overview of XM-1, which is modeled after a conventional bright field microscope is presented in figure 1. Bending magnet radiation from the Advanced Light Source (ALS) provides the illumination source. Radiation from the bending magnet is reflected off of a plane mirror at glancing incidence to filter out higher photon energies. The effective microscope operating range in photon energy is 250 - 900 eV.

The condenser zone plate (CZP), which collects the radiation from the ALS, is used to illuminate the sample. A new CZP constructed using electron beam lithography

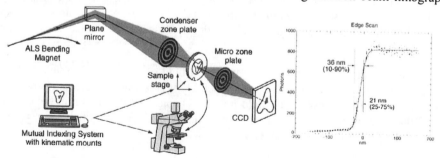

FIGURE 1. a) Optical Layout of the Xm-1 X-ray Microscope b) knife edge resolution test results.

CP507, *X-Ray Microscopy: Proceedings of the Sixth International Conference*,
edited by W. Meyer-Ilse, T. Warwick, and D. Attwood
© 2000 American Institute of Physics 1-56396-926-2/00/$17.00

was installed in October, 1998 (3). The new CZP has a diameter of 9 mm, 41,000 zones, an outer zone width of 55 nm, and much-improved efficiency. The combination of the CZP and a pinhole produces a linear monochromator of moderate spectral resolution by exploiting the known chromatic aberrations of zone plates. Sample illumination wavelength is selected by varying the distance between the CZP and pinhole. Following the pinhole, the beam is brought through a thin window to atmospheric pressure and to the sample holder.

Samples with a thickness of less than 10 microns can be imaged dry, hydrated, or cryogenically fixed. The sample can be mounted between nominal 1000Å thick silicon nitride windows, or on standard TEM grids. The sample holder has a three-point kinematic mount designed to fit both the XM-1 and a custom Zeiss Axioplan visible light microscope (VLM). Sample positions and focus are pre-selected with the VLM, which is mutually indexed with the sample stage of XM-1. X-Y position accuracy is typically 2 microns over a 3mm field with focal accuracy better than 1 micron.

This system is especially useful in reducing total exposure times, as the researcher is able to pre-select individual sample elements with the visible light microscope, followed by automated x-ray imaging with accurate position and focus. Moreover, the researcher is able to view many samples in one session, thus providing a high throughput with hundreds to even thousands of images obtainable per day.

The radiation passing through the sample is collected by the micro zone plate (MZP) and imaged at high magnification onto a 1024 x 1024 pixel x-ray CCD camera. Images are typically magnified 2400x. The pixel size projected back to the sample is typically 10 nm. There is an option to combine several pixels into one by binning the CCD camera, thus reducing the radiation dose and time exposure, for example a factor of 4 with 2 x 2 binning. Exposure times for images with low photon noise are typically on the order of 1 second. Images and microscope parameters are recorded digitally and are available to the user within seconds of exposure.

To first order, the spatial resolution of the microscope is determined by the outer zone width of the MZP (although other factors enter). The present MZP has an outer zone width of 35 nm with 318 zones and a diameter of 45 microns (3). In knife-edge measurements, the 10% to 90% intensity range, which approximates Rayleigh resolution, yields a result of 36 nm (Figure 1). Results are expected to improve in the coming months. A new 25 nm outer zone width MZP is being installed as this article goes to press.

Though the CZP illuminates a 10-micron field of view, it is possible to build images of a larger area using a montage assembly. This automated process builds a larger image based on several sub-field images. Typical images range in dimensions up to 100 x 100 microns. Using cross correlation techniques the smaller images are placed at the proper positions creating a nearly seamless montage.

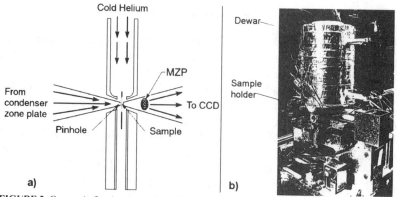

FIGURE 2. Cryogenic fixation system (a) schematic (b) sample stage.

Cryogenic Sample Stage

In order to preserve the structural integrity of biological samples, a cryogenic sample holder has been built. The radiation dose from each exposure at XM-1 is roughly 10^7 Gray which, depending on the sample, can cause changes in the morphology of wet samples after one or two images. In order to keep the samples intact for many images, and also to prevent the formation of ice crystals, the sample is quickly frozen at a rate of about 3000 K/s to a temperature of -130 °C where it is maintained. The freezing of the sample is accomplished by blowing cold helium gas across the sample (figure 2). The helium is cooled as it passes through a dewer filled with liquid nitrogen, and then is directed across the sample window to freeze the sample. Once the sample is frozen, it is able to withstand many exposures. In one experiment a cryogenically fixed 3T3 fibroblast cell was imaged 40 times in the same location (figure 3). At the resolution of this microscope there was no apparent change in the nuclear membrane due to the multiple exposures.

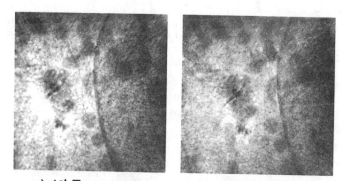

a) 1st Exposure   b) 40th Exposure

FIGURE 3. Cryogenically Fixed 3T3 Cells after a)1st and b)40th Exposures. The thin vertical line is the nuclear membrane which shows no observable modification due to the radiation dose. (with C. Larabell, D. Yager, unpublished)

Spectromicroscopy

As mentioned previously, the sample illumination photon energy is selectable by varying the distance between the condenser zone plate and pinhole. With recent additions to the microscope control software it is now relativity easy to obtain a series of full field sample images, with each image at a different energy. A series of images taken at 100 different photon energies each with a spatial resolution of about 36 nm takes about 20 minutes. In order to measure the spectral resolution of XM-1, the Calibration and Standards Beamline (6.3.2) at the ALS (4) which has a known spectral resolution of $\lambda/\Delta\lambda > 4000$ was used to precisely determine the spectral dependence of absorption of a thin sample of CaF_2 at the Calcium L edge (~350 eV). The same measurement of the sample was also performed with XM-1. The convolution of the high precision 6.3.2 data with spectral bandwidth of XM-1 yields the measured spectra from XM-1. The result of this calculation yielded a spectral resolution of 0.5 eV FWHM indicating a spectral resolving power of $(E/\Delta E)$ approximately 700 at 350 eV. Such a resolution is sufficient to distinguish different elemental species. In some cases it is sufficient to facilitate chemical state identification. It has been shown in a test sample to show that it is possible to distinguish between the different oxidation states of chromium (5).

SCIENTIFIC HIGHLIGHTS

In recent years many exciting scientific studies have been conducted making use of XM-1. In the biological arena, high-resolution protein localization has been demonstrated (6). Gold-labeled antibodies were used to selectively identify proteins and oligonucleotides within a cell. The gold particles were silver-enhanced to form aggregates approximately 50 nm in diameter. These particles were resolved with the XM-1 and a mapping of the location of the target proteins and oligonucleotides were obtained. Some of the proteins and oligonucleotides labeled and imaged are microtubulin, nuclear pore complex, and actin mRNA. An example of the images is shown in figure 4, where tubulin protein labeled within a 3T3 epithelial cell is clearly seen. More recently, cryo-fixed images of a 3T3 cell were obtained as seen in figure 5 (6). In the montage assembly, the nuclear membrane, nucleoli, organelles, and granules are all clearly visible.

In addition, many material and environmental science studies were conducted using XM-1. The microscope has been used to study Alkali-silica reactions in concrete samples used in various water storage dams (7). These reactions can cause swelling and cracking of the dam structure. Reaction gel morphology was studied extensively with XM-1 (figure 6a) The microscope has also been used to observe the distribution of Mn in micronodules produced by biomineralization (8). Making use of the image contrast above and below the Mn L_3 edge provided the mechanism to produce a mapping of the Mn distribution. Moreover the high spatial resolution of the XM-1 was able to reveal a network of very fine needle like shaped sturctures.

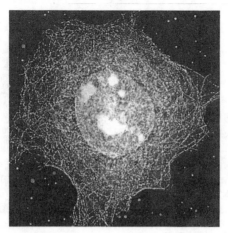

FIGURE 4. Tubulin Network in Epithelial Cell (C. Larabell, S. Lelievre, D. Hamamoto, M. Bissell, A. Nair and W. Meyer-Ilse, submitted for publication)

The microscope was also used to study macromolecular structures of humic substances in aqueous solutions, soils, and in sediments (figure 6b) (9). These studies have shown that the macromolecular structures vary as a function of both solution chemical composition and mineral chemistry. This information is useful in accurately predicting the organo-mineral interactions, C-cycle, and contaminant transport in soils and aquatic systems. The above is just a sampling of studies using XM-1 during the past year. Additional studies involved nanocrystals, chromium-reducing bacteria, and magnetic materials.

Conclusion

A user-friendly, full field, x-ray transmission imaging microscope has been constructed and is in use for a wide variety of scientific studies at a spatial resolution of 36 nm. Users are able to obtain a set of images of their specimens, which can be dry, hydrated, or cryo-fixed. Plans are underway to improve the spatial resolution of the microscope and to develop its spectromicroscopic capabilities.

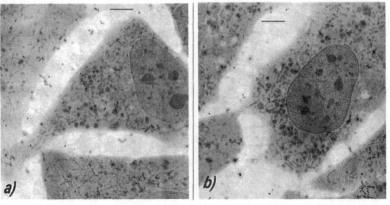

FIGURE 5. Cryogenically Fixed 3T3 Cells (with C. Larabell, D. Yager, T. Shin, these proceedings)

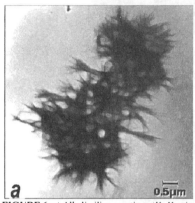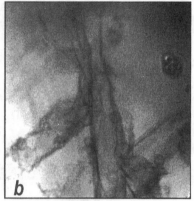

FIGURE 6. a) Alkali-silica reactions (K. Kurtis et al., ref 7) b) Humic substances in aqueous solutions (S. Myneni,et al, ref 9)

Acknowledgments

The authors would like to acknowledge our colleagues of the Center for X-ray Optics, the Life Science Division, and the ALS. In particular, the support from D. T. Attwood, D. Yager, and C. Larabell is to be noted. The US Dept. of Energy and The Office of Navy Research supported this work. AFOSR provided research support for student participation and training. This great instrument is a testament to the hard work, dedication, and insight of Werner Meyer-Ilse. His driving force and light will be deeply missed.

References

1. Meyer-Ilse, W., Medecki H., Jochum, L. Anderson, E., Attwood, D., Magowan, C., Balhorn, R., Moronne, M., Rudolph, D., Schmahl, G., *Synchrotron Radiation News* **8**, 23-33 (1995).

2. Schmahl, G., Rudolph, D., Niemann, B., Christ, O., *Quarterly Reviews of Biophysics* **13**, 297-315 (1980)

3. Anderson, E., Harteneck, B., Olynick, D., "Nanofabrication of X-ray Zone Plates with the Nanowriter Electron-Beam Lithography System", these proceedings

4. Underwood, J.H., Gullikon, E.M., " Beamline for measurements and characterization of multilayer optics for EUV Lithography", in Emerging Lithographic Technologies II, Proceedings of SPIE 3331, pp. 52-61.

5. Denbeaux, G., Johnson, L.E., Meyer-Ilse, W., "Spectromicroscopy at the XM-1", these proceedings

6. Larabell, C., Shin, T., Yager, D., "Localization of proteins and oligonucleotides using soft x-ray microscopy", these proceedings

7. Kurtis, K.E., Monteiro, P.J.M., Brown, J.T., Meyer-Ilse, W., "Investigation of Alkali-silica Reaction by Transmission Soft X-ray Microscopy", these proceedings

8. Rothe, J., Kneedler, E.M., Pecher, K., Tonner, B., Nealson, K.H., Grundl, T., Meyer-Ilse, W., Warwick, T., "Spectromicroscopy of Mn Distributions in Micronododules produced by Biomineralization"

9. Myneni, S.C.B., Brown, J.T., Martinez, G.A., Meyer-Ilse, W., "Imaging of humic substance macromolecular structures in water and soils", Science Magazine, in press.

X-Ray Microscopy Study of Track Membranes and Biological Objects.

I.A.Artioukov[1], V.E.Asadchikov[2], V.I.Gulimova[3], V.E.Levashov[1],
B.V.Mchedlishvili[2], A.N.Kurohtin[4], A.A.Postnov[2], A.V.Popov[4],
S.V.Saveliev[3], I.I.Struk[1] A.I.Vilensky[2], A.V.Vinogradov[1], D.L.Zagorsky[2].

1. Lebedev Physical Institute, RAS, Moscow, Russia.
2. Institute of Crystallography, RAS, Moscow, Russia.
3. Institute of Human Morphology, RAMS, Moscow, Russia.
4. Institute of Earth Magnetism and Radio Wave Propagation, RAS, Troitsk, Russia.

Abstract. The development of two types of X-ray microscopy applying to the organic objects investigation (biological samples and polymer matrix) is reported. Polymer track membranes were investigated using Schwarzchild X-ray microscope with 20 nm wavelength. Pore diameters down to 0.2 μm were clearly imaged. Contact X-ray microscopy at 0.229 nm wavelength was used to obtain clear images of inner structure of native biological samples. High contrast together with the high resolution (about 2-3μm) allowed us to use this method for quantitative analysis of demineralization process taking place in the skeleton of amphibious after several weeks of weightlessness on biosputnik board.

SCHWARZCHILD X-RAY MICROSCOPY OF TRACK MEMBRANES.

Track membranes are created from the polymer film by bombarding with heavy ions. After bombarding the film is etched until required channels appear. The conditions of bombarding and etching defines the shape and sizes of channels together with the pore density. Such membranes were investigated at 20 nm wavelength with 20^X Schwarzchild X-ray objective and laser plasma source equipped with condenser [1]. The source driver was a small scale repetitive laser (0.1J/5ns, second harmonic of 1.06 μm) focused on a W target.

The investigated objects were unregular track membranes made of polyethyleneterephtalate with the pore diameters 1 μm (the matrix thickness is 10 μm) and 0.2 μm and regular track membrane with the pore diameters 0.5 μm and 1 μm period. We obtained the sharp pictures of the filtration channels in all cases (pic.1,2). This images were used to determine pores shape and density distribution.

We must stress that the practically achieved resolution is higher than of a visible light microscope. Moreover, the objects with such a high aspect ratio (up to 50) are very difficult for visible light microscopy study due to the diffraction in the body of the cannel. So, submicron pore diameters in the thick matrix (10-20μm) are opaque for visible light. Better resolution of course can be achieved using electron scanning microscopy, but it still gives information about the modified surface. The Schwarzchild microscope investigates the entire channel and can by used as a powerful instrument to define the rate of pollution of the track membrane when some channels are fouled by the products of filtration.

CP507, *X-Ray Microscopy: Proceedings of the Sixth International Conference,*
edited by W. Meyer-Ilse, T. Warwick, and D. Attwood
© 2000 American Institute of Physics 1-56396-926-2/00/$17.00

CONTACT X-RAY MICROSCOPY

Contact microscopy at 0.229 nm wavelength was applied to study the inner structure of biological samples. The experimental setup was based on the small-angle difractometer AMUR-1 (Institute of Crystallography, RAS, Moscow). Sealed X-ray tube was used as a source and an X-ray film as a detector.

Liofilization through critical point was used to prepare objects for X-ray investigations. This kind of treatment doesn't change the inner structure of the sample and ultrastructure of the cells. The same technology is used to prepare the biological samples for electron microscopy study.

We present the image of *Salamandrela keyserlingii* with different times of exposure (pic.3) One can estimate the "optical density" of different tissues from finger joints to blood system. Due to the monochromatic illumination with the specially selected wavelength of 0.229 nm we can observe high contrast between tissues with close chemical composition and density. The thickness of about 10 μm (blood vessels) is enough to obtain the visible contrast. The use of CCD matrix can bring further improvement of the contrast but with the limited field of view.

Quality of the images and high adsorption of the calcified tissue (bone tissue) in comparison with cartilage allowed us to make a quantitatively investigation of the process of demineralization in triton skeleton after two weeks of weightlessness on biosputnik board.

Several samples *of Pleurodeles waltlii (Amphibia Urodela)* were studied in order to define the influence of space flight on the mineral metabolism. While the spaceflight the control group of tritons developed on the Earth coping the same conditions and events that happened in biosputnik, so all distinctions but weightlessness were compensated. The tritons were mature and all changes had histological character.

On the picture 4 one can see two types of bones: control group and investigated group. These micrographs were studied using Wacom Computer System GmbH – Wacom Ultra Pad equipment. Canvas (v.5.0.2.) software helped us to define perimeters and calculate the square of bone and cartilage tissues. We must stress that absorption picture cannot define strictly the real percentage of two types of tissues. But their rate can be calculated and doesn't depend upon orientation of the bones and time of exposure (the only condition is that all bones must be orientated the same way).

The 27±4% demineralization of space bones was observed. This figure is close to that obtained by histological investigations of the opposite legs of the same animal. But the advantages of contact microscopy are that it is nondestructive and operative way of investigation.

CONCLUSIONS

The high contrast photos of different kinds of objects (frog larvae, early phase embryos, elements of skeleton, track membranes) were obtained by two methods: Schwarzchild microscopy at 20 nm and contact microscopy at 0.229 nm wavelength. The bone tissue demineralization of *Pleurodeles waltlii (Amphibia Urodela)* after 2 weeks of space flight was studied quantitatively. The 27% demineralization was observed. The resolution achieved was 2-3 μm for contact microscopy and at least 0.2 μm for Schwarzschild microscopy.

Both methods have following values:
1. They are nondestructive
2. They investigate the inner structure instead of the surface.
3. They use common inexpensive sources.

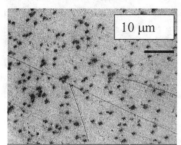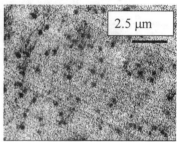

Figure 1. Images of pores in polyethyleneterephtalate obtained by use of Schwarzschild microscope with the working wavelength 20 nm. Images of 1 μm pores (right), images of 0.2 μm pores (left).

Hard X-ray contact microscopy can also provide stereo photos with 2-3 μm resolution.

Assuming the future of these methods in 3D reconstruction of unprepared biological objects and tissues with the 0.229 nm or close wavelength applying it to embryos development and cancer analysis we connect the future success in resolution improvement with the X-ray optics (such as Fresnel zone plates) development. Moscow X-ray optics group is taking steps in this direction [2].

ACKNOWLEDGEMENTS

This work have been supported by ISTC Project # 1051-99.

REFERENCES

1. V.E.Levashov, A.V.Vinogradov. Resonance diffraction efficiency enhancement in sliced multilayers. Appl.Opt., vol.32(7), p.1130-1135, (1993).
2. V.E.Asadchikov, V.I.Beloglazov, A.V.Vinogradov, D.L.Voronov, V.V.Kondratenko, Yu.V.Kopylov, N.F.Lebedev, A.G.Ponomarenko, A.V.Popov, A.A.Postnov, A.I.Fedorenko. "Sliced Zone Plate for Hard X-ray Sources. Manufacturing and Testing." Proc. SPIE. 1997, Vol.3113, pp.384-392.

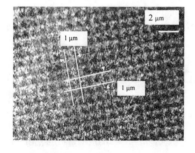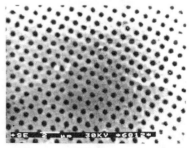

Figure 2. Regular track membrane photo made by Schwarzschild microscope in comparison with those made by electron microscope. Pore diameters are 0.5 μm, period of the structure is 1 μm. Wavelength 20 nm.

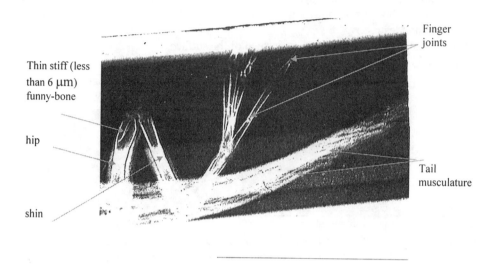

Finger joints

Thin stiff (less than 6 μm) funny-bone

hip

shin

Tail musculature

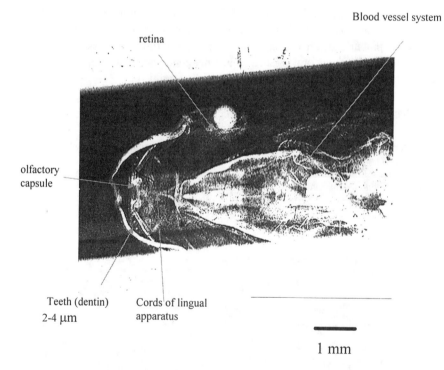

Blood vessel system

retina

olfactory capsule

Teeth (dentin) 2-4 μm

Cords of lingual apparatus

1 mm

Figure3. X-ray contact image of *Salamandrela keyserlingii* with different times of exposure.

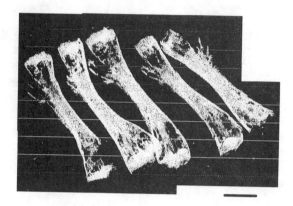

1 mm

Figure4. Humeral bones of *Pleurodeles waltlii* (*Amphibia Urodela*), control group.
Contact photo. Wavelength 0.229 nm.(upper picture)

Humerial bones of *Pleurodeles waltlii* (*Amphibia Urodela*) after two weeks of weightlessness on biosputnik board. Measured demineralization of bone tissue is 27%.(lower picture)

Gadolinium Uptake by Brain Cancer Cells: Quantitative Analysis with X-PEEM Spectromicroscopy for Cancer Therapy

Gelsomina De Stasio [a,b], B. Gilbert[c], P. Perfetti[b], G. Margaritondo[c],
D. Mercanti[d], M. T. Ciotti[d], P. Casalbore[d], L. M. Larocca[e], A. Rinelli[e],
and R. Pallini[e]

[a] *University of Wisconsin-Madison, Department of Physics, Madison WI 53706, USA,*
e-mail: pupa@src.wisc.edu
[b] *Istituto di Struttura della Materia, CNR, Rome, Italy.*
[c] *IPA-EPFL, CH-1015 Switzerland.*
[d] *Istituto di Neurobiologia, CNR, Rome, Italy.*
[e] *Universita' Cattolica del Sacro Cuore, Rome, Italy.*

Abstract. We present the first X-PEEM spectromicroscopy semi-quantitative data, acquired on Gd in glioblastoma cell cultures from human brain cancer. The cells were treated with a Gd compound for the optimization of GdNCT (Gadolinium Neutron Capture Therapy). We analyzed the kinetics of Gd uptake as a function of exposure time, and verified that a quantitative analytical technique gives the same results as our MEPHISTO X-PEEM, demonstrating the feasibility of semi-quantitative spectromicroscopy.

INTRODUCTION

Gadolinium Neutron Capture Therapy (GdNCT)

Gadolinium neutron capture therapy (GdNCT) is a non-invasive alternative therapy for brain cancer, which has never been clinically tested. It is based on two steps: first the patient is injected with a tumor-seeking Gd compound; second, the patient's skull is exposed to thermal neutrons, which induce a short range destructive nuclear reaction ($^{157}Gd(n,\gamma)^{158}Gd$) [1,2]. Gd has a capture cross section for thermal neutrons much higher than any other element, so if Gd can be delivered only to regions of tumor tissue, the neutron capture reaction induces destruction of tumor, leaving neighboring healthy tissue unharmed. Gadolinium compounds selectively accumulate in brain tumor tissues. They are in fact used as tumor contrast-enhancing agents for magnetic resonance imaging, given the large magnetic moment of the Gd^{3+} ion [3].

CP507, *X-Ray Microscopy: Proceedings of the Sixth International Conference,*
edited by W. Meyer-Ilse, T. Warwick, and D. Attwood
© 2000 American Institute of Physics 1-56396-926-2/00/$17.00

EXPERIMENTAL METHODS

Microchemical analysis of glioblastoma cells was performed with the MEPHISTO X-PEEM [4,5] producing images and x-ray absorption spectra [6] from microscopic areas. Spectra can be acquired simultaneously from regions selected on the real time image of the sample surface (the probed depth is on the order of 100 Å). For this work, MEPHISTO was mounted on the HERMON beamline of the Wisconsin Synchrotron Radiation Center. Since the output of the monochromator is smooth around the Gd3d edge, all spectra were normalized dividing by a third-order polynomial fit to the raw data. The normalized spectra depend only on the Gd lineshape and local concentration. This allows us to make comparisons of relative Gd concentrations between different cells or sub cellular structures.

Inductively Coupled Plasma Atomic Emission Spectroscopy (ICP-AES) is a quantitative analysis of aqueous samples that can reach a sensitivity of a few ppb for some elements. The analysis of cell samples is performed by first digesting the 10^6 cells/sample in nitric acid [7].

We cultured cells from a glioblastoma patient that had to undergo surgery independent of our experiment. The (mostly glial) cells were plated on gold coated silicon substrates, and exposed to gadopentetic acid for 0-72 hours. The cell cultures for MEPHISTO analysis were ashed by exposure to UV/O$_3$ to remove carbon and nitrogen, and enhance the local concentration of Gd, which is otherwise not detectable [8].

RESULTS AND DISCUSSION

Figure 1 shows the MEPHISTO micrograph and Gd spectra acquired on glioblastoma cells. These results are representative of about 300 similar regions that we analyzed on a total of 10 cell cultures grown in parallel, exposed to Gd for 0-72 hours.

Figure 2 shows the Gd uptake *vs* exposure time analyzed in MEPHISTO and with ICP-AES. In the MEPHISTO spectra, after normalization, we measured the Gd3d peak intensity at 1172 eV, and calculated the standard deviation from the mean [Gd] value for each exposure time. The results match very closely the curve obtained from ICP-AES: the unexposed samples show no gadolinium signal; the Gd concentration increases with the exposure time, up to 48 hours, then decreases. In the MEPHISTO data of Fig. 2 the normalized Gd 3d peak intensities (arbitrary units) were scaled by a factor of 350,000 to match the ICP data (mg/ml in the original cells). This empirical calibration factor contains all the experimental response contributions (x-ray beam intensity, electron optical transmission function, etc.) to the final video signal. Since the samples studied in MEPHISTO were ashed before analysis, and no independent reference samples of known [Gd] were analyzed, it is not possible to convert the

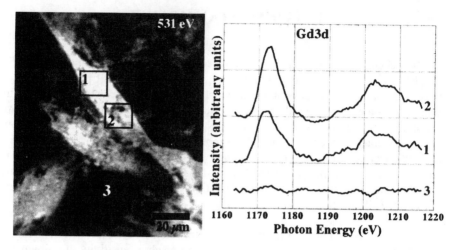

FIGURE 1. Left: MEPHISTO micrograph of a region of glioblastoma cells in culture, exposed to Gd for 48 hours and carefully washed to remove all non-uptaken Gd. The boxes indicate the areas from which the spectra were acquired. Right: Gd3d spectra taken on the box-areas on the image. The 3 spectra were acquired simultaneously. Note the presence of Gd in cells, not on the substrate (region 3), and the higher intensity of Gd spectra on the cell nuclei. The Gd signal from area 2 (nucleus) is clearly stronger than from the flatter, brighter cytoplasm within area 1, and this trend was reproduced across all cell cultures.

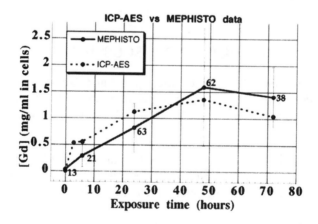

FIGURE 2. Gadolinium uptake curves in glioblastoma cell cultures as followed by Atomic Emission Spectroscopy (ICP-AES) and MEPHISTO spectromicroscopy. Each one of the points represents either the average concentration measure over 10^6 cells in the ICP-AES data or the average over many independent peak intensities of spectra acquired in MEPHISTO on individual cells. The number of cells that contribute to each data point is shown next to each marker. Note the agreement of the two curves within the error bars. [Data from Ref. 2].

MEPHISTO absorption intensities into absolute Gd concentrations. However, the careful normalization procedure enables the relative concéntration comparisons to be made.

The error on the averaged MEPHISTO Gd intensity at each time point is much greater than that for the ICP-AES data. This partly reflects the variance in [Gd] in different cells (lost in the averaged ICP-AES measurement over 10^6 cells) and partly experimental error. The data reported here are the first evidence of semi-quantitative elemental concentration measurements X-PEEM, but other related techniques (XPS, Auger ES) reach accuracies of only 20-50% when quantitative measurements are made.

CONCLUSIONS

X-PEEM spectromicroscopy of Gd in glioblastoma cell cultures, after exposure to Gd for the optimization of GdNCT, demonstrated that [Gd] in cells is maximum after 48 hour exposure, then decreases. This result was verified on independent cell cultures and with two different techniques: quantitative ICP-AES and X-PEEM. The results of these two experiments coincide, demonstrating that X-PEEM is a semi-quantitative analytical technique. The advantage of using spectromicroscopy over other, more accurate, quantitative techniques is the possibility of identifying and localizing elements in subcellular structures. In particular we observed a tendency of Gd to accumulate preferentially in cell nuclei, giving hope for the success of GdNCT.

AKNOWLEDGMENTS

The experiments were performed at the Wisconsin Synchrotron Radiation Center, a facility supported by NSF. We are indebted to Didier Perret for allowing us to use the ICP-AES instrument at the University of Lausanne, and to Mark Bissen for his expert help during the experiments on the SRC-HERMON beamline.

REFERENCES

1. Shih, J. A., Brugger, R. M. "Gadolinium as a Neutron Capture Therapy Agent" in *Progress in Neutron Capture Therapy*, edited by Allen, B. J., Moore, D. E. and Harrington, B. V., New York: Plenum Press, 1992, pp183-186.

2. De Stasio, G., Casalbore, P., Gilbert, B., Mercanti, D., Ciotti, M. T., Larocca, L. M., Rinelli, A., Perret, D., Perfetti, P., Margaritondo, G., and Pallini, R., submitted.

3. Mathur-De Vre, R. "The Structural and Physical Aspects Affecting the Efficiency of Metal Chelates as Contrast Agents: Proton Relaxation Enhancement" in EUR 10986 - *Contrast Agents for MRI Tissue Characterization: Basic Principles and Research Methodology*, edited by de Certaines, J. D.

and Podo, F., Luxembourg: Office for Official Publications of the European Communities, 1988, pp26-42.

4. De Stasio, G., Capozi, M., Lorusso, G.F., Baudat, P.A., Droubay, T.C., Perfetti, P., Margaritondo, G., and Tonner, B.P., *Rev. Sci. Instrum.* **69**, 2062-2067 (1998).

5. De Stasio, G., Perfetti, L., Gilbert, B., Fauchoux, O., Capozi, M., Perfetti, P., Margaritondo, G., Tonner, B. P., *Rev. Sci. Instrum.* **70**, 1740-1742 (1999).

6. Gudat, W., and Kunz, C., *Phys. Rev. Lett.* **29**, 169-173 (1972).

7. E. Andrasi, J. Nadasdi, Zs. Molnar, L. Bezur and L. Ernyei. *Biol. Trace Element Res.*, **26-27**. 691-698 (1990).

8. De Stasio, Gelsomina, B. Gilbert, L. Perfetti, R. Hansen, D. Mercanti, M. T. Ciotti, R. Andres, P. Perfetti, and G. Margaritondo. *Anal. Biochem.* 266:174-180 (1999).

144

XANES of Mammalian Cells in the Soft X-ray Region for the Basis of X-ray Imaging

Atsushi Ito[*], Kunio Shinohara[†] and Yoshinori Kitajima[‡]

[*]*Department of Nuclear Engineering, School of Engineering, Tokai University, Hiratsuka, Kanagawa 259-1292 JAPAN*
[†]*Radiation Research Institute, Faculty of Medicine, the University of Tokyo, Bunkyo-ku, Tokyo 113-0033 JAPAN*
[‡]*Photon Factory, Institute of Materials Structure Science, High Energy Accelerator Research Organization, Tsukuba, Ibaraki 305-0801 JAPAN*

Abstract. To image molecule or element distribution in a cell, XANES of such biomolecules in a cellular environment give direct information about photon energy choosing for imaging. This study describes XANES of dried Chinese hamster ovary cells in the pellet form at the N, O, P and S-K absorption edges. In addition, XANES of dried pellet of nuclear and mitochondrial fraction were measured at the N and O-K edges. At the N-K edge XANES of cells had peaks at the identical energies of isolated DNA and histone, indicating that XANES peaks of biomolecules are insensitive to the cellular environment. The similar results were obtained for the cases of the P-K and S-K edges: XANES profile of cells at the P-K edge is likely to be a mixed pattern of DNA and related molecules including nucleotides, and the profile at the S-K edge could be explained by the superposition of those of SH and SS compounds. However, at the O-K edge, XANES of cells and nuclei exhibited the same peak slightly shifted to the higher energy compared with that in XANES of histone. This result indicates that for molecule or element imaging of a specimen, XANES should be measured under the specimen environment.

INTRODUCTION

Absorption peaks that appear in X-ray Absorption Near Edge Structure (XANES) have been widely recognized as useful for molecule or element imaging in a specimen, because the peaks reflect chemical environment of the constituent elements such as chemical bonding. XANES profiles of biological molecules have been reported: DNA and related compounds at the N-K absorption edge (1), DNA and BSA at the C-K edge (2), DNA and histone at the N-K and the O-K edges (3), Ca compounds at the Ca-L absorption edge (4), and S compounds at the S-K (3, 5) and the S-L edge (3). Biological applications including Ca distribution in bones (6, 7), DNA in a chromosome (2), DNA and protein in a sperm (8), and DNA in a HeLa cell (9) have been performed based on such spectral features. Considering that XANES peaks are likely to be influenced by molecular or elemental environment, XANES peaks to be used for imaging should be determined under the closely related environment to the specimen. Such measurement was very scarce except CHO cells at the C-K edge (10).

In the present study, to obtain XANES spectra of biomolecules in a cellular environment, we measured XANES of whole mammalian cells (cell pellet) and

CP507, X-Ray Microscopy: Proceedings of the Sixth International Conference,
edited by W. Meyer-Ilse, T. Warwick, and D. Attwood

isolated intracellular organelles, and compared them with XANES of biomolecules in thin dry film.

MATERIALS AND METHODS

XANES measurement. XANES at the N-K and O-K edges were measured at BL-11A beamline at the Photon Factory, Tsukuba, Japan with the energy resolution of E/ΔE~1500, while monochromatic X-rays around the P-K and S-K edges were obtained from double Si crystal monochromator with E/ΔE=1500~2000 installed at the beamline of BL-11B.

XANES was measured by the transmission method, i.e. the transmitted X-rays through specimen were detected by silicon photodiode (AXUV-100, International Radiation Detectors Inc., U.S.A.).

Preparation of specimens. Chinese hamster ovary (CHO) cells were used. In the preparation of cell pellet specimen, cell suspension was dropped on a collodion coated EM grid, and then dried in the air. Cellular nuclei and mitochondria were isolated as follows: cells homogenized with Teflon homogenizer were centrifuged at 1,000g. Resulting pellet is mainly composed of nuclear fraction. Supernatant was further centrifuged at 10,000g for 10 min. Mitochondrial fraction is in the pellet. Organelle specimens in the form of pellet were prepared in the same manner as the cell pellet.

RESULTS AND DISCUSSION

Figure 1 shows XANES of cell pellet and nuclei pellet at the N-K edge. For comparison spectra of DNA and histone were plotted in the same figure. The major peak was observed in the both spectra of cells and nuclei at the identical energy to that

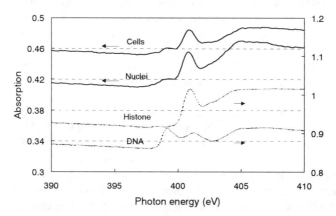

FIGURE 1. XANES of CHO cells, nuclei in dried pellet at the N-K absorption edge. Spectra of DNA and histone were shown for comparison.

of histone. Furthermore minor peak in the lower energy side was clearly noticed, which could be attributable to DNA or DNA related molecules. These results indicate that XANES of major cellular molecules such as proteins and DNA are rather

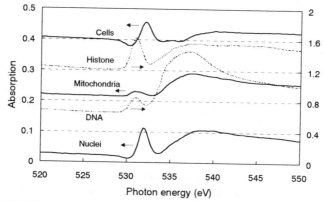

FIGURE 2. XANES of CHO cells, nuclei, and mitochondria in dried pellet at the O-K absorption edge. Spectra of DNA and histone were shown for comparison.

insensitive to the cellular environment around N. On the other hand, as shown in Fig. 2, XANES profiles of cells and cellular organelles at the O-K edge exhibited significant difference from those of biomolecules: the peak was slightly shifted, which is likely produced by the cellular environment, while the spectrum of mitochondria fraction is similar to that of DNA in the peak energies and in the more intense absorption at the higher-energy peak. Although this difference is not well interpreted at present, the results indicate that XANES of biomolecules could be influenced by cellular environment in some case. This is particularly critical in cell imaging.

Figure 3a shows XANES of cells and DNA at the P-K edge. The slight difference of the peak energy was observed. To explore the cause of the difference, spectra of nucleotides were surveyed (Fig. 3b). The peak in DNA was found to be red-shifted compared with those of ATP, ADP and AMP. Consequently the presence of such DNA-related molecules may be reflected to the spectrum of cells.

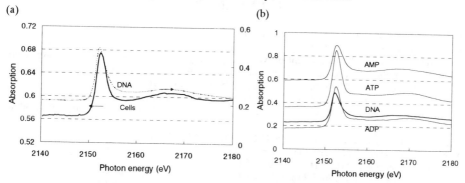

FIGURE 3. (a) XANES of CHO cells in dried pellet at the P-K absorption edge. Spectrum of DNA was shown for comparison. (b) XANES of nucleotides at the P-K absorption edge.

Cellular XANES at the S-K edge was shown in Fig. 4. XANES of glutathione in the reduced form (GSH) and oxidized form (GSSG) were also displayed. As

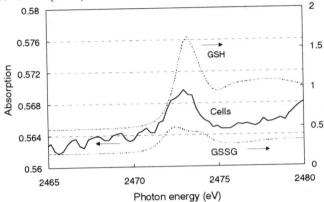

FIGURE 4. XANES of CHO cells in dried pellet at the S-K absorption edge. Spectra of GSH and GSSG were shown for comparison.

discussed in our earlier report (11), GSSG has two peaks; the lower energy one could be assigned to the transition of S $1s \rightarrow \sigma^*$(S-S) and the higher one to S $1s \rightarrow \sigma^*$(S-C), whereas GSH spectrum has a peak probably due to the transition of S $1s \rightarrow \sigma^*$(S-C). The cell spectrum seems to be the superposition of these two spectra.

These profiles of cell spectra could be explained as the superposed pattern of constituent biomolecules, providing that XANES peaks of cells at the P-K and S-K edges are located at the same energy as those of biomolecules.

Finally it should be noted that in addition to the information about the kind of molecules contained in specimen, the molecular composition could be estimated from the recent analysis of the peak height of XANES as attempted by Rompel et al. at the S-K edge (5). Ongoing study on measuring XANES of intracellular areas using contact microscopy should provide the distribution of molecular composition in a cell.

REFERENCES

1. Kirtley, S. M. et al., *Biochim. Biophys. Acta* **1132**, 249-254 (1992).
2. Ade, H. et al., *Science* **258**, 972-975 (1992).
3. Shinohara, K. et al., "Measurement of XANES spectra of biological molecules in the soft X-ray region," in *X-Ray Microscopy and Spectromicroscopy,* edited by J. Thieme, G. Schmahl, E. Umbach, and D. Rudolph, Heidelberg: Springer-Verlag, 1998, pp. III 157-161.
4. Buckley, C. J. et al., "Possibilities for chemical state imaging of calcium compounds," in *X-Ray Microscopy IV,* edited by V. V. Aristov and A. I. Erko, Moscow: Bogorodskii Pechatnik Publishing, 1994, pp. 207-212.
5. Rompel, A. et al., *Proc. Natl. Acad. Sci. USA* **95**, 6122-6127 (1998).
6. Kenny, J. M. et al., *J. Microsc.* **138**, 321-328 (1985).
7. Buckley, C. J. et al., *SPIE* **1741**, 363-372 (1992).
8. Zhang, X. et al., *J. Struct. Biol.* **116**, 335-344 (1996).
9. Ito, A. et al., *J. Synchrotron Radiat.* **5**, 1099-1101 (1998).
10. Zhang, X. et al., *Nucl. Instrum. Methods in Phys. Res.* **A347**, 431-435 (1994).
11. Ito, A. et al., *Photon Factory Activity Report* **14**, 261 (1997).

The Use of Absorption Difference at Distant Wavelengths for Imaging Elements in Mammalian Cells

Atsushi Ito[*], Tetsuya Endo[†], Yoshimasa Yamamoto[‡] and Kunio Shinohara[§]

[*]Department of Nuclear Engineering, School of Engineering, Tokai University, Hiratsuka, Kanagawa 259-1292 JAPAN
[†]Faculty of Pharmaceutical Sciences, Health Sciences University of Hokkaido, Ishikari-Tobetsu, Hokkaido 061-0293 JAPAN
[‡]College of Medicine, University of South Florida, Tampa, FL 33612 U.S.A.
[§]Radiation Research Institute, Faculty of Medicine, the University of Tokyo, Bunkyo-ku, Tokyo 113-0033 JAPAN

Abstract. Intracellular mapping of elements was performed for Cd or Hg loaded mammalian cells. X-ray contact microscopy with an electronic zooming tube were used for cell imaging at the resolution of 0.5 μm. The mapping was obtained by the subtraction between images taken at two different wavelengths. For both elements, pair of the wavelengths at 8.21 nm and 4.48 nm was adopted, because the absorption coefficients of Cd and Hg at 8.21 nm are smaller than those at 4.48 nm, while those of major elements in biological components such as C, N and O exhibit opposite nature. Subtracted data were compared with those for non-loaded cells. The results suggest that the present method is valuable for a new approach to image some kinds of elements such as Cd and Hg that have no absorption edge or have the absorption edge overlapped with those of major elements in this wavelength region.

INTRODUCTION

One of the major advantages of soft X-ray microscopy is mapping of elements in a specimen with high spatial resolution. The common method employed for the mapping is the subtraction of two images taken at the wavelengths below and above the absorption edge of the element. Subtraction using a XANES peak, if available, would be a highly sensitive method for the chemical nature of element (1). However for the heavy elements that have the broad M or N absorption edge, the above subtraction is not always effective. Although X-ray fluorescence analysis in the hard X-ray region is the most popular and sensitive method, the spatial resolution is not enough to detect intracellular distribution in submicron size at the present stage. In addition, with hard X-rays, it is hard to image biological cells by its own components at high resolution.

In this study, we present a new approach to image heavy elements in a cell in the soft X-ray region: we have introduced the subtraction method using a pair of distant wavelengths with different absorption nature of such elements from those of abundant

CP507, *X-Ray Microscopy: Proceedings of the Sixth International Conference,*
edited by W. Meyer-Ilse, T. Warwick, and D. Attwood
© 2000 American Institute of Physics 1-56396-926-2/00/$17.00

elements in a cell. The method was applied to mapping Cd or Hg in a renal tubular epithelial cell line loaded with the element.

MATERIALS AND METHODS

Imaging. X-ray contact microscopy with an electronic zooming tube (2) was employed to image mammalian cells at the resolution of about 0.5 μm over the wide wavelength range from 2.16 nm to 10.88 nm. Monochromatic soft X-rays (E/ΔE~1500) were obtained at the BL-11A beamline at the Photo Factory, Tsukuba, Japan.

Cell preparation. LLC-PK$_1$ cells, a renal tubular epithelial cell line (3), were grown on a collagen coated SiN membrane (a backside of photocathode of the zooming tube) in the presence or absence of Cd or Hg. Cultured cells were then fixed with glutaraldehyde followed by drying in the air.

RESULTS AND DISCUSSION

In the wavelength region tested, no sharp absorption edges usable for subtraction between both sides of absorption edge were found for both Cd and Hg judging from Henke's table for absorption coefficients of elements (4). The M edge of Cd may be useful, but unfortunately the K edge of N is overlapped. The wavelength pair for subtraction having different characteristics in the absorption spectra from major biological elements such as C, N and O was explored. Table 1 proposed one of such candidates, which shows absorption coefficients of C, N, O, Cd and Hg at 8.21 nm and 4.48 nm. It is evident that C, N and O have the larger values at 8.21 nm than at 4.48 nm, while Cd and Hg exhibit the opposite tendency.

TABLE 1. Comparison of mass absorption coefficients at 8.21 nm and 4.48 nm.

Element	Mass absorption coefficient (cm^2/g)	
	8.21 nm	4.48 nm
C	9.94×10^3	1.96×10^3
N	1.75×10^4	3.77×10^3
O	2.79×10^4	6.04×10^3
Cd	4.76×10^3	6.03×10^3
Hg	6.75×10^3	1.43×10^4

These values were cited from the reference (4).

Therefore the ratio image (image at 8.21 nm/image at 4.48 nm) is expected to reflect the distribution feature of Cd or Hg. Figure 1 shows a ratio image of control cells without Cd or Hg. The ratio image (Fig. 1c) showed a similar image of those at 4.48 nm and 8.21 nm: Dense area in the nucleus probably corresponding to nucleolus was clearly observed. On the other hand, for Cd loaded cells the same set of images (4.48 nm image, 8.21 nm image and ratio image) was presented in Fig. 2. Compared

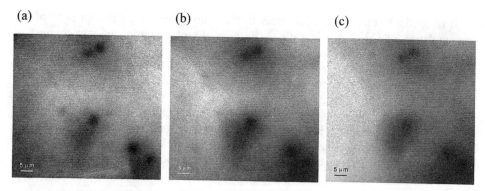

FIGURE 1. Images of LLC-PK$_1$ cells cultured in the absence of Cd and Hg. (a) Image at 4.48 nm. (b) Image at 8.21 nm. (c) Ratio image of (a) and (b) (8.21 nm/4.48 nm).

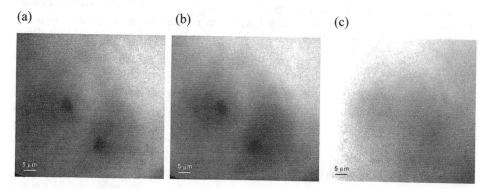

FIGURE 2. Images of Cd loaded LLC-PK$_1$ cells. (a) Image at 4.48 nm. (b) Image at 8.21 nm. (c) Ratio image of (a) and (b) (8.21 nm/4.48 nm).

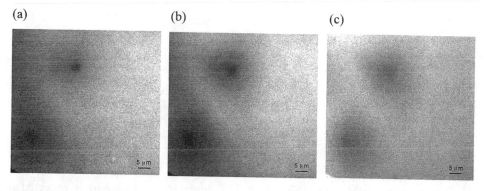

FIGURE 3. Images of Hg loaded LLC-PK$_1$ cells. (a) Image at 4.48 nm. (b) Image at 8.21 nm. (c) Ratio image of (a) and (b) (8.21 nm/4.48 nm).

with control cells, contrast of the dense area in the nucleus became lower. The similar result was obtained with Hg loaded cells (Fig. 3): The dense area in the nucleus was also not so distinct as that in control cells.

It is no doubt that these differences in the ratio images are consequences of the accumulation of Cd or Hg in cells. The possible interpretation of low contrast of nuclear structure observed in heavy element loaded cells would be as follows: 1) non-specific adsorption of heavy elements throughout cells, and 2) selective accumulation of heavy elements in the nuclei, which results in the lower contrast of this region compared with control cells.

Although further analysis remains for the determination of quantitative distribution of these heavy elements, it is proposed that the absorption difference at distant wavelengths is useful to study the distribution of heavy elements in a cell. Soft X-ray contact microscopy with an electronic zooming tube is applicable to this new approach.

REFERENCES

1. Buckley, C. J., *Rev. Sci. Instrum.* **66,** 1318-1321 (1995).
2. Shinohara, K. et al., "X-ray microscopy system with an electronic zooming tube," in *X-Ray Microscopy and Spectromicroscopy,* edited by J. Thieme, G. Schmahl, E. Umbach and D. Rudolph, Heiderberg: Springer-Verlag, 1998, pp. II255-259.
3. Endo, T. et al., *Pharmacol. Toxicol.* **82,** 230-235 (1998).
4. Henke, B. L. et al., *Atomic Data and Nuclear Data Tables* **54,** 181-342 (1993).

X-ray Microimaging of Cisplatin Distribution in Ovarian Cancer Cells

Yasuhiko Kiyozuka[1], Kuniko Takemoto[2], Akitsugu Yamamoto[3], Peter Guttmann[4], Airo Tsubura[1], and Hiroshi Kihara[2]

[1]Department of Pathology II, and [3]Department of Physiology I, Kansai Medical University, 10-15 Fumizono, Moriguchi, Osaka 570-8506, [2]Physics Laboratory, Kansai Medical University, 18-89 Uyamahigashi, Hirakata, Osaka 573-1136, Japan, [4]Univesity Georgia-Augusta at Goettingen, Institute for X-ray Physics, Geiststrasse 11, 37073 Goettingen, Germany

Abstract. X-ray microscopy has the possibility to be in use for elemental analysis of tissue and cells especially under physiological conditions with high lateral resolution. In X-ray microimaging cis-diamminedichloroplatinum II (cisplatin: CDDP), an anticancer agent, which has a platinum atom at its functional center gives sufficient contrast against organic material at sub-cellular level. We analyzed the enhance effect and intracellular distribution of CDDP in human ovarian cancer cells with the transmission X-ray microscope at BESSY, Berlin. Two human ovarian cancer cell lines (MN-1 and EC) were treated with 1 and 10 µg/ml of CDDP for 4 hours and compared with untreated cells. X-ray images of CDDP-treated samples show clearly labeled nucleoli, periphery of the nucleus and mitochondria, in a concentration-dependent manner. CDDP binds to DNA molecules via the formation of intra- or inter-strand cross-links. Higher contrasts at the periphery of nucleus and nucleoli suggest the distribution of tightly packed heterochromatin. In addition, results show the possibility that CDDP binds to mitochondrial DNA. Biological function of cisplatin is not only the inhibition of DNA replication but is suggested to disturb mitochondrial function and RNA synthesis in the nucleolus.

Introduction

X-ray microscopy enables high-resolution analysis of thick biological specimens (5-10 µm) in aqueous conditions (1). Recent advance in x-ray microimaging was achieved by the use of zone plates as imaging elements, which resulted in having a resolving power of less than 40 nm (2). In spite of the excellent contrast without staining in case of X-ray microscopy, a special labeling such as potassium permanganate treatment is often useful for the visualization of the internal cellular membrane systems (3). The staining with heavy metal not only gives an enhancing effect, but makes it possible to observe the intracellular distribution of such substances.

Cis-diamminedichloroplatinum II (cisplatin), one of the most effective agents in cancer chemotherapy, is widely used in the treatment of ovarian cancer patients. Although it is known that CDDP binds to DNA via the formation of intra- and inter-strand cross-links and inhibits DNA replication, detailed intracellular distribution is remained to be elucidated (4). As the CDDP has a platinum atom at its functional center, we thought that the X-ray microscopic analysis will help to add a novel aspect in the intracellular distribution of CDDP.

CP507, X-Ray Microscopy: Proceedings of the Sixth International Conference, edited by W. Meyer-Ilse, T. Warwick, and D. Attwood

In this report, on the basis of a feasibility study (5), we demonstrated novel findings as to the intracellular distribution of CDDP by X-ray microscopy.

Materials and Methods

Cell culture and fixation.

MN-1 (mucinous cystadenocarcinoma) and EC (endodermal sinus tumor) cells established from human ovarian cancer (6,7) were cultured on support foils (8) in Dulbecco's modified Eagle medium supplemented with 10 % fetal bovine serum at 37° C in a 5 % CO^2 incubator in the presence of 1 and 10 µg/ml of CDDP for 4 hours. Untreated control consisted of culture medium alone. Cells were fixed with 2.5 % glutaraldehyde in 0.1 % cacodylate buffer (pH 7.4) for 1 hour, and washed with the same buffer for 3 times followed by a storage in PBS.

X-ray microscopy.

The specimens prepared on the support foils were put in small, hermetically closed envelops filled with PBS and were sent to the Goettingen transmission X-ray microscope at BESSY, Berlin. Samples were stored at 4° C. All investigations were performed within a week after the preparation of the specimens. The support foil with the specimen was put into the environmental chamber (8) which was closed with a cover foil. The chamber was transferred into the transmission X-ray microscope. The liquid layer thickness was then reduced to thinner than 10 µm controlling by a light microscope (Zeiss Axioskop with differential interference contrast mode) which is incorporated in the X-ray microscope. This light microscope was also equipped with a CCD-camera to get light microscope images of the cell under investigation. The specimen was held under atmospheric pressure and at room temperature during the investigation. The micro zone plate used as X-ray objective for imaging had an outermost zone width of 40 nm. The X-ray images were taken at 2.4 nm wavelength using a thinned, backside illuminated CCD as detector (9). Preservation of the fixed cells in the buffer for one week at 4° C caused no ultrastructure change as confirmed under an electron microscope.

Electron microscopy.

Cells were cultured on collagen-coated plastic coverslips (Cell tight C-1 Cell disk, Sumitomo Bakelite Co., Ltd, Tokyo, Japan) and fixed with 2.5 % glutaraldehyde in 0.1 M cacodylate buffer (pH 7.4) and post-fixed with 1 % OsO_4 in the same buffer for 1 hour, respectively. The cells were then washed in distilled water, incubated with 50 % ethanol for 10 min, and block-stained with 2 % uranyl acetate in 70 % ethanol for 2 hours. The cells were further dehydrated with a graded series of ethanol, and were embedded in epoxy resin. Ultrathin sections were stained with uranyl acetate and lead citrate, and observed under a Hitachi H-7000 electron microscope.

Results

X-ray microscopy
Several X-ray images of wet MN-1 and EC cells were put together to obtain a larger field of view (Figs 1 and 2). In the figures of untreated controls (Fig 1a and 2a), nucleus, nucleoli and the cell border can be differentiated. In addition, X-ray dense spherical shaped structures, less than 250 nm in their diameter, were distributed in the cytoplasm of EC cells (Fig 2a). Although X-ray dense nuclear membrane can be detected, the boundary against cytoplasm was not so clear.
On the contrary, images taken in CDDP-treated MN-1 and EC cells clearly showed good enhanced contrast of intracellular organelle, particularly the nucleoli, periphery of the nucleus and mitochondria was enhanced in the concentration-dependent mannner (Figs 1b and 2b). The presence of mitochondria were well defined preferentially at peripheral region of cytoplasm with their rod shaped structure (Fig 3).

Electron microscopy
Types and distribution of organella in X-ray images were confirmed with a comparison of electron microscopy images. MN-1 cells posessed numerous number of lysosome and mitochondria in cytoplasm (Fig 4a). In EC cells (Fig 4b), the numbers of mitochondria and lysosome was lower than MN-1. However, the cluster of spherical bodies with higher electron density than other cytoplasmic component was observed. Monotonous composition of these bodies suggested they are lipid droplets. X-ray dense spherical shaped structures in EC cells were confirmed to be these lipid droplets.

Discussion

It is believed that DNA is the critical target of CDDP and cell death occurs when the nuclear DNA strands are cross-linked (10). Although details of the cytotoxic function of CDDP remained to be elucidated, the determination of the intracellular distribution of CDDP will help to evaluate the functional mechanism. CDDP distribution in cells has been shown by using the X-ray elemental imaging with scanning electron microscopes (11-13). However, there is a considerable disagreement between the report of Khan et al.(11) and of Kirk et al. (12) as to the intracellular distribution of CDDP within the cell. The reason may be attributed to the use of much higher concentration of CDDP or for the use of different types of noncancerous cells. Preliminary, absorption and phase contrasts of DNA and CDDP-DNA complex were calculated based on the report of Schmahl et al. (14). Briefly, the theoretical absorption and phase contrasts were calculated between DNA with 10 nm thickness and compounds consisting of DNA with 10 nm thickness and platinum layers with various thickness instead of cisplatin to simplify the calculation (5). As a result, contrast varied sensitively depending on the platinum thickness. The present study was conducted based on this theoretical calculation.
In general, results in our study by X-ray imaging agree with both of previous observations. Not only the remarkable enhancement (CDDP accumulation) in

nucleolus and nuclear membrane was presented, but the enhancement of cytoplasmic mitochondria was thought to be the evidence for the cytoplasmic distribution of CDDP. DNA is usually packed into chromatin fibers. Chromatin fibers are distributed between relatively open networks of euchromatin and more tightly packed heterochromatin. Frequently, heterochromatin is located in a layer just inside the nuclear envelope and surrounding the nucleolus. Enhance effect of CDDP at nucleolus and nuclear membrane by X-ray microscopy can be induced from the chemical interaction between CDDP and DNA (4), and intracellular distribution of DNA itself. In addition, it is generally accepted that DNA fibers are a regular constituent of mitochondria. Although the DNA fibers visible in the organelle are much thinner than nuclear chromatin fibers, it has the possibility to form adducts with CDDP.

As cells start to remove CDDP-DNA adducts with excision repair system within a few hours after the onset of CDDP treatment (15), an exposure time of CDDP (less than10 µg/ml) was limited up to 4 hours in this study, which allows to keep the initial

distribution of CDDP. X-ray microscopy proved to give an advantage to detect CDDP-adducts in fine structures in such a labeling condition of short time and lower concentration of CDDP.

In accordance with the visual enhance effect of CDDP, the alteration of intensities at nucleoli and nuclear membrane showed an increase parallel to the escalation of CDDP concentration with the intensity profile analysis. However, nucleolus intensity in EC cells was much higher than MN-1 cells, which may be resulted from the phenotypic difference of mitochondrial DNA contents between both cell lines. On the contrary, lipid droplet distributed in the cytoplasm of EC cells was not enhanced with the CDDP treatment as to suggest no interaction between CDDP and lipid component take place. Thus, our finding leads to the conclusion that the cisplatin are selectively distributed to DNA-rich sites in ovarian cancer cells.

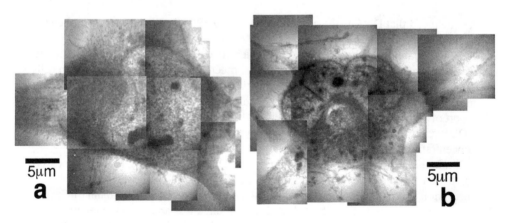

FIGURE 1. X-ray microimages of MN-1 cells. Compared to untreated cells (a), enhanced contrast of intracellular organelle such as the nucleoli and the periphery of the nucleus is demonstrated in cells exposed to 10 µg/ml of CDDP for 4 hours (b).

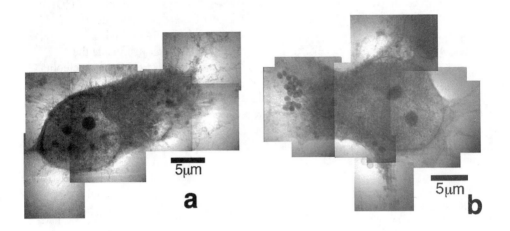

FIGURE 2. X-ray microimages of EC cells with (b) or without (a) the CDDP treatment at the concentration of 10 μg/ml for 4 hours. Compared to MN-1 cells, EC cells contain X-ray dense spherical shaped structure in the cytoplasm.

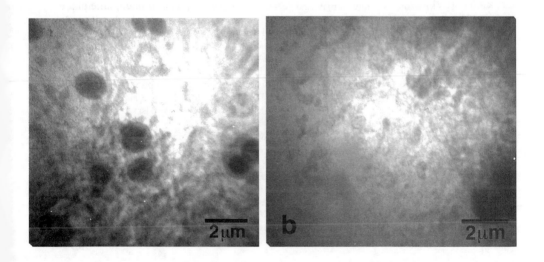

FIGURE 3. X-ray images show the presence of mitochondria preferentially at peripheral region of cytoplasm with their rod shaped structure both in MN-1 (a) and EC (b) cells.

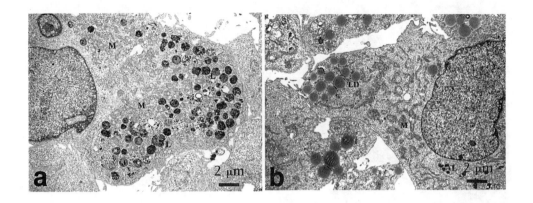

FIGURE 4. Electron microscopy of MN-1 cells (a) presents numerous number of lysosome(L) and mitochondria (M) in cytoplasm. In EC cells (b), the number of mitochondria and lysosome was lower than MN-1. However, the cluster of lipid droplet (LD) was seen.

REFERENCES

1. Kirz, J., Jacobsen, C. , and Howells, M., *Q Rev Biophys* **28**, 33-130 (1995)
2. Schmahl, G., Rudolph, D., Guttmann, P., Schneider, G., Thieme, J., and Niemann, B., *Rev Sci Instrum* **66**, 1282-1286 (1995)
3. Kihara, H., Yamamoto, A., Guttmann, P., and Schmahl, G. *J Electron Spectrosc Relat Phenom* **80**, 369-372 (1996)
4. Huifang, H., Zhu, L., Reid, B.R., Drobny , G.P., and Hopkinst, P.B., *Science* **270**,1842-1845 (1995)
5. Takemoto, K., Kiyozuka, Y., Tsubura, A., and Kihara, H., " *Intracellu localization of cisplatin in ovarian cancer cells by X-ray microimaging*" HAGA' 97 Program & Abstracts, 1997, pp27
6. Yoshida, M., Kiyozuka, Y., Noda, T., Imai, S., and Ichijo, M., *J Jpn Soc Clin Cytol* **32**: 1-8 (1993)
7. Imamura, K., Kiyozuka, Y., Nishimura, H., and Yakushiji, M., *Oncol Rep* **2**, 17-25 (1995)
8. Niemann, B., Schneider, G., Guttmann, P., Rudolph, D., and Schmahl, G., *X-ray microscopy IV*, edited by V.V. Aristov & A.I. Erko, Chernogolovka, Moskow Region: Bogorodski Pechatnik Publishing Company. 1994, pp. 66-75.
9. Wilhein,T., Rothweiler, D., Tusche, A., Scholze, F., Meyer-Ilse, W., *X-ray microscopy IV*, edited by V.V. Aristov & A.I. Erko, Chernogolovka, Moskow Region: Bogorodski Pechatnik Publishing Company. 1994, pp. 470-474.
10. Rosenberg, B., *Cancer* **55**, 2303-2316 (1985)
11. Khan, M. U. A., and Sadler, P. J., *Chem Biol Interact* **21**, 227-232 (1978)
12. Kirk, R. G., Gates, M. E., Chang, C. S., and Lee, P., *Exp Mol Pathol* **63**, 33-40 (1995)
13. Lee, P., Gates, M., Chang, C. S., and Kirk, R. G., *Chin J Physiol* **39**, 205-210 (1996)
14. Schmahl, G., and Rudolph, D., " *Proposal for a phase contrast X-ray microscope*" in *X-ray microscopy*, edited by P.C. Cheng and G.J. Jan, Spring-Verlag Berlin, Heiderberg: 1987, pp. 231-238.
15. Fichtinger-Schepman, A. M. J., van Oosterom, A. T., Lohman, P. H. M., and Berends, F., *Cancer Res* **47**, 3000-3004 (1987)

The X-Ray Microprobe for Studies of Cellular Radiation Response

AG Michette,[1] SJ Pfauntsch,[1] M Folkard,[2] BD Michael[2] & G Schettino[1,2]

[1]*Centre for X-Ray Science, Dept. of Physics, King's College London, Strand, London WC2R 2LS, UK*
[2]*Gray Laboratory, Cancer Research Trust, Mount Vernon Hospital, Northwood, HA6 2JR, UK*

Abstract. An x-ray microprobe, which utilises a small laboratory x-ray source with a carbon target and a zone plate to form a fine focus, has been used for initial studies of cellular radiation response. It is not yet clear whether carbon K x-rays exhibit the hypersensitivity at very low doses for 240 kVp x-rays, which manifests itself predominantly with sparsely ionising radiation. There is some evidence to support the phenomenon of radiation induced bystander effects whereby unexposed neighbours of irradiated cells also exhibit a response.

INTRODUCTION

The microprobe uses a beam of carbon K x-rays (278 eV) to produce controlled clusters of damage to DNA and other structures within individual cells. The x-rays are focused by a zone plate to a spot of a few hundred nanometres diameter. Each x-ray photon produces about 14 ionisation events within a 7 nm range when interacting with biological tissue. This very small range, combined with the spatial resolution offered by the zone plate, allows the creation of clusters of ionisation with sizes comparable to those of critical sub-nuclear structures. Experiments with the microprobe are expected to provide answers to the following questions, among others.

How does clustering of damaged DNA sites affect biological response? Precise control of the spatial dose distribution provides a unique probe of the consequences of localised clusters of energy deposition. Soft x-rays, which deposit their dose over a smaller spatial range than higher energy radiation, are expected to produce more DNA double strand breaks, which are much harder for the cell to repair than single strand breaks are.

How is sensitivity to radiation distributed across the cell nucleus? Some α-particle studies show an apparently even distribution[1] while others indicate that DNA close to the nuclear membrane is most sensitive to damage.[2]

Do effects other than DNA damage contribute to cell death? Some evidence suggests that the cytoplasm is an important target for genotoxic effects;[3] this may have to be taken into account when estimating the risk from exposure to ionising radiation.

Can radiation effects be transmitted from irradiated cells to neighbouring, unirradiated, cells (the bystander effect)?[4]

CP507, *X-Ray Microscopy: Proceedings of the Sixth International Conference,*
edited by W. Meyer-Ilse, T. Warwick, and D. Attwood
© 2000 American Institute of Physics 1-56396-926-2/00/$17.00

THE MICROPROBE

The microprobe is placed on an optical table to minimise vibrations. A three axis micro-positioning stage (accuracy 250 nm per step) supports a cell dish and allows the pre-located biological targets to be aligned precisely at the x-ray focus. An ultraviolet microscope coupled to a CCD camera is used to monitor the targets, which are stained with fluorescent dye. The micro-positioning stage and the CCD camera are controlled by a PC for fast and accurate automatic location and relocation of the cells and for image analysis. Electrons — up to 30 keV and 20 mA — are focused to a 3 μm spot on a graphite target by a water cooled magnetic lens. A silica mirror above the target eliminates the bremsstrahlung component via a 3° grazing incidence reflection, giving a nearly monochromatic carbon K x-ray beam.

The zone plates are made from tungsten, on 100 nm thick Si_3N_4 supports, by electron beam lithography. They are 200 μm diameter, 200 nm outer zone width ($f\sim$9 mm for carbon K x-rays) and are used with a 100 μm diameter axial stop and a 12 μm diameter order selecting aperture (OSA), which also prevents irradiation of cells by unfocused residual background radiation. The first-order efficiency of the current zone plates is typically 7%.

For targeting the cells, it is essential to know precisely the axial (z) and transverse (x–y) positions of the zone plate focus; the zone plate and OSA must be aligned to within a few tens of micrometres in z and a few micrometres in x and y. Since the alignment must be done within the space constraints of the cell positioning stage, conventional micro-manipulators cannot be used. Instead, the stage itself is employed, using the arrangement shown in figure 1. First, the position of the vacuum window is noted using the CCD camera. The zone plate holder is then placed onto the vacuum cap, where it is held lightly by a magnetic sheet. Using an iris diaphragm attached to the stage, the holder is moved in x and y until the zone plate is over the vacuum window, its coordinates are noted and then it is decoupled from the

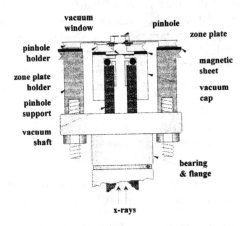

Figure 1. The zone plate and OSA alignment mechanism.

160

diaphragm. This procedure is repeated for the OSA holder to centre the OSA over the zone plate. Finally, the OSA support is moved in z until it is the correct height above the zone plate to block the unwanted diffraction orders.

The location and size of the focus is established by measuring the detector output while scanning a knife edge through the beam at the nominal sample position. The spot diameter at the focus is <500 nm.

LOW DOSE CELL SURVIVAL

A typical count rate incident on the cell dish is ~1200 photons s^{-1}. This corresponds to a dose rate of about $0.1\,Gy\,s^{-1}$ in a typical mammalian cell, averaged over the mass of the cell nucleus. Using the automated cell alignment feature, about 5–6 cells per minute can be irradiated at the lowest doses. After incubation for three days the cells are returned to the microprobe stage for analysis. Using the microscope, each irradiated cell can be revisited, by retrieving stored coordinates, and those that have formed healthy colonies (50 or more cells) are scored as surviving.

Figure 2 shows the percentage of surviving V79 cells as a function of carbon K x-ray dose, compared with (unfocused) aluminium K_α x-rays and 240 kVp x-rays. The lower energy x-rays are clearly more damaging. The inset shows a hypersensitivity at very low doses for 240 kVp x-rays, which manifests itself predominantly with sparsely ionising radiation. It is not yet clear whether carbon K x-rays exhibit this effect.

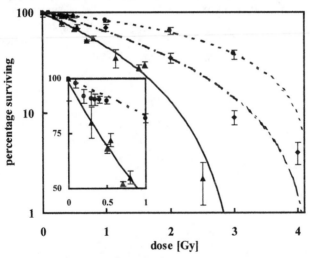

Figure 2. The percentage of surviving V79 cells as a function of carbon K x-ray dose (lower curve), compared with (unfocused) aluminium K_α x-rays (middle curve) and 240 kVp x-rays (upper curve).

RADIATION INDUCED BYSTANDER EFFECTS

Preliminary experiments have been carried out to elucidate the phenomenon of radiation induced bystander effects, whereby unexposed neighbours of irradiated cells also exhibit a response. Single AG01522 human fibroblast cells were exposed to a controlled dose (up to 4 Gy through the nucleus) of focused x-rays, and the induction of micronuclei (indicative of chromosome damage) and apoptosis within the rest of the population (typically ~2000–4000 cells after 24 hours) were measured. Initial results indicate a ~1.5–2 fold increase in both types of damage in unirradiated cells, although the absolute numbers are low (<1%).

ACKNOWLEDGEMENTS

This work was supported by the Biotechnology and Biological Sciences Research Council, UK, grant number E05297. The authors are grateful to Pambos Charalambous, of King's College London, for the zone plates used in this work, and to Peter Anastasi (FasTec Ltd.) for the supply of membranes used for the zone plates and cell dishes.

REFERENCES

1. Raju, M. R. et al., *Radiat Res.* **128**, 204–209 (1991)

2. Cole, A., et al., in *Radiation Biology in Cancer Research*, edited by R. E. Meyn and H. R. Withers, Raven Press: New York, 1980, pp. 33–58.

3. Wu, L-J. et al., *Proc. Natl. Acad. Sci.*, **96**, 4959–4964 (1999)

4. Deshpande, A, et al., *Radiat Res.* **145**, 260–267 (1996)

Visualising Membrane Turnover in Trypanosomes

A.D. Stead[1], T.W. Ford[1], A.M. Page[2], T. Majima[3], H. Shimizu[3] &
T. Tomie[3].

[1]*Biological Sciences, Royal Holloway, Univ. London, Egham, Surrey, UK. TW20 0EX.*
[2]*Biomedical Imaging Unit, Univ. Southampton, General Hospital, Southampton. UK. SO16 6YD.*
[3]*Electrotechnical Lab. Tsukuba, Ibaraki, Japan.*

Abstract. After incubation with colloidal gold, x-ray dense particles accumulated within (or on the surface of) the cells of the trypanosome *Crithidia fasciculata*. These particles appeared to accumulate at the posterior of the cell and it is only when cells were imaged rapidly and after maintenance at 4°C that these particles were seen at the anterior of the cells. It would seem therefore that the internalization of the surface membrane is a very rapid process. Using an alternative photoresist material internal cellular organelles were successfully imaged.

INTRODUCTION

Survival of parasitic trypanosomes in the bloodstream of the vertebrate host depends on their ability to continually "update" their immunological profile by replacing their surface coat with biochemically and antigenically distinct molecules thereby enabling it to evade the hosts immune responses. Ultrastructural studies have shown that markers bound to this surface coat enter the flagellar pocket and are taken up into endocytotic vesicles (1-3).

In the present study the cell surface of the trypanosome *Crithidia fasciculata* was tagged with x-ray dense markers and the fate of these, when bound to the cell surface of the parasites, followed using x-ray microscopy (XRM). The advantage of XRM being the ability to image living material without specimen preparation since membrane turnover, like other cellular processes, is dynamic and may continue during the fixation period needed for EM (4), furthermore it is possible to image the cells within a minute of removal from the culture medium.

MATERIALS AND METHODS

C. fasciculata (clone H56) were grown in a semi-defined culture medium (5) at 27°C; cells attained a concentration of approx.10^8 cells/ml within 3 days and then plateaued but the parasites remained viable for about 2 weeks. Cultures were harvested by centrifugation at 4°C (c.5 x 10^8 cells) and resuspended in 1ml 0.1M HEPES buffer, pH 7.2, and then mixed with 10Fl (1:100 dilution) of cationic gold and sampled either without washing or after washing and resuspension in buffer. For x-ray microscopy 5µl cell samples were

CP507, *X-Ray Microscopy: Proceedings of the Sixth International Conference,*
edited by W. Meyer-Ilse, T. Warwick, and D. Attwood
© 2000 American Institute of Physics 1-56396-926-2/00/$17.00

imaged at various times (5-65min) after treatment using either polymethylmethacrylate (PMMA) or a modified DOW resist material (6) with a 120nm thick silicon nitride window (Fastec, Silverstone, UK) to maintain environmental conditions. Soft x-rays were generated from a yttrium foil target with the sample held 1mm from the target. The laser power was c.300mJ. After exposure resists were cleaned by sonication and developed in 1:1 iso-propanol:MIBK,; optimally developed resists were imaged on a Seiko AFM [7].

RESULTS

Images of unlabelled *C. fasciculata* cells (Fig. 1A) show the conventional choanomastigote morphology characteristic of the genus, with a single anterior flagellum. The cell body of unlabelled cells (Fig. 1A) showed no suggestion of x-ray dense inclusions, however, cells labelled with cationic gold and imaged at 8, 31 and 44mins

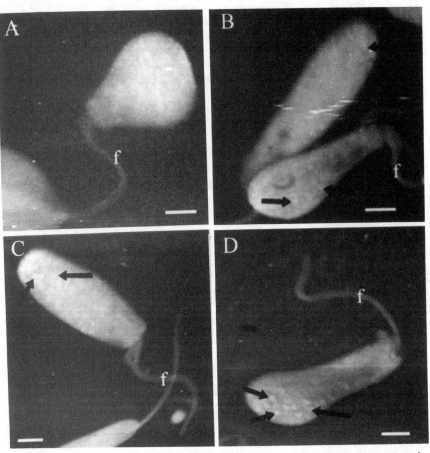

FIGURE 1A-D. AFM images of chemically developed photoresists. A) Control cells not treated with cationic gold. B-D) Labelled with cationic gold and imaged 8, 21 or 44 mins. following treatment, in each case x-ray absorbing inclusions are associated with the cells (arrows) and the single anterior flagellum is visible (f). Scale bars = 2μm.

after treatment (Figs. 1B-D) had a similar morphology to control cells but had large numbers of x-ray dense structures (arrowed) associated with the cells. At earlier sampling times these structures were distributed throughout the cell (Fig. 1B) but at later times they were concentrated towards the cell posterior (Figs. 1C,D). Therefore, when compared to unlabelled cells, the marker appears to be internalised within 10min of it binding to the cells. When cells were incubated at 4^0C, to slow down membrane turnover, and imaged just 4mins after the buffer wash they showed x-ray dense deposits in the region of the flagellar pocket (Fig. 2B). There was also a slight depression in the photoresist images corresponding to the site of the flagellar pocket (Fig. 2A). This feature was seen in a number of cells especially at the earlier incubation times (Figs. 1B). A height profile of a cell from Fig. 2B shows this depression (Fig. 2A).

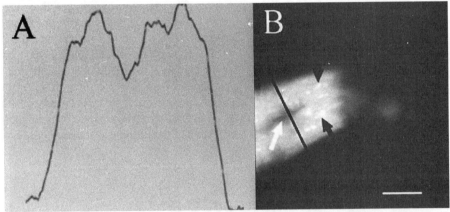

Figure 2A,B. A) Height profile across the image (scan line shown on B) indicating the depression corresponding to the flagellar pocket. B) AFM image of a chemically developed photoresist, the cells were imaged 4mins after treatment with cationic gold at 4°C. The x-ray absorbing inclusions are associated with the anterior of the cell (black arrows) and close to the depression which corresponds to the flagellar pocket (white arrow). Scale bar = 2µm.

Previously (6) we have shown that the chemically amplified DOW photoresist is several orders of magnitude more sensitive than PMMA. However, in the present experiments increased sensitivity was less evident but the discrimination between adjacent cell regions was better. In Fig. 3 the cell outline is clearly seen although areas shaded from the soft x-rays (ie underneath the cells) develop away, ie this is a negative (cf PMMA which is positive) photoresist material. Therefore, the darker (arrowed) areas of these images correspond to areas which were more x-ray absorbing than the surrounding cytoplasm, as such these are probably organelles. Features such as these were much less evident in similar cells imaged with PMMA (Figs.1 & 2) possibly because of the increased sensitivity associated with this resist material.

CONCLUSIONS

The present study has again demonstrated the ability to image living biological

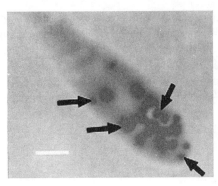

material by SXCM using the very small laser system developed at ETL, however this study has achieved two further goals:

1. The use of the technique to study a dynamic cellular process.

2. Demonstrated the possibility of improving on the PMMA photoresist usually used for SXCM.

The work has also highlighted the difficulty of determining where objects are in the Z plane. The dense aggregations that are a feature of these images could be discrete patches on the outer membrane or ingested particles within an autophagic vacuole. Future work must involve tomographic techniques to allow 3D reconstruction but radiation damage must be avoided, thus the

Figure 3. AFM image of a chemically developed photoresist, in this case using the negative DOW resist material rather than PMMA thus depressions seen by AFM correspond to regions of high x-ray absorbance (arrows). Scale bar = 2μm.

development of cryo imaging facilities is essential. Alternatively two images might be recorded simultaneously using two resist surfaces and two orthogonal soft x-ray sources (8-10).

ACKNOWLEDGEMENTS

This collaboration has been possible by the provision of British Council funding to ADS. The development of the negative photoresist has been partially funded by an EU grant (ERB CHRXCT 940600).

REFERENCES

1. Langreth, S. and Balber, A. E. *J. Protozool.* **22**, 40-53 (1975).

2. Frevert, U., and Reinwald, E. *J. Ultrastr. Res.* **99**, 137-149 (1988).

3. Webster, P. *Eur. J. Cell Biol.* **49**, 303-310 (1989).

4. Kaminskyj, S. G. W., Jackson, S. L., and Heath, I. B. *J. Microsc.* **167**, 153-68 (1992).

5. Shim, H. and Fairlamb, A. H. *J. Gen. Microbiol.* **134**, 807- 817 (1988).

6. Cefalas, A. C., Argitis, P., Kollia, Z., Sarantopoulou, E., Ford, T. W., Stead, A. D., Marranca, A., Danson, C. N., Knott J., and Neely, D. *Applied Physics Letters* **72**, 3258-3260 (1998).

7. Shimizu, H., Tomie, T., Majima, T., Stead, A.D., Ford, T., Muira, K., Yamada, M. and Kanayama, T. X-ray Microscopy and Spectroscopy, Berlin: Springer. pp. I-157-162 (1998).

8. Stead, A. D., Page, A. M., Cotton, R. A., Neely, D., Bagby, R., Miura, E., Tomie, T., Shimizu, S., Majima, T.,Anastasi, P. A. F., and Ford, T. W. Applications of Laser Plasmas II, SPIE Proceedings **2523**, 202-211 (1995).

9. Stead, A. D., Bagby, R., Cotton, R. A., Neely, D., Miura, E., Page, A. M., and Ford, T. W. Ann. Rep. LFC, Rutherford Appleton Laboratory. 49-52 (1995).

10. Stead, A. D., Bagby, R., Neely, D., Page, A. M., Rondot, S., Wolfrum, E., and Ford, T. W. Proceedings XRM'96, II-165-172 (1998).

X-ray Microscopic Studies of Labeled Nuclear Cell Structures

S. Vogt, G. Schneider, A. Steuernagel*, J. Lucchesi*,
E. Schulze†, D. Rudolph, and G. Schmahl

*Institut für Röntgenphysik, Georg-August-Universität Göttingen, Geiststraße 11,
37073 Göttingen, Germany*
* *Biology Department, Emory University, Atlanta, GA, USA*
† *III. Zoologisches Institut - Entwicklungsbiologie, Georg-August-Universität Göttingen,
Humboldtallee 34A, 37073 Göttingen, Germany*
corresponding author: S. Vogt, email: svogt@gwdg.de

Abstract. In X-ray microscopy different proteins are not readily distinguishable. However, in cell biology it is often desirable to localize single proteins, e.g., inside the cell nucleus. This can be achieved by immunogold labeling. Colloidal gold conjugated antibodies are used to mark the protein specifically. With silver solution these are enlarged so as to heighten their contrast. The strong absorption of silver allows easy visualization of the label in the nuclei. In this study male specific lethal 1 protein in male *Drosophila melanogaster* cells was labeled. This protein forms, together with four other proteins, a complex that is associated with the male X chromosome. It regulates dosage compensation by enhancing X-linked gene transcription in males. Room temperature and cryo transmission X-ray microscopic images (taken with the Göttingen TXM at BESSY) of these labeled cells are shown. Confocal laser scan microscopy ascertains the correct identification of the label in the X-ray micrographs, and allows comparison of the structural information available from both instruments.

INTRODUCTION

Immunolabeling is a well established method in visible light and electron microscopy which allows the specific visualization of otherwise indistinguishable proteins inside of cells. In immunofluorescence microscopy fluorophore conjugated antibodies are used to tag the investigated protein. The fluorochrome is then detected in a conventional light or in a confocal laserscan microscope. In electron microscopy colloidal gold conjugated antibodies are used to tag the investigated protein with gold spheres of typically 5-30 nm diameter [1]. When maximum penetration of the antibodies into the sample is desired, smaller gold spheres of only 1 nm diameter are used. To improve visualization of these small spheres, they are often enlarged by a subsequent silver enhancing step, where silver ions in solution are

CP507, *X-Ray Microscopy: Proceedings of the Sixth International Conference,*
edited by W. Meyer-Ilse, T. Warwick, and D. Attwood

reduced onto the metal. The same procedure can be followed to visualize specific proteins for the X-ray microscope. X-ray microscopy is especially well suited to investigate whole, unsectioned cell nuclei, since it has a higher resolution than visible light microscopes, but at the same time is able to image samples of up to 10 μm thickness. In this work we study the distribution of male specific lethal 1 (MSL-1), one of the proteins involved in dosage compensation in *Drosophila melanogaster*. Dosage compensation ensures that males with a single X chromosome have the same amount of most X-linked gene products as females with two X chromosomes. In *Drosophila*, this equalization is achieved by a twofold enhancement of the level of transcription of the X in males relative to each X chromosome in females. The products of at least five genes, maleless (*mle*), male specific lethal 1, 2, and 3 (*msl-1*, *msl-2*, *msl*-3) and males absent on the first (*mof*), are necessary for dosage compensation. The proteins produced by these genes form a complex that is preferentially associated with numerous sites on the X chromosome in somatic cells of males but not of females. Recently, Gu et al. have shown that microscopic techniques serve as powerful tools in studying the sequential assembly of this dosage compensation complex [2]. The exact function of the dosage compensation complex is not known, and higher resolution microscopy studies may contribute to answering this question. This was chosen as a model system to investigate the possibilities of silver enhanced immunogold labeling for transmission X-ray microscopy (TXM). Results from the Göttingen TXM at BESSY will be shown from both room temperature and cryogenic temperature samples, in direct comparison to results from confocal laserscan microscopy (LSM).

MATERIALS AND METHODS

Cells of the male *D. melanogaster* cell line SL2 (ATCC number: CRL#1963) were grown in Schneider's Drosophila medium with 10% fetal calf serum. Support foils were incubated at 25°C with a drop of the above cells in suspension. After ~2 hours they were fixed with 4% formaldehyde in D-PBS for 5-7 minutes.

Rabbit polyclonal antibodies were raised against a glutathion-S-transferase-tag-MSL-1 fusion protein. Nucleotide 1269-3087 of the MSL-1 open reading frame were cloned into the pGEX-2T expression vector. The fusion protein was overexpressed in BL21(DE3), purified on glutathion-S-transferase columns and then used for immunization.

After fixation the cells were extracted with 0.5% Triton X-100 in D-PBS for 15-20 minutes (at 22°C), then washed. Samples were blocked for 120 minutes (at 37°C) with 0.5M NH$_4$Cl and 5% glycine, and for 45 minutes in 10% normal goat serum and 2% bovine serum albumin. They were incubated for 30 minutes at 37°C with the primary antibody (rabbit, against c-MSL-1) and subsequently washed several times. Samples were then incubated with the secondary antibody, either for 30 minutes at 37°C with a fluorochrome (Cy2) conjugated goat-anti-rabbit antibody, or for 8-12 hours at 6-10°C with a 1 nm colloidal gold conjugated F(ab')$_2$ goat-anti-rabbit antibody fragment. Additionally, the later were fixed with fresh 1% glutaraldehyde for 30-60 minutes (at 6-10°C), rinsed in distilled water, silver enhanced for 25-30 minutes, and subsequently rinsed again. When samples were prepared for both fluorescence microscopy and transmission X-ray microscopy (TXM) they were first treated with the colloidal gold conjugated antibody, then with the Cy2 conjugated antibody. The sample was first investigated with the confocal laser scan microscope, then silver enhanced and investigated with the TXM.

ROOM TEMPERATURE TXM OBSERVATIONS

To visualize the label in the TXM colloidal gold conjugated antibodies were used. On the left of Fig. 1, a TXM micrograph of a labeled SL2 cell is shown. The micrograph was taken with the Göttingen X-ray microscope at BESSY in room temperature setup [3] using a zone plate with 40 nm outermost zone width at 2.4 nm wavelength. The MSL-1 label appears as an X-ray dense structure inside the cell nucleus. The cytoplasm (c) surrounding the nucleus (n) has been removed to a large part due to the extraction with Triton X-100. A control experiment was carried out, where the sample preparation was identical to that used for labeled cell, except that the primary antibody (against MSL-1) was omitted. A micrograph of this control sample is shown on the right of Fig. 1. As expected, it does not show any labeling.

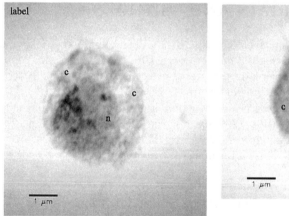 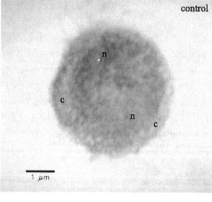

FIGURE 1. TXM micrographs of *Drosophila melanogaster* cells imaged at 2.4 nm wavelength with a zone plate with 40 nm outermost zone width. On the left a MSL-1 labeled cell is shown, the X-ray dense regions show the location of MSL-1 protein. On the right a control sample is shown, where the antibody against MSL-1 was omitted. The cell nuclei are marked with (n), the cytoplasm with (c).

CRYO-TXM AND CONFOCAL LASERSCAN
MICROGRAPHS

To avoid shrinking of the cells due to radiation damage sustained during exposure, experiments at cryogenic temperatures were carried out using the Göttingen X-ray microscope at BESSY in cryo mode [4]. So as to compare the TXM.with immunofluorescence microscopy, the cells were labeled with two sets of secondary antibodies conjugated to two different kinds of markers, to the fluorochrome Cy2 for LSM investigations, and to 1 nm colloidal gold for TXM experiments (after silver enhancement). Micrographs of the same cell are shown from both instruments in Fig. 2. Both images show the same region as labeled. In both microscopes the same regions of high labeling intensity are visible. In the TXM these appear as numerous, well defined, spots of about 100 nm diameter close to each other. There is no labeling in between these spots. In contrast, this observation cannot be made from the LSM micrograph only, since it is not clear whether the fluorescence signal in between the labeled spots is due to the limited resolution or the actual presence fluorochrome.

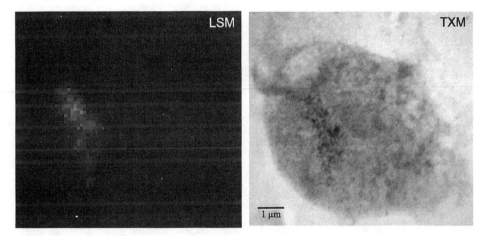

FIGURE 2. The same labeled cell in the Göttingen X-ray microscope at BESSY I in cryo mode (imaged at 2.4 nm wavelength with a zone plate with 40 nm outermost zone width) and in a confocal light microscope (imaged in a Zeiss LSM 510 with an 100x/1.3 oil immersion objective in extended focus mode). The TXM micrograph (left) shows clearly, that there are strongly labeled parts (black regions), in between which there is virtually no label. In the LSM micrograph (right), the same region appears as labeled, but the substructure cannot be resolved.

Resolution present in the TXM and LSM micrographs

To better quantify the difference in resolution present in the confocal laserscan and transmission X-ray micrographs, a spatial frequency analysis is performed on the image data. Both instruments register the images electronically, therefore the image data are directly available as two dimensional arrays containing the image information. So as to avoid artifacts in the Fourier analysis due to image discontinuities at the edges, the image is multiplied with a broad gaussian prior to computing the two dimensional Fourier transform. The resulting two dimensional power distribution in frequency space is integrated circularly over 2π, leading to a one dimensional power spectrum. The different structures in the image contribute to varying degrees to the spectrum. In particular, there are no object structures contributing at higher spatial frequencies than the cutoff frequency associated with the optical system. The only contribution at these higher frequencies is due to shot noise in the image, giving rise to a constant offset at all frequencies. The cutoff frequency is visible in the power spectrum as the sharp bend between constant power at higher spatial frequencies than the cutoff frequency, and varying power

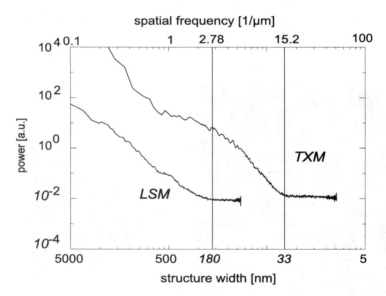

FIGURE 3. Power Spectra of TXM and LSM micrographs. The top x-axis gives spatial frequency in $[1/\mu m]$, the bottom x-axis gives the corresponding widths of the smallest structural information available from a micrograph in nanometers. The y-axis gives the power for each spatial frequency in arbitrary units; the plots have been scaled to fit into the same range. The two vertical lines indicate the cutoff frequency for both micrographs, 15.2 μm^{-1} (TXM) and 2.78 μm^{-1} (LSM), respectively. They correspond to smallest structural information of 33 nm for the TXM and 180 nm for the LSM.

at lower frequencies. To avoid additional contributions by the CCD camera used in the TXM, the power spectra of X-ray micrographs are corrected using flatfield power spectra. The power spectra of LSM and TXM micrographs are shown in Fig. 3. The two vertical lines indicate the cutoff frequencies, 15.2 μm^{-1} for the TXM and 2.78 μm^{-1} for the LSM and the corresponding smallest structure of the micrographs respectively. This is 33 nm for the TXM micrographs, and 180 nm for the LSM micrographs.

DISCUSSION AND OUTLOOK

The results reported here demonstrate that whole cells with immunogold labeled nuclear structures can be investigated advantageously with the transmission X-ray microscope after silver enhancement. The labeling, as visualized with the TXM, is confirmed by confocal laserscan microscopy. While some substructure is visible with the LSM, it is not possible to clearly resolve it. In the TXM these labeled regions appear as numerous spots, about 100 nm across. Since the current limit for the achievable resolution in the TXM is not due to the fundamental diffraction limit, it can be expected that in the future a better resolution of the TXM will be achieved. At the 2D level it may be possible to increase the current spatial resolution limit of 30-40 nm to 10-20 nm. Our aim will be to take tomographic image data of the MSL-1 labeled cells, so as to get high spatial resolution information on the axial configuration of the cell. Further experiments are planned to allow labeling of living cells using microinjection, avoid the silver enhancement, and improve the vitrification process to image labeled cells more closely to the natural state.

ACKNOWLEDGEMENTS

We would like to thank P. Guttman for help with the TXM, J. Herbst for expert technical help, A. Pannuti for constructing the MSL-1 expression vector, and the staff of the BESSY storage ring facility for excellent working conditions. This study was supported by the Deutsche Forschungsgemeinschaft (DFG) under contract number Ru 725/1-1, Ru 725/1-2, Schm 1118/5-1 and Schm 1118/5-2.

REFERENCES

1. Griffiths, G., *Fine Structure Immunocytochemistry*, Berlin: Springer-Verlag, 1993.
2. Gu, W., Szauter, P., and Lucchesi, J. C., *Developmental genetics* **22**,56–64 (1998).
3. Niemann, B., Schneider, G., Guttmann, P., Rudolph, D., and Schmahl, G., In Aristov, V. V. and Erko, A. I., editors, *X-ray Microscopy IV*, Chernogolovka: Bogorodskii Pechatnik, 1994, pp. 66–75
4. Schneider, G., Niemann, B., Guttmann, P., Rudolph, D., and Schmahl, G., *Synchrotron Radiation News*, **8(3)**,19–28 (1995)

Fine Surface Structure of Unfixed and Hydrated Macrophages Observed by Laser-Plasma X-Ray Contact Microscopy

Yoshimasa Yamamoto[1], Herman Friedman[1], Hideyuki Yoshimura[2],
Yasuhito Kinjo[3], Seiji Shioda[4], Kazuhiro Debari[4], Kunio Shinohara[5],
Jayshree Rajyaguru[6], and Martin Richardson[7]

[1]Department of Medical Microbiology and Immunology, University of South Florida College of Medicine, Tampa, FL, USA, [2]Meiji University School of Science and Technology, Kawasaki, Japan, [3]Tokyo Metropolitan Industrial Technology, Tokyo, Japan, [4]Showa University School of Medicine, Tokyo, Japan, [5]Faculty of Medicine, The Tokyo University, Tokyo, Japan, [6]Arnold Palmer Hospital for Children and Women, Orlando, FL, USA, [7]Center for Research and Education in Optics and Lasers, University of Central Florida, FL, USA.

Abstract. A compact, high-resolution, laser-plasma, x-ray contact microscope using a table-top Nd:glass laser system has been developed and utilized for the analysis of the surface structure of live macrophages. Fine fluffy surface structures of murine peritoneal macrophages, which were live, hydrolyzed and not sliced and stained, were observed by the x-ray microscope followed by analysis using an atomic force microscopy. In order to compare with other techniques, a scanning electron microscopy (SEM) was utilized to observe the surface structure of the macrophages. The SEM offered a fine whole cell image of the same macrophages, which were fixed and dehydrated, but the surfaces were ruffled and different from that of x-ray images. A standard light microscope was also utilized to observe the shape of live whole macrophages. Light microscopy showed some fluffy surface structures of the macrophages, but the resolution was too low to observe the fine structures. Thus, the findings of fine fluffy surface structures of macrophages by x-ray microscopy provide valuable information for studies of phagocytosis, cell spreading and adherence, which are dependent on the surface structure of macrophages. Furthermore, the present study also demonstrates the usefulness of x-ray microscopy for analysis of structures of living cells.

INTRODUCTION

Since x-ray microscopy is well suited for high resolution and sensitivity of microscopic analysis of biological specimens without the use of the conventional sample processing such as fixation, which causes dehydration, and staining required for electron microscopy, it enables analysis of fine fragile structures of live and hydrated cells. A compact, high-resolution, laser-plasma, x-ray contact microscope using a table top Nd:glass laser system has been developed and utilized for the analysis of the surface structure of live macrophages. Macrophages play a key role in host defense against a variety of invading microorganisms by phagocytosis, secretion of cytokines, presenting antigens to lymphocytes, and other functions (1). The initial step of the process required for such functions is the recognition of microorganisms by the surface of

CP507, *X-Ray Microscopy: Proceedings of the Sixth International Conference,*
edited by W. Meyer-Ilse, T. Warwick, and D. Attwood
© 2000 American Institute of Physics 1-56396-926-2/00/$17.00

macrophage (2). Therefore, the structural as well as physicochemical property of macrophage surfaces should be critical. To date, most studies on the surface structure of macrophages have been based on electron microscopic techniques, which require specimens to be fixed and stained, meaning dehydrated and non live cells are utilized. In the present study, the surface structure of live and hydrated macrophages was investigated by x-ray microscopy in comparison with ordinary scanning electron microscopy (SEM) as well as light microscopy.

MATERIALS AND METHODS

Macrophages: Peritoneal macrophages were obtained from 8- to 12-week-old female BALB/c mice at 4 days after intraperitoneal injection of thioglycollate broth. The macrophage suspensions in 10 % fetal calf serum (FCS)-RPMI 1640 medium were cultured on a x-ray photo resist (PMMA) supported by a silicon nitride wafer (Fastec Fabrication Services, Ltd., Northants, UK) for 2 hrs at 37°C in 5 % CO_2. Nonadherent cells were removed by washing the resist with warm Hanks balanced salt solution. The resulting monolayers on the resist were incubated in 10 % FCS-RPMI 1640 medium at 37°C in 5 % CO_2 until the monolayers were analyzed by either x-ray microscopy or SEM. In some experiments, the macrophage suspensions were placed on the resist and used for observation by x-ray microscopy without cells adhered on the resist.

X-ray microscope: The laser plasma x-ray contact microscope was utilized for this study. The microscope was in the Center for Research and Education in Optics and Lasers (CREOL), University of Central Florida, Orlando, FL (3). A laser pulse (5ns width, 10-20 J) from a table-top Nd:glass laser system was used for producing plasma. The pulsed laser beam was focused onto Yttrium targets to produce laser plasma x-rays. The sample holder and other elements, except the laser itself, were installed in a vacuum chamber. The specimen was placed on a x-ray photo resist (PMMA) supported by a silicon nitride window. The specimen holder was placed 1 cm away from the target at an angle of 45 degrees from the target.

X-ray image: The photo resist was dissolved in a mixture of methyl isobutyl ketone and isopropyl alcohol to develop the x-ray image. The three-dimensional topological absorption image that resulted from the development was then observed with a differential interference microscope. An atomic force microscope (AFM) was used to reproduce the images from the developed photo resists.

Electron microscope: Macrophage monolayers on a x-ray photo resist were observed by scanning electron microscopy. Using a standard method, the cells were fixed in 2.5 % glutaraldehyde in 0.1 M phosphate buffer. Postfixation was with 1 % OsO4 in 0.1 M phosphate buffer. The cells were dehydrated using ethanol series, then critical point dried and coated with platinum. The cells were observed by a field a field emission type scanning electron microscope (Hitachi S-4700; Hitachi, Tokyo).

Light microscope: Hydrated macrophage monolayers on a glass cover slip without any sample processing were observed by a conventional light microscope.

RESULTS AND DISCUSSION

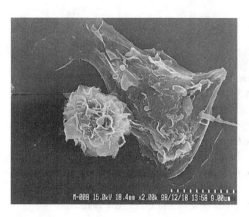

FIGURE 1. SEM image of cultured macrophages. The macrophage monolayers on a photo resist were fixed, stained and dehydrated by an ordinary protocol.

Macrophage surface structures have a pivotal role in the process of phagocytosis as well as in the communication of macrophages with other immune cells, such as lymphocytes. Therefore, precise structures should be studied for understanding macrophage functions, which are critical in the defense system of the host. In order to clarify the fine surface structure of live macrophages, three different methods were utilized in this study for this purpose, i.e. standard light microscopy, electron microscopy and x-ray microscopy. The standard light microscopy is able to show the some structures of cells in medium without any sample processing, but resolution is limited. In fact, some fluffy surface structures of the macrophages were observed in this study, but the resolution was too low to observe fine structures. However, from this observation, it appeared that the macrophage surface is fluffy. In contrast to the light microscope image of macrophages, SEM showed the fine surface structures of macrophages (Fig. 1), but the cells had to be fixed, stained and dehydrated. As seen in the figure, the surface structures of macrophages observed by SEM were rigid and seemed to be shrunk, caused by the sample processing. However, the x-ray images of unfixed- and hydrated-macrophages, i.e. live macrophages in medium, showed fine fluffy surface structures (Fig. 2). The outline of fluffy structure was similar to the image in light microscopy, but was much more clear. Because of high resolution, macrophage surface structures have been mainly studied by electron microscopy. However, the processing of specimens was required for electron microscopy may alter such fragile surface structures of macrophages. In fact, this study showed that the fine fluffy structure of macrophages, which was not observed by SEM, could be demonstrated by x-ray microscopy.

In conclusion, the findings in this study indicate that x-ray microscopy of live cells in medium can reveal fragile structures sensitive to the sample processing required for electron microscopy. The fine fluffy surface structure observed by x-ray microscopy may be involved in attachment and phagocytosis of macrophages.

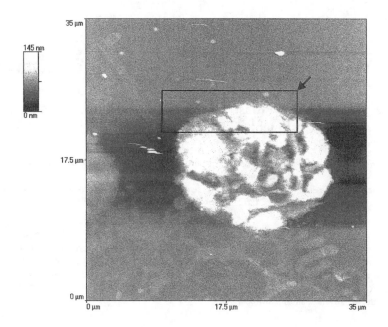

FIGURE 2. X-ray microscopic image of macrophage. The hydrated- and unfixed-macrophage monolayers on a photo resist were examined and the x-ray images observed by an atomic force microscopy. Arrow shows fluffy surface structures.

ACKNOWLEDGMENTS

This work was supported by a grant from the National Institutes of Health (1R21HL60329).

REFERENCES

1. Zwilling, B. S., and Eisenstein, T. K., *Macrophage-Pathogen Interactions*, New York: Marcel Dekker, Inc., 1994.
2. Chakravarti, B., and Chakravarti, D. N., Pathol. Immunopathol. Res., **6**, 316-342 (1987).
3. Richardson, M., Kado, M., Torres, D., Yamamoto, Y., Friedman, H., Rajyaguru, J., and Muszynski, M., SPIE, **3240**, 29-35 (1998).

X-Ray Microscopy Study of Bone Mineralisation

M. Salomé[1], M. H. Lafage-Proust[2], L. Vico[2], D. Amblard[2], B. Kaulich[1], S. Oestreich[1], J. Susini[1], R. Barrett[1]

[1]X-ray Microscopy Beamline ID21, European Synchrotron Radiation Facility, BP 220, F-38043 Grenoble Cedex, FRANCE
[2]Laboratoire de Biochimie et de Biologie du Tissu Osseux, INSERM E 9901, Université Jean Monnet, 15, rue A. Paré, F-42023, Saint-Etienne Cedex 2, FRANCE

Abstract. Transmission spectro-microscopy around the calcium K-edge and fluorescence microscopy were performed respectively on the Transmission X-ray Microscope (TXM) and Scanning X-ray Microscope (SXM) end-stations of ID21 beamline at ESRF, to map the calcium distribution and the Ca/P ratio in bone samples. Preliminary results are presented. The motivation for these experiments is the study of the genetic determinism of bone mineralisation parameters in two different strains of mice.

INTRODUCTION

Bone is a calcified tissue made of a collagen matrix forming a template for calcification. Most of the mineral phase of bone consists of a crystallized calcium phosphate called hydroxyapatite ($Ca_{10}(PO_4)_6(OH)_2$). Several techniques may be used for the study of bone mineralisation. Ash weight and chemical analysis techniques allow a global measurement of bone mineral content, but require the destruction of the sample, consequently losing spatial information. However the study of the mineral content of bone at a microscopic scale is of particular interest, since it can give new insights into remodelling activities, mineralisation process and related mechanical properties. A traditional technique to measure spatial variations of mineral content in bone is microradiography. Other techniques offering higher spatial resolution, such as backscattered electron imaging (BSE), have also been used. Indeed, the grey levels of BSE images have been shown to be significantly related to bone mineral content [1]. Energy dispersive x-ray microanalysis (EDX) [2] allows determination of the elemental composition in well defined regions of the bone sample. At the cellular scale, electron energy loss spectroscopy (EELS) has also successfully been applied to bone study [3]. Fourier-transformed infrared microspectroscopy (FT-IRM) allows mapping of the chemical nature of mineral phases in calcified tissues [4]. However the spatial resolution is limited and the imaging time is slow. Synchrotron radiation infrared microspectroscopy improves both spectral and spatial resolutions [5].

X-ray microscopy with energy tunability, which offers elemental and chemical state sensitivities, is well suited to obtain quantitative mapping of the mineral components of bone. X-ray fluorescence microscopy allows simultaneous elemental quantification and is well suited to Ca/P ratio measurements in bone tissue. In transmission mode,

CP507, X-Ray Microscopy: Proceedings of the Sixth International Conference,
edited by W. Meyer-Ilse, T. Warwick, and D. Attwood
© 2000 American Institute of Physics 1-56396-926-2/00/$17.00

spectroscopic measurements at the calcium absorption edges can be used to map the chemical state and distribution of calcium. Measurements performed on bone samples at the calcium L-edge have been reported [6]. In this contribution, we worked at the calcium K-edge, which offers the advantages of simpler sample environment (since transmission measurements can be performed in air), thicker sample investigation, and higher fluorescence yield.

Recent data show that the peak bone mass acquired at the end of growth is heritable [7], but other bone parameters may also be genetically determined. A study dealing with the genetic determinism of immobilization induced bone loss is in progress at LBBTO, Saint-Etienne, France, with the support of a European Project (ERISTO). Two strains of mice (C3H/HeJ and C57Black/6J), presenting differences in peak bone mass for comparable weight and bone size [8, 9], are involved in this study. Preliminary histomorphometric data show that the level of bone remodeling is different between the two strains when characterized by bone formation rates and mineralisation parameters [10]. A study of mineralisation differences between these mice using spectro-microscopy around the calcium K-edge and fluorescence microscopy of thin histological slices is in progress. Preliminary experiments performed at the ESRF X-ray Microscopy Beamline ID21 are reported in this contribution.

MATERIALS AND METHODS

Sample Preparation

The samples were taken from 18 week-old C3H/HeJ and C57BL/6J mice. The femurs were excised, fixed in buffered formalin for 24h at 4°C, dehydrated in acetone, impregnated in methyl methacrylate for 9 days at -20°C and embedded in polymethyl methacrylate (PMMA) at 4°C. Thin slices (5 or 10 μm thick) were cut using a microtome along the frontal axis. The sample holder consisted of two 100 μm thick stainless steel plates drilled with 2 mm circular windows, between which the sample was stuck. The distal metaphyses of the femurs were centered on the windows.

ID21 Microscopes

Detailed descriptions of the ESRF ID21 beamline [11, 12], the Transmission X-ray Microscope (TXM) [13] and Scanning X-ray Microscope (SXM) [14] are given in other contributions to this conference. We will therefore only give a brief description of the experimental setup. For the spectroscopic measurements performed on the TXM, the condenser used was a gold zone plate with a diameter of 1.2 mm (E. Di Fabrizio, M. Gentili, CNR, Roma, Italy), and the objective lens was a gold zone plate of diameter 64.8 μm (M. Panitz, Institute for X-ray Physics, Göttingen, Germany). The pixel size in the images acquired was 45 nm in the object plane. For the fluorescence measurements performed on the SXM, the focussing lens used was a tantalum zone plate with a diameter of 1.04 mm (H. Kihara, Kansai Medical University, Japan). A 50 μm order selecting aperture was used.

RESULTS

Transmission Spectro-microscopy (on TXM)

The sample imaged was a 10 μm thick slice of femur bone from a C57BL/6J mouse. A series of hundred images were taken around the calcium K-edge (4038 eV), with energy steps of 0.6 eV and 60 s exposure time. The images were flat field corrected, aligned and organized in a three-dimensional array ("energy stack"), allowing quick access to the spectrum of any pixel in the image.

In figure 1, images of a bone trabecula taken at different energies around the K-edge (a) (b) (c) can be seen. These images have been obtained by taking the logarithm of the ratio between the incident and transmitted intensities, and thus correspond to the optical density in the sample. The star-shaped lacuna in the middle of the trabecula is an osteoplast, i.e. a lacuna containing a bone cell called osteocyte. The canaliculi via which the osteocyte is linked to the other bone cells by thin prolongations can also be seen. A calcium map was calculated by subtracting optical density images at 4040.7 eV and 4023.9 eV and dividing by the difference in mass absorption of calcium at those energies. The absorptions of PMMA and collagen were neglected.

In figure 2, the average spectrum over the whole energy stack, in a region of interest of the bone sample is plotted over a reference spectrum of hydroxyapatite, demonstrating the possibility to use spectroscopic information to identify calcium compounds in the sample.

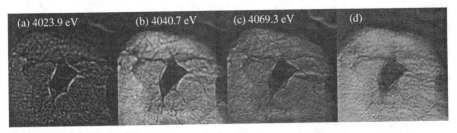

FIGURE 1. (a) (b) (c) Optical density maps of a trabecula in the femur metaphysis of a mouse at different energies. The field of view is 15.7 μm x 15.7 μm. Increasing brightness corresponds to higher values. (d) Calcium map calculated from images (a) and (b).

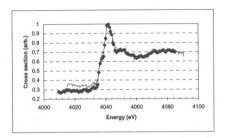

FIGURE 2. Reference x-ray absorption spectrum of hydroxyapatite (plain line) and x-ray absorption spectrum taken through a mineralised area of the trabecular bone sample (dots).

Fluorescence Microscopy (on SXM)

The energy of the impinging beam was set to 5.46 keV, using the double crystal Si monochromator. Fluorescence spectra of bone were recorded and are displayed in figure 3. The strong attenuation of the phosphorus K_α peak (at 2013.7 keV) in air imposes working under vacuum. Energy windows corresponding to the calcium and phosphorus K_α peaks were defined and used to perform elemental maps of a 5 μm thick slice of femur bone from a C57BL/6J mouse (see figure 4). The sample was scanned with 2 μm steps, covering a field of view of 100 μm x 100 μm. The integration time was 5 s/pixel. A Ca/P molar ratio map was calculated after normalisation of the elemental maps. The images obtained show that the degree of mineralisation is higher in the inner part of the trabecula than at the periphery.

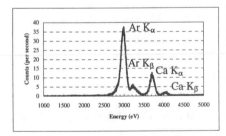 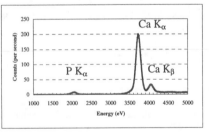

FIGURE 3. Fluorescence spectra of bone in air (left) and under vacuum (right). Parasitic peaks corresponding to the argon of the air can be observed.

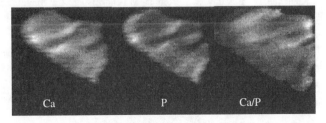

FIGURE 4. Calcium, phosphorus and Ca/P molar ratio maps inside a bone trabecula. The field of view is 100 μm x 100 μm, 2 μm steps.

CONCLUSION

These preliminary experiments have shown the possibility of mapping the mineral content of bone and the Ca/P ratio using fluorescence and spectro-microscopy. Furthermore, an advantage of x-ray microscopy techniques is the possibility to work on the same thin histological slices as conventionally used in histomorphometry, or even directly on polished blocks in fluorescence mode, without staining or decalcification. Further developments will concern the improvement of the accuracy of the quantitative measurements, which is strongly linked to the thickness accuracy and the flatness of the bone slices. We will also investigate the possibility to distinguish between the different calcium compounds involved in bone mineralisation. These techniques will be applied in the framework of a study dealing with the genetic determinism of bone loss.

ACKNOWLEDGEMENTS

The authors are very grateful to L. André, R. Baker, G. Berruyer, A. Beteva, A. Debreyne, F. Demarcq, D. Fernandez, A. Koch, P. Noé, F. Picard, G. Rostaing, F. Thurel, C. Vartanian and H. Witsch for their technical assistance. We also thank N. Laroche (LBBTO) for the preparation of the bone samples.

REFERENCES

1. Bloebaum, R.D., Skedros, J.G., Vajda, E.G., Bachus, K.N., and Constantz, B.R, *Bone* **20:5**, 485-490 (1997).

2. Akesson, K., Grynpas, M.D., Hancock, R.G.V., Odselius, R., and Obrant, K.J., *Calcif Tissue Int* **55**, 236-239 (1994).

3. Bordat, C., Bouet, O., and Cournot, G., *Histochem Cell Biol* **109:2**, 167-174 (1998).

4. Bohic, S., Heymann, D., Pouezat, J.A., Gauthier, O., and Daculsi, G., *C R Acad Sci III* **321:10**, 865-876 (1998).

5. Miller, L.M., Carlson, C.S., Carr, G.L., and Chance, M.R., *Cell Mol Biol* **44:1**, 117-127 (1998).

6. Buckley, C. J., Khaleque, N., Bellamy, S. J., Robins, M., and Zhang, X., "Mapping the organic and inorganic components of bone", in : *X-ray Microscopy and Spectromicroscopy*, edited by J. Thieme et al., Berlin Heidelberg : Springer Verlag, 1998, pp. II-47 – II 55.

7. Slemenda, C.W., Christian, J.C., Williams, C.J., Norton, J.A., and Johnston, C.C. Jr, *J Bone Miner Res* **6**, pp 561-567, (1991).

8. Beamer, W.G., Rosen, C.J., Donahue, L.R., Frankel, W.N., Churchill, J.A., Shultz, K.L., Baylink, D.J., and Pettis, J.L., *Bone* **23**, abst 1058, (1998).

9. Sheng, M.H.C., Baylink, D.J., Beamer, W.G., Donahue, L.R., Rosen, C.J., Lau, K.H.W., and Wergedal, J.E., *Bone* **25:4**, 421-429 (1999).

10. Amblard, D., Lafage-Proust, M.H., Alexandre, C., and Vico, L., *Bone* **23**,abst F424 (1998).

11. Susini, J., and Barrett, R., "The x-ray microscopy facility project at the ESRF", in *: X-ray Microscopy and Spectromicroscopy*, edited by J. Thieme et al., Berlin Heidelberg : Springer Verlag, 1998, pp. I-45 – I 54.

12. Susini, J., Barrett, R., Kaulich, B., Oestreich, S., and Salome, M., "The x-ray microscopy facility at the ESRF : a status report", *this conference*.

13. Kaulich, B., Niemann, B., Rostaing, G., Oestreich, S., Salome, M., Barrett, R., and Susini, J., "The transmission x-ray microscope end-station at the ESRF", *this conference*.

14. Barrett, R., Kaulich, B., Salome, M., and Susini, J., "Current status of the scanning x-ray microscope at the ESRF", *this conference*.

Potential Application of Vanadium Probes for Biological X-ray Microscopy

M.M. Moronne, D.J. Hamamoto, G. Meigs, L. E. Johnson, G. P. Denbeaux, and W. Meyer-Ilse

Lawrence Berkeley National Laboratory, Berkeley, CA, USA

Abstract. Soft x-ray microscopy is now routinely capable of imaging biological specimens with resolutions that are five times better than the best visible light microscopes ($\leq$ 50 nm). However, for biological labeling the only options developed for x-ray microscopy have been silver enhanced gold probes that can be used with both scanning and wide field CCD microscopes, such as XM-1 at the Advanced Light Source (ALS), and luminescent lanthanide probes that necessitate a scanning microscope (SXM). To add to the arsenal of useful x-ray biological probes, we have begun the development of labels that rely on the L-edge absorption lines of vanadium. Vanadium is especially attractive as a biological contrast reagent because it has two strong absorption lines at energies that range from ~512 to 525 eV just below the oxygen K-edge, which makes it an ideal material for imaging in the water window. In this report, we present our initial findings on the application of vanadium for biological labeling. Fixed NIH 3T3 cells grown on silicon nitride windows were incubated with vanadyl sulfate and in some cases basified with triethylamine. After vanadium treatment of the cells, they were thoroughly rinsed and then imaged using XM-1 above and below the vanadium 516 eV resonance. Vanadium staining was clearly visible around and in the cells. These findings suggest that bioconjugated vanadium clusters could provide sufficient x-ray contrast to be used as biological probes.

INTRODUCTION

In the past two decades, cell biology has been revolutionized by the development of fluorescent labels that can accurately target specific biological molecules and report the spatial distribution of cell components ranging from the cytoskeleton to nuclear transcription sites. In addition, the wide variety of emission colors makes it possible to perform simultaneous labeling of distinct cell proteins and to determine their geometric proximity. This information can be especially useful in identifying and quantifying possible functional interactions between specific cell proteins. However, conventional light microscopy using fluorescent probes is limited in resolution to about 200 nm. Although of great utility, in many instances this size scale is simply too large given that the dimensions of most proteins range from 1 to 10 nm.

The most direct approach to increase resolution beyond the visible light limit is to use techniques that can operate at shorter wavelengths. In recent years, soft x-ray microscopy (λ = 2 to 5 nm) has achieved resolutions close to 30 nm [1]. Future improvements are expected to give resolutions approaching 10 nm. However, despite the improved resolution of x-ray microscopes, and the high quality morphological

CP507, *X-Ray Microscopy: Proceedings of the Sixth International Conference,*
edited by W. Meyer-Ilse, T. Warwick, and D. Attwood
© 2000 American Institute of Physics 1-56396-926-2/00/$17.00

detail they can provide, there has been a lack of molecular probes capable of localizing specific cell proteins and other important subcellular targets. To circumvent this problem, labeling agents are needed that can specifically operate using x-ray illumination. In this regard, two approaches have been taken including silver enhanced gold labeling [2], and scanning luminescence x-ray microscopy (SLXM, [3, 4, 5]).

In this report, we describe the first results of soft x-ray contrast enhancement based on vanadium labeling. We show that the vanadium L-edges provide sufficiently narrow resonances, strong x-ray absorption, and favorably positioned water window energies (~512-525 eV) that make it an ideal candidate for biological x-ray labeling. Of particular importance, is the fact that vanadium based probes would be fully compatible with silver enhanced gold labeling making it possible to perform high resolution co-localization studies using CCD based x-ray imaging microscopes such as XM-1 at the Advanced Light Source (ALS) or the similar microscope at BESSY.

METHODS

Vanadium Spectroscopy. Vanadium $L_{2,3}$ edge absorption lines were examined using both dry precipitated samples on a silicon nitride window with the ALS scanning transmission microscope (STXM) on beamline 7.0.1 and in aqueous 1 M $VOSO_4$ solutions sandwiched between silicon nitride windows using the CCD imaging microscope XM-1. The latter instrument provides lower spectral resolution than the undulator driven STXM and there are some differences in the wavelength calibrations, but the results are quite comparable in demonstrating the narrowness of $L_{2,3}$ resonances. The vanadium hydrated oxide precipitate was prepared by adding about 50 mM triethylamine to a 1 M solution of $VOSO_4$ and allowed to react for about ten minutes. This turned the blue vanadyl solution to a turbid slightly gelatinous brownish dispersion that was then collected by centrifugation. The precipitate was resuspended in 18 $M\Omega$ water and re-centrifuged. The washing was repeated and the final precipitate resuspended to approximately 10 mg/ml in water. A few μl of the dispersed precipitate was then applied to a 1000 Å thick silicon nitride window and dried in air.

Cells. NIH 3T3 mouse fibroblasts were cultured in Dulbelco modified Eagles medium containing 10% calf serum (Gibco) at 37°C with 5% CO_2. Specimen substrates consisted of 1000 Å thick silicon nitride windows (3.5x3.5 mm) etched in 100 μm thick 12x12 mm silicon squares. Fibronectin coating was found to greatly improve cell adhesion to silicon nitride windows and was used routinely [5]. In either case, substrates were placed in petri dishes, covered with media, then layered with cells and cultured for 1 to 3 days until a convenient but subconfluent density was obtained. Fixed cell specimens were prepared by first rinsing off the culture media with phosphate buffered saline (PBS) pH 7.4, permeabilized for 30 seconds with 0.5% Triton X-100 in PBS, then fixed with a solution containing 4% paraformaldehyde, 0.2% glutaraldehyde, 0.5% Triton X-100, 5 mM $MgCl_2$, 150 mM NaCl, 1 mM EGTA, and 10 mM phosphate at pH 7.4 for 1 hr. Specimens were subsequently stored in PBS at 4°C until ready for treatment with vanadium. Fixed cell specimens to be labeled with vanadium were first rinsed with 150 mM NaCl then incubated for several hours with 1 M $VOSO_4$. After the vanadyl incubation the cell specimens were rinsed several times with 18 MΩ water. In some samples, after an hour of vanadyl incubation, 10 to 20 mM triethylamine was added inducing hydrolysis of the vanadyl solution. After an additional 10 minute incubation, the specimens were rinsed with water. All vanadium treated silicon nitride cell samples were sandwiched with an opposing silicon nitride window and kept hydrated. All cell imaging was done with XM-1.

RESULTS

Figure 1 compares images taken with the ALS 7.0.1 STXM of the precipitated vanadium hydrated oxide above and below the L-edge resonances. The increase in x-ray absorbance at the higher energy is dramatic attaining 99% reduction of the transmitted signals in thicker areas of the film. Fig. 2 shows the absorption spectra of the precipitate taken with the STXM (at the crosshairs) and a similar spectrum taken with the CCD XM-1 imaging microscope of an aqueous 1 M $VOSO_4$ solution. It can be seen from both spectra that a 3.5-4 eV increase from just below the L-edge resonance (point of maximum transmission), results in about 90% of the full signal decrease corresponding to an optical density of 1.0. A very rough estimate can be made for the attenuation length for vanadium at the L resonance lines by using the signal at 527.5 eV to estimate the thickness of the $VOSO_4$ solution sampled in the XM-1 spectra. Using the tabulated mass absorption coefficients [7], we obtain an average estimate of 2 μm thickness. Using this value and the vanadium concentration of 0.051 gm/ml, we estimate an attenuation length of ~30 nm at the first vanadium L-edge resonance. This value is reasonably consistent with the vanadium attenuation length of 97 nm just past resonance [7].

FIGURE 1. The figure below shows STXM images of precipitated vanadium hydrated oxide taken above and below vanadium $L_{2,3}$ edges (128x128x0.3 μm per pixel). Cross hairs in left image indicate spot used to take spectra in Fig. 2 (left)

Vanadium Precipitate on Si_3N_4 Imaged above and below the $L_{2,3}$ edge

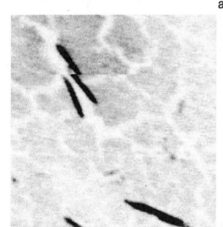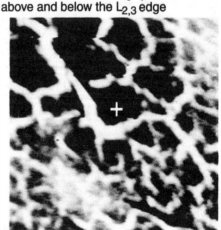

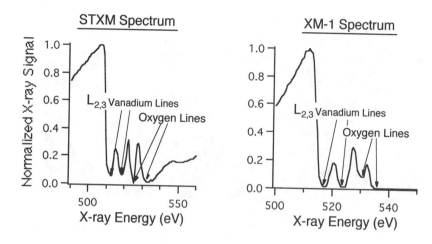

FIGURE 2. The figure above on the left shows the STXM spectrum obtained from the spot marked by crosshairs in Fig. 1. The right hand figure shows a similar spectrum obtained using XM-1 of a 1 M solution of VOSO$_4$.

The figures below show examples of NIH-3T3 fibroblasts incubated with vanadyl sulfate as discussed in METHODS. The vanadyl ion is divalent and as a consequence tends to bind strongly to negatively charged groups common to cell proteins, membranes, and nucleic acids. Figure 3 illustrates a striking increase in contrast of the cell nucleus when imaged at 516 eV compared to 500 eV. At the lower energy, the border of the nucleus is virtually imperceptible, whereas at the higher energy the entire nucleus stands out clearly. Fig. 4 shows a region dense in lamellopodia first imaged at 516 eV, 500 eV, then again at 516 eV.

FIGURE 3. 3T3 cells incubated with VOSO$_4$ then imaged using the CCD imaging microscope XM-1.
Edge of a 3T3 Cell Nucleus Imaged
above and below the Vanadium L$_{2,3}$ Edge

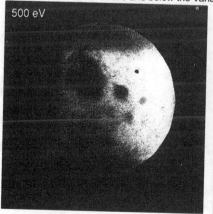
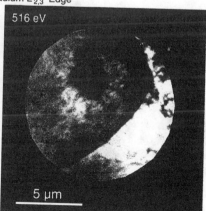

Sequence of Images of Filamentous Region of 3T3 Cell
Treated with VOSO₄ then Basified with Triethylamine

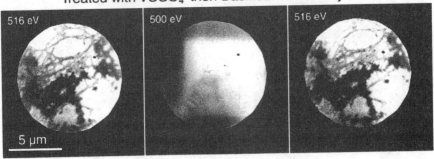

FIGURE 4. 3T3 cells incubated with VOSO₄ then basified with triethylamine. Images were acquired from left to right using XM-1. Dark dense material most likely to be hydrated vanadium oxide crystallites.

Figure 5 shows the contrast enhancement provided by the vanadyl binding to the cytoplasmic edge of a fixed cell. In this pair of images, the change in energy is a mere 3 eV. Nonetheless, the contrast increases at the cell boundary from 29% to 57% (cross, left image).

Vanadium Enhanced Contrast Along Cytoplasmic
Boundary of 3T3 Cell

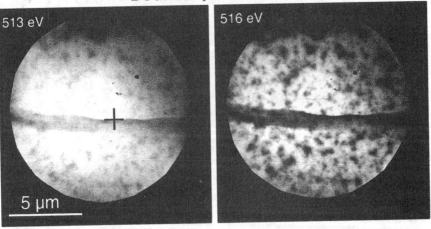

FIGURE 5. 3T3 cells incubated with VOSO₄ and imaged along cytoplasmic edge of lamellopodia. Cross (left image) indicates point used to compare relative contrasts at 513 and 516 eV.

CONCLUSION

It is clear from the data obtained, that vanadium offers the possibility of a second contrast probe to complement silver enhanced gold labeling with CCD-imaging x-ray microscopes. Based on the spectral data collected, the resonant absorption at the L-edge should make it practical to specifically image vanadium particles that are 20 nm or smaller. For labeling purposes, vanadium nanocrystals should be 5 nm or smaller to insure adequate cell penetration. By careful choice of hydrolytic conditions that control vanadium oxide formation, it should be possible to make these small crystals, which can then be derivatized by conventional silane chemistry providing linkers to biological recognition molecules such as antibodies and avidin.

ACKNOWLEDGEMENTS

This paper is dedicated to our dear friend and colleague Werner Meyer-Ilse whose untimely death is a tragic loss for us all.

The authors further acknowledge the support by the US. Department of Energy, Office of Health and Environmental Research, and the Office of Basic Energy Sciences under contract number DE-AC03-76SF00098.

REFERENCES

1. Kirz, J., Jacobsen, C. and Howells, M. (1995). Soft x-ray microscopes and their biological applications. *Quarterly Reviews of Biophysics* 28(1): 33-130.

2. Chapman, H. N., Fu, J., Jacobsen, C., and Williams, S. (1996) Dark field x-ray microscopy of immunogold-labeled cells. *J. Microscopy Soc. Am.* 2(2): 53-62

3. Jacobsen C; Lindaas S; Williams S; Zhang X. (1993) Scanning luminescence x-ray microscopy - imaging fluorescence dyes at suboptical resolution. *J. Microscopy* 172: 121-129.

4. M. M. Moronne, C. Larabell, P.R. Selvin, and A. Irtel von Brenndorff (1994), Development of fluorescent probes for x-ray microscopy. *Microscopy Society of America* 52:48-49.

5. Moronne, M.M. (1999), Development of x-ray excitable luminescent probes for high resolution scanning x-ray microscopy. *J. Ultramicroscopy*, 77:23-36

6. T. Warwick, H. Ade, S. Cerasari, J. Denlinger, K. Franck, A. Garcia, S. Hayakawa, A P. Hitchcock, J. Kikuma, S. Klinger, J. Kortright, G. Meigs, G. Morisson, M. Moronne, S. Myneni, E. Rightor, E. Rotenberg, S. Seal, H-J. Shin, W.F Steele, T. Tyliszczak, and B. P. Tonner (1998). A scanning transmission x-ray microscope for materials science spectromicroscopy at the Advanced Light Source. *Rev. Sci. Instr.* 69(8): 2964-73

7. B.L. Henke, E.M. Gullikson, and J.C. Davis (1993). X-ray interactions: photoabsorption, scattering, transmission, and reflection at E=50-30000 eV, Z=1-92, *Atomic Data and Nuclear Data Tables*, 54 (2): 181-342.

Immunohistochemistry for the MEPHISTO X-PEEM

B. Gilbert[a], M. Neumann[b], S. Steen[b], D. Gabel[b], R. Andres[c], P. Perfetti[d],
G. Margaritondo[a] and Gelsomina De Stasio[e]

a)   Institute of Applied Physics, Swiss Federal Institute of Lausanne, Switzerland
b)   Department of Chemistry, University of Bremen, Germany
c)   Paul Scherrer Institute, Switzerland
d)   Istituto di Struttura della Materia della CNR, Roma, Italy
e)   Department of Physics, University of Wisconsin at Madison, WI 53706

Abstract. Over almost 50 years of its development, the science of immunology has become an indispensable tool for the understanding of the histology of tissues. Antibodies are highly specific probes, so that tissue structure can be interpreted not just by morphological considerations but by association with a wide variety of physiological molecular antigens. The simple concept of an antibody linked to a microscopically dense marker remains constant in each application of the technique, such as the use of fluorescent markers or the incorporation of electron-dense gold colloids for electron microscopy. We describe a new application of immunocytochemistry to x-ray spectromicroscopy: the antibody labeling of tissue structures with nickel precipitates. Regions of positive staining can be seen by the acquisition of nickel distribution maps in the MEPHISTO X-PEEM from the intense Ni L-edge absorption features around 850 eV. The aim of this work is to know the background tissue structures on which the distribution of other relevant elements (such as boron for BNCT) can be mapped. Results are presented showing positive staining by two antibodies in human glioblastoma tissue, anti-Ki-67, a protein found in the nuclei of proliferating cells, and anti-van Willebrandt factor, located in blood vessel endothelia. We show that the criteria for successful staining for optical microscopy are different than for spectroscopic imaging, but useful results can be obtained with careful image treatment.

INTRODUCTION

X-ray spectromicroscopy has the capability of detecting elements and distinguishing chemical states while imaging the specimen under study. There are limits to the use of this technique for complex biological samples, however, as it is usually not possible to differentiate individual proteins from their elemental composition, and generally the determination of protein function is a complicated task. Proteins govern intercellular mechanisms, and the expression of different proteins distinguishes cells and the structures within them. Techniques of staining or labeling with dense markers or fluorescent probes have proven invaluable for both medical analysis and biochemical research by using the natural protein binding affinities to localize and visualize specific targets in samples for optical microscopy. The most specific and versatile probes are antibodies, raised as an immune response against an antigen that could be a protein or other large molecule. Immunohistochemical agents have been conjugated to various

CP507, *X-Ray Microscopy: Proceedings of the Sixth International Conference,*
edited by W. Meyer-Ilse, T. Warwick, and D. Attwood
© 2000 American Institute of Physics 1-56396-926-2/00/$17.00

markers for use beyond conventional optical microscopy, such as fluorophores for confocal optical microscopy, colloidal gold for SEM, and recently lanthanides for luminescent x-ray microscopy [1]. We describe a further application of immunohistochemical staining that incorporates nickel to provide a marker detectable in the X-PEEM due to intense x-ray absorption features at the Ni L-edge. The aim of this work is to correlate the detection of non-physiological elements in tissue sections with cell and tissue structure. More specifically, the experimental therapy for glioblastoma, boron neutron capture therapy (BNCT) requires that the boron isotope ^{10}B be localized in tumor rather than healthy tissue [2 and the references therein]. Following bombardment by thermal neutrons, cell death will occur only in the tumor through the neutron capture and subsequent fission of the boron nucleus. Boron administered to human patients is detectable by X-PEEM in tumor sections [3]; however, it is not clear if boron reaches the nucleus, which is highly desirable for the therapy. One method of addressing this question may be to stain nuclei with nickel in tissue sections that are also analyzed for boron. If subsequent analysis showed a coincidence of nickel and boron areas we could show that boron does reach the nucleus *in vivo*. We present the first results of staining trials with two antibodies: anti-Ki-67, a protein present in the nucleus of proliferating cells, and anti-van Willdebrandt factor, located in blood vessel epithelia.

MATERIALS AND METHODS

Glioblastoma tissue samples were taken from human patients administered with BSH ($B_{12}H_{11}SH$) before surgery for tumor excision [2]. The tumor tissue was fixed overnight in a 10% solution of formaldehyde. It was dehydrated by immersion in baths containing increasing concentrations of ethanol (70%, 96% and 99%, 3 exposures for 30 minutes at each concentration). The tissue was hardened in toluene for 1 hour and finally embedded in paraffin at 60°C. The tissue blocks were microtomed into 7 μm sections and mounted on glass slides (for optical microscopy) or silicon wafer substrates (for MEPHISTO).

Anti-Ki-67 was purchased from Dianova, Germany, anti-van Willebrandt factor from DAKO, Denmark, and used without further purification. In each case the secondary antibodies are conjugated to enzyme-linked streptavidin-biotin complexes so that the method of stain development and nickel incorporation steps are identical (oxidative polymerization of diaminobenzidine by horseradish peroxidase and H_2O_2 in the presence of $NiCl_2$). Following the removal of paraffin with xylene, the tissue sections were rehydrated with increasing concentrations of water in ethanol. Endogenous peroxidase was blocked with 1% H_2O_2 in double distilled water, then the samples were serially incubated in a humid chamber with normal serum, the primary and secondary antibodies, and finally the mixture of diaminobenzidine, H_2O_2, and

NiCl₂. The tissue sections on silicon were ashed in a cold oxygen plasma for 24 hours to remove carbon and increase the relative concentration of trace elements [4].

All samples were analyzed in the MEPHISTO X-PEEM installed on the 10m TGM beamline of Aladdin at the Synchrotron Radiation Center. MEPHISTO acquires micro-XANES spectra with an optimum lateral resolution of 20nm [5] and is used for research in medicine [6], cell biology, materials science and tribology [7]. MEPHISTO was operated in the photoemission mode, acquiring total electron yield photoelectron images as a function of photon energy from the surface of a sample under soft x-ray illumination. The images were manipulated in Adobe Photoshop for Mac to enhance contrast. An intensity threshold function was applied to the nickel distribution maps to remove contribution from the non-specific background.

RESULTS AND DISCUSSION

Figure 1 shows the results of staining blood vessel walls. A nickel distribution map is acquired by digitally subtracting photoelectron images (e.g. Fig. 1 (left)) on and below the Ni 2p 3/2 peak (see spectrum in Fig. 1). Such a map (Fig. 1 (right)) shows that this particular vessel was positively stained with respect to the surrounding tissue, although the contrast between positively stained areas and the non-specific background is low. The major challenge in staining tissue for spectromicroscopy is to optimize the staining protocol to minimize background staining while keeping a detectable nickel concentration in the positively stained areas. Diffuse background nickel staining was detected in all samples and partly removed with digital imaging software.

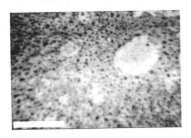 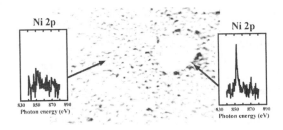

FIGURE 1. Left: Photoelectron micrograph of tumor tissue section stained with nickel against blood vessel endothelia (photon energy = 853 eV, scale bar = 50µm). Right: Nickel distribution map. Inserts: Ni L-edge XANES from stained blood vessel wall and nearby unstained tissue.

Figure 2 shows the results of staining proliferating nuclei. The left image is a sulfur map, the right image the corresponding nickel map. As X-PEEM is surface sensitive (50 - 100 Å) an initial concern was that substantial nickel deposition at stained sites might interfere with the spectroscopy of other elements. There was clearly no problem in imaging physiological sulfur in this sample, however, even at tissue locations stained with nickel. For each section analyzed in MEPHISTO an adjacent tissue section was mounted on glass and stained to act as a reference. There was good qualitative agreement with the results of both staining experiments as observed with optical microscopy and MEPHISTO.

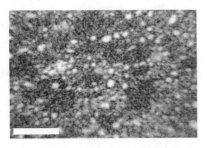 

FIGURE 2. Left: Sulfur distribution (white on black) in tumor tissue section stained with nickel for proliferating nuclei (scale bar = 100 μm). Right: Nickel distribution map (black on white) of the same tissue region.

CONCLUSIONS

We demonstrate that immunohistochemical staining for X-PEEM spectromicroscopy can reveal the location within tissue of specific proteins. We successfully mapped blood vessel walls and nuclei that had been labeled with nickel. The staining was in good agreement with reference sections stained with the same antibodies for the optical microscope, and there was acceptable contrast between areas of positive staining and background. This approach may be especially useful for understanding the location of trace elements within tissue structures.

ACKNOWLEDGEMENTS

Work supported by: the Fonds National Suisse de la Recherche Scientifique, the Consiglio Nazionale delle Ricerche, the Ecole Polytechnique Fédérale of Lausanne, Deutsche Forschungsgemeinschaft, Fonds der Chemischen Industrie and the Biomed Program of the European Commission. We thank Roger Hansen and the staff of the SRC (a national facility supported by the NSF under grant DMR-95-31009) for their help, and Takashi Suda for use of the oxygen plasma oven.

REFERENCES

1. Moronne, M. M., *Ultramicroscopy* **77**, 23-36 (1999).

2. Otersen, B., Haritz, D., Grochulla, F., Bergmann, M., Sierralta, W. and Gabel D., *J. Neuro-Oncology* **33**, 131-139 (1997)

3. Gilbert, B., Perfetti, L., Fauchoux, O., Redondo, J., Baudat, P.-A., Andres, R., Neumann, M., Steen, S., Gabel, D., Mercanti, D., Ciotti, M. T., Perfetti, P., Margaritondo, G., and De Stasio, G., *submitted.*

4. De Stasio, G., Gilbert, B., Perfetti, L., Hansen, R., Mercanti, D., Ciotti, M. T., Andres, R., Perfetti, P. and Margaritondo, G., *Anal. Biochem.* **266**, 174-180 (1999).

5. De Stasio, G., Perfetti, L., Gilbert, B., Fauchoux, O., Capozi, M., Perfetti, P., Margaritondo, G. and Tonner, B. P., *Rev. Sci. Instrum.* **70**, 1740-1742 (1998).

6. De Stasio, G. Gilbert, B., Perfetti, P., Margaritondo, G., Mercanti, D., Ciotti, M. T., Casalbore, P., Larocca, L. M., Rinelli, A. and Pallini, R., in *X-Ray Microscopy: 6th International Conference*, edited by D. Attwood and T. Warwick, AIP Conference Proceedings, New York: American Institute of Physics, 2000, pp 140-144.

7. Canning, G. W., Suominen Fuller, M. L., Bancroft, G. M., Kasrai, M., Cutler, J. N., De Stasio, G. and Gilbert, B., *Tribology Letters* **6**, 159-169 (1999).

MATERIALS AND SURFACE SCIENCES

Morphology and Dynamics of Polymers in Constraint Systems

H. Ade

Dept. of Physics, North Carolina State University, NC27695

Abstract. We review the use of NEXAFS microscopy in transmission and from surfaces to characterize a variety of as-cast and annealed polymer thin films. Special emphasis is placed on systems where the proximity of a surface or an interface effects the polymer morphology or dynamics. We exclude microtomed thin sections of bulk material from the review.

INTRODUCTION

When large molecules such as polymers are confined to geometries that are smaller than a few times the polymer's size, the entropic and energetic consequences of the confining interface can strongly influence polymer characteristics and morphologies. One of the simplest confined geometries is a polymer thin film on a substrate, where the substrate surface provides a rigid and the transition to air a flexible interface. During the last few years, Near Edge X-ray Absorption Fine Structure (NEXAFS) imaging and spectroscopy has been used to study a variety of effects in these constraint systems (1-6). Typically, the quantitative nature of NEXAFS spectra is employed to determine surface composition or to map the mass thickness of the polymer components. The quantitative nature of the NEXAFS data is rather crucial and complements other microscopies, such as Scanning Force Microscopy. In several instances, these complementary microscopies have been used side by side in order to fully illuminate the sample morphology and dynamics (1-5). We will review results from polystyrene/poly(methyl methacrylate) (PS/PMMA) and polystyrene/brominated polystyrene (PS/PBrS) blends and bilayer model systems.

RESULTS

Polymer Blends

The study of phase separation in thin films of binary mixtures is commercially important for the effective production of various coatings and films, including dielectric layers, photographic materials, and paint systems. While films of polymer blends often exhibit more desirable characteristics than individual homopolymers, most blend components are also highly incompatible with each other and will phase

CP507, *X-Ray Microscopy: Proceedings of the Sixth International Conference,*
edited by W. Meyer-Ilse, T. Warwick, and D. Attwood
© 2000 American Institute of Physics 1-56396-926-2/00/$17.00

separate in the melt or in the presence of a common solvent. The degree of separation in blends will greatly effect the resulting morphology, which can have adverse affects on the properties of the resulting film. Conversely, a phase separated film could exhibit more desirable characteristics (7). Tailoring the resulting morphology during preparation could prove very beneficial. Therefore, understanding the variables that affect phase separation morphology and to what degree the kinetics and dynamics are influenced by the presence of interfaces are issues of considerable commercial interest. From a basic science perspective, systematic investigations of variables affecting these processes are important, since a full understanding of blend separation processes such as nucleation and growth, spinodal decomposition and Ostwald ripening are still lacking for polymer thin films.

PS/PBrS blends

Slep *et al.* (3) compared the as-cast and annealed morphologies of PBrS/PS thin film polymer blends as a function of blend composition. Quantitative NEXAFS maps were compared to AFM topographs and complemented with Secondary Ion Mass Spectroscopy (SIMS) depth profiling. The NEXAFS maps where produced by a single value decomposition procedure from transmission images, which, ideally, requires the careful use of undistorted NEXAFS reference spectra of the constituent components. We show the respective reference spectra for the PBrS/PS systems in Figure 1.

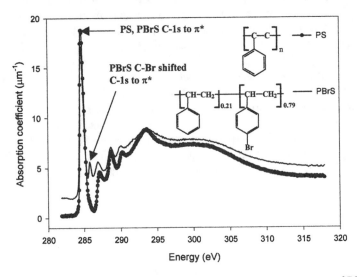

FIGURE 1. Absolute linear absorption coefficients, i.e. NEXAFS reference spectra, of PS and PBrS. Spectra are scaled to Henke data and densities of 1.07 g/cm³ and 1.53 g/cm³ were assumed for PS and PBrS, respectively.

The morphologies observed depend strongly on the blend composition. NEXAFS microscopy of blends with changing compositions showed directly that the morphology changes from droplets to surface holes in a continuous PS layer as the

PBrS concentration increases and when PBrS becomes the majority phase. Fig. 2. illustrates typical NEXAFS microscopy results. The observed morphology for low concentration of PS is contrary to what one would expect from bulk behavior, where the minority phase generally forms dispersed droplets, while the majority component forms the matrix phase. Secondary Ion Mass Spectroscopy (SIMS) data and direct spatially resolved observation of the surface composition with Photoemission Electron Microscopy (PEEM) in PBrS/PS systems showed that the PBrS is encapsulated by PS. The continuity constraint for PS at the surface (see section on bilayers below) and Si interface explained the observed hole morphology in the sample when PS was the minority phase. From the morphologies formed, it could be deduced that 4.6 dyn/cm$<\gamma_{PBrS/Si}<$20.6 dyn/cm, where $\gamma_{PBrS/Si}$ is the PBrS to Si interfacial tension. This indicates that the surface tension for PBrS is approximately 39% polar (3).

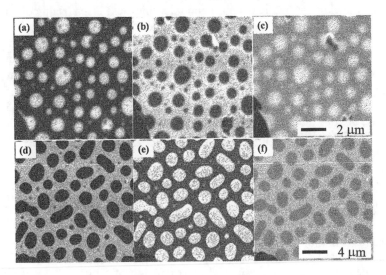

FIGURE 2. NEXAFS maps of 100 nm thick, unannealed 50/50 w/w% PBrS/PS blend (a-c), and 100 nm thick, unannealed 70/30 w/w% PBrS/PS blend (d-f). (a) and (a) are PBrS maps, (b) and (e) PS maps, while (c) and (f) are total thickness maps. (Figure adapted after Figs. 2 and 6 from reference (3))

Dynamics of two dimensional binary fluids: PS/PMMA blends

NEXAFS Microscopy was also used in conjunction with Scanning Force Microscopy (SFM) to observe and characterize the phase separation and coarsening in a thin film blend over a wide range of time scales as these films were annealed. The model polymer blend utilized consisted of monodisperse (Molecular weight Mw=27,000 g/mol, Mw/Mn<1.05) PS and PMMA (Mw=27,000 g/mol, Mw/Mn<1.1). The composition of the phases in the as spun films were determined, the phase separation into smaller domains as well as the development of jagged domains during the early stages of coarsening were observed. These observations were particularly interesting because recent lattice Boltzmann simulations by Wagner and Yeomans (8) have shown that domain growth in binary fluids is a richer phenomenon than hitherto believed and described theoretically. For example, the scaling invariance generally observed for many phase separating systems that are dominated by diffusion breaks down for the phase separation of two-dimensional binary fluids under certain circumstances. The process depends on the relative importance of diffusive to hydrodynamic time scales, i.e. the diffusivity and viscosity of the fluid. For example: i) For diffusive growth domains grow as $R(t) \sim (t-t_o)^\alpha$, and the growth is slow with a growth exponent $\alpha = 1/3$. ii.) Diffusion enhanced collisions provides for faster coalescence, but still with $\alpha = 1/3$. iii) For time scales over which hydrodynamic modes can be excited bulk fluid flow is possible, a process that is faster and leads to a growth exponent of $\alpha = 2/3$. Depending on the viscosities, different morphologies, including jagged ones, evolve. Figure 3 is figure 2 from ref. (2), and illustrates the dynamics and morphologies observed experimentally. It shows PS mass thickness (left column), PMMA mass thickness (middle column) and total thickness maps (right column) of a nominally 143 nm thick 50/50 w/w% PS/PMMA blend annealed for (A-C) 0 min, (D-F) 2 min, (G-I) 10 min, (J-L) 30 min, (M-O) 2 hrs, and (P-R) 1 week. The raw data comprised a set of four to six images for each sample area investigated, including energies corresponding to the C=C 1s to π^* transition at 285.2 eV of the aromatic group of PS, the PS dominated σ^* peak at 295 eV, the C=O 1s to π^* transition of the carbonyl in the PMMA, and the pre and post edge images at 281 and 310 eV respectively. Single value decomposition using densities of 1.07 g/cm^3 for PS and 1.19 g/cm^3 for PMMA resulted in the quantitative composition maps. The observations of the various time domains and the formation of a jagged morphology during coarsening are in qualitative agreement with Wagner and Yeomans.

Confinement induced compatibilization of PS/PMMA blends

To improve the miscibility of polymer blends, copolymer compatibilizers can be added (9). This process is typically limited by the tendency of the compatibilizer to form miscelles in one of the phases and the formation of a micro-emulsion has not been achieved for bulk PS/PMMA blends. Confinement of a blend to a thin polymer film has a significant impact on the process of compatibilization. Samples of a bottom PMMA layer and a top layer of a blend of PS and 30% of a PS-*b*-PMMA diblock copolymer were investigated with NEXAFS microscopy. The loss in configurational entropy as a result of the confinement changed the micellar transition for copolymers (4). When the size of the top layer is comparable to the size of the miscelles, complete miscibility is achieved, resulting in a two-dimensional micro-emulsion. Quantitative thickness maps showed that the film is of uniform thickness and essentially flat (see Fig. 3). Until now, the method of adding copolymer compatibilizers to the blend has only been able to induce complete miscibility in polymer mixtures that are either close to a critical point in their phase diagram or have an attractive interaction between the components. The results by Zhu *et al.* suggest that miscibility can be achieved in thin films irrespective of the type of polymers used. Winesett *et al.* (10) have furthermore investigated the dynamics of the micro-emulsion evolution and found that the growth exponent of the initial growth is 2/3.

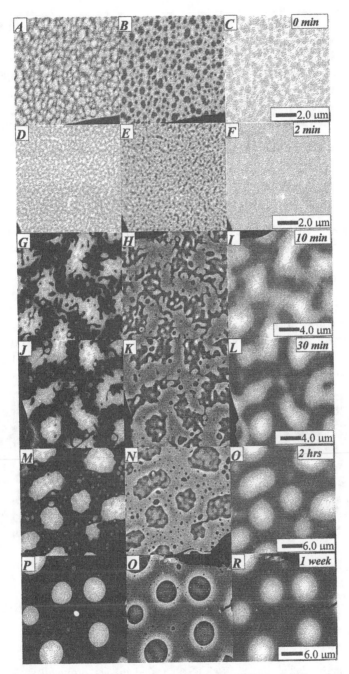

FIGURE 3. PS mass thickness (left column), PMMA mass thickness (middle column) and total thickness maps (right column) of a nominally 143 nm thick 50/50 w/w% PS/PMMA blend annealed for (A-C) 0 min, (D-F) 2 min, (G-I) 10 min, (J-L) 30 min, (M-O) 2 hrs, and (P-R) 1 week. All images are individually scaled for good contrast, with Black = 0 and White = maximum thickness. The maximum thickness of the films increases from 145 nm in Fig. 3A, to 460 nm in Fig. 3R. (From Ref. (2))

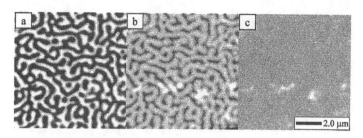

FIGURE 4. Quantitative PS (a), PMMA (b) and total thickness (c) maps of a stable PS/PMMA micro-emulsion in a polymer thin film. (Figure adapted from ref. (4))

Polymer Bilayers

Additional NEXAFS investigations of PBrS/PS model systems focused on the dynamics and morphology formation during the dewetting of a bilayer consisting of a 50 nm thick PBrS film on top of a 30 nm thick PS film (1). These investigations were the first combination of surface (PEEM) and quantitative bulk (STXM) characterization of a polymeric system to assess the three dimensional morphology. As the PBrS is dewetting the PS sublayer, holes are forming randomly and subsequently grow to form Veronoi tesselation patterns. These patterns consist of an interconnected network of spines that eventually break up to form droplets. It was observed from the STXM line profiles that the spines consist of sharply delineated PBrS and are wetted, if not encapsulated, by PS walls (see Fig. 5).

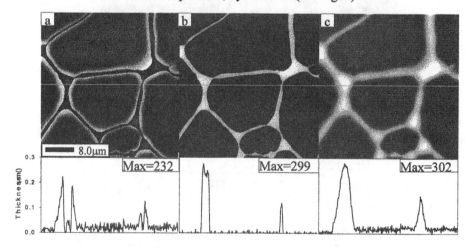

FIGURE 5. Quantitative composition map of dewetting PBrS layers on top of PS layer. (a) PS map, (b) PBrS map, (c) total thickness map.

PEEM studies of a similar PBrS/PS annealed bilayer provided surface NEXAFS spectra from the top 10 nm of the surface from a variety of sample areas including the spines and intersection of the spines. The spectral feature at 286 eV, an energy characteristic of the C-Br shifted C 1s to π^* transition, was not detected from any of the sample areas (see Figure 6). PEEM thus showed unambiguously that the PBrS spines were indeed completely encapsulated. Characterization of the time evolution - from 0 to 11 days of annealing - of the dewetting process with NEXAFS spectroscopy and PEEM furthermore revealed the encapsulation pathway and allowed to distinguish whether diffusion or flow of the polymers are the dominant processes. The results indicated that the encapsulation is delayed until a sufficient number of holes have formed in the PBrS layer through which the PS can penetrate the PBrS film and subsequently flow along the PBrS/air and PBrS/PS interfaces. Additional studies showed that the apparent contact angle at the polymer air interface decreases exponentially with film thickness with a constant that is dependent upon the radius of gyration Rg (a measure of the size of a random coil polymer molecule that increases with the molecular weight of the polymer). NEXAFS data show that the droplets consist of a PBrS core fully encapsulated by PS for substrate thickness greater than Rg while only partial encapsulation is seen for substrates with thickness less than Rg (5).

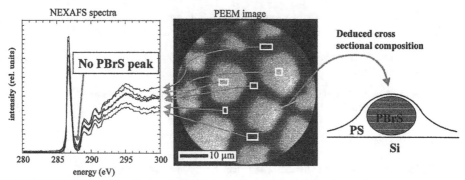

FIGURE 6. PEEM data from dewetted PBrS layer on top of PS layer demonstrating encapsulation of PBrS by PS.

CONCLUSIONS

X-ray microscopy has evolved into a powerful characterization tool for polymer thin films. In particular, the quantitative capabilities afforded by single value decomposition provide a unique complement to scanning probe microscopies. The direct observation of the surface composition with PEEM also provides unique information. It seems entirely justified to be optimistic and to predict that the quantity and breath of applications of NEXAFS microscopy and the concomitant publications will continue to grow and maybe even accelerate in the future.

ACKNOWLEDGEMENTS

Data for Figs. 1-5 acquired with the Stony Brook STXM developed by the group of J. Kirz and C. Jacobsen with financial support from the Office of Biological and

Environmental Research, DOE under contract DE-FG02-89ER60858, and NSF under grant DBI-960-5045. Zone plates are developed by S. Spector and C. Jacobsen of SUNY Stony Brook and D. Tennant of Lucent Technologies with support from NSF under grant ECS-951-0499. Date for Fig. 6 acquired with PRISM at BL8.0 at the ALS. H.A. supported by NSF Young Investigator Award (DMR-945-8060). We thank all co-authors of references 1 through 6 for their contributions.

REFERENCES

1   Ade, H., Winesett, D. A., Smith, A. P., Anders, S., Stammler, T., Heske, C., Slep, D., Rafailovich, M. H., Sokolov, J., and Stöhr, J., Appl. Phys. Lett. **73**, 3773 (1998).

2   Ade, H., Winesett, D. A., Smith, A. P., Qu, S., Ge, S., Rafailovich, S., and Sokolov, J., Europhys. Lett. **45**, 526 (1999).

3   Slep, D., Asselta, J., Rafailovich, M. H., Sokolov, J., Winesett, D. A., Smith, A. P., Ade, H., Strzhemechny, Y., Schwarz, S. A., and Sauer, B. B., Langmuir **14**, 4860 (1998).

4   Zhu, S., Liu, Y., Rafailovich, M. H., Sokolov, J., Gersappe, D., Winesett, D. A., and Ade, H., Nature **400**, 49 (1999).

5   Slep, D., Asselta, J., Rafailovich, M. H., Sokolov, J., Winesett, D. A., Smith, A. P., Ade, H., and Anders, S., Langmuir (submitted) (1999).

6   Winesett, D. A., Ade, H., Sokolov, J., Rafailovich, M., and Zhu, S., Polymer International (in press) (1999).

7   Walheim, S., Schäffer, E., Mlynek, J., and Steiner, U., Nature **283**, 520 (1999).

8   Wagner, A. J. and Yeomans, J. M., Phys. Rev. Lett. **80**, 1429 (1998).

9   Utracki, L. A., *Polymer Alloys and Blends* (Carl Hanser Verlag, Munich, 1990).

10 Winesett, D. A., Ade, H., Gersappe, D., Rafailovich, M., Sokolov, J., and Zhu, S., AIP Proc. of 6th international conference on X-ray microscopy (2000).

Imaging Magnetic Structures With A Transmission X-ray Microscope

P. Fischer[1], T. Eimüller[1], G. Schütz[1], P. Guttmann[2], G. Schmahl[2], and G. Bayreuther[3]

[1]Univ. Würzburg, Inst. f. Experim. Physics IV, Am Hubland, D 97074 Würzburg, Germany
[2]Univ. Göttingen, Inst. f. X-ray Physics, Geiststr. 11, D 37073 Göttingen, Germany
[3]Univ. Regensburg, Universitätsstr. 31, D 93053 Regensburg, Germany

Abstract. The X-ray magnetic circular dichroism (X-MCD), i.e. the dependence of the absorption of circularly polarized X-rays on the magnetization of the absorber exhibits at L-edges of transition metals values up to 25%. This can serve as a huge magnetic contrast mechanism in combination with a transmission X-ray microscope (TXM) to image magnetic domains providing a lateral resolution down to about 30nm. The inherent element-specificity, the possibility to record images in varying external fields within a complete hysteresis loop, the relation of the contrast to local magnetic spin and orbital moments, etc. demonstrate the unique applicability to study the magnetic domain structure in current technical relevant systems like magneto-optics for high density storage media, multilayers for GMR applications or nanostructures for MRAM technology.

INTRODUCTION

The magnetism in systems with reduced dimensionality like thin films is an outstanding issue in view of a fundamental understanding of magnetism on a microscopic scale. In this context selected issues are the occurence of perpendicular anisotropies, the exchange coupling in layered structures or the role of the orbital moment in magneto-optics. On the other hand new effects in ferromagnetic systems like the giant magneto resistance (GMR) to be used in read-out technologies and the field of magneto-electronics (MRAM) where in addition to the charge also the spin of the electrons is taken into account are of increasing importance for current technological developments. Macroscopic magnetic properties, however, are intimately related with the occurence of magnetic domains on a microscopic scale. The interplay between exchange and anisotropy interactions brings the relevant dimensions of magnetic domains into a sub-100nm range. Several techniques to image magnetic domains are available. Among them are electron detecting techniques like scanning electron microscopy with polarisation analysis (SEMPA), photoemission electron microscopy (PEEM) or Lorentz microscopy and optical techniques as scanning near field optical microscopy (SNOM) or Kerr microscopy. Magnetic force microscopy (MFM) detects the stray fields of the magnetic domains with high lateral resolution. A new account to image magnetic structures is the combination of a transmission X-ray

CP507, *X-Ray Microscopy: Proceedings of the Sixth International Conference*,
edited by W. Meyer-Ilse, T. Warwick, and D. Attwood
© 2000 American Institute of Physics 1-56396-926-2/00/$17.00

microscope with the X-ray magnetic circular dichroism as magnetic contrast mechanism (M-TXM). Physical aspects of X-MCD and the basic features of M-TXM will be discussed in the following by selected results obtained with several multilayered and nanostructured Rare Earth-Transition Metal (RE-TM) systems.

X-MCD - A POWERFUL TOOL TO STUDY MAGNETISM

The X-ray magnetic circular dichroism (X-MCD) in core-level absorption detects the magnetization dependence of the absorption coefficient in the vicinity of element-specific absorption edges of circularly polarized radiation. Circularly polarized X-rays are obtainable at synchrotron radiation (SR) sources where either the off-orbit contribution emitted from bending magnets or more recently at third generation SR storage rings asymmetric insertion devices, as helical wigglers and undulators are used. Absorption coefficients are conveniently detected in the transmission mode by counting the incoming and transmitted photon intensity. Due to the limited penetration depth of X-rays in matter, however, the information depth is limited to about 100nm for the soft X-ray regime below 1keV.

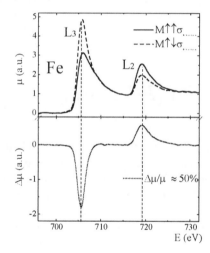

FIGURE 1. Top: Absorption coefficient μ(E) in Fe of circularly polarized photons with the direction of the magnetization M of the sample being parallel (solid) and antiparallel (dashed) to the photon propagation direction. Bottom: Dichroic profile, i.e. difference of the two lines in the top panel which give rise to the magnetic contrast in the magnetic imaging at element specific photon energies,e.g. the L_3 and L_2 edges.

In the absorption process of a circularly polarized photon the spin-orbit interaction in the initial core level and angular momentum conservation causes the photoelectron to act as a spin and orbital polarized probe of the polarisation properties of the unoccuopied levels above the Fermi energy [1]. Thus the spectroscopic X-MCD signal can be related to local magnetic moments. The major attractivity of X-MCD measurements, however, is based on the fact that in principle a separation of local magnetic moments into spin and orbital contributions is possible by applying sum rules [2]. Hence X-MCD is a unique tool to account for the orbital moment, which plays the important role of the coupling the spin to the lattice via spin-orbit coupling.

MAGNETIC TRANSMISSION X-RAY MICROSCOPY

Beyond the pure absorption spectroscopy X-MCD can also be applied to image magnetic domains. A huge magnetic contrast can be obtained in technological relevant transition metals (Fe,Co,Ni) where large dichroic effects with values up to 50% occur at the $L_{2,3}$ edges (see Fig. 1) [3]. The first attempt used a photoemission electron microscope (PEEM) [4]. Recently the combination of X-MCD with the high resolution transmission X-ray microscope (TXM) at a bending magnet station at BESSY I was reported succesfully [5] by the authors.

The optical set-up of the TXM is described elsewhere in detail [6]. To switch to the magnetic imaging mode (M-TXM) circular polarisation is obtained by partially masking the condensor zone plate (CZP) and a small solenoid with fields up to 80mT allows to align the magnetic moments in the sample parallel/antiparallel to the photon propagation direction within a complete hysteresis loop [7].

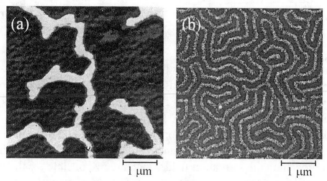

FIGURE 2. M-TXM images taken at the Fe L_3 edge of a 75x(4ÅGd/4ÅFe) layered system prepared onto 325nm polyimid (a) and on a 30nm Si_3N_4 membrane (b) [7,8].

Fig. 2(a) shows a M-TXM image obtained at the Fe L_3 edge of a 75x(4ÅGd/4ÅFe) layered system prepared by magnetron sputtering onto a 325nm polyimid substrate [7]. The illumination time of less than 1 minute needed to record the images is limited by the available flux. An on-line contrast of about 10% allows to distinguish clearly dark and light areas, which can be attributed to magnetic domains, where the direction of the local Fe magnetization points in/out of the paper plane. The corresponding M-TXM image of the same layered system prepared, however, on a 30nm thin Si_3N_4 membrane is shown in Fig. 2(b) [8]. While the structure of the system prepared onto the Si_3N_4 membrane is a self-organized structure, the corresponding domain pattern on the polyimid substrate is dominated by pinning centres, which can be jusitified by taking M-TXM images in varing magnetic fields (see Fig. 3) [9].

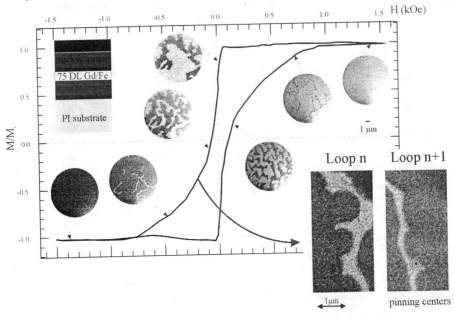

FIGURE 3. M-TXM images taken at the Fe L_3 edge in a multilayered (4ÅFe/4ÅGd)×75) sample prepared on 325nm polyimid in varying applied fields. The role of pinning centres can be studied by recording images in repeated loops [9].

The outstanding property of M-TXM is that the images can be recorded in varying external magnetic fields, which allows a detailed study of the switching of magnetic domains. A typical example is shown in Fig. 3, where M-TXM images have been taken in a (4ÅFe/4ÅGd)×75) multilayered sample prepared onto a 325nm polyimid substrate. Interesting details of the evolution of the microscopic domain structure like the nucleation process can be studied. Magnetically hard worm-like domains could be observed by approaching the saturation field which are responsible for the flattening of the hysteresis profile measured with a vibrating sample magnetometer (VSM). Images taken in repeated loops of the hysteresis cycle allowed to determine unambigously characteristic properties of pinning centres like their density and spatial distribution. The occurence of pinning centres are of particular interest in technological applicatons as they hinder the free movement of domain walls. Generally, the microscopic origin of defects, like inhomogeneities, voids, etc. which have a large impact on the magnetic properties in disordered ferromagnetic systems of technological relevance is at present not completely understood.

Another example demonstrating the capability to image in varying magnetic fields is shown in Fig. 4 where by ion beam etching 1μm×1μm dots could be prepared in a (4ÅFe/4ÅGd)×75) multilayered sample prepared onto a the Si_3N_4 membrane. Both the domain structure within each single dot and the collective switching behaviour can be observed. At present it is unclear whether local variations in the morphology or local variation in the preparation are responsible for the individual switching fields in each dot.

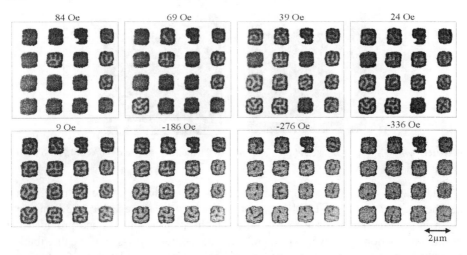

FIGURE 4. M-TXM images taken at the Fe L_3 edge in a microstructured multilayered (4ÅFe/4ÅGd)×75) sample in varying applied fields.

The temporal resolution that can be achieved with the M-TXM is limited by the flux of circularly polarised photons which at present allows to observe dynamical processes on a min time scale. In Fig. 5 magnetic after-effects observed at the Fe L_3-edge in an amorphous $Gd_{25}Fe_{75}$ system, where the penetration of a dark domain into a light domain occurs are clearly visible [7].

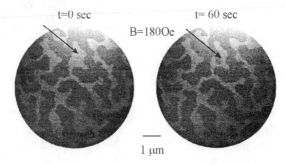

FIGURE 5. M-TXM images taken at the Fe L_3-edge in an amorphous GdFe system within 1 minute [7].

An inherent feature of the M-TXM is the element-selectivity, which allows to address a single element in a multicomponent system. Thus information on the chemical morphology, which influences drastically the behaviour of the global magnetization in the presence of external fields can be obtained. The field dependence of magnetic domains of a (30ÅPt/(4.2ÅCo/14ÅPt) ×30/16ÅPt) multilayered systems recorded at the Co L_3-edge (778eV) are shown in Fig. 6 [11].

close H increasing field

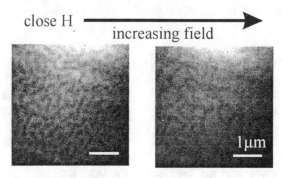

FIGURE 6. M-TXM images taken at the Co L_3-edge in the multilayer system (30ÅPt/(4.2ÅCo/14ÅPt) ×30/16ÅPt) [11].

Quantitative information can be addressed as the amplitude of the dichroic effect, i.e. the contrast in the M-TXM images is related to the average local (3d)-moment given in μ_B (Fig. 7) [10]. Together with the high lateral resolution and provided the resolution function is known, one can in particular address the issue of the distribution of magnetic moments inside the domain wall, which is related to fundamental anisotropy (K_u) and exchange constants (A).

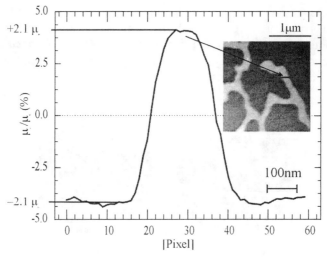

FIGURE 7. Intensity scan of the contrast obtained from M-TXM images of an amorphous GdFe system (1pixel = 13.79nm) [10].

In principle the ratio of M-TXM images taken at L_3- and L_2-edges contains information on the spin and orbital contributions as in spectroscopy. Hence M-TXM is capable to hunt for 2-dim maps of the orbital moment within the resolution provided by the X-ray optics. This is of particular interest, as e.g. in the vicinity of pinning centres the increased coupling of the magnetic moment to defects in the lattice might results in an increased orbital moment. Magnetic images taken at the Fe L_3 and L_2 edge in a (4ÅFe/4ÅGd)×75) multilayered sample prepared onto a 325nm polyimid substrate are shown in Fig. 8 [7]. The reverse in contrast due to the different spin orbit coupling s clearly visible. However the inhomogenous illumination of the images due to the inclined view method hinders at present an accurate quantitative analysis. Although the spectroscopic information which is essential for the applicability of the sum rules to determine spin and orbital moments can be obtained in principle with the M-TXM itself an alternative way is to measure the spectroscopic data at a high resolution spectroscopic beamline with the identical M-TXM sample and to scale images taken at selected energy values to the spectroscopic profile in order to get the microscopic information.

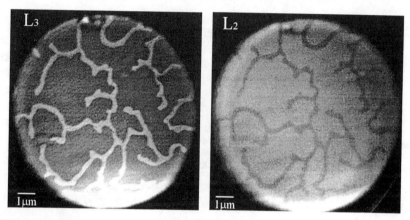

FIGURE 8. M-TXM images of a (4ÅFe/4ÅGd)×75) multilayered sample prepared onto a 325nm polyimid substrate taken at the Fe L_3 (left) and L_2 (right) edge [7].

OUTLOOK

The combination of a high resolution TXM with the X-MCD as huge magnetic contrast mechanism allows to image magnetic domains. The major advantage of this element-specific technique is the possibility to record the images in varying external magnetic fields, which gives information on the magnetization dependent evolution of magnetic domains within a complete hysteresis loop. Quantitative information can be obtained due to the relation of the X-MCD with the local magnetic moment.

A dedicated M-TXM is currently planned to be set-up at a helical undulator at the third generation facility BESSY II in Berlin. High fields (up to 5T) and a temperature variation between 100K and 400K will be provided. The vast field of thin magnetic films with an in-plane anisotropy can be studied by means of tilting the sample. A second branch at this insertion device will house a high resolution monochromatic beamline dedicated to spectroscopic studies of magnetism with X-MCD. Thus the spectroscopic information needed to determine spin and orbital contributions to the local magnetic moment separately can be obtained allowing to study the local distribution of orbital moment. This will give valuable insight into the role of the orbital moment related to the origin of magnetic anisotropies or more general the coupling of the spin to the lattice.

ACKNOWLEDGEMENTS

This work has been funded by the Federal minister of research and technology under the project 05 SC8 WW1 and 05 SL8 MG11.

REFERENCES

1. Schütz, G., Fischer, P., Attenkofer, K., Knülle, M., Ahlers, D., Stähler, S., Detlefs, C., Ebert, H. and deGroot, F.M.F., J. Appl. Phys. 76(10), 6453-6458 (1994).

2. Carra, P., Thole, B.T., Altarelli, M., and Wang, X., Phys. Rev. Lett. 70(5), 694-697 (1993); Thole, B.T., Carra, P., Sette F., and van der Laan, G., Phys. Rev. Lett. 68(12),1943-1946 (1992).

3. Chen, C.T., Sette, F., Ma, Y., and Modesti, S., Phys. Rev. B42, 7262-7265 (1990).

4. Stoehr, J., Wu, Y., Hermsmeier, B.D., Samant, M.G., Harp, G., Koranda, S., Dunham, D., Tonner, B.P., Science 259, 658 (1993).

5. Fischer, P., Schütz, G., Schmahl, G., Guttmann, P. and Raasch, D., Z. f. Physik B 101, 313-316 (1996).

6. Schmahl, G., Rudolph, D., Guttmann, P., Schneider, G., Thieme, J., and Niemann, B., Rev. Sci. Inst. 66(2) 1282 (1995); Niemann, B., Schneider, G., Guttmann, P., Rudolph, D., Schmahl, G., X-ray Microscopy IV, edited by V.V. Aristov and A.I. Erko, Begorodski Pechatnik Publishing Company, Moscow (1995) p 66.

7. Fischer, P., Eimüller, T., Schütz, G., Guttmann, P., Schmahl, G., Prügl, K. and Bayreuther, G., J. Phys. D: Appl. Physics 31(6), 649-655 (1998)

8. Fischer, P., Eimüller, T., Schütz, G., Guttmann, P., Schmahl, G., Köhler, M., and Bayreuther, G., J. Magn. Soc. Japan 23, suppl. S1 205-8 (1999).

9. Eimüller, T., Fischer, P., Schütz, G., Guttmann, P., Schmahl, G., Prügl K., and Bayreuther, G., J. of Alloys and Comp. 286, 20-25 (1999).

10. Fischer, P., Eimüller, T., Schütz, G., Guttmann, P., Schmahl G., and Bayreuther, G., J. of Magn. and Magn. Mat. 198-199, 624-627 (1999).

11. Fischer, P., Eimüller, T., Kalchgruber, R., Schütz, G., Schmahl, G., Guttmann, P., and Bayreuther, G. J. of Synchr. Radiation 6, 688-690 (1999).

Investigation of Alkali-silica Reaction by Transmission Soft X-ray Microscopy

K.E. Kurtis[1], P.J.M. Monteiro[2], J.T. Brown[3], and W. Meyer-Ilse[3]

[1]School of Civil and Environmental Engineering, Georgia Institute of Technology, Atlanta, Georgia
[2] Department of Civil and Environmental Engineering, University of California, Berkeley
[3]Center for X-ray Optics, Lawrence Berkeley National Laboratory, Berkeley, California

Abstract. The soft x-ray transmission microscope XM-1 was used to examine the alkali-silica reaction, a deleterious reaction that produces expansion in some concrete structures. Reactions of silica gel in alkaline solutions (pH $\geq$ 12.4) in the presence of pore solution cations (Na$^+$, Ca^{++}) and a chemical additive (CaCl$_2$) were observed, and the morphology of the reaction products were examined to investigate the mechanisms of expansion and expansion control. From the investigation of the effect of pore solution cations, it was found that reactive silica combined with alkalis present in the pore solution to produce a reaction gel capable of swelling, while the reaction of silica in the presence of calcium ions alone resulted in the formation of a non-swelling product – calcium silicate hydrate. Reaction of the silica gel in the presence of CaCl$_2$ also resulted in the formation of calcium silicate hydrate.

INTRODUCTION

The alkali-silica reaction (ASR) is a deleterious reaction that occurs in concrete between reactive silicates present in some aggregate and alkalis present in the concrete pore solution which are contributed by the cement during hydration. The pore solution in concrete is highly alkaline with pH typically measuring 12.5-13.1 when produced with low alkali cement, but the pH may measure 13.5-13.9 when produced with high alkali cement. The alkali-silica reaction product is an alkali-silicate gel which is believed to expand by swelling. The expansion can result in cracking and loss of stiffness, strength, and impermeability of concrete. While a general knowledge of the chemistry of the reaction exists, the fundamental mechanism of expansion associated with the reaction is not understood. With increasing alkali contents in cement and diminishing aggregate resources, motivation exists to develop an improved understanding of the mechanisms of expansion associated with the alkali-silica reaction [1]. In addition, research has shown that some chemical additives, including LiCl and CaCl$_2$, in certain doses may control the expansion, providing a means to arrest or limit damage in existing structures experiencing ASR [2-5]. However, the mechanisms by which expansion may be controlled by these chemical additives are also not understood. This lack of fundamental understanding of the effect of these additives is one obstacle to their use in existing structures to control expansion and damage by ASR.

CP507, X-Ray Microscopy: Proceedings of the Sixth International Conference,
edited by W. Meyer-Ilse, T. Warwick, and D. Attwood
© 2000 American Institute of Physics 1-56396-926-2/00/$17.00

Study of expansion associated with the alkali-silica reaction in concrete will enhance understanding of the damage mechanisms and will provide more dependable means for predicting performance and for limiting damage. However, in the past, investigations of the reaction product morphology and the mechanisms of expansion have been limited by the lack of appropriate characterization techniques. Because damage to concrete is caused by expansion believed to be generated by swelling of the gel, removal of water from the gel, as required by most existing high-resolution techniques, will introduce artifacts in the specimens. Thus, sample preparation requirements may limit the usefulness of these techniques for in situ examination of reaction products and reactions over time.

Transmission soft x-ray microscopy, a microscopy technique, not previously applied to concrete technology, was used to study the alkali-silica reaction with the objective of improving understanding of the mechanisms of the alkali-silicate gel expansion. Transmission soft x-ray microscopy is a high-resolution technique where reactions occurring in wet samples can be imaged over time at standard pressure, making this technique advantageous for the study of hydrated systems such as those encountered when examining ASR. The transmission soft x-ray microscope XM-1, built and operated by the Center for X-ray Optics (CXRO) and located at beamline 6.1.2 of the Advanced Light Source (ALS) in the E.O. Lawrence Berkeley National Laboratory (LBNL), was used to examine morphology of alkali-silica reaction products in this investigation. (Further information about XM-1 can be found in this volume [6].) Transmission soft x-ray microscopy was used to study the effects of pore solution cations (Na^+, Ca^{++}) and a chemical additive which have been shown to affect expansion ($CaCl_2$) on the morphology of an alkali-silicate gel.

EXPERIMENT

Samples consisted of slurries of various solutions containing chemical grade silica gel (60-200 mesh). Samples were prepared according to the pessimum proportion ($SiO_2/Na_2O=3$) where the ratio of silica to alkali concentrations has been shown to produce the greatest expansion through silica dissolution and repolymerization where alkalis are incorporated into the product gel [7,8]. To study the influence of calcium ions and alkalis on the ASR gel, the ground gel was exposed to three solutions: (a) saturated $Ca(OH)_2$, (b) 0.05M NaOH, and (c) 0.7M NaOH. To examine the effect of some chemical additives on the ASR gel from FURNAS Dam, the ground gel was exposed to two solutions: (a) 0.7M NaOH (control/background) and (b) 0.7M NaOH + 0.1M $CaCl_2$. The pH of the saturated $Ca(OH)_2$ measured 12.4, and the pH of the other solutions measured 12.6.

Pessimum proportion samples were observed periodically by preparing a wet chamber containing a 2 or 3 µl drop of the gel/solution slurry mounted between two silicon nitride films, each about 100 nm thick. To prevent the windows from collapsing together or breaking, commercially available polystyrene beads of 6µm nominal diameter were inserted between the windows with the sample. Sample preparation and experiment design are described in greater detail in [9-11].

RESULTS

To establish a basis for comparison, the silica gel was observed "dry", or in the absence of solution. X-ray images of the dry gel (Figure 1) show that the particles are of irregular shape, with clearly defined edges, and with no evidence of organized internal structure [9-11].

After exposure the sodium hydroxide solution, edges of the silica gel particles are not as clearly defined as when examined dry. While the interior or central regions of the original particles remain dense, the edges of the particles are more transparent to x-rays. These indications of dissolution and repolymerization are consistent with the mechanisms described in the literature [7,8,12], where dissolution is initiated at the silica or silicate surface by hydrolization

$$Si\text{-}O\text{-}Si + H_2O \rightarrow Si\text{-}OH \cdots OH\text{-}Si \qquad (1)$$

and hydroxyl attack

$$Si\text{-}OH + OH^- \rightarrow \quad Si\text{-}O^- + H_2O \qquad (2)$$

Areas of decreased density located at the perimeter of the particles suggest that the silica network, originally dense and less pervious, has become less ordered, allowing the ions in the surrounding solution to permeate more freely into the silicate network. Because of low diffusivity in the sample (as in concrete), the area near the site of dissolution may reach a saturation concentration for silica at a given pH value. As a result, the silica repolymerizes, incorporating sodium and potassium cations into the alkali-silicate gel. The repolymerized gel, seen in Figure 2 just above the original surface of the silica gel particle, appears to be significantly less dense than the original silicate, suggesting a mechanism for expansion due to attack of the silicate structure by alkalis.

In contrast to the products formed in the sodium hydroxide solution, spherulitic or near-spherulitic dendritic structures are produced from the reaction of the silica gel in the saturated calcium hydroxide solution and in the sodium hydroxide solution containing the chemical additive calcium chloride (0.7M NaOH + 0.1M CaCl$_2$). These structures, such as shown in Figure 3, bear close resemblance to calcium silicate hydrate (C-S-H[1]) of the sheaf-of-wheat morphology described in the literature [13-15]. C-S-H is the notation in cement chemistry for the range of compounds with the general formula CaO$_x$·SiO$_2$·H2O$_y$, where x and y vary over a wide range. C-S-H is the primary strength-giving phase in portland cements, and late age formation of C-S-H is

---

[1] In cement chemistry the following oxide abbreviations are used:
C: CaO; S: SiO$_2$; A: Al$_2$O$_3$, S̄: SO$_3$, H: H$_2$O

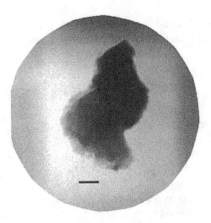

Fig·re 1. X-ray image of chemical grade silica gel. The image was taken with a 10.000 s exposure time with a beam current of 165.6 mA at an original magnification of 2400x. scalebar = 1μm. 60918023

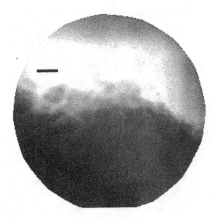

Figure 2. X-ray image of the pessimum proportion of chemical grade silica gel in 0.05M NaOH. The image was taken with a 30.000 s exposure time with a beam current of 180.5 mA at an original magnification of 2400x. scalebar = 1μm. 60918012

often encouraged (e.g., use of pozzolanic materials) as impermeability will be improved. It is therefore believed that the formation of this product during the alkali-silica reaction would not produce expansion or damage.

In a previous investigation [9], reaction of siliceous material in a solution of saturated calcium hydroxide solution with an addition of 0.1M NaOH, the resulting product included both lath-like structures, similar to the dendrites in the sheaf-of-wheat morphology, and repolymerized gel. Comparison of the results from the previous investigation [9,10] with results of this investigation, shows the effect of

Ca^{++} concentration in solution. When poorly soluble calcium-containing additives are used or when soluble additives are used in small amounts, fewer calcium ions are available for reaction with dissolved silica species to produce a non-swelling C-S-H. Instead, dissolved silica species will repolymerize as alkali-silicate gels which may or may not contain adsorbed calcium ions. These results suggest that the concentration of calcium ions in the pore solution, which is dependent upon the solubility of the chemical additive and the percent addition, relative to the concentration of silica species in solution is an important parameter for effective control of expansion associated with ASR.

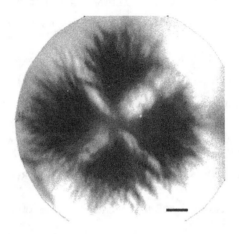

Figure 3. X-ray image of chemical grade silica gel after 20 minutes in saturated Ca(OH)$_2$ solution. The image was taken with a 90.000 s exposure time with a beam current of 381.1 mA at an original magnification of 2400x. scalebar =1μm. 61106105

CONCLUSION

From the investigation of the effect of pore solution cations, it was found that reactive silica combined with alkalis to produce a reaction gel of decreased density – a reaction which appears to result in expansion. The reaction of silica in the presence of calcium ions alone resulted in the formation of a non-swelling product, i.e. calcium silicate hydrate. In the investigation of the effect a chemical additive, a structure resembling calcium silicate hydrate formed in a solution containing sodium hydroxide and calcium chloride, suggesting that the relative concentration of calcium ions to silica species in solution may influence expansion.

ACKNOWLEDGEMENTS

The first-named author wishes to acknowledge financial support by the National Science Foundation graduate research fellowship. Research at XM-1 is supported by the United States Department of Energy, Office of Basic Energy Sciences under contract DE-AC 03-76SF00098.

REFERENCES

1. Idorn, G.M. Concrete Progress: From Antiquity to Third Millennium, Thomas Telford (1997).

2. McCoy, W.J. and Caldwell, A.G. (1951) "A New Approach to Inhibiting Alkali-Aggregate Expansion", *J. ACI*, **47**, 693-706.

3. Prezzi, M.; Monteiro, P.J.M.; and Sposito, G. (1997) *ACI Materials Journal*, **94**, N. 1, 10-17.

4. Prezzi, M.; Monteiro, P.J.M.; and Sposito, G. (1998) *ACI Materials Journal*, **95**, N1, 3-10.

5. Ramachandran, V.S. "Alkali-aggregate expansion inhibiting admixtures", *Cem. Concr. Comp.*, **20**, N 2-3, 149-161 (1998).

6. W. Meyer-Ilse, E. H. Anderson, W. Bates, G. Denbeaux, L.E. Johnson, and A. Lucero, "The High Resolution X-ray Microscope, XM-1", *Proc. XRM99*, (1999).

7. Dent Glasser, L.S. & Kataoka, N. "The Chemistry of 'Alkali-Aggregate' Reaction", *Cem.and Concr. Res.*, **11**, 1-9 (1981).

8. Dent Glasser, L.S. & Kataoka, N. "On the Role of Calcium in the Alkali-Aggregate Reaction", *Cem.and Concr. Res.*, **12**, 321-331 (1982).

9. K.E. Kurtis, P.J.M. Monteiro, J.T. Brown, and W. Meyer-Ilse, "Imaging of ASR gel by Soft X-ray Microscopy", *Cement and Concrete Research*, V28:411-421 (1998).

10. K.E. Kurtis, P.J.M. Monteiro, J.T. Brown, and W. Meyer-Ilse, "Analysis of Deterioration Products Developed in Large Concrete Dams by High Resolution Transmission Soft X-ray Microscopy", *Journal of Microscopy*, in press

11. K.E. Kurtis, "Transmission Soft X-ray Microscopy of the Alkali-Silica Reaction", UC Berkeley Dissertation (1998).

12. Iler, R.K. The Chemistry of Silica, John Wiley & Sons, New York (1979).

13. Rashed, A.M. "The Microstructure of Air-entrained Concrete", UC Berkeley Dissertation (1989).

14. Williamson, R.B. "Constitutional Supersaturation in Portland Cement Solidified by Hydration", *J. Crystal Growth*, **34**, 787-794 (1968).

15. Zampini, D.; Shah, S.P.; and Jennings, H.M. "Early Age Microstructure of the Paste-Aggregate Interface and Its Evolution", *J. Materials Res.*, **13**, N.7, 1888-98 (1998).

Spectromicroscopy Of Catalytic Relevant Processes With Sub-Micron Resolution

S. Günther[1], F. Esch[1], L. Gregoratti[1], M. Marsi[1], M. Kiskinova[1],
U.A. Schubert[2], P. Grotz[2], H. Knözinger[2], E. Taglauer[3],
E. Schütz[4], A. Schaak[4], R. Imbihl[4]

[1] Sincrotrone Trieste, Area Science Park, I-34012 Basovizza, Italy
[2] Institut für Physikalische Chemie, Ludwig Maximilians Universität München, Butenandstr. 5-13, 81377 München, Germany
[3] Max-Plack-Institut für Plasmaphysik, Boltzmannstrasse 2, 85748 Garching, Germany
[4] Institut für Physikalische Chemie und Elektrochemie, Universität Hannover, Callinstr. 3 - 3a, 30167 Hannover, Germany

Abstract. The capabilities of the Scanning Photo Electron Microscope (SPEM) at ELETTRA as a unique probing tool in the field of catalysis and surface science are illustrated presenting results of two recent investigations. The lateral resolution and the high surface sensitivity of the SPEM has enabled imaging the initial steps of the spreading processes of MoO_3 crystals on an alumina support surface, a model system of a catalyst used in petrochemistry. In the second study the local adsorbate coverage inside a pulse of a chemical wave occurring in the catalytic $NO + H_2$ reaction on a Rh(110) single crystal surface has been determined. The microscope was used to monitor the sample surface in situ during the reaction and thus characterizing a temporal and spatial inhomogeneous system. The so-called excitation cycle of the pulse formation has been verified and the adsorbate gradient inside a chemical wave was measured.

INTRODUCTION

The capability of SPEM to provide core and valence level spectral information from sub-micron areas of solid surfaces and adlayers offers the unique possibility of investigating small, catalytic relevant phases in their stationary and – in limits- in their dynamical state. In the following we will illustrate this with the help of two examples, the preliminary steps of spreading of a supported model catalyst and the evolution of chemical waves of adsorbates on a metal single crystal surface during reactions.

EXPERIMENTAL

In the SPEM at ELETTRA, the photon beam provided by an undulator and subsequently monochromatized is demagnified to a sub-micrometer diameter spot by a zone plate optical system. The emitted photoelectrons (PEs) are collected by a 100 mm hemispherical analyzer, mounted at $70°$ with respect to the sample normal and the incident photon beam.[1] The microscope can work in imaging mode collecting photoelectrons with chosen kinetic energy while scanning the sample and in spectroscopy mode with the beam focused to a feature selected from the maps. All photoemission measurements in this study were performed with a photon energy at $\approx$ 600 eV, a lateral resolution of 0.15 μm and an energy resolution of 0.4 eV.

CP507, X-Ray Microscopy: Proceedings of the Sixth International Conference,
edited by W. Meyer-Ilse, T. Warwick, and D. Attwood
© 2000 American Institute of Physics 1-56396-926-2/00/$17.00

SPREADING OF ALUMINA SUPPORTED MO-BASED MODEL CATALYSTS

Supported MoO_3 based catalysts are used in petrochemistry.[2] For example using these catalysts for Hydrodesulfurization (HDS) reactions sulfur is removed from raw oil fractions and "clean" hydrocarbons are obtained. Industrial supported catalysts consist of support particles that provide a high surface area per gram material, which is desirable because the active surface of the catalyst should be proportional to its reactivity. Very often Al_2O_3 particles are used. Onto this support surface the active substance of the catalyst is deposited (in our case Mo oxide). Industrially this is achieved by an impregnation technique: a defined amount of an aqueaous amoniumheptamolybdate solution is deposited onto the support powder, which absorbs the solution. After drying in air at 770 K a surface heptamolybdate species is stabilized on the Al_2O_3 support.[3] It has been shown that alternatively this state of the catalyst can be reached by a solid-solid wetting procedure. The support powder and the active substance (in form of small MoO_3 crystals) are physically mixed. After annealing in humid oxygen the MoO_3 crystals dissolve and Mo-oxide species homogeneously wet the support surface. The preparation technique by solid-solid wetting is expected to produce catalysts with sufficient quality avoiding waste-water problems inherent in impregnation procedures.[3]

The dissolution of crystalline MoO_3 and the wetting of the support was first observed in 1984.[4] Since then these type of spreading processes were intensively studied.[2, 5, 6, 7] Nevertheless the active mechanism accounting for the transport of Mo-species from the crystalline MoO_3 to the spread phase wetting the support is still unknown. We have studied these phenomena using the surface sensitivity and the lateral resolution of the SPEM at ELETTRA. The system MoO_3- Al_2O_3 was chosen because it represents a model system for a so-called monolayer type catalyst, where after the preparation procedure a 1 ML thick Mo-oxide wets the Al_2O_3 support.[2] To obtain an easier geometry necessary for microscopy the support particles were replaced by a flat polished Al foil covered by a 20 nm thin polycrystalline γ- Al_2O_3 film. The thin oxide film was used to diminish charging under the illuminating x-ray beam to about 2 eV on the support.[8] The MoO_3 crystals were placed onto the support foil by a suspension of MoO_3 crystals in methanol.[8] This preparation of the model surface with well separated MoO_3 crystals allowed to allocate a specific crystal with the SPEM. This model catalytic system was ex-situ treated under different conditions (temperature and gas environment) After each treatment the sample was transferred back into the SPEM and the changes of the already defined and characterized area were followed. To image the spreading processes we collected the Mo 3d photoelectrons emitted from the flat support while scanning the sample. Due to a mixture of charging and photon-induced-reduction the Mo 3d peak energy of the spread phase is slightly energy shifted with respect to one obtained from the MoO_3 crystals.[8] Accounting for this energy shift and subtracting the background signal an image with a gray scale proportional to the Mo 3d intensity is obtained. It reflects the amount of Mo species in the spread phase. To reduce topographic artefacts and normalize to the photon flux each image was divided by an Al 2p image of the same area. After this procedure a chemical map of the surface is obtained. Details of the procedures are discussed elsewhere.[9, 10] Fig. 1 shows chemical maps of an area with four MoO_3 crystals taken before and after temperature treatment in oxygen with increasing duration.

 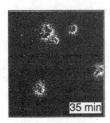 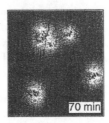 

FIGURE 1. Spreading of the MoO_3 clusters after annealing in dry oxygen at 630 K. Image size: $(60 \times 60) \, \mu m^2$.

It is clearly seen that each MoO_3 crystal has released material that wets the support surface. The lateral resolution as well as the high surface sensitivity of the SPEM enable these first measurements of the initial steps of spreading. The loss of Mo species from the MoO_3 crystals was quantified taking Mo 3d and Al 2p spectra at increasing distance from an isolated MoO_3 crystal. The Mo 3d/Al 2p intensity ratio can serve as a measure of the amount of spread Mo-oxide material on the support. Fig. 2 shows the gradients corresponding to the images in Fig. 1.

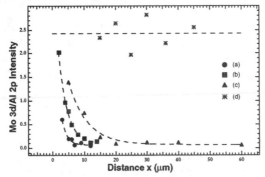

FIGURE 2. Gradient around a MoO_3 crystal after temperature treatment in dry oxygen at 630 K: a) 35 min, b) 70 min, and c) 170 min. Complete spreading after annealing d) 270 min at 720 K in air.

The stars show the Mo 3d/ Al 2p ratio corresponding to a completely spread crystal, where the support should be covered by one monolayer (ML). Obviously all gradients measured in the initial steps of the spreading are well below 1 ML. This finding shows that the favored, so-called unrolling carpet mechanism of the spreading process seems unlikely. In the unrolling carpet mechanism Mo species are assumed to be mobile only on top of MoO_3 and of the already wet support. Once the clean support is reached the Mo species are trapped. Thus, in the special surface diffusion process a carpet of a thickness of > 1ML is unrolled around each MoO_3 crystal. Since this coverage is not reached according to the describe experiment the unrolling can be questioned as an active transport mechanism during spreading.[11]

CHEMICAL WAVES IN THE NO + H₂ REACTION ON RH(110)

The interaction of NO with Rh surfaces is of great interest for the development of the automotive catalytic converter.[12] Especially the catalytic reaction of NO + H_2 on a Rh(110) crystal surface is an important system where the formation and the dynamical properties of chemical waves and the adsorption layers involved were intensively studied.[13-16] Mertens et al. proposed an excitation cycle of this system that triggers a pulse of a chemical wave.[17] The aim of the experiments described below was to confirm this model which will be described with the help of Fig. 3.

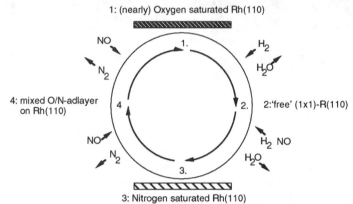

FIGURE 3. The excitation cycle that triggers a pulse in the catalytic NO+H_2 reaction on Rh(110).

On the starting oxygen-covered surface (state 1) H_2 is not able to adsorb, except in the vicinity of a surface defect which will trigger the pulse. There hydrogen reacts with the adsorbed oxygen forming H_2O, which immediately desorbs, producing a largely adsorbate "free" surface (state 2). On this "free" surface H_2 and NO can adsorb and dissociate. Since oxygen is further reacted-off by H, N accumulates leading to a N-saturated surface (state 3). On the N-covered surface NO can still adsorb and dissociate leading to an increase of oxygen and thus an increase of the total adsorbate coverage (state 4). As will be explained below, this leads to the destabilization and desorption of nitrogen and the surface becomes enriched with oxygen until the initial state 1 is reached. The sequence described is called the excitation cycle of the system, which triggers the conversion of the initially oxygen-covered surface (1) in to its reduced state (3) and then back to the original state (1). The described cycle can be regarded as realistic because it is known that the states 1, 2, 3 and 4 can be prepared in a laterally homogeneous way in separate experiments. Under the present reaction conditions an oxygen coverage of 0.66 ML is easily reached, producing an ordered c(2x6) structure, whereas to reach higher coverages requires very high exposure time.[18] The N-saturated surface of 0.5 ML (state 3) shows a (2x1) structure.[15, 16] On this surface additional oxygen can be adsorbed leading to a mixed N+O adlayer (state 4) where the N atoms become destabilized.[16, 19] The destabilized nitrogen desorbs already at a substrate temperature < 500 K, i.e. below the reaction temperature in the described experiment. This accounts for the conversion of the nitrogen-saturated surface to a mixed N+O covered surface and finally to its initial state (1).[16, 19] To prove the described model a pulse was imaged by SPEM collecting the O 1s and to the N 1s core level photoelectrons, respectively (see Fig. 4a). To obtain the displayed images the traveling wave had to be slowed down virtually in order to match its traveling velocity with the acquisition time of the microscope. Therefore a slow offset ramp was added to the scanning of the image, so that the illuminating x-ray beam followed the traveling pulse. After calibrating the gray scale of the image to the one corresponding to the stationary oxygen-saturated or nitrogen-saturated state of the surface the local coverage can be calculated. The result is shown in Fig. 4b.

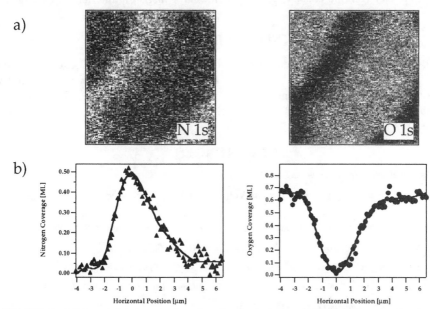

FIGURE 4. a) N 1s and O 1s images of a pulse of a chemical wave. b) Calibrated gray scales obtained from the images show the equivalent adsorbate coverage and thus the correct gradient inside the pulse.

The results in Fig. 4b clearly confirm that the proposed coverages of the excitation cycle (Fig. 3) are reached within the pulse. Furthermore, we observe simultaneously a steep increase of the N coverage, followed by a slow N-decrease and a steep decrease and slow increase of the O coverage. The measured adsorbate gradients are compatible with the intuitive expectation and with computational simulations of the pulse.[20] By changing the kinetic energy during acquisition of an image from the N 1s to the O 1s core level it can be measured whether there is a spatial shift between the decrease of the oxygen and the increase of the nitrogen coverage. This shift can be detected only, if the Rh(110) surface is adsorbate-free (state 2) over a wide spatial area inside the pulse. Within our resolution limit of $\approx$ 1 μm, which is mostly influenced by the acquisition time (the pulse is moving), rather than by the lateral resolution of the instrument, this is not the case. Therefore the adsorbate-free zone cannot exceed a lateral area of 1 μm. The described experiment is the first one where the lateral resolution of a SPEM has been used to quantify the adsorbate coverage in-situ in a dynamical reaction–diffusion system. This provided a consistent explanation of the excitation cycle of the $NO+H_2$ reaction system.[20]

CONCLUSION

The two examples described here show that SPEM can successfully be used to image catalytic relevant processes. The investigations are hindered by three main restrictions: the measurements have to be carried out in vacuum, sometimes the interaction with the highly focused illuminating x-ray beam causes problems and finally, mostly only model systems can be investigated instead of real catalysts in order to maintain a defined geometry. Despite these restrictions, which are inherent to all types of electron microscopy, SPEM provides new insight in the field of catalysis.

ACKNOWLEDGEMENTS

We would like to thank the entire staff of ELETTRA, especially Diego Lonza and Gilio Sandrin for their excellent technical support. This work was financially supported by an EC grant under contract ERBCHGECT920013, the Deutsche Forschungsgemeinschaft (SFB 338) and by Sincrotrone Trieste SCpA.

REFERENCES

1. Casalis et al., *Rev. Sci. Instrum.*, **66**, 4870 (1995).

2. Knözinger, H. and Taglauer, E., *Catalysis* Vol. 10, Cambridge, The Royal Society of Chemistry, 1993, pp. 1-40.

3. Koranyi, T. I., Paal, Y., Leyrer, J., and Knözinger, H., *Appl. Catal.* **64**, L5-L8 (1990). 231.

4. Liu, Y., Xie, Y., Li, C., Zou, Z., and Tang, Y., *J. Catal. (China)* **5**, 234 (1984).

5. Leyrer, J., Margraf, R., Taglauer, E., and Knözinger, H., *Surf. Sci.* **201**, 603-623 (1988).

6. Kisfaludi, G., Leyrer, J., Knözinger, H., and Prins, R., *J. Catal.* **130**, 192-201 (1991).

7. Mestl, G. et al., *Langmuir* **12**, 1817-1829 (1996).

8. Günther, S. et al., *J. Phys. Chem. B* **101**, 10004-10011 (1997).

9. Marsi, M. et al., *Journ. Electron. Spect.* **84**, 73-83 (1997).

10. Günther, S., Kolmakov, A., Kovac, J., Kiskinova, M., *Utramicroscopy* **75**, 35-51 (1998).

11. Günther, S., et al., *J. Chem. Phys.*, submitted.

12. Thomas, J. M., Thomas W. J., *Principles and Practice of Heterogeneous Catalysis*, Weinheim, VCH-Verlag, 1997, pp. 576-590.

13. Mertens, F., and Imbihl, R., *Nature* **370**, 124-126 (1994).

14. Mertens, F. and Imbihl, R., *J. Chem. Phys.* **105 (10)**, 4317-4322 (1996).

15. Gierer, M., Mertens, F., Over, H., Ertl, G., and Imbihl, R., *Surf. Sci.* **339**, L903-L908 (1995).

16. Comelli, G. et al., *Surf. Sci. Rep.* **32 (5)**, 165-231 (1998).

17. Mertens, F., Schwegmann, S., and Imbihl, R., *J. Chem Phys.* **106 (10)**, 4319-4326 (1997).

18. Over, H., *Prog. Surf. Sci.* **58 (4)**, 249-376 (1998).

19. Kiskinova, M., et al., *Appl. Surf. Sci.* **64**, 185-196, (1993).

20. Schaak, A., et al., *Phys. Rev. Lett.* **83 (9)** 1882-1885 (1999).

Application Of X-Ray Microscopy In Food Science
Investigation Of High Pressure Affected Bacterial Spores

Susanne Mönch[1], Volker Heinz[1], Peter Guttmann[2], Dietrich Knorr[1]

1 TU Berlin, Department of Food Biotechnology and Process Engineering, Königin – Luise Str. 22,
   14195 Berlin, Germany
2 University Georgia – Augusta at Göttingen, Institute for X – Ray Physics, Geiststr. 11,
   37073 Göttingen, Germany

Abstract. Using the Göttingen transmission X-ray microscope at BESSY the effect of different pressure and temperature levels during the high hydrostatic pressure (HP) treatment was investigated. At 150 MPa and temperatures up to 50°C the triggering of germination was observed by standard microbiological methods with *Bacillus subtilis* spores. Increasing the temperature to 70°C at the same pressure level killed the spores without any indication of germination. By X-ray microscopy images it could be shown that the typical disintegration of the protoplast is inhibited. This suggests that the enzymic reaction pathway is possibly affected under specific pressure temperature conditions.

INTRODUCTION

The application of high hydrostatic pressure (HP) for food preservation has been extensively investigated by several research groups for many years (review [1]). Similar to the wide-spread thermal methods of preservation the less invasive HP-treatment can kill vegetative microorganisms which might spoil the food or bear health hazards for the consumers. However, bacterial endospores are more resistent to pressure.

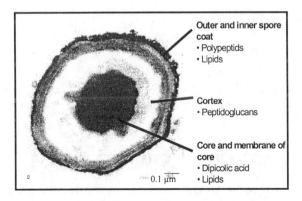

FIGURE 1. Scanning electron microscopy and composition of a spore from *B. stearothermophilus* [2]

CP507, *X-Ray Microscopy: Proceedings of the Sixth International Conference*,
edited by W. Meyer-Ilse, T. Warwick, and D. Attwood
© 2000 American Institute of Physics 1-56396-926-2/00/$17.00

The formation of bacterial spores (Fig. 1) represents a unique strategy of survival of some bacterial species. Sporeformers surviving the most food preservation methods because of their high resistance against heat, radiation and chemical agents (Fig. 2).

The effect of pressure in a range from 100 - 700 MPa on food borne microorganisms should be examined. X-ray microscopy [3] provides a powerful tool to investigate microorganisms in their natural wet environment. Especially, the identification of the inactivation mechanism of bacterial endospores could benefit from the analysis of X - ray microscopy images.

The germination process is triggered by nutrients, chemical agents or enzymes. After activated germination specific enzymes decompose the cortex. Corematerial (dipicolic acid and Ca^{++}) is released (Fig. 3) and an influx of water occurs. The spore loses its typical resistance against heat, radiation and chemical agents. This process can also be triggered by high pressure [4]. After germination the inactivation of the rehydrated spore by temperature or pressure treatment is facilitated.

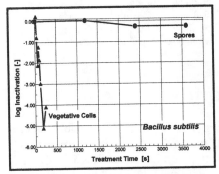

FIGURE 2. Comparison between lethal effect of high pressure on vegetative cells and spores (parameters: 300 MPa, 30°C)

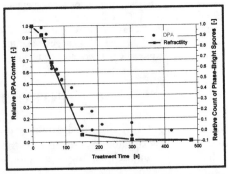

FIGURE 3. Release of dipicolic acid during high pressure treatment (parameters: 150 MPa, 33 °C)

Assumption

High pressure in combination with temperature triggers the germination process. The spore loses its typical heat resistance. Subsequently, the germinated spore is inactivated by the continued action of pressure and temperature treatment.

MATERIALS AND METHODS

Endospores were prepared by surface occulation of Nutrient-Agar in addition of 10 mg $MnSO_4$/L plates with a growing culture of *Bacillus subtilis*, followed by incubation at 37°C for 4 days. The spores were then harvested and suspended in distilled water, concentrated by centrifugation at 4000 g, washed twice with 70% Ethanol and resuspended in Ringer's solution.

For high pressure experiments an aliquot of spore solution with a concentration of 10^9 spores/ml were transfered in a NUNC cryo tube vial which was placed in the multi vessel high pressure unit [5]. Immediately after the pressure cycles, probes were transfered on ice to the X-ray microscope and to the analysis of viable, germinated and inactivated spores. In addition, the release of the spore specific dipicolic acid was measured, which is an indication for the onset of germination.

The X-ray microscopy images were taken with the Göttingen transmission X-ray microscope (TXM) at BESSY I in Berlin [3]. For the investigation of wet specimens an environmental chamber is used. Details of the handling is given elsewhere [e.g. 3, 6]. For an X-ray image exposure times in the range of several seconds are needed. Using two seconds exposure time the dosage given to the spores is in the range of 10^5 - 10^6 Gy.

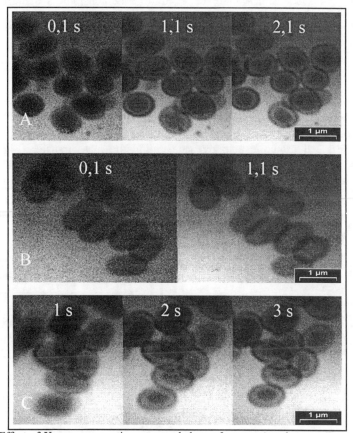

FIGURE 4. Effect of X-ray exposure time on morphology of spores; A – dormant, B – germinated (30°C, 150 MPa), C – inactivated (70°C, 150 MPa)

The effect of X-ray dosage on the structure of spores is negligible (Fig. 4) compared to the effect of high pressure. First short exposure has reduced the movement of the spores in the environmental chamber. Therefore, the second exposure or even longer first exposure times produces sharper images without changing the visible structure of the spores. This effect is the same on dormant, germinated or inactivated spores. The structural differences between these three types of spores are clearly visible (Fig. 4)

The number of viable, germinated and inactivated spores were estimated from colony counts (details see [4]). By phase contrast light microscopy the state of the spores can be assessed. Phase bright spores are dormant and phase dark are germinated.

The release of DPA (2,6-pyridinedicarboxylic acid, dipicolic acid) a typical mass transport process occurring during germination, was detected by the kinetics of accumulation in the medium (details see [4]). The DPA concentrations are indicated relative to the suspension's spore concentration. Dimensionless quantification was achieved by relating the released DPA with the average DPA content of a single spore.

RESULTS AND DISCUSSION

The detection of surviving, germinated or inactivated spores (Fig. 5) shows a clear pressure and temperature optimum of the germination at 30°C and 150 MPa and an inactivation optimum at 70°C and 150 MPa. Increasing temperature up to 40°C speeds up the release of dipicolic acid by accelerating the mass transfer (Fig. 6) due to triggering the germination process. Above 40°C no further acceleration is observed suggesting that an effective transport limitation occured in the surrounding layers of the core.

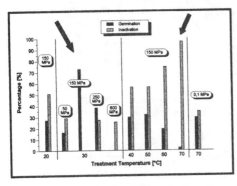 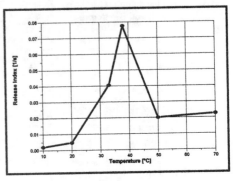

FIGURE 5. Germination and inactivation after different pressure and temperature combinations

FIGURE 6. Release of dipicolic acid as function of treatment temperature during pressure treatment at 150 MPa

Phase contrast light microscopy images yielded differences in the appearance of the spores after pressure treatment at 30°C and 70°C. Whereas complete germination was observed at 30°C, the whole spore population turned to phase dark (Fig.7, middle top). At 70°C the appearence of the spores is similar to dormant spores (Fig.7, right top).

The analysis of X-ray microscopy images suggested a physiological explanation for the anomal temperature response during pressure induced germination. The image of germinated spores show a typical density loss of the core. Similar to dormant spores the core of inactivated spores appears to have a comparably high density. The typical degradation of the core during germination is inhibited.

Considering the X-ray images and the portion of viable spores it is suggested that enzymatic reactions which are involved in triggering the germination are destroyed during pressure treatment of 150 MPa and temperature of 70°C.

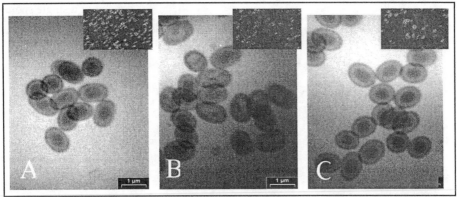

FIGURE 7. Phase contrast microscopy (top) and X-ray microscopy images from spores of *Bacillus subtilis;* A – dormant, B – germinated spores (30°C, 150 MPa), C – killed spores (70°C, 150 MPa)

Conclusion

Pressure treatment at 150 MPa and temperature below 40°C triggers the germination process. Increasing the temperature to 70°C at the same pressure level destroys the enzyme system which is responsible for the degradation of the cortex during germination. Hence, this type of spore inactivation is based on a one step mechanism which does not necessarily involve germinative events.

ACKNOWLEDGEMENTS

The biological part of the work presented here has been supported by the EC - Grant FAIR 1 - CT96 - 1175. The X - ray microscopy part of the work has been supported by the Federal Minister of Education, Science, Research and Technology, BMBF, Bonn, under contract number 05SL8MG11.

REFERENCES

1. Knorr, D., *Current Opinion in Biotechnology* **10**, 485-491 (1999)

2. Vidal, D. R. et al., *Microbios* **92**, 7-18 (1997)

3. Schmahl, G. et al., *Naturwissenschaften* **83**, 61-70 (1996)

4. Heinz, V., and Knorr, D., "High pressure germination and inactivation kinetics of becterial spores" in *High Pressure Food Science, Bioscience and Chemistry*, edited by N.S. Isaacs, Cambridge: The Royal Society of Chemistry, 1998 pp. 435-441

5. Arabas, J. et al., „New Technique for Kinetic Studies of Pressure-Temperature Induced Changes of Biological Materials" in Advances in High Pressure Bioscience and Biotechnology, edited by H. Ludwig, Berlin, Heidelberg New York: Springer, 1999 pp. 537-540

6. Methe, O. et al., *Journal of Microscopy* **188**, 125-135 (1997)

X-ray Spectromicroscopy Studies of Polymer Microstructure

A.P. Hitchcock[1], T. Tyliszczak[1], I. Koprinarov[1], H. Stöver[1], W.H. Li[1],
Y.M. Heng[1], K. Murti[2], P. Gerroir[2], J. R. Dutcher[3], K. Dalnoki-Veress[3]
and H.W. Ade[4]

1. BIMR, McMaster University, Hamilton, ON, Canada L8S 4M1
2. Xerox Research Centre of Canada, Mississauga, ON, Canada L5K 2L1
3. Physics, University of Guelph, Guelph, ON, Canada N1G 2W1
4. Physics, North Carolina State University, Raleigh, NC 27695 USA

Abstract. Scanning transmission X-ray microscopy (STXM) has been used to study the chemical composition and morphology of spontaneous and artificially generated polymer microstructures. Examples include quantitative studies of the internal structure of polymer microspheres, photoconductive thin film structures, and free-standing confined polystyrene thin films.

INTRODUCTION

Soft X-ray microscopy is emerging as a powerful tool for polymer analysis at sub micron scale. Scanning transmission X-ray microscopy (STXM) provides NEXAFS spectroscopy at high spatial resolution and energy selective imaging. The latter provides sensitivity to many sample properties such as local chemical composition, orientation, adsorbate–substrate relationships [3], etc. The intrinsic power of STXM spectromicroscopy is further enhanced by sophisticated acquisition and analysis methodologies which simultaneously exploit the spatial and spectral domains, such as linescans, composition mapping based on singular value decomposition [4], and inversion of image sequences using multivariate statistical analysis or regression procedures. This report illustrates current capabilities of STXM using three examples from recent polymer microanalysis at the BL 7.0 STXM located at the Advanced Light Source.

RESULTS

Quantitative Analysis of Polymer Microspheres

Figure 1 illustrates the large contrast changes that occur in variable energy imaging of core-shell polymer microspheres prepared by precipitation polymerization [5]. The core of these microspheres is DVB55 (55% divinylbenzene / 45% ethylvinyl-benzene), while the shell, prepared in a separate precipitation polymerization step, consists of a copolymer of DVB55 and EGDMA (ethyleneglycoldimethacrylate) where the EGDMA content ranges from 10 - 100%. Sets of images were inverted using singular value decomposition (SVD) [4] to generate maps of the spatial distributions of the poly-DVB55 and poly-EGDMA components.

CP507, X-Ray Microscopy: Proceedings of the Sixth International Conference,
edited by W. Meyer-Ilse, T. Warwick, and D. Attwood
© 2000 American Institute of Physics 1-56396-926-2/00/$17.00

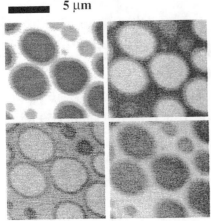

5 μm

FIGURE 1 STXM images at 285, 288.2, 288.4 and 305 eV of thin sections of microspheres composed of a DVB55 core and a shell which is 70 v/v% EGDMA/30% DVB55. The ~5 μm diameter microspheres are embedded in an epoxy which is chemically similar to EGDMA. Changes of ~0.2 eV in the region of the strong π^*(C=O) peak readily distinguish the shell and epoxy components.

Since optical density, $OD = \log_{10}(I_0/I) = \alpha\rho t$ is directly proportional to mass thickness (ρt), linear algebra can be used to convert images into equivalent thickness or composition maps. In principle, the decomposition problem can be expressed as a matrix equation $A\mathbf{x}=\mathbf{d}$, where $\mathbf{x}$ is a vector describing the unknown distribution (ρt), $\mathbf{d}$ are the measured images (converted to OD scale), and A is the matrix of absorption coefficients [6]. The absorption coefficients for poly-DVB55 and poly-EGDMA at 281.0, 285.1, 288.3 and 305.0 eV were obtained from the near-edge absorption spectra of the pure materials, extracted from the microsphere sample with a shell of 100% EGDMA. The model OD spectra were converted to mass absorption scales by matching the spectrum below 282 eV and above 310 eV to the curves for the elemental compositions [6].

The isolated composition maps and the resulting quantitative chemical analysis of the shell regions are presented in Figure 2. We have found the most reliable way to estimate the shell composition (the parameter of interest) is through the ratio of the poly-DVB55 signal in the shell to that of the poly-DVB55 in the core.

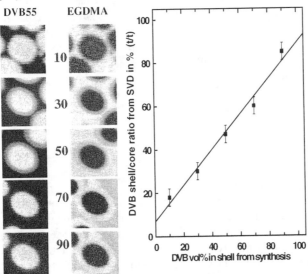

FIGURE 2 (*left*) DVB55 and EGDMA component images for 10 - 90 v/v % EGDMA shell composition, derived from SVD analysis of images at 281.0, 285.1, 288.3, and 305.0 eV. The intensity of the shell region shows changing concentration of each species. The epoxy matrix is bright in the EGDMA map due to strong spectral similarity. (*right*) EGDMA in shell derived from the DVB55 composition maps. The core is assumed to be 100% DVB55 while the shell is assumed to consist of only DVB55 and EGDMA.

This corrects for variable thickness of the microtomed samples, and also reduces sensitivity to the porosity of the samples (and thus extent of in-diffusion of the epoxy resin), while at the same time reflecting the composition of the as-made microsphere. Attempts to quantify the composition of these ~200-300 nm wide shells using point or line spectra were much less successful than the SVD procedure on account of positional instability and the curved interface. In principle application of a 3-component rather than a 2-component SVD analysis would map the internal porosity. However, the similarity of the OD of the epoxy and the EGDMA at the energies studied, prevented clean separation of these two components.

Analysis of a Photoconductive Thin Film Structure

STXM studies of a test photoconductive thin film structure provided by Xerox Research Center of Canada illustrates the power of polymer STXM relative to analytical electron microscopy. The goal of the analysis was to investigate the degree of spatial uniformity of a critical N-containing component present in a protective polycarbonate (PC) capping layer above the image sensing layer. Figure 3 presents an image at 407 eV of the film, a N 1s linescan, and an analysis of the film uniformity. This single 5 minute measurement showed that the (~10%) N-content of the polycarbonate layer was relatively uniform (±20%). By comparison, although it provided higher spatial resolution images, it was not possible to obtain a satisfactory analysis of the spatial distribution of the critical component by either energy loss or X-ray emission spectroscopy in a state-of-the art analytical TEM.

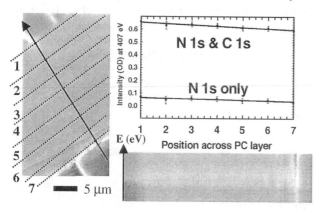

FIGURE 3 (*left*) STXM image of a photoconductive thin film structure at 407 eV. From bottom to top the layers are: a poy(ethlylene-terephthalate) base, a conductor, a pigment layer, a polycarbonate (PC) cap with additive, and an epoxy. (*centre, lower*) **linescan** - optical density as a function of E and position along the indicated line. (*upper*) total and N 1s OD at 407 eV, obtained by integrating signals in the indicated regions.

Fundamentals of Polymer Self Organization

Currently there is much interest in understanding the fundamental processes of self-assembly on the nanometer scale. When free-standing, confined homo-polymer films are annealed close to or above their glass transition temperature (T_g) highly organized patterns are formed. The driving force for this self-assembly is postulated to be long range Van der Waals or dispersion forces which lower the total energy of the system by reducing the distance between two, mutually attractive, confining layers [7].

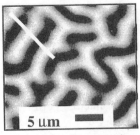

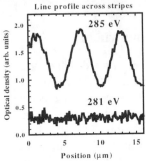

Line profile across stripes

FIGURE 4 (up) Optical density image (285 eV) of a free standing 75 nm thin film of polystyrene (pS) confined between two 70 nm films of SiOₓ.
(low) Intensity at 285 eV (relative pS thickness) across the indicated line, compared to the intensity at 281 eV across the same line.

Quantitative models have been proposed. We have studied a 60-70 nm layer of polystyrene (pS) coated on each side with a silicon suboxide layer. STXM images reveal the same morphology as seen optically [7]. The quantitative and chemically explicit NEXAFS response (Fig. 4) allowed us to measure the OD and thus the pS thickness throughout the pattern. The pS thickness varies sinusoidally, with a change of ~60% from thin to thick regions. STXM measurements of the degree of lateral segregation of the pS in such films are being used to test theoretical models of the pattern formation.

SUMMARY

STXM spectromicroscopy is rapidly developing into a very powerful tool for quantitative polymer microanalysis. Its good spatial resolution and high chemical sensitivity give it unique advantages relative to other techniques. In the near future, improvements in zone plate technology (see elsewhere in these proceedings), and improved access via completion of several soft X-ray STXM facilities currently under construction will make polymer STXM even more powerful and more widely available.

ACKNOWLEDGEMENTS

X-ray microscopy carried out at the Advanced Light Source (supported by DoE under contract DE-AC03-76SF00098). Research supported financially by NSERC (Canada). We thank Tony Warwick, George Meigs and others of the BL 7 STXM team for their extremely competent assistance.

REFERENCES

1. Ade, H., *Trends in Polymer Science* **5**, 58-66 (1997).
2. Ade, H., *Exp. Meth. Phys. Sci.* **32** 225-262 (1998).
3. Hitchcock, A.P., Tyliszczak, T., Heng, Y.M., Cornelius, R., Brash, J.L., and Ade, H. *these proceedings*
4. Zhang, X., Balhorn, R., Mazrimas, J., and Kirz, J., *J. Struc. Biol.* **116**, 335 (1996); Osanna, A., Jacobsen, C. Kirz, J., Kaznacheyev, K., and Winn, B., *these proceedings*.
5. e.g. see Li, W.-H., Li, K., Stöver, H.D.H., *J. Polym. Sci., Polym. Chem.* **37**, 2295 (1999).
6. Henke, B.L., Gullikson, E. M. and Davis, J C., *Atom. Data Nucl. Data Tables*, **54**, 181 (1993).
7. K. Dalnoki-Veress, K., Nickel, B.G., and Dutcher, J.R., *Phys. Rev. Lett.* **82**, 1486 (1999).

X-ray Spectromicroscopy Studies of Protein-Polymer Interactions

A.P. Hitchcock[1], T. Tyliszczak[1], Y.M. Heng[2], R. Cornelius[2],
J.L. Brash[2], H. Ade[3], S. Anders[4], A. Scholl[4] and F. Nolting[4]

1. BIMR, McMaster University, Hamilton, ON, Canada L8S 4M1
2. Pathology, Chem. Eng., McMaster University, Hamilton, ON, Canada L8S 4L7
3. Physics, North Carolina State University, Raleigh, NC 27695 USA
4. Advanced Light Source, Berkeley Lab, Berkeley, CA, 94720 USA

Abstract. Identification of the sites of protein binding on laterally heterogeneous polymer surfaces can help understand mechanisms of bio-passivation, which in turn has practical impact on developing improved polymers for medical uses, such as blood contact applications. We have used scanning X-ray transmission microscopy and photoelectron emission microscopy to explore the sensitivity of these X-ray microscopies to detection of human serum albumin adsorbed onto the surface of two different polyurethane polymers. Both techniques appear capable of the requisite monolayer sensitivity.

INTRODUCTION

Increasingly medicine uses artificial devices to replace or reinforce diseased body parts or to replace defective body functions. One example is hemodialysis which uses devices made from synthetic polymers. Key problems here and in the many other situations where blood is in contact with artificial surfaces - heart-lung bypass for open heart surgery, artificial heart valves, heart assist devices, arterial grafts,

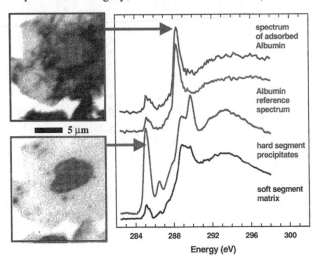

Figure 1 (*upper*) 288.2 eV image of polyurethane thin film exposed to a solution of 5 mg/mL of albumin. (*lower*) 285.1 eV image of the same region. (*right*) Spectra of polymer components (soft and hard regions of the same polymer sample prior to albumin deposition); reference spectrum of an albumin film on a formvar coated TEM grid; and protein spectrum extracted from a ~1 μm² region of the protein coated polyurethane by subtraction of appropriately weighted contributions of the polymer component spectra.

Labels in figure: spectrum of adsorbed Albumin; Albumin reference spectrum; hard segment precipitates; soft segment matrix; 5 μm; Energy (eV); 284 288 292 296 300

CP507, X-Ray Microscopy: Proceedings of the Sixth International Conference,
edited by W. Meyer-Ilse, T. Warwick, and D. Attwood
© 2000 American Institute of Physics 1-56396-926-2/00/$17.00

intravascular stents, involving thousands of patients daily worldwide - are activation of blood coagulation, thrombosis, and the immune system. These effects are known to be initiated by interactions of blood proteins with the surface of the material which is often a polymer. The goal of our research is to develop surfaces which prevent or minimize these phenomena [1-3]. Many of the most promising materials have chemically differentiated surfaces. We are exploring the utility of scanning X-ray transmission microscopy (STXM) and various surface-sensitive X-ray microscopies - photoelectron emission microscopy (PEEM), scanning photoelectron microscopy SPEM, and total electron yield detection in STXM (TEY-STXM) for characterizing phase segregation at polymer surfaces, and for investigating specificity of cell and protein interactions with these surfaces. Here we outline progress using STXM and PEEM for spatially resolved analysis of polymer-protein interactions at the monolayer level. Studies were performed at BL 7.0 and BL 7.3 of the Advanced Light Source.

RESULTS

STXM of Albumin on a Phase Segregated Polyurethane

Initial experiments used STXM with transmitted light detection to study protein *deposited* on the surface of a polyurethane with micron-scale phase segregation. Human serum albumin (HSA) was deposited from an un-buffered aqueous 5 mg/ml solution onto the surface of an ~80 nm polymer thin film section. C 1s and N 1s STXM images and spectra of the sample were examined before and after the exposure. Figure 1 shows images of the same area of the protein coated polymer at 285.1 and 288.3 eV. After subtraction of the spectrum of the underlying polymer, C 1s spectra of the albumin coated regions are in good agreement with the C 1s reference spectrum of albumin. This shows that NEXAFS readily differentiates the three chemical components - protein, hard segment (MDI-urea) and soft segment (polyether) signals - and thus is able to locate proteins relative to an underlying phase segregated polymer, as long as adequate quality data can be acquired from monolayer levels.

Encouraged by the clear spectroscopic answer, we next explored detection limits. C 1s and N 1s spectral regions were studied; in STXM C 1s has better sensitivity. As shown in Fig. 2, STXM can easily detect a monolayer of protein adsorbed onto a Si₃N₄ membrane, where there is no polymer background. Fig. 3 shows detection of HSA down to few layer, possibly monolayer, level, deposited on a typical thickness of polymer (80-100 nm). In this case, regression analysis of image sequences was found to be helpful in locating and measuring the amount of protein. These results demonstrate that conventional mode STXM has the potential to monitor selectivity in adsorption of protein on laterally heterogeneous polymers, and thus has considerable promise for biomaterial surface studies.

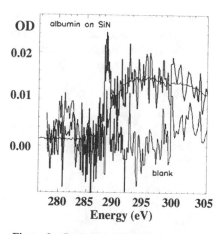

Figure 2. Spectrum of albumin deposited on Si₃N₄ membrane. The C 1s optical density (OD) of 0.01 corresponds to a thickness of ~ 1 monolayer.

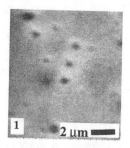
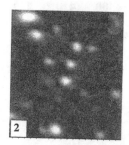
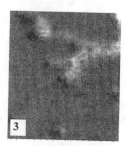

Figure 3 Component maps (1 = matrix, 2 = PIPA fillers, 3 = albumin) derived from regression analysis of a sequence of 86 STXM images (280-315 eV) from a polyurethane with aromatic filler particles (PIPA) which had been exposed to a 1 mg/ml solution of human serum albumin for a few minutes. In each map, the intensity of a given pixel is proportional to the local concentration. The spectra for the 3 components are given in Fig. 1. The faintest detectable albumin signal is an equivalent OD of 0.1, corresponding to only a few layers of protein.

PEEM of albumin on an aromatic-rich polyurethane

PEEM is very sensitive to topography and frequently has problems with charging for insulating samples like polymers. In addition, the technique can not be readily extended to wet samples (unless an inverted geometry is used [4]). However, the intrinsic surface sensitivity of electron detection should given much better sensitivity to low levels of protein adsorbed on polymers, and thus we have explored its applicability to this problem. Figure 4 illustrates N 1s regime PEEM detection of human serum albumin adsorbed onto a poly(tetramethyleneoxide)-ethylenediamine-methylenediisocyanate polyurethane (PTMO-ED-MDI) under conditions known to give monolayer levels on a polyurethane. Although the protein cannot be identified in any single image, the use of regression analyses of image sequences readily maps the protein at very low levels. Further work is needed to quantify the image sequence analysis approach for PEEM, especially since the strength of PEEM signals depend on the work function [5], which may differ from the pure albumin reference to the albumin adsorbed on a polyurethane.

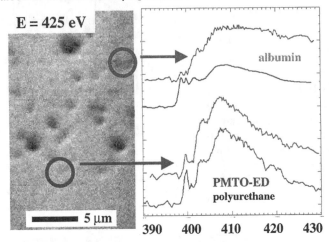

Figure 4. (left) Image at 425 eV of a PMTO-ED-MDI polyurethane with monolayer adsorbed HSA protein.
(right) Comparison of localized spectra extracted from a N 1s image sequence to those of model spectra recorded with PEEM from pure polymer and pure protein, solution cast onto a Si wafer.

SUMMARY

STXM and PEEM spectromicroscopy have been shown to provide useful analytical sensitivity to protein adsorption on polymers at the monolayer level. This is only achieved if one uses these techniques in full spectromicroscopy mode - ie by exploiting both the spectral and spatial aspects of the data. Further work is underway to convert the present demonstration experiments into quantitative tools, and then to apply them to systematic studies aimed at understanding fundamental aspects of protein-polymer interface selectivity and thereby helping to optimize the performance of blood contact polymers.

ACKNOWLEDGEMENTS

X-ray microscopy carried out at the Advanced Light Source (supported by DoE under contract DE-AC03-76SF00098). Research supported financially by NSERC (Canada). We thank Tony Warwick, George Meigs, Rick Steele, and others of the BL 7 team for their extremely competent assistance. We thank Billy Loo for providing helpful discussion regarding the quantitative spectroscopy of proteins.

REFERENCES

1. Santerre, J.P. and Brash, J.L., *Macromolecules* **24**, 5497 (1991) ; ibid, *J. Biomed. Mater. Res.* **26**, 1003 (1992) ; ibid, *Biomaterials* **13**, 1103 (1992); ibid, *J. Appl. Polymer Sci.* **52**, 515 (1994).
2. Skarja, G.A.; Brash J.L., *J. Biomed. Mater. Res.*, **3** , 439 (1997).
3. Woodhouse, K.A.; Weitz, J.I; Brash, J.L., *Biomaterials* **17**, 75 (1996).
4. De Stasio, G. ; Gilbert, B.; Perfetti, L.; Nelson, T.; Capozi, M.; Baudat, P.-A.; Cerrina, F.; Perfetti, P.; Tonner, B.P.; Margaritondo, G., *Rev. Sci. Inst.* **69**, 3106 (1998).
5. Gilbert, B.; De Stasio, G. these proceedings.

X-Ray Digital Microlaminography
for BGA and Flip-Chip Inspection

Alexander Sassov and Filip Luypaert

SkyScan, Aartselaar, Belgium

Abstract. X-ray laminography allows getting local depth information from big flat objects like PCBs and electronic assemblies, which cannot be reconstructed by tomographical approach. According to the needs in high-resolution inspection for electronic and micromechanic industries an X-ray microlaminography system has been developed. This instrument based on a new approach for the x-ray geometry with a minimum of moving parts and a digital extraction of depth information about all layers during one object turn.

INTRODUCTION

Laminographical methods were developed initially for medical application as a "non-computerized" layer-by-layer visualization of human body [1]. In this case an inclined initial X-ray beam project image of specific layer of object to the detector surface with defocusing of other layers during synchronous coplanar rotation of object and photoplate.

|In medical area this method recently completely changed to computerized tomography because of advantages in image sharpness and possibilities to get analytical information about local densities. In the same time all computerized tomographical (CT) methods have one strong limitation. During investigation an object should be completely displaced inside field of view. This requirement can be satisfied in medical applications, but most electronic devices and boards are the planar structures and cannot be completely rotated around axle in the device plane. By another hand big multilayer assemblies cannot be transmitted by X-ray through the direction in parallel to the object surface. Both listed CT limitations not exist in the classical laminography scanning geometry. For laminography the internal 3D-structure can be visualized layer by layer during rotation around axle orthogonal to the object surface. Combination of classical laminography approach with modern digital acquisition technique allows improving spatial resolution and image quality is the standard font and layout for the individual paragraphs.

INDUSTIAL LAMINOGRAPHY

Classical approach for industrial coplanar rotational laminography is shown in Fig.1[2]. An X-ray source and X-ray detector rotate synchronously around the testing

CP507, *X-Ray Microscopy: Proceedings of the Sixth International Conference,*
edited by W. Meyer-Ilse, T. Warwick, and D. Attwood

point in the specimen to produce sharp image from one corresponding layer in the object plane. All images during rotation integrated in X-ray CCD-camera. In real instruments the emission point can be turned by the physical rotation of the X-ray source or by the circular scanning of electron beam through the target inside the source. To change projection plane for visualization in different depths, the camera should be shifted up or down. This classical approach for industrial laminography realized in the instruments from the HP-4Pi company [3].

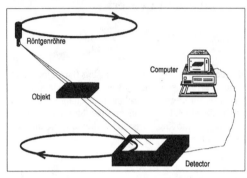

FIGURE 1.
Coplanar Rotational Laminography

Other equipment existing in the market from Digiray company [4] based on the "reverse geometry" with raster scans of emission point inside the X-ray tube and data acquisition by several pointer detectors. These approach more close to X-ray stereo imaging and cannot obtain enough good depth resolution. There are also several other systems with linear movement (1-dimensional) through the conical beam [5]. In this case usable depth and spatial resolution can be achieved for specifically oriented parts of object only.

All recently commercially available X-ray systems for laminography have a spatial resolution in the range of hundred of microns. The spatial resolution mainly limited in accuracy of complicated coplanar movements of two heavy parts of the system. Because of modern multilayer PCBs, flip-chips, BGA-connections etc. contains structures and soldering points of 10-20 microns in size, the spatial resolution in existing systems is not enough for such applications.

There are several main limitations for spatial resolution in the listed laminography systems. In the x-ray source the spatial resolution limited by focal spot size. In the sources with power of hundreds watts the spot size cannot be smaller than 0.5 mm because of heat dissipation in the target. Modern technology open possibilities to build the microfocus X-ray sources with spot size of <10 microns. But such sources have the power in the range of 2-10 watts. That means the corresponding acquisition time should be increased to get reasonable signal-to-noise ratio.

Other limitation for spatial resolution can be found in the detector. Limited number of pixels in the camera array can be a reason for pure resolution in the case of big field of view. For example, if field of view should be 10 by 10 mm with camera division 512x512 pixels the pixel size will be approximately 20 microns. To improve relation of field of view to the spatial resolution the megapixel sensors can be used.

But a main limitation for spatial resolution can be found in complicated coplanar mechanical movements of object, camera and source. All displacements and rotations for heavy x-ray sources and cameras should be done with accuracy better than spatial resolution for all possible object positions. In the case of big size assembles and PCBs it's difficult to avoid vibrations, axle play and object non-planarity during testing.

DIGITAL MICROLAMINOGRAPHY

Digital acquisition technique allows improving several important details in laminographical methods. Two-dimensional digital X-ray camera can collect images to computer memory. That means camera rotation can be changed to the image rotation into the memory by special program. Image rotation in the memory can be done very accurate and, in compare to coplanar mechanical camera movement, don't need complicated and precision mechanics. Main principles for digital multilayer microlaminography are shown in the Fig.2. During object rotation different planes inside the object produce images with rotations around different points into detector. If all projection images during one object turn will be collected into the memory the multilayer separation can be done in once just by re-rotation of initial images around corresponding points.

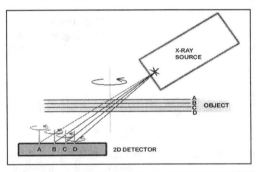

FIGURE 2. Principles of digital multilayer laminography

With digital technique we don't need to rescan object for visualization from another depth plane. In this case all acquired images can be rotated around corresponding points for simultaneous separation for all layers of the object structure. Digital image processing during depth separation also allows correcting most geometrical disalignments and distortions in acquired data. Using this approach the Digital Microlaminography system SkyScan-1080 has been developed. Digital microlaminography allows minimized mechanical movements and reach very good spatial resolution. In the same time the system allows to visualized full 3D-structure just after one fast scan with powerfull post-processing and visualization possibilities.

DESKTOP MICROLAMINOGRAPHY SYSTEM

SkyScan-1080 is a compact desktop microlaminography instrument (Fig.3). As an x-ray source we use a microfocus sealed X-ray tube (Hamamatsu, Japan) operated at 80kV with <8um spot size, integrated with power supply. Source with 130kV sealed tube can be supplied as an option. Images acquired by immovable intensified X-ray

CCD-camera with 560x768 or 1280x1024 pixels. This camera have internal compensation of thermal noise accumulation for acquire images with integration time up to 10 sec. Object stage can hold the objects up to 150x150 mm in size with precision rotation by stepping motor. For microlaminographical depth visualization any 5x5 mm area inside 150x150mm object stage can be selected. Operator can do this selection with looking to the object surface by visual camera or by programming of coordinates from the reference point. One object turn with collecting information about 3D internal structure and separation of 20-80 layers during scanning takes 0.5-5 minutes. System control and depth separation done by internal industrial single or dual Pentium III computer operated under MS Windows NT. Pixel size in any final image of the separated layer inside the object is 10 microns. Depth separation can be done with step size in the range of 10 microns to 2mm.

FIGURE 3. Digital desk-top microlaminography system SkyScan-1080

Powerful library of functions for post-processing allows improving image quality and depth separation. Nonlinear (logarithmically) restoration in initial images is necessary for correct independent separation of different layers. Big library of filters (Laplas, Sobel, median, unsharp mask high frequency...) can be applied to improve image quality and depth separation inside the object. Another set of functions (emboss, color transformation, zoom...) improve visualization possibilities for detection of small defects.

After processing in microlaminography system we will get a set of separated depth sections inside the object. This set can be shown on to the screen layer by layer. Special program inside the software package allows creating realistic 3D image of the object internal structure. After 3D rendering one can select the surface properties for the objects (color, reflection, diffusion...), illumination properties (color, direction, kind of the source...) and move / rotate / zoom / pan the 3D scene onto the screen. Operator can even cut this model of the internal 3D microstructure in any alternative direction or fly through the internal microstructure.

APPLICATION EXAMPLES

Main application areas for high-resolution microlaminography are the electronic assemblies and devices (PCBs, BGA, Flip-Chip...), multilayer structures (laminates), precision mechanics (watch...) etc. We will show several application examples for using microlaminography system SkyScan-1080 for inspection of electronic devices and assemblies.

Fig.4 shows investigation of multilayer PCB. Left image is a conventional microradiographical image where information from all 4 layers mixed together. Microlaminograpohy allows separation of information from different PCB layers, as shown in the right side of the Fig.4.

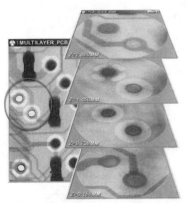

FIGURE 4. Microlaminography of a multilayer PCB

Microlaminography investigation for BGA-inspection is shown in the Fig.5. Left image shows conventional microradiography information. Right images obtained from microlaminography system. In this case the set of separated cross sections through BGA under IC shown as the realistic 3D images with different cuts by software. We can easily found several voids in balls and check all 3D details in the interconnections under IC.

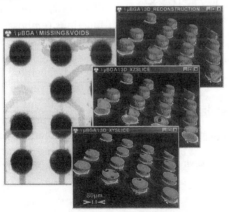

FIGURE 5. Microlaminography application for BGA inspection

Last example shown in Fig.6. demonstrate possibilities of microlaminography for Flip-Chip inspection. Transmission X-ray image, shown in the left side, not allows making any defect inspection for such devices. In the right separated layers one can easily see all details in interconnection under the Flip-Chip body as well as several voids with 10-30 micron in size.

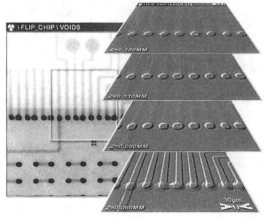

FIGURE 6. Microlaminography application for Flip-Chip inspection

CONCLUSIONS

New approach in digital microlaminography was used for high-resolution layer-by-layer separation of internal microstructure in any place of the big planar objects (multilayers, PCBs etc.), which cannot be reconstructed by computerized tomography because of limited possibilities in rotation. Depth and lateral spatial resolutions in the micron range can be reached. Microfocus X-ray sources and digital acquisition techniques with nonrotatable cameras allow improving performances.

REFERENCES

1. G.T.Herman, Image Reconstruction from Projections, Academic Press, 1980

2. U.Ewert et al,. In: *International Symp. on Computerized Tomography for Industrial Applications*, Berlin, June 1994, pp.148-159

3. Hewlett-Packard Press Reliases. In: web-site http://www.hp48.com/pressrel/feb97/25feb97b.htm

4. Digiray® Digital X-Ray Systems. In: web-site http://www.digiray.com

5. M.Maisl et al. In: *International Symp. on Computerized Tomography for Industrial Applications*, Berlin, June 1994, pp.226-233

Microscopy Of Thin Polymer Blend Films Of Polystyrene And Poly-n-butyl-methacrylate

T. Schmitt[1,2 §], P. Guttmann[3], O. Schmidt[2], P. Müller-Buschbaum[4], M. Stamm[5 *], G. Schönhense[2], and G. Schmahl[3]

[1] Max-Planck-Institut für Polymerforschung, Ackermannweg 10, D-55128 Mainz, Germany
[2] Institut für Physik, Johannes Gutenberg-Universität Mainz, Staudinger Weg 7, D-55099 Mainz, Germany
[3] Institut für Röntgenphysik, Georg-August Universität Göttingen, Geiststraße 11, D-37073 Göttingen, Germany
[4] Technische Universität München, Physik Department E13, James Frank Str. 6, D-85747 Garching, Germany
[5] Institut für Polymerforschung Dresden e.V., Hohe Str. 6, D-01069 Dresden, Germany
§ present address: Department of Physics, Uppsala University, Ångström Laboratory, Box 530, S-75121 Uppsala, Sweden
* to whom correspondence should be addressed
e-mail stamm@ ipfdd.de

Abstract. The structure of thin polymer blend films of polystyrene (PS) and poly-n-butyl-methacrylate (PnBMA) was examined with Transmission X-Ray Microscopy (TXM), Scanning Force Microscopy (SFM), X-Ray Photoemission Electron Microscopy (X-PEEM) and Optical Microscopy (OM). Thin films were prepared by spin casting of a toluene solution of the polymer mixture onto silicon wafers retaining the native oxide. Depending on blend composition and annealing conditions smooth films with and without holes or films with well pronounced surface features (ribbons or islands) were produced. By TXM measurements a high lateral resolution study of the as cast and the annealed polymer blend samples was performed. The contrast in TXM is due to different absorption of x-radiation of the used polymers and due to variations in thickness. With X-PEEM the lateral distribution of the two polymers near the surface was mapped by employing the characteristic Near Edge X-ray Absorption Fine Structure (NEXAFS) spectra of the polymers. The TXM technique is a microscopic method integrating over the total film thickness, whereas the X-PEEM technique is a highly surface sensitive method. TXM and X-PEEM are therefore complementary methods which provide important information on the structure of thin polymer blend films additional to the standard techniques SFM and OM.

INTRODUCTION

Thin polymer films are of increasing importance for high-tech applications in fields like electronics, optics, biotechnology and paper industry. Typical industrial products range from various surface coatings to multicomponent paints. By the synthesis of new polymers optimized materials can be created that serve specific properties required for technological applications. An alternative to this is the much less cost intensive blending of existing polymers having already well defined physical properties. In general high molecular weight polymer mixtures are incompatible because of their extremely small entropy of mixing. Variation of composition and

CP507, X-Ray Microscopy: Proceedings of the Sixth International Conference,
edited by W. Meyer-Ilse, T. Warwick, and D. Attwood
© 2000 American Institute of Physics 1-56396-926-2/00/$17.00

preparation conditions may therefore result in different morphological structures arising from decomposition of the polymer blend films [1, 2, 3]. Microscopy of these structures is of special interest since they are closely related to physical material properties [4, 5]. In the present study thin film blends of the weakly incompatible polymers polystyrene (PS) and poly-n-butyl-methacrylate (PnBMA) were examined with Transmission X-Ray Microscopy (TXM) [6], Scanning Force Microscopy (SFM) and X-Ray Photoemission Electron Microscopy (X-PEEM) [7]. An in-situ study on the kinetics of phase separation of thin blend films of PS and PnBMA employing optical microscopy (OM) during annealing above the glass transition temperature of both polymers will be reported elsewhere by the authors.

SAMPLE PREPARATION

Thin film blend samples were cast on native oxide covered silicon substrates by spin-coating (1950 rpm, 30 sec) from a toluene solution of a binary polymer blend. Depending on blend composition and annealing conditions smooth films with and without holes or films with well pronounced surface features (ribbons or islands) were produced. Windows with a thickness of 100 nm and a diameter of 3 mm in the substrates used for the TXM investigations were etched from the backside prior to film preparation. The used polymers had the following molecular weights: $M_w = 68.3$ kg/mol, $M_n = 66.0$ kg/mol for PS and $M_w = 87.0$ kg/mol, $M_n = 84.5$ kg/mol for PnBMA. For the X-PEEM measurements a different PS sample was used with $M_w = 139.4$ kg/mol and $M_n = 134.4$ kg/mol.

TRANSMISSION X-RAY MICROSCOPY

The contrast in Transmission X-Ray Microscopy (TXM) is due to different absorption of x-radiation of the used polymers and due to variations in thickness. TXM images of 120 nm thick films of different polymer content (molar fraction) were taken at 2.4 nm wavelength with the Göttingen TXM at BESSY [6]. The samples were situated in air under normal pressure while performing the image acquisition. The first row in Fig. 1 shows TXM images of as cast films of different polymer content. These pictures display with high lateral resolution the different developing decomposition features originating from the mutual incompatibility of both polymers and due to the influence of substrate and air interface. At a polymer content of 40% of PS round droplets with different sizes can be seen. A ribbon like structure with substructures is viewed at 45% PS, whereas we see holes with and without substructures at 50% PS and 55% PS, respectively. In addition we examined equivalently prepared thin blend film samples of symmetric composition (50% molar fraction of each) annealed in a vacuum oven at different times at T = 130∘C. TXM images of four different annealing states are shown in Fig. 3. During the annealing process the structures first grow in lateral sizes (30 min), then form connected network structures (3 h and 9 h) and finally they build islands (25 h).

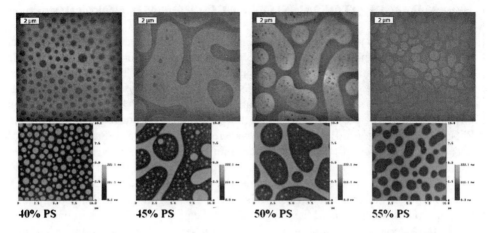

FIGURE 1. TXM images at λ = 2.4 nm (upper row) and SFM images (lower row) of the same as cast samples of thin film blends of PS and PnBMA with a thickness of d = 120 nm and different compositions (molar fractions of PS).

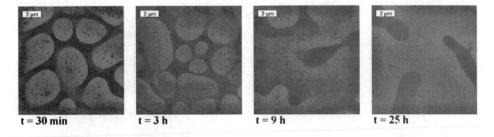

FIGURE 2. TXM images at λ = 2.4 nm of equivalently prepared thin films of mixtures of 50% PS and 50% PnBMA (molar fractions) with a thickness of d = 120 nm after different annealing times at T = 130°C.

SCANNING FORCE MICROSCOPY

In order to compare the TXM images with SFM, images in tapping mode with a Nanoscope IIIa (Digital Instruments) were made on the same samples as used for the TXM investigations. The SFM images on films with different polymer content are shown in Fig. 1 in the lower row. In these pictures the bright features correspond to structures with higher film thickness. These structures represent the dark features in the TXM images (Fig. 1, upper row) because of higher absorption of sample positions with higher film thickness. The comparison between SFM and TXM images visualizes that the main contrast in TXM is determined by the topography of the film.

X-RAY PHOTOEMISSION ELECTRON MICROSCOPY

With X-PEEM [7] the lateral distribution of the two polymers near the surface was mapped by employing the characteristic Near Edge X-ray Absorption Fine Structure (NEXAFS) spectra of the polymers. A 96 nm thick PS/PnBMA blend film with a PS mass fraction of 30% was examined by recording the secondary electron yield which is emitted from the surface after illumination with x-rays (c.f. Fig. 3, right). These measurements of the polymer spectra and the X-PEEM images were performed at the PM-3 monochromator at BESSY [7]. First we recorded the NEXAFS spectra (Fig. 3, left) on the C K-edge of two homopolymer thin film samples. Images were taken at 285.2 eV and at the background below the absorption edge (284 eV). The peak (285.2 eV) corresponds to the C=C 1s to π^* transition of the aromatic group of PS. Hence there is no aromatic group in PnBMA, this peak is missing in the PnBMA absorption spectrum. Thus, subtraction of the image taken at the peak from the background image yields the lateral distribution of PS. The bright ribbon features in the resulting difference image (Fig. 3, right) represent the PS rich phase, which is embedded in the sea of the PnBMA rich phase. The weak contrast and resolution in this PEEM image is due to the segregation of PnBMA to the air interface. In the spectrum of the thin blend film of the polymer mixture (Fig. 3, left) it can be seen that the sharp peak at 285.2 eV is damped in comparison to the homopolymer film of PS.

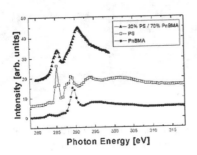

FIGURE 3. NEXAFS spectra of the C K-edge of PnBMA, PS and of a thin blend film with a mass fraction of 30% PS and 70% PnBMA (left). PEEM difference image I(285.2 eV) – I(284 eV) of a thin blend film with a mass fraction of 30% PS and 70% PnBMA with a thickness of 96 nm (right).

CONCLUSIONS

Different microscopic techniques for the determination of the structure of polymer blend films were compared and the weakly incompatible polymer blend system PS / PnBMA was investigated as a model system to some detail. It can be concluded that the different techniques are largely complementary with respect to spatial resolution, depth sensitivity and ease of use. While OM gives first indication of phase behavior and can be used also in-situ for kinetic experiments, its resolution is limited to typically 1 micrometer. SFM provides a much better lateral resolution and offers with different scan modes also the possibility of materials contrast in addition to surface topography. X-PEEM is very surface sensitive, and information about surface segregation of one component can be obtained, while TXM also gives information on the blend composition at larger depth at high spatial resolution.

The investigated blend system reveals lateral phase separation after spin coating, where structures are similar to those reported previously for other polymer blend systems [2, 3, 8]. Structures can be expected to be frozen-in non-equilibrium structures, which develop during a complicated thermodynamic pathway during the spin coating process. Depending on blend composition spherical or ribbon like structures develop. As an interesting feature again small structures are observed in the larger islands (see e.g. [8]). From a depth sensitive analysis it is concluded that PnBMA is enriched at the surface. Besides the lateral phase separation also the film thickness varies over the film as a function of local composition. Further investigations are necessary to follow the structure evolution during annealing, where as a competing effect also dewetting may set in. The combination of different microscopic techniques provides, however, a means to obtain and separate different aspects of the phase separation process.

ACKNOWLEDGEMENTS

This work has been funded by the Max-Planck-Gesellschaft (MPG) and the German Federal Minister of Education, Science, Research and Technology (BMBF) under contract number 05SL8UM10 (X-PEEM) and 05SL8MG11 (X-ray microscopy). We gratefully acknowledge the help of V. Scheumann (MPI für Polymerforschung) in performing the SFM measurements.

REFERENCES

1. Sung, L., Karim, A., Douglas, J. F., and Han, C. C., *Physical Review Letters* **76**, 4368-4371 (1996).

2. Affrossman, S., et al., *Macromolecules* **29**, 5010-5016 (1996).

3. Affrossman, S., O'Neill, S. A., and Stamm, M., *Macromolecules* **31**, 6280-6288 (1998).

4. Ade, H., "NEXAFS and X-Ray Linear Dichroism Microscopy and Applications to Polymer Science", in *X-Ray Microscopy and Spectromicroscopy*, edited by J. Thieme, G. Schmahl, D. Rudolph, and E. Umbach, Berlin, Heidelberg: Springer Verlag, 1998, pp. III-3 – III-13.

5. Ade, H., et al., *Applied Physics Letters* **73**, 3775-3777 (1998).

6. Schmahl, G., et al., *Naturwissenschaften* **83**, 61-70 (1996).

7. Swiech, W., et al., *Journal of El. Spec. and Rel. Phenomena* **84**, 171-188 (1997).

8. Gutmann, J. S., Müller-Buschbaum, P., and Stamm, M., *Faraday Discuss.* **112**, 285-297 (1999).

Reorganization Of Clusters In Cluster-Cluster-Aggregation

O. Vormoor, J. Thieme

Georg-August-Universität Göttingen, Institut für Röntgenphysik,
Geiststraße 11, 37073 Göttingen, Germany
Corresponding author: O. Vormoor (ovormoo@gwdg.de)

Abstract. X-ray microscopy and scattering experiments give different experimental results when determining the fractal dimension of colloidal clusters. A two dimensional simulation model is presented to describe reorganization of clusters during the aggregation process. Considering this reorganization gives a rapidly increasing fractal dimension from $d=1.53$ to $d=1.61$. This result implies that the fractal dimension of the clusters is mainly affected by reorganization of small sub clusters.

INTRODUCTION

Aggregation processes, i. e. the mechanism of cluster formation from single particles, are important for many areas of science, e. g. precipitation processes in aqueous media. The resulting clusters can be characterized by their fractal dimension d [1], which is a measure for their surface.

Observation of colloidal systems with X-ray microscopy shows, that the fractal dimension d of the clusters is $d<2$ and increases with growing cluster size [2], [3]. In contrast scattering experiments measure a decreasing d with growing cluster size while d remains $d>2$.

This paper presents an aggregation model that considers a reorganization of the clusters during the aggregation process to describe the results won by X-ray microscopy.

THEORY

The interaction between colloidal particles is mainly caused by two forces, first the repulsive electrostatic force, which is depending on the ion concentration c_{ion} in the dispersion, and second the attractive van der Waals force. The resulting interaction between two colloidal particles is shown in figure 1 for various ion concentrations. At the so called critical coagulation concentration (*c.c.c.*) repulsive and attractive forces cancel each other except for very short distances (curve 4).

CP507, *X-Ray Microscopy: Proceedings of the Sixth International Conference,*
edited by W. Meyer-Ilse, T. Warwick, and D. Attwood
© 2000 American Institute of Physics 1-56396-926-2/00/$17.00

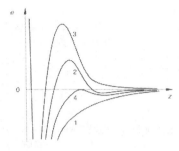

FIGURE 1. (from [4]) The resulting interaction energy e per area between two colloidal particles with their distance z. The curves 2 and 3 show the case $c_{ion} < c.c.c.$, curve 4 $c_{ion} = c.c.c.$ and curve 1 the case $c_{ion} > c.c.c.$

SIMULATION MODEL

The simulation model developed to describe the aggregation process is based on the cluster-cluster-aggregation (CCA) [5]. All particles are statistically placed in the simulation space and perform a simultaneous random walk. If two particles collide, they stick together and form a new bond. This algorithm is repeated until all monomers form a single cluster or a certain time interval without a collision is exceeded.

A potential wall between two particles, i. e. the peaks of the curves 2 and 3 in figure 1, is simplified to a box function in the simulation model. This reduces with growing peak height the sticking probability of two hitting clusters.

For particles with attractive potentials reorganization of the clusters during the aggregation process may play an important role. To consider this procedure a reorganization step is introduced after a collision between two clusters. The bond formed at the last collision is an elastic point in the newly formed cluster. Around this bond the sub clusters may rotate to form a second bond between them – see figure 2. Due to numerical implementation and performance this reorganization step is assumed to be complete before this cluster forms a new bond.

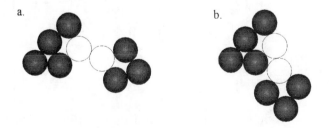

FIGURE 2. Model of the reorganization of the clusters. The highlighted balls mark the newly formed bond: a. shows the cluster directly after the bond formation, b. shows it after the reorganization has occurred.

To model a varying strength of the attractive forces a reorganization coefficient α is introduced and varies between 0 and 1. With N as the total number of monomers in the dispersion only clusters with size smaller than αN reorganize.

RESULTS

A number of simulations with different monomer concentrations and different N have been performed to ensure a good statistics. Simulations without a box potential and without reorganization show, that the fractal dimension d of the resulting clusters is nearly independent of the monomer concentration c (figure 3a.). So d can be accepted as $d=1.52-1.53$.

Modeling a repulsive potential gives similar results: d is nearly constant (figure 3b.) when the height V of the repulsive potential is varied over a wide range.

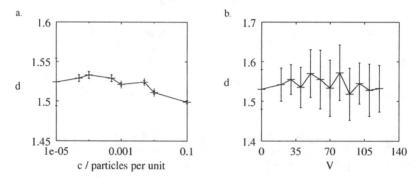

FIGURE 3. The box dimension d of the resulting clusters in two dimensions, which is nearly constant in both cases: a. as a function of the monomer concentration c and b. for varying height of the repulsive potential V. The larger errors in plot b. result from averaging over several clusters in a simulation.

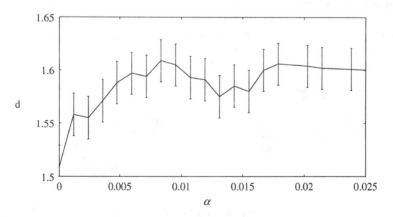

FIGURE 4. The fractal dimension d increases rapidly from $d=1.52-1.53$ to $d=1.6$, if a reorganization of the clusters is considered. The dip at about $\alpha=0.015$ has to be interpreted as a statistical error.

a. 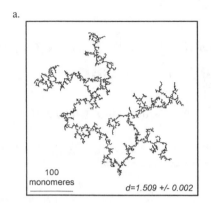 b.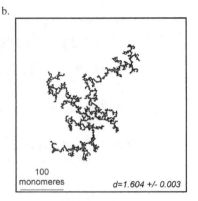

FIGURE 5. Two resulting clusters of simulations, a. shows a cluster without reorganization ($\alpha=0$, $c_{ion}=c.c.c.$) and b. shows a cluster with complete reorganization ($\alpha=1$, $c_{ion} > c.c.c.$).

The results of simulations considering the reorganization as described above are shown in the figures 4 and 5. d increases rapidly for $\alpha < 0.01$ and remains constant until total reorganization is reached.

DISCUSSION

The results of the simulations show, that the fractal dimension is mainly affected by the reorganization of small clusters during the aggregation process. Reorganization of large sub clusters, which will be slow compared to the aggregation process itself, is negligible. Considering the above described reorganization of the clusters, the developed aggregation model is in good agreement with the experimental X-ray microspoy results [2][3].

REFERENCES

1. Mandelbrot, B. B., *Die fraktale Geometrie der Natur*, Birkhäuser, 1991.

2. Thieme, J. and Niemeyer, J., *Geol. Rundsch.* **65**, 2503-2504 (1996)

3. Ruprecht, A., Diploma Thesis, University Georgia Augusta at Göttingen, 1998.

4. Mögel, H. J., *Grenzflächen und Kolloide*, Spektrum Akademischer Verlag 1993

5. Meakin, P. *Phys. Rev. Lett.* **51**, 1119-1122, 1983

Analytical X-Ray Microscopy Using Laboratory Sources: Method And Results

J.M. Wulveryck[a], S. Odof, J.M. Patat and D. Mouze

DTI/LASSI, Faculté des Sciences, BP1039, 51687 Reims Cedex 2, France
[a]*Electronic mail : jm.wulveryck@univ-reims.fr*

Abstract. The implementation of X-ray imaging techniques (X-Ray Projection Microscopy or X-Ray Microtomography) for analytical purpose with laboratory equipment is not easily obtainable, due to the use of polychromatic X-ray sources. To overcome this drawback, we propose here a mathematical technique based upon the calculation of the X-ray emission spectra and the knowledge of the spectral response of the camera. For example, we show that the composition of a ternary sample can be deduced from two images measurements recorded from two primary radiation. The accuracy of the method is discussed.

INTRODUCTION

X-ray microscopy, including x-ray microradiography and x-ray microtomography, is a standard method for specimen imaging. The contrast in the images depends both on the local elemental composition and density of the specimen. Nevertheless, when a laboratory source is used, i.e. a polychromatic radiation, these techniques are not easily usable to obtain a quantitative information on the specimen composition. In this paper, we show that it is possible to overcome this disadvantage *without filtering* provided that both the x-ray spectra and the spectral response of the imaging device are known. As for the knowledge of the x-ray spectra, instead of the tedious recording of numerous experimental spectra, we have opted for the development of a calculation freeware, whereas the spectral response of the image sensor has been measured experimentally[1].

X-RAY SPECTRA SIMULATIONS

X-ray spectra simulations are mainly used in Electron Probe Micro Analysis (EPMA) and X-Ray Fluorescence (XRF). Our freeware is a derivation of the well known ZAF method applied in EPMA. The calculus are adapted to the experimental conditions particular to XRM, i.e. to pure metallic target in transmission geometry (see fig. 2).

CP507, *X-Ray Microscopy: Proceedings of the Sixth International Conference,*
edited by W. Meyer-Ilse, T. Warwick, and D. Attwood
© 2000 American Institute of Physics 1-56396-926-2/00/$17.00

The calculation of both line and continuum spectra due to electron impact is usually done by either of the two following methods : analytical models[2,3] (the simplest and least expensive), and the Monte Carlo modeling[4,5]. As for us, we have chosen the analytical models for its short computed time. The flexibility of this program is such that the users entirely control the operating conditions.

The parameters which can be governed by the user concern :

- The target : composition and thickness

- The electron gun : accelerating voltage (from 5 to 100kV), current beam, and acquisition time

- The geometrical parameter : incidence angle, take-off angle, detection solid angle

The reliability of simulated x-ray spectra depends mainly on the accuracy of the interaction cross sections used in the calculation. X rays mainly originated from electron impact ionization of inner shells and from bremsstrahlung emission; thus the cross sections used to simulate these processes should be the most accurate available. To give an element of discussion about the computation accuracy, figure 1 shows an example of some discrepancies between ionization cross sections, for the nickel.

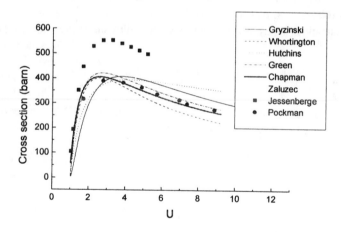

FIGURE 1. Some ionization cross sections formula (solid lines) and experimental data points for the nickel.

X-RAY IMAGE SENSOR

In order to evaluate the spectral sensitivity of our x-ray imaging camera, a monochromatic x-ray laboratory source was needed. This was obtained by using a wavelength dispersive x-ray fluorescence spectrometer as a monochromator. Our x-ray image sensor is made of a scintillator screen optically coupled to a CCD camera.

The results obtained from the measurements are presented in an other paper. These results must be compared with that we could expect with such an instrument. The only element which governs the *spectral efficiency* of the x-ray camera concern the scintillating layer. In this respect, the composition of this layer, its thickness and the number of visible photons emitted per absorbed x-ray photon are of the utmost importance. On the contrary, once the x-ray to visible light conversion is carried out, the other part of the camera are unimportant for the spectral efficiency in the x-ray range.

DESCRIPTION OF THE QUANTITATIVE FORMALISM

To understand the problem of quantitative analysis in projection x-ray imaging, we must refer to figure 2. In fact, the incident X-ray radiation characterized by its spectral distribution, $g(E)$, will be partially absorbed when it go through the specimen. Finally, this transmitted spectral distribution will be detected by the image sensor with a given spectral efficiency, $\varepsilon_{CCD}(E)$. So, we can write the following relation :

$$I_{CCD} = i_{DC} + \int_0^{E_0} g(E)\varepsilon_{CCD}(E)\exp\left[-\mu_e(E)t_e\right]dE \qquad (1)$$

where I_{CCD} is the recorded intensity for one pixel of the CCD, i_{DC} is the dark current, and the exponential term accounts for the specimen attenuation.

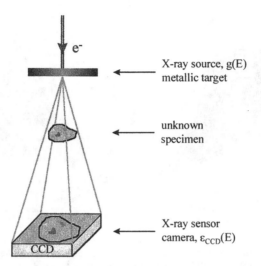

FIGURE 2. Projection x-ray microscopy using laboratory source. The polychromatic radiation, enlighten the specimen. The X-ray image is projected on the x-ray sensor camera which is characterized by its spectral response ε_{CCD}.

Since we know *a priori* the incident spectral distribution, $g(E)$, and the spectral response of our image sensor, $\varepsilon_{CCD}(E)$, and since the dark level is easily measurable, the only term playing a role in the previous relation is the attenuation part. This one is a function of the local composition in the specimen. If, the thickness is known, the attenuation part is solely linked to the elementary concentrations in the specimen.

$$I_X = I_{CCD} - i_{DC} = \int_0^{E_0} g(E)\varepsilon(E)\exp\left\{-t_e \times \sum_{i=1}^n c_i\rho_i \times \sum_{i=1}^n c_i\left[\frac{\mu_i}{\rho_i}(E)\right]\right\}dE \qquad (2)$$

Therefore, to resolve the problem, we have to establish the curve giving the theoretical output signal (in ADU unit) as function of the elementary mass concentrations.

For example, in the case of a ternary alloy $Fe_xCr_yNi_{1-x-y}$, i.e. a stainless steel, using only one radiation, a given signal level corresponds to a set of elementary mass concentrations. On the other hand, the use of another radiation (giving another signal level for the same pixel) leads to a second set of concentrations. The intersection between this two sets of data enables us to determine the elementary mass concentrations. Moreover, for a ternary compound, only two concentrations are independents, the third concentration being deduced from the relation $C_A + C_B + C_C = 1$. In the figure 3, the characteristics curves obtained with an iron target (left side) and a copper target (right side) are shown.

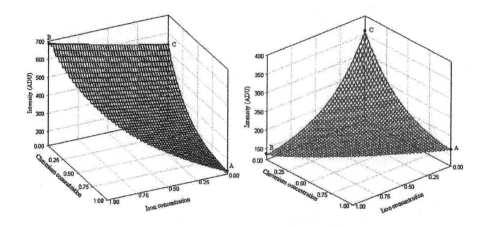

FIGURE 3. Intensity - elemental concentrations curves obtained under the same operating condition (accelerating voltage : 30kV, primary current : 100nA, solid angle : 10^{-5}str., acquisition time : 100s). Between these two curves only the targets are different, left side : iron (5µm); right side : copper (5µm).

CONCLUSION

In a first step, we have shown that quantitative analysis of a specimen in projection x-ray imaging is possible even if the x-ray source is polychromatic. To achieve this result, we have had to decrease the number of unknown in the problem, e.g. the incident spectral distribution and the spectral response of the detector device. Remain the question of the inaccuracy of the proposed formalism. In fact, we consider honestly that this inaccuracy is close to 20%, but this is mainly due to the inaccuracies on the calculated spectra (~10%) and on the measured spectral response of our image sensor (~5-10%). So, if you could determine experimentally the incident spectra and more precisely the spectral response, the inaccuracy should be smaller (up to 5-10%).

REFERENCES

1. Wulveryck, J. M., and Mouze, D., *Review of Scientific Instruments*, **70**, 9, pp. 3549-3553 (1999).

2. Reed, S. J. B., *Review of Physics in Technology*, **2**, 92 (1971).

3. Gilfrich, J. V., Burkhalter, P. G., Whitlock, R. R., Warden E. S., and Birks, L. S., *Analytical Chemistry*, 43, 7, pp. 934-936 (1971).

4. Acosta, E., Llovet, X., Coleoni, E., Riveros, J. A., and Salvat, F., *Journal of Applied Physics*, **83**, 11, pp. 1-12 (1998).

5. Ding, S. J., Shimizu, R., and Obori, J., *Scanning Microsc. Suppl.*, **7**, 81 (1993).

Elemental Mapping with an X-Ray Fluorescence Imaging Microscope

Kimitake Yamamoto, Norio Watanabe, Akihisa Takeuchi, Hidekazu Takano,Tatuya Aota, Masaru Kumegawa, Takuji Ohigashi, Ryuichi Tanoue, Hiroki Yokosuka, and Sadao Aoki

Institute of Applied Physics, University of Tsukuba, 1-1-1,Tennoudai, Tsukuba, Ibaraki, 305-8573, JAPAN

Abstract. An X-ray fluorescence (XRF) imaging microscope with a Wolter-type grazing-incidence mirror as an objective was constructed at the beamline 39XU of SPring-8 (8GeV, 70mA) at Japan Synchrotron Radiation Institute. The monochromatic undulator X-rays in the energy range of 6-10keV were used to produce XRF of a specimen. The microscope system was set normal to the incident beam to reduce elastic scattering from a specimen and to improve signal/background ratio. The two-dimensional elemental mappings of a test specimen (Cu, Ni, Fe wires) and inclusions (Fe, Co, Ni) in a synthesized diamond could be obtained by utilizing the absorption edges of the corresponding elements.

1. INTRODUCTION

X-ray fluorescence microscopes have been recently developed for trace-element analysis. One of them is an X-ray fluorescence imaging microscope. It needs only a moderate position control system and can record an image in a reasonably short time. In some cases, a real time observation is also possible. In the previous paper, we demonstrated that X-ray fluorescence could be successfully imaged with a Wolter-type mirror [1]. As the mirror has no chromatic aberration, all the X-rays above the critical wavelength can be imaged simultaneously. In order to obtain an image of a specific element, an absorption edge of the element concerned can be utilized.

In this paper, we show the X-ray optical system constructed at the beamline 39XU of the SPring-8 first. Then X-ray fluorescence images of metal wires are given. Application of the imaging to a synthesized diamond is also given.

2. OPTICAL SYSTEMS

We built an X-ray fluorescence imaging microscope by using the Wolter mirror at the SPring-8. A schematic diagram of the experimental arrangement at the beamline 39XU is shown in Fig.1.

The incident X-ray path between the upstream end of the experimental hutch and the slit, and the X-ray fluorescence path between the specimen and the detector were evacuated.

CP507, *X-Ray Microscopy: Proceedings of the Sixth International Conference,*
edited by W. Meyer-Ilse, T. Warwick, and D. Attwood
© 2000 American Institute of Physics 1-56396-926-2/00/$17.00

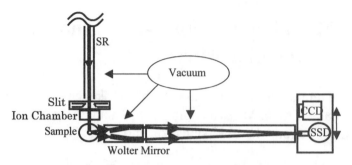

FIGURE 1. Schematic diagram of experimental arrangement.

The area detector was an X-ray sensitive charge-coupled device (CCD) camera (HAMAMATSU, TI, TC-215). The number of pixels is 1000×1018 and each pixel area was 12μ m $\times 12 \mu$ m. The detector was cooled by a Peltier cooler below 253K to reduce the dark current. And an energy dispersive detector which was used was Ge solid state detector (SSD).

3. EXPERIMENTAL RESULTS

3.1 X-ray fluorescence image of a specific element

Three kinds of metallic wires (which were Cu, Ni and Fe and diameters of Cu and Ni wires are 25 μ m, respectively and Fe wire is 100 μ m) were used to evaluate the performance of the microscope. Figure 2 (a) shows the visible light image of the wires. X-ray fluorescence images of the wires were taken by changing the excitation energy of the incident X-ray. Figure 2 (b) shows the X-ray fluorescence image of the wires at an excitation energy of 9.000 keV. Figures 2 (c) and 2 (d) show the X-ray fluorescence images which were obtained with the excitation energy of 10eV above the absorption edge of each element. The exposure time is 1 minutes. The results show that the selective excitation of a specific element can be made by using an absorption edge.

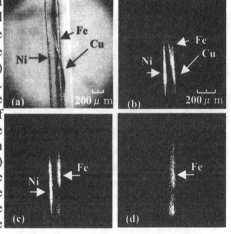

FIGURE 2. (a) A visible light image of the Cu, Ni and Fe wires, (b) the X-ray fluorescence image at 9.000 keV, (c) 8.343 keV and (d) 7.122 keV.

3.2 Application to a synthesized diamond

To show the practical applicability of the microscope, a synthesized diamond was analyzed. The diamond which was produced by the solvent method includes some metallic particles such as Fe, Co and Ni and so on [2]. Characterization of these particles is very important for the synthesis of the diamonds. Figure 3 (a) shows the visible light image of the synthesized diamond. Figure 3 (b) shows the x-ray fluorescence image and its spectrum, which were obtained with the excitation energy of 20eV above the absorption edge of the Ni element. Figures 3 (c)-(e) show the x-ray fluorescence images of the synthesized diamond and their spectra, which were obtained with the excitation energy of 20eV below the absorption edges of the Ni, Co and Fe elements. It took 20 minutes for each exposure of the x-ray fluorescence images, and 5 minutes for each measurement of the spectra. Figures 3 (f)-(h) show the x-ray fluorescence images and their spectra, which were obtained by subtracting the adjacent data. Because of the statistical errors or for some other reasons, there were negative values in their spectra. These subtracted X-ray fluorescence images show that the elements of Ni, Co and Fe were similarly distributed in the inclusions.

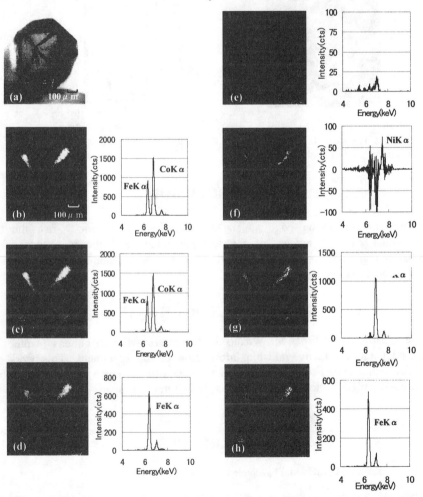

FIGURE 3. (a) A visible light image of the synthesized diamond, (b) X-ray fluorescence images and their spectra at 8.353keV, (c) 8.313keV, (d) 7.689keV, (e) 7.092keV, (f) X-ray fluorescence images and their spectra which were obtained by subtracting the adjacent data of Ni, (g) Co and (h) Fe.

261

3.3 Evaluation of elemental composition of the inclusions in a synthesized diamond

Figure 4 shows the X-ray fluorescence image of some inclusions in a synthesized diamond at the excitation energy of 8.343 keV. Incident X-rays which were adjusted by the slit were respectively irradiated on some inclusions and the X-ray fluorescence generated from the inclusion was measured. Figure 5 shows the elemental weight ratio of the inclusions which were estimated by the standardized intensity of the X-ray fluorescence, which was cobalt of No.3. According to this graph, the total weight of the elements of No.3 is about 3 times heavier than that of the other inclusions, but the elemental composition ratio of the inclusions are almost the same.

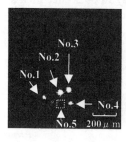

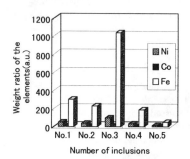

FIGURE 4. The X-ray fluorescence image of the inclusions in a synthesized diamond.

FIGURE 5. The elemental weight ratio of the inclusions in a synthesized diamond.

4. CONCLUSION

Using the absorption edge as an excitation monochromatic X-ray, the two-dimensional element images of the test specimens and the inclusions of the synthesized diamond were obtained. And elemental composition of the inclusions in a synthesized diamond could be evaluated.

ACKNOWLEDGEMENTS

We greatly thank Prof. M. Wakatsuki for offering the specimens (synthesized diamonds) and for many valuable information of the specimens. This work was partially supported by the Grants-in-Aid for Scientific Research No. 11305011 from the Ministry of Education, Science, Sports, and Culture of Japan.

REFERENCES

1. Aoki,S., Takeuchi, A., Ando,M. J.Synchrotron Rad. 5,1117-1118 (1998).

2. Wakatsuki,M., "Synthesis Researches of Diamond," in Materials Science of the Earth's Interior, edited by I. Sunagawa, Tokyo, Terra Science Publishing Company, 1984, pp. 351-374.

A X-ray Microscope for Stored Energy in Single Grains of Cold-Rolled Steel

M. Drakopoulos[1], I. Snigireva[1], A. Snigirev[1], O. Castelnau[2],
T. Chauveau[2], B. Bacroix[2], C. Schroer[3], and T. Ungar[4]

[1] *ESRF, BP220, F-38043 Grenoble, France*
[2] *LPMTM-CNRS, Université Paris-Nord, av. J.B. Clément, F-93430 Villetaneuse, France*
[3] *II. Physikalisches Institut, RWTH, D-52056 Aachen, Germany*
[4] *Inst. for General Physics, Eötvös University Budapest, P.O.B. 323, H-1445 Budapest, Hungary*

Abstract. A new set-up for X-ray microdiffraction has been developed on the ESRF beamline ID22. This set-up allows microscopic characterization of materials in diffraction mode, the size of the focussed beam being only of a few microns. This facilitates the measurement of material quantities as average size of the coherently diffracting volume, local dislocation density, residual stress, local fluctuation of the residual stress, and intragranular misorientation from single grains of a polycrystalline material. The first application on an IF-Ti steel after different thermo-mechanical treatments (recrystallization, cold-rolling, annealing) is presented.

INTRODUCTION

The deformation of a polycrystalline material is heterogeneous at the scale of grains and subgrains, owing to the elastic and plastic anisotropy of grains and to the interaction between neighbored grains. As plastic deformation proceeds, the density of dislocations increases by several orders of magnitude with respect to the local strain (intra-crystalline hardening), the local misorientation between dislocations cells and subgrains increases, and long range internal stresses develops. A stored energy is associated to the lattice distortion created by the dislocation structure. The local gradient of this stored energy is an important parameter (but rarely determined) for the understanding of recrystallization processes.

Dislocation density in metals deformed at large plastic strain can be best estimated by X-ray diffraction line profile analysis (1). In a classical x-ray diffraction experiment on a polycrystalline material having a typical grain size of a few microns, a beam with a cross-section of several hundred micrometers diffracts on all those grains, which by chance fulfil the diffraction condition in terms of crystal orientation and plane spacing. The measured line profile then will contain an averaged information about the stored energy in those grains, and any microscopic resolution is lost.

In this work we present a method, where the diffraction comes from a specified single grain and therefore reveals the local strain- and dislocation-state of the

CP507, *X-Ray Microscopy: Proceedings of the Sixth International Conference,*
edited by W. Meyer-Ilse, T. Warwick, and D. Attwood
© 2000 American Institute of Physics 1-56396-926-2/00/$17.00

polycrystalline material under investigation. In combination with a scanning-mode a 2-dimensional map of the dislocation-density can be obtained.

EXPERIMENTAL

The experiment was carried out at the ESRF beamline ID22, particularly dedicated to micro-fluorescence, micro imaging, and micro-diffraction. The set-up is shown in Fig. 1. The experimental method is basically a single-crystal diffraction set-up with a focused beam. The plane of diffraction is the horizontal plane. The sample is mounted on a xyz-scanning stage to enable mapping. This sample stage is placed in the centre of the Eulerian cradle of a four-circle diffractometer. The diffracted beam is recorded with a 2-dimensional gas-filled detector, which is mounted on the 2-Theta circle. Resolution and dynamic range of such a detector match the requirements to measure quantitative diffraction profiles.

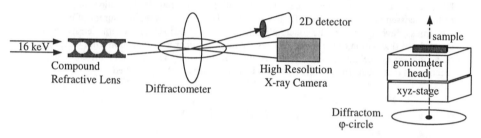

FIGURE 1. Schematic draw of the experimental set-up

A monochromatic beam with photon energy of 16 keV and a relative bandwidth of 1.3×10^{-4} was produced with a Si 111 monochromator, higher order wavelengths have been cut off with an flat Si-mirror at an glancing angle of 0.11 degree. The beam was focussed with a parabolic Compound Refractive Lens (CRL) (2) at a distance of 1276 mm from the lens. The spot size at that distance was $13 \times 3\ \mu m^2$ (H×V). The use of the CRL leads to a gain of flux of about a factor 350 as compared to the unfocussed beam. The divergence of the focused beam is given by the effective aperture of the CRL and is around 5×10^{-4} degree. Both, bandwidth and divergence are small enough not to contribute to significant instrumental broadening.

The focus is positioned exactly into the diffractometer centre by means of a high-resolution x-ray camera (3), which is placed after the diffractometer. The diffractometer centre is determined using a polystyrene sphere of 40 μm diameter as sample. Phase contrast from the transparent sphere gives enough contrast to visualise the position of the sphere in the direct beam with micrometer precision (Fig. 2b).

The main difficulty along such an experiment is the positioning of a particular grain into the focused beam and orientating of the crystal planes afterwards. Both, positioning and orientating should be accomplished fast to ensure an efficient mapping procedure. Here we have established two procedures. First, the positioning is done with the help of a direct calibration as follows. The sample is fitted with three markers (polystyrene spheres), which are placed around the region of interest. Then an image is taken with an optical microscope, showing the individual grains as well (Fig 2a). The position of those markers in the unfocused x-ray beam is recorded, again using the high-resolution camera (Fig. 2b). The so obtained three co-ordinates serve for a transformation between sample-stage co-ordinates and the digitised image co-ordinates. Now, the positioning can be accomplished by image-display software. Second, the crystallographic orientations of the grains are determined by means of electron-back-scattering before in the lab. The obtained orientation map is directly translated into commands to the diffractometer, which then can be controlled by image-display software as well. Besides comfortable working, the procedure is efficient and has the advantage, that microscopic information obtained by x-rays can be easily compared with other microscopic images such as electron micrographs.

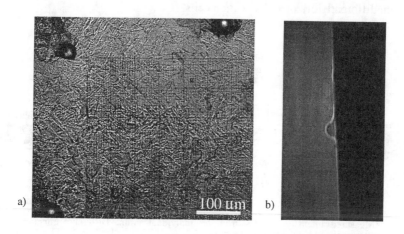

FIGURE 2. a) Image of sample and markers taken with optical microscope (the mesh on top is a gold grid deposited after deformation). b) Image of sample and one marker taken with x-ray high-resolution camera. The markers are visible at the diffractometer angle $\omega = 0$. A similar sphere was used to define the diffractometer centre.

Because dislocations are vectored quantities, a diffraction profile from a single reflection would not transport enough information to reveal the dislocation density. In this work we recorded for each grain the diffraction profiles from 8 reflections (planes {110}, {200}, {211}, {220}, {310}, {222} and {400}). In ten days about 350 line profiles have been recorded from 45 different grains.

RESULTS

The material studied in this work was Interstitial Free Titanium killed steel (IF-Ti) with a grain size of around 30 micrometers. Three specimens were studied after

265

different treatment. The first sample (IF60) was hot rolled and has re-crystallised completely. The second sample (IF61) was cold rolled up to 26% relative deformation, and the third one (IF63) was cold rolled up to 26% and then annealed at 550°C for 10 minutes. After treatment, the specimens were mechanically and chemically polished down to half the initial thickness. Final etching revealed the grain boundaries.

Intra-granular misorientation

The vertical distribution of diffracted intensity gives an indication of the lattice misorientation within the diffraction volume. The simultaneous study of local misorientation and line profiles is particularly interesting because both are the result of the dislocation structure. The 2-dimensional detector records both effects simultaneously; the horizontal scale analyses the profile and the vertical scale the misorientation (Fig. 3). Each horizontal section of the detector records the line profile coming from differently oriented dislocation cells.

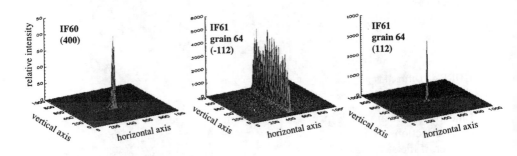

FIGURE 3. Measured intensity profiles from an individual grain. Different reflections from recrystallized steel (IF60) and cold-rolled steel (IF61).

Fig. 3 shows three examples of reflections measured on re-crystallised (IF60) and deformed (IF61) specimens. On IF60, reflections are sharps and local misorientations are always small, of the order of 0.1 degree, indicating that grains have a well-defined contour and a well-defined lattice orientation. Some grains show a very particular behaviour with an excessively small misorientation within the diffracting volume. This indicates a sharp fragmentation of the grain (Fig. 3, right).

Strain anisotropy

Line profiles are essentially measurements of the average displacement of atoms in the direction parallel to the diffraction vector k. Because the displacement field created by dislocations is strongly anisotropic, not all dislocations will contribute to the line profile broadening. From the analysis of a single line profile, one can only estimate the "apparent dislocation density" $\rho^*_{(hkl)}$. The real dislocation density can be obtained by applying a correction coefficient $C_{(hkl)} \propto \rho*_{(hkl)} / \rho$, denoted "contrast factor" of the dislocations (4), (5). The contrast factors contain average information about nature and orientation of the dislocation structure. For intermediate deformation as considered here, we expect that only a few slip systems have been activated. This can be easily verified by superimposing several line profiles of the same {hkl} family and the same grain (Fig. 4).

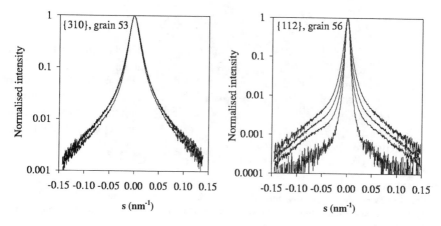

FIGURE 4. Diffraction profiles from one {hkl}-family superimposed ($s = 1/d_{hkl}$). Two different grains of deformed steel (IF61).

Line profiles from the same {hkl}-family may have show similar shape, as the {310} reflections of grain 53 (Fig 4, left). This shows isotropic strain distribution. However, if the strain anisotropy is very strong, as for grain 56 (Fig 4, right), the lines profiles become (hkl) dependent. These observations offer the possibility to develop adequate models about the distribution of strain and dislocations within the grains on the base of x-ray diffraction experiments.

Dislocation density

To calculate the dislocation density, a modified Williamson-Hall is used. The full-width half-maximum ΔK of the line profiles is decomposed into the contribution of the size D of the coherently diffracting volume and the contribution of the elastic strain linked to the distortion of the lattice by the dislocations:

$$\Delta K = 0.9/D + (\pi A b^2/2)^{1/2} \rho^{1/2} |k| C^{1/2} , \tag{1}$$

with b being the norm of the Burgers vector, ρ the dislocation density, and A a normalization coefficient. It should be noted that in the case of a dislocated crystal, the proper scaling factor of the FWHM is $|k|C^{1/2}$ instead of $|k|$ as in the original Williamson-Hall procedure. The contrast factors are determined assuming a random distribution of the dislocations on all possible slip systems. In Fig. 5, the obtained dislocation densities are shown in comparison to the dislocation density on non-deformed steel. One new finding is that the grains with almost the same lattice orientation can exhibit very different dislocation densities, up to a factor of 5.

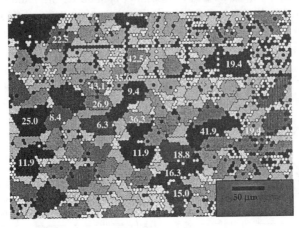

FIGURE 5. Microscopic dislocation density map of cold-rolled steel (sample IF61). The dislocation density is marked by numbers in relative units (dislocation density of non-deformed steel = 1). The gray-scaled image is an orientation map obtained with electron back-scattering. Single grains appear in different grey values.

CONCLUSION

A new set-up for X-ray microdiffraction experiment has been developed. It allows any local characterization of the material such as given by the line profile analysis, residual stress and its fluctuation within the diffracting volume, local misorientation, etc. The first application concerns IF-Ti steel specimens after different thermomechanical treatments (recrystallization, cold-rolling, annealing). Both misorientation and line profile data show that dislocations are not randomly distributed on all possible slip systems. From our first evaluation, we observe a large fluctuation of the dislocation densities from grain to grain. The dislocation density is not clearly correlated to the lattice orientation. Deeper data analysis is actually under progress.

REFERENCES

1. Ungár, T., Groma, I., Wilkens, M., *J. Appl. Cryst.* **22**, 26-34 (1989)

2. Lengeler, B., Schroer, C. Richwin, M., Tümmler, J., Drakopoulos, M., Snigirev, A.

   Snigireva, I. *Appl. Phys. Lett.*, **74**, 3924-3926 (1999)

3. Koch, A., Raven, C., Spanne, P., Snigirev, A., *J. Opt. Soc. Am.*, **A15(7)**, 1940-1951 (1998)

4. Wilkens, M., *Phys. Stat. Sol.* (a) **104**, K1-K6 (1987)

5. Ungár, T., Tichy, G., *Phys. Stat. Sol.*, *Phys. Stat. Sol. (a)*, **171**,425-434 (1999)

Polycrystalline Metal Surfaces Studied by X-ray Photoelectron Spectro-Microscopy

A.W.Potts[1], G.R.Morrison[1], S.R.Khan[1], L.Gregoratti[2] and M.Kiskinova[2]

1) Department of Physics, King's College London UK
2) Sincrotrone Trieste, 32014 Trieste, Italy

Abstract. The scanning photoelectron microscope (SPEM) on beam line 2.2 at the Elettra synchrotron produces small spot XPS spectra from a sub-micron radiation microprobe. It is also capable of producing surface images in terms of the energy resolved photoelectron signal. This microscope has been used to study oxidation on polycrystalline tin and lead surfaces and the variations in reactivity between different crystallite surfaces. The diffusion of gold and silver films on polycrystalline metal surfaces has also been followed.

INTRODUCTION

Following the development of third generation synchrotrons with high brightness undulator beam lines it is now possible to undertake X-ray photoemission studies with high spatial resolution. A number of such synchrotrons now have beam lines with photoelectron microscopes (1,2). We have used the scanning photoelectron microscope on beam line 2.2 at the Elettra synchrotron in Italy to study the processes of oxidation and diffusion taking place on polycrystalline metal surfaces. The microscope has allowed us to image surfaces in terms of both metallic and chemically shifted core level photoelectron signal with a spatial resolution of better than 1μm. It has also enabled us to carry out small spot spectroscopy on core levels at various stages in the oxidation process with an energy resolution of better than 1 eV. In this way we hope to examine the effects of surface inhomogeneity on surface reactions and to determine the extent to which studies on single crystal surfaces can be generalized to the more common polycrystalline surface. For oxidation we have studied lead and tin surfaces at various oxygen doses. Preliminary results on our work on Sn have already been published (3) so we shall concentrate here on the lead study.

To examine the effects of surface diffusion on polycrystalline nickel surfaces sub-atomic layers of silver and gold were first deposited. The surface distributions of the Ag and Au were then recorded by imaging the surfaces in terms of the Ag 3d and Au 4f photoelectron signals before and after the samples had been annealed.

CP507, X-Ray Microscopy: Proceedings of the Sixth International Conference,
edited by W. Meyer-Ilse, T. Warwick, and D. Attwood
© 2000 American Institute of Physics 1-56396-926-2/00/$17.00

OXIDATION

Experimental

Lead targets were prepared from high purity commercial samples by pressing metal discs onto a glass optical flat. The surfaces were then given a light chemical etch and the grain boundary structure checked with an optical microscope. Before imaging in the microscope the samples were cleaned by argon ion bombardment in the sample preparation chamber and were shown to be free from carbon contamination by Auger spectroscopy. The metal samples were then loaded into the microscope and brought to focus by imaging the surfaces in terms of the Pb 4f photoelectron signal, varying the target position until the sharpest image was obtained. Samples were dosed in-situ with molecular oxygen at pressures of 10^{-6} mbar. The base pressure used during imaging was $\sim 10^{-10}$ mbar.

An example of an image of an oxidized lead surface is shown in figure 1. This surface has been exposed to 1000L (Langmuir) of oxygen and was recorded in terms of the chemically shifted lead 4f peak. Although the chemical shift in this peak is relatively small (~ 0.9 eV) it is possible by setting the energy window of the microscope to the low kinetic energy side of the $4f_{7/2}$ feature to produce an image which is sensitive to the oxide distribution.

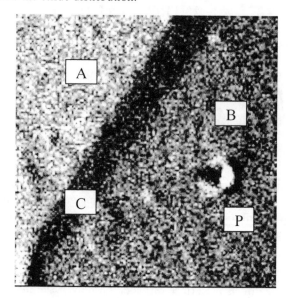

FIGURE 1. 64μm x 64μm image of an oxidized lead surface imaged in terms of PbO Pb $4f_{7/2}$ electrons.

In figure 1 the bright area (A) corresponds to a surface grain structure which is rapidly oxidized while the dark area (B) corresponds to a surface which is oxidized more slowly. The raw image is of course sensitive to surface topography. While this can be compensated for to some extent by subtraction of images recorded in terms of secondary electron signal (3) a more reliable measure of surface composition is via small spot spectroscopy.

The detection geometry of the microscope affects the contrast from surface topography. The focused radiation strikes the surface normally while the energy analyzer accepts photoelectrons only from angles close to 20° to the surface leading to characteristic shadowing from surface irregularities. In figure 1 electrons are being collected from the right of the image. Thus the dark band C between areas A and B is interpreted as a sloping boundary between the two grains. Small spot spectroscopy on this boundary in fact shows that it has similar oxide sensitivity to region A (Table 1). The feature P showing a shadow to the left and bright spot to the right is interpreted as small spherical lead particle on the surface possibly formed by the sputtering process. The clear shadow to the left of the object enables the particle height to be estimated at around 3μm. The shadow to the right of this feature suggests that the particle sits in a small pit formed during the sputtering process.

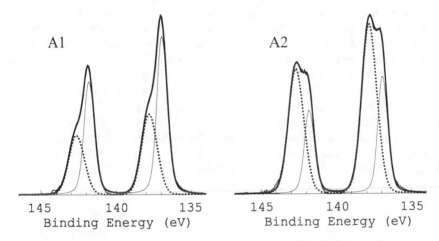

FIGURE 2. Pb 4f spectra for region A for 500L (A1) and 1000L (A2) doses of oxygen .

The oxide shift is clearly seen in the small spot spectra. (Figure 1) The spectra have been fitted using one pair of Doniach-Sunjic functions (4) for the metalic Pb 4f doublet , and another pair for the chemically shifted peaks (shown as the dashed lines in figure 2.) The oxide shift was measured to be 0.88eV. The fraction of oxidized lead atoms present in the surface layer sampled by the radiation is then the area of the oxide doublet divided by the total area of the doublet (oxide + metal) The percentage of oxidized lead atoms calculated in this way for areas A, B and C and from a small spot spectrum of particle P is shown in table 1.

TABLE 1. % of oxidized surface lead atoms for different oxygen doses.

Feature/Area	500L	1000L
A	29	56
B	15	33
C	28	54
P	26	44

It is clear from these results that region C is similar to region A, its low intensity in fig.1 being due to topography rather than oxide concentration. All spectra apart from that on P show a doubling in oxide with a doubling of dose. It is probable that the discrepancy with the particle measurement is simply due to a small error in the positioning of the microprobe for the measurements with different oxygen doses.

DIFFUSION

The process of diffusion on a polycrystalline nickel film has been investigated by evaporating sub-monolayer amounts of silver and gold onto commercial nickel foils. These were then heated to approximately 620 K for 10 minutes. The surfaces were imaged in terms of Ag 3d or Au 4f signal before and after heating. Ag is essentially insoluble in Ni at 620 K while Au shows considerable solubility (5). This results in different behaviour for the two surfaces. For Ag coated Ni, Ag atoms appear to collect at surface irregularities which are not necessarily grain boundaries. For the Ni surface, imaged in terms of Au 4f$_{7/2}$ electrons, Au atoms did not diffuse to surface irregularities but appeared to diffuse into the bulk of the metal at a rate depending on the surface structure of the particular crystallite. This lead to contrast between areas where Au diffusion into the bulk occurred at different rates leading to different surface Au concentrations. The effect is apparent in figures 3A and 3B which show the same Au coated surface before and after annealing. Surface topography complicates the images but the effect of diffusion is clear.

A   **B**

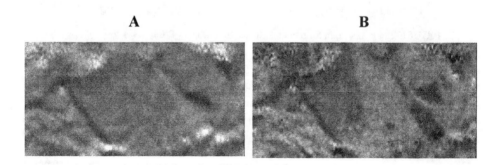

FIGURE 3 64μm x 32μm Au 4f image of a Ni surface A, before annealing and B, after annealing.

ACKNOWLEDGEMENTS

We are grateful to Sincrotrone Trieste ScpA for allowing us access to this facility and for providing support. The work was supported by an EC grant under contract ERBCHGECT920013.

REFERENCES

1. Casalis,L., Gregoratti,L., Kiskinova,M., Margaritondo,G., Braz Fernandes,F.M., Silva,S., Morrison,G.R., and Potts,A.W. *Surface and Interface Analysis* 25,374-379 (1997).

2. Ko,C-H., Kirz,J., Ade,H., Johnson,E., Hulbert,S., and Anderson E., *Rev.Sci.Instrum.* **66**, 1416-1418 (1995).

3. Potts,A.W., Morrison,G.R., Gregoratti,L., Gunther,S., Kiskinova,M., and Marsi,M., *Chem. Phys. Lett.* **290** 304-310 (1998)

4. Hüfner,S., *Photoelectron Spectroscopy 2nd Ed.* ,Chapter 4, Springer, 1996.

5. Hansen,M., *Constitution of Binary Alloys*, McGraw-Hill Book Company , 1958

Compatibilization Dynamics in Highly Immiscible Polymer Blends

D.A. Winesett, D. Gersappe*, M. Rafailovich*, J. Sokolov*, S. Zhu*, H. Ade

Department of Physics, North Carolina State University, Raleigh, NC 27695-8202
*Department of Materials Science and Engineering, SUNY@StonyBrook, Stony Brook, NY 11794.

Abstract. The morphology of incompatible polymer blends are often stabilized by the addition of block copolymers that ideally will localize to the polymer-polymer interface and reduce the interfacial tension. However, the effectiveness of adding copolymer is significantly reduced by the tendency of the diblock to form micelles that become trapped within one of the phases. Recent theoretical and experimental results show that using compatibilizers in confined physical geometry will reduce the configurational entropy of the diblock and make it energetically more favorable for the diblock to locate to the interface [1]. Dynamics studies with Scanning Transmission X-ray Microscopy (STXM) show two regimes in the dynamical process, where growth regimes are characterized by growth exponents, α, where $R(t) \sim R^{\alpha t}$. The first growth regime consists of round micelle-like domain formation and relatively fast growth ($\alpha=2/3$) of these structures within the PS layer. The second regime results in a relatively stable bi-continuous domain formation with slow growth ($\alpha=1/20$).

INTRODUCTION

Polymer films are important in numerous technological applications such as paint systems, adhesives, photolithographic printing, magnetic disk coatings, and index-matched optical coatings. Unfortunately, the macromolecular nature of polymer chains significantly reduces the entropic gain during mixing such that different polymers are usually immiscible. Hence, thin film blends will phase separate or dewet during preparation which creates large agglomerations of like-phases and dramatic surface modulation which adversely affect the electrical and mechanical properties of the film. To prevent phase separation, many polymer systems are compatibilized with the addition of block copolymers that ideally will localize to the interface and reduce the interfacial tension [2]. However, the effectiveness of adding copolymer is significantly reduced by the tendency of the diblock to form micelles that become trapped within one of the phases [3], reducing the desired compatibilizing effect of the copolymer. Recent theoretical and experimental studies have shown that using compatibilizers in a confined physical geometry can alter this micellar transition in block copolymers to achieve miscibility in thin films even of highly immiscible polymers [1]. As the diblock layer thickness was reduced below the equilibrium bulk micelle size, the constraining effect of the thin film reduces the configurational entropy substantially and it becomes energetically more favorable for the diblock to

CP507, *X-Ray Microscopy: Proceedings of the Sixth International Conference,*
edited by W. Meyer-Ilse, T. Warwick, and D. Attwood
© 2000 American Institute of Physics 1-56396-926-2/00/$17.00

locate to the interface, thus facilitating miscibility. We have extended this investigation to characterize the growth exponent of the compatibilization process.

EXPERIMENTAL

We examined model systems of monodisperse polystyrene (PS) and poly(methyl methacrylate)(PMMA) purchased from Polymer Laboratories. 80 nm thick films of PMMA (M_w=296,000 g/mol, M_w/M_n < 1.1) were prepared by spin casting from toluene onto HF stripped silicon surfaces. 30 nm PS (PMMA (M_w=200,000 g/mol, M_w/M_n < 1.05) films with 30% added PS-b-PMMA diblock copolymer were prepared separately by spinning from toluene onto a glass microscope slide and floated onto the PMMA layer. Samples were annealed in a vacuum oven (10^{-3} Torr) at 180°C for times ranging from 30 minutes to 2 weeks. After quenching to room temperature, the films were removed from the annealing surface by dissolving the silicon with NaOH and floating off the films in a bath of distilled water, with subsequent transfer onto an electron microscopy (EM) grid. Films on EM grids were examined with the Stony Brook Scanning Transmission X-ray Microscope (STXM) at beamline X1A at the National Synchrotron Light Source [4]. A series of transmission X-ray micrographs from the same sample area were acquired of each sample at energies coinciding with characteristic Near Edge X-ray Absorption Fine Structure (NEXAFS) absorption peaks of the constituent components [5]. Utilizing a singular value decomposition procedure [6] we extract mass thickness maps [7] of each component in the sample image areas.

RESULTS AND DISCUSSION

Figure 1 shows compositional PS maps of identically prepared bilayer samples annealed for a) 5 hours, b) 10 hours, c) 15 hours, d) 96 hours, e) 168 hours, and f) 336 hours. Initially (Fig.1a) we observe the formation of micelle-like structures with PMMA cores with an average domain size of 160 nm. These structures are much larger than the thickness of the top confining layer of 30 nm and the equilibrium bulk micelle size of 56 nm [8]. Therefore, these formations likely have already undergone significant growth and dynamics before reaching this stage, with the diblock located at the PS-PMMA interface. Dynamic Secondary Ion-Mass Spectrometry on these samples (data not presented) supported this conclusion. Further annealing (Fig. 1b) produces larger individual domains (now approximately 220 nm) and indications of the growth mechanism. The larger connected domains appear to be forming through collisions of the smaller, round micelle-like domains. Annealing for 5 more hours (Fig. 1c) results in a nearly bi-continuous structure nearing its equilibrium state. Annealing 96 hours (Fig. 1d) yields refinement and slight growth, but the morphology remains similar.

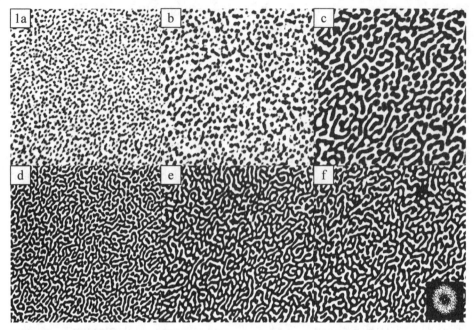

FIGURE 1. PS compositional maps acquired from identically prepared bilayer samples annealed for a) 5 hours, b) 10 hours, c) 15 hours, d) 96 hours, e) 168 hours, and f) 336 hours (inset = $|FFT|^2$ of 336 hr. sample). Fig a-c area = 10 μm^2, Fig d-f = 20 μm^2.

Annealing for very long times (Fig. 1e,f) shows little agglomeration, indicating that the film had been stabilized as the copolymer has saturated the interface. We ascertained the location of the block copolymer after microemulsion formation by imaging a sample that had been annealed for 168 hours and washed with cyclohexane to remove the PS phases (Figure 2a). At the photon energy utilized (285.3 eV, PS $\pi^*_{C=C}$) any styrene content will appear dark while methyl-methacrylate is completely transparent. The dark halos around the domains of this washed sample thus indicate that the PS-b-PMMA diblock migrated preferentially to the interface during emulsification.

We have quantified the growth of the morphology in these samples in the following way. The absolute value of the two-dimensional Fourier transform ($|FFT|^2$) of each sample exhibits a doughnut like pattern (see inset to Fig. 1). We determined the average radius of this pattern for each of the PS maps and plotted on a normalized log-log scale the radius against the annealing time (Fig. 2b). The initial micelle growth is very fast, with a growth exponent of $\alpha=2/3$. An exponent of 2/3 indicates a hydrodynamic growth regime and mechanism [9]. However, we expected these domains to be developing via a diffusion/collision-like behavior at the interface, which should exhibit a growth exponent of 1/3 [10]. We currently have no explanation for this discrepancy. They point to some interesting physics in the dynamics of the microemulsion formation that we currently don't understand sufficiently. After the formation of the bi-continuous microemulsion, the phase structure is stable exhibiting very little morphological development, with small growth exponent that can be approximated with $\alpha=1/20$.

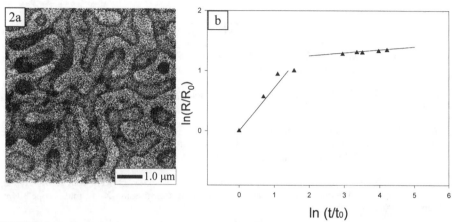

FIGURE 2. a) PS $\pi^*_{C=C}$ image acquired at 285.3 eV from a 7 day annealed bilayer sample washed with cyclohexane to remove all PS. b) Normalized plot of domain size vs. time for identical bilayer samples annealed for various times. Lines are fits to the data with slopes of $\alpha=2/3$ and $\alpha=1/20$, respectively.

CONCLUSIONS

The compatibilization of blends of PS and PMMA using a diblock PS-*b*-PMMA copolymer in the confined geometries prepared by us occurs in at least two distinct stages. First, micelle-like domains with PMMA cores form in the PS layer and migrate to the interface due to the energetic penalty of micelle formation in confined geometry. As the diblock localizes to the PS-PMMA interface it reduces the interfacial tension and a relatively stable two-dimensional microemulsion is formed. This two-dimensional microemulsion phase exhibits no dewetting or surface modulations signifying the compatibilization of the two phases. The initial micelle-like domains at the interface grow with a characteristic exponent $\alpha=2/3$ which is typically associated with a hydrodynamic growth mechanism.

AKNOWLEDGEMENTS

H. Ade and D. A. Winesett are supported by NSF Young Investigator Award DMR-9458060. M. H. Rafailovich, J. Sokolov and S. Zhu are supported by NSF DMR-9732230 (MRSEC Program) and DOE-SG02-93-ER45481. Data acquired with the Stony Brook STXM at the NSLS developed by the group of Janos Kirz and Chris Jacobsen at SUNY Stony Brook, with support from DOE (DE-FG02-89ER60858) and NSF (DBI-9605045). The zone plates were developed by S. Spector and C. Jacobsen of Stony Brook and D. Tennant of Lucent Technologies Bell Labs, with support from the NSF (ECS-9510499). The NSLS is supported by the Office of Basic Energy Sciences, Energy Research, Department of Energy.

REFERENCES

1   S. Zhu, Y. Liu, M. H. Rafailovich, J. Sokolov, D. Gersappe, D. A. Winesett, and H. Ade, "Confinement Induced Miscibility in Polymer Blends", Nature **400**, 49 (1999).

2   K. Shull, "Interfacial Phase-Transitions in Block Copolymer Homopolymer Blends", Macromolecules **26**, 2346 (1993).

3   W. C. Hu, J. T. Koberstein, J. P. Lingelser, and Y.Gallot, "Interfacial Tension Reduction in Polystyrene/Poly(Dimethylsiloxane) Blends by the Addition of Poly(Styrene-B-Dimethylsiloxane)", Macromolecules **28**, 5209 (1995).

4   C. Jacobsen, S. Williams, E. Anderson, M. T. Brown, C. J. Buckley, D. Kern, J. Kirz, M. Rivers, and X. Zhang, "Diffraction-limited imaging in a scanning transmission X-ray microscope", Opt. Commun. **86**, 351 (1991).

5   H. Ade, "Compositional and orientational characterization of polymeric materials with x-ray microscopy", Trends Polym. Sci. **5**, 58 (1997).

6   X. Zhang, R. Balhorn, J. Mazrimas, and J. Kirz, "Mapping and measuring DNA to protein ratios in mammalian sperm head by XANES imaging", J. Struc. Biol. **116**, 335 (1996).

7   H. Ade, D. A. Winesett, A. P. Smith, S. Qu, S. Ge, S. Rafailovich, and J. Sokolov, "Phase Segregation in Polymer Thin Films: Elucidations by X-ray and Scanning Force Microscopy", Europhys. Lett. **45**, 526 (1999).

8   A. N. Semenov, "Theory of Diblock-Copolymer Segregation to the Interface and Free-Surface of a Homopolymer Layer", Macromolecules **25**, 4967 (1992).

9   H. Furukawa, "Role of Inertia in the Late-Stage of the Phase-Separation of a Fluid", Physica A **204**, 237 (1994).

10 H. Tanaka, "A New Coarsening Mechanism of Droplet Spinodal Decomposition", Hournal of Chemical Physics **103**, 2361 (1995).

Precision micro-Xanes of Mn in corroded roman glasses

A. Simionovici[a], K. Janssens [b], A. Rindby[c], I. Snigireva[a], A. Snigirev[a]

[a]ESRF, BP 220, 38043 Grenoble, FRANCE
[b]MITAC, Univ. of Antwerp, BELGIUM
[c]Chalmers Univ., Gøteborg, SWEDEN

Abstract. The highest spatial resolution µ-Xanes experimental results to date were obtained on an archeological glass sample containing Mn. Both the fluorescence (SIXES) and absorption collection modes were used to record maps of the elemental distribution throughout the surface corrosion layer. By using two excitation energies near the Mn threshold at 6.5 keV, direct speciation maps were obtained. The analysis was carried out using the ESRF, ID 22 microbeam, with a 3 x 5 μm^2 beamspot.

INTRODUCTION

In recent studies of archeological samples of cultural and historical interest, the stress is laid on the use of multiple analysis methods to fully characterize the structure and chemistry of various regions of the artifacts [1]. This implies a localized analysis, with a spatial resolution on the order of a few microns, in order to elucidate the structure of small samples. As a further complication, the samples are often precious and unique artifacts and the major requirement in such cases is the guarantee of a non-destructive analysis, capable of going beyond the surface layers, to analyze in-depth the samples. The necessity of manipulating relatively large objects on the scale of a few centimeters, without any special preparation precludes the use of specialized analyses such as electron microscopy. The only probe which fulfills all the above requirements is the synchrotron microprobe, capable of delivering intense beams of X-ray photons to spots of a few microns in surface, penetrating several hundred microns inside the objects and requiring no special sample preparation. The analysis is performed in the air, using a precise scanning movement of the sample in front of a focused beam.

The studies of archeological samples using Synchrotron Radiation span several methods of investigation such as as fluorescence, absorption, diffraction and other imaging techniques. Sometimes, it is necessary to establish the provenance of artifacts and there the detection of fluorescence of trace elements establishes the fingerprint of specific geographical sites. The absorption spectroscopy is employed to assess the state of

CP507, *X-Ray Microscopy: Proceedings of the Sixth International Conference,*
edited by W. Meyer-Ilse, T. Warwick, and D. Attwood

conservation of particular materials or to distinguish original from add-on materials, used by past restoration techniques. Finally, diffraction techniques are used to analyze metal alloys from samples of historical interest. A review of micro-analitical investigations of archeological samples of similar interest can be found in [2].

EXPERIMENT

Experimental set-up at the ESRF, ID22 beamline

The ID 22 beamline is dedicated to micro X-ray spectroscopy: fluorescence, absorption, fluorescence tomography, as well as imaging: phase contrast topography, tomography, holography and diffraction. The microprobe, presented in figure 1 consists of a focusing lens, a sample holder with 6 degrees of freedom, allowing the scan of the sample in X Y Z θ (and φ ψ for fluo-tomography) in front of the focused beam. Beam intensity monitors are placed before and after the sample for precise monitoring. They are either ionization chambers - for low energy beams – or PIN diodes. A system of pinholes of varying sizes (5, 10 20 µm) is located before the sample and serves as OSA (Order Sorting Aperture) for the FZP beams which feature an intense zero order transmitted beam. A SiLi, solid state detector records the fluorescence at 90° (to minimize Compton/Rayleigh scattering) with respect to the incident beam. A CCD X-ray camera of high resolution (0.5 µm) is located behind the sample, allowing to record phase contrast images of the samples and directly align it in the beam.

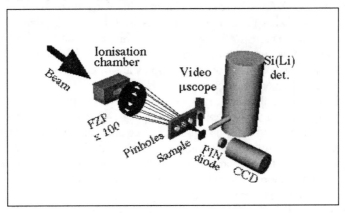

Figure 1. Experimental setup of the micro-spectroscopy apparatus on beamline ID22 at the ESRF

The optics of the beamline consists of a horizontally deflecting flat mirror of extremely low micro-roughness (< 1.5 A) and slope error (< 1.5 µrad), designed to suppress higher undulator harmonics and decrease the total heatload on the monochromator. Following is a double crystal, fixed exit Kohzu monochromator, LN$_2$ cooled which uses two sets of crystals: Si 111 and 311. The beam is delivered by a standard ESRF undulator of 42 mm period and 1.65 m length.

Results

The sample analyzed was a glass fragment from Qumran, in the Jordan valley in Israel. This glass was buried for 2000 years in a soil sediment containing water. The exposure to the environment produced a large corrosion body at the surface of the glass where Na and C cations were leached out. Mn oxides form as dark precipitates at the surface of the exposed zone and facilitate the advance of the corrosion layer by difusing in the crevices of the glass.

The experiment was performed using both modes: fluorescence, recorded by the SiLi detector which was taking energy spectra and integrating over the Mn K$_\alpha$ region and absorption, by recording incident/transmitted flux in the ionization chamber, respectively PIN diode. A windowless ionization chamber was used for the I$_o$ measurement so as not to disturb the beam too much. The energy scans of the monochromator were taken between 6.52 and 6.59 keV, with 0.7 eV/step, using the Si 111 crystals and an acquisition time of 5 sec/point was used. The spectra obtained from Mn metal and KmnO$_4$ as well as two different zones rich in MnO, respectively MnO$_2$ from the glass sample are presented in figure 2.

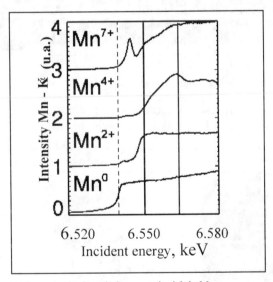

Figure 2. XANES spectra from standards and glass sample rich in Mn.

Fluorescence maps of a region 300 x 300 µm were recorded, at two different energies: E = 6.564 keV and E = 6.550 keV. Owing to the fact that the Mn^{4+} oxidation state features a prominent crest (white line) at E = 6.564, and a very small absorption at E = 6.550 kev, while the Mn^{2+} is almost constant at both energies (above edge) it is possible to obtain an enhanced contrast of the varying Mn absorption oxidation state. By subtracting from the map obtained at higher energy the one at low energy, normalized, we obtain a map of Mn fluorescence intensity, highly enriched in Mn^{4+}. This is the so-called SIXES speciation method which allows direct estimation of the oxidation states of samples containing strongly differing Xanes spectra of same element.

The maps obtained in the fluorescence scans of the glass sample are presented in figure 3. We obtained simultaneously concentration maps of several elements present in the glass, such as: Si, Cl, K, Ca, Ti and Mn.

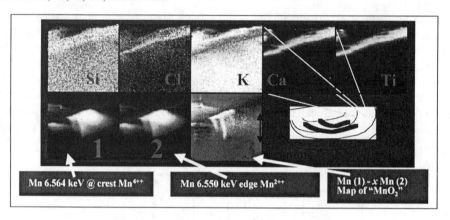

Figure 3. Maps of the elemental concentrations in the glass sample

By performing image processing corrections on the two elemental maps of the Mn concentration (maps 1 and 2 in figure 3) a map of the normalized difference is obtained (map 3).

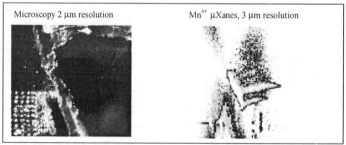

Figure 4. Micro-Xanes map of the Mn^{4+} oxidation state vs. a visible micrograph

282

Maps 1 and 2 do not differ visibly, however, taking the normalized difference produces a striking contrast, revealing the V-shaped region enriched in Mn^{4+} oxidation state. The results of the image subtraction are presented in figure 4 in comparison with a visible microscopy of the same region.

CONCLUSION

The SIXES collection mode was applied to a Mn-rich glass sample of archeological interest. By using the microprobe at ID22, a direct map of the Mn^{4+}-enriched regions in the sample was obtained. This represents a time gain of a few hundred times, which would have been otherwise necessary, had the regular Xanes scanning collection mode been used. Moreover, the spatial resolution recorded (3 x 3 μm) was the highest ever obtained for such Xanes maps, in a relatively short time. Detailed results of this investigation as well as other related studies of archeological artifacts will be presented elsewhere.

ACKNOWLEDGEMENTS

The authors are grateful to M. Gentili and E. Di Fabrizio, IESS Rome, for the loan of the FZP lens array used in the experiment.

REFERENCES

1. Ch. Lahanier, G. Amsel, Ch. Heitz, M. Menu, H. H. Andersen, Nucl. Instr. Meth. B 14, 1 (1986)

2. F. Adams, A. Adriaens, A. Aerts, I. De Raedt, K. Janssens, O. Schalm, J. Anal. Atomic Spectrom. 12, 257 (1997)

3. K. Janssens, G. Vittiglio, I. De Raedt, A. Aerts, B. Vekemans, L. Vincze, F. Wei, I. Deryck, O. Schalm, F. Adams, A. Knochel, A. Simionovici, A. Snigirev, X-Ray Spectrometry, to be published.

Grain Orientation Measurement of Passivated Aluminum Interconnects by X-ray Micro Diffraction

Chang-Hwan Chang[1], B.C. Valek[2], H.A. Padmore[3], A.A. MacDowell[3], R. Celestre[3], T. Marieb[4], J.C. Bravman[2], Y.M. Koo[6], J.R. Patel[3,5]

[1]Research Institute of Industrial Science & Technology (RIST), Pohang 790-600, Korea
[2]Departmant of Materials Science and Engineering, Stanford University, Stanford, CA, 94305
[3]Advanced Light Source, Lawrence Berkeley National Laboratory, Berkeley, CA 94720
[4]Intel Corporation, Santa Clara, CA
[5]SSRL/SLAC, Stanford University, Stanford, CA 94309
[6]Department of Materials Science and Engineering, Pohang University of Science and Technology, Pohang 790-600, Korea

Abstract. The crystallographic orientations of individual grains in a passivated aluminum interconnect line of 0.7-μm width were investigated by using an incident white x-ray microbeam at the Advanced Light Source, Berkeley National Laboratory. Intergrain orientation mapping was obtained with about 0.05° sensitivity by the micro Laue diffraction technique.

INTRODUCTION

The steady, continuing trend of miniaturization of electronic components in integrated circuits has placed stringent demands on interconnects between the circuit elements. Smaller interconnect cross-sections lead to extremely high current densities. The performance of these interconnects (usually Al-Cu or Cu) is of major concern to the Semiconductor Industry (1). Failure can arise from two main causes. First, as the dielectric encapsulated (passivated) interconnect cools from its deposition temperature, the metal will contract more that the silicon substrate and passivation layer, placing the metal under tensile stress orders of magnitude greater than its yield stress. Some dislocation motion will occur to relieve this stress, but the encapsulation prevents further relaxation. Stresses in an Al line can be as high as 400 MPa. Some relaxation also occurs by vacancy flow and this can result in stress voids that reduce the cross section of the line. The second failure mode, and by far the most important, is electromigration in active circuits (2). Atoms in the metal interconnects are acted on by a force due to the high electron current density and are actually moved from the cathode end of the line to the anode. Over time in a passivated interconnect, the motion of atoms down the line leads to compressive stresses building at the anode and tensile stresses at the cathode. Void formation can result due to the depletion of material at the cathode, leading to open circuit failure. Since electromigration is a diffusive process, stress gradients in the line (3,4), as well as the grain boundary structure and orientation (5), influence the time until failure of the interconnect.

CP507, X-Ray Microscopy: Proceedings of the Sixth International Conference,
edited by W. Meyer-Ilse, T. Warwick, and D. Attwood
© 2000 American Institute of Physics 1-56396-926-2/00/$17.00

It is important, therefore, to understand these failure mechanisms in a fundamental manner. So far the tools to accomplish this in a meaningful, quantitative way on individual lines have not been available. The key experiments required are (a) grain orientation measurements of individual grains in lines and (b) accurate strain measurements in individual grains along a line with and without current flow in the line. The only tool that can accomplish both these objectives on passivated interconnects is x-ray diffraction using suitably prepared x-ray beams. With the availability of Modern Third Generation high brilliance synchrotron source, we have for the first time the capability of producing micron and submicron x-ray beam with sufficient intensity to do meaningful orientation and strain measurements on individual interconnect lines.

EXPERIMENTAL

We have used bend magnet radiation from the synchrotron source at the Advanced Light Source. At the source point, the size is $300 \times 30 \ \mu m^2$ FWHM (horizontal and vertical) and is imaged with demagnifications of 300 and 60 respectively by a set of platinum-coated, elliptically bent, Kirkpatrick-Baez (K-B) focusing mirrors. Imaged spot sizes on the sample are about a micron in size. Photon energy is either white or monochromatic (energy range 6-14 keV), generated by inserting a pair of Si(111) channel-cut monochromator crystals into the beam path. A property of the four-bounce monochromator is its ability to direct the monochromatic primary beam along the same direction as the white radiation. Thus, the sample can be irradiated with either white or monochromatic radiation while maintaining the focal spot on the sample. White radiation is chosen for Laue experiments that determine crystallographic orientation and monochromatic radiation is chosen for d-spacing measurements in strain determination of single grains in the metal line. We will only

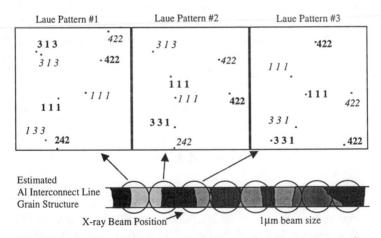

FIGURE 1. Three Laue patterns taken at 1 μm intervals along the Al interconnect line. The estimated Al grain structure is shown below the Laue patterns with x-ray beam position indicated. For each x-ray beam position, the left most grain is indexed in italic type, and the right most grain is indexed in bold type.

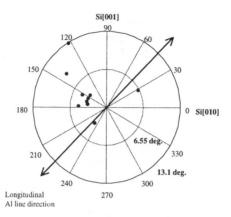

FIGURE 2. Al(111) orientations of aluminum grains (circular dots) relative to longitudinal line direction and silicon substrate.

discuss our initial results involving crystallographic orientation determinations.

The focused x-ray microbeam measurements were performed on an aluminum line deposited to 0.75-μm thickness and 0.7-μm width on an oxidized silicon substrate. The line was passivated with a one micron thick layer of SiO₂. The sample was mounted on a 0.05-μm resolution x-y translation stage, and x-ray microbeam measurements were made by translating along an interconnect line in a 1-μm step scan. Laue patterns were collected using white radiation and an x-ray CCD camera. The exposure time was 1 second and sample-to-CCD distance was 19.6 mm. Details of the experimental arrangement are described elsewhere (6).

The origin on the CCD detector array was determined by moving the CCD camera in the radial direction from the sample and recording the silicon Laue patterns at various distances from the sample. The origin was determined on the CCD where the lines drawn through the succession of the same Laue spots intersected. All aluminum spot positions were coordinated to the origin and indexed by using an automated indexing computer code. From the indexation of the aluminum pattern as well as the silicon crystal substrate, the orientation of aluminum grains was determined in an accuracy of ~0.05°.

RESULTS

Figure 1 shows the results of a grain orientation scan in the passivated aluminum interconnect wire. The background silicon Laue pattern is constant as the line is translated because the silicon is a single crystal substrate. The aluminum Laue pattern was obtained following digital subtraction of the silicon pattern from the silicon and aluminum pattern (7). In the schematic diagram of the grain structure as shown in Figure 1, it is assumed that the wire has a bamboo or near-bamboo structure. The positions of the grain boundaries in the interconnect were estimated from the intensities of the Laue spots. The out of plane misorientation of adjacent aluminum grains ranges from 0.5° to 10°. Figure 2 shows Al(111) orientations of the aluminum grains relative to the longitudinal aluminum line direction as well as the silicon substrate. The circle at 6.55° indicates that (111) orientations of most grains are within this range. The longitudinal line direction lies along [011] direction of the silicon substrate crystal. The majority of misorientation angle ranges between 3° and 4° and one grain shows a large angle of ~13°. Initial work shows that the in-plane grain orientation is random.

CONCLUSION AND FUTURE DEVELOPMENT

We have demonstrated that x-ray micro-diffraction is capable of determining the crystallographic orientation of individual grains in passivated metallic interconnect lines. The orientation mapping is performed by collecting the Laue patterns from individual grains along the length of the line and using a computerized indexing code. Beyond the work of orientation mapping, the requirement is to measure the d-spacing of various aluminum planes to determine the stress and strain state of individual grains along the length of the aluminum interconnect line. This question is currently being addressed with a specially designed, high absolute accuracy diffractometer.

ACKNOWLEDGEMENTS

This work was supported by the Director, Office of Basic Energy Sciences, Materials Sciences Division of the US Department of Energy, under Contract no. DE-AC03-76SF00098. J.R. Patel would like to thank John Carruthers of Intel for his support of this work.

REFERENCES

1. Ryan J.G., Geffken R.M, Poulin N.R., Paraszczak J.R., *IBM Journal of Research and Development* **39** (4), pp. 371-381 (1995)

2. Lloyd J.R., *Journal of Applied Physics* **69** (11), pp. 7601-7604 (1991)

3. Blech I.A. and Tai K.L. *Applied Physics Letters* **30** (8), pp. 387-389 (1977)

4. Korhonen M.A., Borgesen P., Tu K.N., Li C., *Journal of Applied Physics* **73** (8), pp. 3790-3799 (1993)

5. Attardo M.J. and Rosenberg, R., *Journal of Applied Physics* **41** (4), pp. 2381-2386 (1970)

4. MacDowell A.A., Celestre R., Chang C.H., Franck K., Howells M.R., Locklin S., Padmore H.A., Patel J.R., and Sandler R., *SPIE Proceedings* **3152**, 1998, pp. 126-133.

5. Chang C.H., MacDowell A.A., Thomson A.C., Padmore H.A., and Patel J.R., AIP Conference Proceedings **449**, New York: American Institute of Physics, 1998, pp. 424-426.

ENVIRONMENTAL AND SOIL SCIENCES

Charge state mapping of mixed valent iron and manganese mineral particles using Scanning Transmission X-ray Microscopy (STXM)

K. Pecher[1], E. Kneedler[2], J. Rothe[3], G. Meigs[4], T. Warwick[4], K. Nealson[1], B. Tonner[5]

[1]CalTech/JPL, Pasadena, CA
[2]Surface/Interface Inc., Sunnyvale, CA
[3]Institut für Nukleare Entsorgungstechnik, Forschungszentrum Karlsuhe, Germany
[4]Advanced Light Source, LBNL, Berkeley, CA
[5]Dept. of Physics, University of Central Florida, Orlando, FL

Abstract. The interfaces between solid mineral particles and water play a crucial role in partitioning and chemical transformation of many inorganic as well as organic pollutants in environmental systems. Among environmentally significant minerals, mixed-valent oxides and hydroxides of iron (e.g. magnetite, green rusts) and manganese (hausmanite, birnessite) have been recognized as particularly strong sorbents for metal ions. In addition, minerals containing Fe(II) have recently been proven to be powerful reductants for a wide range of pollutants. Chemical properties of these minerals strongly depend on the distribution and availability of reactive sites and little is known quantitatively about the nature of these sites. We have investigated the bulk distribution of charge states of manganese (Mn (II, III, IV)) and iron (Fe(II, III)) in single particles of natural manganese nodules and synthetic green rusts using Scanning Transmission X-ray SpectroMicroscopy (STXM). Pixel resolved spectra (XANES) extracted from stacks of images taken at different wave lengths across the metal absorption edge were fitted to total electron yield (TEY) spectra of single valent reference compounds. Two dimensional maps of bulk charge state distributions clearly reveal domains of different oxidation states within single particles of Mn-nodules and green rust precipitates. Changes of oxidation states of iron were followed as a result of reductive transformation of an environmental contaminant (CCl$_4$) using green rust as the only reductant.

INTRODUCTION

An understanding of geochemical and biochemical processes in the environment, such as biomineralization or redox transformations of pollutants on minerals, requires knowledge about the physical structure as well as the thermodynamic and electronic properties of materials involved. One inherent feature of such interactions is the spatial inhomogeniety of the chemical processes involved, which lead to a variety of different products and heterogeneous phases. In order to be able to understand reaction kinetics and mechanisms therefore, it is necessary to characterize specific binding sites or preferred sites of reaction. Although a wide range of modern analytical, diffraction and spectroscopic techniques are available to the geochemical community, there are few

CP507, X-Ray Microscopy: Proceedings of the Sixth International Conference,
edited by W. Meyer-Ilse, T. Warwick, and D. Attwood
© 2000 American Institute of Physics 1-56396-926-2/00/$17.00

that combine submicron spatial resolution and chemical specificity. In particular, there is a clear need for a sensitive, element-specific site and valence probe (1).

Scanning Transmission X-ray Microscopy (STXM) of transition metal compounds offers exactly this combination (2, 3, 4). It is the only x-ray absorption technique which allows for samples to be analyzed in their fully hydrated, wet state. Since the early 90′s, the number of applications of x-ray absorption spectroscopy (XAS) has increased as a consequence of improved synchrotron assisted spectral resolution (5) and better theoretical interpretation of the resulting spectra (6, 7, 8). 2p XAS of 3d transition metals is based on dipole allowed, bound-state electron transitions from the core 2p level to empty 3d states. Because of the large Coulomb interaction between these two levels, the dipole transition energies i.e. position of lines in the spectra, and their occurrence probabilities i.e. the intensity of lines in the spectra, depend on the local electronic structure of the absorbing ion. Contrary to hard x-ray absorption spectra, core-hole lifetime broadening is small, resulting in sharp multiplet structures and the $2p_{3/2}$ (L_3) and $2p_{1/2}$ (L_2) spectral parts are clearly separated by core-hole spin-orbit interaction. Thus, analysis of the $L_{2,3}$ absorption structure provides information primarily about the oxidation state and upon further analysis, information about site symmetry, spin state, and crystal field splitting of the absorbing transition metal ions.

Although we are currently applying several theoretical models to explain the spectral features of our samples (9), it is convenient in this study to do a more "chemical approach to XANES" (x-ray absorption near edge structure), i.e. utilizing the XANES features as "fingerprints" and trying to reveal information by comparing XANES spectra of the samples investigated with a series of known and well characterized reference compounds (10). In this article we demonstrate the capabilities of STXM to map out domains of different charge states of transition metals in single particles of synthetic green rusts and natural Mn nodules. We also discuss the comparability of x-ray absorption spectra taken with different detection methods, i.e. transmission vs. total electron yield (TEY).

MATERIALS AND METHODS

All chemicals used (CCl_4, $FeCl_2$ (anhydr.), $MnSO_4$, Mn(III)-oxide, Mn(IV)-oxide) were of >99.99% purity and were used as purchased from Aldrich and Fluka. Magnetite (Fe_3O_4) was purchased from CERAC (Stock.No. I-1061, Lot No. 3268-A) and Goethite (α-FeOOH) was purchased from Bayer AG (Bayferrox 910). Green rust is a mixed valent, greenish to blue colored Fe(II,III) hydroxide and is structurally related to the pyroaurite group of double metal hydroxides (11, 12). A aqueous suspension of the sulfate form of green rust used in this study was kindly provided by Hansen (13). All minerals used were proven to be crystallographically clean by powder x-ray diffraction. Samples of Mn nodules were obtained from sediment samples collected from Green Bay (Lake Michigan).

All x-ray absorption measurements were done at the Advanced Light Source on beam line 7.0.1. The spatial resolution of STXM varied between 150-200 nm depending on the zone plates used and ambient vibrations of the scanning stages. Instrumental details are given elsewhere (14, 15, 16). An aliquot of 1-2 μl of aqueous suspension of particles was sandwiched between two Si_3N_4-membranes and the membrane sandwich was fixed onto an aluminum sample holder with tape and high vacuum grease. Samples of air sensitive material (green rust) were prepared under inert gas atmosphere (glove box), transported to STXM in an airtight jar and mounted to the scanning stage as fast as possible. In operation, the microscope is constantly purged with He, thereby avoiding any chance of contamination by oxygen. Spectra were extracted from stacks of images acquired over the range of photon energies of the metal absorption edges after careful alignment of these sequential images. Computer programs used for this analysis were provided by C. Jacobsen and A. Hitchcock. Extraction of single valent charge components was done by a multiple linear regression routine which matched the spectral vector of each pixel of a stack to a linear combination of single valent reference spectra. This analysis takes about 30 seconds on a 266 MHz Pentium II for a stack of 150 images (= photon energy points), each 150 by 150 pixels. Beam damage was not found to be relevant for the samples presented in this study.

TEY-spectra were taken from reference compounds pressed into indium foil onto stainless steel pucks. The absorption signal was monitored by collecting the total current from the sample as a function of excitation energy. The beam intensity (I_0) was recorded as the current from a gold mesh upstream. The spectra were divided by I_0 and normalized as described for each spectrum in the text. A spherical grating monochromator with slits set at 30 μm corresponds to a resolving power of 0.1 eV at the Mn-edge and 0.2 eV at the Fe-edge. All reference materials (except magnetite) were crushed to fine powders in an achate mortar prior to sample preparation. Air sensitive samples ($FeCl_2$ (anhydr.), green rusts) were kept under nitrogen in a glove box (<0.1 ppm O_2). These samples were loaded onto the pucks in the glove box and transferred to the beamline inside of a high vacuum transfer suitcase without any contact to air. Energy calibration was done prior to each experimental run by matching the $L_{2,3}$ maxima of Goethite or Mn(IV)-oxide to predetermined fixed values.

RESULTS AND DISCUSSION

Comparison of TEY and STXM Spectra

Although TEY and counting of transmitted photons constitute two completely different detection principles, spectra of both methods (Fig. 1c) are well comparable in terms of spectral lines and relative intensities. Spectra have been normalized to an edge jump of one and background subtracted (integral background) to adjust both intensity scales i.e. I/I_{ref} of TEY-measurements and optical density scale ($\ln(I_0/I)$) of transmission measurements. This is of particular importance, as both methods probe different parts of the sample, STXM being bulk sensitive and TEY probing the uppermost 50-150 Å depending on sample roughness and the angle of the incident photon beam. Figure 1 shows results of STXM and TEY together with a raster electron microscopy image of a sample of Goethite (α-FeOOH).

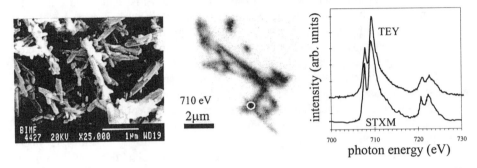

Figure 1. Raster electron microscopy image (a), STXM image (b), and comparison of TEY- and STXM-spectra (c) of a synthetic Goethite (∝-FeOOH, Bayferrox, 910). The circle in (b) labels the area where the spectrum in (c) was taken.

α-FeOOH is one of the thermodynamically most stable, naturally occurring iron oxide minerals, and therefore no transformations are to be expected during storage of the material in air or aqueous suspension. The batch of α-FeOOH we used, consisted of acicular long needles, 0.3-1.5 μm in length and 0.1-0.2 μm in diameter. Single needles are clearly differentiable in figure 1b, even though the edges look blurred, demonstrating the limits of STXM as it has been used in this study.

A different result is obtained when comparing spectra of magnetite (Fe_3O_4), which has an inverse spinel structure with Fe(II) occupying octahedral sites and Fe(III) occupying both tetrahedral and octahedral sites (Fig.2b). The magnetite has not been ground for TEY and STXM measurements. The two large peaks at the L_3-edge correspond to contributions from Fe in the Fe^{2+} and Fe^{3+} charge state. The ratio of both states varies between TEY and STXM spectra and among STXM spectra taken at different areas of the agglomerated particles in figure 2a. We attribute the apparently lower Fe(II)/Fe(III)-ratio in the TEY spectra to a thin oxidized surface layer which could be probed by TEY but got diluted in the bulk sensitive transmission measurement. Fe_3O_4 is frequently non-stoichiometric in which case it has a cation deficient Fe(III) sublattice (17). A variable stoichiometry might explain the observed spatial variation of the Fe(II)/Fe(III)-ratio in the STXM spectra. As spatial resolution was not sufficient to clearly separate out single particles (Fig. 2a), we cannot determine whether the found heterogeneity is inter or intra particle specific. Thus, if we want to measure TEY spectra as models for STXM, we need to ensure that the surface layer is not different from the bulk composition of the sample. This can be done for polycrystalline samples by careful grinding and avoidance of reoxidation of freshly exposed surfaces of sensitive samples by working under inert gas.

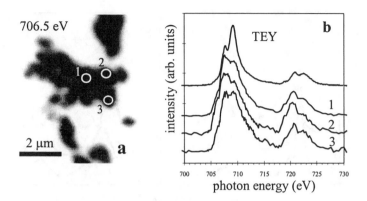

FIGURE 2. STXM image (a) and comparison of TEY and STXM spectra (b) of Fe_3O_4; numbers denote areas where spectra were taken.

TEY Spectra of Single Valent Mn and Fe Compounds as Models for Charge State Mapping of Mixed Valent Samples

The oxidation states of metal cations in homovalent Mn model compounds have been shown to dominate the L-edge spectral features when no change from high to low spin state occurs upon changes in the ligand environment (21). Within these constraints and neglecting influences from different minor ligand field strengths, L-edge spectra can be used as quantitative valence state probes (1). There are no systematic studies of Mn and Fe L-edge spectra in transmission. However, TEY spectra have been published for a variety of Fe and Mn compounds (6, 18, 19, 20). Using TEY spectra as models for transmission experiments, we have to (i) adjust the intensity scales of TEY relative to transmission measurements as mentioned previously, and (ii) properly weigh the single valent model spectra to each other, so that mixed valent samples can be modeled correctly. Spectra of all reference compounds have been normalized accordingly. Our basic assumption to correctly set the intensity scale for a set of reference compounds is the total intensity being proportional to the number of valence holes (22). Crocombette et al. (23) determined a theoretical ratio of $L_{2,3}$ edge areas of $Fe(III)_{oct}/Fe(II)_{oct}=1.24$. We used this value to normalize the Fe(II) and Fe(III) compounds in figure 3. For lack of literature data, we used the nominal value of empty valence d-orbitals in Mn^{2+} (=5), Mn^{3+} (=6), and Mn^{4+} (=7) to calibrate the relative intensities of the Mn reference spectra as shown in figure 3. These spectra were used as input to fit single valent reference compounds to stack data of mixed valent samples.

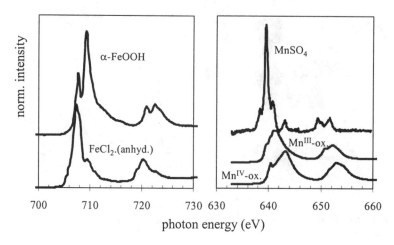

FIGURE 3. Edge jump normalized and area weighted TEY spectra of single valent Fe and Mn reference compounds.

Charge State Mapping of Mixed Valent Fe and Mn Compounds

Figure 4 shows charge state maps of a sample of SO_4-green rust. The gray scale of the maps is directly comparable in terms of mass contribution of single valent components to the overall spectrum. Bright regions correspond to higher concentration. We see irregularly shaped particles where Fe(II) is dominating (area 2 in figure 4) with corners and edges which look more oxidized (area 1). We are currently investigating whether this more oxidized material is part of the green rust crystals or whether it constitutes a separate, probably amorphous phase.

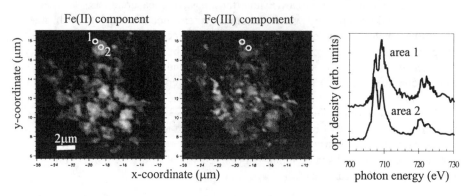

FIGURE 4. Spatial distribution of single valent iron components in SO_4-green rust and XANES spectra (STXM stack mode) extracted from two different regions.

The expected ratio of Fe(II):Fe(III) in this type of green rust is around 2:1 (24). Taking the ratio of both charge state maps in figure 4 and zooming in to a single particle of SO_4-green rust, we get the result shown in figure 5. The upper left part of the zoomed in image coincides with area 1 in figure 4.

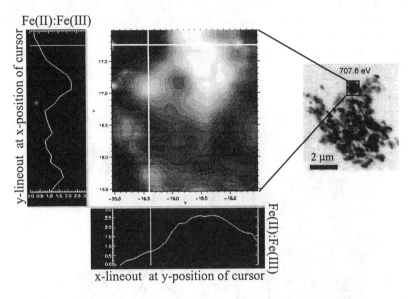

Fe(II):Fe(III)

y-lineout at x-position of cursor

707.6 eV

2 μm

Fe(II):Fe(III)

x-lineout at y-position of cursor

FIGURE 5: Spatial distribution of Fe(II)/Fe(III)-ratio in a single particle of SO_4-green rust.

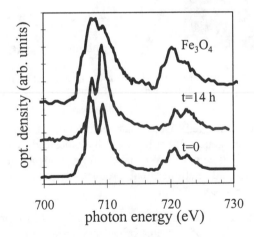

Fe_3O_4

t=14 h

t=0

FIGURE 6. Sample averaged XANES spectra of SO_4-green rust before (t=0) and after (t=14 h) reaction with CCl_4. A spectrum (STXM stack mode) of Fe_3O_4 is shown for comparison.

The Fe(II)/Fe(III)-ratio increases to about 2-2.5 moving from this spot along the x- and y-lineouts but does not stay constant throughout the entire particle. By taking the ratio of the extracted charge state maps, we eliminate any possibility of introducing artifacts due to varying particle thickness or background (25). This heterogeneous distribution of domains of charge states within a single crystallite is a new feature of green rusts. Without a diffraction microscope with similar spatial resolution we can only speculate about the consequences of this discovery on crystallographic features. However, this finding has some very important implications on the reactivity of organic and inorganic chemicals with green rusts. A single particle of green rust needs to be looked at as a multi-site reaction complex and it is not only the morphology (kinks, steps, interlayers) that might determine the reactivity of specific sites. The distribution and exact ordering of charge state domains is expected to have a big influence as well.

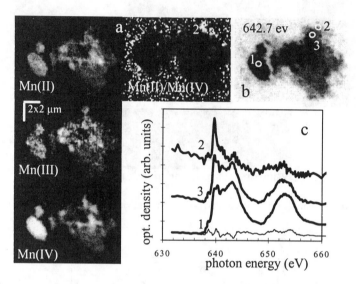

FIGURE 7. Charge state maps of Mn in a sample of Mn-micronodules (a), STXM image (b), and XANES spectra averaged over three different areas (c). The numbers correspond to the labeled areas in (b).

Product formation during reductive dehalogenation of CCl_4 on green rust was also investigated. In a recent paper (26) it has been hypothesized that Fe_3O_4 is one of the major products of SO_4-green rust. We have strong evidence that Fe_3O_4 is only a transient intermediate and that the final product is a more oxidized compound, probably maghemite (Fig. 6).

The final example is a continuation of a study about Mn biominerals started by Rothe et al. (2). Mn-micronodules are formed as a result of respiratory activity of bacteria in lake sediments and little is known about the reaction mechanisms and the composition of nodules on a submicron scale. Using conventional wet chemical analysis, researches have found that the average oxidation state of Mn in these precipitates varies around 4 (27). We found a considerable amount of Mn(II) within our sample, both uniformly dispersed throughout the micronodule and as an almost pure precipitate (labeled 2 in Fig. 7). The Mn(II)-precipitate (upper right corner of Fig. 7a) becomes nicely visible when looking at the Mn(II)/Mn(VI)-ratio map. Also shown in figure 7c is a plot of the fit residuals of the region labeled 1, i.e. the amount of spectral intensity not accounted for by a linear combination of the three model components. The residuals tend to have a little structure independent of the region the spectrum was taken. We are currently investigating whether this residual structure is due to spectral component not being represented by the used models. The Mn(III)-component contributes least to the overall spectral intensity, and whether there is a significant amount of Mn(III) at all and is currently under investigation.

CONCLUSIONS AND OUTLOOK

STXM combines submicron resolution and high resolution x-ray absorption spectroscopy. This, together with the possibility to look at samples in their wet, fully hydrated state makes it a very powerful tool to gain new insights in composition and features of environmentally important minerals and particles. Mapping of charge state domains in reactive particles, like green rusts, is likely to lead to a better understanding of transformation pathways and uptake mechanisms of environmental pollutants. Therefore, we are currently trying to extent the capabilities of STXM to study dynamic, more dilute systems as well. Following chemical reactions in situ with STXM despite the constraints of soft x-ray penetration depth, is a challenge we are taking up as part of an upgrade to the existing version of the instrument.

ACKNOWLEDGEMENTS

This work was supported by grants from DOE Division of Materials Sciences FG02-98ER45688 and DOE NABIR FG02-97ER62474. Part of the beamtime for this research has been granted under proposal ALS-00056. Clarissa Drummer (Bayerisches Geoinstitut) prepared the REM images. We appreciate the comments of D. McCubbery who carefully reviewed the manuscript.

REFERENCES

1. Schofield, P.F.; Henderson, C.M.B.; Cressey, G.; der Laan, G.van *J. Synchroton Rad.* **1995**, 2,93-98.

2. Rothe, J.; Kneedler, E.M.; Pecher, K.; Tonner, BP.; Nealson, K.H.; Grundl, T.; Meyer-Ilse, W.; Warwick, T. *J. Synchroton Rad.* **1999**, 6,359-361.

3. Droubay, T.; Mursky, G.; Tonner, B.P. *J. Electron Spec.* **1997**, 84,159-169.

4. Tonner, B.P.; Droubay, T.; Denlinger, J.; Meyer-Ilse, W.; Rothe, J.; Kneedler, E.; Pecher, K.; Nealson, K.; Grundl, T. *Surf. Interf. Anal.* **1999**, 27,247-258.

5. de Groot, F.M.F.; Fuggle, J.C.; Thole, B.T.; Sawatzky, G.A. *Phys. Rev. B* **1990**, 42,5459-5468.

6. de Groot, F.M.F. *J. Electron Spectrosc. Relat. Phenom.* **1994**, 67,529-622.

7. der Laan, G.van; Kirkman, I.W. *J. Phys.: Condens. Matter* **1992**, 4,4189-4204.

8. Cressey, G.; Henderson, C.M.B.; der Laan, G.van *Phys. Chem. Minerals* **1993**, 20,111-119.

9. Droubay, T.; PhD Thesis: University of Wisconsin at Milwaukee, in preparation.

10. Behrens, P. *Trends in analytical chemistry* **1992**, 11,237-244.

11. Allmann, R. *Chimia* **1970**, 24,99-108.

12. Brindley, G.W.; Bish, D.L. *Nature* **1976**, 263,353.

13. Koch, B.C.; Hansen, H.C.B. 16th World Congress of Soil Science - Proceedings, 10-26. 8., Montpellier, France **1998**,poster presentation.

14. Warwick, T.; Franck, K.; Kortright, J.B.; Meigs, G.; Moronne, M.; Myneni, S.; Rotenberg, E.; Seal, S.; Steele, W.F.; Ade, H.; Garcia, A.; Cerasari, S.; Denlinger, J.; Hayakawa, S.; Hitchcock, A.P.; Tyliszczak, T.; Kikuma, J.; Rightor, E.G.; Shin, H-J.; Tonner, B.P. *Rev. Sci. Instruments* **1998**, 69,2964-2973.

15. Warwick, T.; Ade, H.; Hitchcock, A.P.; Padmore, H.; Rightor, Ed.G.; Tonner, B.P. *J. Electron Spec.* **1997**, 84,85-98.

16. Ade, H. *Experimentals Methods in the Physical Sciences* **1998**, 32,225-262.

17. Cornell, R.M.; Schwetmann, U. The iron oxides.; VCH: Weinheim, 1996.

18. Grush, M.M.; Muramatsu, Y.; Underwood, J.H.; Gullikson, E.M.; Ederer, D.L.; Perera, R.C.C.; Callcott, T.A. *J. Electron Spectrosc. Relat. Phenom.* **1998**, 92,225-229.

19. Grush, M.M.; Chen, J.; Stemmler, T.L.; George, S.J.; Ralston, C.Y.; Stibrany, R.T.; Gelasco, A.; Christou, G.; Gorun, S.M.; Penner-Hahn, J.E.; Cramer, S.P. *J. Am. Chem. Soc.* **1996**, 118,65-69.

20. Chen, J.G. *Surface Science Reports* **1997**, 30,1-152.

21. Cramer, S.P.; de Groot, F.M.F.; Ma, Y.; Chen, C.T.; Sette, F.; Kipke, C.A.; Eichhorn, D.M.; McKee, V.; Mullins, O.C.; Fuggle, J.C. *J. Am. Chem. Soc.* **1991**, 113,7937-7940.

22. Thole, B.T.; der Laan, G.van *Phys. Rev. B* **1988**, 38,3158-3171.

23. Crocombette, J.P.; Pollak, M.; Jollet, F.; Thromat, N.; Gautier-Soyer, M. *Phys. Rev. B* **1995**, 52,3143-3150.

24. Hansen, H.Ch. **1999**, pers. communication

25. Koningsberger, D.C.; Prins, R. X-ray Absorption.; John Wiley & Sons: New York, 1988; p 108.

26. Erbs, M.; Hansen, H.C.B.; Olsen, C.E. *Environ. Sci. Technol.* **1999**, 33,307-311.

27. Murray, L.W.; Balistrieri, L.S.; Paul, B. *Geoch. et Cosmoch. Acta* **1984**, 48,1237-1247.

Studies of Colloidal Systems in Soils with X-Ray Microscopy

J.Thieme[1], C.Schmidt[1], J.Niemeyer[2]

[1] *Institut für Röntgenphysik, Georg-August-Universität Göttingen,*
Geiststraße 11, 37073 Göttingen, Germany
[2] *Fachbereich VI - Bodenkunde, Universität Trier, 54286 Trier, Germany*

Abstract. Soil is a part of the environment where colloidal systems play an important role when it comes to the description of its properties. A great variety of chemical reactions within these systems occur in an aqueous phase. Due to its ability to image specimen directly in their natural aqueous environment, X-ray microscopy is very suitable tool for the study of these systems. Man-made contamination of soils, especially organic contaminations, can be removed with thermal clean-up techniques. These techniques influence strongly the shape of the colloidal systems within the treated soils. X-ray microscopy studies of an agriculturally used soil and a forest soil give examples of this influence.

INTRODUCTION

Soil is the animated part of the top crust of the earth. Its limits are the underlying rocks, the vegetation of top and the atmosphere. The lithosphere, the hydrosphere and the atmosphere penetrate it. Due to this penetration soil is a very heterogeneous medium. Its fabric consists of a pore system, which determines the inner surface of the soil. The radii of the pores range from mm to sub-µm. Pores with radii of some µm or less are the most frequent and therefore form the largest part of the inner surface. This inner surface affects many important soil parameters as the turnover of nutrients, the adsorption of toxicants, the water storage capacity and the water flow [1].

This microstructure of soils is built up by colloids. Two types of colloidal particles can be found in soils, inorganic and organic particles. Representatives for inorganic particles are mainly clays or particles from quartz, from iron oxides, and from aluminum oxides. Organic colloidal particles are bacteria, fungi, humic substances, xenobiotica etc. Nearly all chemical reactions occur in an aqueous phase [2]. Therefore, there is a strong need for techniques, that allow the investigation of soil colloids directly in their aqueous environment with sub optical resolution.

X-ray microscopy is a tool that is very well suited to fulfill this task. This has been shown already elsewhere [3-7]. First, it supplies the investigator with a sub optical resolution needed for studying interactions in the colloidal range. Second, due to a natural contrast between water and other substances, it provides the ability to study samples in aqueous media. This natural contrast has its origin in the different absorption and phase shift of X-rays by water and other substances.

CP507, *X-Ray Microscopy: Proceedings of the Sixth International Conference,*
edited by W. Meyer-Ilse, T. Warwick, and D. Attwood

Figure 1 shows the linear absorption coefficient μ of water, the clay mineral smectite and a protein as an organic substance as a function of wavelength. The differences in μ are clearly visible. The x-ray image in figure 2 illustrates the differences in absorption of particles and water. Here, a flock of soil colloids from a dystric cambisol in water is imaged. The random association of particles of different shapes can be seen clearly.

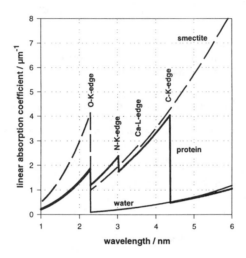

FIGURE 1. The linear absorption coefficient μ is plotted as a function of wavelength for three different substances, i.e. water, the clay mineral smectite and a protein as an organic molecule. The important absorption edges of oxygen, nitrogen, calcium, and carbon are indicated.

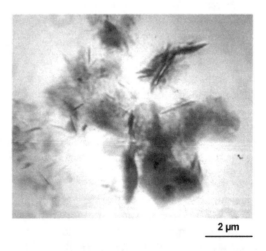

FIGURE 2. X-ray microscopy image of a flock of soil colloids from a dystric cambisol.

MICROSTRUCTURE OF SOILS BEFORE
AND AFTER THERMAL CLEAN-UP

Man made contaminations of soils consist mainly of organic substances as oil, lubricants or polyaromatic hydrocarbons, and heavy metals as lead or cadmium. Many decontamination procedures exist to extract the introduced substances from the soil [8]. To remove organic contaminations among others thermal clean up procedures are used, especially for lubricants. Here, temperatures of several hundred degrees are introduced to the soil, burning away all organic substances including the organics natural in soils. Organic colloids that take part in the building process of the micropores are burnt away. In addition, the water that is stored in the spaces between the colloids is removed completely. Apparently changes in the microstructure of the soil treated this way might occur. Therefore, X-ray microscopy studies with two different types of soil have been done to visualize these changes. These experiments have been performed with the X-ray microscope of the Institute for X-ray Physics at the electron storage ring BESSY in Berlin [9].

An agriculturally used soil and a forest soil, both contaminated with lubricants and polyaromatic hydrocarbons from a factory site nearby have been studied. In all cases presented here a 1% dispersion (weight to weight) of the soil sample in distilled water has been prepared. The microstructure of the flocks formed by the colloids has been imaged with the X-ray microscope. All images have been taken at the X-ray wavelength $\lambda = 2.4$ nm making use of the lowest possible absorption of X-rays in water within the water window. The figures 3a and 3b show the porous structure of flocks formed by the soil colloids in the agriculturally used soil; the figures 4a and 4b show a similar appearance of flocks from the forest soil. The strong absorbing, round particle in figure 4a is probably a drop of the contaminating lubricants.

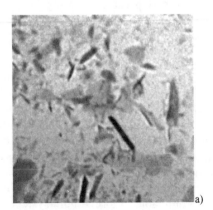 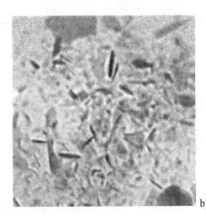

a) b)

FIGURE 3. Porous structure of flocks formed by soil colloids in an agriculturally used soil, the size of the images is 6 x 6 μm.

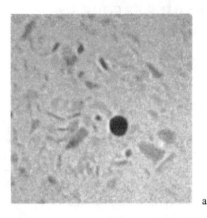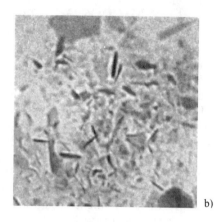

a)   b)

FIGURE 4. Porous structure of flocks formed by soil colloids in a forest soil, the size of the images is 6 x 6 μm.

These contaminated soils have been treated in a thermal clean up procedure with different temperatures, 200°C, 400°C and 600°C. Lubricants are very difficult to remove with other clean-up treatments as biological with bacteria or chemical with detergents. After the thermal treatment a 1% dispersion of this substrate in distilled water has been prepared, one for each temperature and soil. The flocks formed by the colloidal particles have been imaged with the X-ray microscope. Figures 5a, 5b, and 5c show particles from the agriculturally used soil, figures 5d, 5e, and 5f from the forest soil. In all images a drastic change of the structure formed by the colloidal particles is apparent. The small particles visible in figures 3 and 4 do not appear in figure 5. Instead, large particles can be seen on the images, which overlap each other partly. These particles are probably thin stacks of clay particles. The regular shaped dense structures, which are also visible on the images, are edge-on views of the large clay stacks. Due to sheet-like structure of the clay particles the absorption is much higher edge-on than face-on.

The explanation for this difference in appearance is of course the thermal treatment. Organic substances have been burnt away from the soils. These substances very often serve as glue between colloidal particles building up an open and porous structure in untreated soils. The lack of organics in the thermally treated samples leads to a collapse of these structures. This collapse is supported by the complete removal of water. Even in a naturally dry sample there is still some water in the sub-μm pores and between the sheet-like colloids. Introducing the very high temperatures of the thermal clean-up treatment means to remove this water too. Therefore, stacks of colloids appear which are very dense packed. They are only able to form a pore system with larger radii when put again into an aqueous environment. This drastic change of the pore system influences the soil parameters mentioned above. The inner surface of this soil is reduced drastically, so the storage capacity for substances like nutrients is reduced. In addition, the water storage capacity is reduced; vice versa the ability of water to flow through these pores is increased.

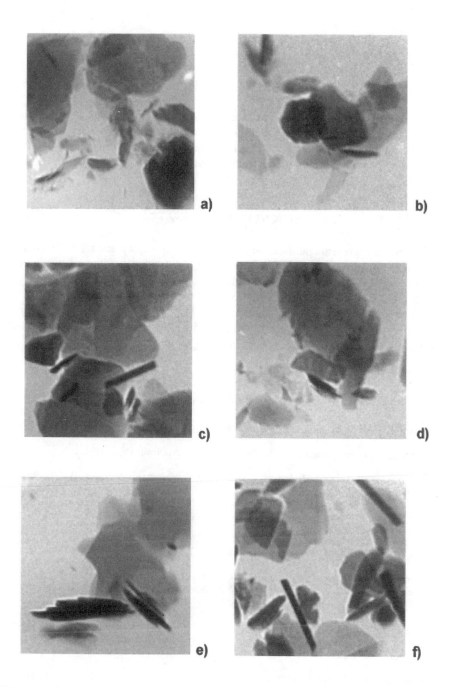

FIGURE 5: X-ray microscopy images of thermally treated soils. The agriculturally used soil is imaged in figures 5 a – 5 c, the forest soil in figures 5 d – 5 f. The size of the images is 6 x 6 μm.

CONCLUSION

The results of these X-ray microscopy studies show, that thermal clean-up techniques change the shape of colloidal structures in soils very strongly, thus influencing important soil parameters. They show too, that X-ray microscopy is a very useful tool for studying colloidal systems in soils.

ACKNOWLEDGEMENTS

This paper represents publication no. 98 of the Priority Program 546 "Geochemical processes with long-term effects in anthropogenically-affected seepage- and groundwater". Financial support was provided by Deutsche Forschungsgemeinschaft. In addition, this work was supported by the Deutsche Bundesstiftung Umwelt under contract number 03149, and the Federal Ministry of Education, Science and Technology, BMBF, under contract number 05 644 MAG. We would like to thank the staff of BESSY for providing excellent working conditions.

REFERENCES

1. Scheffer, F., and Schachtschabel, P., *Lehrbuch der Bodenkunde*, Stuttgart: Enke, 1992.

2. Sparks, D.L., *Environmental Soil Chemistry,* San Diego: Academic Press, 1995.

3. Thieme, J., Guttmann, P., Niemeyer, J., Schneider, G., David, C., Niemann, B., Rudolph, D., and Schmahl, G., *Nachr Chem Tech Lab* **40**, 562-563 (1992)

4. Thieme, J., Schmahl, G., Rudolph, D., and Umbach, E. (Eds.), *X-Ray Microscopy and Spectromicroscopy*, Heidelberg: Springer, 1998.

5. Thieme, J., Niemeyer, J., *Progr Colloid Polym Sci* **111**, 193-201 (1998)

6. Niemeyer, J. and Thieme, J., "Reactions of Clay Particles in Aqueous Dispersions Studied by X-Ray Microscopy", in *Synchrotron X-ray Methods in Clay Science,* edited by R. B. Schulze, J.W. Stucki, and P.M. Bertsch, Boulder: The Clay Mineral Society, 1999, pp. 208-219.

7. Neuhäusler, U., Abend, S., Ziesmer, S. Schulze, D., Stott, D., Jones, K., Feng, H., Jacobsen, C., Lagaly, G., "Soft X-ray spectromicroscopy on hydrated colloidal and environmental science samples", this volume

8. Weber, H., and Neumaier, H. (Eds.), *Altlasten – Erkennen, Bewerten, Sanieren*, Heidelberg: Springer, 1993.

9. Niemann, B., Schneider, G., Guttmann, P., Rudolph, D., and Schmahl, G., "The new Göttingen X-Ray Microscope with Object Holder in Air for Wet Specimens", in *X-Ray Microscopy IV*, edited by V.V. Aristov and A.I. Erko, Chernogolovka: Bogorodski Pechatnik, 1994.

Probing Chemistry Within the Membrane Structure of Wood with Soft X-ray Spectral Microscopy

George D. Cody

Geophysical Laboratory, Carnegie Institution of Washington,5251 Broad Branch Rd., Nw Washington, DC 20015

Abstract. Scanning Transmission Soft X-ray spectral microscopy on Carbon's 1s absorption edge reveals the distribution of structural biopolymers within cell membrane regions of modern cedar and oak. Cellulose is extremely susceptible to beam damage. Spectroscopic studies of beam damage reveals that the chemical changes resulting from secondary electron impact may be highly selective and is consistent with hydroxyl eliminations and structural rearrangement of pyranose rings in alpha-cellulose to hydroxyl substituted γ pyrones. A study of acetylated cellulose demonstrates significantly different chemistry; principally massive decarboxylation. Defocusing the beam to a 2 µm spot size allows for the acquisition of "pristine" cellulose spectra. Spectral deconvolution is used to assess the distribution of lignin and cellulose in the different regions of the cell membrane. Using the intensity of the hydroxylated aromatic carbons 1s-π^* transition, the ratio of coniferyl and syringyl based lignin within the middle lamellae and secondary cell wall of oak, an angiosperm can be determined.

INTRODUCTION

Advances in X-ray focusing techniques coupled with brilliant, synchrotron derived, soft X-ray sources now make it possible to interrogate bioorganic structure for functional group distributions at spatial resolutions approaching 50 nm (1). Scanning Transmission (Soft) X-ray Microscopy (STXM) instrumentation has recently been used to explore the geochemical evolution of organic matter entrained in sedimentary rocks (2). In this report, micro C-XANES is used to characterize the organic chemistry within structurally differentiated regions of the membranes of tracheid cells in recent woods, the possible extension of these methods to the structure of ancient woods is straightforward.

Tracheid cells in the xylem tissue of mature trees are constructed with an unusually high degree of chemical and structural differentiation. This differentiation is defined both by variations in the distribution of specific structural biopolymers such as lignin, hemicellulose, and cellulose, as well as variations in the structure of polysaccharides microfibrils (3). The high level of differentiation reflects the multiple functions required of the tracheid cell wall, e.g. conduction, storage, and support.

CP507, *X-Ray Microscopy: Proceedings of the Sixth International Conference,*
edited by W. Meyer-Ilse, T. Warwick, and D. Attwood
© 2000 American Institute of Physics 1-56396-926-2/00/$17.00

Chemical differentiation within the tracheid cell wall of modern cedar (Fig. 1 left) is as follows; first, there is the middle lamellae, a thin, lignin rich region that defines the border of each individual cell; adjacent to this is the thicker and relatively lignin poor S1 region of the secondary cell wall; This is followed by a thick lignin rich layer, and, innermost, a structurally complex, lignin depleted layer. Note that this last layer exhibits branched structures similar to helical cavities commonly observed (3) using transmission electron microscopy. These last two layers constitute the S2 region of the secondary cell wall.

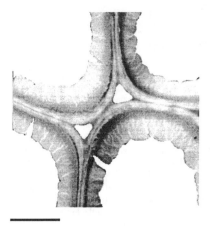

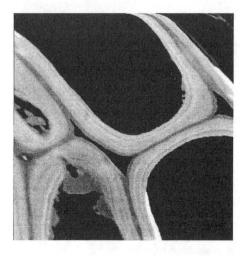

4 µm

FIGURE 1 (left) X-ray image of Red cedar (a gynmosperm), (right) X-ray Image of Oak (an angiosperm). These images were acquired with monochromator tuned to 285.1 eV where absorption contrast is based on intensity of aromatic carbon 1s-π* transition. As only the biopolymer lignin contains aromatic carbon both images are "maps" of the distribution of lignin. Note that the image of Oak (right) the tracheid cells are filled with epoxy, an aromatic polymer, hence the strong absorption. Considerable fine structure is evident within both cell membranes.

The oak cell membrane structure (Figure 1 right) exhibits distinctly different membrane texture than that of cedar, although the general differentiation, i.e. lignin rich middle lamellae and polysaccharide rich secondary cell wall, persists. The biochemistry of these two woods is also significantly different. Cedar is a member of the subdivision gymnosperms, derivative of the most ancient vascular plants whose origins are traced back nearly 400 million years ago. Oak is a member of the angiosperm subdivision, the most recent member of the tracheophyta division. The earliest angiosperms appeared ca. 75 million years ago. At the biochemical level, these two divisions differ most strikingly in the chemistry of their lignin. Gymnosperms are composed of lignin predominantly composed of the monomers derived from coniferyl alcohol, a dihydroxy aromatic molecule. Lignin in angiosperms is composed of a mixture of monomers derived from coniferyl alcohol and monomers derived from syringyl alcohol, a trihydroxy aromatic molecule.

Once angiosperms were established, they rapidly out competed gymnosperms and are currently the dominant vascular plant species. Clues to the success of angiosperms might be found in the function of lignin. While it is observed that lignin plays an important structural role in supporting the mass of large vascular plants, it may also serve another role of protecting the plant from attack by fungal microorganisms. Given the apparent dual functionality of lignin, i.e. both as a structural element and an anti-microbial protector, it appears possible that the lignin chemistry in angiosperms might be variable throughout the cell wall. In other words, lignin in the middle lamellae might serve a structural function, whereas lignin in the secondary cell wall serves a predominantly protective function. Early work by Goring using UV microscopy suggested that the secondary cell wall might be enriched in syringyl based lignin [see references in (3)]. The strength of their conclusions, however, were undermined by the non-specific nature of the UV-visible spectra. C-XANES on the other hand is extremely specific in terms of quantitatively assessing the degree of aromatic hydroxylation. Using the micro-spectroscopic capabilities of the scanning transmission X-ray microscope, it should be possible to ascertain precisely whether the lignin chemistry in angiosperms varies across the cell membrane.

The Problem Of Soft X-ray Beam Damage On The Quantitative Analysis of Tracheid Cell Membranes using C-XANES

Although micro C-XANES clearly has enormous potential for revealing chemical differentiation within the membrane structure of plants, the predominant biopolymer cellulose may be highly susceptible to beam damage. Basically, photoemitted electrons with energies above ~ 10 eV are capable of inducing fragmentation reactions. Polyhydroxylated molecules, such as cellulose and other polysaccharides will be particularly susceptible to loss of hydroxyls and subsequent structural rearrangement. Lignin, by virtue of its aromatic character is expected to be much more resistant to beam damage.

It turns out the beam damage issues may easily be overlooked for the following reasons. In order to obtain the best contrast between lignin and polysaccharides, imaging is usually done at ~285 eV where the 1s-π* transitions of aromatic carbon clearly define lignin rich and poor regions within the membrane, e.g. Figure 1. However, imaging at 285 eV will not reveal beam damage if there are no losses or gains of the organic functionality responsible for absorption.

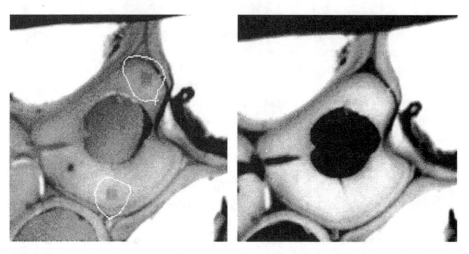

FIGURE 2 (left) X-ray image of Oak cell membrane acquired at 286.6 eV. The white circles highlight small rectangular1 μm x 1 μm regions that were "burned" in by irradiation (500 ms dwell) at 289.5 eV (left, upper) and 285 eV (left lower). (right) the same region imaged at 285.5 eV after "burning" of regions.

As an example of this, two regions (1.0 x 1.0 μm^2) were "burned" into the secondary cell wall of oak tracheid cell membrane. In one case, the monochromator was tuned to 289.5 eV (corresponding to the principal absorption band of polysaccharides); the other case the monochromator tuned to 285.1 eV (corresponding to one of the principal absorption bands of lignin). The damaged regions are clearly seen in Figure 2(left). The dark spot to the left of the lower region corresponds to a region where repeated C-XANES spectra were acquired. In Figure 2(right) the same image was then acquired at 285.5 eV. Note, that at this energy no evidence of beam damage is observed.

Sequential C-XANES spectra from the same region of oak's secondary cell wall reveals the chemistry associated with the beam damage (Figure 3, left). The losses at 289.5 eV correspond to a loss of secondary alcohol functionality in the polysaccharides. The gain in intensity at 286.6 eV corresponds to an enol, i.e. an oxygen substituted olefin. Note that no protonated or alkylated olefin is formed. This is why the intensity at ~ 285 eV is invariant with beam damage (e.g. Figure 2 (right)). Potential fragmentation paths are shown in Figure 3 (right). Both the cyclic ketone and the pyrone suggest that an absorption in addition to the enol at 286.6 eV is required. Examination of Figure 3 (left) reveals growth of spectral intensity at ~ 288.5 eV, possibly consistent with the ketone, although this energy is closer to that expected for a carboxylic acid group. The growth of absorption at 286.6 eV is a problem if the degree of ring hydroxylation is to be determined as hydroxylated aromatic carbon has its absorption band at 287.2 eV [see Figure 3 (left)] and will overlap with the enol.

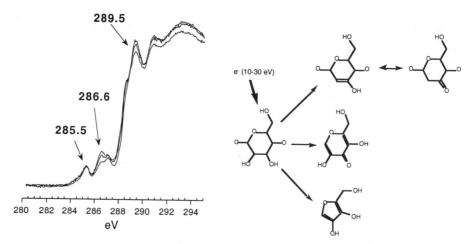

FIGURE 3 (left) Sequential micro C-XANES of Oak secondary cell wall. Note that intensity at 289.5 eV (cellulose secondary alcohol) decreases while intensity at 286.6 (enol) increases: there is also an increase in intensity at 288.5 eV. Intensity at 285.5 eV does not change indication that non-oxygen substituted olefins are formed. (Right) Proposed reactions of cellulose upon electron impact from 10 to 30 eV electrons, the upper pathway involves elimination of water followed by tautomerization to the cyclic ketone; the middle pathway involves cleavage of the acetal carbon and loss of hydrogen, the lower pathway involves loss of CO and hydrogen.

A systematic study varying dwell times, exit slit widths, and beam spot size was performed to find the best method to obtain C-XANES spectra. It was found that defocusing the X-ray beam to 2μm and keeping the dwell below 50 ms effectively minimizes beam damage. Having to increase the beam spot size is a bit of a sacrifice. In the case of the oak secondary wall, this does not present an significant problems, however, in studies of other membrane structures, one should not expect cell walls any where near this thickness. Two possible solutions exist; either micro C-XANES spectra should be obtained at cryogenic temperatures or sequential spectra should be acquired. With sequential spectra it should be possible to deconvolute the spectra and extrapolate back to t=0 exposure time. The chemistry evident in Figure 3 appears straightforward and this method should work.

In the case of the Oak secondary cell wall, defocusing was a possibility. Micro C-XANES spectra of both the secondary cell wall and the middle lamellae are presented in Figure 4. The most obvious differences in the chemistry between these two regions is in the lignin to polysaccharide ratio. The middle lamellae is nearly 100% lignin, where as the secondary cell wall is predominantly composed of polysaccharide (note the pronounce absorption at 289.5 eV indicative of the presence of polysaccharides). The ratios of the intensities of the aromatic and hydroxylated aromatic 1s-π* transitions (highlighted) can be used to determine the fraction of coniferyl and syringyl monomers within the lignin. Quantitation is achieved through comparison of model compound studies of phenol, catechol, and gallol (unpublished results).It is evident that lignin within the secondary cell wall contains more of the syringyl monomer, confirming and quantifying the earlier conclusions of Goring (3). Now that it is established that variations in lignin chemistry can be distinguished spatially, a number of applications are in progress. Perhaps the most noteworthy is an investigation into potential disproportional metabolism of lignin in the secondary cell wall by fungal micro-organisms.

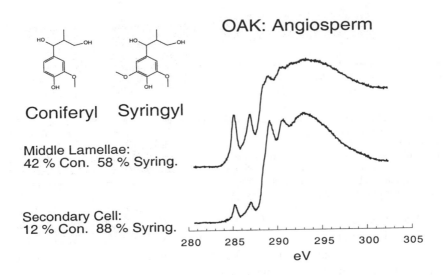

OAK: Angiosperm

Coniferyl Syringyl

Middle Lamellae:
42 % Con. 58 % Syring.

Secondary Cell:
12 % Con. 88 % Syring.

280   285   290   295   300   305
eV

ACKNOWLEDGMENTS

Help, encouragement, and advice from Chris Jacobsen, Janos Kirz, and Susan Wirick is gratefully acknowledged. All experiments were performed at the X1A beam line, a DOE supported facility.

REFERENCES

1. C. Jacobsen et al. *Optics Communications* **86**, 351--364 (1991).

2. G. D. Cody et al., *Energy Fuels* **9**, 75 (1995); G. D. Cody et. al. *Int. Journ. Coal Geol.* **32**, 69 (1996); G. D. Cody et al. *Org. Geochem* **28**, 441 (1998).

3. E. Sjöström, *Wood Chemistry: Fundamentals and Applications* (Academic Press, New York, 1993).

Association Of Particles And Structures In The Presence Of Organic Matter

C. Schmidt[1], J. Thieme[1], U. Neuhäusler[1,2], U. Schulte-Ebbert[3], G. Abbt-Braun[4], C. Specht[4], C. Jacobsen[2]

[1] *Institut für Röntgenphysik, Georg-August-Universität Göttingen, Geiststr. 11, D-37073 Göttingen*
[2] *Department of Physics and Astronomy, SUNY Stony Brook, Stony Brook NY 11794-3800, USA*
[3] *Institut für Wasserforschung GmbH Dortmund, Zum Kellerbach 46, D-58239 Schwerte*
[4] *Engler-Bunte Institut , Universität Karlsruhe, Engler-Bunte-Ring 1, D-76128 Karlsruhe*

Abstract. The influence of organic matter, the humic substances, on the precipitation behavior of the iron in groundwater aquifers was studied. Size and structure of these associations are determined using fractal geometry. The humic substances were characterized spectroscopically. For differentiation of organic and inorganic substances, the contrast at the carbon K-absorption edge was used. Spectra were taken at different spots of the samples to investigate, whether there are differences in the spectra with and without iron.

INTRODUCTION

The size of most of the particles which can be found in aquifers are in the colloidal range that means between 1nm and 1µm. With soft X-ray microscopy at a resolution down to 30 nm we are able to investigate those particles in their natural aquatic state and under atmospheric pressure. In the energy range of the water window, we can use the natural contrast between the surrounding water and these particles for imaging.

A large fraction of organic matter in aquifers are the so-called "humic substances". Most of the organic Carbon on earth is bonded in those substances. They are anionic polyelectrolytes and can act as complexing agent for metals. In this way, they serve as carriers of substances in natural water and play an important role in the environment [1]. In this paper we focus on the interaction of iron with humic substances first on precipitation processes in groundwater and then on the usage of iron as flocculation agent for humics.

ANAEROBIC GROUNDWATER WITH ORGANIC MATTER

In colloidal systems, chemical and physical reactions are enhanced due to the large surface of the colloidal particle. When studying properties of natural colloidal systems, one major issue is the question, whether the shape of the associations formed can be described morphometrically. Fractal geometry can be used for describing the size and

CP507, *X-Ray Microscopy: Proceedings of the Sixth International Conference,*
edited by W. Meyer-Ilse, T. Warwick, and D. Attwood
© 2000 American Institute of Physics 1-56396-926-2/00/$17.00

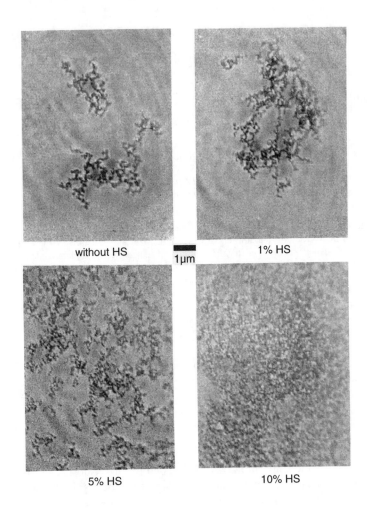

without HS   1µm   1% HS

5% HS   10% HS

FIGURE 1. When anaerobic groundwater gets in contact with oxygen, small single aggregates of iron containing particles occur . The formation of larger aggregates begins, influenced by the organic matter. When adding different amounts of a humic substance (1 to 10 % HS) to the dispersion, the associations grow bigger and denser with increasing organic content. With a high content (10% HS) of humic substances also networks appear.

In groundwater, iron is an abundant cation. In anaerobic groundwater aquifers it is present in a reduced form as a bivalent cation. When this groundwater gets in contact with oxygen, the bivalent iron cation is oxidized to a trivalent state; insoluble iron compounds are formed in consequence. The images in figure 1, taken with the Göttingen transmission X-ray microscope at BESSY [4], show aggregates of such particles in originally anaerobic groundwater after oxidization. First, small single aggregates of iron containing particles occur. For studying the formation of these structures in the presence of organic matter, different amounts of a humic acid were added to the dispersion. Larger aggregates and also networks appear due to the amount of the organic matter. The iron particles may act as nuclei for additional aggregation processes. The determination of the fractal dimension with box-counting shows that the with increasing organic content the fractal dimension increases significantly from $D_f = 1,60$ without organic matter to $D_f = 1,75$ for 10 % organic matter that means that the associations not just grow bigger but also become denser. That means that available surface becomes smaller, which in an important parameter for the transportation of substances.

Spectroscopical Characterization Of Humic Substances

A humic substance extracted from a soil near Göttingen, has been analyzed spectroscopically in a dry state as well as in an aqueous dispersion in comparison to a dry sample of a synthetic fulvic acid. The NEXFAS spectra, taken at the NSLS with the Stony Brook STXM at the carbon-K-absorption edge, are shown in figure 2 [5],[6]. Differences between the samples can be clearly seen.

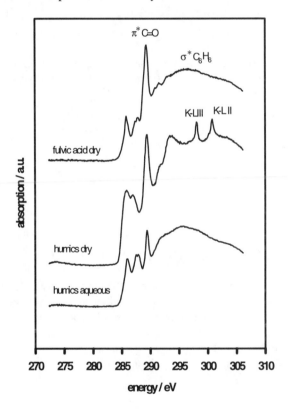

FIGURE 2. NEXAFS-spectra of a synthetic fulvic acid, dry, and a colloidal humic substance extracted from a soil, a calcareous aquic vermudoll, dry and in aqueous dispersion.

Both, the colloidal humic substance and the fulvic acid are substances of heterogeneous chemical composition. Therefore, uncertainties may arise in this assignment of the resonances (tab. 1). But some resonances can be certainly assigned. The resonance at 288 eV is the π * resonance of the CO-double bond of the carboxyl group [7]. The fact that the fulvic acid has more acidic groups than the humic acids, can be seen in the NEXAFS spectra, the resonance of the CO-double bond of the carboxyl group is more prominent in the spectra of the fulvic acid than in the spectra of the humic acid. So, there is a possibility to distinguish between these two substances spectroscopically. The broad peak above 290 eV can be assigned to resonances of the aromatic groups of the molecules. These resonances are higher for the humic acid, which have a more aromatic character than the fulvic acid.

Also differences between the spectra of the aqueous and dry sample of the humic acid can be observed. The absorption peaks 298.1 and 300.8 eV can be assigned to the L_{III} - and L_{II} - edges of potassium. The ionic bonding of the potassium to the humic substance leads to shifts in the positions of these peaks compared to edge energies tabulated in the Henke-data [8]. These peaks do not occur in the spectra of the hydrated humic acid which may be caused by dilution. The potassium-cations were solved in the aqueous suspension and diluted, so they can not be detected anymore.

TABLE 1. Energies of the absorption peaks (fig. 2) and their assignment

Energy/ eV	Resonances and absorption edges
285,9	$\pi_1^* C_6H_6$
287,0	3s CH_3, CH_2
287,9	3s CH_3, CH_2
288,4	$\pi^* C=O$ $\pi_2^* C_6H_6$
292,0	$\sigma^* C-C$
293,9	$\sigma_1^* C_6H_6$
298,1	K L_{III} (294,6 eV)
300,8	K L_{II} (297,3 eV)

FLOCCULATION OF HUMIC SUBSTANCES WITH IRON

After the spectroscopic characterization of the humic substances, the flocculation of those substances with iron cations was studied. Iron chloride ($FeCl_3$) was added as flocculation agent in a dispersion of humic acid. This agent is used for removing organic matter in sewage. Through the addition of metal cations, the charge of humic substance is reduced and the solubility decreases. The formation of flocs begins. Figure 3 shows those flocs dried on a silicon nitride window.

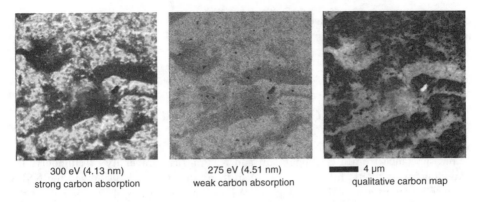

| 300 eV (4.13 nm) | 275 eV (4.51 nm) | ▬ 4 µm |
| strong carbon absorption | weak carbon absorption | qualitative carbon map |

Figure 3. STXM images of a dry sample of a humic acid which is flocculated with iron taken near the *K* absorption edge of carbon. The carbon map (*right*) is calculated by dividing the left image, taken at an energy with strong absorption for carbon, by the center image, taken at weak absorption. The greyscale in the right image indicates the carbon amount.

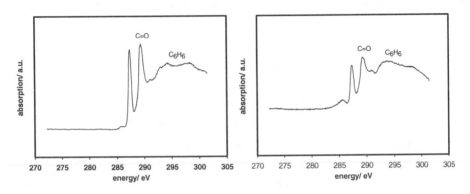

FIGURE 4. NEXAFS-spectra of a humic acid without (*left*) and with iron (*right*), taken at the C-absorption edge.

Taking images above and below the carbon K-absorption edge with the Stony Brook STXM at the NSLS lets one observe the distribution of carbon in the sample (right image figure 3). The dark dots in the center image (taken at an energy for weak absorption for carbon) are iron particles, since iron is quite absorptive at this energy.

For the spectroscopic characterization of the flocculation process, NEXAFS-spectra of this humic acid with and without iron were taken (figure 4). There is a significant change between both spectra in the relations of the peak height of this resonance for the CO double bond and the aromatic peak. The height of the resonance peak of the CO double bond decreases compared to the broad aromatic peak above 292 eV. It is known, that the carboxyl ions play a prominent role in the complexation of cations by humic acids, but there are also some evidence in IR spectra that the oxygen of the CO double bond of carboxyl-group can be involved in the metal-ion binding [9]. This

317

could be the reason for the decrease of the resonance of the CO double bond at 289 eV in the spectra.

ACKNOWLEDGEMENTS

We would like to thank J. Niemeyer, University of Trier, for the preparation of the humic substance from a Göttingen soil and K. Pranzas, University of Hamburg, for providing the synthetic fulvic acid. We also thank the staff of BESSY and the NSLS for excellent working conditions and Sue Wirick from Stony Brook for the help at the beamline. This work represents publication no. 20 in the Priority Program 546 "Geochemical processes with long-term effects in anthropogenically-affected seepage- and groundwater". Financial support was provided by Deutsche Forschungsgemeinschaft within the Priority Program 546, the Deutsche Bundesstiftung Umwelt, under contract number 03149 and the Federal Ministry of Education, Science, Research and Technology, BMBF, under contract number 05 644 MAG and by a fellowship for Ph.D. research studies (U.N.) from German Academic Exchange Service (DAAD).

REFERENCES

1. Stevenson, F.J., Humus chemistry, John Wiley & Sons Inc, 1994

2. Ruprecht, A., Diploma Thesis, Georg-August-Universität Göttingen, 1998.

3. Thieme, J. and Niemeyer, J., *Geol. Rundsch.* **65**, 2503-2504 (1996)

4. Niemann, B., Schneider , G., Guttmann, P., Rudolph, D. and Schmahl, G. " The new Göttingen X-Ray microscope with object holder in air for wet specimens" in *X-Ray Microscopy IV*, edited by V.V. Aristov and A.I. Erko, Proceedings of the 4th International Conference, Chernogolovka, Russia 1994, pp. 20-24

5. Jacobsen, C. Williams, S., Anderson, E., Brown, M.T., Buckley, C.J., Kern, D., Kirz, J., Rivers, M., and Zang, X., *Optics Communications* **86**, 351-364 (1991)

6. Neuhäusler, U., Jacobsen, C., Schulze, D.G, Stott, D. and Abend, S, *J. Synchrotron Rad.,* in press (1999)

7. Ishii, I. and Hitchcook, A.P., J. *Electron spectroscopy and related Phenomena* **46**, 55-84 (1988)

8. Henke, B.L., Gullikson, E.M., Davis, J.C., *Atomic Data and Nuclear Data Tables* **54**, 181-342 (1993)

9. Piccolo A. and Stevenson F.J., *Geoderma*, **27**, 195 (1981)

Use Of The High-Energy X-ray Microprobe At The Advanced Photon Source To Investigate The Interactions Between Metals And Bacteria

K. M. Kemner,[1] B. Lai,[1] J. Maser,[1] M. A. Schneegurt,[2] Z. Cai,[1]
P. P. Ilinski,[1] C. F. Kulpa,[2] D. G. Legnini,[1] K. H. Nealson,[3] S. T. Pratt,[1]
W. Rodrigues,[1] M. Lee Tischler,[4] W. Yun[5]

1. Argonne National Laboratory, Argonne, IL 60439, USA
2. University of Notre Dame, Notre Dame, IN 46556, USA
3. Jet Propulsion Laboratory, Pasadena, CA 91109, USA
4. Benedictine University, Lisle, IL 60532, USA
5. Lawrence Berkeley National Laboratory, Berkeley, CA 94720, USA

Abstract. Understanding the fate of heavy-metal contaminants in the environment is of fundamental importance in the development and evaluation of effective remediation and sequestration strategies. Among the factors influencing the transport of these contaminants are their chemical speciation and the chemical and physical attributes of the surrounding medium. Bacteria and the extracellular material associated with them are thought to play a key role in determining a contaminant's speciation and thus its mobility in the environment. In addition, the microenvironment at and adjacent to actively metabolizing cell surfaces can be significantly different from the bulk environment. Thus, the spatial distribution and chemical speciation of contaminants and elements that are key to biological processes must be characterized at micron and submicron resolution in order to understand the microscopic physical, geological, chemical, and biological interfaces that determine a contaminant's macroscopic fate. Hard X-ray microimaging is a powerful technique for the element-specific investigation of complex environmental samples at the needed micron and submicron resolution. An important advantage of this technique results from the large penetration depth of hard X-rays in water. This advantage minimizes the requirements for sample preparation and allows the detailed study of hydrated samples. This paper presents results of studies of the spatial distribution of naturally occurring metals and a heavy-metal contaminant (Cr) in and near hydrated bacteria (*Pseudomonas fluorescens*) in the early stages of biofilm development, performed at the Advanced Photon Source Sector 2 X-ray microscopy beamline.

ENVIRONMENTAL RESEARCH

Chemical contamination of soil and groundwater is a universal problem of immense complexity and great global concern. Sources of contamination include past and present agricultural and industrial activities, operations at national defense sites, and mining and manufacturing processes. Assessment of thousands of hazardous waste sites in the United States alone (including over 1,200 on the National Priority List) has identified the presence of an array of toxic substances. These include heavy metals (such as Pb, Cr, As, Zn, Cu, Cd, Ba, Ni, and Hg), radionuclides (including U, Pu, Sr, Cs, Co, and Tc), and potentially hazardous anions such as arsenate, chromate, and selenate.[1] The

CP507, *X-Ray Microscopy: Proceedings of the Sixth International Conference,*
edited by W. Meyer-Ilse, T. Warwick, and D. Attwood
© 2000 American Institute of Physics 1-56396-926-2/00/$17.00

restoration of soils and groundwaters that have been contaminated by combinations of these stable and radioactive substances (e.g., mixed wastes) presents significant scientific and engineering challenges, because the interactions among the contaminants are unknown.

THE NEED FOR SUBMICRON HIGH-ENERGY X-RAY MICROPROBES IN ENVIRONMENTAL AND MICROBIOLOGY RESEARCH

A current focal point of molecular environmental science involves the pathways, products, and kinetics of chemical reactions of contaminant species with inorganic and organic compounds, plants, and organisms in the environment.[1] These reactions often occur at aqueous solution-solid interfaces and can have many different results. The contaminant can be precipitated from the solution to the solid interface, transformed into a different species, incorporated into a solid phase, or released from the solid surface into the solution. Such interfacial reactions play a very important role in the transport and dispersal of toxic species in soils and natural waters. Therefore, discovering what is occurring at these interfacial surfaces is key to understanding the bioavailability of many contaminants. Despite this importance, these surfaces and their associated chemical reactions are not well understood. Consequently, little is known about the mechanisms by which biota, in particular microorganisms, determine the speciation, forms, reaction rates, and distribution of contaminants in soils and groundwater.

The heterogeneity of most environmental samples makes their study very difficult. Because environmental samples are almost always hydrated, high-energy X-rays having the ability to penetrate water are very useful. In addition, it can be valuable to probe both sides of the interfaces in these heterogenous samples to elucidate transformations that result in the movement of the contaminant across the interface. Thus, the smallest possible probe is needed to analyze the homogeneous region on either side of the interface selectively. These requirements make the use of micron and submicron X-ray beams advantageous. One difficulty encountered in investigating the contaminant-microbe interface is sample heterogeneity. For instance, in such a system, the metal contaminant may be bound in a variety of ways: (1) in solution, (2) to extracellular material, (3) within cell membrane regions, or (4) within the bacterium. To study the spatial distribution and chemical speciation of a contaminant metal at the microbe-metal interface and thus to elucidate the interactions occurring at this interface, the dimension of the X-ray probe must allow the vast majority of the X-rays to be positioned at the contaminant-microbe-mineral interface. The size of most bacteria is approximately 1 μm. Therefore, to investigate the speciation and spatial distribution of elements associated with bacteria, the dimensions of the X-ray probe must be smaller than 1 μm. Finally, although soft X-ray microprobes have considerable utility for studying biological samples,[2] hard X-ray microprobes provide improved fluorescence yields, better penetration of hydrated samples, and access to the K edges of third-row and heavier elements. Many of these elements are important nutrients, micronutrients, and environmental contaminants.

Third-generation X-ray sources such as the Advanced Photon Source (APS), where our experiments were performed, provide an increase in brilliance of approximately three orders of magnitude compared to second-generation synchrotron X-ray sources. In addition, advances in microfabrication technologies have resulted in X-ray phase zone plates[3] with spatial resolution better than 0.20 µm and focusing efficiency better than 33%. The combination of the increased brilliance of X-ray beams provided by the APS and improved zone plate fabrication technology provides unique capabilities in X-ray microscopy and spectromicroscopy.

THE USE OF X-RAY MICROPROBES FOR INVESTIGATING THE ROLE OF BACTERIA AND THEIR ORGANIC EXUDATES IN CONTAMINANT TRANSPORT

The objectives of our studies are (1) to determine the spatial distribution and chemical speciation of metals near bacteria-geosurface interfaces and (2) to use this information to identify the interactions occurring near these interfaces among the metals, mineral surfaces, and bacterially produced extracellular materials under a variety of conditions. We have used hard X-ray phase zone plates to investigate the spatial distribution of 3d elements in a single hydrated *Pseudomonas fluorescens* bacterium adhered to a Kapton film. The zone plate used in these microscopy experiments produced a focused beam of cross section 0.15 µm^2 and had an effective focal length of 12.5 cm at 10.0 keV. The samples were mounted on a computer-controlled XYZ piezo stage at 10 degrees to the incident beam, thus negligibly affecting the X-ray footprint on the sample in the horizontal dimension. The intensity of the fluorescence radiation from the sample was monitored by a single-element solid-state detector that enables efficient detection of fluorescent X-rays with energies greater than 1.5 keV. The elements were mapped by scanning this sample in 0.15-µm steps through the focused monochromatic X-ray beam and integrating the selected Kα fluorescence for 5 sec/pt. The total data collection time was approximately 6 hours.

Figure 1 shows results of the X-ray microprobe measurements, qualitatively indicating the spatial distributions of Cr, K, and Ca in and near a hydrated *Pseudomonas fluorescens* bacterium, adhered to a Kapton film at ambient temperature, that was exposed for 6 hours to Cr 1000 ppm in solution. Observation of these images indicates that monitoring the spatial distribution of the K and Ca Kα fluorescent radiation coming from the sample enabled identification of the rod-shaped *Pseudomonas fluorescens*, as well as the extracellular exudes associated with it. In addition, comparison of the distribution of Cr with that of K or Ca indicates that the majority of the Cr in this sample was associated extracellularly. Because most of the Cr remained outside the cell, the Cr(VI) was probably not actively metabolized. Finally, although these results demonstrate the utility of imaging hydrated bacteria at ambient temperature, a cryostat might be required to quick-freeze the samples in future spectromicroscopy studies in order to reduce the effects of radiation damage.

SUMMARY

We have demonstrated the utility of X-ray microbeams, particularly those produced by hard X-ray phase zone plates, for investigating biological and environmental systems. Specifically, we have illustrated the use of submicron hard X-ray beams (0.15 µm) for determining the spatial distribution of metals in a hydrated bacterium that was exposed to Cr 1000 ppm in solution for six hours. The further development of these techniques for such applications promises to provide unique opportunities in the field of microbiology and environmental research.

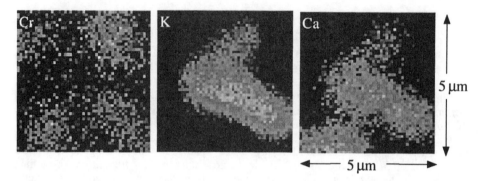

FIGURE 1. Elemental maps of a hydrated *Pseudomonas fluorescens* bacterium treated with Cr(VI) solution. See text for further details.

ACKNOWLEDGMENTS

This work was supported by the U. S. Department of Energy, Office of Science, Office of Basic Energy Sciences and Office of Biological and Environmental Research (NABIR Program), as well as internal Argonne National Laboratory (LDRD) funds. Additional support was from the Center for Environmental Science and Technology, University of Notre Dame. Use of the Advanced Photon Source was supported by the U.S. Department of Energy, Office of Science, Office of Basic Energy Sciences, under Contract No. W-31-109-Eng-38. We thank K. W. Germino and M. A. Mundo for assistance with image analysis and sample preparation during this work.

REFERENCES

1. *Molecular Environmental Science: Speciation, Reactivity, and Mobility of Environmental Contaminants*, Report of DOE Molecular Environmental Science Workshop, Airlie Center, Virginia, July 5-8, 1995.

2. Kirz, J., Jacobsen, C., and Howells, M., *Quart. Rev. Biophys.* **28**, 33-130 (1995).

3. Lai, B., Yun, W., Legnini, D., Xiao, Y., Chrzas, J., Viccaro, P. J., White, V., Bajikar, S., Denton, D., Cerrina, F., Difabrizio, E., Gentili, M., Grella, L., and Baciocchi, M., *App. Phys. Lett.* **61**, 1877-1879 (1992).

Soft X-ray Spectromicroscopy On Hydrated Colloidal And Environmental Science Samples

U. Neuhäusler [1,2], S. Abend [3], S. Ziesmer [3], D. Schulze [4], D. Stott [5], K. Jones [6], H. Feng [6], C. Jacobsen [1], G. Lagaly [3]

[1] *Department of Physics and Astronomy, State University of New York, Stony Brook, NY 11794-3800, U.S.A.,* [2] *Institut für Röntgenphysik, Universität Göttingen, Geiststraße 11, 37073 Göttingen, Germany,* [3] *Institut für Anorganische Chemie, Universität Kiel, Olshausenstraße 40-60, 24098 Kiel, Germany,* [4] *Agronomy Department, Purdue University, West Lafayette, IN 47907-1150, U.S.A.,* [5] *U.S. Department of Agriculture, Agricultural Research Service, West Lafayette, IN 47907-1196, U.S.A.,* [6] *Department of Applied Science, Brookhaven National Laboratory, Upton, NY 11973-5000, U.S.A.*

Abstract. Various aqueous colloidal systems (solid-stabilized oil/water emulsions, clay mineral systems with organics and contaminated estuarine sediment) have been studied with a wet specimen chamber based on silicon nitride windows designed to be used with the Stony Brook scanning transmission X-ray microscope (STXM) [3] at X-1A beamline [8] at the NSLS. The samples were imaged with sub-100 nm spatial resolution at photon energies within the 'water window', employing carbon-K- and calcium-L-absorption edge contrast.

INTRODUCTION

Soft X-ray microscopes offer especially favorable contrast mechanisms for studying colloidal systems. Unlike in electron microscopy, samples can be easily examined in transmission in a hydrated state at atmospheric pressure without any pretreatment and with a currently approximately tenfold higher spatial resolution than achievable in visible light microscopy. The term 'colloidal' covers systems with particle sizes ranging between 1 nm and 1 µm, so many systems that are not accessible by light microscopy techniques can be studied in an X-ray microscope. The 'water window' energy range between the K absorption edges of carbon and oxygen (284 – 543 eV) was employed for good intrinsic amplitude contrast in hydrated samples and the carbon-K and calcium-L absorption edges were used to highlight compounds containing these elements in the sample.

Figure 1 shows the wet specimen chamber used for the experiments [5]: The aqueous sample is sandwiched as a thin layer of 1 to 2 µm between two 100 nm thin silicon nitride windows, made of 200 µm thick silicon wafers. For carbon edge experiments, the thickness of the sample layer has to be in this range because the 1/e-absorption length for water is 2 µm at 280 eV while it is 10 µm for 540 eV. A 25 µm thin shim metal support allows the placement of the sample as close as 200 µm downstream of the order sorting aperture. Design and dimension of the wet cell are determined by restrictions in space and mass as imposed by the microscope setup. The wet cell is sealed by an o-ring so the sample can be kept stable for many hours.

CP507, *X-Ray Microscopy: Proceedings of the Sixth International Conference,*
edited by W. Meyer-Ilse, T. Warwick, and D. Attwood
© 2000 American Institute of Physics 1-56396-926-2/00/$17.00

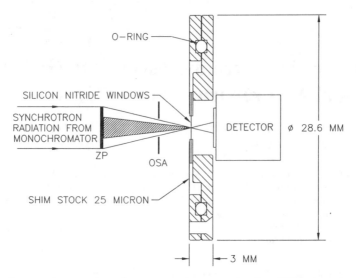

FIGURE 1. Cross section overview of the wet specimen chamber for STXM. Note that the wet cell dimensions are not to scale in respect to zone plate (ZP), order sorting aperture (OSA) and detector.

EMULSIONS STABILIZED BY SOLID COLLOIDS

Emulsions are systems of non-mixable liquids dispersed in each other. They are not stable over time unless they are stabilized. Colloidal particles such as clay minerals can act as stabilizers in emulsions [6] and can help to partially or completely substitute classical surfactant based emulsifiers, which is desirable from an environmental and toxicological point of view. The paraffin oil-water emulsions examined with STXM have been stabilized only by a clay mineral (sodium montmorillonite, Wyoming) and calcium/aluminum layered double hydroxide (LDH) without any additional surface active agents [1].

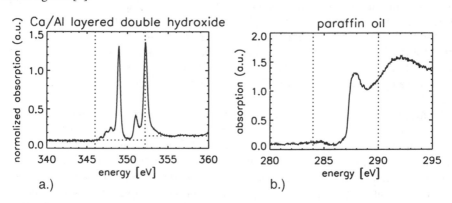

FIGURE 2. X-ray absorption spectra of (a) Ca/Al layered double hydroxide (suspended in water) near the calcium-L-absorption edge and (b) paraffin oil (bulk sample) near the carbon-K-absorption edge. The photon energies used in the images shown in Fig. 3 and 4 are marked with vertical dashed lines.

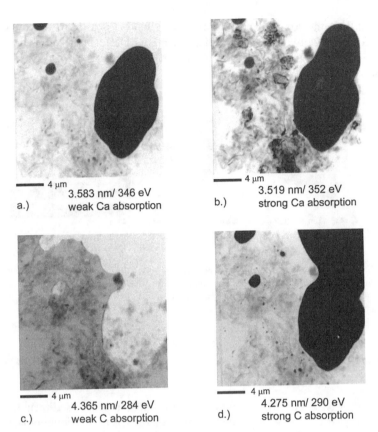

a.)   3.583 nm/ 346 eV
weak Ca absorption

b.)   3.519 nm/ 352 eV
strong Ca absorption

c.)   4.365 nm/ 284 eV
weak C absorption

d.)   4.275 nm/ 290 eV
strong C absorption

FIGURE 3. STXM images of an oil-water-emulsion, containing 10 % v/v paraffin oil and equal amounts of montmorillonite and layered double hydroxide (overall content of solids 1 % w/w), diluted 1:10 with water. The photon energies used for imaging are indicated in the spectra in figure 2. (a) Calcium very weakly absorbing, the clay mineral is visible. (b) Calcium strongly absorbing, LDH is visible. (c) Carbon weakly absorbing, the paraffin oil droplet is transparent. (d) Carbon strongly absorbing, the oil droplet appears black. The images were taken in the sequence a,b,d,c. During the experiment, the oil droplet remains at a fixed position, where heterocoagulate is caging the oil droplet (lower part of the image), but can disperse in regions without stabilizing envelope (upper part). The images have 300 x 300 pixels and 70 nm pixel size. The data acquisition time was 10 (20) min for images taken near the calcium (carbon) edge.

Due to the interaction between negatively charged clay mineral and positively layered double hydroxide, heterocoagulate networks form that surround and cage the oil droplets in the emulsion and prevent them from coalescing and forming separate oil/water phases. Solid stabilized emulsions have also been studied with a TXM [7], however without any spectral information. We mapped and distinguished the two liquid phases oil and water near the carbon absorption edge and the two solid compounds (sodium montmorillonite and Ca/Al-LDH) near the calcium absorption edge and supported the theoretical model of this stabilization process [4]. While the heterocoagulate network extends rather far away from the oil-water interface for the emulsion shown in figure 3, the situation is different for the emulsion shown in figure 4: Only one solid stabilizer (namely LDH) is used and the colloids do not form a heterocoagulate network but are located right in the oil-water-interface, thus stabilizing the emulsion.

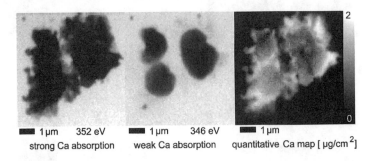

| 1 µm   352 eV | 1 µm   346 eV | 1 µm |
| strong Ca absorption | weak Ca absorption | quantitative Ca map [µg/cm^2] |

FIGURE 4. STXM images of an emulsion that contains only Ca/Al-LDH as a solid compound. The concentration of oil and solid stabilizer is the same as in for the emulsion in figure 3. For high energy resolution monochromators and low calcium concentration, it is possible to use resonances (here at 352 eV) for quantitative mapping [2] (right image).

CARBON CONTAMINANTS IN NY HARBOR SEDIMENT

In the port of New York/New Jersey, 4 million cubic meters of sediment have to be dredged each year from navigational channels. A large fraction is contaminated with PCBs, PAHs, pesticides etc., so there is a need for remediating this material. Differential imaging across the carbon absorption edge shows carbon compounds in different forms (figure 5).

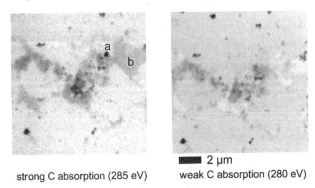

2 µm

strong C absorption (285 eV)   weak C absorption (280 eV)

FIGURE 5. Carbon mapping in a contaminated marine sediment from NY harbor (Newtown Creek), suspended in water. Carbonaceous material can be found in dense spots, most likely tar or biological material (a) and extended liquid phases, most likely oil (b). A carbon coating of sediment particles as it has been assumed, could not be shown on length scales of the available spatial resolution.

XANES spectra of carbon in the contaminated sediment were compared to spectra from sediment that underwent a cleaning procedure developed by Biogenesis Enterprises to find out about the effectiveness of the cleaning technique. However, there are several possible conclusions that can be drawn from that data. Unlike well defined laboratory systems from colloid chemistry, natural environmental science samples are usually very heterogeneous and the conclusions must be supplemented by information obtained by other techniques.

STUDIES OF HYDRATED CLAY-ORGANIC INTERACTIONS

Agronomy: Kaolinite – Polyacrylamide Aggregates

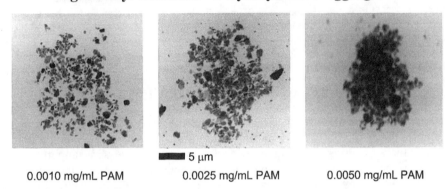

| 0.0010 mg/mL PAM | 0.0025 mg/mL PAM | 0.0050 mg/mL PAM |

FIGURE 6. Fully hydrated aggregates consisting of 2 - 0.2 µm kaolinite clay mineral particles at different concentrations of polyacrylamide (PAM), suspended in an aqueous solution of 0.01 M CaCl$_2$. The structures become more and more dense for increasing PAM concentration.

Polyacrylamide (PAM) is a flocculant that is added to irrigation water so as to reduce erosion by 94 % and increase irrigation water uptake by 15 % in loam soils. We visualized the flocculation of a KGa-1 kaolinite clay suspension at various concentrations of polyacrylamide (figure 6). The results are in good agreement with the concentrations needed for field applications. X-ray microscopy provides a very direct way of visualizing the effects of the flocculant.

Colloid Chemistry: Sodium Montmorillonite - TMA-PEG Aggregates

The structures of aggregates formed by negatively charged sodium montmorillonite clay mineral and polyethylenglycol (TMA-PEG) with a molecular weight of 20000 (containing positively charged trimethylammonium groups) were studied: Dense and on the length scale of the spatial resolution homogenous clay mineral-PEG aggregate networks form that are surrounded by an excess liquid phase of TMA-PEG (figure 7). Only in the aggregate-water interface, aggregate structures can be seen (right image in figure 7).

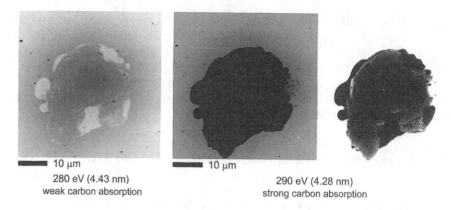

| 280 eV (4.43 nm) | 290 eV (4.28 nm) |
| weak carbon absorption | strong carbon absorption |

FIGURE 7. STXM images across the carbon-*K*-absorption edge of an aggregate formed by an aqueous dispersion of 0.1-% sodium montmorillonite and 10 g/l polyethylenglycol (TMA-PEG). The right image is the center image scaled to enhance the contrast for dark image regions.

ACKNOWLEDGEMENTS

I would like to thank my colleagues at Stony Brook, Sue Wirick for her help at the beamline, all co-authors of this paper for the fruitful collaboration and Prof. G. Schmahl for supporting my visit at Brookhaven. This work was supported by a fellowship for Ph.D. research studies (U.N.) from German Academic Exchange Service (DAAD), by a travel grant from Boehringer Ingelheim Fonds for basic medical research (S.A.) and by the Office of Biological and Environmental Research U.S. DoE under contract DE-FG02-89ER60858.

REFERENCES

1. Abend, S., Bonnke, N., Gutschner, U., and Lagaly, G., *Colloid Polym. Sci.* **276**, 730-737 (1998)

2. Buckley, C.J., Bellamy, S.J., Zhang, X., Dermody, G., and Hulbert, S., *Rev. Sci. Instrum.* **66** (2), 1322-1324 (1995)

3. Jacobsen, C., Williams, S., Anderson, E., Browne, M.T., Buckley, C.J., Kern, D., Kirz, J., Rivers, M., and Zhang, X., *Optics Communications* **86**, 351-364 (1991)

4. Neuhäusler, U., Abend, S., Jacobsen, C., and Lagaly, G., *Colloid Polym. Sci.* **277**, 719-726 (1999)

5. Neuhäusler, U., Jacobsen, C., Schulze, D.G., Stott, D., and Abend, S., *J. Synchrotron Rad.,* in press (1999)

6. Pickering, S.U., *J. Chem. Soc.* **97**, 2001-2021 (1907)

7. Thieme, J., Abend, S., and Lagaly, G., *Colloid Polym. Sci.* **277**, 257-260 (1999)

8. Winn, B., Ade, H., Buckley, C., Howells, M., Hulbert, S., Jacobsen, C., Kirz, J., McNulty, I., Miao, J., Oversluizen, T., Pogorelsky, I., and Wirick, S., *Rev. Sci. Instrum.* **67(9)**, 1-4 (1996)

The Use of the Box-Counting Dimension for Characterization of the Aggregation of Colloidal Hematite

A. Ruprecht, J. Thieme

Institute for X-Ray Physics, University Georgia August at Göttingen,
Geiststraße 11, 37073 Göttingen Germany

Abstract. Studying properties of natural colloidal systems, one major issue is the question, whether the shape of the associations formed can be described morphometrically. Therefore, X-ray microscopy experiments have been performed with hematite particles, which have been forced to coagulate by adding increasing amounts of sodium sulfate as a coagulation agent. An important figure is the critical coagulation concentration c.c.c., where the repulsive potential around these colloidal particles is matched in its range with the attractive van der Waals potential. To calculate the fractal dimension of the associations in the X-ray micrographs is one way to characterize the structures formed. The method used here is the box-counting algorithm. The calculated box-counting dimensions show that the associations have an increasingly denser shape when crossing the c.c.c..

INTRODUCTION

Studying properties of natural colloidal systems, one major issue is the question, whether the shape of the associations formed can be described morphometrically. In nature associations of iron containing colloids can be found for example at the transition from anaerobic to aerobic groundwater. Insoluble compounds occur; new particles are formed in consequence. These particles associate to larger structures due to the chemical conditions of the environment. Size and structure of these associations determine the surface available for interactions with that environment. These interactions are among others adsorption, desorption, and the colloid bound transport of nutrients and toxicants.

For a better understanding of the coagulation processes, three approaches have been carried out: X-ray microscopy experiments with dispersions of hematite particles, presented here and in [1], modeling of the association of particles with repulsive and attractive potentials [2], and X-ray microscopy experiments with formerly anaerobic groundwater, where particle formation occur [3].

CP507, *X-Ray Microscopy: Proceedings of the Sixth International Conference,*
edited by W. Meyer-Ilse, T. Warwick, and D. Attwood
© 2000 American Institute of Physics 1-56396-926-2/00/$17.00

EXPERIMENTS

Hematite particles are iron containing colloidal particles that could be synthesized in the laboratory and have been used as a model system for the experiments described here [4]. Stable dispersions of hematite particles have been forced to coagulate by adding increasing amounts of sodium sulfate as a coagulation agent. With increasing concentrations of sodium sulfate the repulsive electrostatic potential around the particles is reduced in its range. Thus, the stability of the dispersion is reduced, coagulation gets more and more important. Sedimentation takes place in ever-decreasing times. An important figure hereby is the critical coagulation concentration c.c.c., where the repulsive potential is matched in the range with the attractive van der Waals potential [5]. Especially the range around this figure has been scrutinized.

The diameter of the colloidal hematite particles was measured by electron microscopy. The particles show diameters of about 100 nm (see figure 1). Increasing amounts of sodium sulfate were added to a 1% hematite dispersion in water thus realizing different states of coagulation of the particles. Samples of these dispersions were examined with the X-ray microscope of the Institute for X-ray Physics at the electron storage ring BESSY in Berlin [6]. Figure 2 shows some examples of these pictures at different concentrations of sodium sulfate.

FIGURE 1. Electron microscopic picture of hematite particles.

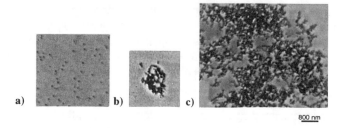

800 nm

FIGURE 2. X-ray microscopic pictures at different concentrations of Na_2SO_4: a) original dispersion b) sodium sulfate concentration of 0.109 millimol/l c) sodium sulfate concentration of 0.155 millimol/l

The method used to calculate the fractal dimension is the box-counting algorithm. Since this method is only an approximation to the fractal dimension the result is called box-counting dimension. To calculate the box-counting dimension of the associations the pictures have to be binarized in a first step. As all the pictures have been taken in phase contrast mode, halos around the structures might occur. These halos can cause problems at the binarization. Because of this a morphometrical correction, which suppresses the halo, is used (see chapter 2). Figure 3 shows binarized pictures of typically shaped associations at different concentrations of sodium sulfate.

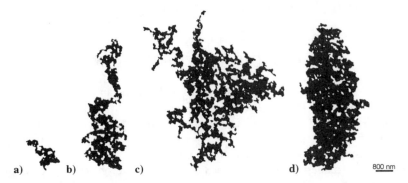

FIGURE 3. Binarized picture of a typical shaped associations at sodium sulfate concentrations of a) 0.105 millimol. / l , b) 0.112 millimol / l, c) 0.155 millimol / l, d) 0.197 millimol / l

IMAGE PROCESSING

The first processing step is the morphometrical correction. It consists of a dilation followed by erosion [7]. The dilation shrinks dark structures and broadens bright structures in the picture. The erosion works the other way around and restores the remaining structures to their original gray-level value. If the parameters were chosen that way, that the dark structures that represent the hematite particles are lost in the first step, the calculated picture contains only the bright halo. By subtracting this calculated picture from the original picture the halos are almost completely removed (see figure 4).

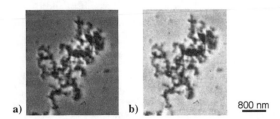

FIGURE 4. Original X-ray microscopic image (a) and morphometrical corrected image (b) of the same association of hematite particles.

To binarize a picture a threshold value must be found that defines whether a point is colored black or white. The binarized picture varies with this value. If the threshold is shifted to a value of brighter color the structures in the binarized picture broaden. This is used to find the correct value. The correct threshold value is found if single particles or structures with a width of one particle equal the width of the measured particle size of 100 nm. Most binarized pictures show single particles, parts of other associations or single black points from background noise. To calculate a correct value of the box-counting dimension these things must be erased from the binarized pictures. Particles or associations were treated as separated from each other if there is a gap larger than one particle diameter between them.

BOX-COUNTING DIMENSION

The fractal dimension is usually a noninteger number and can be used to describe the structures of a fractal object. The range of values is the same as the topological dimension. The fractal dimension D_f can be interpreted as follows: The more the fractal object fills the space like a plane ($D_f = 2.0$) than like a line ($D_f = 1.0$), the higher is D_f. To measure the fractal dimension one has to cover the object with balls of size ε and find the smallest covering. Instead of covering the object with balls one can also cover with boxes. A simple way to do this is to cover with a lattice of spacing d and determine the number of nonempty boxes N. This is called the box-counting algorithm. In a log(N) versus log(d) plot the box-counting dimension D_b is the negative slope. To find the smallest covering all different positions of the association to the lattice are examined for each lattice and the lowest number of counted boxes chosen. Because of their finite size physical fractals have a lower and an upper cutoff length scale. The lower length is equal to the size of a single particle and the upper length is equal to the size of the whole fractal object. For correct calculations the values of d must be between these cutoff lengths.

At high sodium sulfate concentrations some hematite associations were larger then the picture field of the x-ray microscope. Since the margin of the picture cut the association, the shifting of the grid can cause problems. An extension of the box-counting algorithm prevent this problems: If a nonempty box is cut by the margin the number of counted boxes is not raised by one but only by the percentage of the box that is inside the picture. Table 1 shows a comparison of the two algorithms for a part of a straight line, a plane and a sierpinski carpet [8]. The calculations were made with the program XDimension [9].

TABLE 1. Calculated box-counting dimensions of subimages of objects.

	Without margin correction	With margin correction
1000 pixel section of a straight line	D=0.951	D=1.000
1000x1000 pixel section of a plane	D=1.903	D=2.000
750x600 pixel section of a sierpinski carpet (D=1.893)	D=1.811	D=1.890

Figure 5 shows the calculated box-counting dimensions of the scrutinized hematite associations. The calculated box-counting dimensions show that the associates have an increasingly denser appearance when crossing the c.c.c..

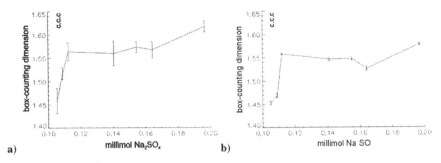

a)

b)

FIGURE 5. Averaged box-counting dimensions. The box-length ranges from a) particle size to the size of the smallest association, b) particle size to half of the association size.

ACKNOWLEDGEMENTS

The work presented here has been supported by the Federal Minister of Education, Science, Research and Technology, BMBF, Bonn, under contract number 05 644 MAG. We thank the staff of BESSY for excellent working conditions. We also thank PD Dr. J. Niemeyer, University of Trier for the hematite dispersions.

REFERENCES

1. Thieme, J., and Niemeyer, J., *Geol Rundsch,* **85,** 852-856 (1996)

2. Vormoor, O., and Thieme, J., "Reorganization of clusters in cluster-cluster-aggregation", this volume

3. Schmidt, C., Thieme, J., Neuhäusler, U., Schulte-Ebbert, U., Abbt-Braun, G., Specht, C., Jacobsen, C., "Association of particles and structures in the presence of organic matter", this volume

4. Schwertmann, U. and Cornell, R. M., *Iron Oxides in the Laboratory*, VCH, 1991.

5. e.g. Sparks, D.L., *Environmental Soil Chemistry*, Academic Press, 1995.

6. Niemann, B., Schneider, G., Guttmann, P., Rudolph, D., and Schmahl, G., "The new Göttingen X-Ray Microscope with Object Holder in Air for Wet Specimens", in *X-Ray Microscopy IV*, edited by V.V. Aristov and A.I. Erko, Chernogolovka: Bogorodski Pechatnik, 1994.

7. Haralick, Sternberg, Zhuang, *IEEE Transitions on Pattern analysis and Machine Intelligence,* PAMI-**9(4)**, 532ff (1987).

8. e.g. in Mandelbrot, B.B., *The Fractal Geometry of Nature,* Freeman , 1977.

9. Program "Xdimension", programmed by A. Ruprecht.

INSTRUMENTATION DEVELOPMENT

Characterization of Normal-Incidence Water-Window Multilayer Optics

G. A. Johansson, M. Berglund, L. Rymell, Y. Platonov[*], and H. M. Hertz

Biomedical and X-Ray Physics, Royal Inst. of Technology, SE-100 44 Stockholm, Sweden
*) *Osmic Inc., 1788 Northwood Drive, Troy, Michigan 48084-5532 USA*

Abstract. Spherical normal-incidence multilayer mirrors are attractive alternatives as condensers in compact water-window x-ray microscopes. In this paper we discuss the properties of such multilayer mirrors and the requirements when they are combined with a line-emitting soft x-ray source.

INTRODUCTION

Recently we demonstrated compact water-window x-ray microscopy with sub-100 nm resolution.[1] This microscope is based on a droplet-target laser-plasma source, a normal-incidence spherical multilayer condenser, a micro zone plate for the high-resolution imaging and a back-illuminated thinned CCD detector. One important observation during the development of the microscope is that the choice of condenser optics is critical for ease of alignment and exposure times. Normal-incidence multilayer optics is an attractive alternative,[2] despite severe demands as regards reflectivity, layer uniformity, and spectral response. Such optics provide high numerical aperture, ease of alignment and wavelength selectivity. Furthermore, the technology of fabrication shows promise for significant advances, making future multilayer condensers potentially very efficient.

EXPERIMENTS AND DISCUSSION

Our present compact X-ray microscope utilizes an 100 Hz ethanol-droplet-target laser-plasma source emitting characteristic line-emission at $\lambda=3.37$ nm.[3] The radiation is collected by a normal-incidence spherical mirror and focussed on the sample, as is shown in Fig. 1. The W/B$_4$C condenser mirror has a radius of curvature of 343 mm, and a diameter of 58 mm. The 200 bi-layers of W/B$_4$C are deposited by rf sputtering.

FIGURE 1. The experimental arrangement for the compact x-ray microscope.

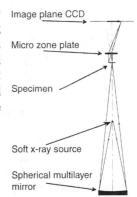

Image plane CCD

Micro zone plate

Specimen

Soft x-ray source

Spherical multilayer mirror

CP507, *X-Ray Microscopy: Proceedings of the Sixth International Conference,*
edited by W. Meyer-Ilse, T. Warwick, and D. Attwood
© 2000 American Institute of Physics 1-56396-926-2/00/$17.00

The fabrication of the mirror for this condenser arrangement is complicated. Not only is high reflectivity for short-wavelength normal-incidence radiation necessary. In addition, the 2d-spacing must exactly match the fixed emission line from the source. For the present mirrors, the FWHM of the reflectivity curve is typically 0.036 nm ($\lambda/\Delta\lambda \approx 100$). For these mirrors the peak of the reflectivity curve must be within 0.01 nm from the λ=3.37 nm emission wavelength to utilize more than 80% of the peak reflectivity. Finally, this accuracy in d-spacing must be achieved over the full area of the mirror. Figure 2 shows one-dimensional measurements of reflectivity for the 3 mirrors produced so far. The measurements are performed with the laser-plasma source at λ=3.37 nm. It is clear that the reflectivity varies considerably both within and between the mirrors. One possible explanation is that the d-spacing is not uniform.

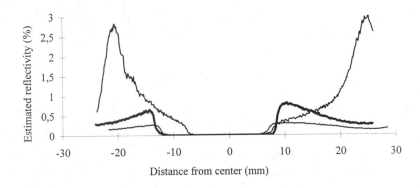

FIGURE 2. Reflectivity measurements of the three condenser mirrors. The center is obscured by a beam block.

In order to produce high-quality multilayers, an accurate and fast method for d-spacing measurements is necessary to give rapid feedback to the manufacturing process. For this purpose we are constructing a compact instrument that allows accurate in-house measurements of d-spacing via Bragg's law.

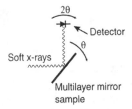

FIGURE 3. Schematic arrangement for measurements of d-spacing.

A 10 Hz laser-plasma x-ray source, which is based on a ethanol liquid-jet target, produces high-brightness, line-emission soft x-ray radiation in the water window.[4] A narrow x-ray beam from the source is incident on the multilayer sample. The multilayer sample is rotatable in a range of 90 degrees and the reflected x-ray radiation from the sample is measured by a detector rotating with twice the angle of the sample (cf. Fig. 3). The detector is based on a highly sensitive silicon photodiode and a low-noise integrating amplifier connected to a computer-based measuring system. A second detector monitors the initial source intensity. Due to the narrow line width of the source (typically $\lambda/\Delta\lambda \approx 700$), no monochromatization of the incident x-ray radiation is necessary for accurate measurements of d-spacing The spectral resolution $\lambda/\Delta\lambda$ of the reflectivity measurements depends on incident angle and is predicted to be in the range of 100-500 for relevant angles. Figure 4 shows a preliminary measurement of a first-order W/B_4C multilayer sample. The reflected intensity is due to the $\lambda=2.847$ nm emission line from the source.

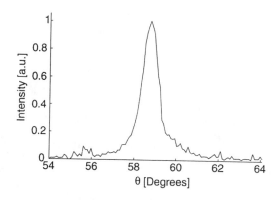

FIGURE 4. Reflectivity curve of first-order multilayer mirror.

ACKNOWLEDGEMENTS

We authors thank F. Eriksson and J. Birch for stimulating discussions.

REFERENCES

1. M. Berglund, L. Rymell, M. Peuker, T. Wilhein, and H. M. Hertz, "Compact Water-Window Transmission X-Ray Microscopy", submitted to *J. Microscopy*

2. H. M. Hertz, L. Rymell, M. Berglund, G. A. Johansson, T. Wilhein, Y. Platonov, and D. Broadway, "Normal-incidence condenser mirror arrangement for compact water-window x-ray microscopy", *SPIE Vol.* **3766**, to appear.

3. L. Rymell and H. M. Hertz, *Opt. Commun.* **103**, 105 (1993).

4. L. Malmqvist, L. Rymell, M. Berglund, and H. M. Hertz, *Rev. Sci. Instrum.* **67**, 4150 (1996).

A Microscope for Hard X-Rays Based on Parabolic Compound Refractive Lenses

C. G. Schroer*, B. Lengeler*, B. Benner*, F. Günzler*, J. Tümmler*, M. Drakopoulos[†], T. Weitkamp[†], A. Snigirev[†], and I. Snigireva[†]

*2. Physikalisches Institut, RWTH Aachen, D-52056 Aachen, Germany
[†]European Synchrotron Radiation Facility ESRF, BP220, F-38043 Grenoble, France

Abstract. We describe refractive x-ray lenses with parabolic profile that are genuine imaging devices, similar to glass lenses for visible light. They open considerable possibilities in x-ray microscopy, tomography, microanalysis, and coherent scattering. Based on these lenses a microscope for hard x-rays is described, that can operate in the range from 2 to 60keV, allowing for magnifications up to 50. At present, using aluminium lenses, it is possible to image an area of about 300μm in diameter with a resolving power of 0.3μm. Using beryllium as a lens material, the resolution can be increased below 0.1μm. The microscope allows to image opaque samples without destructive sample preparation and without the need of a vacuum chamber. It is particularly useful for in situ studies of wet samples, like biological and geological specimens. Imaging in both absorption and phase contrast is possible.

INTRODUCTION

There is a growing demand for x-ray microscopy techniques in both fundamental science and technology, in particular in bio-medicine, earth and environmental sciences, microelectronics, and material science. Soft x-ray full field microscopy techniques are very advanced and many dedicated soft x-ray microscopes are operational worldwide (see the many soft x-ray microscopy contributions to XRM99). It is only in recent years that hard x-ray imaging optics based on diffraction and refraction have become available and full field image transfer has been demonstrated [1–6]. Recently, we have introduced aluminium *compound refractive lenses* (CRLs) with parabolic profile and rotational symmetry that are *genuine imaging devices* like glass lenses in visible light [7,8]. They are free of spherical aberration, and have excellent imaging properties. They allow the construction of a high resolution hard x-ray full field microscope [7]. They are also very useful for scanning microscope applications (see contribution 160 to XRM99).

The hard x-ray microscope based on CRLs is complementary to other types of microscopes and is particularly suited for nondestructive imaging of opaque samples

CP507, X-Ray Microscopy: Proceedings of the Sixth International Conference,
edited by W. Meyer-Ilse, T. Warwick, and D. Attwood
© 2000 American Institute of Physics 1-56396-926-2/00/$17.00

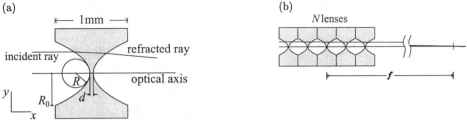

FIGURE 1. Schematic sketch of a parabolic compound refractive lens. The individual lenses (a) are stacked behind each other to form a CRL (b).

that do not tolerate sample preparation. There is no vacuum required in the beam path and a sample can be kept in its natural, e. g., wet, environment. Currently, with aluminium lenses, energies from about 10 to 60keV can be used for imaging, and a lateral resolution of about 300nm can be obtained over a field of view of about 300μm, with magnifications between 10 and 50.

In the future it is planned to use more transparent lens materials, such as beryllium, to reduce the absorption inside the lens. This increases the effective aperture of the lens and the resolution of the setup. It also improves the transmission and allows for shorter exposure times, reducing radiation damage in the sample and allowing for time resolved imaging. Beryllium lenses with an aperture of 1mm can have efficiencies of up to 40% and operate at energies as low as 2keV. With Be lenses, the resolution of the microscope can be pushed below 100nm [8].

The large depth of field of this microscope (see below) allows to acquire sharp projection images of objects up to several millimeters in thickness, but requires combination with tomographic techniques when longitudinal resolution is required.

The microscope can operate in both absorption and phase contrast and examples of both are given below. Using phase contrast at high x-ray energies, objects can be imaged with high contrast and a minimum of radiation damage, since absorption in the sample can be kept low. It is planned to build a full field x-ray microscope open to users at beamline ID22 of the ESRF.

COMPOUND REFRACTIVE LENSES

The main physical constraints to build refractive optics for hard x-rays are weak refraction and strong absorption of x-rays in matter. The weak refraction can be accounted for by using both a strong curvature ($R = 0.2$mm, see Fig. 1(a)) of the lens surfaces and a stacking of many lenses (N typically ranging from 20 to 150) behind each other forming a compound refractive lens (Fig. 1(b)). To avoid strong absorption, the lenses must be as thin as possible ($d \approx 10\mu$m in Fig. 1(a)) and made from low Z materials, such as Be, B, C, or Al. Since the refractive index of hard x-rays in matter is smaller than one, focussing lenses have concave shape. The parabolic profile with rotational symmetry avoids spherical aberration.

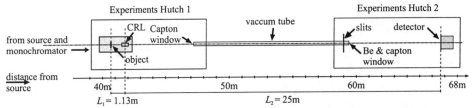

FIGURE 2. Hard x-ray full field microscopy setup at beamline ID22 of the ESRF.

The lenses described here are made from polycrystalline aluminium and have been designed and manufactured at the University of Technology in Aachen, Germany. The single lenses (Fig. 1(a)) are made by a pressing technique and are aligned in a stack (Fig. 1 (b)) of variable length on two high-precision shafts. For a detailed description and discussion see [6–8]. Due to the modularity of the CRLs a lens can be assembled to meet the specific requirements of a given experiment. Assembly and adjustment of the lens take no more than 15 minutes. The straight optical path allows for easy alignment of the setup.

The focal distance of a CRL is given by $f = R/(2\delta N)$ ($n = 1 - \delta - i\beta$ index of refraction) and ranges from 0.4m to 2.5m in typical applications. Due to the stronger absorption in the outer parts of the lens, a CRL has an effective aperture D_{eff} smaller than the geometric aperture R_0 (Fig. 1(a)).[1] For aluminium lenses D_{eff} typically ranges from 100 to 250μm.

CRLs are robust and withstand the white beam of an ESRF undulator source. Therefore, they can be used in "pink" beam (radiation from one undulator harmonic ($\Delta E/E = 10^{-2}$)) increasing the flux up to two orders of magnitude with respect to the monochromatic beam. They might be suited to be used together with future free-electron laser sources.

FULL FIELD X-RAY MICROSCOPE USING CRLS

In the full field x-ray microscope described here, the sample is illuminated from behind by monochromatic hard x-rays. The lens is placed a distance L_1 slightly larger than its focal distance f behind the object and transfers an enlarged image of the object onto a two dimensional position sensitive detector at a distance $L_2 = L_1 f/(L_1 - f)$ behind the lens. A scintillator based high resolution ccd-camera (FRELON [9]) or high resolution x-ray film is used as a detector. Both detectors have about 1μm lateral resolution. As for visible light optics, the magnification of this microscope is given by the ratio of L_2 and L_1 ($m = L_2/L_1$ typically 10 to 50) and the lateral resolution is given by $\Delta_{\text{lat}} = 0.75L_1\lambda/D_{\text{eff}} = 0.38\lambda/NA$ for incoherent illumination and is typically in the range of a few 100nm. For coherent illumination the resolution is slightly worse ($\Delta_{\text{lat coh}} = \sqrt{2}\Delta_{\text{lat incoh}}$). Here, NA is the numerical aperture $NA = D_{\text{eff}}/(2L_1)$, which is typically in the range of 10^{-4} for

---

[1] See [8] for a detailed description of D_{eff} and its dependence on lens parameters.

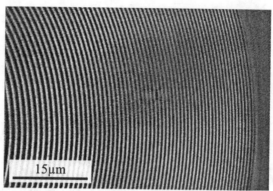

FIGURE 3. X-ray micrograph of a Fresnel zone plate at 23.5keV (magnification 12). The outermost zones of the zone plate (smallest structures to the right) have a width of about 0.3μm.

this type of setup. Highest resolution is obtained for largest NA, which requires $L_1 \approx f$, i. e., large magnifications. We have measured the lateral resolution of the microscope by a knife edge technique [7] and found good agreement with the theoretical predictions. The depth of field is given by $\Delta_{\text{long}} = \Delta_{\text{lat}}/NA$ and is in the millimeter range. For a detailed discussion of the theoretical aspects of this microscope see [8].

Fig. 2 shows a typical setup for the x-ray microscope at beamline ID22 of the ESRF. Sample and lens are placed in the first experiments hutch and the image is recorded by a detector in the second experiments hutch. The total length of the setup is about 25m. In between the two hutches, the beam is passed through an evacuated tube with two capton and one polished beryllium window.

Fig. 3 depicts the x-ray micrograph of a gold Fresnel zone plate (thickness 1.15μm) on silicon nitrate (thickness 100nm) at $E = 23.5$keV fabricated at IESS in Italy. The zone plate has a diameter of 200μm and is composed of 169 zones. The outermost zone (width 0.3μm) is clearly resolved. The CRL ($N = 62$) used in the microscope had a focal distance $f = 1.65$m. With $L_1 = 1.79$m a sharp image of the FZP was formed at a distance $L_2 = 21.44$m from the lens on high resolution x-ray film. This setup has a magnification of 12 and a lateral resolution of $\Delta_{\text{lat}} = 0.34\mu$m. The image was recorded on high resolution x-ray film with about 1μm lateral resolution. At a magnification of 12 the smallest structures visible on the x-ray film are about 3μm in size. Thus the resolution is not limited by that of the film.

The high lateral coherence of x-rays from third generation undulator sources allows for imaging in phase contrast mode [11,12] by slightly defocussing the sample (see Fig. 4). Like for Garbor inline-holography the contrast is produced by refraction effects in the sample. Weakly absorbing objects can be imaged with high contrast and a minimum of radiation damage. Depending on the defocusing

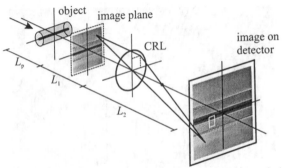

FIGURE 4. Setup of the phase contrast full field microscope. The sample is slightly defocused by L_p and illuminated by partially coherent x-rays form a third generation undulator source.

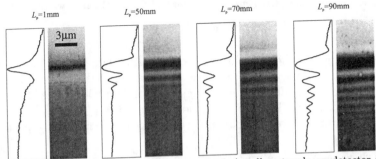

FIGURE 5. Detail of the phase contrast micrograph (small rectangle on detector in Fig. 4) of a tungsten borate-boron interface inside a boron fibre with tungsten borate core for different defocusing distances L_p.

distance L_p, the object can be imaged in the edge contrast or Fresnel regime.

We have used a boron fibre with tungsten borate core as a test object for phase contrast imaging. We used an CRL ($N = 50$) to construct an x-ray microscope operating at $E = 14.4$keV ($f = 0.72$m, $L_1 = 0.75$m, $L_2 = 18.0$m, magn.= 24, $\Delta_{lat} = 430$nm). Fig. 5 shows the image of the edge of the tungsten borate core for different distances L_p. The transition from the edge contrast regime (small L_p) to the fresnel regime (large L_p) can be clearly observed. A quantitative evaluation of these results together with an evaluation of the resolution will be published elsewhere.

Fig. 6 shows two detail x-ray micrographs of an insect imaged at $E = 23.5$keV in the same geometry as the FZP above, but with a slight defocusing. The images were recorded with a ccd-camera (FRELON II [9]). Strong contrast can be observed from this pure phase object. Due to the large depth of field, the full sample is imaged sharply. To obtain longitudinal resolution, the imaging needs to be combined with tomographic techniques.

We would like to thank H. Schlösser for his excellent work in manufacturing the

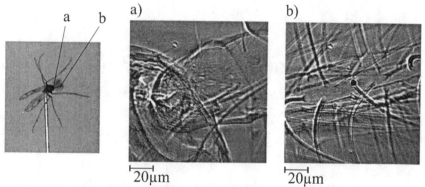

FIGURE 6. X-ray micrograph of an insect at $E = 23.5\text{keV}$ (magnification 12). CRLs allow for imaging pure phase objects illuminated by coherent x-rays.

pressing tools for the lenses, and J.-M. Rigal for his excellent technical support during the measurements made at beamline ID22 of the ESRF.

REFERENCES

1. A. Snigirev, I. Snigireva, P. Bösecke, S. Lequien, I. Schelokov, *Opt. Commun.* **135**, 378 (1997)
2. B. Lai, et al., *Rev. Sci. Instrum.* **66**, 2287 (1995).
3. A. Snigirev, V. Kohn, I. Snigireva, B. Lengeler, *Nature* **384**, 49 (1996).
4. P. Elleaume, *Nucl. Instrum. Methods Phys. Res.* **A 412**, 483 (1998).
5. A. Snigirev, V. Kohn, I. Snigireva, A. Souvorov, B. Lengeler, *Appl. Opt.* **37**, 653 (1998).
6. B. Lengeler, J. Tümmler, A. Snigirev, I. Snigireva, C. Raven, *J. Appl. Phys.* **84**, 5855 (1998).
7. B. Lengeler, C. Schroer, M. Richwin, J. Tümmler, M. Drakopoulos, A. Snigirev, I. Snigireva, *Appl. Phys. Lett.* **74**(26), 3924 (1999).
8. B. Lengeler, C. Schroer, J. Tümmler, B. Benner, M. Richwin, A. Snigirev, I. Snigireva, M. Drakopoulos, *J. Synchro. Rad.* **6**(6), 1153 (1999).
9. Labiche, J. C., Segura-Puchdes, J., van Brusel, D., Moy, J. P., *ESRF Newslett.* **25**, 41 (1996).
10. Spanne, P., Raven, C., Snigireva, I., Snigirev, A., *Phys. Med. Biol.* **44**, 741 (1999).
11. Gureyev, T. E., Raven, C., Snigirev, A., Snigireva, I., Wilkins, S. W., *J. Phys. D: Appl. Phys.* **32**, A160-165 (1999).

Development of an X-Ray Projection Microscope using Synchrotron Radiation

Kunio Shinohara[*], Atsushi Ito[†], Toshio Honda[‡], Hideyuki Yoshimura[§] and Keiji Yada[¶]

[*]*Radiation Research Institute, Faculty of Medicine, the University of Tokyo, Bunkyo-ku, Tokyo 113-0033 JAPAN*
[†]*Department of Nuclear Engineering, School of Engineering, Tokai University, Hiratsuka, Kanagawa 259-1292 JAPAN*
[‡]*Department of Information & Image Sciences, Faculty of Engineering, Chiba University, Chiba, Chiba 263-0022 JAPAN*
[§]*Department of Physics, School of Science and Technology, Meiji University, Kawasaki, Kanagawa 214-8571 JAPAN*
[¶]*Tohoku University, Sendai, Miyagi 980-8577 JAPAN*

Abstract. Projection microscopy using synchrotron radiation as a light source has been developed. The system consists of a Fresnel zone plate to produce a small point source, an aperture (pinhole), a specimen holder, and a back-illuminated CCD camera. With this system, we obtained the images of EM-grids and dried mammalian cells. The resolution was estimated to be around 1.3 μm from the intensity change of 10 to 90% in the image of the edge pattern of an EM-grid.

INTRODUCTION

Projection microscopy has unique advantages over other types of X-ray microscopy. Representative features include focus-free optics, wide view area and easy-to-zoom-up optics. As a projection microscope, presently available system uses a point source of fluorescent X-rays from a metal target illuminated with a focused electron beam (1, 2). In such a system, however, available wavelength is limited. In the present study, we have developed a projection microscope system using synchrotron radiation a tunable light source.

OPTICAL LAYOUT OF PROJECTION MICROSCOPE

The optical layout of the microscope system was illustrated in Fig. 1. The system was composed of a Fresnel zone plate for producing a small point source, an aperture at the focal point of the zone plate and a back-illuminated CCD camera. It was separated into two parts: a high vacuum chamber with about 10^{-7} Torr including zone plate and a low vacuum chamber with about 10^{-3} Torr, where a pinhole, specimen, and

CP507, *X-Ray Microscopy: Proceedings of the Sixth International Conference,*
edited by W. Meyer-Ilse, T. Warwick, and D. Attwood

CCD camera were arranged. Both chambers were separated by SiN window of 200 μm thickness and 1 mm x 1 mm in size.

Monochromatic X-rays from BL-12A, a bending magnet beamline, at the Photon Factory, Tsukuba, Japan, were focused on the aperture by a Fresnel zone plate of 0.625 mm diameter with 80 nm of the outermost zone width made of 400 nm thick Tantalum (NTT Advanced Technology Corp., Japan). A series of pinholes ranging from 1 to 20 μm in their diameter, made of stainless steel of 13 μm thickness, were mounted on a stage that is movable two dimensionally with 10 nm precision (Sigma Koki, Japan). A specimen holder was set immediately downstream of the pinhole; the distance (d) between specimen and the pinhole can be changed from 10 mm down to the order of 10 μm. Enlarged image was captured by a back-illuminated CCD camera with 512 x 512 pixels and 12.3 mm square in size (C4880-30-26WS, Hamamatsu Photonics K.K., Japan).

In addition, silicon photodiode (AXUV-100, International Radiation Detectors, Inc., U.S.A.) can be inserted for the alignment of the zone plate and the SiN window along the incident beam.

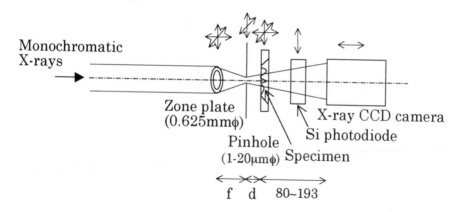

FIGURE 1. Layout of projection X-ray microscope. f is distance between zone plate and pinhole depending on the wavelength. d denotes distance between pinhole and specimen that should be adjusted from 10 mm down to the order of 10 μm.

RESULTS AND DISCUSSION

Figure 2 shows images of EM reference grid with 400 mesh (H7; Graticules, U. K.). The condition of observation was as follows: wavelength at 1.5 nm, pinhole with 5 μm diameter, and distance of 193 mm between the focal point and the CCD camera. The magnifications described in the figure were calculated from the sizes of bars and spacings in the grid. For the most enlarged image (Fig. 2c) the magnification was estimated based on the distance between the pinhole and the specimen which is

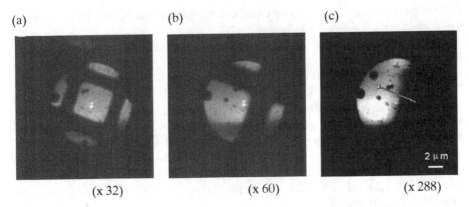

(a)   (b)   (c)

(x 32)   (x 60)   (x 288)

FIGURE 2. Images of H7 reference grid with 400 mesh at various magnifications described below the pictures. A white line across the edge of grid in (c) is for the line profile of photon intensity that is analyzed in Fig. 3.

measured by the reading of an installed distance setting micrometer. To estimate spatial resolution of the system, the intensity profile along the vertical line across the edge drew in Fig. 2c was shown in Fig. 3. The distance required for the change from 10 % to 90 % in the intensity jump was estimated to be 1.33 µm. Other line profiles chosen from the same figure showed the values of 1.67 and 1.25 µm.

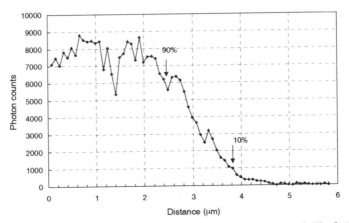

FIGURE 3. Line profile of photon intensity along the line shown in Fig. 2(c).

To check that the present system is consistent with the geometry of optics, the size of imaging area was measured under the various conditions such as different pinhole size and different diffraction order. Table 1 compared the theoretical sizes of imaging areas on the CCD (the 7th column) and the measured diameters of the imaging areas (the 6th column). The both values were very close except the case of the 2nd order diffraction. The difference in this case could be partly ascribed to the vague boundary of the area due to the low photon intensity.

TABLE 1. Size of imaging area.

wavelength (nm)	Diffraction order	Pinhole size (μmφ)	Distance A[*] (mm)	Distance B[*] (mm)	Distance C[*] (mm)	Distance C[*] (theoretical)
1.5	1	5	38.35	193	3.67	3.15
1.5	1	10	39.3	193	3.63	3.07
1.5	2	20	13.8	193	12.8	8.74

[*] Distance A and B denote the distance between zone plate and pinhole and that between pinhole and CCD, respectively. Distance C means the diameter of image on CCD camera.

Preliminary result of biological specimen was shown in Figure 4. The image of Chinese hamster ovary (CHO) cells was taken in a dry state with the magnification of 64. Although the image is not so clear, a nucleus in the cell surrounded by a white dotted outline can be recognized.

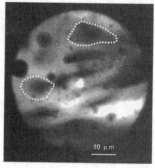

FIGURE 4. Image of dried CHO cells at the magnification of 64.

Future development should be done in the following points:
1) Improvement of resolution that could be achieved by the use of pinhole of the smaller size or coherent illumination to the zone plate
2) Development of image processing method to correct image blurring due to Fresnel diffraction
3) Development of stereo imaging

ACKNOWLEDGMENTS

This work was partly supported by a Grant-in-Aid for Scientific Research (A) from the Ministry of Education, Science, Sports and Culture.

REFERENCES

1. Yada, K. and Takahashi, S. "Projection X-ray microscope observation of biological samples," in *X-Ray Microscopy in Biology and Medicine*, edited by K. Shinohara, K. Yada, H. Kihara and T. Saito, Tokyo/Berlin: Japan Scientific Societies Press/Springer-Verlag, 1990, pp. 193-202.
2. Yoshimura, H. et al., "Contrast enhancement and three-dimensional computed tomography in projection X-ray microscopy" in this proceedings (in press).

Principal Component Analysis for Soft X-ray Spectromicroscopy

Angelika Osanna and Chris Jacobsen

X-ray Microscopy Group, Department of Physics and Astronomy, State University of New York at Stony Brook, Stony Brook, NY 11794-3800

Abstract. A principal component analysis approach to extract thickness maps from spectromiscroscopic image stacks is described and illustrated with a model calculation.

INTRODUCTION

In recent years, XANES (X-ray Absorption Near-Edge Structure) mapping has become a very powerful method in x-ray microscopy [1]: reference spectra of chemical or biological components in a sample provide a means to enhance the contrast of components one is interested in, by selectively imaging at energies that give strong contrast, such as π^* peaks in a XANES spectrum. From the differences in contrast of a set of images, a spatial map of a sample component can be extracted. A commonly used procedure is to use singular value decomposition to obtain maps of two or more components. The analysis is based on the following relation between the optical densities D obtained from the sample, and the absorption coefficients μ and the mass thicknesses t of the different components in the sample:

$$D_{np} = \mu_{n1}t_{1p} + \mu_{n2}t_{2p} + \mu_{n3}t_{3p} + \cdots \tag{1}$$

where μ_{nm} is the absorption coefficient for component m at energy E_n, and t_{mp} is the mass thickness of component m. The index p is the pixel number in an image, $p = i_{\text{cols}} + n_{\text{cols}} \cdot i_{\text{rows}}$, with a maximum of $P = n_{\text{cols}} \cdot n_{\text{rows}}$. The optical densities are calculated by normalizing the microscope image I with the incident flux I_0,

$$D_{np} = D_p(E_n) = -\ln \frac{I_p(E_n)}{I_0(E_n)}. \tag{2}$$

Equations (1) and (2) are nothing other than an expression of Beer's law, which describes the absorption of x rays in matter:

$$I(E) = I_0(E) \cdot e^{\sum_{m=1}^{M} \mu_m t_m}. \tag{3}$$

CP507, *X-Ray Microscopy: Proceedings of the Sixth International Conference,*
edited by W. Meyer-Ilse, T. Warwick, and D. Attwood
© 2000 American Institute of Physics 1-56396-926-2/00/$17.00

Equation (1) can be written as a matrix equation

$$
\begin{pmatrix}
D_{11} & D_{12} & \cdots & D_{1P} \\
D_{21} & D_{22} & \cdots & D_{2P} \\
\vdots & \vdots & \ddots & \vdots \\
D_{N1} & D_{N2} & \cdots & D_{NP}
\end{pmatrix}
=
\begin{pmatrix}
\mu_{11} & \mu_{12} & \cdots & \mu_{1M} \\
\mu_{21} & \mu_{22} & \cdots & \mu_{2M} \\
\vdots & \vdots & \ddots & \vdots \\
\mu_{N1} & \mu_{N2} & \cdots & \mu_{NM}
\end{pmatrix}
\cdot
\begin{pmatrix}
t_{11} & t_{12} & \cdots & t_{1P} \\
t_{21} & t_{22} & \cdots & t_{2P} \\
\vdots & \vdots & \ddots & \vdots \\
t_{M1} & t_{M2} & \cdots & t_{MP}
\end{pmatrix}
$$

$$
\text{or} \qquad \mathbf{D}_{(N \times P)} = \mu_{(N \times M)} \cdot \mathbf{t}_{(M \times P)}. \tag{4}
$$

This means that we have STXM images D_n at $E = E_n$ energies, $n = 1, 2, \ldots N$. We expect our sample to have m components with thicknesses t_m, $m = 1, 2, \ldots M$. To separate M components, we need at least M images, so that the number of energies should be $N \geq M$. In general, the system of equations (4) will be overdetermined, and a large number of images will give better statistical significance of the resulting thickness maps. To solve (4), we will need to multiply with $\mu^{-1}_{(M \times N)}$ from the left:

$$
\mu^{-1}_{(M \times N)} \cdot \mathbf{D}_{(N \times P)} = \mu^{-1}_{(M \times N)} \cdot \mu_{(N \times M)} \cdot \mathbf{t}_{(M \times P)} = \mathbf{t}_{(M \times P)}. \tag{5}
$$

Matrix inversion is conveniently done using singular value decomposition (SVD). SVD is usually available as a black-box routine in mathematical software and is described for example in [2].

The SVD analysis outlined above assumes that we know ahead of time the absorption spectra $\mu_{(N \times M)}$ of all M components in the sample. However, frequently we will be investigating samples where we hope spectromicroscopy will tell us how many and which statistically independent components are in the data set, without specifying the components *a priori*. One method for spectromicroscopy (see *e.g.* [3]) involves the acquisition of images of the sample at many (closely spaced) energies. We would like to extract spectra from such a "stack" of images and find a set of basis images that correspond to the different chemical components in the sample, similar to the maps from reference spectra described above. This type of statistical analysis was first developed in the social sciences and is called Principal Component Analysis (PCA) or Factor Analysis. It was first applied to x-ray spectroscopy by Wasserman [4].

PRINCIPAL COMPONENT ANALYSIS

The following discussion is based on [4] and [5]. To analyze a set of N images with P pixels, D_{np}, we form a data matrix $\mathbf{D}_{(N \times P)}$ by putting the images into the matrix rows. From this, we calculate the *covariance matrix* $\mathbf{Z}$ by multiplying $\mathbf{D}$ with its transpose, $\mathbf{D}^{\mathrm{T}}$:

$$
\mathbf{Z}_{(N \times N)} = \mathbf{D}_{(N \times P)} \cdot \mathbf{D}^{\mathrm{T}}_{(P \times N)}. \tag{6}
$$

Each element of $\mathbf{Z}$ is the inner product of two image "vectors", representing how similar the images are to each other. On the other hand, we can consider a stack

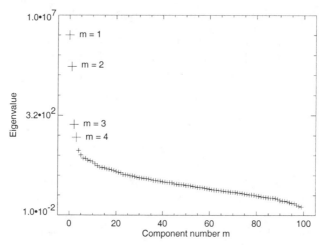

FIGURE 1. The eigenvalues of the covariance matrix for a model stack. In this case, we decide that only the first four eigenimages are of significance. The model calculation is described below.

as a set of P spectra (one for each pixel), and each spectrum consists of N data points. We see that we can express all information contained in the data matrix with the eigenvalues and eigenvectors ("eigenspectra") of $\mathbf{Z}$. The eigenvalues of $\mathbf{Z}$ give the relative importance of each component and the eigenspectra associated with the most significant eigenvalues are called the *principal components*. The number of principal components can be determined from the eigenvalues and it will be equal to the number of physical components in the sample. Only those eigenvalues significantly larger than the rest (see Fig. 1) represent non-noise components m, and this reduced set has $m = 1, 2, \ldots \bar{M}$. For a successful factor analysis, it is necessary that the data matrix contains significantly more images than the number of expected principal components, or that each spectrum in the stack is measured at more energies (or wavelengths) than any reasonable guess of the number of components.

Next, we want to obtain a set of eigenimages for the stack. We form a matrix $\mathbf{C}_{(N \times M)}$ that contains the m principal components in its columns; since the principal components are (normalized) eigenvectors of $\mathbf{Z}$, $\mathbf{C}$ is an orthonormal matrix. The matrix

$$\mathbf{R}_{(\bar{M} \times P)} = \mathbf{C}^{\mathrm{T}}_{(\bar{M} \times N)} \cdot \mathbf{D}_{(N \times P)} \tag{7}$$

will contain the $\bar{M}$ eigenimages in its rows. On the other hand, we can use the eigenimages and principal components to "reconstruct" the data matrix and compare it to the original data set:

$$\mathbf{D}'_{(N \times P)} = \mathbf{C}_{(N \times \bar{M})} \cdot \mathbf{R}_{(\bar{M} \times P)}. \tag{8}$$

It is important to repeat here that we are using only the components with the $\bar{M}$ largest eigenvalues out of the complete set of eigenvectors of $\mathbf{Z}$. If we have indeed retained all significant components in $\mathbf{C}$, then $\mathbf{D}'$ will agree with $\mathbf{D}$ within the experimental error, $\mathbf{D}' = \mathbf{D}$. A discrepancy will indicate that we need to include one or more additional eigenvectors as principal components. It is important, however, to not use more principal components than necessary, since insignificant eigenvectors will just reproduce experimental errors (see below). The number of significant principal components $\bar{M}$ gives an upper limit on the number of spectroscopically different, physical components in the data set.

There is a bit of mathematical freedom in how to approach PCA. It is equally valid to form the data matrix by putting the spectra into the columns of $\mathbf{D}$, resulting in an $(P \times P)$ covariance matrix. If $P < N$, this will require less computing power and time. In general, one will always arrange the dimensions of the matrices $\mathbf{D}$, $\mathbf{C}$ and $\mathbf{R}$ such that the computing efforts are optimized and it can be shown that either choice is equally valid and yields equivalent results [5].

It should be pointed out that the eigenvectors of $\mathbf{Z}$ are *abstract* components and usually have no direct physical meaning. By "abstract", we mean that these components simply describe natural groupings of the data, but not necessarily in the way we might best understand them. Therefore, we need to find a transformation that leads from the principal components and eigenimages to the spatial maps of the sample components, in terms of reference spectra.

In Eq. (4), we decomposed our data into a set of thickness maps t and absorption coefficients μ as

$$\mathbf{D}_{(N \times P)} = \mu^{\text{physical}}_{(N \times M)} \cdot \mathbf{t}^{\text{physical}}_{(M \times P)}. \tag{9}$$

On the other hand, with the principal component analysis, we have shown that our data can be represented with a set of (abstract) eigenimages and eigenspectra. To compare these two approaches, let us change the names of two entities in Eq. (8):

$$\mathbf{D}_{(N \times P)} = \mu^{\text{abstract}}_{(N \times \bar{M})} \cdot \mathbf{t}^{\text{abstract}}_{(\bar{M} \times P)}. \tag{10}$$

Equating Eq. (9) with (10), we see that the abstract and the physical decomposition of $\mathbf{D}$ are connected through a "rotation" $\mathbf{F}_{(M \times M)}$:

$$\mu^{\text{abstract}} \cdot \mathbf{t}^{\text{abstract}} = \mu^{\text{abstract}} \cdot \mathbf{F} \cdot \mathbf{F}^{-1} \cdot \mathbf{t}^{\text{abstract}} = \mu^{\text{physical}} \cdot \mathbf{t}^{\text{physical}} \tag{11}$$

$$\text{or} \quad \mu^{\text{physical}} = \mu^{\text{abstract}} \cdot \mathbf{F} \quad \text{and} \quad \mathbf{t}^{\text{physical}} = \mathbf{F}^{-1} \cdot \mathbf{t}^{\text{abstract}}. \tag{12}$$

Therefore, we can conclude that the (physical) thickness maps of the sample components are obtained as linear combinations of the (abstract) eigenimages.

In order to determine the transformation matrix $\mathbf{F}$, we now have to come up with some assumptions about the components contained in our sample. The matrix $\mu^{\text{physical}}_{(N \times M)}$ contains the absorption spectra of sample components in its rows. We can

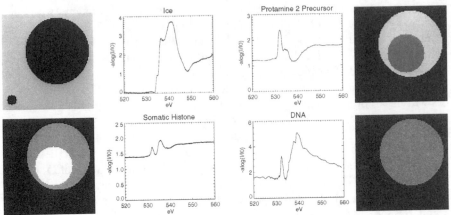

FIGURE 2. The reference spectra used to generate the model data set, with the respective maps next to them.

test if a component is present in the sample by trying to fit a reference spectrum to the principal component spectra: a spectrum of a physical component $\mu^{physical}_{(N \times M)}$ can be represented as a linear combination of principal components $\mu^{abstract}_{(N \times M)}$, with the fit coefficients obtained through a least-squares fit. The transformation matrix **F** contains the fit coefficients for each component reference spectrum in its rows. Details on how to best obtain a least-squares fit to the principal component spectra can be found *e.g.* in [2].

When comparing PCA and the procedure of obtaining maps from reference spectra using SVD described in the introduction, one sees that the two approaches are very much complementary and as closely connected as the sides of a coin: SVD does a least square fit to the data where the number of components is an input parameter, based on a model for the sample. PCA *determines the number of factors (components)* and lets the user come up with a suitable model.

MODEL CALCULATION FOR PCA

The steps of principal component analysis described above have been tested with a model data set. When generating this data set, we used four reference spectra at the oxygen edge that are relevant to a spectromicroscopy project currently under investigation at Stony Brook. The spectra are shown in Fig. 2 and we denote them as μ_{ice} (ice), μ_{p2p} (protamine 2 precursor), μ_{sh} (somatic histone) and μ_{dna} (DNA). The model maps t_{ice}, t_{p2p}, t_{sh} and t_{dna}, simulating an object consisting of DNA and two proteins embedded in ice, are also shown in Fig. 2. Random noise of the order of a few percent was added to each map. A model STXM stack was generated by calculating optical densities at 100 different energies E_n,

$$D(E_n) = \mu_{ice} \cdot t_{ice} + \mu_{p2p} \cdot t_{p2p} + \mu_{sh} \cdot t_{sh} + \mu_{dna} \cdot t_{dna} \tag{13}$$

TABLE 1. The fit coefficients for the spectra in Fig. 4

	m = 1	m = 2	m = 3	m = 4
ice	16.2909	-7.50436	0.139019	-0.149354
protamine 2 precursor	15.3866	3.49711	1.84571	-1.40350
somatic histone	16.6577	2.64456	0.0470331	-1.68316
DNA	26.8258	-0.681556	-2.42283	4.28298

and they can be converted into transmission images according to eq. (3) using a suitably chosen incident flux spectrum $I_0(E_n)$. The data matrix is formed as described and the covariance matrix (6) is decomposed into eigenvalues and eigenvectors. Fig. 1 shows the eigenvalues in a semi-logarithmic plot. The first four eigenvalues appear significantly larger than the rest.

Fig. 3 shows on the left the first five eigenvectors. The first four are chosen as principal components; the fifth one shows that the uncertainties increase most where the reference spectra overlap most. This will, in case of significant overlap, make it difficult to separate the components. The corresponding eigenimages are shown in the right column of Fig. 3.

Fitting the reference spectra to the principal components gives the fit coefficients listed in Table 1 and as we can see from Fig. 4, the spectra are reproduced very nicely up to noise-induced inaccuracies. From the fit coefficients, a (4×4) transformation matrix $\mathbf{F}$ is formed and the eigenimages are transformed into the thickness maps (see eq. (12)). The resulting maps are shown in Fig. 5 and we can see that they agree very well (within the noise error) with the original maps.

REFERENCES

1. See *e.g.* H. Ade, X. Zhang, S. Cameron, C. Costello, J. Kirz, and S. Williams, *Science* **258**, 972 (1992); C. J. Buckley, *Review of Scientific Instrumentation* **66**, 1318 (1995); X. Zhang, R. Balhorn, J. Mazrimas, and J. Kirz, *Journal of Structural Biology* **116**, 335 (1996); and numerous papers in these proceedings.
2. W. H. Press, S. A. Teukolsky, W. V. Vetterling, and B. P. Flannery, *Numerical Recipes in C*, Cambridge University Press, 1992.
3. C. J. Jacobsen, C. Zimba, G. Flynn, and S. Wirick, Soft x-ray spectroscopy from sub-100 nm regions. *Journal of Microscoscopy*, in press.
4. S. R. Wasserman, *Journal de Physique IV France* **7** (vol. C2), 203 (1997).
5. E. R. Malinowski and D. G. Howery, *Factor Analysis in Chemistry*. John Wiley and Sons, 1998.

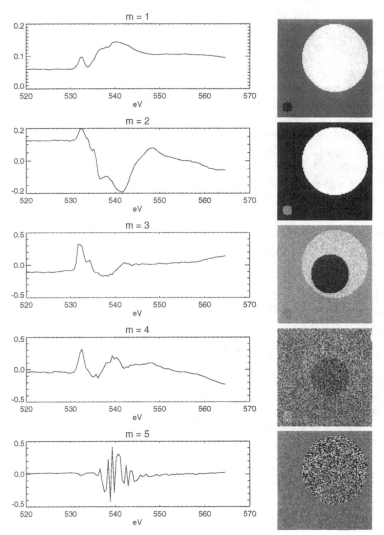

FIGURE 3. The first 5 eigenspectra (on the left) with the corresponding eigenimages. The first four eigenspectra were chosen as principal components.

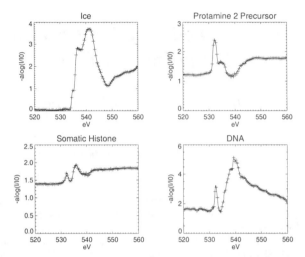

FIGURE 4. Fitting the reference spectra to the principal components. The reference spectra points are shown as a solid line, while the PCA fits are are indicated as +.

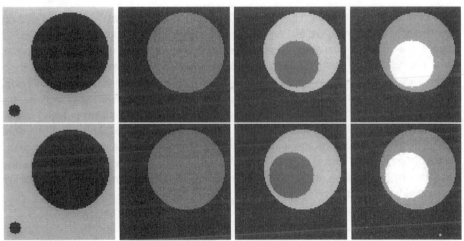

FIGURE 5. In the top row are the "original" model maps and below are the maps obtained from the rotation of the eigenimages (left to right: ice, DNA, protamine 2 precursor and somatic histone).

Evaluation Of The Detection Efficiency Of An X-Ray Image Sensor

J.M. Wulveryck[a] and D. Mouze

DTI/LASSI, Faculté des Sciences, BP1039, 51687 Reims Cedex 2, France
[a]Electronic mail : jm.wulveryck@univ-reims.fr

Abstract. This work deals with the measurements of the spectral efficiency of an x-ray camera by a straightforward method, using laboratory x-ray sources. By mean of a fluorescence wavelength dispersive spectrometer (WDS), we carry out measurements in a large energy range, typically from a few keV to tenths of keV. The camera consists of a slow scan transfer device (CCD) optically coupled to a scintillator screen (Y_2O_2S:Eu). The spectral responsivity of the x-ray imaging detector - in ADU/keV units - is discussed. In particularly, the assumption of a linear conversion from incident x-ray energy to visible photon energy is confirmed by the experiment.

INTRODUCTION

In the field of non-destructive testing, x-ray imaging, and particularly x-ray absorption microscopy (XAM) plays a major role. In this method, the contrast in the images is a mixed combination of density and compositional information. More often, x-ray projection microscopy make use of laboratory sources namely microfocus x-ray generators. Unfortunately, because these sources emit polychromatic radiation, quantitative microanalysis leads to errors in the specimen composition. To overcome this drawback, one must master both the physical phenomenon allowing observation, i.e. the x-ray source, and the features of the imaging device. The latter is more particularly the subject of this paper. Previous papers have been published on the subject from authors making use of synchrotron radiation source [1,2], radioactive x-ray sources [3] or wavelength dispersive spectrometer [4,5]. The present measurements are performed with this third technique.

CP507, *X-Ray Microscopy: Proceedings of the Sixth International Conference,*
edited by W. Meyer-Ilse, T. Warwick, and D. Attwood

EXPERIMENTAL FEATURES AND MEASUREMENTS

X-Ray Sensor Camera

The first stage of the x-ray camera consists in a scintillating screen which converts the incident x-ray photons into optical ones. It is optically coupled to a CCD sensor to carry out the conversion of optical photons into electron-hole pairs. The charge output is read out, amplified, and digitized by mean of the processing electronic unit.

The main characteristics of the detector assembly was as follows :

- Scintillating screen

   material : P22 red phosphor (Y_2O_2S:Eu), surface density : 6.4 mg/cm^2

- Optical coupling

   35 mm f/1.2 lenses, magnification factor M = 0.71

- CCD arrays

   model EEV P86231, format 385 $\times$ 578, pixel size 22 μm $\times$ 22 μm, dark current < 0.1 e$^-$/pixel.s, full frame transfer, cooling : -40°C (3 Peltier stages)

- A/D precision :

   16 bits

Though such a coupling is reputed to provide lower throughput than optical fibre coupling, it offers significant advantages for some applications such as microtomography : variable magnification allowing an adjustment of the field of view, separation of the phosphor screen assembly from the camera head.

Experimental Set-up

In order to evaluate the spectral sensitivity of our x-ray imaging camera, a monochromatic x-ray laboratory source was needed. This was obtained by using a wavelength dispersive x-ray fluorescence spectrometer as a monochromator. In conventional use, a fluorescence spectrometer comprise an x-ray tube, i.e. a polychromatic source, to enlighten the unknown specimen, an analyser crystal to resolve secondary x-rays emitted from the sample and a gas-flow proportional counter or scintillation counter for photons counting. For our application, pure elements was used as specimen, acting as secondary targets. In this case, the spectral distribution of the emitted photons consists in characteristic fluorescence lines of the pure element on a very low background. This lines are resolved by the analysing crystal, thus a monochromatic radiation is obtained. Measurements was performed by substituting successively the x-ray camera and a lithium-drifted silicon detector for the counter.

This allows us to use the Si(Li) detector as a standard one and to make an estimation of the count rate to evaluate the true x-ray flux incident on the camera (see next section).

The x-ray generator we used as primary source was a 1.5kW gold anode tube. The drift of the stability of this generator was estimated by previous measurements to be less than 2% for an exposition time of one hour (the stability defects must be taken into account as a systematic error in our measurements). The analyser crystal used was LiF(200). In our experiments nine secondary targets were used giving eleven worth working x-ray lines, the weak lines being unexploitable for imaging.

Experimental Measurements

The knowledge of the main structural features of the Si(Li) detector allow us to evaluate precisely its efficiency for the detected x-ray lines. These features concern the thickness of the different layers which the detector is made of, from these thicknesses it is possible to evaluate the part of the x-ray flux efficiently absorbed. For our detector - model Edax PV9755/02 - these characteristics was the following : beryllium window, t_{Be}=8μm; gold contact, t_{Au}=200nm; silicon dead layer, $t_{Si.dl}$=100nm; active area, $t_{Si.i}$=3mm.

Eleven images have been acquired in the same operating conditions as the previous measurements. The acquisition time was one hour for each of the images. To obtain a correct evaluation of the signal in the image, the measurements was made by averaging the output signal of the camera in a selected area with the same surface as the Si(Li) active area. Each selected areas grouped about ten thousands pixels.

The weak flux issued from the x-ray secondary sources has entailed long acquisition times (# 1 hour). This make necessary a dark current correction procedure. This was obtained by acquiring ten images without x-ray enlightenment. The averaging of these images was used as a dark image. It was then subtracted from each of the eleven raw images to obtain resultant net images. To evaluate the accuracy of each measurements three images was recorded and corrected from the dark current. This allowed us to evaluate the standard deviation σ attached to these measurements.

The CCD signal is given in analog-to-digital converter units (ADU). It concerns the mean signal registered in one pixel per unit time. Assuming a normal distribution, we have calculated the probable error (confidence level = 50%) = 0.67σ.

DISCUSSION

The aim of the measurement is to evaluate the signal level in a CCD pixel versus the x-ray energy incident on the corresponding imaged area of the scintillating screen (i.e. *scintillating screen pixel*). Thus if E^X_{ph} is the x-ray photon energy for a given line, the incoming x-ray energy E^X_{in} on a s.c. pixel at the entrance of the camera can be evaluated by the relation :

$$E_{in}^{X}(keV) = E_{ph}^{X}R\frac{I_{SiLi}}{\varepsilon_{SiLi}} \qquad (1)$$

where R is the surface ratio between the x-ray camera pixel and the Si(Li), and I_{SiLi} is the output count rate from the detector. Then the efficiency of the x-ray camera is :

$$\varepsilon_{CCD}(ADU/keV) = \frac{I_{CCD}}{E_{in}^{X}} \qquad (2)$$

The results obtained from this relation and the measurements are plotted in the figure 1. These results must be compared to what we could expect with such an instrument. The only element which governs the *spectral efficiency* of the x-ray camera concern the scintillating layer. In this respect, the composition of this layer, its thickness and the number of visible photons emitted per absorbed x-ray photon are of the utmost importance. On the contrary, once the x-ray to visible light is carried out, the other part of the camera are unimportant for the spectral efficiency in the x-ray range.

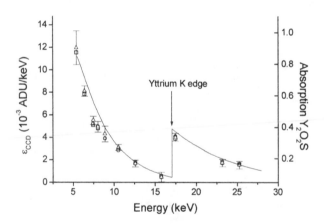

FIGURE 1. Plot of the efficiency of the x-ray camera, ε_{CCD} (10^{-3} ADU/keV), versus energy. On the right axis we plot the absorption curve of the x-ray in the scintillating screen.

So, if we assume a linear relationship between the x-ray energy of a photon absorbed in the scintillating layer and the number of photons emitted in the visible light range, the responsivity of the camera, defined as the ratio of the signal output to the incident x-ray energy, would fit the absorption curve of the x-rays in the scintillating screen.

We have shown in this work that measurements of the efficiency of an x-ray sensor camera can be easily performed with a simple laboratory equipment. As expected, the measured responsivity of the x-ray camera, ADU/keV units, fits well with the theoretical energy conversion efficiency curve of x-rays to visible photons in the phosphor screen, which confirms a linear energy conversion process in the phosphor material.

REFERENCES

1. Prigozhin, G. Y., Woo, J., Gregory, J. A., Loomis, A. H., Bautz, M. W., Ricker, G. R., and Kraft, S., *Optical Engineering*, **37**, 10, pp 2848-2854 (1998).

2. Koch, A., *Nuclear Intruments and Methods in Physics Research A*, **348**, pp. 654-658 (1994).

3. Shepherd, J. A., et al., *Optical Engineering*, **36**, 11, pp 3212-3222 (1997).

4. Hashimotodani, K., et al., *Review of Scientific Instruments*, **69**, n°11, 3746-3750 (1998).

5. Naday, I., Westbrook, E. M., Westbrook, M. L., Travis, D. J., Stanton, M., Phillips, W. C., O'Mara, D., and Xie, J., *Nuclear Instruments and Methods in Physics Research A*, **334**, pp.635-640 (1994).

A Soft-X-Ray Imaging Microscope With Multilayer-Coated Schwarzschild Optics

M. Toyoda,* Y. Shitani, M. Yanagihara, T. Ejima, M. Yamamoto
and M. Watanabe

Research Institute for Scientific Measurements, Tohoku University
2-1-1 Katahira, Aoba-ku, Sendai 980-8577, Japan
*Present address: Nikon Corp. 1-6-3 Nishi-ohi, Shinagawa-ku, Tokyo 140-8601, Japan

Abstract. We constructed a soft-X-ray imaging microscope based on a multilayer-coated Schwarzschild objective. The Schwarzschild objective was designed to have a 50 x magnification and a numerical aperture of 0.25. The mirrors of the objective were coated with a Mo/Si multilayer to reflect the Si L emission. The overall throughput of the objective was 14% at a peak wavelength of 13.3 nm. The 5-μm wide stripe of SiO_2 lithographically patterned was observed under irradiation with an electron beam of 1 μA.

INTRODUCTION

Soft X-ray microscopy has been expected to observe biological samples under wet environment in the water window region with a high spatial resolution. In imaging soft X-ray microscopy, transmitted soft X-rays or those scattered by a sample are focused by soft X-ray optics.[1] On the other hand, Aoki et al.[2] constructed an imaging X-ray emission microscope using Wolter type I optics. However, no imaging type emission microscope has been so far constructed in the soft X-ray region. In this spectral region we observe K emission of the light elements. Owing to the development in fabrication of multilayer coatings, image formation has been possible in the soft X-ray region. The multilayer exhibits a narrow reflection band because of the interference effect.[3] We can design a multilayer mirror to selectively reflect the characteristic soft X-rays of an element to be studied. A multilayer-coated Schwarzschild objective forms an image of a distribution map of the specific element in terms of the characteristic soft X-rays over a large area with a spatial resolution higher than 100 nm as described later. It is also able to provide images of elements distributed in buried layers because of the long mean-free-path of the soft X-rays, which should be compared with photoelectrons. Besides, a Schwarzschild objective has a great merit that the numerical aperture is large to collect the soft X-ray emission. We have constructed a soft-X-ray imaging microscope based on multilayer-coated Schwarzschild objective to observe biological samples in the water window region. In order to demonstrate the possibility of the microscopy as a first step, we have recently succeeded in forming an image of SiO_2 pattern of 5 μm wide by imaging the Si L emission.

CP507, *X-Ray Microscopy: Proceedings of the Sixth International Conference,*
edited by W. Meyer-Ilse, T. Warwick, and D. Attwood
© 2000 American Institute of Physics 1-56396-926-2/00/$17.00

SCHWARZSCHILD OBJECTIVE

In this imaging microscope the soft X-ray emission from a sample produced with an electron beam or synchrotron soft X-rays is focused on a position-sensitive detector using a Schwarzschild objective. From the practical requirements we designed that the Schwarzschild objective had a 50 x magnification and a numerical aperture of 0.25 and that the distance between the sample and the detector was 1 m. By means of ray tracing we finally optimized the distance between the centers of curvature of the concave and the convex mirrors to be 0.69 mm.[4] The ideal resolution was found to be 30 nm on the optical axis.

The mirrors were produced from fused quartz. Their shape error was about $\lambda/10$ (λ = 632.8 nm), and the surface roughness was 0.3 nm. We coated the two mirrors with a Mo/Si multilayer[5] to reflect the Si L emission using a magnetron sputter system. The two mirrors were mounted to a holder to form a Schwarzschild objective using an autocollimator.[6] The overall throughput of the objective was measured using a soft-X-ray reflectometer equipped with a laser-produced-plasma source[7] to be about 14% at a wavelength of 13.3 nm (Fig. 1).

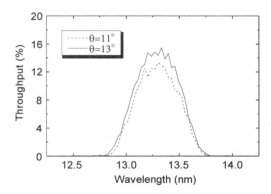

FIGURE 1. Spectral throughput of the Scwarzschild objective measured at angles of incidence of 11° and 13° .

EVALUATION OF THE MICROSCOPE

In order to examine the optical alignment of the microscope, we observed a test sample using visible light in the atmosphere. We mounted a cross hair ruled on a glass to the sample holder and a CCD camera of a 15-μm resolution to the detector holder. By illuminating the cross hair from backside with a tungsten lamp, we observed the image projected on a TV monitor (Fig. 2a). Judging from the profile linescan across the cross hair (Fig. 2b), we ascertained that the resolution of the system was higher than 1 μm, that is, fairly close to the diffraction limit. This result means that the optical alignment for the microscope as well as the Schwarzschild objective was exactly carried out.

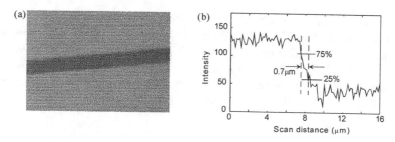

FIGURE 2. a: Transmission image of a cross hair sample, and b: a profile linescan across it.

In order to estimate the performance of the microscope for soft X-rays, we prepared a test sample. It was lithographically patterned with stripes of SiO_2 and W, both 5 μm wide and 50 nm thick, on a Si wafer. Figure 3a shows its SEM image, where the dark area is SiO_2, and the light area is W. The image shows the edge area of the stripe pattern. Figure 3b shows the Si L emission image of the sample obtained for the same area as that shown in Fig. 3a under irradiation with an electron beam of 1 μA accelerated to 2.5 kV. The footprint of the electron beam on the sample was confirmed to be 0.40 x 0.36 mm^2. In this time about 10^5 photons/sec came to the detector. It was recorded for 6 min. The Si L emission emitted from the Si wafer under the W layer may be partially observed. As is shown, the stripe pattern was obviously observed. This is the first result obtained by a soft X-ray imaging microscope.

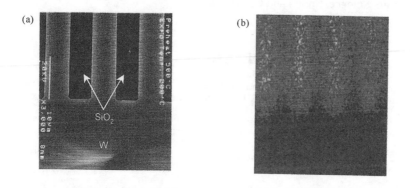

FIGURE 3. a: SEM image of a test sample, and b: its Si L emission image.

Judging from the image shown in Fig. 3b the resolution of the microscope is about 2 or 3 μm. This resolution is lower than what was expected from the result of the visible light test (Fig. 2). The most probable reason is the resolution of the detector. From the magnification of the microscope and the nominal resolution of the detector,

1 μm is resolvable on the object. However, 1 μm may be overestimation. Merely 2 or 3 μm would be expected for the resolution from the detector and the magnification. We need a resolution higher than that of the CCD camera, 15 μm, at least. We should employ an X-ray film or an imaging plate to record the image with high resolution. We should also make the microscope free from the vibration. This improvement will be indispensable for further studies towards a higher resolution.

SUMMARY

We have constructed a soft-X-ray imaging microscope based on a multilayer-coated Schwarzschild objective. The spatial resolution of the microscope was estimated by observing 5-μm wide stripes of SiO_2 lithographically fabricated. The pattern was obviously observed under irradiation with an electron beam of 1 μA. This is the first observation of an image made by focusing soft X-ray emission. This microscope will be used for observation of mineralogical samples as well as biological samples.

ACKNOWLEDGEMENTS

We would like to thank Y. Fuda for making the microscope apparatus. We also thank I. Tanaka for producing the Schwarzschild mirrors. Prof. T. Matsuura is also acknowledged for preparing the test sample. This work was also supported in part by Grant-in-Aid for Scientific Research (B) (Contract No. 10555012) from the Ministry of Education, Science, Sports and Culture, Japan.

REFERENCES

1. For example, Iketaki, Y., Horikawa, Y., Mochimaru, S., Nagai, K., Atsumi, M., Kamijou, H., and Shibuya, M., *Opt. Lett.* **19**, 1804-1806 (1994); Murakami, K., Oshino, T., Nakamura, H., Ohtani, M., and Nagata, H., *Appl. Opt.* **32**, 7057-7061 (1993).

2. Aoki, S., Takeuchi, A., and Ando, M., *J. Synchrotron Rad.* **5**, 1117-1118 (1998).

3. Spiller, E., *Vacuum Ultraviolet Spectroscopy I*, eds. Samson, J.A., and Ederer, D.L., (San Diego, Academic Press, 1998) Chap. 14.

4. Toyoda, M., Master's thesis, Faculty of Engineering, Tohoku University 1999 (in Japanese).

5. Nomura, H., Mayama, K., Sasaki, T., Yamamoto, M., and Yanagihara, M., *Proc. SPIE* **Vol. 1720**, 395-401 (1992).

6. Horikawa, Y., Mochimaru, S., Iketaki, Y., and Nagai, K., *Proc. SPIE* **Vol. 1720** 217-225 (1992).

7. Nakayama, S., Yanagihara, M., Yamamoto, M., Kimura, H., and Namioka, T., *Physica Scripta* **41**, 754-757 (1990).

Instrumentation Advances And Detector Development With The Stony Brook Scanning Transmission X-ray Microscope

M. Feser, T. Beetz, M. Carlucci-Dayton, C. Jacobsen

Department of Physics and Astronomy, State University of New York at Stony Brook, Stony Brook NY 11794-3800, USA

Abstract. Driven by the requirements of new x-ray microscopy instrumentation the Stony Brook microscopy beamline X-1A has undergone considerable evolution [1]. The room temperature scanning transmission X-ray microscope (STXM) has been completely redesigned improving performance, ease of use and compatibility with other experiments. We present the highlights of the new design, the available detectors and the result of early tests of this new microscope.

INTRODUCTION

In this paper we describe the features of the redesigned room temperature STXM operating at beamline X1 of the National Synchrotron Light Source at Brookhaven National Laboratory. It succeeds a STXM [2] that has seen a lot of use during the last decade and has become a valuable tool for scientists addressing problems in biological, environmental, polymer science and other fields. In contrast to the old STXM, where most of the work has been done at the K-absorption edge of carbon, the new STXM extends that work to the K-absorption edges of nitrogen and oxygen, since it is designed to operate in a closed Helium atmosphere. Matched sets of zone plates (ZP) [3] and order sorting apertures (OSA) can be interchanged rapidly to accommodate for different working distance / resolution requirements. Three detectors can be mounted simultaneously on a platform that is motorized and computer controlled. For maximum signal to noise detection a continuous flow proportional counter is employed. A multi channel silicon detector with segmentation is under development that is matched to the geometry of a STXM. This detector has the potential to record not only the "bright field" (absorption map), but simultaneously record the "differential phase contrast" and "dark field" information that are present in the detector plane. Since many experiments at X1 share the beamlines, the new STXM is designed to allow rapid interchange with a cryo STXM [4] located downstream of the STXM.

CP507, *X-Ray Microscopy: Proceedings of the Sixth International Conference,*
edited by W. Meyer-Ilse, T. Warwick, and D. Attwood
© 2000 American Institute of Physics 1-56396-926-2/00/$17.00

SEALED VACUUM CHAMBER WITH HELIUM ATMOSPHERE

The K-absorption edges of oxygen and nitrogen are within the energy range of beamline X1. For high quality imaging and spectroscopy, oxygen and nitrogen have to be removed from the beam path. This is accomplished by putting all microscope components into a vacuum chamber (see Fig. 1). Since some microscope components are not suited for operation in vacuum, the chamber is evacuated and filled with Helium with sensitive equipment turned off during the evacuation period. First tests have shown that with one evacuation-refill cycle we can get the air content down to less than 0.1%, making quantitative oxygen and nitrogen edge spectroscopy work possible. Oxygen and nitrogen are abundant in samples of interest in biology, material and polymer science.

 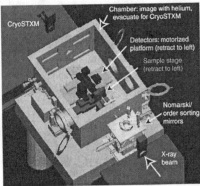

FIGURE 1. Left: The STXM installed at the beamline. **Right:** Schematic View. The vacuum vessel features a Plexiglas top and 3 view ports for accessibility. Evacuation and refilling with Helium takes about 10 minutes. A glove allows access to the interior of the chamber without breaking the seal to make sample changes possible. The chamber is kinematically mounted with 20 μm positioning capability (translation and tilt). An additional vacuum chamber located upstream of the STXM holds a set of 4 mirrors that are used for order sorting and Nomarski differential interference contrast illumination. The Nomarski experiment is done in collaboration with F. Polack (Centre Universitaire, Orsay, France) and D. Joyeux (Institut d'Optique, Orsay, France) [5].

MECHANICAL DESIGN

For diffraction limited imaging in a STXM it is essential to eliminate vibrations in the mechanical connection between the ZP and the sample. Wiggling of the probe on the sample spoils the spatial resolution. The microscope is mechanically isolated from the experimental floor by an air table. The arm that connects the ZP with the sample is kept as light and stiff as possible to minimize the amplitude of remaining vibrations. Fig. 2 shows the mounts of the ZP and OSA. Little or no adjustments have to be made when switching between matched pairs.

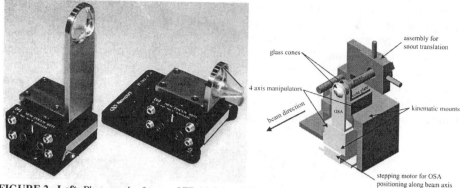

FIGURE 2. Left: Photograph of a set of ZP (right) and OSA (left) mounts. The ZP and OSA are glued to the end of the glass cones, which are connected to 4 axis manipulators on kinematic mounts. Once the units holding the ZP / OSA are aligned in the microscope, they can be removed and put back without loosing the alignment. Many ZP / OSA pairs can be prepared to match the working distance / resolution requirements of different experiments. We are using circular glass cover slips with laser drilled holes as OSAs. **Right:** Perspective view of the ZP / OSA units mounted in the microscope. The glass cones fit into each other, the ZP is stationary and the OSA is motorized along the beam axis to accommodate for changes in focal length as the X-ray energy is tuned.

ELECTRONICS AND SOFTWARE

The microscope electronics can acquire simultaneously 8 channels of analog and 6 channels of pulse data at 16 Bit precision. Multi-channel detectors can be used to increase the total flux capability or extract information of the spatial distribution of X-ray intensity in the detector plane.

All electronics are running on a general purpose interface bus (GPIB) and communicate with a PII class PC equipped with a GPIB ISA bus card running Linux. The software controlling the microscope electronics is written in C++ making use of freely available toolkits for file I/O (Unidata NetCDF), GPIB communication (Linux Lab Project Driver), graphical user interface (Troll Tech Qt), compilation and programming (GNU make, GNU GCC).

X-RAY DETECTORS

For maximum signal to noise detection a continuous flow multi wire gas proportional counter will be used. The counter operates with a 16 ns shaping time and has shown linear response to soft X-ray photons for countrates up to 1 MHz for one channel. This detector is not position sensitive and integrates over the intensity in the detector plane.

For high count rate experiments one can afford to accept the electronic noise of a charge integrating detector, since it adds only very little total noise compared to the photon shot noise of the experiment. A novel charge integrating detector with segmentation is being developed. Soft X-ray photons are being absorbed beneath a shallow P/N junction in high resistivity silicon and create a charge proportional to the energy of the X-ray photon. The charge is integrated over a time window that is matched to the dwell time per pixel of the scanning microscope. Custom built low noise electronics generate a voltage proportional to the collected charge per time and segment that is read by the microscope electronics. The detector will be very attractive for count rates in excess of 100 kHz. The segmentation (see Fig. 3) is designed to pick up information about the spatial distribution of x-rays in the detector plane. Dark field contrast and Nomarski differential phase contrast are two applications of this scheme.

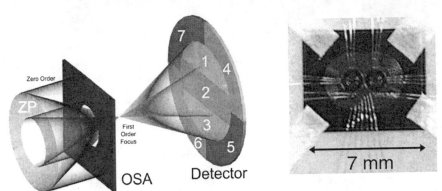

FIGURE 3. Left: The detector has 7 active segments (right). Segments 1-3 are matched to the bright field cone of light coming from the zone plate (left). The pinhole (OSA) selects the light focused to the positive first order. Segments 4-7 of the detector can be used for dark field and differential phase contrast measurements. **Right:** The silicon chip mounted on a PC board for first tests. Two detector structures are on each chip. The extra detector can potentially be used to subtract low frequency noise pickup.

The microscope has a detector platform for three detectors that can be positioned accurately and reproducibly with stepping motors. Switching between the visible light microscope (used for sample inspection and alignment) and X-ray detectors is computer controlled.

HIGH RESOLUTION SCANNING STAGES – FIRST TESTS

For large area scans stepping motors with encoders that offer 0.2 µm resolution and good reproducibility are used. For high resolution scans we employ a piezo actuated flexure stage with capacitance micrometers in closed loop feedback (PI model 731.20). Using the full 100 µm of travel the minimum step size of this stage is 5 nm using a 16 bit digital to analog converter. A scan of a test pattern (see Fig. 4) demonstrates the high spatial resolution and operation without distortions.

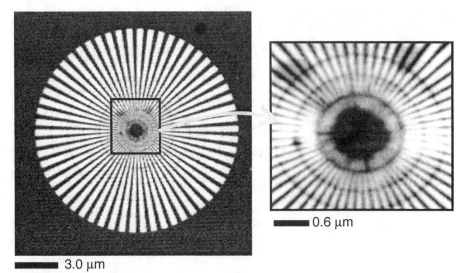

■■■ 3.0 μm

■■■ 0.6 μm

FIGURE 4. First images of a test pattern taken with a 80 μm diameter, 30 nm outermost zone width zone plate at a photon energy of 535 eV. **Left:** The large overview scan demonstrates the orthogonality and linearity of the fine scanning stage. **Right:** A high resolution scan of the inner part of the test pattern. The images have been recorded with a non-optimized detector that did not allow diffraction limited imaging and thus do not represent the resolution limit of the zone plate used.

ACKNOWLEDGEMENTS

We want to thank S. Wirick for her support at beamline X1A and J. Kirz for his input into the design, advice and support. S. Wang, A. Stein and T. Oversluizen have been involved in the early stages of development of the microscope. U. Neuhäusler designed the visible light microscope that is used in the STXM. D. Joyeux and F. Polack initiated the fruitful collaboration on Nomarski differential interference contrast. The gas proportional counter was built in collaboration with G. Smith and B. Yu (Instrumentation Division Brookhaven National Laboratory). The integrating silicon detector with segmentation is being developed in collaboration with P. Rehak and G. DeGeronimo (Instrumentation Division Brookhaven Nation Laboratory). We gratefully acknowledge the support of the U.S. Department of Energy for support under grants DE-FG02-89ER60858 and DE-FG02-96ER14655.

REFERENCES

1. Rarback, H., et al., *Journal of X-ray Science and Technology* **2**, pp. 274-296 (1990). See also Winn, B., et al., "Considerations for a soft x-ray spectromicroscopy beamline" in *Optics for High-Brightness Synchrotron Radiation Beamlines II*, edited by L. E. Berman and J. Arthur, SPIE proceedings vol. 2856, Bellingham, Washington: Society of Photo-Optical Instrumentation Engineers (SPIE), 1998, pp. 100-109.

2. Jacobsen, C., et al., *Optics Communications* **86**, pp. 351-364 (1991). See also Zhang, X., et al., *Nuclear Instruments and Methods in Physics Research A* **347**, 431-435 (1994).

3. Spector, S., et al., *Journal of Vacuum Science and Technology B* **15**, 2872-2876 (1997).

4. Maser, J., et al., *Journal of Microscopy*, in press. Also see Maser, J., et al., "Development of a cryo scanning x-ray microscope at the NSLS" in *X-Ray Microscopy and Spectromicroscopy*, edited by J. Thieme et al., Conference Proceedings XRM96, Springer Verlag, 1998.

5. Polack, F., "A Wavefront Profiler As An Insertion Device For Scanning Phase Contrast Microscopy" in *X-Ray Microscopy and Spectromicroscopy*, edited by J. Thieme et al., Conference Proceedings XRM96, Springer Verlag, 1998, II pp. 201-205. See also Polack, F., these proceedings.

Performance of the CAMD Microprobe utilizing a Kirkpatrick-Baez mirror system

Nicholas Mölders[1], Paul J. Schilling[2], Jon M. Schoonmaker[1]

[1] Ctr. for Advanced Microstructures and Devices, Louisiana State University, Baton Rouge, LA 70806
[2] Mechanical Engineering Department, University of New Orleans, New Orleans, LA 70148

Abstract. A Kirkpatrick-Baez micro-focusing system has been implemented at the CAMD X-ray microprobe beamline. The system utilizes gold-coated float glass mirrors at grazing incidence angles that can be varied up to 15 mrad. Dynamically controlled 4-point bending mechanisms are used to approximate elliptical surfaces. The mirrors were characterized for slope error and roughness considerations with regard to the effective focusing based on the source characteristics and geometry. Ordinary float glass blanks were used after a screening process based on slope error. The measured focused spot size compares well with results of ray-tracing and simple geometric optics calculations. The focused spot size is ~ 18.8 μm x 7.0 μm (σ), with the limiting factor being the size of the CAMD source. Using such a mirror system accommodates operation in both monochromatic and polychromatic ('white') modes. The system therefore allows performance of micro-fluorescence measurements using polychromatic light and micro-XANES measurements over an energy range of 2000 eV – 12000 eV. Characterization of the focusing system and tests of the spectro-microscopy and micro-spectroscopy capabilities are presented.

INTRODUCTION

Due to the spectral photon flux of the CAMD storage ring in the low-energy region (E_c=1660 eV at 1.3 GeV) (1,2), and the scarcity of spatially-resolved x-ray absorption near-edge structure spectroscopy (micro-XANES) capabilities in the energy region around 2000 eV to 3000 eV (3), the CAMD microprobe was optimized to perform measurements in this region. We chose to use an achromatic Kirkpatrick-Baez (KB) mirror focusing system (4-7), because at low incident photon energies mirror systems are very beneficial permitting large grazing angles and hence a large acceptance. Furthermore, mirrors offer the advantage of being achromatic allowing (*i*) the use of high intensity polychromatic (white) x-ray beam to perform spatially-resolved elemental determination via x-ray fluorescence spectroscopy (micro-XRF) and (*ii*) the use of monochromatic x-ray beam for spatially-resolved chemical speciation via x-ray absorption near-edge structure spectroscopy (micro-XANES). To minimize the cost of the focusing system, we decided to use inexpensive float glass blanks as the substrate for our x-ray mirrors. However, the performance of the focusing system is ruled by the quality of the float glass mirror blank, which is determined by the intrinsic rms slope error, and the rms surface roughness of the float glass blank. Therefore, we describe in this paper the initial (before coating) and final slope error screening process of the float glass blanks using a *Continental Optics* long-trace-profiler LTP II. Thereafter, we

CP507, X-Ray Microscopy: Proceedings of the Sixth International Conference,
edited by W. Meyer-Ilse, T. Warwick, and D. Attwood

discuss the performance of the focusing system considering source size, demagnification, and the impact of inexpensive float glass mirrors. However at first, we would like to give a brief description of the microprobe end-station.

LAYOUT OF THE END-STATION

The incident white or monochromatic x-ray beam passes through a motorized 4-jaw entrance slit, defining the beam size, which illuminates the vertical and horizontal focusing mirrors of the Kirkpatrick-Baez (KB) system. After the KB focusing system, the beam travels through a 1" long ion chamber and impinges on the sample. The sample, with a maximal size of 45 mm x 15 mm, is mounted to the sample holder, which is located on the motorized goniometric sample stage assembly under 45° to the incident x-ray beam. The entire end-station, including the KB system and sample stage, is enclosed in a Plexiglas helium chamber. To collect data in transmission a photo-diode is positioned behind the sample in the direct beam. Fluorescence data, on the other hand, is acquired with an energy-dispersive germanium detector, which is also coupled to the helium atmosphere. To obtain an optical microscope image of the sample, a long-working distance microscope is pointed through a view port under 90° to the sample (8).

CHARACTERIZATION OF THE FLOAT GLASS MIRROR BLANKS AND PERFORMANCE OF THE MICROPROBE

To be able to take full advantage of the synchrotron source brightness, the integrated slope error budget σ_{Slope}, given as the sum of the intrinsic slope error and other sources of slope error due to heat load, sag etc., has to be smaller than the vertical source size σ_z, viewed from the position of the mirror (6). This can be expressed as

$$\sigma_{Slope} < \frac{\sigma_z}{f_1},\qquad(1)$$

where f_1 is the distance from the source to the center of the mirror. In the case of the CAMD microprobe set-up with a vertical source size σ_z of 187 µm and f_1 of 9.09 m, we obtain an integrated slope error budget σ_{Slope} 20.5 µrad. To determine the intrinsic rms slope error of the float glass mirror blanks, we used a long-trace-profiler (LTP II) (9,10). 22 mirror blanks were ordered from *Abrisa Industrial Glass* and were measured before and after coating along a 90 mm strip of each mirror. Data were collected from both sides of the mirror blanks. Figure 1 depicts schematically the optical principle of the LTP II. Light from the laser diode (LD) is sent into the beam splitter (BS 1) and 2 prisms (SP, AP), so that 2 parallel, coherent beams emerge downward towards the surface of the float glass mirror (FGM). The beams are reflected from the float glass mirror's surface (FGM) and hit the beam splitter (BS 2), which directs the beams into the Fourier transform lens (FTL). The FTL projects the interference pattern of the beams, onto the linear photo-diode array detector (LPAD). In addition, an extra beam pair is produced in beam splitter 2 (BS 2). It will be reflected onto a reference mirror (RM) and directed through the beam splitter (BS 2) into the Fourier transform lens (FTL). In this way, a 2nd interference pattern is produced and is also detected in LPAD. The 2nd beam pair is used to correct the slope data of the FGM surface for laser beam instability and sag of the beam, on which the optical head (indicated by the dashed box around the main optical system) is moving along while measuring the FGM (11,12).

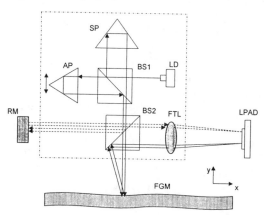

FIGURE 1. Schematic of the optical system of the Long-Trace-Profiler (LTP II) and the float glass mirror blank under test.

PERFORMANCE OF THE MICROPROBE USING FLOAT GLASS MIRRORS AND SUMMARY

The data analysis of the rms slope error measurements involved removal of a 0th and 1st order polynomials, representing the tilt and curvature of the mirror blank, which can be compensated for by the design of the focusing system via dynamically bending the mirror. The result yields the intrinsic rms slope error, which ranged between 9.5 μrad and 70 μrad. Furthermore, the results indicate that each mirror blank has a concave and a convex side. However, so far no correlation between surface sign and the float glass mirror side, which was in touch with the liquid tin during the manufacturing process, could be established. Based on the current geometry and the resulting large radii of curvature (~ 200 m), it is favorable to use the convex (negative sign) surface of the mirror, since this curvature will be reversed during the focusing procedure by applying a bending moment at either side of the mirror. The best mirror with an intrinsic slope error of σ=9.5 μrad was installed as the vertical focusing mirror. The horizontal focusing mirror has an intrinsic slope error of σ=10.7 μrad. Experimental knife-edge and pinhole scans performed through the focal spot yield horizontal and vertical focal spot sizes of 18.8 μm x 7.0 μm. These numbers are in close agreement with the theoretical horizontal and vertical focal spot size of 13.6 μm x 5.3 μm based on a source size of 831 μm x 187 μm (σ_x, σ_z) at 1.3 GeV and a horizontal and vertical demagnification of 61 x 35. Unfortunately, surface roughness measurements of the float glass mirrors were neither conducted before nor after the coating. However, the smooth Gaussian intensity profile obtained by conducting a 2D pinhole scan through the focal spot indicates no scattering out of the focal spot. Therefore, we conclude that the surface roughness is very small and has no significant impact on the performance of the float glass mirror. Hence, inexpensive float glass mirrors proved to be very effective for the CAMD microprobe project.

ACKNOWLEDGEMENTS

We thank Peter Eng and Mark Rivers, GSE/CARS, University of Chicago, for their collaboration on the Kirkpatrick-Baez focusing system and for their continuous support throughout the project; Mark Petri and Len Leibowitz, Argonne National Laboratory for the initiation of the project and the procurement of the instrumentation; This work was supported by Argonne National Laboratory and the State of Louisiana through the Center for Advanced Microstructures and Devices

REFERENCES

1. Craft, B. C., Findley, A. M., Findley, G. I., Scott, J. D. and Watson, F. H., *Nuclear Instruments and Methods in Physics Research B* **40/41**, 379-383 (1989).

2. Stockbauer, R. L., Ajmera, P., Poliakoff, E. D., Craft, B. C. and Saile, V., *Nuclear Instruments and Methods in Physics Research A* **291**, 505-510 (1990).

3. Susini, J., Barrett, R., Kaulich, B., Oestreich, S. and Salome, M., "The X-ray Microscopy Facility at the ESRF: a status report," presented at the 6th International Conference on X-Ray Microscopy, 1999.

4. Kirkpatrick, P. and Baez, A., *Journal of the Optical Society of America* **38**, 766-774 (1948).

5. Kirkpatrick, P., Baez, A. and Newell, A., *Physical Review* **73**, 535-536 (1948).

6. Yang, B. X., Rivers, M. L., Schildkamp, W. and Eng, P. J., *Review of Scientific Instruments* **66**, 2278-2280 (1995).

7. Eng, P. J., Newville, M., Rivers, M. L. and Sutton, S. R., "Dynamically Figured Kirkpatrick Baez X-Ray Micro-Focusing Optics," presented at the X-Ray Microfocusing: Applications and Technique, 1998.

8. Mölders, N., Moser, H. O., Saile, V. and Schilling, P. J., *Scientific Report* FZKA 6314, 1999.

9. Takacs, P. Z. and Qian, S., United States of America (1989).

10. Takacs, P. Z., Feng, S. K., Church, E. L., Qian, S. and Liu, W., "Long trace profile measurements on cylindrical aspheres," presented at the Advances in Fabrication and Metrology for Optics and Large Optics, 1989.

11. Irick, S. C., McKinney, W. R., Lunt, D. L. J. and Takacs, P. Z., *Review of Scientific Instruments* **63**, 1436-1438 (1992).

12. Irick, S. C., *Review of Scientific Instruments* **63**, 1432-1435 (1992).

X-RAY FLUORESCENCE PROJECTION MICROSCOPY IN A SEM : FIRST RESULTS

D. Erre, H. Jibaoui and J. Cazaux

DTI EP 120 CNRS, BP 1039, Université de Reims
51687 REIMS Cedex 2, France
E-mail : damien.erre@univ-reims.fr

Abstract. A simple arrangement is proposed which permits recording of X-ray fluorescence images (non analytical) of surfaces and interfaces inside a conventional SEM equipped with a two dimensional detector and a special mechanical set-up.

INTRODUCTION

X-Ray Fluorescence Projection Microscopy (XRFPM) consists in acquiring global images by X-ray fluorescence emitted from a specimen surface irradiated with a divergent X-ray beam. At the present stage of development only non analytical images are obtained. Compared with another conventional imaging technique by X-ray fluorescence (i.e. images acquired from the scanning of the sample across the beam) [1], the advantage of the present technique is the rather short acquisition time : less than ten minutes for a surface of 1 to 5 mm^2.

The work presented in this paper shows the feasibility of this method which provides chemical and topographical contrast in images.

The basic instrument is a scanning electron microscope which has been transformed into a X-ray projection microscope [2]. With the aim to perform XRFPM experiments the main modification concerns the relative position of the X-ray source and of the specimen with respect to the cooled CCD camera.

PRINCIPLE AND PERFORMANCE

The Principle of XRFPM consists in irradiating a specimen with an X-ray divergent beam issued from a point source and in collecting the X-ray fluorescent beam, through a pinhole (principle of the stenope) with an X-ray sensitive CCD camera, the image plane of which being parallel to the surface of the specimen (Fig. 1).

CP507, *X-Ray Microscopy: Proceedings of the Sixth International Conference,*
edited by W. Meyer-Ilse, T. Warwick, and D. Attwood

From geometrical argument, the direct magnification G of the obtained images is easily evaluated (Fig. 2). It is given by the expression :

$$G = A'B'/AB = b/a$$

Where "a" is the specimen to pinhole distance and "b" the pinhole to detector distance.

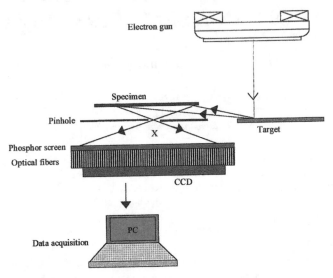

FIGURE 1. Schematic arrangement of the fluorescence X-ray microscope.

In the plane of the specimen, the lateral resolution r_o of this technique is given by the expression :

$$r_o = \frac{r_i}{G} = d\left(1 + \frac{1}{G}\right)$$

in which d is the diameter of the pinhole, and r_i is the intrinsic resolution of XRFPM.

The value r_o is function of two parameters : the direct magnification G and the diameter of the pinhole d. We must notice that this expression remains valid as far as r_i is larger than the lateral resolution of detector (for the laboratory instrument: 60 μm). For example, if a=1 mm, b=9 mm and d = 10 μm, the lateral resolution is $r_o \approx 11$ μm. On the other hand, the region of the specimen that can be imaged is function of the direct magnification.

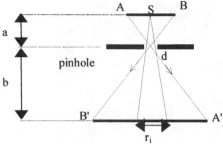

FIGURE 2. Schematic arrangement to calculate the lateral resolution of the XRFPM.

RESULTS

The figure 3 shows an image of a cooper grid obtained by X-ray fluorescence projection microscopy. This sample was chosen to illustrate the feasibility of this technique with our laboratory instrument.

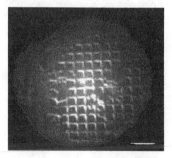

300µm

FIGURE 3. XRFPM image obtained with a cooper gate. Operating conditions : target Mo; electron beam E_o=30 keV and I_o=1µA; acquisition time 20 min; direct magnification G=3; pinhole diameter d=10 µm.

Application of XRFPM to the topography

When the chemical composition of the sample is homogeneous, the image obtained by XRFPM is only sensitive to the topography, the surface being irradiated with a low-angle incident beam (3° to 10°). This phenomenon is illustrated in Fig 4. The image is that of a pure nickel electrochemical deposit. The same surface observed by SEM presents craters of 100 to 200 µm in diameter.

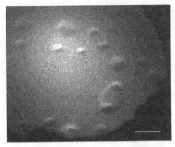

200 μm

FIGURE 4. Topographical XRFPM image of a nickel deposit. Operating conditions : target Zn; electron beam E_o=30 keV and I_o=1μA; acquisition time 20 min; direct magnification G=3; diameter of the pinhole d=10 μm.

Application of XRFPM to the tomography

With this technique, in-depth profiling is possible since the penetration depth can be controlled by changing the incidence angle (obtained from a slight shift of the point X-ray source). The image difference between two images acquired with two positions of S (two depths z and z+dz) permits to give the map of the layer (thick dz) in volume (Fig. 5).

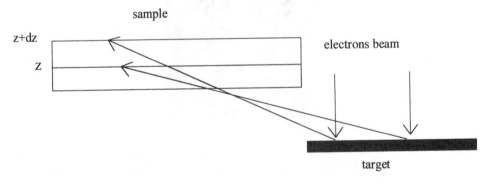

FIGURE 5. Schematic arrangement for the implementation of tomography in XRFPM.

In Fig. 6, the images of copper precipitates in an aluminum matrix have been obtained by XRFPM with different shift of the X-ray point source.

The imaged specimen region is about 3 mm². In this region there are 4 bright patches (regions surrounded A, B, C and D) corresponding to copper precipitates. These patches present different intensities which may be attributed to the difference in the in-depth position or to the difference in the dimension (volume) of each detail. The result is obtained by shifting the source S increasing the grazing angle from 1° to 3°. The corresponding irradiation depth are 60 μm, 80 μm and 90 μm. Indeed in I_3 the point D disappeared, showing that its the depth of this precipitate is greater than 60 μm but the depths of the precipitate corresponding to the points A, B and C are less than 60 μm.

These recent results illustrate the feasibility of the tomography but are not quantitative for the moment.

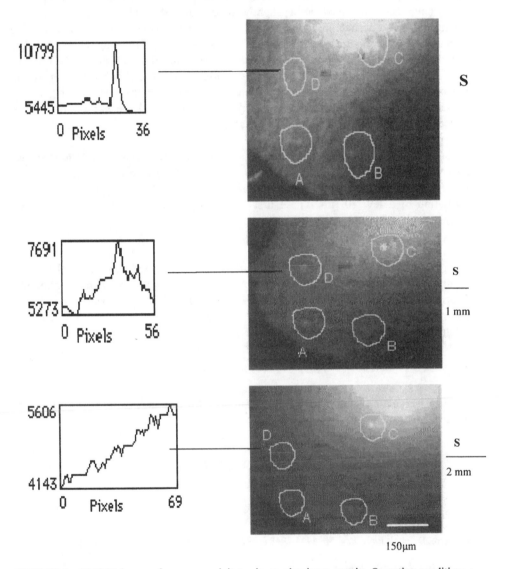

FIGURE 6. XRFPM image of copper precipitates in an aluminum matrix. Operating conditions : target Mo; electron beam E_o=30 keV and I_o=1μA; acquisition time 40 min; direct magnification G=3; diameter of the pinhole d=10 μm.

CONCLUSION AND PROSPECTS

In this work, the first experimental results obtained in XRFPM have been reported. They have been obtained in a slightly modified SEM. The same experimental arrangement also permits to acquire global fluorescence images by the pinhole imaging technique. But at the present time, these fluorescence images obtained by irradiating a sample surface with a divergent X-ray beam are non analytical. This could be done by means of two methods : the use of filters (thin foils) [3] or the change of the energy of the primary radiation. With these solutions the contrast in the image owing to the presence of an element in a matrix could be enhanced and the surface or volume concentration of this element could be measured.

Compared with the other classical imaging techniques using X-ray fluorescence (scanning of the specimen followed by elementary reconstruction of images) , this set up requires acquisition time of only tens of minutes for a 2 mm^2 region with a lateral resolution of a few 10 µm. These first results prove the feasibility of this method delivering chemical and topographical contrast in images.

Since the images are obtained in the digital form they can undergo any mathematical treatment. From the signal measured it should be possible to obtain quantitative information on the local composition of a sample. This work we expect to perform in the near future opens a way for several interesting applications in particular in electrochemical systems analysis.

ACKNOWLEDGMENTS

The authors would like to thank Mouze D. for help during the realization of the experiments.

REFERENCES

1.   Janssens, K., et al, *J. Anal. At. Spectrom.* **9**, 151 (1994).

2.   Erre, D., et al, *Phys. Rev.* **56**, 4944-4948 (1997).

3.   Cazaux, J., *X-Ray Spectrom.* **28**, 9-18 (1999).

Total Reflection X-Ray Microscopy in a SEM : 1. Principle and Performance

H. Jibaoui, D. Erre and J. Cazaux

DTI EP 120 CNRS, BP 1039, Université de Reims
51687 REIMS Cedex 2, France
E-mail : hussein.jibaoui@univ-reims.fr

Abstract. Total Reflection X-ray Microscopy permits to obtain topographic images of surfaces and interfaces. The principle and performance of this new microscopy are described and discussed. The experimental arrangement is installed inside a conventional SEM equipped with a CCD camera and with a special mechanical set-up . Some images illustrating the performance are also shown.

INTRODUCTION

The refractive index of all the materials is less than the unity in the x-ray domain. Then, x-rays are reflected from all material surfaces irradiated at glancing angles less than the critical angle. Benefits of this physical fact are taken in x-ray reflectometry operated at around the critical angle for obtaining information on the roughness of the investigated surface and on density changes in the direction perpendicular the this surface. X-ray reflectometry is widely used in modern technologies for the control of the mirror quality, or of the microelectronic components as well as the study of the metal corrosion and of the multilayer systems[1-5].

The same physical fact may be used for obtaining total reflection images of the irradiated surfaces but, to the authors' knowledge and excepting their previous contribution to this technique [6,7], the corresponding microscopy, (Total Reflection X-Ray Microscopy :TRXRM), has never been suggested in the past .The present paper reports mainly on the principles and performance of this microscopy, most of the recent experimental results being given in the next article [8]. This new technique permits to acquire rapidly (a few seconds) digital images of a specimen surface or interface, the visualized area is of around 1-2 mm², the lateral resolution is in the micron range and the vertical resolution is in the nm range The basic instrument is a SEM which has been previously transformed into a X-ray projection microscope: the x-ray source results from the electron bombardment of a thin foil target. In order to perform TRXRM experiments, the main changes concern the relative position of the X-ray source and of the specimen with respect to the cooled CCD camera.

CP507, *X-Ray Microscopy: Proceedings of the Sixth International Conference,*
edited by W. Meyer-Ilse, T. Warwick, and D. Attwood

PRINCIPLE

The principle of TRXRM is the following one : a flat specimen is irradiated with a divergent x-ray beam issued from a point source and the specularly reflected X-rays are then collected with a CCD camera, parallel to the specimen.

The x-ray images have been obtained from the total reflection phenomenon and the region of the specimen that can be imaged corresponds to incidence angles θ smaller than the critical angle θ_c given by :

$$\theta_c(rad) = 2.28\,10^{-3}\lambda\left[\rho\frac{(Z+f')}{A}\right]^{1/2}$$

with λ : X-ray wavelength (Å) ; ρ : density of material (g/cm^3) ; Z : atomic number; A: average atomic mass; f' is the real part of the average atomic scattering factor. The value of θ_c ranges from approximately 0.1 to 100 mrad, depending on the wavelength and the type of materials.

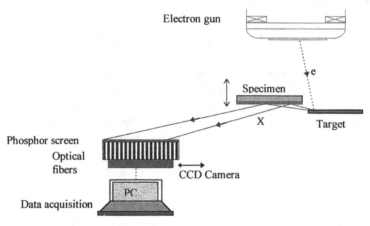

FIGURE 1. Schematic arrangement of the total reflection X-ray microscope.

PERFORMANCE

From geometrical arguments and like in shadow X-ray microscopy, the direct magnification G of the obtained images is easily calculated. It is given by (see fig 2a) :

$$G = \frac{\overline{A'B'}}{\overline{AB}} = (a+b)/a$$

with $a = \overline{S'H}$ and $b = \overline{HO}$.

In the specimen plane, the geometrical lateral resolution $r_g(/\!/)$ is also given by:

$$r_g(//) = s + (d - s)/G$$

But in TRXRM, there is, in addition , a vertical resolution given ,$r(\perp)$, by (Fig.2.b) :

$$r_g(\perp) = r_g(//)\,\theta\,/\,2$$

with s :size of the point source (1 - 2 μm); d: lateral resolution size of the detector (60 μm ; Fig 2a).

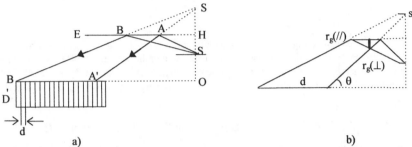

a)   b)

FIGURE 2. a) Geometrical configuration of TRXRM. b) Geometrical configuration used to evaluate the horizontal and vertical resolutions. E : specimen, D : two dimensional detector, S : point source, S' : virtual point source, AB : imaged area of the specimen and A'B' : image on the detector.

There are two manners for evaluating the minimal detectable difference of height Δh_{min} of TRXRM . One is deduced from the minimum contrast detectable in the image (Fig. 3) :

$$MN = \Delta h_{min} = [(a + b)/2]\left(1 - \sqrt{1 - \frac{3}{\sqrt{N_{sat}}}}\right)$$

where N_{sat} : number of photons saturating the CCD camera.

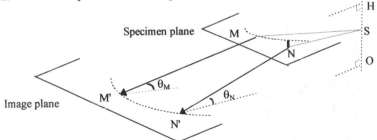

FIGURE 3. Geometrical configuration to calculate the difference of height Δh_{min} on surface by a contrast variation.

The other manner is only related the lateral resolution in the plane of the camera (Fig.4.a) :

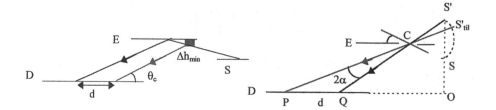

FIGURE 4. a) Geometrical configuration to evaluate the difference of height Δh_{min} on surface from geometrical arguments. b) Configuration used to evaluate the angular sensitivity of TRXRM: When the plane of a detail is rotated of the angle α, the reflected intensity is missing in the initial direction CP and it is deflected of 2α in the CQ direction. The corresponding well and bump can be detected when PQ>d.

The sensitivity for the detection of tilted zones, α, on the surface is given by (Fig.4.b) :

$$\alpha = \theta^2 \, \overline{PQ} / (\overline{OS} - \theta \, \overline{PQ})$$

and the minimal detectable tilt, α_{min}, is obtained with $\overline{PQ} = d$ within the approximation $\overline{OS} \gg \theta \, \overline{PQ}$

$$\alpha_{min} \approx d \, \theta^2 / \overline{OS}$$

Finally the performance may also be characterized by the sensitivity to detect a mass density changes on the surface. Such a superficial density variation $\Delta \rho$ causes a change in the critical angle value and then the transition region (between the non-reflection and the total reflection) is shifted from $\overline{T_1 T_2}$ which is given by (see Fig.5):

$$\overline{T_1 T_2} = \overline{OS'} \left[\frac{1}{\theta'_c} - \frac{1}{\theta_c} \right]$$

The minimum detectable density variation corresponds to $\overline{T_1 T_2}$ when it is equal to the lateral resolution of detector $\overline{T_1 T_2}_{\,min} = d$.

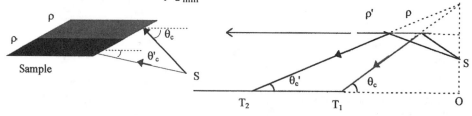

FIGURE 5. Geometrical configuration to evaluate the shift of the transition region caused by a superficial density change.

RESULTS AND DISCUSION

The specimen is now a bulk specimen set parallel at the CCD detector and the X-ray source results from the electron bombardment of a bulk target set below the specimen plane (like in Fig .2.a with a=50 μm and b=550 μm). This target is made of a pure copper block and a 20 keV electron beam generates mainly CuLα photons (λ=13.34 Å) in addition to the bremsstrahlung. The scanning of the electron beam on the target permits to change easily the angle of incidence of the emitted X-rays on the specimen surface. We present two images illustrating the potential of TRXRM.

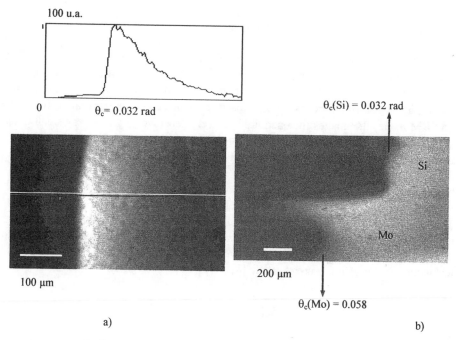

a)   b)

FIGURE 6. a) TRXRM image of a silicon crystal with a perfectly flat surface : (target : Cu; I_o =0.8 μA; E_o=20 keV; acquisition time : t=30 seconds ; direct magnification G=12); top part : intensity profile. b) TRXRM image of a sample make up of two twin regions one made of silicon and the other of molybdenum : (target : Cu; I_o =0.6 μA; E_o=18 keV; acquisition time : t=40 seconds ; direct magnification G=12).The x-ray source is situated on the left of each image.

Figure 6.a shows a TRXRM image of a silicon crystal with a perfectly flat surface obtained with CuLα radiation. The corresponding critical angle is $θ_c$= 0.032 rad.
The intensity profile (see the top left of fig 6) corresponds to the illumination of the CCD camera .From left to right one observes, first, an absence of reflection ; next ,the transition zone preceding the critical angle. Beyond the maximum, a monotone decrease of the illumination (as $θ^3$) corresponds to the change of the solid angle of collection [6]. It may be observed that the local variations in intensity correspond to

topographical variations of the specimen combined to the surface heterogeneities of the phosphor converter (which are in fact responsible of the contrast observed on fig.6a).

Figure 6.b presents a TRXRM image of a sample make up of two homogeneous regions but having different densities : Si (ρ= 2.33 g/cm^3) and Mo (ρ= 10.2 g/cm^3). This image is also obtained with Cu Lα radiation. It shows the shift between the two reflection regions related to the different critical angles. The transition region of Mo is set before the one of Si, because θ_c(Si)=0.032 rad<θ_c(Mo)=0.058 rad. The corresponding shift may be easily measured, $T_1 T_2$, on this type of image: it corresponds here to 320 pixels (7 mm for G=12).

CONCLUSION

The principle and the performance of the TRXRM have been reported. They have been obtained in a slightly modified SEM. The same experimental arrangement also permits the acquisition of Kossel patterns in the reflection mode from single crystals irradiated with hard X-ray photons at incidence angles larger than θ_c [6]. In the TRXRM mode, the images then acquired are very sensitive to a few very small details isolated on a flat surface and then they provide an information different from that, the mean roughness, obtained from x-ray reflectivity experiments. The images are in a digital form and then easy to proceed [7]. The lateral resolution is only a few microns but the resolution normal to the surface is about a few tens nm or less, while the minimal angular deviation that can be detected is a small fraction of θ_c.

This new technique seems to us very promising because of its potential applications in the surface and near interfaces studies mainly in the case of liquid/solid interface topography where efficient techniques are rather scarce. Some applications of this technique are shown on the next article [8].

REFERENCES

1 Parratt, LG., *Physical Review* **95**, n°2, 359-369 (1954).

2 Nevot, L., Croce, P., *Revue de Phys. Appl.* **15**, 761-779 (1980).

3 Naudon, A., Chihab, J., Goudeau, P., Minault, J., *J. Appl. Crystallogr.* **22**, 460-464 (1989).

4 De Boer, DKG, *Phys. Rev. B* **49**, 5817 (1994).

5 De Boer, DKG., Leenaers, AJG., Van den Hoogenhof, WW., *X-ray Spectrometry* **24**, 91 (1995).

6 Erre, D., Jibaoui, H., and Cazaux, J., *Journal de Physique IV* **C7**, 393-398 (1996).

7 Erre, D., Jibaoui, H., and Cazaux, J., *X-Ray Microscopy and Spectromicroscopy*, Thieme, Schmal, Rudolph and Umbach (Eds). Springer-verlag, II 237-II242 (1998).

8 Jibaoui, H., Erre, D., and Cazaux, J., this volume.

Total Reflection X-Ray Microscopy in a SEM : 2. Application to Surfaces and Interfaces

H. Jibaoui, D. Erre and J. Cazaux

DTI EP 120 CNRS, BP 1039, Université de Reims
51687 REIMS Cedex 2, France
E-mail : hussein.jibaoui@univ-reims.fr

Abstract. Some applications of Total Reflection X-Ray Microscopy (TRXRM) are given. It is demonstrated that this new imaging technique permits to acquire rapidly (a few seconds) digital images related to the topography of the irradiated surfaces. An important illustration is the direct imaging of slightly buried solid/solid interfaces.

INTRODUCTION

Total Reflection X-Ray Microscopy (TRXRM) is a new microscopy providing images of surfaces and interfaces [1]. The illustrations of its applications on a solid surface concerns the direct imaging of scratches (lateral dimension of a few μm), of agglomerates (height of a few hundreds nm), and of slightly tilted region (tilt angle of about 10 μrad). The direct imaging of solid/solid interfaces is also shown and the constraints to be satisfied by the specimen are established.

The instrument is a laboratory equipment based on a scanning electron microscope which was first transformed into a X-ray projection microscope. To perform TRXRM experiments, the main changes concern the relative position of X-ray source and of specimen with respect to the cooled CCD camera [1].

PRINCIPLE

The principle of TRXRM has been given in the previous contribution to this meeting [1] (see also Fig.1 for an illustration.). It consists in irradiating a flat specimen at grazing incidence with a divergent X-ray beam issued from a point source and the specularly reflected X-rays are then collected with a CCD camera, parallel to the surface of the specimen [1-3]. The X-ray imaging is based on the total reflection phenomenon and the region of the specimen that can be imaged corresponds to incidence angles θ smaller than critical angle θ_c (typically $\theta_c < 100$ mrad).

The performance of this technique in our laboratory instrument is a few μm for lateral resolution $r_g(//)$, a few 10 nm for vertical resolution $r_g(\perp)$, a few ten nm for

CP507, *X-Ray Microscopy: Proceedings of the Sixth International Conference,*
edited by W. Meyer-Ilse, T. Warwick, and D. Attwood
© 2000 American Institute of Physics 1-56396-926-2/00/$17.00

minimal topographical change in height, a few μrad for minimal detectable tilt on the surface [1].

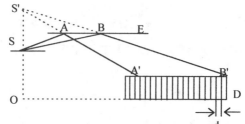

FIGURE 1. Schematic arrangement of the total reflection X-ray microscope. E : specimen, D : two dimensional detector, S : point source, S' : virtual point source. AB : imaged area of the specimen, A'B' : image on the detector.

APPLICATION OF TRXRM TO SOME SOLID SURFACES

Fig. 2 shows the TRXRM image of Ag surface obtained with CuLα radiation (λ = 13.34 Å). A scratch of lateral dimension 50 μm is present on this surface and this scratch gives rise to the oblique dark band (absence of reflection) situated, on the right, in the region of total reflection (bright area). The black area on the left corresponds angles of incidence larger than the critical angle.

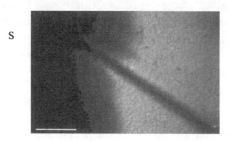

100 μm

FIGURE 2. TRXRM image of Ag surface which presents a scratch defect. Experimental conditions : target : Cu ; I_o = 0.4 μA; E_o = 17 keV ; acquisition time t = 20 s ; direct magnification G=10. The position of the X-ray source is S on the top left.

Also obtained with the CuLα radiation, Fig.3 shows the TRXRM image of a Si surface on the top of which, there are two small clusters (agglomerates) of height ranging in a few 100 nm. In the reflection image, these two details give rise to the two tip-shape shadows (on the right; the X-ray source ,S, being on the left).

390

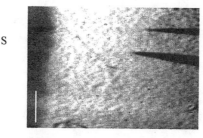

80 μm

FIGURE 3. TRXRM image of a silicon surface with two small details (clusters) leading to the two shadows seen on the right. Experimental conditions : target : Cu ; I_o =0.8 μA ; E_o = 20 keV ; acquisition time t = 40 s ; direct magnification G=12. The X-ray source, S, is on the left.

Obtained again with the CuLα radiation, Fig. 4 corresponds to the TRXRM image of a Ag surface. The contrast changes (indicated by A, B, C, D) on this image corresponds to topographical variations related to height changes on the specimen surface. From considerations on the minimum contrast being detectable, the difference in height can be estimated from [1] :

$$\Delta h_{AB} = [(a+b)/2] \left[1 - \sqrt{\frac{I_A}{I_B}}\right]$$

where Δh_{AB} is the difference of height between A and B, I_A and I_B respectively the intensities measured on the areas A and B.

For example : Δh_{BA}~7 μm, Δh_{BD}~5 μm and Δh_{BC}~7μm.

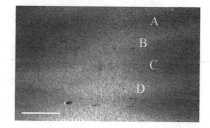

FIGURE 4. TRXRM image of a silver surface related to topographical changes. Experimental conditions : target : Cu ; I_o = 0.9 μA ; E_o = 20 keV ; acquisition time t = 120 s ; direct magnification G=10. The X-ray source, S, is on the left.

The image of a silicon surface having a slightly tilted region (illumination from the left) is shown on Fig.5 (CuLα radiation). When the angle of tilt, α, increases the local incident angle θ (but keeping it below its critical value) a brighter region followed by a darker one are expected [2]. The darker region (Q) corresponds to a deficit in intensity in the initial direction because of the beam deflection by a tilted zone and the same beam deflection reinforces the brightness of another region (P). From simple geometrical arguments, the angle of tilt, α, may be easily deduced from the measurement of the distance $\overline{PQ}$ as obtained from the recorded intensity profile (see Fig.5b) :

$$\alpha\,(\text{rad}) = \frac{\theta\ \overline{PQ}}{b - \overline{PQ}}$$

Here (Fig.5a) the angle of tilt is estimated to be at about 0.6 mrad.

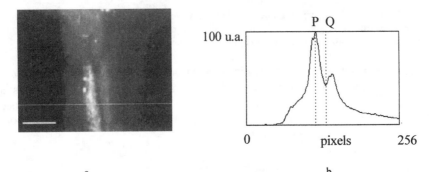

FIGURE 5. a) TRXRM image of a silicon crystal with a tilted zone defect. Experimental conditions : target : Cu ; $I_o = 0.6\ \mu A$; $E_o = 15$ keV ; acquisition time t = 60 s ; direct magnification G=12. The X-ray source, S, is on the left. b) : Profile taken across the image a).

APPLICATION OF TRXRM TO SOLID/SOLID INTERFACES

The large penetration depth of X-rays permits to consider positively the possible imaging of solid/solid interfaces. For total reflection experiments the additional constraint is that the critical angle at the vacuum/specimen surface, θ_{cs} must be less than the critical angle, θ_{ci}, at this solid/solid interface. This angular constraint may be converted into a refractive index constraint or in a mass density constraint (via the well known eq. relating these different parameters : see this eq. in ref.[1]) :

$$\rho_s < \rho_i \Leftrightarrow \theta_{cs} < \theta_{ci}$$

In this situation and for a layered specimen, two distinct regions are expected in the image plane (see Fig.6) :

i) the region which corresponds to : $\theta_{cs} < \theta < \theta_{ci}$. The incident X-ray beam penetrates into the top layer before being reflected by the interface with the second layer.

ii) the region where $\theta < \theta_{cs}$. The incident X-ray beam is only reflected by the surface. In the object plane the extent of imaged zone of the interface, T_sT_i, can be estimated from geometrical arguments :

$$T_sT_i = \frac{T'_s\,T_i}{G} = OS'\left[\frac{1}{\theta_{ci}} - \frac{1}{\theta_{cs}}\right]$$

Moreover it is also possible to evaluate the thickness of the layer (from simple geometrical considerations or from the absorption law applied to the first layer).

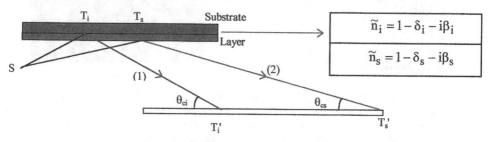

$$\tilde{n}_i = 1 - \delta_i - i\beta_i$$

$$\tilde{n}_s = 1 - \delta_s - i\beta_s$$

FIGURE 6. Schematic principle of TRXRM applied to the solid/solid interface imaging.

Fig.7 (left) illustrates these above points. It is a TRXRM image (CuLα rad.) of a sample made up of 3 layers Al/resin/Au (see Fig.7 right). The two distinct region are easily identified on this image. The first in dark grey ($T'_iT'_s$) on the left hand side corresponds to the buried resin/Au interface (absorption in the Al + resin. layer). The second in light grey ($T'_sT'_o$) corresponds to the total reflection on the Al surface.

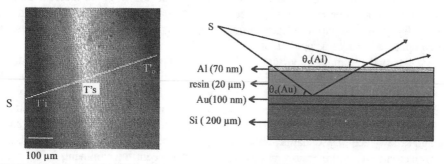

Al (70 nm)
resin (20 μm)
Au (100 nm)
Si (200 μm)

100 μm

FIGURE 7. Left : TRXRM image (obtained with CuLα) of a sample made up of 3 layers Al/resin/Au. This layered structure is shown on the right. Experimental conditions : target : Cu ; I_o =1 μA ; E_o = 25 keV ; acquisition time t = 120 s ; direct magnification G=10. S indicates the source position on the left.

To obtain the TRXRM image shown on Fig.8 (left), the specimen being imaged was similar to that imaged on Fig.7 but small details have been added on the gold/resin interface (see Fig.8 right). The first reflection region on the left (T_iT_s) corresponds to the resin/Au interface. It is striated by the shadowing effects of the details set at the interface. Their height is estimated to be of around 500 nm.

No contrast effect can be identified on second reflection region which appears very bright and which corresponds to the total reflection on the Al surface.

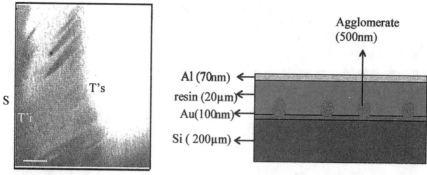

FIGURE 8. Left : TRXRM image (CuLα radiation) of a sample made up of 3 layers Al/resin/Au interface with some agglomerates (defects) sitting on the gold interface: see the cross-section of the specimen on the right. Experimental conditions : target : Cu; I_o = 1 μA ; E_o = 25 keV ; acquisition time t = 120 s; direct magnification G=20. The X-ray source, S, is on the left.

CONCLUSION

The Figures 2 to 5 have demonstrated the possibility of surface imaging by TRXRM. The Figures 7 and 8 have demonstrated that the imaging of interfaces buried below an opaque coating, 20 microns thick, is also possible. Over techniques such near field microscopy the advantage of TRXRM for the surface imaging is its speed of operation for identifying an isolated detail on very flat and large surfaces, a domain where X-ray reflectometry cannot be used. Conversely, the use of TRXRM is of limited interest for the investigation of surfaces showing many topographic details because of the complicated images being then obtained.

It seems that the most important applications of TRXRM are related to the direct imaging of deeply buried solid/solid interfaces, a domain where there are strong needs and where the techniques to be used, are very scarce. We hope that the present work opens a new way for many interesting applications in particular in electrochemical systems and in the investigation of multilayer systems. As the exposure time is short enough (a few seconds), dynamical phenomena could be observed too and the (missing here) microanalytical dimension may be given by the recent development, in the same basic equipment, of X-ray fluorescence microscopy at grazing incidence [4].

REFERENCES

1   Jibaoui, H., Erre, D., and Cazaux, J., this volume.

2   Erre, D., Jibaoui, H., and Cazaux, J., *X-Ray Microscopy and Spectromicroscopy*, Thieme, Schmal, Rudolph and Umbach (Eds), Springer-verlag, II237-II242 (1998).

3   Erre, D., Jibaoui, H., and Cazaux, J., *Journal de Physique IV* **C7**, 393-398 (1996).

4   Erre, D., Jibaoui, H., and Cazaux, J., this volume.

Sealed Cell For In-Water Measurements

K. Kaznacheyev, A. Osanna, B. Winn.

Department of Physics and Astronomy, SUNY, Stony Brook, NY 11794, USA.

Abstract. We describe here the use of a very simple sealed cell to carry out absorption spectroscopy measurements on hydrated thin films, particular for oxygen K edge studies, where the sample thickness is limited by about 0.5 microns due to the strong water absorption. The cell is small enough to be mounted on a standard TEM holder, and has been used in the vacuum chamber of the Stony Brook cryo STXM. An application includes the measurement of the Si_3N_4 wetting properties and spectroscopy of amino acid solutions. Comparison with frozen hydrated glycine shows that special care must taken to avoid the drying of submicron water layer during sample preparation, since precipitation may take place and influence XANES spectra.

INTRODUCTION

Soft x-ray microscopes offer unique capabilities for studies of hydrated specimens and wet specimen chambers of different design have already been used in transmission x-ray microscopes [1] for the energy below oxygen K-edge through several μm-thick water films. When crossing the oxygen absorption edge, hydrated specimens should be no more than about 0.5 μm thick to obtain high quality absorption spectra. In addition, the sample must be held in a vacuum or in an atmosphere with less than about 0.01% oxygen over a path length of several mm so as to avoid additional absorption structures. Such experimental requirements offer challenges to oxygen edge spectroscopy of liquids. We describe here the use of a very simple sealed cell to carry out absorption spectroscopy measurements on hydrated thin films. The cell may be mounted on a standard TEM holder, and was used in the vacuum chamber of the Stony Brook cryo STXM. This approach has made it possible to obtain oxygen near-edge spectra of water, and to study wetting phenomena with a particular focus on the drop shape. In studies of glycine in solution, we have also learned of complications (e.g., the drying of sub-μm water layers may lead to precipitation of the solute), and the precautions that should be taken to avoid them.

EXPERIMENTAL SETUP

The cryo Scanning Transmission X-ray Microscope (STXM) is operated at the National Synchrotron Light Source (NSLS) at Brookhaven National Laboratory [2]. It is capable of imaging biological and other specimens with a spatial resolution of about 60 nm in present experiments. The beamline was recently upgraded to increase the energy resolving power to 2000-4000 for the energy range 250-700 eV. Modified

CP507, *X-Ray Microscopy: Proceedings of the Sixth International Conference,*
edited by W. Meyer-Ilse, T. Warwick, and D. Attwood
© 2000 American Institute of Physics 1-56396-926-2/00/$17.00

cryo-TEM specimen holders allow one to maintain specimen at temperature between 90 K and 320 K. The specimen holder is inserted into a JEOL 1000 TEM-type rotational airlock for quick transfer of the sample into the experimental chamber, where the vacuum is at the 10^{-6} torr level. The size of the Si chip ($2*2$ mm^2) was chosen to fit into the existing holder. For surface wetting studies, the Si_3N_4 windows were not treated beyond the KOH etch, acid rinse, and water rinse steps of the window fabrication process. Several different window sizes all with 100-nm thick Si_3N_4 etched membrane were tried. The best performance was found for small windows ($200*200$ μm^2), but bigger ones ($500*500$ μm^2) are also robust enough to withstand the vacuum forces. The choice of the glue used for sealing the wet cell is mostly determined by vacuum and cryo conditions; we have found that Torr Seal (Varian Vacuum Products) works well, though clear fingernail polish can also be used with reduced vacuum requirements. Water drop as small as possible (about 0.2-0.5 μL mostly determined by capillary forces) was placed on one Si_3N_4 window. The other Si chip was gently pressed upon the lower one and aligned. External water was removed by filter paper. The cell was sealed and left to dry for between 10 minutes and an hour, depending on the glue used. Afterwards, the cell was examined under the optical microscope to check the water film thickness, and then imaged in the cryo STXM.

MECHANICAL STRAIN IN SEALED CELL

Before we start to discuss the data, let us briefly consider the mechanical properties of thin wall cell. We are interesting to answer the following questions: (i) whether the membrane will break, (ii) how big is the mechanical deformation/ displacement under the vacuum forces.

Let us consider a more simple case of circular membrane with radius R, which is clamped around the rim and do not stress preliminary to the load. According to Timoshenko [3], for small deflection we may neglect of shearing force acting due to the stretching of the middle plane of the plate, and we have deformation equal to:

$$z = \frac{q}{64D}\left(R^2 - r^2\right)^2; \quad z_{max} = \frac{qR^4}{64D}.$$ Here, q is the external pressure, r is the radius of point of interest, and D is the flexural rigidity of the plate defined as $D = Eh^3 \Big/ \left[12(1-v)^2\right]$, where E is Young's modulus, h is the thickness of the plate, and v is Poisson's ratio. Maximum stress occurs at the rim and is equal to $\sigma_{max}^{bending} = \frac{3}{4}q\frac{R^2}{h^2} = \frac{1}{v}\sigma_{max}^{twisting}$. For a very thin plate (membrane), where the deflection of the plate become large in comparison with h, and hence the shearing force due to the stretching of the middle plane become dominant, the maximum deviation is $z_{max} \approx 0.66R\sqrt[3]{qR\Big/Eh}$ for $v=0.25$ so it do not proportional anymore to the load but varies as a cube root of the external pressure. The maximum of the tensile stress

happens at the center of the membrane and is equal to $\sigma_{max}^{tensile} \approx 0.42 \sqrt[3]{q^2 E R^2 / h^2}$. In an actual case the values of maximum stress and deformation are somewhere between above extremes. Using the suggested parameters for Si_3N_4, $q=100$kPa, $\sigma_{max}=20$ GPa, $E=310$ GPa, $\nu=0.25$, and $h=100$ nm, the membrane is expected to break whenever $R>20$ µm. Experimentally we had found that windows larger than $800*800$ µm are likely to be broken during vacuum evacuation, whereas $500*500$ µm membranes are robust enough to survive the vacuum forces. We may also estimate the size of the window which bends under the vacuum force but does not change the distance between two windows substantially. Let us suppose that the deformation is negligible if $z_{max}<n\lambda_0/4\sim0.2$ µm, so that the interference picture viewed by the optical microscope do not change significantly, this leads to $R\sim10$ µm. But even for the bigger window size such as $200*200$ µm, the water film thickness within 10 µm of the border of the window remain unchanged.

WETTING PROPERTIES OF SI_3N_4

The second question to be addressed is: what is the natural thickness of a water layer squeezed between two Si_3N_4 membranes and sealed? Two long range forces to consider are: an attractive van der Waals force and repulsion due to the interaction of spontaneously charged planar surface through the water (so named double-layer forces). Because the pressure built by van der Waals forces between two planar surface is inversely proportional to the cube of layer separation, at long distance the double-layer repulsion is greater and produces disjoint pressure $p(d) = 2\varepsilon\varepsilon_0 (\pi kT / ze)^2 \, 1/d^2$, where d is the plate to plate separation, ε_0 and ε are dielectric constants of the media (water), k is Boltzmann's constant, T is the absolute temperature, and z and e are the valence and charge of the electrolyte used (water) [4]. Therefore the natural thickness of water layer may be as big as micron. Also a second minimum on a potential curve is possible for a smaller separation, so a thick water layer may be unstable.

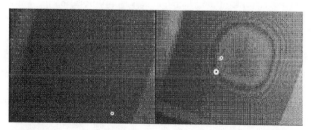

FIGURE 1. Two optical microscope images taken with the time delay of 50 s show how the phase separation is built in thin water film placed between two Si_3N_4 windows. The field of view is 800 (horizontal)*400 (vertical) µm². Note, what the black lines correspond to the destructive interference and so they correspond to water thickness difference $2h= \lambda$ or $h\sim0.23$µm.

We have found that a water layer sealed between two Si_3N_4 membranes has a tendency of forming a bubble. This bubble starts to grow from the center of the cell, pushes water toward the edges (Fig.1), and leaves tiny droplets in the middle. The size of the drop left depends on how clean the membrane is, and in our case varied from 5 μm to 0.5 μm.

STABILITY OF MICRODROPS

For micro drops, where the capillary length is much smaller then $k^{-1} = \sqrt{\gamma / g\rho}$, where ρ is the liquid density and g the gravitational acceleration, gravity is negligible compare with other forces. For pure water at 20 C°, the capillary length is equal to 2.7 mm. So, one may expect that drops with a characteristic size in the micron range do not move at all. This is not exactly correct. A series of x-ray images of the same region taken with a time difference 10 h are presented on Fig.2(a-c). The main contribution is the scanner drift, so the relative positions of the drops remain mostly unchanged. But still, while some drops are stable, others disappear or merge.

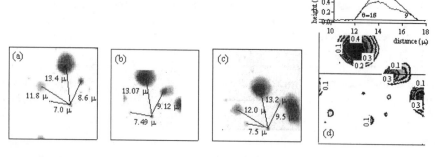

FIGURE 2. (a-c) Three x-ray images of the same region taken with time difference of 10 h. The energy of x-ray is set to 538.8 eV, where the water drops strongly absorbs. The acceleration due to the gravity is in the plane of the picture and pointed down. Over such long time the main motion is due to X-Y stage drift, but the relative distance of the droplets, as indicated, is very much unchanged. (d) The x-ray image with droplet contour profiles (line separation is 0.1 μm), based on the water x-ray attenuation depth. On the top, the droplet line profile clearly shows the deviation from spherical cap shape, which is and would be expected in the case of thermodynamic equilibrium.

As one can see from Fig.2(b) the drop, the second from the top, will move toward its lower right neighbor. The contour plot of this drop is shown on Fig.2(d), with a line profile superimposed. It reveals a nice tail (precursor film) formed, and the negative curvature in the vicinity of this tail suggests that the drop is thermodynamically unstable.

DRYING OF A THIN LAYER OF SOLVENT

Due to the small thickness of the water layer, it may be completely dried in the range of tens of seconds. This process may be accompanied by precipitation of the products dissolved in water. As a result, there is a new morphology (contrast) due to such phase separation. For this particular example, we had measured the drying dynamics of glycine (*Gly*) solution. A 1.3M water solution of *Gly* was mixed with *HCl* to shift its *pH* to pK_a value (*pH* of solution was measured to be ~ 2.38). Two different types of the samples were prepared. For 'dry' samples, a drop of the solution was placed on Si_3N_4 membrane and naturally dried. For 'cryo-fixed' samples, a #400 formvar coated *Au* grid was dipped in the solution. The excess water was drained from the grid by filter paper gently pressed upon the edge of the grid. The grid was left to dry for 20 sec and afterwards was quickly propelled into liquid ethane to preserve its hydration state [2].

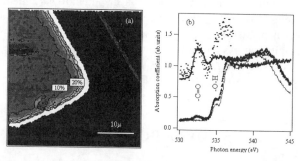

FIGURE 3. The contour plot of *Gly* concentration in water for 'cryo-fixed' sample, extracted from x-ray images of the same sample taken at 532.5 and 535 eV. As one can see, the *Gly* near the mesh bar is completely dry and its contribution quickly decreases towards the center of the mesh. The spectrum for the center part of the mesh (triangles) was measured to extract the spectra of *Gly* in solution. The extracted spectra of *Gly* in solution (dots) looks different from the spectrum of dry *Gly* (solid line).

In order to extract the relative contribution of water and *Gly*, two images of the same mesh, which contains thin layer of *Gly* in water solution, were taken at 532.5 and 535 eV. The images were decomposed for *Gly* and water contributions, based on the relative absorption strength of dry *Gly* layer and pure ice (thick and thin solid lines on Fig. 3(b)). Due to the difference in the spectra, it was possible to determine the relative thickness of water and *Gly*. Fig.3(a) shows a relative concentration of *Gly* in water across the mesh. As one can see, *Gly* is pulled toward the edges and forms a dry 0.5 μm thin film at the border. The spectrum of the dry sample differs from the spectrum of solution, measured at the center of the mesh, as well as from *Gly* contribution, determine by subtraction of the ice spectrum from spectrum of the solution. Due to an extremely small thickness of the water film, special care must be taken to prevent it from drying, because it may lead to precipitation of the solvent products. The XANES spectra of these precipitates may differ from the XANES spectra of the material in solution.

CONCLUSION

We have described the use of a very simple sealed cell to carry out absorption spectroscopy measurements on hydrated thin films, particularly for oxygen K edge studies. The cell consists of two Si chips with Si_3N_4 windows etched in them, squeezed and sealed. The whole cell is small enough to be mounted on a standard TEM holder, and has been used in the vacuum chamber of the Stony Brook cryo STXM. Windows larger than 800*800 μm are likely to be broken during vacuum evacuation, whereas 500*500 μm membranes are robust enough to survive the vacuum forces. The cell with Si_3N_4 windows (and pure water inside) has a tendency for phase separation with a vapor bubble formed in the middle and the water layer pulled toward the edges. The x-ray 1/e attenuation length for water is as small as 340 nm for 537 eV, so the drying of such thin water layer happens in the range of seconds. In the case of the solvents this may lead to the precipitation. As a result, a new morphology/ contrast due to the phase separation may result. In addition the XANES spectra of the solution may be different from the spectrum of the precipitate, possibly leading to misinterpretation of the data.

ACKNOWLEDGEMENTS

Data taken using the X-1A beamline developed by the group of Janos Kirz and Chris Jacobsen at SUNY Stony Brook, with support from the Office of Biological and Environmental Research, U.S. DoE under contract DE-FG02-89ER60858, and the NSF under grant DBI-9605045. The authors particular acknowledge the technical help of Michael Feser and Sue Wirick from Stony Brook.

REFERENCES

1. U. Neuhäusler *et al.*, A specimen chamber for soft x-ray spectromicroscopy on aqueous and liquid samples, submitted to *Journal of Synchrotron Radiation*, 1999 and citations therein.

2. J. Maser *et al.*, Soft x-ray microscopy with a cryo STXM: I. Instrumentation, imaging, and spectroscopy, submitted to *J. Microscopy* (1999).

3. Timoshenko S., *Theory of Plates and Shells*, New York: McGraw-Hill book company, 1940.

4. Israelachvili J., *Intermolecular and Surface Forces,* San Diego: Academic Press, 1991.

An Imaging Zone Plate X-ray Microscope Using a Laser Plasma Source

Hiroaki Aritome, Masato Fujii, Goh Fujita

Research Center for Materials Science at Extreme Conditons, Osaka University,
Toyonaka, Osaka 560-8531 ,Japan

and

Kunio Shinohara

Radiation Research Institute, Faculty of Medicine, The University of Tokyo,
Bunkyo-ku, Tokyo 113-0033, Japan

Abstract. An imaging x-ray microscope using x-rays from a carbon plasma generated by a high repetition rate, moving slab glass laser, an objective zone plate and an ellipsoidal condenser mirror with multilayer coating is described. The 3.37 nm line emission from a carbon plasma is filtered and focused by the ellipsoidal condenser mirror and used as a light source. A magnified image of a zone plate and dried HeLa cells are obtained.

INTRODUCTION

Plasmas of low-atomic number materials such as carbon and nitrogen emit intense lines in the soft x-ray region.[1] These lines have very narrow bandwidth and are good x-ray sources for the zone-plate microscope which requires the illumination of very narrow bandwidth smaller than the reciprocal of the zone number. The laser plasma x-ray source is also of particular interest because there is a possibility to observe a living biological specimen before the influence of the radiation damage occurs by its pulse nature. A laboratory size, imaging x-ray microscope using a Nd-YAG laser was reported previously.[2] To obtain an image by a single x-ray pulse, a glass laser has advantage because of its larger laser energy. A laboratory size, imaging x-ray microscope using a high-repletion-rate Nd-glass laser is described.

OPTICAL ARRANGEMENT

The optical arrangement of our x-ray microscope is shown in Fig. 1. The drive of the plasma source is a moving slab Nd –glass laser (wavelength: 1.05 μ m) with an energy of 5 J and the pulse duration of 10 ns.[3] The repetition rate is 3 Hz. The laser

CP507, *X-Ray Microscopy: Proceedings of the Sixth International Conference,*
edited by W. Meyer-Ilse, T. Warwick, and D. Attwood
© 2000 American Institute of Physics 1-56396-926-2/00/$17.00

beam was focused by a lens onto a carbon sheet target. To provide a fresh surface, the target was rotated around the normal to the target surface while being translated parallel to the surface during the laser operation. The condenser mirror is a part of an ellipsoid with multilayer coating. The incident angle is 71 degrees and the numerical aperture is 0.024. The multilayer coating is designed to have the peak reflectivity at 3.37 nm. The multilayer is composed of 100 pairs of $NiCr/V_2O_5$ with a thickness period of 5.42 nm and a thickness ratio of 1:1. The calculated reflectivity and spectral resolution are 19% and 28, respectively. The reflectivity and the spectral resolution measured by synchrotron radiation was 11% and 25, respectively. After reflection by the condenser, the carbon 1s-2p line (3.37 nm) is filtered.

The objective is a Ta zone plate. The outermost zone width is 70 nm. The focal length and the numerical aperture at a wavelength are 1.57 mm and 0.024, respectively.

A filter (200 nm-thick aluminum on 200 nm-thick poly-p-xylene) was placed between the plasma source and the condenser mirror to eliminate debris and stray light from the plasma source.

Images were detected by a back-illuminated charge coupled device (CCD).[4] The pixel size of the CCD is 22 μ m x 22 μ m.

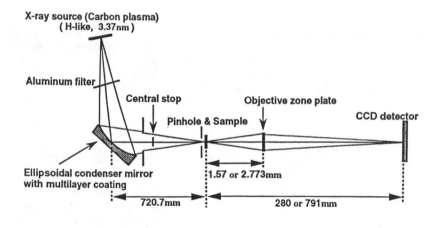

FIGURE 1. Optical arrangement of the x-ray microscope.

X-RAY IMAGING EXPERIMENT

To investigate the spatial resolution of our microscope, the x-ray image of another zone plate was taken as a specimen. The image of taken with 225 laser puses with a laser energy of 2 J. The magnification is 500. The obtained image is shown in Fig. 2. The zone of the width of 90 nm is clearly resolved which is close to the expected calculated value.

Figure 3 shows the x-ray image of dried HeLa cells taken with 26 pulses with laser energy of 5.5 J. The magnification is 95. The microvillus (approximately 0.5 μ m wide) is resolved.

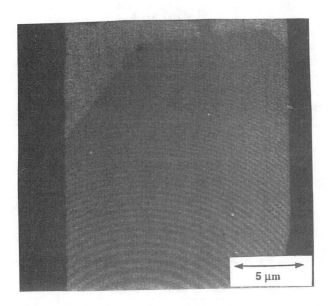

FIGURE 2. X-ray image of Ta zone plate. Magnification is 500 (225 laser puses with energy of 2 J). 90 nm structure is resolved.

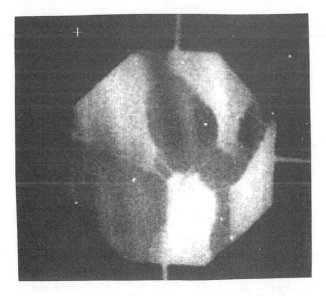

FIGURE 3. X-ray image of HeLa cell. Magnification is 95 (26 laser puses with energy of 5.5 J).

SUMMARY

A laboratory size, x-ray microscope was developed using x-rays from a carbon plasma generated by a high-repetition-rate glass laser. A magnified image of 90 nm structure is obtained by multiple x-ray pulses. The present results show that it is possible to obtain an image with resolution better than 50 nm by a single x-ray pulse by improving the x-ray brightness and x-ray optical components. Such an x-ray microscope would give additional information about biological specimen not obtained so far.

REFERENCES

1. G. Zeng, H. Daido, T. Togawa, M. Nakatsuka, S. Nakai, and H. Aritome, J. Appl. Physics 69, 7460-7464 (1991).
2. S. Nakayama, K. Haramura, G. Zeng, H. Daido, M. Nakatsuka, S. Nakai, N. Katakura, H. Nagata and H. Aritome, Jpn. J. Appl. Phys. 33, L1280-L1282 (1994).
3. H. Aritome, K. Haramura, H. Sekiguchi, H. Hara, and T. Mochizuki, OSA Proceedings on Soft X-Ray Projection Lithography, Vol.18, 138-141 (1993).
4. H. Aritome, S. Nakayama, G. Zeng, H. Daido, M. Nakatsuka, S. Nakai, M. Sakurai, and K. Yamashita, Proc. SPIE 1741, 276-279 (1992).

A Modified Electron Multiplier X-Ray Detector For Synchrotron and Laser Plasma Sources

C.J. Buckley[a], G. Dermody[a1], N. Khaleque[a2], A. G. Michette[a],
S. J. Pfauntsch[a], Z. Wang[a3] I.C.E. Turku[b] , W. Shaik[b] and R. Allott[b]

[a]*Department of Physics, King's College London, Strand, London WC2R 2LS, U.K.*
[b]*Rutherford Appleton Laboratory, Chilton, Didcot, Oxfordshire, OX11 0QX, U.K.*

Abstract. A 129EM electron multiplier has been modified to detect soft x-rays by coating the first dynode with CsI. The prototype detector was tested at the X1a beamline of the NSLS synchrotron source at Brookhaven Lab and at the laser plasma source at the Rutherford Lab. Initial tests have shown that the detector type is suitable for use in soft x-ray microscopy.

INTRODUCTION

The detection of x-rays is typically achieved by photon interactions that produce charged particles which are collected and their number amplified to form an electronic signal. Ideally, the interaction of the x-ray photon with the sensitive volume of the detector will produce many charged particles resulting in an electronic signal which is considerably greater than the electronic noise in the system. Soft x-rays have energies in the range 200 to 700eV and will generally produce a small number of charged particles. Where small numbers of photons are involved, the detector will typically need an intrinsic amplification process in order to produce an electronic signal of sufficient magnitude.

Here we introduce a modified electron multiplier tube as a detector for both photon counting and pulse height analysis applications in x-ray microscopy. Initial tests on synchrotron and laser plasma sources indicate that the prototype has some pulse-height non-linearity, but is both useful and competitive with existing scanning soft x-ray microscope detectors.

A Modified Photomultiplier Tube as a Soft X-Ray Detector

Channel electron multipliers are inexpensive and an be used for the detection of soft x-rays. However, their poor dynamic range and linearity performance[1,2] makes them unsuitable as detectors for x-ray microscopy using either synchrotron or laser plasma

---

[1] Now at DERA Fort Halstead, Sevenoaks, Kent, UK.
[2] Now at RSI, Crowthorne, Berkshire, UK.
[3] Visiting KCL from Changchun Institute of Optics and Fine Mechanics, Chinese Academy of Sciences.

CP507, *X-Ray Microscopy: Proceedings of the Sixth International Conference*,
edited by W. Meyer-Ilse, T. Warwick, and D. Attwood
© 2000 American Institute of Physics 1-56396-926-2/00/$17.00

sources. Photo multiplier tubes overcome pulse height variation effects suffered by channel electron multipliers by having focussed dynode systems which provide for linear gains and insignificant pulse height spread. These tubes are linear over several decades of input level[3]. To take advantage of these properties, an Electron Tubes Ltd 129EM photo-electron multiplier tube was modified at King's College London for use as a soft x-ray detector. The first dynode was coated with 100nm of caesium iodide which acted as a photo-emitter and the dynode chain then provided amplification of the photoelectron signal. Figure 1 shows a schematic of the arrangement. The prototype detector was tested at the scanning x-ray microscope facility at the X1a synchrotron beamline in Brookhaven National Lab[4] and also at the laser plasma x-ray source at the Rutherford Laboratory[5]. These tests and the results are presented below.

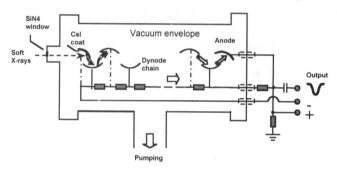

FIGURE 1. A schematic diagram of a 129EM photomultiplier modified to detect soft x-rays

Electron Multiplier Tests At The X1a Beam Line Of The NSLS

The electron multiplier assembly was positioned on the scanning microscope at the X1a beam line (see figure 2), where monochromatic x-rays (400eV) were focussed by a zone plate to provide a flux of ~5×10^6 photons per second.

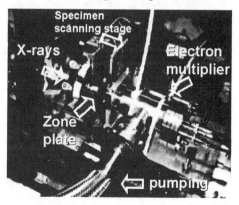

FIGURE 2. The electron multiplier x-ray detector mounted on the scanning x-ray microscope at the X1a beamline of the NSLS.

The detector signal was fed into a discriminator via a times 50 charge amplifier. The output of the discriminator was recorded and displayed via Camac based a multi-channel scaler and VAX computer system. The power supply leads to the amplifier were unscreened resulting in a 20mV electronic noise pick-up at its output. To overcome this, the discriminator was set to threshold out signals below this level.

Results of Synchrotron Tests

The desirable characteristics of a detector on a quasi-continuous source are that it is efficient and is linear within the desired photon rate range and has near zero dark count. In the case of the scanning soft x-ray microscope at the X1a undulator beamline, the maximum diffraction-limited incident count rate was about 5×10^6 photons per second at an x-ray energy of 400eV. The linearity of the detector was checked by recording the count rate for different settings of the beam line exit slit and at each setting of the slit, the beam was attenuated by a fixed amount by replacing the helium with air. This provided an attenuation factor of approximately 58. Any saturation or dead-time effects will show up as a swing to the attenuated axis on a plot of the unattenuated beam versus the attenuated The results are shown in figure 3.

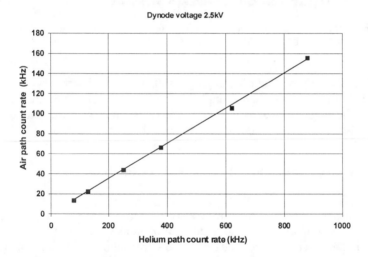

FIGURE 3. Test of linearity for the 129EM on the X1a beam line at the NSLS. The count rate was recorded for a range of exit slit settings while the incident flux was alternately attenuated by a fixed path in air and helium.

The detector showed good linearity up to a count rate of almost 0.9 MHz. This count rate was the maximum count rate that was obtainable with a -2.5kV dynode voltage. However, prior tests of the incident flux with a single wire proportional counter indicated that the maximum incident flux was in excess of 2.5 MHz. Observations on an oscilloscope indicated that there were a considerable number of pulse heights below 20mV. Also, the efficiency of the CsI will have been less than

407

unity. It is likely that these factors acted together to produce the reduced maximum count rate.

The detector was operated at -2.5 kV with a discriminator level of 20mV to provide a linear detection system with zero dark count (set by electrical pick-up noise) with the X1a scanning x-ray microscope. Figure 4 shows an image of a wood composite sample obtained with this microscope and detector combination.

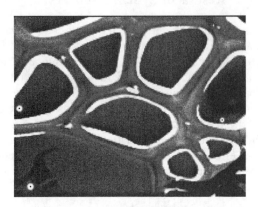

Figure 4. An x-ray transmission image of a wood-composite section, formed using the X1a scanning transmission x-ray microscope with the modified 129EM detector. The x-ray energy was 400eV.

The modified 129EM produced encouraging results when used with a synchrotron source. It is likely that with a suitable amplifier, such as the LeCroy 612AM (NIM configuration), the detector would give a count rate which was limited by the multipliers pulse width of 15nS. This should provide a count rate in excess of 10 MHz.

Electron Multiplier For Pulsed X-Ray Sources

For use with a pulsed x-ray source, a detector should produce an output signal which is consistently proportional to the number of photons which is incident on the detector. Here, in the absence of non-linearity effects, the number of discernable grey levels will be a function of the number of photons incident on the detector (Poissonian statistics) and the detector efficiency. I.e., if N photons are incident on the detector which has an efficiency ε, the number, ξ, of discernable grey levels will be:

$$\xi = \sqrt{\varepsilon N}$$

The pulsed x-ray source at the Rutherford laboratory[5] was used to test the electron multiplier x-ray detector. The electron multiplier detector was mounted outside the source chamber as shown in figure 5.

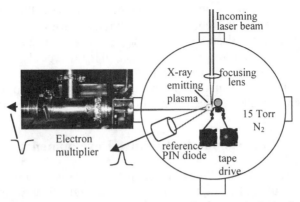

FIGURE 5. Schematic/photo of laser-plasma x-ray source & detectors at the Rutherford Laboratory.

The output of the laser plasma x-ray source varies considerably with each pulse and thus a large-area reference PIN diode is included in the chamber to normalise the smaller detected signals on the outside of the chamber. Charge pulses from both the 129EM and the PIN diode were measured and stored via a LeCroy 2249A module, a Camac crate and a PC. Most of the energy filtered x-ray output from the plasma is in the line at 3.37nm. The PIN captures on average 10^8 photons per pulse. Based on this signal, and taking into account the filters in front of the 129EM and the solid angle it subtends to the source, we calculate that approximately 2000 photons on average per pulse were incident on the entrance window. The 129EM was operated at -2.5 kV.

Results tests with the Laser Plasma X-Ray Source

The output of the laser plasma was recorded for approximately 30,000 laser plasma pulses. The reference diode and electron multiplier data pairs were sorted by reference diode signal magnitude and a ratio of electron multiplier to reference diode was plotted against reference diode value. This is shown in figure 6.

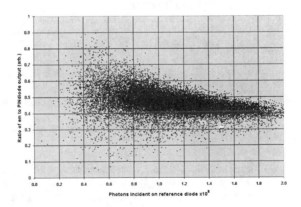

Figure 6. The ratio of the output of the 129EM to that of the PIN reference diode for a range of x-ray pulse intensities taken from 30,000 laser plasma generated soft x-rays.

The ratio of the 129EM output to the reference diode should be a constant with a statistical variation about this constant which improves with increasing signal level (i.e. with increasing laser plasma x-ray output). However, the data plotted in figure 6 shows a 30% drop in ratio over nearly a decade of x-ray input level. This is almost certainly due to inter-dynode voltage drop at higher charge-pulse levels[6]. The effect can be seen both in the drop in the ratio and also in the asymmetry of the spread about the mean - which is smaller on the higher ratio side due to saturation effects.

Conclusions And Future Developments

The output current of the 129EM in this application was a substantial fraction of the dynode chain current leading to charge starvation at the final dynodes. It is likely that this was the main reason for the non-linearity in the laser plasma experiments. Also, this gain non-linearity produced the small pulse heights obtained with the synchrotron experiments which in part limited the maximum count rate. The tests on this prototype have give a strong indication that a modified design will produce a soft x-ray detector with desirable attributes. A detector based on a 143EM is currently under development. This detector will have a dynode chain current which will be an order of magnitude larger that that of the 129EM at the same voltage. Also, the system will have higher gain and external voltage stabilisation on the on the final dynodes together with access to the first to second dynode potential for maximising the devices efficiency.

Acknowledgements

The authors would like to thank Sue Wirick, Chris Jacobsen and Janos Kirz for assistance with and use of their x-ray microscope at the NSLS. Thanks also go to Sarah Huntington for assistance during tests with the Rutherford Laboratory x-ray source. We would also like to acknowledge the fact that a detector based on a phosphor and a PMT is being pursued in parallel by Tony Warwick and colleagues at LBL.

References

1. Buckley, "X-Ray Detectors", in *X-Ray Science and Technology*, edited by A.G. Michette and C.J.Buckley, IOP publishing, 1993, pp 207-253.

2. J. Adams and B. Manley, *IEEE Transactions on Nuclear Science*, NS-13, pp 88-99.

3. Electron Tubes Ltd. "Photomultipliers and accessories", Technical literature 1998.

4. M.Feser et. al. "X-ray microscopy and microscope development at Stony Brook" - These proceedings.

5. A.G. Michette et. al. " Status of the King's College Laboratory x-ray microscopes" - These proceedings.

6. C. Delaney and E. Finch, "Radiation Detectors", *Oxford University Press*, 1992.

X-Ray Microscopes At BESSY II

P. Guttmann, B. Niemann, J. Thieme, U. Wiesemann,
D. Rudolph and G. Schmahl

*University Georgia-Augusta at Göttingen, Institute for X-Ray Physics,
Geiststraße 11, 37073 Göttingen, Germany*

Abstract. The undulator U41 at BESSY II will be used as source for X-ray microscopes. An overview of the X-ray microscopy area is presented. After finishing the construction phase a transmission X-ray microscope, a scanning transmission X-ray microscope and an X-ray test chamber will be available. The transmission X-ray microscope will allow investigations with high lateral resolution at moderate energy resolution while the scanning transmission X-ray microscope will allow high energy resolution at moderate lateral resolution of the same specimen.

INTRODUCTION

One of the requirements for X-ray microscopy is the need for high brilliance of the source. The new synchrotron radiation source BESSY II in Berlin-Adlershof is an electron storage ring of the third generation optimized for the operation with insertion devices [1]. Especially for the purpose of X-ray microscopy using wavelengths within the so called water window (2.34 nm to 4.38 nm) and spectromicroscopy an undulator U 41 for a low beta section at BESSY II has been built [1,2].

FIGURE 1. Status of the X-ray microscopy area at BESSY II in July 1999

CP507, *X-Ray Microscopy: Proceedings of the Sixth International Conference*,
edited by W. Meyer-Ilse, T. Warwick, and D. Attwood
© 2000 American Institute of Physics 1-56396-926-2/00/$17.00

X-RAY MICROSCOPY AREA AT BESSY II

The status of the X-ray microscopy area as in July 1999 is shown in Fig.1. The X-ray microscopes will share the beam time at the undulator U 41 with a BESSY beamline for spectromicroscopy. Therefore, a switching mirror unit is installed in front of the beamlines, as shown in Fig.2. For radiation safety the direct beam will not be used. The beamline for the X-ray microscopes is set up at the left side downstream and splitted into separate beamlines for three instruments described in the following. Only one of these instruments can be supplied with the X-ray beam at a time. A preparation laboratory equipped with different light microscopes and several instruments which will be necessary for the applications of X-ray microscopy in the fields of e.g. biology, colloidal physics, environmental and soil science will be available directly at the X-ray microscopy aera (see Fig.1).

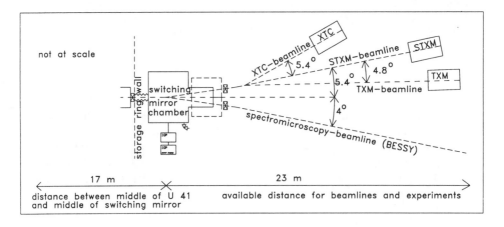

FIGURE 2. The schematic arrangement of the X-ray microscopy beamlines at BESSY II.

The Transmission X-Ray Microscope (TXM)

The first system built up is a transmission X-ray microscope (TXM) providing high lateral resolution with moderate energy resolution. This is an improved version of the TXM at BESSY I [3]. The schematic optical setup of this instrument is shown in Fig.3. The first two water cooled plane mirrors M1 and M2.1 will be coated with nickel to suppress the higher harmonics of the undulator. Therefore, the thermal load of the first element of the following monochromator will be reduced strongly. A new condenser concept has to be incorporated because the highly collimated X-ray beam of the undulator can not be used with a conventional zone plate linear monochromator, which was used with dipol radiation at BESSY I previously. For this purpose, a dynamical aperture synthesis with a high numerical aperture condenser-linear monochromator will be realized by using a fixed off-axis transmission zone plate (OTZ), an adjustable plane mirror and two rotating small plane mirrors [4]. This system will produce a non-rotating source image and a one dimensionally dispersed, spatially fixed spectrum of the source in the object plane. The monochromaticity will be improved considerably. Due to the higher flux of about a factor of 60 compared to BESSY I shorter exposure times will be possible. The CCD camera will be movable in the beam direction to allow various X-ray magnifications. Therefore, it will become possible to make an high resolution image of a specimen detail or to make an overview image of a great part of the specimen. Besides X-ray microscopy at room temperature cryo X-ray microscopy and X-ray tomography will be possible.

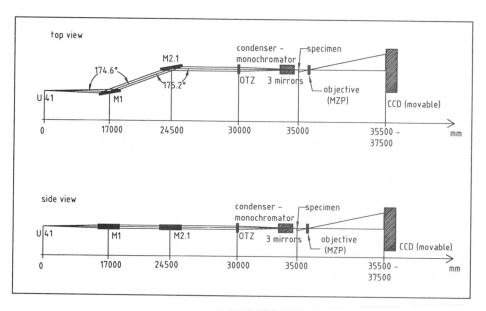

FIGURE 3. Schematic optical setup of the transmission X-ray microscope (TXM). M1 and M2.1 are plane mirrors. OTZ is the off-axis transmission zone plate which works together with the three mirrors as condenser-monochromator [4]. The X-ray microscope objective is a micro zone plate MZP.

The Scanning Transmission X-Ray Microscope (STXM)

A new scanning transmission X-ray microscope (STXM) is under development [5]. The STXM provides high energy resolution with moderate lateral resolution. The STXM uses the gain in spatially coherent photons/sec in the scan spot of a factor 10^4 due to the higher brilliance compared to BESSY I. This will reduce the integration time per pixel considerably. The schematic optical setup of this instrument is shown in Fig.4. The monochromator principle using a water cooled plane mirror M2.2 and a plane grating G2.2 was already tested at BESSY I [6]. It is planned to use the same specimen holder for the STXM as it is used for the TXM. Therefore, it will become possible to do investigations of a specimen or specimen detail with both instruments. Quantitative analysis and elemental mapping will be done with the STXM whereas structural studies will be done with the TXM.

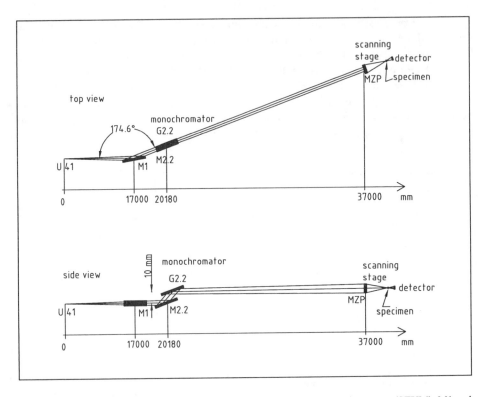

FIGURE 4. Schematic optical setup of the scanning transmission X-ray microscope (STXM). M1 and M2.2 are plane mirrors. G2.2 is a plane grating which acts together with M2.2 as monochromator [6]. MZP is a micro zone plate which produces the scan spot. The scanner is described elsewhere [5].

The X-Ray Test Chamber (XTC)

The X-ray test chamber (XTC) is a very versatile instrument which is necessary for the calibration of detectors and optical elements for X-ray microscopy, e.g. zone plates built in our institute [7, 8]. The schematic optical setup is shown in Fig.5. The monochromator of the STXM will be used for this instrument, too. A plane mirror M2.3 will be used to deflect the beam in the XTC beamline. The XTC itself will be transferred from BESSY I to BESSY II. Several types of detectors are available. Switching between them can be done without venting the system. A micro channel plate (MCP) with a phosphor screen converting the X-ray photons into visible light can be used to align the optical elements which shall be tested. For the determination of the lateral not resolved intensity of the X-rays a windowless photodiode is incorporated. A thinned, back-side illuminated CCD camera can be utilized to determine the intensity of the X-rays with lateral resolution, e.g. for the radially resolved determination of zone plate efficiencies [7].

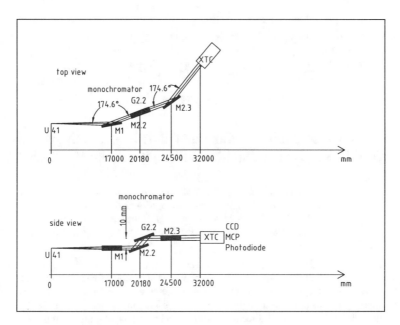

FIGURE 5. Schematic optical setup of the X-ray test chamber beamline. M1 and M2.2 are plane mirrors. G2.2 is a plane grating which acts together with M2.2 as monochromator [6]. A CCD camera, a micro channel plate (MCP) or a photodiode can be used as detector.

ACKNOWLEDGEMENTS

This work has been funded by the German Federal Minister of Education, Science, Research and Technology (BMBF) under contract number 05SL8MG1 1.

REFERENCES

1. BESSY homepage: http://www.bessy.de

2. Guttmann, P. et al., "The X-ray microscopy project at BESSY II" in *X-Ray Microscopy and Spectro-microscopy*, edited by J. Thieme et al., Springer-Verlag Berlin Heidelberg, 1998, pp. I-55 - I-64

3. Schmahl, G. et al.., *Rev. Sci. Instrum.* **66**, 1282-1286 (1995).

4. Niemann, B. et al., "The condenser-monochromator with dynamical aperture for the TXM at an undulator beamline at BESSY II", this volume

5. Wiesemann, U. et al., "The new scanning transmission X-ray microscope at BESSY II", this volume

6. Irtel von Brenndorff, A. et al., *J. Synchrotron Rad..* **3**, 197-198 (1996).

7. Peuker, M., "Nickel zone plates for soft X-ray microscopy ", this volume

8. Hambach, D., "High numerical aperture zone plates using higher orders of diffraction", this volume

Phase-Contrast X-Ray Microscopy with A Point Focus X-Ray Source

Kengo Takai, Takashi Ibuki, Kazushi Yokoyama, Yoshiyuki Tsusaka, Yasushi Kagoshima, Junji Matsui and Katsuhito Yamasaki*

Factory of Science, Himeji Institute of Technology
3-2-1 Kouto, Kamigori, Ako, Hyogo, 678-1297, Japan
Department of Radiology, Kobe University
7-5-2 Kusunokicho, Chuo-ku, Kobe, Hyogo, 650-0017, Japan

Abstract. Phase-contrast x-ray microscopy has been performed using a point focus x-ray source. The method is essentially the projection microscopy, but the contrast enhancement due to the refraction is obtained by taking a certain sample-detector distance large enough to spatially resolve phase gradients in the x-ray beam. The spatial resolution was measured to be 2.5 μm in horizontal and 2.2 μm in vertical directions. A phase contrast image of a glass tube was taken, and the contrast enhancement was clearly observed in boundary regions. The phase contrast images of a butterfly and a mosquito were also taken. It was confirmed that this method was useful to observe samples composed of light elements such as insects.

INTRODUCTION

Imaging with hard x-rays is one of the most important diagnostic tools in materials and biological science. When soft tissues are to be observed, one can obtain very poor contrast because of their low absorption in the hard x-ray region. On the other hand, phase-contrast imaging[1] can enhance the contrast in boundary regions of the density distribution even with hard x-rays. Therefore, the phase-contrast imaging can be applied to observe weakly absorbing samples, namely, carbon-based biological systems. In this paper, the experimental system and some results of the phase-contrast x-ray microscopy with a point focus x-ray source are presented.

EXPERIMENTAL SETUP

A point focus x-ray source equipped with a LaB_6 filament and double-focusing electromagnetic lenses is used in our phase-contrast microscopy. The target can be exchanged easily, and a silver foil with a thickness of 5 μm was used in the experiment. The acceleration voltage and the target current were 40 kV and 2 μA, respectively, which corresponded to 8×10^{-2} W. The experimental setup is shown in Fig. 1. The sample was mounted on XYZ pulse-motor-driven stages, and a detector was placed apart from the sample in order to observe a phase contrast phenomenon. As an image detector, a CCD camera or an x-ray film was used. Their specifications

CP507, *X-Ray Microscopy: Proceedings of the Sixth International Conference,*
edited by W. Meyer-Ilse, T. Warwick, and D. Attwood
© 2000 American Institute of Physics 1-56396-926-2/00/$17.00

are shown in Fig. 1. The magnification, M, is defined by $M = L_d / L_s$, where L_d and L_s are source-detector and source-sample distances, respectively. The typical L_s was about 15 mm. The large magnification can cover the poor spatial resolution of the CCD camera.

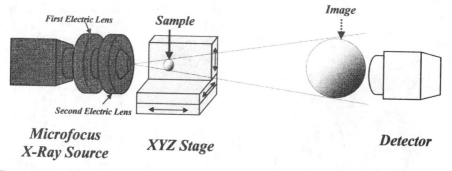

Figure 1. Experimental Setup of Phase-Contrast X-Ray Microscopy with Point Focus X-Ray Source (RIGAKU New-µFlex). A 5-µm-thick-silver foil was used as a target. X-ray films used are FUJI-IX 150 (0.9 µm resolution). The CCD camera (SX-TE/CCD-1024SB: Princeton Instruments, inc.) has 1024 × 1024 pixels with the pixel pitch of 24 µm × 24 µm.

RESOLUTION TEST

The spatial resolution was measured by using a tungsten slit and a CCD camera. Absorption contrast images of the slit were taken and the edge response profiles are shown in Fig. 2(a). The exposure time was 2 minutes and the magnifications were 134 (horizontal case) and 107 (vertical case). The profiles of the slit images were differentiated and fitted to the Gaussian distribution. The spatial resolution was evaluated to be 2.5 µm in horizontal and 2.2 µm in vertical directions.

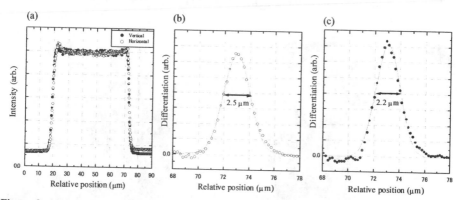

Figure 2. (a) Intensity Profiles of the Images of the Tungsten Slit. Differentiations of the intensity profiles are also shown in (b) horizontal and (c) vertical directions. Circles and dots are differential data and solid lines are fitted curves to the Gaussian distribution. The FWHM are 2.5 µm and 2.2 µm, respectively.

APPLICATIONS

As an example of the phase-contrast x-ray microscopy, an image of a glass capillary was taken with the CCD camera. An outside radius and an inside radius of the capillary were 300 μm and 100 μm, respectively. The magnification and exposure time were 28.3 times and 90 sec, respectively. The image is shown Fig 3(a). The contrast enhancement was clearly observed near edges in the intensity profile (b).

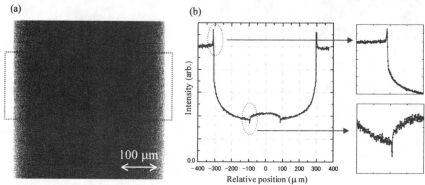

(a)   (b)

Figure 3. (a) X-Ray Micrograph of Grass Capillary. (b) Horizontal Intensity Profile of (a). The profile was derived from vertical integration inside a rectangular area surrounded by a dotted line in (a). Contrast enhancement was clearly observed at the boundaries.

Next, images of insects were taken by our microscopy system. One of the most important advantages of the phase-contrast imaging is that the high contrast can be obtained even with the low-absorption-objects. Figure 4(a) shows the phase-contrast images of a butterfly. This image was taken with the x-ray film. The magnification and exposure time were 34 times and 30 minutes, respectively. The curled straw of butterfly is a good example exhibiting the phase-contrast phenomenon.

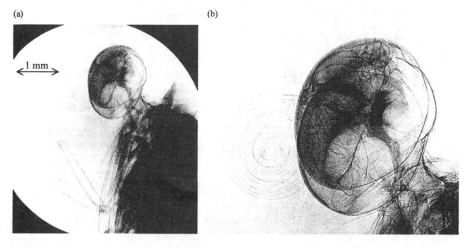

(a)   (b)

Figure 4. (a) X-Ray Micrograph of Butterfly and (b) Magnification View of Butterfly's Head.

A phase-contrast image of a mosquito was also taken with the x-ray film. The magnification and exposure time were the same as those for the image of butterfly. The complicated structures of the joints and the head were clearly observed. Figures 4 and 5 demonstrate the applicability of the present phase-contrast x-ray microscopy to imaging samples composed of light elements.

(a)   (b)

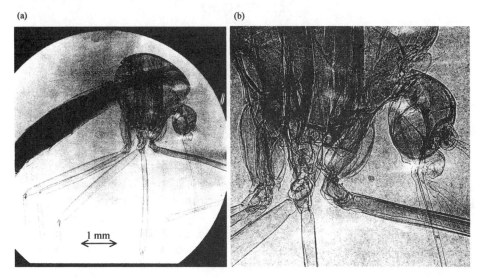

Figure 5. (a) X-Ray Micrograph of Mosquito and (b) Magnification View of Mosquito's Body.

ACKNOWLEDGEMENTS

This work was supported in part by Grant-in-Aid for Scientific Research (Contract No. 09357008 and 10670849) from the Ministry of Education, Science, Sports and Culture, Japan.

REFERENCES

1. Wilkins, S.W., Gureyev, T.E., Gao, D., Pogany, A., and Stevenson, A.W., Nature **384**, 335-338 (1996)

The King's College Laboratory Scanning X-Ray Microscope

AG Michette, CJ Buckley, SJ Pfauntsch, NR Arnot, J Wilkinson,
Z Wang[¶], NI Khaleque[†] and GS Dermody[‡]

Centre for X-Ray Science, Dept. of Physics, King's College London, Strand, London WC2R 2LS, UK

Abstract. The King's College laboratory scanning x-ray microscope has been used on the laser plasma source at the Rutherford Appleton Laboratory, which has been extensively characterised. Resolution and contrast in the initial images are limited by electrical noise in the detector system.

INTRODUCTION

The concept of the scanned source x-ray microscope was discussed at the previous x-ray microscopy conference in Würzburg.[1] Since then the x-ray source at the Lasers for Science Facility (LSF), Rutherford Appleton Laboratory (RAL), has been extensively characterised, and the microscope (figure 1) has been built and used for initial imaging. The results are encouraging but highlight the need for improvements.

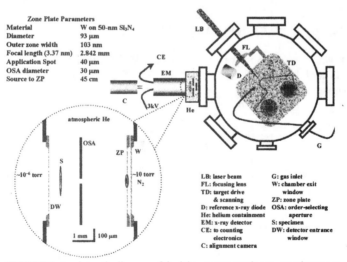

Zone Plate Parameters

Material	W on 50-nm Si₃N₄
Diameter	93 μm
Outer zone width	103 nm
Focal length (3.37 nm)	2.842 mm
Application Spot	40 μm
OSA diameter	30 μm
Source to ZP	45 cm

LB: laser beam
FL: focusing lens
TD: target drive & scanning
D: reference x-ray diode
He: helium containment
EM: x-ray detector
CE: to counting electronics
C: alignment camera

G: gas inlet
W: chamber exit window
ZP: zone plate
OSA: order-selecting aperture
S: specimen
DW: detector entrance window

FIGURE 1. A schematic diagram of the laboratory scanning x-ray microscope.

[¶] Visiting scientist from Changchun Institute of Optics and Fine Mechanics, China
[†] Now at: Research Systems International (UK) Ltd, Crowthorne, Berkshire RG45 6LS, UK
[‡] Now at: DERA Fort Halstead, Sevenoaks, Kent TN14 7BP, UK

CHARACTERISTICS OF THE LASER PLASMA X-RAY SOURCE

The RAL laser system uses a spatial multiplexer to give a train of eight 7 ps pulses over 20 ns at up to 100 Hz. The average energy per pulse at ~20 mJ at λ=248 nm gives an irradiance of ~2×10^{19} W cm^{-2} when focused to a diameter of ~10 µm.

A plasma formed from Mylar is an intense emitter in the water window and at slightly shorter wavelengths. The emission is mainly from H- and He-like carbon and oxygen, resulting in a line-dominated spectrum with some recombination radiation but very little bremsstrahlung. Previous work has shown that the lines have $\lambda/\Delta\lambda \gg 1000$ for similar Mylar plasmas.[2] A low pressure gas in the source chamber prevents debris damage and acts as a spectral filter to give a quasi-monochromatic spectrum suitable for zone plate microscopy. The source emission has been characterised for a range of nitrogen pressures (up to 5 torr m) and for 6.7 torr m of helium, using a pinhole camera and a transmission grating spectrometer, both using characterised soft x-ray film.[3]

The pinhole camera images (figure 2) were fitted to two-dimensional distributions using the Pearson VII peak function,[4] a computationally inexpensive approximation to a Voigt function. In one dimension it is $f(x) \propto [1+\{(2(x-x_0)\surd(2^{1/M}-1))\}^2]^{-M}$, where x_0 is the mean, w is the FWHM and $f(x)$ tends to a Gaussian (Lorentzian) as $M\rightarrow\infty(0)$. The elongation of the image is due partly to the camera angle and the laser beam shape, but mostly to a subsequently corrected misalignment of the multiplexer resulting in a double source. The fitted FWHM spot sizes, after correcting for the camera angle, are ~27×16µm and ~27×19µm, with relative intensities of ~1:2, separated by ~25µm. This source size allows a microscope resolution of ~0.2µm.

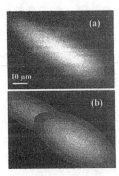

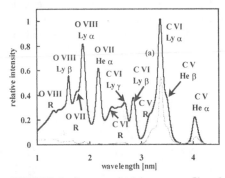

FIGURE 2. (a) The emission distribution measured with the pinhole camera. (b) The fit to the two components caused by multiplexer misalignment.

FIGURE 3. Spectra of Mylar targets filtered with (a) helium and (b) nitrogen, showing the monochromating effect.

Mylar and metal spectra have been obtained. Gold and silver give reasonably flat spectra in the water window and could be used in a tunable source for elemental mapping, but the low intensities mean that a source upgrade will be required for this development. Figure 3 shows Mylar spectra filtered by 6.7 torr m of helium and 4 torr m of nitrogen. Contributions from higher diffraction grating orders have been removed to leave just the first order. The lines were fitted to Pearson VII distributions and the recombination edges (R) to triangles convolved with Lorentzians. The flux in the nitrogen filtered 3.37 nm line has been measured to be ~5000 photons per laser shot at the zone plate focus.

INSTRUMENTATION FOR THE X-RAY MICROSCOPE

All vibration susceptible components were mounted on cork pads, generally reducing vibrations to <0.1 μm but at times amplitudes of ~0.6 μm were observed, making the microscope essentially unusable. A PC controls the microscope, and acquires and displays images, which are normalised by reference diode data to take account of fluctuating plasma emission.

Thin tape targets give little debris and present a fresh surface for every laser shot. The target drive incorporates motions for source scanning, which would be difficult with liquid or gas targets. Tape 12 mm wide gives several tracks by moving the tape across the beam, allowing acquisition of large images. The lens, positioned centrally in the ±150 μm range of constant water-window emission, and target are scanned together in steps down to 1 μm in two orthogonal directions. The movement is demagnified ~156× by the zone plate so that a step size of ~15 μm allows an image resolution of ~0.2 μm (Shannon sampling). Oscillations of the tape surface along the beam direction must be ≪15 μm to prevent blurring; measured oscillations are ≪1 μm. Specimens are mounted on a motor driven translation stage to place the specimen at the focus, to locate areas of interest and for coarse scanning.

The LSF x-ray laboratory is electromagnetically noisy, due to the complicated laser system needed to offer the wide range of laser beam properties required in a multi-user facility. The photodiode originally planned as the detector could not be shielded from this noise and at the same time cooled to reduce the dark current. An electron multiplier was therefore modified for soft x-ray detection.[5] This was less sensitive to noise but did not totally eliminate it

CHARACTERISATION AND USE OF THE MICROSCOPE

The zone plate and OSA are optically aligned with the chamber exit window. The OSA is finely aligned by scanning the x-ray beam across a pinhole to give a characteristic donut image. The focus is then found by scanning the beam across an edge and fitting the change in the normalised detector signal to a Gaussian. This gave an edge resolution of ~0.4 μm, a factor of two worse than expected due to the electrical noise.

Figure 4 shows an image of unstained myofibrils on an electron microscope finder grid, obtained by scanning the specimen stage in 2 μm steps with four shots per pixel. A dense clump of myofibrils can be seen at upper centre, with a less dense clump extending downwards from the curl of the 6. Figure 5 shows two myofibrils with the characteristic 2.4 μm repeat, extending to the right from a grid bar. This image was obtained by scanning the source with 16 shots per pixel and a step size of 100 μm, giving a pixel size of 0.64 µm. The resolution and contrast of all the images obtained so far are limited by the effects of electromagnetic noise.

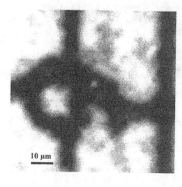

FIGURE 4. (left) A low resolution image of unstained myofibrils, obtained by scanning the specimen.
FIGURE 5. (above) A higher resolution image of myofibrils, obtained by scanning the source.

FUTURE WORK

The next stage in the work is to continue the development of the microscope as a user facility. This will require several improvements, including better vibration isolation and a detector with 10× larger dynode chain current. With external voltage stabilisation on the final dynodes and adjustable voltage between the first and second dynodes, this will need less external amplification. Alternatively, the elimination of the effects of electromagnetic noise will allow the use of a photodiode detector as originally intended.

ACKNOWLEDGEMENTS

The development of the laboratory x-ray microscope was supported by the Engineering and Physical Sciences Research Council, UK, under grant GR/K73381. The support of scientific and technical staff at the LSF is gratefully acknowledged.

REFERENCES

1. Michette, A. G., et al. in *X-Ray Microscopy and Spectromicroscopy*, edited by J. Thieme et al., Berlin Heidelberg: Springer-Verlag, 1998, pp. II-87–II-91

2. Turcu I. C. E., et al., *Proc. SPIE* **2015**, 243–260 (1994)

3. Tallents, G. J., et al., *Proc SPIE* **3157**, 281–290 (1997)

4. Elderton, W. P., and Johnson N. L., *Systems of Frequency Curves*, London: Cambridge University Press, 1969

5. Buckley, C. J., et al., *Proc SPIE* **3449**, 208–214 (1998)

High-resolution X-ray imaging and tomography at the ESRF beamline ID 22

T. Weitkamp[a], M. Drakopoulos[a], W. Leitenberger[a], C. Raven[a],
C. Schroer[b], A. Simionovici[a], I. Snigireva[a] and A. Snigirev[a]

[a]European Synchrotron Radiation Facility (ESRF)
B. P. 220, 38043 Grenoble, France

[b]Rheinisch-Westfälische Technische Hochschule (RWTH)
52056 Aachen, Germany

Abstract.
    At the ESRF micro-fluorescence, imaging and diffraction (μ-FID) beamline ID 22, a microimaging and tomography setup has been operational for several months. The coherence properties of the high-energy X-ray undulator beam at ID 22 make the setup especially suited for phase-contrast tomography including possible holographic reconstruction, but it is suited for absorption tomography too. A fast-readout, low-noise CCD camera makes time-resolved imaging possible. Recent developments in magnifying X-ray optics such as Compound Refractive Lenses (CRL) and Fresnel Zone Plates (FZP) open up the field of magnified-X-ray imaging with a resolution of less than 300 nm. Imaging techniques using a "pink beam", i. e. a beam with limited monochromaticity obtained by filtering one harmonic from the undulator spectrum, can increase flux in intensity-limited experiments.

The coherence properties of X rays from third-generation synchrotron radiation (SR) sources open up new fields of coherent X-ray imaging. At the ESRF undulator beamline ID 22, where SR in a spectral range between 5 and 70 keV is available, a microimaging and tomography station has been developed and put in user mode during the past two years.

The instrument consists of a single-crystal scintillator coupled to a cooled CCD detector by a visible-light microscope. The CCD is a 2048×2048-pixel, 14-bit, fast-readout low-noise (FReLoN 2000) model developed at ESRF [1]. The minimum effective pixel size of the system is 0.35 μm.

The setup is used on a routinely basis for both absorption and phase-contrast imaging and tomography. In studies using absorption contrast, the fixed-exit design of the monochromator at the beamline and the possibility to move the undulator gap during scans make it possible to scan the photon energy during an image series. First tests of tomographic absorption imaging on nuclear fuel particles at different energies around

CP507, *X-Ray Microscopy: Proceedings of the Sixth International Conference,*
edited by W. Meyer-Ilse, T. Warwick, and D. Attwood
© 2000 American Institute of Physics 1-56396-926-2/00/$17.00

the absorption L edge of Uranium have been carried out; a detailed analysis of these data is in progress.

For low-absorbing samples, phase-contrast imaging can yield better results than conventional absorption radiography and tomography. These samples include most biological samples as well as other organic materials. Phase contrast is generated in an in-line geometry using a coherent beam by the propagation of a distorted wavefront downstream from the sample. In the near field, phase contrast results in an edge enhancement in the radiograph (and, consequently, in a tomographic reconstruction) [2,3]. Figure 1 shows examples of phase-contrast tomograms and projection images obtained with the presented setup.

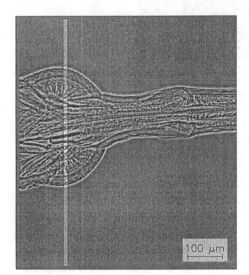 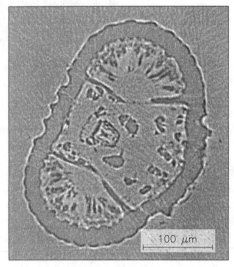

FIGURE 1. Phase-contrast projection image (*left*) and tomogram (*right*) of the head of a weevil, a beetle with a beaklike rostrum. The black rectangle in the projection indicates the area, in the region of the insect's eyes, where the reconstructed tomographic slice was taken. For this reconstruction, 1250 projections were taken over an angular interval of 180 degrees.

For the quantitative retrieval of the phase shift distribution of a wavefront after transmission through a weakly or non-absorbing sample, different methods based on an iterative [4] and a transport-of-intensity equation [5] approach, all using data acquired in the near field, have been tested with very promising results [3,5].

The fast readout of the FReLoN camera opens up the field of time-resolved imaging both in absorption and phase-contrast mode. It was thus possible for the first time to image the dendritic growth of an alloy solidification front [6] (Figure 2). The opacity of the materials, the time scale of the process, and the required spatial resolution had so far prevented the in-situ observation of such processes.

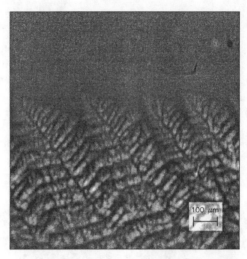

FIGURE 2. Real-time image of dendritic solidification of an Sn-Pb alloy melt. The image is part of a time series taken at a repetition rate of 2 frames per second and with an exposure time of 0.1 second. The X-ray energy was 25 keV. Since the sample transmitted only 10 per cent of the incoming intensity, the image makes use of only 0.5 per cent of the dynamic range of the FReLoN camera, whose low readout noise is therefore crucial for this kind of time-resolved studies (Collaboration with R. Mathiesen and L. Arnberg, NTNU Trondheim, Norway).

The spatial resolution in hard X-ray imaging at third-generation synchrotron light sources is generally limited by the detection system. The ultimate resolution limit of a system using conversion to visible light is the diffraction limit of the visible-optics microscope. The actual limit to resolution, however, is determined by X-ray scattering in the scintillator screen, which limits the resolution to about 1 micrometer [7].

The recent development of lenses for hard X rays [8–10] with good imaging quality makes it possible to overcome these limitations by building a true hard X-ray microscope. The feasibility of such a device using Fresnel zone plates (FZP) has been shown by Lai et al [10]. At ID 22, hard X-ray microscopes have been tested and are ready for use that use either FZPs or compound refractive lenses [9] (CRLs). These devices produce images that show no distortions and need no background correction; structures of 0.3 μm thickness can easily be resolved (see Figure 3). Magnified phase-contrast imaging is possible. The long focal lengths of lenses for the hard X-ray regime —typically between one and two meters for photon energies between 10 and 25 keV— require access to the beam at positions far apart from each other. Beamline ID 22 with its two experimental hutches meets this requirement.

To overcome the problem of insufficient flux in hard X-ray microscopy and time-resolved imaging yet maintain some coherence, the beam can be band-limited only by filters absorbing low-energy photons and by total reflection on a plane mirror surface, which cuts off the high-energy part of the spectrum, leaving only a few or, ideally, a single undulator harmonic. First experiments on coherent pink-beam imaging show that

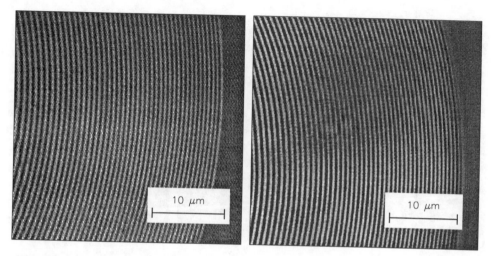

FIGURE 3. X-ray micrographs of the outermost region of a Fresnel zone plate, taken with two different X-ray microscopes. *Left:* Image taken with a zone-plate microscope at a photon energy of 14 keV. The object-to-lens and lens-to-image distances are 1.9 and 24 m respectively, with a focal length of the zone plate of 1.85 m and a magnification of 12.6. *Right:* Image taken with a CRL microscope of similar geometry, but at an energy of 23.5 keV. The outermost zone width of the object is 0.3 μm.

although, as should be expected, interference phenomena are weaker in pink-beam mode than with a conventionally monochromatized beam, clear phase-contrast outline images can still be obtained (Figure 4).

CONCLUSIONS

The microimaging and tomography instrument at the ESRF beamline ID 22 is a routinely-working user facility where absorption and phase-contrast imaging and tomographic experiments with a spatial resolution of down to less than 1 μm, in hard X-ray microscopy even down to less than 0.3 μm, can be carried out. A fast-readout, low-noise FReLoN CCD detector system allows users to perform time-resolved studies of otherwise experimentally inaccessible phenomena such as nonequilibrium thermodynamic processes in opaque materials with a temporal resolution of less than 0.1 s and a granularity of 0.5 s. For intensity-limited studies such as time-resolved imaging or hard X-ray microscopy, the use of a pink beam instead of a monochromatic one is a way of increasing intensity yet maintaining possibilities of coherent imaging.

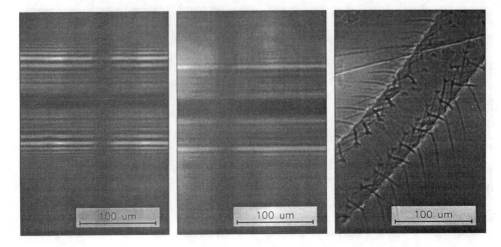

FIGURE 4. Monochromatic versus pink-beam coherent imaging: the two frames on the left show images of the same object, made of two crossed 100-μm-thick boron fibers with a 15-μm-thick tungsten core, taken with monochromatic 15-keV radiation (*left*) and with a pink beam of roughly the same energy (*center*). The interference fringes from diffraction on the sample boundaries are weaker in the pink-beam image, but still visible. Note that due to the larger horizontal source size and resulting poorer coherence in that direction, interference fringes can only be observed that arise from diffraction in vertical planes. A pink-beam image of a spider's leg (*right*) is an example of pink-beam phase contrast imaging of biological samples.

ACKNOWLEDGMENTS

The zone plate used in the FZP microscope was provided by the Advanced Photon Source through a collaboration with the Center for X-ray Lithography (U. of Wisconsin-Madison) and the Istituto di Elettronica dello Stato Solido (IESS-CNR, Italy).

REFERENCES

1. Labiche, J. C., Segura Puchades, J., and Brussel, D. C., and J. Moy, *ESRF Newsletter 25*, 41–43 (1996)
2. Spanne, P., Raven, C., Snigireva, I., and Snigirev, A., "In-line holography and phase-contrast microtomography with high energy X rays," *Phys. Med. Biol.* **44**, 741–749 (1999)
3. Raven, C., *Microimaging and Tomography with High-Energy Coherent Synchrotron X rays,* Aachen: Shaker Verlag, 1998
4. Kohn, V. G., "The method of phase retrieval of a complex wavefield from two intensity measurements applicable to hard X rays," *Physica Scripta* **56**, 14–19 (1997)

5. Gureyev, T. E., Raven, C., Snigirev, A., Snigireva, I., and Wilkins, S. W., "Hard X-ray quantitative non-interferometric phase-contrast microscopy," *J. Phys. D: Appl. Phys.* **32**, 563–567 (1999)

6. Mathiesen, R., Arnberg, L., and Weitkamp, T., to be published.

7. Koch, A., Raven, C., Spanne, P., and Snigirev, A., "X-ray imaging with submicrometer resolution employing transparent luminescent screens," *J. Opt. Soc. Am.* **15**, 1940–1951 (1998)

8. Snigirev, A., Kohn, V., Snigireva, I., and Lengeler, B., "A compound refractive lens for focusing high-energy X-rays," *Nature* **384**, 49–51 (1996)

9. Lengeler, B., Schroer, C. G., Richwin, M., Tümmler, J., Drakopoulos, M., Snigirev, A., and Snigireva, I., "A microscope for hard X rays based on parabolic compound refractive lenses," *Appl. Phys. Lett.* **74**, 3924–3926 (1999)

10. Lai, B., Yun, W., Xiao, Y., Yang, L., Legnini, D., Cai, Z., Krasnoperova, A., Cerrina, F., DiFabrizio, E., Grella, L., and Gentili, M., "Development of a hard X-ray imaging microscope", *Rev. Sci. Instrum.* **66** (2), 2287–2289 (1995)

The New Scanning Transmission X-Ray Microscope at BESSY II

U. Wiesemann, J. Thieme, P. Guttmann, B. Niemann, D. Rudolph, G. Schmahl

Georg-August-Universität Göttingen, Institute for X-Ray Physics, Geiststr. 11,
37073 Göttingen, Germany
Corresponding author: U. Wiesemann (uwiesem@gwdg.de)

Abstract. A new scanning transmission X-ray microscope (STXM) is under construction and will be operated at the undulator U41 of the BESSY II storage ring. In the monochromator without entrance or exit slits, a plane grating with variable line density is used. A spectral resolution of 2200 to 6500 is expected in the energy range of 680 eV to 200 eV. The sample is located in air so it can be exchanged easily. The photon rate in the focus will be of the order of 10^9 / s at a monochromaticity of 3000. A pnCCD capable of handling this rate will be used as a detector.

INTRODUCTION

Scanning transmission X-ray microscopes obtain images by scanning the sample with a scan spot generated by a Fresnel zone plate. The illumination of the zone plate has to be monochromatic and spatially coherent in order to get a diffraction limited focal spot size, so a source of high spectral brilliance is needed. Existing scanning microscopes are used in a wide variety of applications, including environmental science, materials science and biology. A summary of these applications can be found in [1]. A new STXM will be installed at the BESSY II storage ring. The schematic of the microscope is shown in Fig. 1. Planned research activities are among others in environmental and soil sciences, water chemistry and geochemistry.

THE SOURCE

The new microscope will be operated at the undulator U41 which is installed at a low beta section of the BESSY II storage ring [2]. The undulator covers a photon energy range of 172 eV to 596 eV in the first and 518 eV to 1340 eV in the third harmonic. Its spectral brilliance is about 10^{18} photons / s mm^2 mrad^2 0.1% BW. The size of the source is $\sigma_x = 84.3$ µm in the horizontal and $\sigma_y = 21.7$ µm in the vertical direction. The microscope is located at a distance of 37 m from the undulator. Due to the small source size and the long distance between the zone plate and the source, a zone plate with a diameter of 200 µm is illuminated coherently, so there is no need to form a smaller source with a pinhole.

CP507, *X-Ray Microscopy: Proceedings of the Sixth International Conference,*
edited by W. Meyer-Ilse, T. Warwick, and D. Attwood
© 2000 American Institute of Physics 1-56396-926-2/00/$17.00

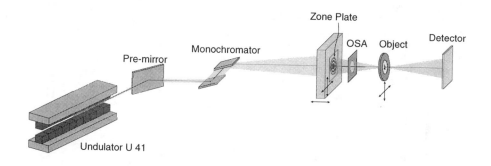

FIGURE 1. Schematic of the Scanning Transmission X-Ray Microscope

THE MONOCHROMATOR

The monochromator is very similar to the one used in an earlier STXM at BESSY I [3]. It consists of a pre-mirror and a plane grating (Fig. 2). To prevent a loss of monochromaticity due to the divergence of the beam, the line density of the grating is varied. The diffraction order -1 is used because of its high dispersion and the reduction of the vertical beam diameter. To acquire a spectrum of a small wavelength range, only the grating is turned. To adjust the angle of incidence when scanning over a larger range, the mirror is moved. It is suspended excentrically under the axis of the grating, so it does not have to be shifted along the beam axis. The undulator emits radiation in higher harmonics in addition to the desired wavelength. A part of this unwanted radiation is diffracted by the grating and the zone plate in higher orders of diffraction and also illuminates the sample. To prevent this spectral pollution, the mirror is coated with nickel which has a low reflectivity at shorter wavelenghts.

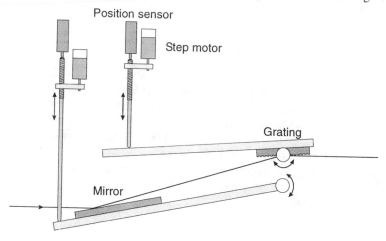

FIGURE 2. Side View of the Monochromator.

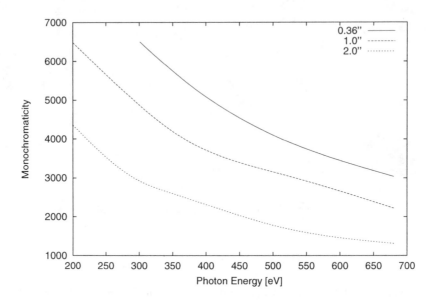

FIGURE 3. Expected Monochromaticity for Different Slope Errors of the Grating with a Fixed Exit Angle of 0.89°.

The small source size of the undulator U41 and the variable line density of the grating make it possible to have a setup without entrance or exit slits. This keeps the alignment procedure. Ray tracing simulations show that monochromaticities of several thousands can be achieved (Fig. 3). The measured slope error of the grating substrate is 0.36''. Even at higher slope errors, the spectral resolution is still sufficient for NEXAFS measurements and to illuminate the zone plate with a monochromaticity higher than the number of zones. Due to the limited area of the grating illuminated by the beam, the monochromaticity is limited to about 6500.

THE MICROSCOPE UNIT

The sample is located in air and thus can be exchanged easily. It is possible to remove the detector vacuum chamber and insert a visible light microscope for optical alignment, sample inspection and focusing. The object can be moved with step motors to obtain a large, low resolution X-ray image and to move it to an interesting location for a high resolution scan. The high resolution scan is performed by moving the zone plate in vacuum by a piezo driven X-Y flexure stage, which can be shifted along the beam axis for focusing. An order sorting aperture (OSA) is used together with a central stop to prevent light of the 0th order of diffraction of the zone plate from reaching the sample (Fig. 4).

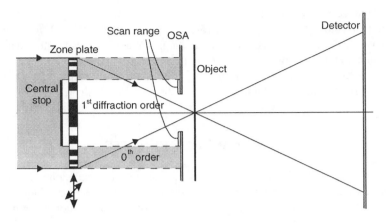

FIGURE 4. Schematic Optical Setup

TABLE 1. Parameters of the Micro Zone Plates

Outermost zone width dr_N	Radius r_N	Number of Zones N	Focal length at 190 eV	Focal length at 700 eV
40 nm	100 µm	1250	1.23 mm	4.52 mm
30 nm	100 µm	1667	0.92 mm	3.39 mm

The OSA is identical with the exit window. It does not have to be scanned together with the zone plate. This setup restricts the area for a high resolution scan to approximately 20×20 µm^2. In order to have comfortable working distances between zone plate and the object, large zone plates with long focal lengths are used (Tab. 1). The total length of the beam path in air is only a few hundred micrometers. The estimated photon rate in the focus is about 10^9 / s. The scanning motion of the zone plate, the object stage, the detector, the monochromator and the vacuum system will be controlled by computers running the Linux operating system with the RTLinux real time extension.

THE DETECTOR

The transmitted photons are detected using a back illuminated pn-charge coupled device (pnCCD) developed by the MPI für Extraterrestrische Physik [4]. A whole line of the CCD is read out in parallel to allow very fast readout with low noise. The gate structures on the CCD chip are isolated from the light sensitive silicon by pn-junctions instead of SiO$_2$, which improves the radiation hardness. The detector is read out in a continuous mode, where one CCD line is read every 25 µs. If an area of 10×10 pixels is illuminated, the detector is able to detect 10^9 photons / s with a noise level below 10 electrons per CCD pixel.

433

An advantage of the position resolution of the pnCCD is the option to use it as a configured detector [5]. It is possible to reconstruct images with different contrast mechanisms from the data obtained in a single scan. One can implement phase contrast imaging by taking the intensity difference between the left and the right detector part as the signal. By counting only the photons outside of the bright field cone, a dark field image is generated.

ACKNOWLEDGEMENTS

We would like to thank the X-ray microscopy groups at Berkeley and Stony Brook for helpful discussions. This project is a collaboration with the Chair of Water Chemistry, University of Karlsruhe and the Institute for Water Chemistry and Chemical Balneology, TU München.

This work has been funded by the German Federal Minister of Education and Research (BMBF) under contract number 02 WU 9893/0.

REFERENCES

1. *X-Ray Microscopy and Spectromicroscopy*, edited by J. Thieme, G. Schmahl, D. Rudolph and D. Umbach, Berlin: Springer Verlag, 1998.

2. Guttmann, P., Niemann, B., Thieme, J., Wiesemann, U., Rudolph, D., Schmahl, G., "X-Ray Microscopes at BESSY II", this volume.

3. Irtel von Brenndorff, A., Niemann, B., Rudolph, D., and Schmahl, G., *J. Synchrotron Rad.* **3**, 197-198 (1996).

4. Strüder, L., Bräuninger, H., Meier, M., Predehl, P., Reppin, C., Sterzik, M., Trümper, J., Cattaneo, P., Hauff, D., Lutz, G., Schuster, K., Schwarz, A., Kendziorra, E., Staubert, R., Gatti, E., Longoni, A., Sampietro, M., Radeka, V., Rehak, P., Rescia, S., Manfredi, P. F., Buttler, W., Holl, P., Kemmer, J., Prechtel, U. and Ziemann, T. , *Nucl. Instrum. Methods Phys. Res. A* **288**, 227-235 (1990).

5. Morrison, G. R., and Niemann, B., "Differential Phase Contrast X-Ray Microscopy", in [1], I-85 – I-94.

Combined near-field STXM with cylindrical collimator and AFM

R E Burge (1,2), M T Browne (2), P Charalambous(2), and X-C Yuan(1,2)

(1) Cavendish Laboratory, University of Cambridge, UK
(2) Physics Department, King's College London, UK

Abstract. The SNXM [1] was first installed at the European Synchrotron Research Facility in November 1998. It has been designed for water-window operation at a spatial resolution of about 10nm and in its final form will comprise a Zone Plate focusing X-rays onto a cylindrical collimator 10 - 20 nm in diameter, made by drilling an AFM tip, with its exit aperture within a few nm of the specimen surface. The operation of the microscope may be loosely defined to be in near-field in analogy with the optical SNOM e.g.[2]. Point to point resolution equal to the collimator diameter is expected for specimens up to 200 nm thick. The collimator to surface separation is also monitored by the AFM scanning tip. Simultaneous signals are available from X-ray transmission and surface topography. A progress report is given limited by the current availability of high energy X-rays (3-6Kev).

INTRODUCTION

The progress of the work has been influenced by the time scale of the making available at the European Synchrotron Research Facility, Grenoble, of the X-ray beam-line itself and particularly X-ray wavelengths within the so-called water window. The latter facility, which is crucially important to establish the potential of the near-field microscope, i.e. X-ray energies from about 250 to 500 eV, is expected to be available soon at the ESRF.

CONSTRUCTION OF THE MICROSCOPE

A schematic diagram of the microscope is shown in Figure 1. The essential new component, as compared with a conventional X-ray scanning microscope, is the AFM tip pierced with a nanometre scale diameter hole to pass X-rays to the specimen maintained a few nanometres from the tip aperture. The hole in the tip may be in the range of 10 to 50 nm in diameter, held a few nanometres from the specimen so that the scanning probe at the specimen will be close in diameter to the tip aperture.

CP507, X-Ray Microscopy: Proceedings of the Sixth International Conference,
edited by W. Meyer-Ilse, T. Warwick, and D. Attwood
© 2000 American Institute of Physics 1-56396-926-2/00/$17.00

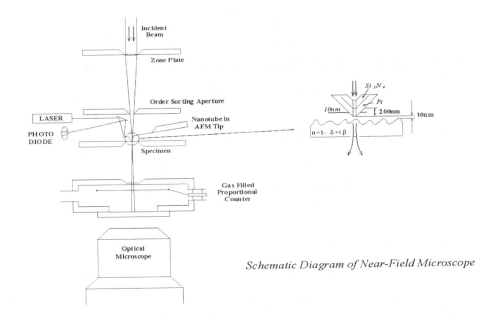

FIGURE 1. Schematic diagram of the SNXM

The main new problems in making the microscope concern tip piercing and tip mounting and alignment. The Zone Plate alignment is carried out using a piezzo X-Y stage, range 50 x 50 μm, mounted on a compound flexure stage, with 8 mm Z-range, which itself is kinematically mounted on three ministeppers acting vertically to give coarse X-Y scans with a range of 2 x 2 mm. The Zone Plate Z-movement is required for focusing over a range of wavelengths for XANES (NEXAFS) measurements. The Zone Plate is followed downstream in the X-ray beam by the Order Sorting Aperture (OSA), mounted on a blade attached to a nanostepper X-Y-Z stage. The specimen is mounted on a piezzo X-Y stage mounted kinematically on three ministeppers acting vertically. The AFM tip assembly is fitted between the OSA and the specimen. The specimen stage can be transferred to a similar kinematically mounted optical system of which the scanning stage is linked to the movement of the SNXM specimen stage and can be used to select a specimen area of interest prior to X-radiation. The optical system can be used in stand-alone mode as a SNOM.

It is necessary to align the Zone Plate so that the focussed spot is centered on the fixed AFM tip, and the intensity transmitted through the collimating hole forms our probe. It is also necessary to incorporate an optical system to detect deflection of the tip cantilever. The specimen is moved by a piezzo transducer up to the tip, until it gives a pre-determined deflection. The Z distance traveled by the specimen is recorded, and is used to form topographic images of the surface sample.

The X-ray detector for transmitted radiation is currently a PIN diode coupled to a current amplifier followed by an integrating A/D converter. This design allows the

436

simultaneous recording of X-ray transmission and topographic images of the same area of the specimen.

Development of Cylindrical Collimator (drilled AFM tip)

The lowest photon energy so far made available is 3 KeV; such a high energy is not conducive to the use of a tube collimator. This is because of the difficulty of fabricating a "long hole" such that photons not travelling through the hole are almost totally absorbed and kept away from the specimen. At water window energies the aperture (say 10 nm in diameter with parallel sides fabricated through the AFM tip) needs to be about 250 nm long in gold or other heavy metal, which gives a not unreasonable problem in ion beam/chemical fabrication to solve. At 3 – 5 KeV the aperture needs to be several μm long, which is not practical at the 10 to 50 nm diameter level, but may be at the 200 nm level.

In collaboration with FEI (Europe), using a Focussed Ion Beam facility, we have so far produced tips pierced with holes at the 50 nm level of diameter. In Figure 2 is shown, for illustration a commercial pyramidal AFM tip. initially fabricated in Si_3N_4, here coated with 0.5 μm of tungsten, with an 80 nm diameter drilled hole at the tip.

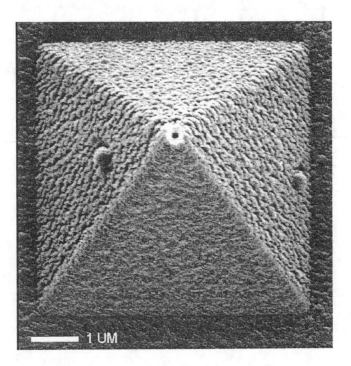

FIGURE 2. Micrograph of an AFM tip, coated with 500 nm of Tungsten, and pierced with an 80 nm hole, using a focussed Ion Beam facility.

We have not as yet imaged with X-rays through a collimator, mainly because of the unsuitable X-ray energies available to us, but we are hoping to do so in the near future. We did however collect a number of ordinary STXM images using the condenser ZP as an objective, while at the same time a displaced AFM tip allowed the collection of contact mode images of adjacent areas. The displacement which was of the order of 20 µm was necessary to allow the X-ray beam through to the specimen.

IMAGES: X-RAY TRANSMISSION AND SURFACE TOPOGRAPHY

The current stage of performance of the microscope may be demonstrated by results shown in Figures 3 and Figures 4. The images show regions of a Tungsten Zone Plate, prepared for high energy X-ray imaging and used as a specimen. The range of zone widths (250 – 500 nm) and the large thickness of the zones (700 nm), made this an ideal test sample for X-ray imaging, in view of the high photon energy during the experiment.

Figures 3(a,b) illustrate an image and a line scan perpendicular to the zones in the central region of a Tungsten Zone Plate gained by using the AFM tip in contact mode. The zone widths were of the order of 300 – 400 nm. The triangular profile of the fringes in the line scan are not a true representation of the sample, but arise from the inability of the tip to penetrate the full depth of the zones because of its physical dimensions (see fig. 2). Figures 4(a,b) show complementary X-ray transmission and AFM contact images of the outermost zones of the same ZP. The AFM tip easily resolves the zone separation of 250 nm with a depth resolution of 15nm; the estimated lateral resolution is of the order of 10 nm. The X-ray micrograph was taken at 3 KeV, and the imaging ZP had an outermost zone width of 120 nm.

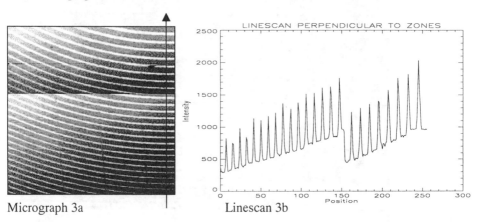

Micrograph 3a   Linescan 3b

FIGURE 3. Left, AFM image of Tungsten zones, 700 nm deep. Right, Linescan perpendicular to the zones, showing good lateral and depth resolution in the contact image.

 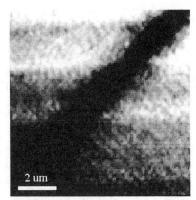

Micrograph 4a   Micrograph 4b

FIGURE 4. Left, AFM image of 250 nm Tungsten zones. Right, 3 KeV X-ray image of same area

CONCLUSIONS

Sufficient development has been conducted to show that cylindrical collimators with diameters less than 50 nm can be made. The most serious consequence of delay in the availability of water window energies is that we have not yet had the chance to test our cylindrical collimators with soft X-rays. We await the opportunity for the simultaneous acquisition of a topographic image and a transmitted X-ray image from the same area. We are optimistic that the design specifications of the SNXM could improve significantly on the 30 to 50 nm resolution so far achieved with Zone Plate objectives, and at the same time provide topographic information about the sample.

It is already clear that the harnessing of combined AFM and X-ray transmission is a significant advance. This dual image signal facility has not been made available before in soft X-ray imaging. The AFM signal gives, without beam damage, an image scanning facility at a resolution at or below that of the X-ray microscope; the tip function has been shown to be remarkably robust.

ACKNOWLEDGEMENTS

We are grateful for financial support to the Leverhulme Trust and the Paul Instrument Fund administered by the Royal Society.

REFERENCES

1. Burge, R. E.,Yuan, X. C., Browne, M. T., Charalambous, P., *Ultramicrscopy* **69,** 259-278 (1997)

2. Pohl, D. W., Courjon, D., Near Field Optics, Dordrecht, Kluwer, 1992

The Condenser-Monochromator With Dynamical Aperture Synthesis For The TXM At An Undulator Beamline At BESSY II

B.Niemann, P.Guttmann, D.Hambach,
G.Schneider, D.Weiß, G.Schmahl

Institute for X-Ray Physics (IRP), University Georgia Augusta at Göttingen
Geiststraße 11, D-37077 Göttingen, Germany

Abstract. The Göttingen transmission X-ray microscope at the low emittance electron storage ring BESSY II will use the concept of dynamical aperture synthesis [1] for the condenser-monochromator. The concept is well suited as a condenser, as it can match the undulator U41 to the TXM objective and has many other advantages, too. It can use an off-axis transmission zone plate with comparatively wide zones of low aspect ratio, which therefore can be produced with almost theoretical efficiency. It will deliver a monochromaticity of 1000 to 3000. As the numerical aperture of *any* existing micro objective zone plate can be matched, the achievable resolution increases to the theoretical limit. Phase contrast imaging is possible with annular phase plates of extremely small width, a fast switching from amplitude contrast to phase contrast is possible. Stereo imaging with arbitrary stereo axis will be possible without tilting the object.

1. INTRODUCTION

The transmission X-ray microscope (TXM) installed at the electron storage ring BESSY I uses the X-rays delivered by a bending magnet. In this TXM the object illumination is performed with an on-axis condenser zone plate (CZP) operating as a linear condenser-monochromator, which illuminates the object quasi-monochromatically in critical illumination. However, the small source size obtained at BESSY I [2, 3] leads to an illuminating spot size in the object region of about 7 µm FWHM in diameter. If larger objects have to be imaged, a scanning of the illuminating spot over the object, which in turn reduces the monochromaticity unacceptably or stitching of a series of images is required.

In the near future TXMs will run routinely at electron storage rings with undulator sources of ever reduced source size and divergence, e.g. at BESSY II in Berlin. Therefore critical illumination with an on-axis CZP linear condenser-monochromator [4] would produce an even smaller illuminating spot size. Furthermore, only a CZP with a relatively large central stop can deliver a sufficiently quasi-monochromatic object illumination. Therefore a CZP such as KZP7 [5] - which has got a central stop of about 4 mm diameter - could only be illuminated partially and asymmetrically, as the highly collimated undulator beam is only a few mm in diameter. An asymmetrical transfer function of the TXM would result. If on the other hand - at a given required

CP507, X-Ray Microscopy: Proceedings of the Sixth International Conference,
edited by W. Meyer-Ilse, T. Warwick, and D. Attwood
© 2000 American Institute of Physics 1-56396-926-2/00/$17.00

monochromaticity and *numerical aperture* (NA) of the beam illuminating the object - an on-axis CZP linear condenser-monochromator with a CZP not larger than the X-ray beam diameter would be used at these highly collimated undulator sources, a very small CZP with a reduced focal length would be required. An even smaller illuminating spot size would result, aggravating the problems mentioned above. However, the problems will be overcome by using the concept of dynamical aperture synthesis (DAS) [1] including an off-axis transmission zone plate (OTZ) and a pair of rotating mirrors which was already proposed earlier [6,7]. This concept will be used in the TXM at the X-ray undulator source U41 at BESSY II.

2. THE OFF-AXIS TRANSMISSION ZONE PLATE LINEAR MONOCHROMATOR

Figure 1 illustrates the concept of an OTZ linear monochromator, which will be used at BESSY II. An OTZ with an area 2 x 4 mm² in size, located at R = 26 mm from the zone plate center completely accepts the central beam of the strongly collimated undulator radiation at 30 m distance from the source. As the OTZ is very small and

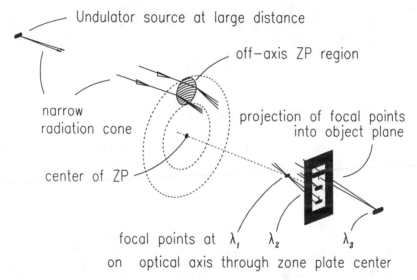

FIGURE 1. Schematic of an off-axis zone plate linear monochromator. A small off-axis region of a zone plate, which also can be regarded as a focusing grating, produces a series of focal spots on the optical axis. The projection of the focal spots into the object field is also sketched and results in a one-dimensionally dispersed spectrum of the source.

the zone structures therein only represent a short arc, it is more appropriate to regard it as a "focusing grating", which vertically disperses the radiation into a stripe-shaped area in the object plane, located 5 m behind the OTZ. The vertically dispersed spectrum, which is actually transferred to the object plane by the complete optical set-up according to fig. 3 will extend some hundred µm in height and is of very homogeneous intensity in this vertical range. The source will be demagnified 6x by the

OTZ. The monochromatic source image will extend $\varsigma_{y,image,fwhm} = 9$ µm in vertical and $\varsigma_{x,image,fwhm} = 33$ µm in horizontal direction (electron beam coupling: 1%, Twiss correction: included). It is - compared to previous TXMs - much larger in horizontal width. Therefore the homogeneity of the object illumination is also significantly improved in horizontal direction, perpendicular to the direction of dispersion.

The obtainable monochromaticity only depends on the extension of the - commonly elliptical - source in the direction of dispersion. If the direction of the smallest source diameter is oriented parallel to the direction of dispersion, best monochromaticity is achieved. For comparison, in an on-axis CZP monochromator the monochromaticity is determined by the two-dimensional extension of the source, because the radiation is dispersed in *two* dimensions, which also leads to a comparatively inhomogenous intensity distribution with a strong maximum in the center. This means that an OTZ linear monochromator is superior to any on-axis CZP linear monochromator.

Applying the equation for the monochromaticity $M = \lambda/\Delta\lambda = R/d$ [4], with $R = 26$ mm and an object field diameter of e.g. $d = 26$ µm we get $M \approx 1000$. This field can be subdivided of regions of $\varsigma_{y,image,fwhm} = 9$ µm height with $M \approx 3000$. $M = 1000$ is sufficient to use long focal length micro zone plates (MZPs) with as much as 1000 zones. The OTZ has 4.28 m focal length. For 2.4 nm radiation and at $R = 26$ mm its zones have got 198 nm mean width (zone number: 65722), which can easily be electroplated in nickel with the optimum aspect ratio of ≈ 1.2. Thus the OTZ will show almost theoretical diffraction efficiency in the 20% range.

3. MATCHING A TXM TO AN UNDULATOR

Figure 2 illustrates how the OTZ linear condenser-monochromator can be used to perform a dynamical aperture synthesis (DAS) with the help of a rotating condenser

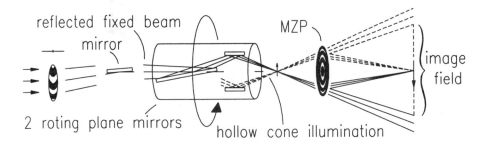

FIGURE 2. Schematic of the RK-TXM including a rotating condenser matching the NA of the MZP. The small source size and the well collimated beam of the undulator source are of great advantage, as they allow to realize a high performance optical set-up with comparatively small optical elements.

(RK), which contains two rotating mirrors [6]. The inclination angle of the second rotating mirror performs the matching of the required numerical aperture. As the OTZ is only a few mm in diameter and the focal length is several m, the angular diameter of the illuminating beam generated by the OTZ has got an angular width of less than 1 mrad diameter. Therefore very short rotating mirrors can be used, if they are located close to the spot illuminating the object.

The DAS of a *hollow cone* illumination has several advantages. The image field can be increased at least by a factor of two compared with the TXM at BESSY I using a KZP7, as the shadowed region inside the illuminating solid angle is increased significantly. The depth of focus increases at least by a factor of two compared to full cone illumination as the angular sequence of the individual images is recorded with very narrow illuminating condenser beams. The image will be free of coherent noise even with coherent X-ray beams, as the principle of time-sequential recording and superposition of images destroys the edge ringing present in the individual images. As the angular diameter generated by the OTZ is less than 1 mrad in diameter, the minimum width of the phase ring for phase contrast is smaller than 1 µm at 1 mm focal length of the micro objective. This value is much smaller than the width required with KZP7 at BESSY I. Therefore the halo present in phase contrast images is reduced.

Stereo imaging is possible without tilting the object. For this purpose two exposures are taken while the mirrors are rotated from ß .. ß+180°, and ß+180° .. ß+360°, where ß is the arbitrarily selected stereo axis; thereby the two images are taken at two different oblique illumination angles which are required to get a pair of stereo images, though the stereo angle is very small and the images always contain phase information

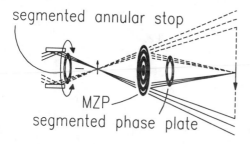

FIGURE 3. Phase contrast arrangement for the RK-TXM. A segmented annular phase plate is located behind the MZP objective. Phase resp. amplitude contrast imaging is obtained if the direct beam can only pass through the phase plate segments resp. cannot pass this segments. Chopping of the beam is achieved with the help of a fixed segmented annular stop behind the rotating mirrors, which only transmits the radiation in the appropriate angular positions of the rotating mirrors. Switching from phase to amplitude contrast is achieved by rotating the segmented stop a certain amount.

of the object. It is known from the literature that an image obtained with oblique illumination (angle: µ) in the amplitude contrast imaging mode will produce an image which contains both amplitude and phase information of the object. Solely the intensity-superposition of images obtained at µ and -µ cancels the phase information and a pure amplitude contrast image will remain. This behavior is fundamental for incoherent imaging. Köhler illumination e.g. uses oblique plane waves illuminating an object, and the angle of obliqueness is changed in a statistical manner, which is determined by the location of the radiating atom of the thermal source; the images are sequentially integrated on the detector. The rotating condenser system generates oblique, nearly plane (though sligthly convergent) waves - it just puts the angles of obliqueness in a well defined row. During imaging the detector integrates the intensities of all subsequent images and the read-out is again the average over all images.

4. IMAGING PERFORMANCE

A theoretical model describing the image performance of a TXM using a rotating condenser was established [8]. For this purpose it was assumed, that protein fibers embedded in water or ice are imaged in a TXM with an X-ray objective of about 25 nm Rayleigh resolution. The protein fibers have a square cross-section with the edge length d, are infinitely long and are arranged to form a grating like structure with a period of 2d. In Fig. 4 the upper graph shows the contrast as it is generated by the fiber-object and observable *directly behind it*. It is plotted as a function of the parameter 1/2d. The other graphs show the contrast obtainable in the *image* plane for different condenser set-ups using the given X-ray micro objective.

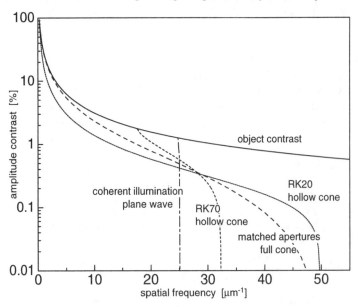

FIGURE 4.
Image contrast achieved with different condenser arrangements when imaging a model protein grating at 2.4 nm wavelength with a micro objective of about 25 nm Rayleigh resolution. Best resolution is obtained with RK20. A rotating condenser RK20 (RK70) just matches a numerical aperture of a micro objective with a smallest, outermost zone width of 20 nm (70 nm) at the regarded wavelength.

For comparison the imaging quality is also shown for other commonly used illumination conditions. If on-axis plane waves illuminate the object - no rotating condenser arrangement is necessary - we get coherent imaging. The contrast can be retained fully up to the cut-off frequency of 25 LP/μm. If the X-ray source is incoherent and a full-cone condenser with an aperture matched to the X-ray objective is used the cut-off frequency is extended to 50 LP/μm. However, the contrast transfer rapidly decreases with increasing spatial frequency, as it is common for incoherent imaging systems. The contrast transfer at high spatial frequencies can be increased significantly if the rotating condenser RK20 is employed - however, the price which is paid is a further reduction of the contrast transfer at lower spatial frequencies. Nevertheless, for high resolution imaging application these lower spatial frequencies are of less importance.

When the rotating condenser RK70 is employed, the spatial frequency transfer is linear up to moderate spatial frequencies of about 18 LP/μm, therefore this imaging condition is superior if quantitative data have to be restored from the images. However the cut-off frequency is reduced to 33 LP/μm.

5. SUMMARY

The monochromator-condenser presented here realizes the concept of dynamical aperture synthesis with an off-axis transmission zone plate and a pair of rotating mirrors. It takes full advantage of the increased brilliance of modern X-ray sources and can ideally match these sources to a TXM, compare also [9]. It has already been stated since long that the introduction of undulator sources introduces a dramatic progress for scanning transmission X-ray microscopes. This paper explained that this is true also for transmission X-ray microscopes.

6. ACKNOWLEDGMENTS

This work has been founded by the German Federal Minister of Education, Science and Technology (BMBF) under contract number 05SL8MG11.

7. REFERENCES

1. "Partially filled, synthetic aperture imaging synthesis: coherent illumination", in *The new physical optics notebook*, edited by Reynolds, De Velis, Parrent, Thomson, SPIE, pp.536 - 548 (1990)

2. D.Rudolph, G.Schneider, P.Guttmann, G.Schmahl, B.Niemann and J.Thieme: "Investigations of Wet Biological Specimens with the X.ray Microscope at BESSY", in *"X-Ray Microscopy III"*, edited by A.G.Michette, G.R, Morrison, C.J.Buckeley, Berlin Heidelberg, Springer Verlag,pp. 392-396 (1992)

3. G.Schneider and B.Niemann: "Cryo X-ray microscopy Experiments with the X-Ray Microscope at BESSY", in: see ref. 2, pp I/25-34

4. B.Niemann, D.Rudolph and G.Schmahl: Opt. Com., Vol. 12, pp. 160-163 (1974)

5. M.Hettwer and D.Rudolph: "Fabrication of the X-Ray Condenser Zone Plate KZP7", in: see ref. 2, pp. IV/21-26

6. B.Niemann: "High numerical aperture X-ray condensers for transmission X-ray microscopes", in: see ref. 2, pp.IV/45-55

7. B.Niemann: international patent application, PCT DE 97 00033, 10.1.1997

8. G.Schneider: "Cryo X-ray microscopy with high spatial resolution in amplitude and phase contrast", Ultramicroscopy, Vol. 75 , pp. 85-104 (1998)

9. S.Oestreich and B.Niemann: "Design of a condenser for an X-ray microscope at a low-ß section at the ESRF", in: see ref. 2, pp. IV/77-82

Transmission X-ray Microscopy with 50nm Resolution in Ritsumeikan Synchrotron Radiation Center

K. Takemoto(1), A. Hirai(2), B. Niemann(3), K. Nishino(2), M. Hettwer(3), D. Rudolph(3), E. Anderson(4), D. Attwood(4), D.P. Kern(5), Y. Nakayama(2) and H. Kihara(1)

(1)Physics Laboratory, Kansai Medical University,Uyamahigashi, Hirakata, Osaka, Japan
(2)Department of Physics, Ritsumeikan University, Noji-Higashi, Kusatsu, Shiga, Japan
(3)Forschungseinrichtung Röntgenphysik, Georg-August-Universität, Geiststraße, Göttingen, Germany
(4)Center for X-ray Optics, Lawrence Berkeley National Laboratory, Berkeley, California, USA
(5)Institut für Angewandte Physik, Universität Tübingen, Auf der Morgenstelle , Tübingen , Germany

Abstract. Transmission X-ray microscope at Rits SR Center at Ritsumeikan University has been upgraded to get a higher performance. Its optical configuration allows a continuous wavelength change throughout water window region, 2.2−4.3nm. The present paper describes results observed with the high-performance X-ray microscope, and compared the images taken at two different wavelengths, 2.4nm and 3.2nm. Its achieved resolution with each wavelength was below 50nm. Several specimens such as latex spheres of 0.23μm diameter, diatoms of 0.1μm inner structures could be clearly resolved.

1. DEVELOPMENT OF X-RAY MICROSCOPY

We have been developing a transmission X-ray microscope at Ritsumeikan (Rits) SR Center [2, 3]. It was transferred from UVSOR (Okazaki, Japan), where it was initially installed [4], and has been operating since 1996. At present, including our X-ray microscope, four X-ray microscope stations operate in the world [5-7].

The storage ring at Rits SR Center is in operation under a single circular superconducting magnet with an orbital of 1m diameter. Electrons are injected by a 150MeV microtron. The storage ring is usually operated at 575MeV with 300mA. The source emits photons with energy range from about 0.1eV to 9keV. The lifetime of the electron beam in the storage ring is more than 5h as of May 1998. Its beam size is 0.28 $\times 2.6mm^2$ (2σ) [8].

In 1997, in order to improve the resolution, several changes have been done.

A higher-performance condenser zone plate (CZP) was introduced. This new CZP is a Göttingen KZP 7 type [9]. The new CZP was fabricated by holographic lithography, the specifications of which are listed in Table 1. Its groove efficiency and absolute efficiency at 2.5nm were measured to be 7.5% and 3.6%, respectively.

A SiC plane mirror with a grazing incidence angle of 40mrad was installed at the upstream, which cuts off X-rays of higher energy, thus its protecting the CZP from the

CP507, *X-Ray Microscopy: Proceedings of the Sixth International Conference,*
edited by W. Meyer-Ilse, T. Warwick, and D. Attwood
© 2000 American Institute of Physics 1-56396-926-2/00/$17.00

heat load and suppressing higher order diffraction images. As a result of mounting the SiC mirror, the energy about 900eV above is cut-off, and the heat load decreases to about 50%.

A thick Al mask (2mm diameter, about 4mm thickness) was set at the downstream immediately after the CZP to eliminate zero-th order diffraction of X-rays.

2. OPTICAL SETUPS

The present optical configuration of the X-ray microscope is the same as a conventional one [10]. It consists of two parts: a condensing part and an imaging part [4]. They are in vacuum conditions. Si_3N_4 films of $0.2 \times 0.2 \mu m^2$ and $0.1 \mu m$ thickness shield the vacuum part from atmospheric pressure. Samples are mounted at atmospheric pressure between the two vacuum parts.

The X-rays reflected by the SiC plane mirror are condensed and monochromatized by a CZP through a pinhole of $20\mu m$ diameter, and then beamed at a specimen in air. The transmitted photons are enlarged by an OZP to form an image on a CCD camera (Astromed, U.K.). The CCD chip is liquid nitrogen-cooled, thinned and back illuminated with 512×512 pixels of $24\mu m$ size. Images are focused with the first-order diffraction and observed in the shadow of the central stop of the CZP.

A wavelength in water window region, $2.2 - 4.3nm$, is used for imaging. Usually, two specific wavelengths, $\lambda=2.4nm$ and $3.2nm$ were used. These wavelengths were chosen as a little below oxygen and nitrogen absorption edges, respectively.

Although a distance from the SR source is fixed at 7265mm, it is necessary to change a distance between the CZP and the pinhole depending on an observation wavelength. At $\lambda=3.2nm$, a focal length is 151mm, and at $\lambda=2.4nm$, 202mm. Similarly, the OZP is adjusted to have the best focusing position. The distance between the OZP and a Si_3N_4 window at $\lambda=3.2nm$ is narrower than that at $\lambda=2.4nm$, and its are usually about sub-mm.

The CZP is a Göttingen KZP7 type[9]. The OZP was fabricated at IBM/LBNL[11]. Their specifications are also tabulated in Table 1. Combination with the CZP and the OZP, allows the achievable resolution of 50nm.

TABLE 1. Specifications of zone plates.

	CZP	OZP
Diameter (μm)	9000	50
Number of the zones	41890	277
Outermost zone width (nm)	53.7	45
Zone material	Ge (300nm thick)	Ni (130nm thick)

RESULTS AND DISCUSSION

A resolution of the X-ray microscope was experimentally evaluated by observing a Cu mesh (#2000). In Fig. 1, images taken at wavelengths of 2.4nm and 3.2nm are shown. An X-ray magnification at 2.4nm is 990 and that at 3.2nm is 655. Their resolutions were estimated to be 49nm (25−75%) and 73nm (20−80%) at 2.4nm and 46nm (25−75%) and 56nm (20−80%) at 3.2nm from the intensity gradient of the knife-edge of the mesh, respectively. In order to take an image with the same degree contrast at each wavelength, an exposure time was different, 60s at 2.4nm and 10s at 3.2nm.

A Ge test pattern was observed at λ=3.2nm and 2.4nm, as shown in Fig. 2. As shown in the figures, periodic lines and spaces between the first and the second annulus can be detected. The observable minimum lines and spaces were calculated to be 52nm[10]. These are in good agreement with the values of resolutions estimated by knife-edge analyses.

Latex spheres (Dow Chemical Co.) of 1.09μm diameter and 0.23μm diameter were observed and are shown in Figs.3(a) and (b), respectively. As seen in the figure, the latex spheres show good contrast. Especially, a network structure of latex spheres of 0.23μm diameter demonstrates that the X-ray microscope is capable of observing specimens of organic materials with reasonable contrast. In Fig. 4, images of a part of a diatom are shown. The image in Fig. 4(b) was digitally enlarged, and is shown in Fig. 4(b). Fine structures of 0.1μm size in the diatom cell can be clearly observed. These observations indicate that biospecimens such as subcellular structures could be analyzed *in vivo*.

4. CONCLUSION

A transmission X-ray microscope with zone plates has been operating at BL-12 of Ritsumeikan SR Center. With the newly introduced CZP, we could obtain the resolution below 50nm from both of knife-edge analyses at wavelengths of λ=3.2nm and 2.4nm. From images of a test pattern, we could distinguish 52nm lines and spaces. Several specimens such as latex spheres of 0.23μm diameter and diatoms of 0.1μm inner structures could be clearly resolved. Both wavelengths of λ=3.2nm and 2.4nm are applicable to observe biospecimens such that subcellular structures could be analyzed *in vivo*.

ACKNOWLEDGEMENTS

We thank Dr. P. Guttmann (University of Göttingen), who kindly offered the use of the Siemens test pattern, and Mr. Y. Shimanuki (Tsurumi University), for diatom preparation.

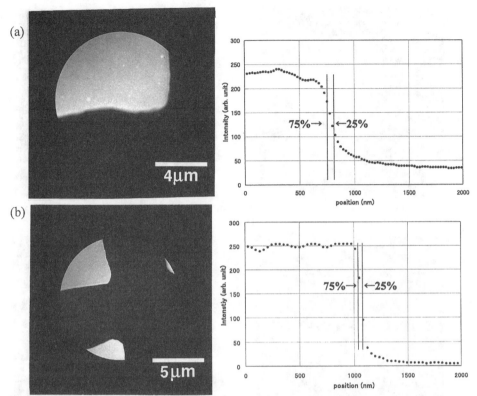

FIGURE 1. X-Ray images of Cu #2000 mesh and knife-edge analyses. (a) At a wavelength of 2.4nm. Exposure time is 60s. The knife edge analyses show the resolution to be 48.7nm at 25−75% and 72.8nm at 20−80%. (b) At a wavelength of 3.2nm. Exposure time is 10s. The knife edge analyses shows the resolution to be 46.3nm at 25−75% and 56nm at 20−80%.

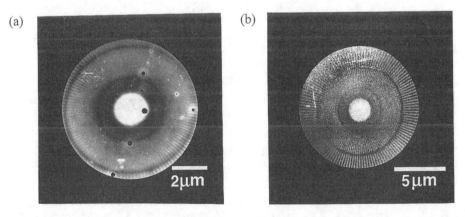

FIGURE 2. X-Ray images of a test pattern. (a) A wavelength is 2.4nm. Exposure time is 60s. (b) A wavelength is 3.2nm. Exposure time is 10s.

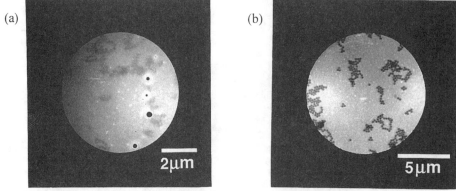

FIGURE 3. X-Ray images of a latex spheres of 0.23µm diameter taken at a wavelength. (a) A wavelength is 2.4nm. Exposure time is 60s. (b) A wavelength is 3.2nm. Exposure time is 10s.

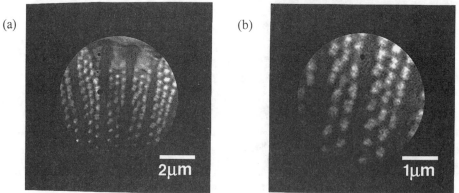

FIGURE 4. X-Ray image of diatom. (a)A wavelength is 2.4nm. Exposure time is 60s. (b) Digitally enlarged image of (a).

REFERENCES

1. Hirai, A., Watanabe, N., Takemoto, K., Nishino, K., Anderson, E., Attwood, D., Kern, D., Shimizu, S., Nagata, H., Aoki, S., Nakayama, Y., and Kihara, H., "X-Ray Microscopy and Spectromicroscopy", eds. J. Thieme, G. Schmahl, D. Rudolph and E. Umbach, Springer-Verlag, Heidelberg, 1998, pp.I-123-I-128.

2. Hirai, A., Takemoto, K., Nishino, K. , Watanabe, N., Anderson, E., Attwood, D., Kern, D., Hettwer, M., Rudolph, D., Aoki, S., Nakayama, Y., and Kihara, H., *J. Synchrotron Rad.* **5** 1102-1104 (1998).

3. Watanabe, N., Hirai, A., Takemoto, K., Shimanuki, Y., Taniguchi, M., Anderson, E., Attwood, D., Kern, K., Shimizu, S., Nagata, H., Kawasaki, K., Aoki, S., Nakayama, Y., and Kihara, H., "X-Ray Microscopy and Spectromicroscopy", eds. J. Thieme, G. Schmahl, D. Rudolph and E. Umbach, Springer-Verlag, Heidelberg, 1998, pp.I-65-I-70.

5. Meyer-Ilse, W., Medecki, H., Brown, J. T., Heck, J. M., Anderson, E. H., Stead, A., Ford, T., Balhorn, R., Petersen, C., Magowan, C., and Attwood, D., "X-Ray Microscopy and Spectromicroscopy", eds. J. Thieme, G. Schmahl, D. Rudolph and E. Umbach (Springer-Verlag, Heidelberg, 1998), pp.I-3-I-12.

6. Wang, J., Kagoshima, Y., Miyahara, T., Ando, M., Aoki, S., Anderson, E., Attwood, D., and Kern, D., *Rev. Sci. Instrum.* **66** 1401-1403 (1995).

7. Medenwaldt, R., and Uggerhøj, E. E.: *Rev. Sci. Inst.* **69** 2974-2977 (1998).

8. Iwasaki H., Nakayama Y., Ozutsumi K., Yamamoto Y., Tokunaga Y., Saisho H., Matsubara T., and Ikeda S., *J. Synchrotron Rad.* **5** 1162-1165 (1998).

9. Schmahl, G., Rudolph, D., Niemann, B., Guttmann, P., Thieme, J., Schneider, G., David, D., Diehl, M., and Wilhein, T. *Optik* **93** 95-102 (1995)

10. Hirai, A., Takemoto, K., Nishino, K., Niemann, B., Hettwer, M., Rudolph, D., Anderson, E., Attwood, D., Kern, D. P., Nakayama, Y., and Kihara, H. *Jpn. J. Appl. Phys.* **38** 274-278 (1999).

11. Anderson, E., and Kern, D. "X-Ray Microscopy III", eds. A.G. Michette, G.R. Morrison, and C.J. Buckley, Springer-Verlag, Berlin, 1992, pp.75-78 (1998).

Configured Detector System For STXM Imaging

W J Eaton[1], G R Morrison[1], N R Waltham[2]

1. Dept. of Physics Kings College, Strand, London WC2R 2LS, UK
2. Space Science Dept. Rutherford Appleton Laboratory, Chilton, Didcot OX11 0QX, UK

Abstract. A configured detector system based on an 80×80 element x-ray sensitive CCD array has been developed to replace the conventional transmitted x-ray detector used in the scanning transmission x-ray microscope. This from of detector allows a flexible choice of imaging modes to be made simultaneously from only a single scan of the sample. Details of the theoretical benefits expected, and the hardware implementation, are described.

INTRODUCTION

The vast majority of x-ray images recorded to date have contrast that derives from variations in the local absorption characteristics of the sample, but, at energies significantly above 1 keV, imaging by phase contrast techniques is increasingly attractive, since the real part of the refractive index decrement δ can be larger than the absorption index β, assuming a refractive index of the form n=1-δ-iβ. The feasibility of phase contrast imaging has already been demonstrated using a variation of the Zernike method (1), and with harder x-rays using other techniques that exploit the high coherence of the illumination (2). The use of a multi-element detector in the STXM geometry provides a very flexible form of imaging; in addition to the incoherent brightfield and darkfield signals, antisymmettric detector combinations yield a differential phase contrast (DPC) signal (3), and the use of a first-moment configuration results in a system with an ideal linear response (4).

In this paper we describe the development of a CCD-based configured detector system for the STXM that will allow a number of different imaging modes to be realised simultaneously. In the STXM, the transmitted x-ray detector can be treated as if it is in a plane conjugate to the objective lens aperture, and so the complex amplitude in this plane $\Psi(\mathbf{k},\mathbf{r}_0)$ is related by a Fourier transform to the complex amplitude $\psi(\mathbf{r},\mathbf{r}_0)$ in the specimen exit plane. The vector $\mathbf{r}_0$ denotes the position of the x-ray probe on the sample, and $\mathbf{k}$ is a position in the detector plane. The STXM image signal is then given by

$$s(\mathbf{r}_0) = \int |\Psi(\mathbf{k},\mathbf{r}_0)|^2 R(\mathbf{k}) d\mathbf{k}, \qquad (1)$$

and since the detector response function $R(\mathbf{k})$ for the CCD detector can be made an arbitrary function of $\mathbf{k}$, a wide range of different imaging modes are possible simply by choosing $R(\mathbf{k})$ appropriately. Since it is the function $|\Psi(\mathbf{k},\mathbf{r}_0)|^2$ that represents the

CP507, X-Ray Microscopy: Proceedings of the Sixth International Conference,
edited by W. Meyer-Ilse, T. Warwick, and D. Attwood

data collected experimentally, the choice of imaging modes can be made after the experimental data have been collected, although for user convenience it is helpful to have a few 'standard' modes available on-line to allow real-time inspection of the sample. Assuming that the optical system is isoplanatic, so that the incident wave on the sample can be written as $\psi_0(\mathbf{r} - \mathbf{r}_0)$, and that the sample can be described by a complex amplitude transmittance $(1 - a(\mathbf{r}))\exp i\phi(\mathbf{r})$, then the complex amplitude at the specimen exit plane can be written as

$$\psi(\mathbf{r}, \mathbf{r}_0) = (1 - a(\mathbf{r}))\exp i\phi(\mathbf{r})\psi_0(\mathbf{r} - \mathbf{r}_0).$$

Choosing $R(\mathbf{k})=1$ for all $\mathbf{k}$ as the detector response in equation (1) produces an incoherent brightfield (absorption contrast) image signal

$$s(\mathbf{r}_0) = (1 - a(\mathbf{r}_0))^2 \otimes |\psi_0(\mathbf{r}_0)|^2, \qquad (2)$$

where $\otimes$ denotes the convolution operation. Any anti-symmetric choice of $R(\mathbf{k})$ will produce a differential phase contrast image. Of particular interest is the choice $R(\mathbf{k})=\mathbf{k}$, so that the image signal is given by the first moment of the transmitted intensity distribution (3), since then the image signal can be written in the form

$$s(\mathbf{r}_0) = (1 - a(\mathbf{r}_0))^2 \nabla\phi(\mathbf{r}_0) \otimes |\psi(\mathbf{r}_0)|^2 / 2\pi. \qquad (3)$$

For weakly absorbing samples, the term $a(\mathbf{r}_0) \to 0$, and the first-moment detector configuration therefore produces a signal that depends linearly on the phase gradient of the sample transmittance.

HARDWARE IMPLEMENTATION

The chosen method for recording $|\Psi(\mathbf{k}, \mathbf{r}_0)|^2$ is to use an 80×80 element array CCD camera as the detector. This is a back-face-thinned frame-transfer CCD (EEV Ltd., model CCD39-01), that has an active area of 1.92 mm × 1.92 mm. It is used in conjunction with fast readout electronics developed by the Space Science group at the CLRC Rutherford Appleton Laboratory as part of a project to develop adaptive optics for applications in astronomy (5). The CCD detector presently provides 14-bit data, and modifications to the CCD control electronics allow the detector system to interface to a PC based 16-bit digital framestore (Imaging Technology IC-DIG). Figure 1 shows a block diagram of the hardware elements and communications links. The hardware and software have been designed so that they can be interfaced to any STXM scanning system with a minimum number of changes to the STXM control software; the CCD based detector is intended as a direct replacement for the more conventional transmitted x-ray detector on the STXM.

The STXM specimen stage is raster scanned in the conventional manner, and the CCD detector, placed in the $\mathbf{k}$ plane, collects one frame of CCD data for each position in the image raster. The detector control PC allows the user to define the format of CCD frame that is read out, rebinning the frame to smaller array sizes when the full 80×80 frame is not required, and to change the frame integration time. This involves

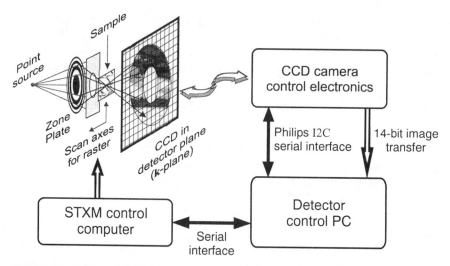

FIGURE 1. CCD based STXM detector showing the hardware elements and communications links.

changing the CCD control waveforms and control programs that are stored in the CCD camera's local memory, and is accomplished via the high-speed serial link (Philips I2C interface) shown in Figure 1. In general, a single raster scan can generate much larger volumes of data than the STXM control computer would normally expect, so the detector control PC stores these data for later analysis, while at the same time carrying out some simple real-time processing of the data to provide an immediate form of image signal that can be passed to the STXM control computer for display in the usual manner. By default, the image signal forwarded to the STXM control system is formed by summing all CCD pixels for each point in the raster, giving the incoherent brightfield image signal described by equation (2). It is straightforward to change this default to provide, for example, a differential phase contrast signal, or a darkfield image signal, for display by the STXM control system. Since the volume of image data forwarded to the STXM display is now much smaller, this is conveniently achieved using serial communications. This serial link is bi-directional, allowing essential parameters describing the format of the STXM raster scan to be passed to the detector control PC.

The control software running on the detector control PC has been written in C/C++, and provides a user-friendly GUI to modify the configuration of the CCD control electronics, and control the detailed operation of CCD frame capture, image storage, and real-time processing of image data prior to transfer to the STXM control computer. The serial communication between the two computers involves the use of a small library of C function calls that must be added to the STXM control program, so that the incorporation of the CCD detector system requires only small changes to existing STXM control software.

IMAGE MODELLING AND ANALYSIS

It is also possible to use the detector control PC for more sophisticated analysis and display of the image data. The same suite of programs can be used for the analysis of simulated image data, that are also generated by IDL routines designed to emulate the complete sequence of operations carried out by the STXM when forming an image. Both real and simulated data are stored in a common format.

$R(\mathbf{k})=1$ $R(\mathbf{k})=k_v$ $R(\mathbf{k})=k_x$

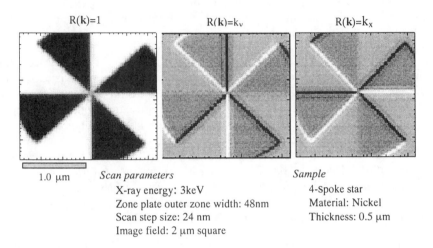

1.0 µm

Scan parameters
X-ray energy: 3keV
Zone plate outer zone width: 48nm
Scan step size: 24 nm
Image field: 2 µm square

Sample
4-Spoke star
Material: Nickel
Thickness: 0.5 µm

Figure 2. The results of image simulations for three different detector configurations. The first is the incoherent brightfield image. The other two images show orthogonal components of the image signal produced by a first moment detector response function $R(\mathbf{k}) = (k_x, k_y)$

The simulation program suite was written in the IDL™ (RSI Inc.) programming language and consists of four separate parts. The first manages a database of optical constants provided by the Henke tables (5). From these data optical properties for compound materials can be produced and their respective β and δ value for the complex refractive index displayed. A second program simulates the geometry of the sample, and allows the user to choose the material and associated optical properties, and overall physical dimensions, such as thickness. The resultant array can be stored in a file for use by the third member of the suite, which allows the user to define STXM scan parameters to be used in the image-scanning simulation. These include: x-ray energy, zone plate outer zone width (and hence the focus spot size), sample step size and defocus. The program will carry out the necessary scan procedure by performing the mathematical transforms from sample plane to detector plane. The result is a series of arrays representing each CCD frame of data that is stored on disc in a format compatible with real experimental data. Finally a suite of analysis routines can be used to provide image signals based on a variety of different detector response functions $R(\mathbf{k})$. The results of such a simulation is shown in figure 2 where the absorption image, and two orthogonal components of the differential phase contrast image expected from a first moment detector configuration, are displayed along with the properties of the sample and the conditions of the scan.

455

CONCLUSIONS

There are a number of benefits expected from the use of the configured detector system described in this paper. The most important is that the user can choose the detector response function $R(k)$ that is most appropriate for the sample composition and incident beam energy being used. This may be an absorption contrast, phase contrast, or darkfield mode, or it may be better to use a combination of a number of different modes to extract the maximum possible information from the sample. In any case, all such signals are available simultaneously, from a single scan of the sample, so that the dose to the sample is minimised.

A detector system of the type described has only recently become technically feasible, since it requires two enabling technologies. The first is the existence of high brightness x-ray beamlines for imaging experiments, providing enough signal spread over the detector plane to give reasonable signal to noise ratios from each element in the detector plane. The second is the availability of fast-readout low-noise, x-ray sensitive CCD's, since it is essential to be able to read out the full CCD frame in only a few milliseconds. Both these conditions must be met to ensure that the acquisition time for a full STXM image raster is not unacceptably long. A further spin-off from recent developments in computer technology is the ready availability of low-cost mass storage devices, so that it is now quite feasible to consider routinely acquiring individual image data sets that are several 100MBytes in size.

One additional point worth noting is that the CCD detector system described here may solve a potential problem created by the latest generation of high brightness x-ray sources, in that the level of signal on single-element detectors is now too high to allow photon counting methods to be used. By spreading the signal over several thousand CCD pixels, each of which is capable of counting x-ray photons, it should be possible to extend significantly the dynamic range of signal for which a linear, photon-counting response can be achieved.

Simulation software suggests that the use of a CCD as a configured detector will produce a useful range of phase and absorption contrast signals, if suitable detector response functions are applied to the data set. However, the stored image data sets also offer the possibility of applying methods such as Wigner deconvolution (6), in which the interference of x-rays in the detector plane are used to provide image resolution beyond the optical limitations.

The hardware necessary to use the detector system described here is now in the final stages of testing prior to its first use on a STXM beamline. The modular nature of both the hardware and software design is intended to allow the system to be fitted to almost any existing STXM, but in the first instance it will be used on ID 21 at the ESRF Grenoble, where the availability of high brightness at photon energies up to about 4 keV should be ideal for the initial tests of the system, using an STXM designed and built at King's College (7).

ACKNOWLEDGEMENTS

The authors wish to acknowledge the support of the UK EPSRC for the provision of a research studentship (for WJE), and are grateful to the Space Science department, CLRC Rutherford Appleton Laboratory, and to EEV Ltd, for the assistance provided with the CCD control electronics.

REFERENCES

1. Schmahl, G., Rudolph, D., Schneider, G., Guttmann, P., Niemann, B., Optik, **97**, 181 (1994)

2. Snigirev, A., Snigireva, I., Kohn, V., Kuznetsov, S. Schelokov, I., Rev. Sci. Instrum., **66**, 5486 (1995)

3. Morrison, G.R., Chapman, J.N., Optik, **64**, 1-12 (1983)

4. Waddell, E.M., Chapman, J.N., Optik, **54**, 83-96 (1979)

5. Henke, B.L., *et al., Atomic Data and Nuclear Data Tables.* **66**, 181-342 (1993).

6. Chapman, H N., *Ultramicroscopy* **66**, 153-172 (1996)

7. Burge, R E *et al., Ultramicroscopy* **69**, 259-278(1997)

Current Status Of The Scanning X-ray Microscope At The ESRF

Ray Barrett, Burkhard Kaulich, Murielle Salomé, Jean Susini

X-ray Microscopy Group, European Synchrotron Radiation Facility, BP220, 38043 Grenoble Cedex, France

Abstract. A short description of the Scanning X-ray Microscope of the ESRF ID21 X-ray microscopy beamline is given and the consequences of the relatively wide operating energy range discussed. The current capabilities of the instrument are demonstrated through images and spectra recorded from a variety of pilot experiments, including X-ray fluorescence imaging, microdiffraction and XANES measurements.

INTRODUCTION

The Scanning X-ray Microscope (SXM) is located on the ID21 X-ray microscopy beamline (on a low-beta straight section of the ESRF storage ring). The beamline [1], which also houses a full-field Transmission X-ray Microscope [2] on a separate end-station, operates under UHV conditions. The SXM endstation is served by two high resolution monochromators: a double crystal system (with optional upstream mirror pre-focussing) and a plane grating monochromator permitting a relatively wide operating energy range of 0.2-7keV using tuneable, undulator-type insertion devices. This energy range offers access to many interesting absorption edges both for materials science and biological applications but necessitates a certain versatility of the mechanical design. The microscope design is particularly adapted for X-ray fluorescence and spectro-microscopy (e.g. XANES) investigations [3,4]. The instrument, is now fully installed and is operating in the energy range 2–7keV.

DESIGN FEATURES

The microscope (shown in Figure 1) is designed to use primarily zone plate focussing optics. A sealed enclosure allows operation in air, vacuum (10^{-6}mbar), or controlled gas environments. Within the microscope chamber, the sample position is fixed along the beam direction. This strategy allows any off-axis detectors (notably the fluorescence detector) to be aligned to a fixed position regardless of the focussing optics. The chromatic nature of the zone plates and the spectral span requires that the instrument be capable of accommodating a wide range of focal lengths. In most instances this chromaticity also obliges a dynamic compensation of the focal length during energy scans if the X-ray probe is to remain optimally focussed. Effort has

CP507, *X-Ray Microscopy: Proceedings of the Sixth International Conference,*
edited by W. Meyer-Ilse, T. Warwick, and D. Attwood

been therefore been made to develop high-quality mechanical translation stages that minimise the movement defects (straightness, roll, yaw and pitch) in order to limit movement of the focussed probe on the sample during energy scans. Whilst the residual defects remain greater than those necessary to accomplish 'single pixel' energy scans at the highest resolutions offered by current zone plate technology, they are sufficiently good to perform scans at the sub-micron scale and limit the post-acquisition image realignment necessary for 'energy stack' type imaging. The zone plate (ZP) and order selecting aperture (OSA) are alignable in 3 directions and are in the microscope enclosure. Immediately upstream of the ZP, a moveable window separates the beamline vacuum from the microscope environment.

The sample is scanned on a two level translation stage: stepper motor driven stages give coarse translation ranges of ±5mm whilst a capacitively-encoded two-axis piezo-electric driven x-y flexure stage gives a fine scan range of ±50µm. The instrument is intended especially for spectro-microscopic applications (XANES, EXAFS, XRF).

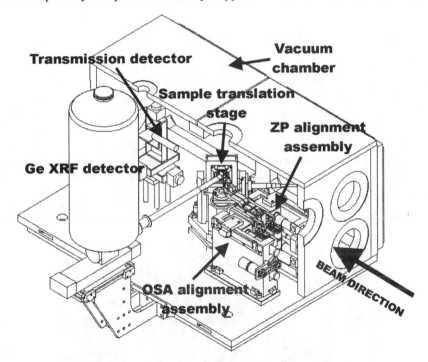

FIGURE 1. An oblique, 'cut-away' view of the SXM. The vacuum exit window from the beamline (which is immediately upstream of the zone plate) is not shown for clarity.

SXM CONTROL SYSTEM & SOFTWARE

SpecGUI, a Unix application running as an overlayer of the ESRF standard Spec control application, provides the top-level control interface of the instrument. This approach offers access to a scripting language allowing rapid and straightforward programmation of repetitive tasks and scan/image series. The user interface accesses the various hardware devices either directly or via distributed device servers (OS9, UNIX, Windows 95) using Ethernet. The software integrates both SXM and beamline control (including undulator gaps and beam-shutter). Fast image acquisition is controlled by a routine running under OS9 in a VME environment and allows pixel dwell times <1ms. On-line image display is performed by the ESRF DIS program and allows basic image analysis/processing operations (line profiles, zooms, colour map modification) to be performed during image acquisition. An IDL based image analysis program (MIAM) has been developed for manipulation of the acquired images (background subtraction, stack alignment, statistical measurements...).

APPLICATIONS

Absorption Contrast Imaging

The parallel detection offered by micro-probe scanning techniques means that the absorption signal can be collected simultaneously with, for example, a fluorescence signal thus allowing direct correlation of several complementary images. The use of a higher energy probe has a number of consequences: for a given ZP the NA varies inversely with the energy and consequently the depth of field increases, thus relaxing the constraints on specimen thickness for well focussed images. However, since the absorption lengths increase with energy, away from absorption edges, for a given sample the contrast becomes reduced. Coupled with dose considerations for some materials [5], this means that it may be preferable in many instances to use differential phase contrast detection techniques in the 'keV' energy range [6]. Figure 2 shows an absorption contrast image of a gold test pattern.

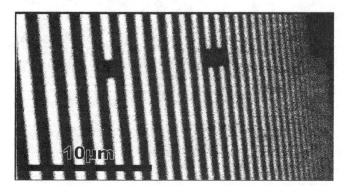

FIGURE 2. Absorption contrast image (500x250pixels, 10ms dwell-time) of a 350nm thick gold test pattern at 4keV. 150nm structures are resolvable at the right hand-side of the image. (Collaboration with M.Panitz & A.Duevel, FE Röntgenphysik, Göttingen).

Microfluorescence Imaging

Detection of the fluorescent X-rays excited by the primary focussed probe using an energy-resolving detector provides a means of characterising the chemical composition of the sample under study. Such methods offer benefits in spatial resolution and/or sensitivity compared with 'competitor' techniques such EDX, PIXE or SIMS and are readily applicable to bulk samples. By using appropriate energy 'windowing' around the fluorescence emission peaks it is possible to map the local distribution of several elements simultaneously. Figure 3 shows a series of images from a Ag stained salivary gland cell of a bloodworm larva obtained with transmitted light microscopy and by fluorescence yield imaging with X-ray excitation.

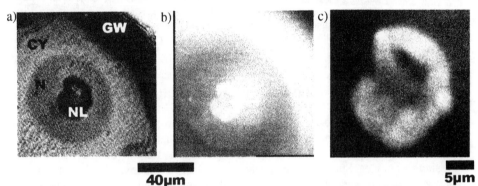

FIGURE 3. b) and c) Silver L-series fluorescence yield images of the nucleolus (NL) of a salivary gland cell of a chironomus thummi larva (bloodworm) after silver staining. Other cell structures are visible in the images (gland wall: GW, cytoplasm :CY, nucleus: N) The primary beam energy was 5 keV. Granular structures are visible within the nucleolus in the higher magnification image. a) A transmitted light micrograph (TLM) of the same cell is shown for comparison. (Collaboration with M. Gromova & M. Robert-Nicoud, IAB, Grenoble).

XANES Imaging

The value of XANES type imaging in Carbon bearing systems has been widely demonstrated in the soft X-ray régime, for example in the identification of different components in biological materials [7] and the direct imaging of phase distributions in polymer composite blends. [8]. Such chemical-state sensitive imaging can be extended to the higher energy range and finds applications in the materials, biological and environmental sciences.

In the case of chromium the presence of a strong pre-peak at 5.995keV in their K-edge absorption spectra allows the detection of Cr^{6+} bearing compositions using XANES imaging. By virtue of this effect it is possible to detect and image the heterogeneous distribution of Cr valence states by collecting image series at different strategic energies around the absorption edge. Figure 4 shows an image series of this type collected from an anti-corrosion Cr conversion coating on an Al substrate.

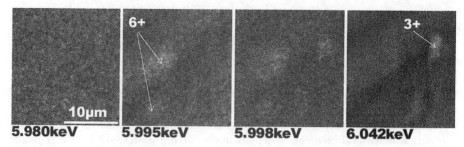

6+   3+

10µm

5.980keV   5.995keV   5.998keV   6.042keV

FIGURE 4. The images and absorption spectrum above show the evolution of the Cr Kα fluorescence signal for various probe energies in the Cr K-edge XANES region from a thin (~600nm) chromate conversion coating on an Al test coupon. Due to the large intensity variations between images the intensity scales are adjusted individually for each image. Quantitative intensity measurements allow the identification of the chemical state ($Cr^{3+/6+}$)of the different inhomogeneities. The 2 diagonally aligned structures originate from the rolling process of the native Al sheet. It is currently unclear whether they are visible due to topographic effects or due to an interruption of the chromate coating.

Microdiffraction Imaging

The possibility to use large acceptance zone plates to focus 'high' (>5keV) energy X-rays allows forward-scattering measurements to be made from weakly scattering systems with considerably higher spatial resolution than available from other focussing techniques. Such experiments are of interest for the investigation of polymer systems where an understanding of crystalline micro-texture can provide useful insights into the effects of different processing techniques and deformation mechanisms. Figures 5 and 6 show diffractograms acquired from a 12 µm diameter Kevlar fibre.

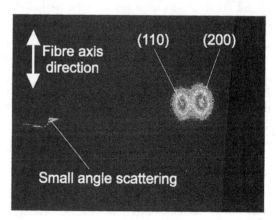

Fibre axis direction   (110)   (200)

Small angle scattering

FIGURE 5. Diffraction pattern recorded from the central zone of a 15µm diameter poly(p-phenylene terephthalamide) [Kevlar] fibre using a 1.5µm diameter focussed beam at 6keV. The pattern was recorded using an intensified CCD camera with 60s exposure time. (Collaboration with M. Müller & C. Riekel, ESRF).

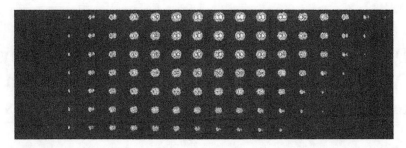

FIGURE 6. Detail of the 110 (left) and 200 (right) reflexions from a series of diffraction patterns recorded from different positions over the fibre. The displacement of the fibre between patterns is 1μm horizontally and 0.1μm vertically. The relative intensity evolution of the 110 and 200 reflexions demonstrates the texturing within the fibre (the 200 planes lying mostly radially and the 110 preferentially aligned tangentially).

ACKNOWLEDGEMENTS

The zone plates essential for the successful operation of the SXM came from a variety of sources. We are grateful to the various suppliers: M.Panitz, G.Schneider & G. Schmahl of FE Röntgenphysik, Göttingen, P. Charalambous of Dept. of Physics, Kings College, London, E. di Fabrizio & M.Gentili of Elettra, Trieste & IESS, Rome and H. Kihara & K. Takemoto of Kansai Medical University, Hirokata.

It is a pleasure to thank L. Andre, R. Baker, G. Berruyer, S. Blanchard, A. Debreyne, P. Noe, F. Picard, G. Rostaing, F. Thurel and C. Vartanian for their invaluable technical assistance and patience during the various stages of this project.

REFERENCES

1. Susini J.et al., "The X-ray Microscopy Facility at the ESRF: A Status Report", *this conference.*

2. Kaulich et al., "The Transmission X-ray Microscope End-station at the ESRF", *this conference.*

3. Susini, J. and Barrett, R., "The X-ray Microscopy Facility Project at the ESRF" in *X-ray Microscopy and Spectromicroscopy,* edited by J.Thieme et al., Springer-Verlag 1998, pp 45-54.

4. Barrett, R. et al., "The Scanning Microscopy End-station at the ESRF X-ray Microscopy Beamline" in *X-Ray Microfocusing: Applications and Techniques*, edited by I. McNulty, Proc. SPIE **3449**, 80-90 (1998).

5. Kirz, J. et al., *Quart. Rev. Biophys.*, **28**, 33-130 (1995)

6. Morrison, G. and Niemann, B., "Differential Phase Contrast X-ray Microscopy" in *X-ray Microscopy and Spectromicroscopy,* edited by J.Thieme et al., Springer-Verlag 1998, pp I 85-94.

7. Zhang, X. et al., *J. Struct. Biol.* **116**, 335-344 (1996)

8. H. Ade et al., *Science* **258**, 972 (1992)

Dynamical Coherent Illumination for X-ray Microscopy at 3rd Generation Synchrotron Radiation Sources: First Results with X-Rays at the Ca-K Edge (4 keV)

S. Oestreich*, G. Rostaing, B. Niemann[†], B. Kaulich, M. Salomé, J. Susini and R. Barrett

European Synchrotron Radiation Facility (ESRF), BP 220, F-38043 Grenoble, France
[†]*Forschungseinrichtung Röntgenphysik, Geiststr. 11, D-37073 Göttingen, Germany*
**Current address: FOM-Institute for Plasmaphysics Rijnhuizen, PO-Box 1207,*
NL-3430 BE Nieuwegein, The Netherlands, email: oestreic@rijnh.nl

Abstract. Dynamical coherent illumination is a way to deal with the illumination problems which emerge from the use of low emittance sources at recent synchrotron radiation sources for a transmission X-ray microscope (TXM). An illumination system, which uses two rotating mirrors to provide dynamical coherent illumination has been realized and tested at the TXM at ESRF's ID 21 Beamline. First results are presented and compared to results obtained with partial coherent illumination.

INTRODUCTION

In a full-field microscope, best imaging properties are achieved when the object is illuminated incoherently. Approximately, the illumination can be considered as incoherent, if the size of a coherence area in the object is smaller than the resolution element of the objective. This is realized by matching the numerical apertures (NAs) of condenser and objective. A smaller NA of the condenser increases the size of the coherently illuminated area and thus leads to partially coherent illumination. This causes a loss of spatial resolution and, in case of a high degree of coherence, artifacts, which appear due to coherent noise, and to the spatial distribution of the object's phase shift.

In a transmission X-ray-microscope (TXM), diffractive optical elements are used. Therefore the illumination has to protect the part of the detector in which the magnified image is formed from direct light which is not diffracted by the micro zone plate used as objective. Therefore, a condenser with a central obstruction is commonly used. Illumination with an annular condenser pupil can not only achieve high spatial resolution out of a low emittance beam, but also enhances the transfer of high spatial frequencies compared to a filled circular pupil condenser (1, 2).

CP507, *X-Ray Microscopy: Proceedings of the Sixth International Conference,*
edited by W. Meyer-Ilse, T. Warwick, and D. Attwood
© 2000 American Institute of Physics 1-56396-926-2/00/$17.00

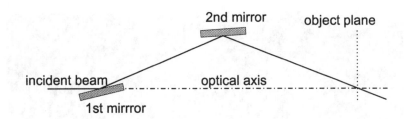

FIGURE 1. Sketch of the two mirror system.

Oblique illumination allows a large image field, but the images show artifacts, which are dependent on the orientation of the object details. Achieving a large image field size with using common illumination schemes thus is difficult on a recent SR-radiation source without a loss in imaging properties. The artifacts of oblique illumination disappear if the oblique illumination is made circular symmetric by means of rotating the illuminating, oblique beam around the optical axis. In this scheme, a highly coherent beam is used to achieve incoherent illumination by means of moving it and integrating the image over a long time; therefore the name "dynamical coherent illumination" (3). With dynamical coherent illumination formed by rotating oblique illumination, absorption contrast as well as Zernike phase contrast can be used (4).

EXPERIMENTAL SETUP

The experiments were performed at the TXM at ESRF's ID21 Beamline (5); it uses an undulator source, a Si-crystal monochromator (6), and a condenser zone plate. A micro zone plate with 70 nm outermost zone width made of Au (7) served as objective and phosphor-coupled CCD camera as detector.

The setup described in Fig. 1 and Ref. (8) was used to achieve rotating oblique illumination: Two plane mirrors are placed in the convergent beam behind the condenser zone plate. The reflecting surfaces of the mirrors are facing towards each other. The first mirror deviates the incoming beam from the optical axis, the second mirror, which is displaced by several mm from the optical axis, reflects it back into the object field with the angle that corresponds to the desired numerical aperture. Because a double reflection is used, the illuminating spot does not rotate with the mirrors. During one exposure, the system has to do an entire number of revolutions.

The system is comparatively insensitive to translational and tilt tolerances, because the mirrors reflect in opposite directions and the reflecting surfaces are nearly parallel and nearly compensate tilts and translations. Nevertheless, the small mirror cross-section and the requirements on the coplanarity of the surface normals require a cautious alignment process. A photograph of the actual system is shown in Fig. 2.

FIGURE 2. Photograph of the 2-mirror system.

RESULTS AND DISCUSSION

With the setup described above, X-ray microscopy images have been obtained at a photon energy of 4 keV both with and without rotating mirrors. Without the rotating mirrors, the direct beam from the condenser zone plate has been used to illuminate the sample. In this case, the image is not centered on the optical axis. With rotating mirrors the direct light reaches the detector outside the image field, and is absorbed by a circular aperture directly in front of the detector. Fig. 3 shows both images, corrected to the intensity distribution incident on the object. The image field without dynamical coherent illumination is smaller, and the background intensity is much less homogenous compared to the image obtained with the rotating mirrors. First images acquired with dynamical coherent illumination show some artifacts in the nearly vertical edge of the test-object, but not on the edge rectangular to it. These artifacts appear due to alignment problems, which result in the fact that the illuminating beam disappears for some angular positions of the rotation, and thus lead to conditions similar to oblique illumination

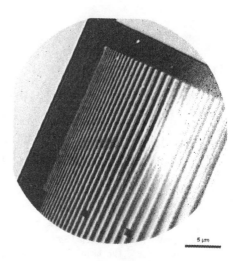

FIGURE 3. X-ray microscopy image (E=4 keV) of a test grating with smallest structures of 70 nm, obtained without dynamical coherent illumination (left) and with dynamical coherent illumination (right).

ACKNOWLEDGEMENTS

The authors thank Prof. Dr. G. Schmahl for his constant encouraging support, M.-Panitz for the supply of the test object and the micro zone plate, the detector group and the technical staff of the ESRF.

REFERENCES

1. Horikawa, Y. *Optik* **95**, 119-134 (1993)

2. Jochum L. and Meyer-Ilse W. *Applied Optics* **34**, 4944-4950 (1995)

3. Cronin D. J. and Smith A. E. *Optical Engineering* **12**, 50-55 (1973)

4. Cronin D. J., De Vlies J. B. and Reynolds G. O. *Optical Engineering* **15**, 276-178 (1976)

5. Kaulich B., Niemann B., Rostaing G., Oestreich S., Salome M. Barrett R. and Susini J. (this conference)

6. Oestreich S., Kaulich B. and Susini J. *Rev Sci Instr* **70**, 1921-1925 (1999)

7. Panitz M., Schneider G., Peuker M., Hambach D., Kaulich B., Susini J. and Schmahl G. (this conference)

8. Oestreich S. and Niemann B. "Design of a condenser for an X-ray microscope on a low beta section undulator source at the ESRF," in :Thieme et al (Eds.) *X-ray Microscopy and Spectromicroscopy*, Springer 1998.

Development of X-ray Photoelectron Microscope with an X-ray Laser Source

Tadayuki Ohchi[1,*], Naohiro Yamaguchi[1], Chiemi Fujikawa[1], Tamio Hara[1], Katsumi Watanabe[2], Ibuki Tanaka[2] and Masami Taguchi[2]

[1] *Toyota Technological Institute, 2-12-1 Hisakata, Tempaku, Nagoya 468-8511, Japan*
[2] *ULVAC-PHI Inc., 370 Enzo, Chigasaki, Kanagawa 253-0084, Japan*

Abstract. We have constructed an x-ray photoelectron microscopic system with an x-ray laser as an x-ray source. The lasing line is the Li-like Al 3d-4f transition at 15.47 nm where the recombining Al plasma is used as the x-ray laser medium. The beam from the x-ray laser cavity was then focused by using a Schwarzschild mirror coated with Mo/Si multilayers. The x-ray beam size with a diameter less than 0.5 µm and the estimated photon number of about 2 x 10^6 photons/shot into the spot were achieved.

INTRODUCTION

X-ray photoelectron spectroscopy (XPS) is a useful technique to analyze the chemical composition, chemical states of surface materials. Recently spatial resolved XPS is actively studied at many synchrotron radiation facilities (1-4). In laboratory scale, x-ray tube and laser-produced plasma x-ray source (5-7) are used as an x-ray source. Such x-ray sources still provide not enough photon flux to observe surface characteristics in a small area. One of the x-ray sources realizing microscopic analysis is an x-ray laser. Because x-ray lasers have high brilliance, narrow spectral width and small divergence, an XPS with such an x-ray source would be able to analyze materials with high spatial and spectral resolving powers.

We have developed a tabletop x-ray laser pumped by a YAG laser. To realize such an x-ray laser, we proposed plasma production by the irradiation of pulse-train lasers (8, 9). The use of the pulse-train laser is an effective method to obtain high gain media of x-ray laser through the recombination process. In our recent experiment, we observed the amplification of the two Li-like Al lines (10.57 nm and 15.47 nm) using the pulse-train YAG laser with an input energy of only 1.5-2 J/cm (10). The gain-length product for the Li-like Al 3d-4f transition at 15.47 nm was about 3.5 and maintained for about 1 ns. Moreover, we have performed cavity experiments using two multilayer x-ray mirrors for 15.47nm transition. The results show clear enhancement in the output signal (11).

---

* Present Address: National Institute of Materials and Chemical Research, 1-1 Higashi, Tsukuba, Ibaraki 305-8565, Japan. E-mail Address: ohchi@nimc.go.jp.

CP507, *X-Ray Microscopy: Proceedings of the Sixth International Conference,*
edited by W. Meyer-Ilse, T. Warwick, and D. Attwood
© 2000 American Institute of Physics 1-56396-926-2/00/$17.00

In this paper, we describe our x-ray photoelectron microscopic system with an x-ray laser as an x-ray source. We equipped a microfocus beamline using a Schwarzschild optics. The beam size was determined according to the knife-edge method. The experimental results showed that the x-ray beam size was about 0.45 μm in diameter and the estimated photon number of the microbeam was about 2 x 10⁶ photons/shot.

EXPERIMENTAL SETUP

An x-ray photoelectron microscopic system is schematically shown in Fig. 1. The pulse-train YAG laser whose wavelength was 1064 nm was focused onto an Al tape target, which was made by coating aluminum foil with 10 mm wide and 10 μm thick on a polyethylene-terephthalate tape of 0.1 mm thickness. The focal line was 10-mm long and 50-μm wide. The pulse-train consisted of 16 pulses of 100-ps pulse-width with the interpulse time of 200 ps (10). The intensity of the second half of the pulse-train was reduced to less than one fourth of the first one. By use of the tape target, an operation with 10 Hz repetition rate was possible in our x-ray laser system. An x-ray laser cavity consisted of two Mo/Si multilayer mirrors that were a flat mirror and a concave mirror (R=60mm) (11). These mirrors were installed 50 mm apart with the plasma placed between them. The flat mirror had an output coupler with a square orifice of 100 μm x 100 μm. This orifice was set at 0.3 mm from the target surface, where the highest enhancement was obtained. A Schwarzschild mirror (Olympus Optical Co.) coated with Mo/Si multilayers for 15.47 nm x-ray was set on the beamline at a distance of about 1 m from the orifice (12). The designed degree of demagnification is 224 on the image plane.

Samples were set on a 4-axis manipulator with a high-precision piezo stage (Physik Instrumente GmbH & Co., Model P762-20). The electron analyzer (ULVAC-PHI Inc., Model 1600C) was a spherical capacitor analyzer with mean diameter of 279.4 mm.

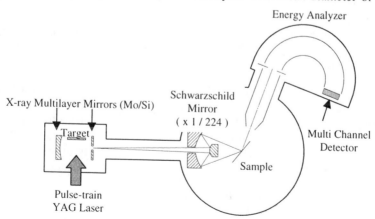

FIGURE 1. Schematic of the x-ray photoelectron microscopic system.

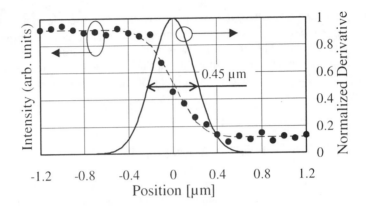

FIGURE 2. The intensity profiles of the microbeam. The raw data at each knife-edge position was obtained by only one shot. The dashed curve and solid curve are the fitted curve to the error function and its derivative, respectively.

The electron detector was a microchannel plate (MCP) with 16 channels.

RESULTS AND PRESENT STATUS

First, the size of microbeam was measured using the knife-edge method (12). X-rays through the Schwarzschild mirror were detected by an MCP with phosphor screen. Figure 2 shows the characteristic curve of the knife-edge response at the best focus condition. The dashed curve represents the fitting curve of the raw data to the error function and the solid curve shows its derivative. The beam size, which was the full width at the half maximum of this solid curve, was about 0.45 μm. It agrees well with the expected spot size in this focusing optics.

Next, we estimated the photon flux on the spot. From the result of the x-ray cavity experiment (11), the photon flux from the cavity at the wavelength of 15.47 nm is measured to be about 6×10^{12} photons/shot/str. Taking into account the acceptance solid angle (3.9×10^{-6} str) and the reflectivity of the mirrors, we can estimate the flux on the spot to be about 2×10^{6} photons/shot.

At present, we have combined the x-ray optical system with the electron energy analyzer. Then we are going to align the axes of x-ray optics and electron analyzer system. Moreover steps to reduce electrical noise caused by electromagnetic induction from the YAG laser system must be taken.

CONCLUSIONS

We have constructed an x-ray photoelectron microscopic system with an x-ray laser. In the microbeam experiment, the focusing of x-ray beam into a spot of about 0.45 μm has been achieved. The photon flux on the spot was estimated to be about 2×10^6 photons/shot. The advantage of the use of a pulse x-ray source is that the photon flux comes into a sample in the very short time. It is possible to obtain the information of the instantaneous state of the sample surface by a single-shot irradiation. But at the present status, the output power is still not enough for various applications. We are planning to extract significant x-ray laser power, by using more efficient cavity geometry, increasing the plasma length, and optimizing x-ray lasant conditions. So by improving the performance of the x-ray laser and by introducing a fast detector, one-shot detection of photoelectrons from a microscopic area would be possible.

ACKNOWLEDGEMENTS

This work is supported in part by a Grant-in-Aid for Scientific Research (Contract No. 09750053) from the Ministry of Education, Science, Sports and Culture and by X-ray Laser Research Consortium, Toyota Technological Institute.

REFERENCES

1. Ng, W., Ray-Chaudhuri, A.K., Liang, S., Singh, S., Solak, H., Welnak, J., Cerrina, F., Margaritondo, G., Underwood, J.H., Kortright, J.B. and Perera, R.C.C., *Nucl. Instrum. Methods Phys. Res.* **A347**, 422 (1994).

2. Ade, H., Ko, C.H. and Anderson, E., *Appl. Phys. Lett.* **60**, 1040 (1992).

3. Voss, J., *J. Electron Spectros. Relat. Phenom.* **84**, 29-44 (1997).

4. Johansson, U., Zhang, H. and Nyholm, R., *J. Electron Spectros. Relat. Phenom.* **84**, 45-52 (1997).

5. Aoki, S., Ohchi, T., Sudo, S., Nakajima, K., Onuki, T., and Sugisaki, K., *Jpn. J. Appl. Phys.* **32**, L1574-L1576 (1993).

6. Ohchi, T., Aoki, S. and Sugisaki, K., *J. Electron Spectros. Relat. Phenom.* **80**, 37-40 (1996).

7. Kondo, H., Tomie, T. and Shimizu, H., *Appl. Phys. Lett.* **72**, 2668 (1998).

8. Hara, T., Ando, K., Negishi, F., Yashiro, H. and Aoyagi, Y., *X-ray Lasers*, York, UK: IOP Publishing, 1990, p. 263.

9. Hirose, H., Hara, T., Ando, K., Negishi, F. and Aoyagi, Y., *Jpn. J. Appl. Phys.* **32**, L1538 (1993).

10. Yamaguchi, N., Hara, T., Fujikawa, C. and Hisada, Y., *Jpn. J. Appl. Phys.* **36**, L1297 (1997).

11. Yamaguchi, N., Hara, T., Ohchi, T., Fujikawa, C. and Sata, T., *Jpn. J. Appl. Phys.* **38**, 5114-5116 (1999).

12. Ohchi, T., Yamaguchi, N., Fujikawa, C. and Hara, T., *J. Electron Spectros. Relat. Phenom.* **101-103**, 943-947 (1999).

A Hard X-Ray Scanning Microprobe for Fluorescence Imaging and Microdiffraction at the Advanced Photon Source*

Z. Cai, B. Lai, W. Yun[#], P. Ilinski, D. Legnini, J. Maser
and W. Rodrigues[%]

Experimental Facilities Division, Argonne National Laboratory, Argonne, IL 60439
[#] *Advanced Light Source, LawrenceBerkeley National Laboratory, Berkeley, CA 94720*
[%]*Department of Physics & Astronomy, Northwestern University, Evanston, IL 60208*

Abstract. A hard x-ray scanning microprobe based on zone plate optics and undulator radiation, in the energy region from 6 to 20 keV, has reached a focal spot size (FWHM) of 0.15 μm (v) x 0.6 μm (h), and a photon flux of $4x10^9$ photons/sec/0.01%BW. Using a slit 44 meters upstream to create a virtual source, a circular beam spot of 0.15 μm in diameter can be obtained with a photon flux of one order of magnitude less. During fluorescence mapping of trace elements in a single human ovarian cell, the microprobe exhibited an imaging sensitivity for Pt (L_α line) of 80 attograms/$μm^2$ for a count rate of 10 counts per second. The x-ray microprobe has been used to map crystallographic strain and multiquantum well thickness in micro-optoelectronic devices produced with the selective area growth technique.

The recent availability of high-brilliance synchrotron radiation sources[1] and x-ray Fresnel zone plate microfocusing optics with high spatial resolution and high focusing efficiency[2] has made possible the creation of a new tool for material characterization on micron and submicron length scales. A hard x-ray microprobe (HXRM) that combines microfocusing capabilities with x-ray sensitivity to trace element distributions, crystallographic strain, and the ability to penetrate deep in a specimen, has been developed for high-resolution fluorescence mapping and microdiffraction at the Advanced Photon Source (APS) at Argonne National Laboratory.

The HXRM utilizes the radiation from the high-brilliance source generated from an electron beam of 7 GeV in the APS storage ring and a 3.3-cm-period undulator (APS undulator A). The energy of the radiation can be tuned from 3.2 to 45 keV by a combination of varying the undulator gap and selecting among the first, third, and fifth harmonics of the undulator. The undulator has been optimized so that continuity in brilliance is achieved when tuning from one harmonic energy to the next.

The HXRM is installed in a dedicated beamline (2-ID-D) specifically designed and developed for x-ray microscopic applications. The beamline was specially designed to achieve conservation of the source brilliance, selectivity of energy bandwidth, and the capability of reducing the effective source size. The beamline uses a windowless operation between the front end and the beamline to avoid degradation of wavefront due

CP507, *X-Ray Microscopy: Proceedings of the Sixth International Conference,*
edited by W. Meyer-Ilse, T. Warwick, and D. Attwood
© 2000 American Institute of Physics 1-56396-926-2/00/$17.00

to vacuum barriers. The nominal emittance of the beam in the APS storage ring is 8.2×10^{-9} m-rad. Given a one-percent coupling between the vertical and horizontal emittance and beta functions of 14 m in horizontal and 6 m in vertical, the FWHM source size of the photon beam at the center of the straight section is 52 μm in vertical and 790 μm in horizontal. In order to preserve the brilliance, we use a water-cooled grazing incidence (0.15°) horizontally deflecting mirror (1.2 meters long) as a first optical component. The grazing incidence of the mirror considerably reduces the power density on the reflecting surface and thus improves the performance of the first component. The cutoff of the reflected energy provided by the mirror considerably reduces both total radiation power and heat flux incident on downstream optical components, thus enhancing their performance. The horizontal deflection geometry takes advantage of less stringent requirements in slope error for avoiding brilliance degradation in the horizontal direction because the emittance in the horizontal direction is two orders of magnitude larger than that in vertical direction. Three reflecting surfaces (Si, Rh, and Pt) cover x-ray energies up to 40 keV. Use of the mirror also provides suppression of high-order harmonics of the undulator radiation, which is particularly important for high-energy storage rings like that of the APS. Either a double-crystal monochromator (DCM) or a double-multilayer monochromator (will be installed soon) located 64 to 65 meters from the source is used to monochromatize the mirror-reflected undulator beam. The combination of the large monochromator-to-source distance and the power filtering of the mirror reduces the maximum power density at normal incidence at the monochromator to less than 13.3 W/mm², thirteen times less than that of a beamline without mirror filtering and with the monochromator located in the first optical enclosure. When a proper mirror coating is used, the maximum heat flux on the first crystal is less than 2.3 W/mm². A monolithic, U-shaped, and water-cooled Si(111) crystal, with less than 2 microradians of tangential slope error over the footprint of the FWHM of the undulator radiation, is used as the first crystal for the DCM. It can handle a peak heat flux of 5W/mm² before the beam brilliance degradation becomes severe.

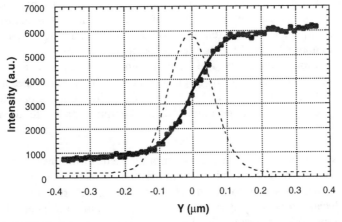

Fig 1. Resolution test of a zone plate at E = 8 keV using a Cr knife edge scan. Dots are measured fluorescence intensity, solid line is the fitting curve, and the dashed line is the derivative of the fitted curve. The FWHM beam size is measured to be 150 nm.

The phase zone plates we have developed for the HXRM have first-order focusing efficiency from 15% to 40% within the x-ray energy region from 6 keV to 20 keV.[3] The thickness of the gold zone plates ranges from 1.5 μm to 3.5 μm, and the focal length ranges from 5 cm to several meters at 8 keV. The outermost zone width, which determines the spatial resolution of the focusing optics, has reached 100 nm. Future development of the phase zone plates will improve the spatial resolution and extend the capability of the HXRM toward higher energy. In the meantime, x-ray energies higher than 20 keV can be achieved by employing a multiple-zone-plate stacking technique. The HXRM is located in the experiment station 71.5 meters away from the source. For a zone plate with a focal length of 10 cm at 8 keV and a diameter of 150 μm, we have a demagnification factor of 710. In order to coherently illuminate the zone plate in the horizontal direction so that a horizontal spot size limited by the outermost zone width of the zone plate is obtained, a water-cooled and adjustable slit located 28.3 meters from the center of the insertion device straight section is used to reduce the effective horizontal source size.

The vertical size of the focal spot generated from a zone plate of 10-cm focal length at 8 keV is shown in Fig. 1. The profile was obtained by scanning a 100-Å-thick Cr knife edge across the focal spot and measuring its fluorescence intensity. The FWHM spot size was obtained by first fitting the fluorescence intensity profile with error functions and then taking derivative of the fitted profile. Using a Si(111) monochromator, we have obtained a focal spot size (FWHM) of 0.15 μm (v) x 1.0 μm (h)[4] and a photon flux of 4×10^9 photons/sec at the focal spot and thus a photon flux density gain of 15,000. A circular beam spot 0.15 μm in diameter can be achieved by creating a virtual source (horizontal only) at 43.5 meters upstream of the zone plate, with an order of magnitude less flux in the focal spot.

The HXRM consists of a zone plate, a platinum order-sorting aperture (5-30 μm) for selecting the first-order focused beam, and a sample holder assembly that allows three orthogonal translations and a full rotation around a vertical axis (theta rotation). The translation stages on the sample assembly provide a minimum step size of 0.1 μm. A peizo-driven XY stage on the assembly provides finer scanning steps. The sample assembly also contains chi and phi segments for angular positioning of samples when performing diffraction experiments. The HXRM is equipped with a Ge energy-dispersive detector (a 13-element detector will be installed soon) for fluorescence measurements and a CCD or NaI detector mounted on the 2-theta arm for diffraction measurement. In the following, we describe two experiments using the microprobe to demonstrate its capability.

Cisplatin is an anticancer agent that is currently used for the treatment of testicular, ovarian and other tumors. The effectiveness of the agent sometimes is limited because the cancer cells develop resistance to this drug. There have therefore been intensive efforts over the last decade to develop derivatives, which maintain anticancer activity against cisplatin-resistant cells. A group[5] in La Trobe University, Australia, has recently identified a cisplatin derivative (Pt103) that has good activity against cisplatin-resistant cells in culture. In order to quantitatively determine the effectiveness of the Pt103 agent,

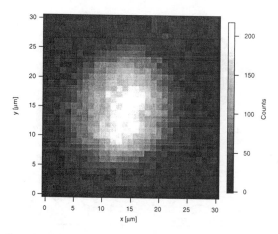

Fig. 2. X-ray fluorescence image of Pt distribution in an ovarian cancer cell treated with cisplatin. The specimen was provided by D. Phillips' group in La Trobe University, Australia.

x-ray fluorescence mapping of platinum's L_α-line emission has been used to image ovarian cancer cells with and without development of cisplatin resistance, at increasing times of exposure to cisplatin and Pt103, respectively. Figure 2 shows a platinum image of an ovarian tumor cell treated with cisplatin. The level of detection sensitivity of platinum is crucial in determining the concentration of platinum in cells that have developed resistance and been exposed to cisplatin for a short time. There is an overlap between the spectrum of the Pt-L_α line and the Cu-K_β, Zn-K_α and Zn-K_β lines. Both Cu and Zn are trace elements in the cell. Therefore, the platinum count rate at each pixel of the image in Fig. 2 was obtained through a careful fitting of the spectrum among the spectra of Cu, Zn, and Pt.[6] The microprobe exhibits sensitivity for the Pt-L_α line of 80 attograms/μm^2 with a count rate of 10 counts per second. Our results show that the x-ray microprobe detection of platinum in tumor cells is many orders of magnitude better than that of a proton microprobe, and the sensitivity now available is suitable for studies of platinum distribution in tumor cells subjected to clinical dose of platinum anticancer agents.

An electroabsorption modulator laser (EML) array for telecommunication requires sophisticated integration of multiquantum well (MQW) lasers, modulators, and optical waveguides on different regions of the same wafer. Integration is possible in the quaternary alloys $In_{1-x}Ga_xAs_{1-y}P_y$ because of the metal-organic-chemical-vapor deposition (MOCVD) technique of selective area growth (SAG). While the group III precursors in MOCVD readily bond to a free InGaAsP surface, they will not stick to a SiO_2 surface. Therefore, the group III precursors just above a thin oxide mask patterned onto the InGaAsP can then diffuse to the free-InGaAsP surface in the vicinity of the oxide, leading to an enhancement of the epitaxial growth rate. By the appropriate choice of an oxide mask pattern, SAG allows precise control of the spatial variation of multilayer thickness, composition, and crystallographic strain on micron length scales. In order to understand the details of the growth process, we have measured the crystallographic

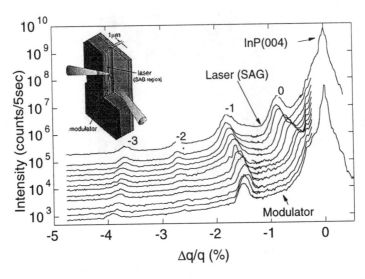

Fig. 3. X-ray microdiffraction along the multilayer growth direction in a fully processed EML device measured with a 0.5 μm x-ray spot. The position of the x-ray beam is scanned from the thickness-enhanced MQW laser region to the modulator region where there is no growth enhancement.

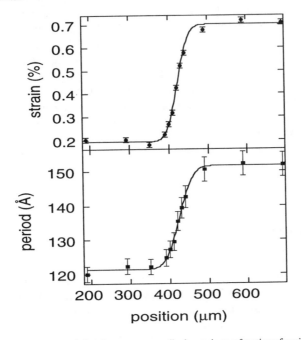

Fig. 4. MQW period and average perpendicular strain as a function of position.

476

strain and multilayer thickness of as-grown InGaAsP multilayer device material produced with SAG. The microprobe was needed because of the ultrasmall volume of material in the active region of optoelectronic device. Figure 3 shows a series of microdiffraction scans taken from an EML device with an active mesa of MQW about 1 μm wide (see the inset in Fig. 3). The principal features in Fig. 3 are the InP(004) substrate Bragg peak and the MQW superlattice peaks labeled by order. The position of the zero-order peak gives the average strain perpendicular to the layer of the MQW. The separation between the adjacent superlattice peaks is proportional to the inverse of the MQW period. In the Fig. 4, we plot the MQW period and average perpendicular strain, which clearly show that the MQW material varies smoothly from the laser to the modulator.

Since the commissioning of the HXRM in December 1997, the microprobe has been used for the development of many x-ray microfocusing-based techniques and a broad range of scientific applications. They include high-spatial-resolution fluorescence microscopy,[6-8] diffraction microscopy,[9-12] microspectroscopy,[7] fluorescence correlation spectroscopy,[13] and fluorescence tomography.[14] In addition, we are currently carrying out experiments of transmission microscopy and transmission microtomography.

*This work was supported by the Department of Energy, Basic Energy Sciences, Office of Science, under Contract No. W-31-109-ENG-38.

REFERENCES

1. G. K. Shenoy, P. J. Viccaro, and D. M. Mills, Argonne National Laboratory Report, ANL-88-9.

2. B. Lai, W. Yun, D. Legnini, Y. Xiao, J. Chrzas, P. Viccaro, V. White, S. Bajikar, D. Denton, F. Cerrina, E. Di Fabrizio, M. Gentili, L. Grella, and M. Baciocchi, *Appl. Phys. Lett.* **61**, 1877(1992).

3. W. Yun, B. Lai, A. Krasnoperova, E. Di Fabrizio, Z. Cai, F. Cerrina, Z. Chen, M. Gentili, E. Gluskin, *Rev. Sci. Instrum.* **70**, 3537 (1999).

4. W. Yun, B. Lai, Z. Cai, J. Maser, D. Legnini, E. Gluskin, Z. Chen, A. Krasnoperova, Y. Valdimirsky, F. Cerrina, E. Di Fabrizio, M. Gentili, *Rev. Sci. Instrum.* **70**, 2238 (1999).

5. D. Phillips et al., unpublished results.

6. P. Ilinski, B. Lai, Z. Cai, W. Yun, D. Legnini, T. Talarico, M. Cholewa, D. Phillips et al., to be submmited.

7. W. Yun, S. Pratt, R. Miller, Z. Cai, D. Hunter, A. Jarstfer, K. Kemner, B. Lai, H-R. Lee, D. Legnini, W. Rodrigues, C. Smith, *J. Synchrotron Rad.* **5**, 1390 (1998).

8. K. Kemner, W. Yun, Z. Cai, B. Lai, H-R. Lee, J. Maser, D. Legnini, W. Rodrigues, J. Jastrow, R. Miller, S. Pratt, M. Schneegurt, C. Kulpa Jr., *J. Synchrotron Rad.* **6**, 639 (1999).

9. Z. Cai, W. Rodrigues, P. Ilinski, D. Legnini, B. Lai, W. Yun, E. Isaacs, K. Lutterodt, J. Grenko, R. Glew, S. Sputz, J. Vandenberg, R. People, M. Alam, M. Hybertsen, L. Ketelsen, *Appl. Phys. Lett.* **75**, 100 (1999).

10. H. Solak, Y. Vladimirsky, F. Cerrina, B. Lai, W. Yun, Z. Cai, P. Ilinski, D. Legnini, W. Rodrigues, *J. Appl. Phys.* **86**, 884 (1999).

11. H.-R. Lee, D. Kupperman, W. Yun, Z. Cai, W. Rodrigues, *Rev. Sci. Instrum.* **70**, 175 (1999).

12. G. Wong, Y. Li, I. Koltover, C.R. Safinya, Z. Cai, W. Yun, *App. Phys. Lett.* **73**, 2042 (1998).

13. J. Wang, A.K. Sood, P.V. Satyam, Y. Feng, X. Wu, Z. Cai, W. Yun, S.K. Sinha, *Phys. Rev. Lett.* **80**, 1110 (1998).

14. M. Naghedolfeizi, J. –S. Chung, G. E. Ice, W. Yun, Z. Cai, B. Lai, *Mat. Res. Soc. Symp. Proc.* **524**, 233 (1998).

Spectromicroscopy at the XM-1

Greg Denbeaux[1,2], Lewis Johnson[1], Werner Meyer-Ilse[1]

[1]Center for X-ray Optics, LBNL, Berkeley, CA 94720
[2]Duke University, Free Electron Laser Lab, Durham, NC 27708

Abstract. The XM-1 x-ray microscope was built to obtain high-resolution transmission images from a wide variety of thick (< 10 micron) samples. Modeled after a conventional full-field microscope, XM-1 makes use of zone plates for the condenser and objective elements. Recent work has enabled the microscope to be used for spectroscopic imaging as well. The bandwidth of light on the sample is limited by a linear monochromator which is formed by the combination of a condenser zone plate (CZP) and a pinhole at the sample plane. This combination gives a good spectral resolution which has been measured to be $\lambda/\Delta\lambda = 700$. This is high enough to be able to distinguish between different elements and even some chemical states on the same scale as the spatial resolution of the instrument which is 36 nm. The measured spectral resolution and the calculated spectral resolution will both be shown.

INTRODUCTION

The XM-1 x-ray microscope is a full field imaging microscope (1). Radiation produced by a bending magnet at the Advanced Light Source (ALS) is collected by a "large" (9 mm diameter) fresnel condenser zone plate (CZP) which projects the light through a pinhole and illuminates the sample. The photon energy which illuminates the sample can be changed within 250 and 900 eV. The light that passes through the sample is imaged through a high precision objective micro zone plate (MZP) (45 μm diameter) and magnified onto a CCD camera. The field of view of the microscope is 10 μm with a spatial resolution of 36 nm.

The monochromator is composed of the CZP and a pinhole about 100 μm from the sample plane and is used to control the sample illumination wavelength. Zone plates have chromatic aberrations, so a change in the distance between the CZP and the sample changes the energy which is focused onto the sample plane. In order to uniformly illuminate the object, the CZP is uniformly scanned to fill the area being imaged. Since the spectral resolution is highest at the center of the spot from the CZP, reducing the area of the CZP scanning motion increases the spectral resolution with a corresponding decrease in the field of view. The pinhole is at a location very near to the sample plane compared to the CZP to sample plane distance (200 mm), so the aperture of the monochromator system is often limited by the size of the sample plane being imaged. The main function of the pinhole is to reduce background light in the system, thus improving spectral purity. During spectroscopic imaging with XM-1, the MZP and CCD camera also move in order to form a proper image at each energy.

CP507, *X-Ray Microscopy: Proceedings of the Sixth International Conference,*
edited by W. Meyer-Ilse, T. Warwick, and D. Attwood
© 2000 American Institute of Physics 1-56396-926-2/00/$17.00

In order to obtain a spectrum we take multiple full field images, each at a different photon energy. Each image is a two dimensional array representing the transmitted photon flux through the sample. The combined data set stacks the individual images along an axis representing photon energy. As long as each image is aligned to within the spatial resolution, then the spectra for each single pixel of the image is found in the energy direction of the data set. This alignment can be difficult, as there can be unwanted off axis motion from the movement of the MZP and CCD. This motion could cause a shift in the projected image on the CCD camera.

IMAGE ALIGNMENT

In order to combine imaging with spectral information, each image at different energies must be aligned. Then each pixel in the series of images at different energies represents the spectral information at that location of the image. Images at different energies with XM-1 require motion of the micro zone plate along the optical axis to properly focus at the chosen energy. The motion of the MZP allows for the possibility of a component of the motion to be transverse to the optical axis and cause a corresponding shift to the image relative to the previous image. Two ways to compensate for this are to use the autocorrelation of the images to determine the relative shift between them or to measure the actual shift of the micro zone plate relative to the optical axis. For either of these methods, software can be used to automatically shift each image into alignment. The autocorrelation method can be used without any additional hardware, but the images need to contain distinct features for the method to work properly. A lightly absorbing sample without distinct features may not always align properly using this technique. In order to avoid this problem, four capacitive micrometers were installed on XM-1 to accurately measure the location of the MZP. With this information, subsequent images could be automatically shifted into alignment with software. The system was designed to measure the location of the MZP to within an accuracy of a few nanometers, but initial calibrations of the system have only been accurate to within 70nm rms. Further testing and calibration should be able to reduce this error to better than the spatial resolution of 36nm.

CALCULATED SPECTRAL RESOLUTION

A numerical simulation was performed to calculate the expected spectral resolution of our system. We have assumed that the CZP has no aberrations other than the known chromatic aberrations, and that the bending magnet illumination is spatially incoherent. These calculations have been performed with no scanning of the CZP. The CZP-to-sample distance was set at 200 mm for which the energy in focus at the sample plane is 500 eV. The condenser zone plate has 41,000 zones and a 9-mm diameter. There is a 3-mm diameter central stop which obstructs the undiffracted light from reaching the sample.

The convolution of the point spread function of the XM-1 condenser zone plate with the source distribution from the ALS bending magnet yields the field distribution at the sample. The nominal size of the ALS source at the bend magnet is $\sigma = 53$ µm horizontal and $\sigma = 44$ µm vertical (2). For a given wavelength, λ, the lens plane field at a point (x, y) from a point source in the source plane with coordinates (ε, η) is (3,4):

$$u(x, y; \varepsilon, \eta) = \frac{1}{i\lambda z_1} \exp(\frac{ik}{2z_1}[(x - \varepsilon)^2 + (y - \eta)^2]) \qquad (1)$$

Where z_1 is the distance from the source to the CZP lens, 17 meters, and the wavenumber, k, is $2\pi/\lambda$. Just past the lens plane, the field of the CZP can be modeled as a simple thin lens:

$$u'(x, y; \varepsilon, \eta) = u(x, y; \varepsilon, \eta)P(x, y)\exp(\frac{-ik}{2f}(x^2 + y^2)) \qquad (2)$$

Where the annular pupil function $P(x, y)$ is 1 between the central obstruction (3 mm diameter) and the lens aperture (9 mm diameter) and is 0 elsewhere. The exponential term is the impulse response of a lens with focal length, f. The field at the image plane (u, v) due to a point source at the source plane is given by the convolution of u' with the point spread function of free space.

$$u_{image}(u, v; \varepsilon, \eta) = \frac{1}{i\lambda z_2} \iint u'(x, y; \varepsilon, \eta)\exp(\frac{ik}{2z_2}[(u - x)^2 + (v - y)^2])dxdy \qquad (3)$$

where z_2, the distance from the lens to the image plane, is 200 mm for this calculation. The convolution with the source can be performed with numerical integration techniques for a series of different photon energies and corresponding different focal lengths from the CZP. The result of these calculations is the illumination at the sample plane for each photon energy. Figure 1 shows the calculated spectral distribution of the illumination near 500 eV on the sample, averaged over areas of $4\mu m^2$, 9 µm^2 and 16 µm^2. Using ΔE as the FWHM of these distributions, the spectral resolution calculations yield $E/\Delta E = 1400$ within a 4 µm^2 spot, $E/\Delta E = 900$ within a 9 µm^2 spot, and $E/\Delta E = 700$ within a 16 µm^2 spot. Since the best spectral resolution is located at the center of the CZP spot on the sample, the choice of a small area CZP scan sacrifices the field of view for high spectral resolution.

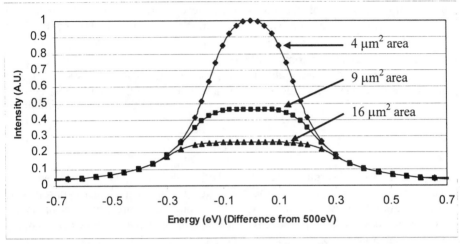

Figure 1. Average intensities at energies near 500 eV for 4 μm², 9 μm², and 16 μm² areas.

MEASURED SPECTRAL RESOLUTION

The measured spectrum of a sample in an instrument is given by the convolution of the actual sample spectrum with the instrument bandwidth. The spectrum of a thin sample of CaF_2 at the calcium L edge (~350 eV) was measured in both XM-1 and the Calibration and Standards Beamline (6.3.2) at the ALS. The Calibration and Standards beamline has a measured spectral resolution of $E/\Delta E$ greater than 4000 (5). The absorption in XM-1 was measured over a 9 μm² imaging area. Both the measured spectrum of CaF_2 from XM-1 and from 6.3.2 are shown in Figure 2.

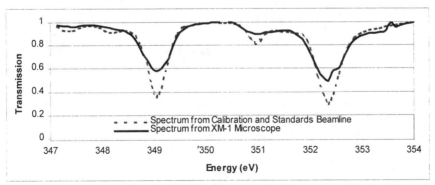

Figure 2. Measured Spectra of CaF_2 by XM-1 microscope and Calibration and Standards Beamline, 6.3.2.

The distribution of the bandwidth of XM-1 was simulated with a gaussian distribution and was convolved with the high spectral resolution data from the Calibration and Standards Beamline. The parameters of the gaussian distribution were adjusted until the result of the convolution very nearly matched the measured spectrum from XM-1. The result was a distribution which has a FWHM of 0.5 eV at 350 eV. This yields a measured spectral resolution for XM-1 for this 9 μm^2 field of view is $E/\Delta E = 700$. This is comparable to the $E/\Delta E = 900$ which was calculated for this field of view. Such bandwidth is sufficient to distinguish between different elements and some chemical states on the spatial resolution of the microscope.

RESULTS

We have observed absorption peaks from the K edges of C,N,O and the L edges of Ca,Ti,V,Cr,Mn,Fe. An example of the absorption features that we have observed is shown in Figure 3. This is the absorption for chromium in three different oxidation states. The increase in oxidation state increases the binding energy and can be seen by the shift to higher energy of the absorption peaks.

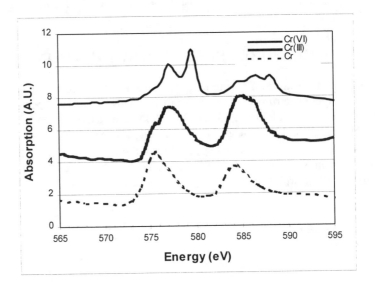

Figure 3. Spectra from Cr, Cr^{3+}, and Cr^{6+}. The shift in energies of the peak absorption is large enough to be distinguished with this spectral resolution.

Cr^{6+} compounds are often soluble and toxic while Cr^{3+} compounds are typically insoluble and therefore not as toxic. There are bacteria which are known to reduce the more toxic Cr^{6+} to the less toxic Cr^{3+}, but there is a lot not known about the process. The combination of a high spatial resolution of 36 nm with the spectroscopic resolution capable of distinguishing the different chemical species will allow us to address this in future research.

The combined capabilities of a high spatial resolution of 36 nm, moderate spectral resolution of $\lambda/\Delta\lambda = 700$, and the ability to image thick, wet samples makes XM-1 a unique tool for many scientific applications. We plan to continue to develop and to actively exploit these features in the future.

ACKNOWLEDGMENTS

The authors would like to acknowledge our colleagues of the Center for X-ray Optics, the Life Science Division, and the ALS, especially D. T. Attwood, E. Gullikson, K. Goldberg, and P. Naulleau. This work is funded by the U.S. Dept. of Energy office of Basic Energy Science, the U.S. Navy, Office of Naval Research under grant N00014-94-1-0818 and the U.S. Army Research office under grand DAAH04-96-1-0246. This great instrument is a testament to the hard work, dedication, and insight of Werner Meyer-Ilse. His driving force and light will be deeply missed.

REFERENCES

1. Meyer-Ilse, W., et. al., "The High Resolution X-Ray Microscope, XM-1," to be published in these proceedings.

2. Private communication with H. Padmore (ALS).

3. Goodman, J., *Introduction to Fourier Optics*, New York: McGraw-Hill, 1996.

4. Born, M., and Wolf, E., *Principles of Optics*, Cambridge: Cambridge University Press, 1980.

5. Underwood, J.H., et al., "High-resolution, high-flux, user friendly VLS beamline at the ALS for the 50-1300eV energy region," *Journal of Electron Spectroscopy and Related Phenomena*, **92**, 265-272 (1998).

X-ray Microscopy in Aarhus

Erik Uggerhøj, Joanna V. Abraham-Peskir

ISA, Institute for Storage Ring Facilities, University of Aarhus, DK-8000, Denmark
E-mail: ugh@ifa.au.dk

Abstract. The Aarhus imaging soft X-ray microscope is now a busy multi-user facility. The optical set-up will be described and project highlights discussed. a) Metal-induced structural changes in whole cells in solution. The effects of aluminium, copper, nickel and zinc on protozoa investigated by using a combination of light microscopy, confocal scanning laser microscopy and X-ray microscopy. b) Botanical studies by X-ray microscopy used to compliment electron microscopy studies. c) Sludge morphology and iron precipitation in Danish freshwater plants by combining X-ray, scanning electron and transmission electron microscopy.

THE AARHUS X-RAY MICROSCOPE

The full-field imaging X-ray microscope (XM) at ISA, Aarhus, Denmark (1) is now a busy multi-user facility. The Aarhus XM plays a significant role in a variety of research projects in areas of botany, colloidal chemistry, food science, medical biology and microbiology.

An object is illuminated by focussing soft X-rays from the Aarhus electron storage ring (ASTRID). Generally, X-rays with a wavelength of 2.4 nm are used, so that transmission through water is high. A condenser zone plate and pinhole act together to produce monochromatic soft X-rays in the water window and a micro zone plate images the sample onto a thinned, back illuminated, Peltier-cooled charge coupled device (CCD) camera with an array of 1024 x 1024 pixels. The condenser zone plates were produced at the Institute for X-ray Physics, Göttingen, by holographic recording (2). The micro zone plates were produced by electron lithography in Göttingen (3). The optimal resolution is around 30 nm. Images are stored digitally for subsequent analysis.

Wet samples are sealed in a specially constructed chamber between two silicon wafers thinned to approximately 150 nm in the central part. The liquid layer can be adjusted via attached syringes. Washed polymer beads (Dynospheres, Plano, Marburg, Germany) with a diameter of 5 µm are used as spacers to maintain a liquid layer sufficiently thick to allow motility within the sealed chamber and prevent collapse of the silicon foil supports. Cells remain in their ambient solution throughout imaging. Typical exposure times are 4 - 25 seconds.

CP507, *X-Ray Microscopy: Proceedings of the Sixth International Conference,*
edited by W. Meyer-Ilse, T. Warwick, and D. Attwood

METAL-INDUCED STRUCTURAL CHANGES IN PROTOZOA

It was found that the flagellated protozoon *Chilomonas paramecium* was sensitive to elevated levels of copper in the external medium (4). XM revealed ultrastructural detail of whole cells in solution (Figure 1a), compartments had formed and ejectosomes were released in response to toxic levels of the free metal ion of copper. Changes in the pellicle thickness were quantified.

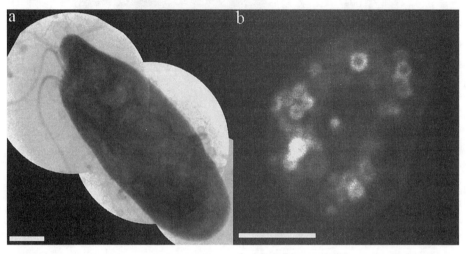

FIGURE 1. *Chilomonas paramecium.* (a) Aarhus XM image of a control cell. (b) Confocal scanning laser microscopy image of a cell loaded with Newport Green and then exposed to aluminium. Fluorescence shows localisation of aluminium into annular-shaped structures. Scale bars = 2 µm.

This work was extended to study the toxic effects of aluminium and zinc on *C. paramecium* (5). A combination of conventional LM, XM and confocal scanning laser microscopy was used. A fluorescent dye, Newport Green (Molecular Probes), was used to localise metals after uptake and accumulation into the cell. Aluminium was accumulated into previously unreported annular-shaped structures (Figure 1b). The pattern of localisation was different for aluminium and zinc. XM studies gave additional information on the nature of the compartments and further ultrastructural details not seen by LM.

BOTANICAL STUDIES

Viscin threads are a unique kind of pollen-connecting thread thought to aid pollen dispersal (Figure 2a). A detailed comparative study combining XM and the existing literature on thread morphology was made (6). Nine plant species (7 Onagraceae and 2 Ericaceae) were investigated and compared to previous electron microscopy studies.

The pollen grain wall has three domains of different chemical compositions. The exine is composed mainly of sporopollenin, a biopolymer with high chemical, physical and biological resistance. The chemical composition is not known in detail. Highly purified exines from Typha angustifolia L. pollen were solubilised, the solubilisate fractionated, and re-aggregated (7). The re-aggregated sporopollenin-like materials were analysed by SEM, TEM and XM. High structural congruence to the initial material was shown and additional distinct substructures were revealed (Figure 2b).

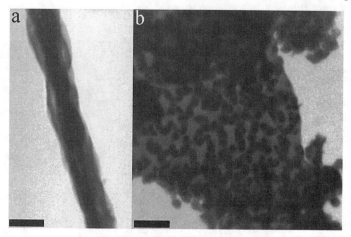

FIGURE 2. Aarhus XM images. (a) Compound viscin thread of *Boisduvalia subulata*. (b) Sporopollenin. Scale bars = 2 μm.

IRON PRECIPITATION IN DANISH FRESHWATER PLANTS

The iron and manganese content in Danish groundwater is generally high, so that precipitation is necessary before the water enters the freshwater supply. Groundwater of South Western Jutland, Denmark, has a high content of iron. Precipitation of the excess iron can be achieved chemically (oxidation by air) or biologically (oxidation by bacteria). Biological precipitation is economically more favourable, as it results in far less sludge than after chemical precipitation.

Sludge containing iron oxide precipitates was obtained from three freshwater plants, Astrup, Forum, and Grindsted. These were investigated to determine the effect of chemical or biological precipitation on the sludge morphology (8). Astrup was the only freshwater plant in Denmark designed to precipitate iron biologically. Figure 3a shows sludge from Astrup containing exopolymers produced by iron bacteria and particles of iron. Forum was designed to precipitate iron chemically and the sludge had a homogenous morphology (Figure 3b). Grindsted was designed to precipitate iron chemically. Figure 3c shows sludge from Grinsted containing exopolymers produced by iron bacteria, demonstrating that iron had been precipitated biologically.

Very recently ISA has become involved in the coming Mars missions. For these projects a key question is: Did bacteria take part in the Fe^{2+} to Fe^{3+} transformation on Mars? For such investigations the Aarhus XM will be a key instrument.

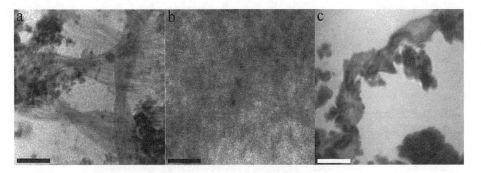

FIGURE 3. Aarhus XM images of sludge from the three different water plants. (a) Astrup. Shows the network of exopolymers produced by bacteria such as *Gallionella,* to which iron particles are attached. (b) Forum. The sludge morphology is homogeneous, though the colloidal nature of the sample is easily recognised. (c) Grindsted. The sludge contains exopolymers and iron particles. Scale bars = 2 μm.

ACKNOWLEDGEMENTS

We thank all collaborators for their interest and continuing support. Robin Medenwaldt and Stuart Lunt for providing valuable assistance.

REFERENCES

1. Medenwaldt, R., and Uggerhøj, E., *Rev. Sci. Instum.* **69**, 2974-2977 (1998).

2. Schmahl, G., Rudolph, D., Guttmann, P., and Christ, O., "Zone Plates for X-ray Microscopy," in *X-ray Microscopy*, edited by G. Schmahl and D. Rudolph, Springer Series in Optical sciences 43, Proceedings of the International Symposium, Göttingen, Fed. Rep. of Germany, 1983, pp. 63-74.

3. David, C., Medenwaldt, R., Thieme, J., Guttmann, P., Rudolph, D., and Schmahl, G., *J. Opt.* **23**, 6 (1992).

4. Abraham-Peskir, J. V., *Europ. J. Protistol.* **34**, 51-57 (1998).

5. Wendt-Larsen, J. Diploma Thesis. Institute of Physics, Odense University, Denmark (1998).

6. Abraham-Peskir, J.V., Searle, R., and Medenwaldt, R., *Grana* **36**, 343-346 (1998).

7. Thom, I., Grote, M., Abraham-Peskir, J., and Wiermann, R., *Protoplasma* **204**, 13-21 (1998).

8. Søgaard, E. G., Medenwaldt, R., and Abraham-Peskir, J. V., *Water Res.* (in press).

X-Ray Speckle Correlation Interferometer

Rachel Eisenhower and Gerhard Materlik

Hamburger Synchrotronstrahlungslabor (HASYLAB) at Deutschen Elektronen-Synchrotron (DESY), Notkestrasse 85, D-22603 Hamburg, Germany

Abstract. Speckle Pattern Correlation Interferometry (SPCI) is a well-established technique in the visible-light regime for observing surface disturbances. Although not a direct imaging technique, SPCI gives full-field, high-resolution information about an object's motion. Since x-ray synchrotron radiation beamlines with high coherent flux have allowed the observation of x-ray speckle, x-ray SPCI could provide a means to measure strains and other quasi-static motions in disordered systems. This paper therefore examines the feasibility of an x-ray speckle correlation interferometer.

INTRODUCTION

Speckle Pattern Correlation Interferometry (SPCI) is a well-established technique in the visible-light regime [1]. As with holographic and Moire interferometry, SPCI provides a full-field, non-destructive method of deducing an optically opaque object's motion from the appearance of interference fringes. Visible region SPCI is performed by using optical components to split, scatter, and recombine an incident laser beam such that the coherent addition of the sample's speckle wavefield and a reference wavefield (which can also be a speckle pattern) is achieved. The resulting superimposed wavefield is sensitive to relative phase changes between the sample and reference wavefields, hence the overall sensitivity to sample motion is greatly increased compared to that of a single-beam speckle field.

Two main geometries exist for performing visible-region SPCI. The first is a Michelson interferometer type, in which one of the split beams scatters from the sample to be studied, and the other beam scatters off a stationary reference sample. This type of interferometer is most sensitive to motion perpendicular to the sample surface ("out-of-plane" motion). The second main geometry is usually called "dual illumination", referring to the illumination of one sample by two beams, each coming from a different direction (but having originated from the same laser source). This geometry is most sensitive to in-plane motion.

Finding a suitable geometry for performing x-ray SPCI poses a real challenge due to the special nature of x-ray optical components. Mirrors can be used for beam

CP507, X-Ray Microscopy: Proceedings of the Sixth International Conference,
edited by W. Meyer-Ilse, T. Warwick, and D. Attwood
© 2000 American Institute of Physics 1-56396-926-2/00/$17.00

splitting and recombination, but must be of extremely high quality in order to preserve beam coherence. Crystal interferometers could perform splitting, but are too limited in angular acceptance to recombine speckle fields. Placing two samples side-by-side such that each intercepts a lateral portion of the beam is feasible, but does not provide much room for acting upon only one of the samples. The geometry that is investigated in this paper is illustrated in Figure 1. It has also been tested experimentally [2], with encouraging initial results. It consists of two samples placed one after another such that each intercepts the entire lateral size of the beam. The idea is that enough of the upstream sample will pass through the downstream sample to enable the fields scattered from each to combine. In order for coherent combination of the fields to occur, all scattering must take place within the coherence volume defined by the product of the lateral and longitudinal coherence lengths.

CALCULATIONS

As shown in [2], the net wavefield at point P as shown in Figure 1 is given by

$$\psi(\vec{R}) = \frac{\mu e^{ikR}}{4\pi R} \left[\sum_{j=1}^{N} e^{-i\vec{Q}_1 \cdot \vec{r}_{1j}} f(Q_1) \sum_{k=1}^{M} e^{-i\vec{Q}_2 \cdot \vec{r}_{2k}} f(Q_2) \right] \tag{1}$$

where the r vectors point to discrete scatterers, of which there are N in the first sample and M in the second. Assuming identical sorts of scatterers in each sample, the initial intensity recorded at P is given by

$$|\psi_{initial}(\vec{R})|^2 = \frac{2f^2(\theta)}{R^2} \left[1 + \sum_{j=1}^{N} \sum_{k=1}^{M} \cos(\vec{Q}_1 \cdot \vec{r}_{1j} - \vec{Q}_2 \cdot \vec{r}_{2k}) \right] \tag{2}$$

Let the motion of sample 2 be described by $\vec{r}_k \rightarrow \vec{r}_k + \vec{d}_k$. Then, the difference between post-motion and pre-motion intensities is given by

$$|\psi_{final}(\vec{R})|^2 - |\psi_{initial}(\vec{R})|^2 =$$
$$\frac{2f^2(\theta)}{R^2} \left[1 + \sum_{j=1}^{N} \sum_{k=1}^{M} \cos(\vec{Q}_1 \cdot \vec{r}_{1j} - \vec{Q}_2 \cdot (\vec{r}_{2k} + \vec{d}_k)) - \cos(\vec{Q}_1 \cdot \vec{r}_{1j} - \vec{Q}_2 \cdot \vec{r}_{2k}) \right] \tag{3}$$

This difference is shown in Figure 2 for the case of a $0.5\mu m$ vertical translation of sample 2. Other paramaters used in the simulation were an x-ray energy of 8 keV, a 1.5 m origin-detector distance, a viewing area of $2mm \times 2mm$, and an incident beamsize of $7\mu m \times 7\mu m$. For simplicity, the coherence volume was assumed infinite and single-scattering was assumed.

The location of the correlation fringes can be explained by considering Equation 3. For those observation positions P for which $\vec{Q} \cdot \vec{d}_k = 2n\pi, n = 0, 1, 2, ...$

for each k, then the difference in intensity is zero. For the case of a bulk translation of the sample such as the one simulated, $\vec{d}_k = \vec{d}$ for all scatterers, hence $\vec{Q} \cdot \vec{d} = 2n\pi$ can readily be solved for the fringe locations. For example, for the simulated motion shown in Figure 2, $\vec{Q} \cdot \vec{d} = \frac{4d\pi}{\lambda} \sin\theta\cos\theta = 2n\pi$ gives fringe positions of 0, $\pm 0.46\,mm$, $\pm 0.92\,mm$,, which agree with those shown in Figure 2. Another example of a bulk translation is shown in Figure 3, in which sample 2 is moved longitudinally; i.e. parallel to the incident beam direction by $1.3\,mm$. Note that this amount of motion is quite a bit larger than that of the vertical translation, illustrating that this geometry for performing x-ray speckle correlation interferometry is much more sensitive to in-plane motion than out-of-plane.

If the sample motion is not a simple bulk translation, but rather a stretch or rotation, then $\vec{d}_k$ is not a constant vector for all scatterers. However, the fringe locations can still be deduced if one assumes that the displacement extrema occur at the edges of the sample. In Equation 3, the $\vec{Q}_{1,2} \cdot \vec{r}_{1j,2k}$ terms describe the speckle, while the $\vec{Q}_2 \cdot \vec{d}_k$ terms describe the fringe locations. For each detector position, the value of $\phi_k = \vec{Q}_2 \cdot \vec{d}_k$ will span a range of values, starting at zero. The fringes will be located at those positions for which the maximum value of ϕ_k is $2n\pi$, since $\int_0^{2\pi} \cos\phi_k = 0$ and $\int_0^{2\pi} \sin\phi_k = 0$. Hence, $\vec{Q}_2 \cdot \vec{d}_{k,max} = 2n\pi$ can be solved for fringe locations.

DISCUSSION

This paper shows that it is theoretically possible to perform x-ray speckle correlation interferometry by summing the coherent small-angle scattering from two samples in series. The method is particularly well-suited for hard x-rays, since re-absorption of the upstream sample's speckle field in the downstream sample is minimized. In addition, the fringe spacings become smaller with increasing energy. What this particular geometry for performing x-ray SPCI could provide is highly accurate measurements of sub-micrometer in-plane motions of disordered materials. This would provide more accurate measurements of piezoelectric parameters, as an example, as well as a tool to answer questions about directions of movement at the onsets of phase changes. In order to have sensitivity to out-of-plane motion, an alternative geometry is proposed, namely, the grazing-incidence small-angle scattering from two closely-placed samples. This geometry has been previously used to interfere two coherent beams from Fresnel mirrors [3], and would be more suitable for soft x-rays.

REFERENCES

1. Jones, R., and Wykes, C., *Holographic and Speckle Interferometry*, Cambridge: Cambridge University Press, 1983.

2. Eisenhower, R., Gehrke, R., Materlik, G., Drakopolous, M., Simionovici, A., and Snigirev, A., *J. Synch. Rad.* **6**, 1168-1173 (1999).
3. Fezzaa, K., Comin, F., Marchesini, S., Coisson, R., and Belakhovsky, M., *J. of X-Ray Science and Technology,***7**, 12-23 (1997).

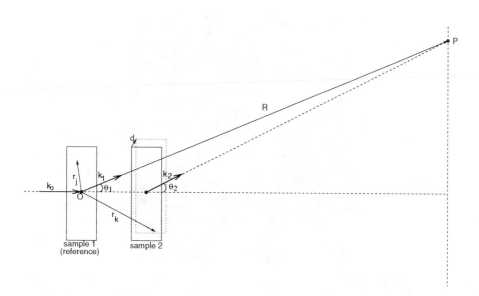

FIGURE 1. Scattering geometry for the x-ray speckle correlation interferometer.

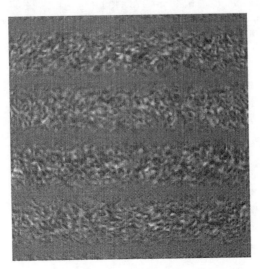

FIGURE 2. The difference between speckle patterns before and after a $0.5\,\mu m$ vertical translation of sample 2. The image size is $2\,mm \times 2\,mm$, at a distance of $1.5\,m$ from the center of the upsteam sample. The direct-beam location is at the center of the image.

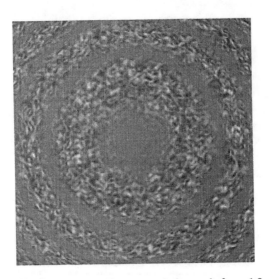

FIGURE 3. The difference between speckle patterns before and after a $1.3\,mm$ motion of sample 2. The image size is $2\,mm \times 2\,mm$, at a distance of $1.5\,m$ from the center of the upsteam sample.

Soft X-ray Holography at NSRL

Shiping Jiang*   Yuxuan Zhang   Xinyi Zhang   Shaojun Fu
Andong Xia   Xiangdong Xu   Yilin Hong

*National Synchrotron Radiation Laboratory, University of Science and Technology of China, Hefei,
230029, P. R. China*

Abstract.   Soft X-ray Microscopy station is one of the first constructed facilities at NSRL. The main instrument of the beamline is a monochromator which can provide the monochromized X-rays varying from 2.0nm to 5.0nm. Our holographic image employs the X-rays, so called "water window", and experimental setup is arranged in a Gabor in-line way. Holograms are recorded on the photoresist polymethyl methacrylate (PMMA) and readout with ordinary light microscopes or atomic force microscopes (AFMs). They are reconstructed using optical or numerical methods. So far the holograms of some samples and their reconstructed images are acquired. Those samples include minute granules, cobweb and so on. The estimated resolution of the reconstructed images reaches the level of submicron. It is limited by the coherence of the X-rays used for these experiments. And it is possible to improve the resolution in the future.

INTRODUCTION

The Synchrotron Radiation (SR) source at Hefei National Synchrotron Radiation Laboratory (NSRL) was operated and available for experiments in the end of 1991. The soft X-ray microscopy facility is one of the five firstly constructed beamlines and stations at NSRL. We have developed soft X-ray contact imaging technique and constructed a prototype of transmission scanning microscope since 1991. The second version of scanning microscope will be in operation in the near future. We are also carrying out other types of X-ray imaging studies, e.g. imaging X-ray microscopy and holography. In this paper, some results of Gabor in-line soft X-ray holography at NSRL are reported.

SOFT X-RAY SOURCE

Hefei SR facility is an electron storage ring with the energy of 800MeV. The characteristic wavelength of its radiation is about 2.4nm, which is at the region of 'water window', so it is very suitable for the study of biological specimens. Coherent

---

* correspondence e-mail: spjiang@ustc.edu.cn

CP507, *X-Ray Microscopy: Proceedings of the Sixth International Conference,*
edited by W. Meyer-Ilse, T. Warwick, and D. Attwood
© 2000 American Institute of Physics 1-56396-926-2/00/$17.00

X-ray is obtained by filter the broad spectrum and limit the divergence of the SR. The white SR light is limited by a series of apertures and then monochromized by a linear monochromator (1), which consists of a Fresnel zone plate and a pinhole. It covers the waveband from 1.97nm to 5.40nm, and the spectrum width is about 0.15nm. The photon flux at the exit of the monochromator is about 10^9 photons per second, and the diameter of the pinhole is 30 microns. At a distance of 700mm from the pinhole, the lateral coherent width is about 70 microns and the coherent length is about 0.07 micron.

X-RAY HOLOGRAPHIC EXPERIMENTAL SETUP

Optical setup for X-ray holography experiments are arranged in a standard Gabor geometry (Figure 1). The distance between the exit of the monochromator and the experimental samples is 700mm. The samples and the recording medium for the holograms, photoresist of polymethyl methacrylate (PMMA), are fixed on the same sample-holder so as to reduce the influence of oscillation. The experiments are carried out both in vacuum and atmosphere. The monochromatic X-rays may exit from vacuum to a helium chamber or air through a Si_3N_4 film window, the thickness of which is 120nm and the area is $0.3 \times 0.3mm^2$. The samples used in our experiments include metal grid, dust granule, cobweb and so on, and the dimension of these samples is one or several microns. The distance between the samples and the recording material is varying from 0.4mm to 10mm, which satisfy the far field condition. The exposure time varied from several tens of minutes to several hours, according the experimental conditions.

RECONSTRUCTION OF THE HOLOGRAMS

After exposure and development, the photoresist is etched by the X-rays, and its height represents the local intensity of the X-rays. The hologram is recorded on the photoresist in form of the topography. It must be readout by microscope before it can

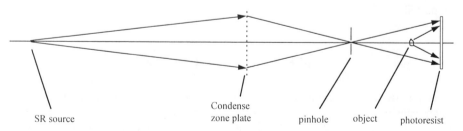

FIGURE 1. The experimental setup for Gabor in-line soft X-ray holography.

be reconstructed. Optical microscope, electron microscope and scanning probe microscope (SPM) can be used. Because of the relatively low resolution of optical microscope, it is not suitable when the spatial frequency of the interference fringe is very high. The fluctuation of the fringe is usually several hundreds angstrom and it must be coated with a heavy metal layer in order to increase the contrast in electron microscopy. Reading out the hologram by SPM such as atom-force microscope (AFM) is more convenient, better results is easy to get.

The hologram can be copy to transparent film and reconstructed by optical method, or it can be digitized from the film or directly obtained by AFM, then reconstructed by numerical method. Modern digital signal processing technique provide the ability to correct the aberration and distortion during the recording and reading out procedure, besides this, extra noise can be avoid which is difficult in optical method. We have tried the numerical method as well as the optical method. The reconstruction process includes the modulation and propagation of a planar wave, which can be describe by Fresnel-Kirchhoff diffraction integral:

$$\psi(x,y) = \iint t(\xi,\eta) \frac{\exp(ikr)}{i\lambda r} d\xi d\eta$$

where the $\psi(x,y)$ and $t(\xi,\eta)$ represent the complex amplitude in the image plane and optical transmissivity in the hologram plane respectively. In numerical method, this formula can be discrete and be used to compute the propagation of the wave.

SOME RESULTS

We have tested a lot of samples, and tried several techniques in hologram readout, including LM and AFM. The holograms are successfully reconstructed using numerical method (2,3). Here are some X-ray holography results at NSRL.

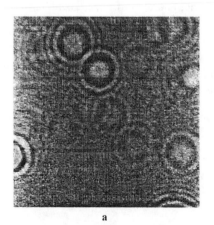 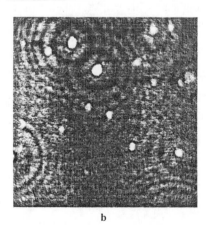

a   b

FIGURE 2. The X-ray hologram (a) and its reconstruction image (b) of dust granules.

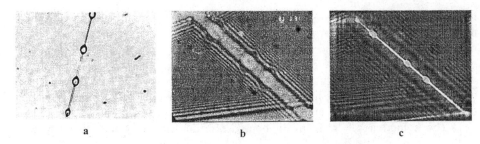

FIGURE 3. The original object (a), the X-ray hologram (b) and its reconstruction image (c) of cobweb.

Figure 2 shows the X-ray hologram and its reconstruction image of some dust granules. The wavelength for recording is 2.3nm, and the distance from the sample to the photoresist is 400μm. The hologram is read out by AFM, and reconstructed using numerical method. Its area is 15×15 μm^2. The size of the smallest particle is less than 0.3μm.

In Figure 3 some biological samples were used. The original objects are some cobwebs (Figure 3a). The recording wavelength is 3.2nm and the distance is 10mm. The hologram is examined by an ordinary microscope and copied to a film, then digitized by a microdensitometer and reconstructed using a computer. The ball-like objects and the thread are clearly reconstructed (Figure 3c). The diameter of the wire is 1.2μm or so.

ACKNOWLEDGMENTS

The authors would like to thank Dr. Zhu Jijun and Prof. Shen Yaochun of Southeast University for their help in AFM imaging, and also to Porf. Chen Jianwen, an NSRL user, for his cooperation.

REFERENCE

1. Zhang, Y. W., Xu, C. Y., Xu, X. L. et al., *Physical Scripta*, **41,** 422-424(1990)
2. Zhang, Y. X., Jiang, S. P., Fu, S. J. et al., *Acta Optics Sinica*, **17**, 1599-1600(1997)
3. Jiang, S. P., Zhang, Y. X., Zhang, X. Y. et al., *Chinese Journal of Laser*, **B8**, 81-84(1999)

Fourier-Transform Holography With Coherent Hard X-Rays

W. Leitenberger and A. Snigirev

European Synchrotron Radiation Facility (ESRF), BP 220, F- 38043 Grenoble - CEDEX, France.

Abstract. Experimental results of recording Fourier-transform holograms using 14keV synchrotron radiation are presented. A Fresnel zone plate was used as a beam splitter to realize a submicron reference point source by its first diffraction order and at the same time a object wave by the transmitted zero order beam. The holograms were recorded by a high resolution CCD-camera and were numerically reconstructed afterwards.

INTRODUCTION

Experiments with Fourier transform holography (FTH) have already been already successfully done using visible light [1,2] and soft X-rays [3,4,5]. The aim of the presented experiments was to show the possibility of microscopic holography with multi-keV X-rays. We describe experiments with FTH at the ESRF undulator beamline ID18. The main advantage of the FTH technique compared to other microscopic technique is the large fringe spacing in the hologram and therefore low requirements to the lateral resolution of the detector. The resolution of the method is determined by the size of the reference wave and the aperture of the hologram. For the experiment the good coherence properties of the undulator radiation are very useful.

THEORY

In FTH a reference source is placed in the plane of the object. The phase of the wave diffracted by the object can be reconstructed from its interference with the reference wave. The reference point source is formed by the first order diffracted beam of a Fresnel zone plate which is used as a beamsplitter and at the same time the object wave is delivered by the transmitted zero order beam. The resulting interference pattern consists of low constant modulations, in analogy to the double slit experiment. The fringe distance D_f is given by $D_f = \lambda x / s$, where λ is the wavelength, x the object to hologram distance and s the reference-to-object distance. If the object is placed close to the reference point and the hologram is recorded in the far field, the intensity distribution in the hologram plane corresponds to the Fourier transform of the trans-mission function of the object. Therefore the object can be reconstructed by the inverse Fourier transform of the hologram.

In order to observe interference effects some requirements on the coherence of the incident wave are necessary. The largest possible object-to-reference distance depends

CP507, *X-Ray Microscopy: Proceedings of the Sixth International Conference,*
edited by W. Meyer-Ilse, T. Warwick, and D. Attwood

on the maximum path difference between the object wave and the reference wave up to what they can interfere. This distance is the longitudinal coherence length $l_l = \lambda^2/\Delta\lambda$ determined by the monochromaticity $\Delta\lambda/\lambda$ of the beam. For the experimental conditions of $\lambda=0.86\,\text{Å}$ and a Si 111 double crystal monocromator with $\Delta\lambda/\lambda=10^{-4}$ we get a value of $l_l=0.6\,\mu m$. From the object distance x follows the largest possible object to reference distance $s_{max} = \sqrt{l_l x}$. In the experimental set-up this value was $s_{max}=1.3\,mm$. In order to coherenly illuminate the object the incident wave must have a constant phase. The maximum distance in the object plane over which the phase difference of the beam is smaller than one wavelength is described by the transverse coherence length $l_t=D\,\lambda/d_s$. It is given by the undulator source size d_s and the source distance D. From the source size of the undulator beam of $54\,\mu m$ vertical and $812\,\mu m$ horizontal and a source distance of $D=56$ m follows the transversal coherence length of $88 \times 6\,\mu m^2$ (vert. × hor.). For enhancement of the horizontal coherence properties the horizontal beam size was decreased by a slit at 23m upstream of the sample which was closed to $25\,\mu m$ width. It gives a secondary source with a horizontally coherence length of $79\,\mu m$. The maximum observable fringe distance is obviously given by the aperture of the hologram which is determined by the numerical aperture of the zone plate delivering the reference wave.

EXPERIMENTS

The 'beam splitting' Fresnel zone plate (FZP) had the following charactristic parameters: height of the individual gold zones 1.15 µm, radius of the central zone of 7.8µm, width of the outermost zone 0.3µm. It was made by X-ray lithography on a SiN substrate. The FZP has 200 zones giving a full aperture of 200µm. At 14.4keV the focal spot of the FZP was at f=0.72 m and the diffraction efficiency was 8 %. The geometry of the setup is schematically shown in Fig.1. The object is directly illuminated by the transmitted beam (zero order diffracted beam). At the same time the reference wave is formed by the first order diffracted beam. The object under investi-

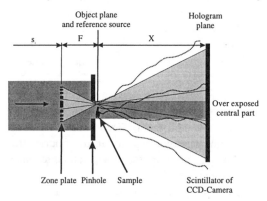

FIGURE 1. Experimental set-up for Fourier transform holography

gation is placed close to the reference source. The zero and higher than first order diffracted portions were removed by a pinhole placed 5 mm before the sample. The hologram was detected at a distance of x=2.7 m behind the sample using a high resolution CCD camera. The effective pixelsize in the arrangement was 1 μm and the exposure time for each hologram was two minutes. The X-ray hologram was converted into light by a thin scintillator screen and the image was magnified by a microscope and recorded on a CCD-camera.

Fig.2 shows the hologram which was observed using a vertical wire of 5μm diameter as a sample. The sample has a distance of 5 μm from the reference source. The central hologram part is overexposed to the direct beam. The concentric circles are the hologram of the 50 μm pinhole and the hyperbola-shaped interference fringes on both sides of the overexposed central part are the hologram of the wire itself. The fringe distances are different for the left and the right side, 21 μm and 68 μm, respectively, which corresponds to object distances of 3.4 μm and 11 μm. The large fringe spacing belongs to the wire side which is close to the reference beam wheras the small fringe spacing belongs to the opposite side of the wire.

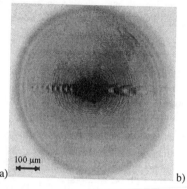

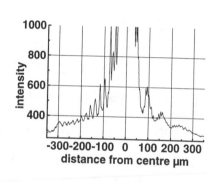

a)   b)

FIGURE 2. a) Fourier transform hologram of a 5 μm tungsten wire. The wire is vertically oriented and placed 5 μm beside the focal spot of the Fresnel zone plate. The width of the circular hologram is 680 μm (black is high intensity)   **b)** Horizontal intensity profile through the centre of image a)

As a more complex test object a gold grid with 15 μm grid spacing and 3.5 μm width of the grid lines was used. The fine grid was adjusted in a way that the reference wave could pass an empty square of it. A 100 μm pinhole was used for background supression and the reference source was about 10 μm from its border. The numerical reconstruction was done after flatfield correction and background subtraction. The influence of the zero order radiation which passes the pinhole in the central part of the hologram was diminished by numerically cutting all values above a certain treshold value.

The result of the inverse Fourier transform is shown in Fig. 3b. The two slightly overlapping circles represent the image and the corresponding twin image of the 100 μm pinhole with the grid inside. The region of overlap of the image with the twin image corresponds to the distance of the reference source to the edge of the pinhole. From the effective pixel size of the camera and the pixel size in the reconstruction which was determined from the known object size follows a effective demagnification by a factor of 0.9.

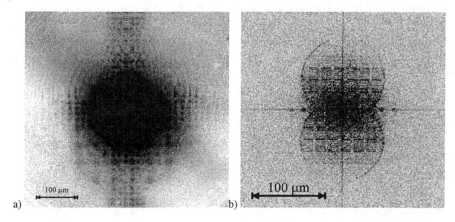

a)   b)

FIGURE 3. **a)** Fourier transform hologram of of a grid structure. The part of $0.5 \times 0.5 \text{mm}^2$ which corresponds to 900×900 pixels was used for reconstruction. **b)** numerical reconstruction of the hologram in a) by applying inverse Fourier transformation. The 500×500 pixels central part is shown corresponding to 280 µm in the object plane.

SUMMARY AND OUTLOOK

Fourier transform holograms with hard X-rays recorded at an ESRF undulator beamline were successfully numerically reconstructed. Due to the geometry of the experiment especially the relatively small distance between sample and hologram plane there was no magnification effect and the final resolution was below the resolution of the used detector. For the near future, experiments are planned that allow larger magnification and investigation weak absorbing phase objects.

ACKNOWLEDGEMENTS

The authors would like to thank E. Di Fabrizio and M. Gentili for providing the Fresnel zone plates. For support during the experiments we thank the following persons: C.Raven, E.Boller, P.Cloetens, W.Ludwig, M. Di Michiel, J.C. Labiche, T.Weitkamp, A.Chumakov (ESRF); R.Mathiesen (NTNU Trondheim); C.Schroer, F.Günzler (RWTH-Aachen). One of the authors (wl) was supported by a Marie Curie Fellowship of the European Commission.

REFERENCES

1. Winthorp, J.T. and Worthington, C.R., Phys. Lett. **15**, 124 (1965).

2. Stroke, G.W., Appl. Phys. Lett. **6**, 201, (1965).

3. Aoki, S., Ichihara, Y. and Kikuta, S., Jap. J. Appl. Phys. **11**, 1857 (1972).

4. Reuter, B. and Mahr, H., J. Phys. E. **9**, 746 (1976).

5. McNulty, I., Kirz, J., Jacobsen, C., Andersen, E., Howells, M. and Kern,D. Science, **256**, 1009 (1992).

Experimental Investigation of Soft X-ray Diffraction

Xiangdong Xu, Yilin Hong, Tonglin Huo, Shipiing Jiang,
Xiaobin Shan, Shaoguang Yang and Shaojun Fu

National Synchrotron Radiation Laboratory, USTC, Hefei, 230029, China

Abstract. The experimental study on soft x-ray diffraction using gold test objects has been carried out with synchrotron radiation at NSRL. Some images, which correspond to 0、2 and 80 mm away from a specimen, were recorded as relief pattern on photoresist with use of 3.2nm x-rays from U12A bemline. The experimental result shows that patterns vary with the distance between specimen and photoresist detector, which is in excellent agreement with the theoretical one.

INTRDUCTION

Diffraction imaging is one of the numerous imaging methods in soft x-ray microscopy. Its interest lies in its being potentially the high-resolution form, and in its special ability to obtain information regarding 3-D structure in a single exposure [1]. In the early part of 1980s, Sayre[2,3] suggested that the use of longer-wavelength x-rays from synchrotron sources might permit the recording of the diffraction pattern of a general microscopic specimen, and that this might in turn open the way to detailed imaging of structures, which do not occur in nature in accurately duplicated form. The first patterns of this type were obtained from single small diatoms.

In this paper we mainly describe the experimental aspects and the preliminary result of soft x-ray diffraction with gold test objects at SXM station at NSRL.

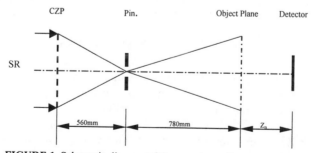

FIGURE 1. Schematic diagram of the experimental arrangement

CP507, *X-Ray Microscopy: Proceedings of the Sixth International Conference,*
edited by W. Meyer-Ilse, T. Warwick, and D. Attwood
© 2000 American Institute of Physics 1-56396-926-2/00/$17.00

EXPERIMENT

The experiments were conducted on U12A beamline at NSRL, which source and beamline have been described [4]. The experimental arrangement is shown schematically in Fig. 1. It consists of a condenser zone plate (CZP), a pinhole and a specimen. In our case, x-rays from the source are monochromatized to a wavelength of 3.2 nm and condensed on the pinhole (30μm) by a CZP, a specimen was placed 780 mm downstream from the pinhole.

Optical system alignment is one of the most important steps. We might judge whether it is alignment by means of fluorescent pattern emitted by soft x-ray impinging on a phosphor coated on the sample holder or view window. When a CZP and a pinhole are coaxial the pattern is an annular shaped image (Fig.2,), which symmetry is no variety when CZP is moved along axial. On the contrary, the pattern is asymmetric annular shaped one, which could rotate or change into two splitting patterns when a CZP moved along axial.

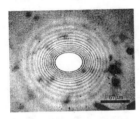
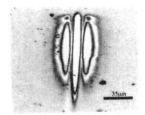

FIGURE 2. Intensity distribution in object plane   **FIGURE 3.** Diffraction of the double slits

We used a positive photoresist (PMMA) as a detector. PMMA is spun at 800 rpm onto silicon substrates, resulting in ~ 80 nm films. The films were baked at 170℃ for 1h. The object-detector packages were aligned to the x-ray beam in a sample chamber $(1.2\times10^{-3}Pa)$, and diffraction patterns were recorded over exposure times of 15~40 minutes. After that the PMMA was developed by immersion in a mixture of 1 part methyl isobutyl ketone to 3 parts isopropyl alcohol for 10~30s, briefly rinsed in isopropyl alcohol, and dried with a filtered jet of nitrogen gas. The patterns could be measured by TEM and AFM for higher resolution. But, as preliminary study we use only a light microscope to observe.

RESULTS AND DISCUSIONS

Two types of objects were used in our experiments. One is a double slits whose width and the spacing are 2.5μm and 10μm, respectively; the other is a 2-D array with 9μm hole diameter and 6μm spacing. It is very difficult to align the double slits to the x-ray beam by naked eyes, then we obtained only a diffraction image (Fig. 3) under the condition of $Z_0=80$mm. However, it is very convenient to align to the larger 2-D array specimen $(3\times3mm^2)$ to the x-ray beam so that we recorded a large number of soft x-ray images at various distances away from the specimen for understanding the characteristics of soft x-ray diffraction imaging. As has been described above, the x-ray diffraction patterns are magnified using a light microscope.

Fig. 4 shows the soft x-ray contact image of array ($Z_0=0$). Its characteristic is no diffraction phenomenon and is only a replica of the specimen. The circular contour in the figure is a standard EM grid (#300) supported specimen, and the central part reveals the silicon substrate because of the intensity distribution of the incident x-rays.

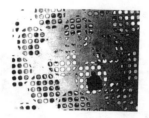

FIGURE 4. Contact image of the array

Fig. 5 and Fig. 6 show observed and their calculated images obtained at the distances of 2、80mm away from the specimen. It is easy to see that these images are obviously different from the soft x-ray contact one, that is, there is an obvious diffraction phenomenon. As the distance between the detector and the specimen increasing, The intensity modulation signal becomes weak and the scattered one becomes strong, dimming the original characteristic outline. In Fig. 5 we might see that the original feature except some diffraction patterns adding to it because of a short distance away from the specimen where the intensity modulation signal stronger than the scattered one. However, we couldn't see that the original feature in Fig. 6 because of a long distance away from the specimen where the intensity modulation signal weaker than the scattered one, resulting in the original characteristic outline vague.

As emulsion film, the photoresist film has also a exposure threshold which can break bonds on the polymer backbone of the resist by secondary electrons generated by x-ray that is absorbed in the resist and the initial x-ray photon, and thus lower the molecular weight of the exposed area. With immersion in the solvent, the exposed areas are dissolved at a much higher rate than are the unexposed ones, so that the incident irradiant pattern is transferred into a surface relief pattern. Under the condition of Z_0 =2mm, the measured image (Fig. 5(a)) agrees well with the calculated (Fig.5(b)) except the parallel fringes among the arrays were not recorded simultaneously because of low intensity. Based on the previous exposure conditions, we consciously increased exposure dose to record the weak diffraction fringes. As shown in Fig. 6(a), the weak fringes have been recorded and agree well with Fig. 6(b), and disappearance of strong patterns is due to the overexposure, resulting in the resist was dissolved out.

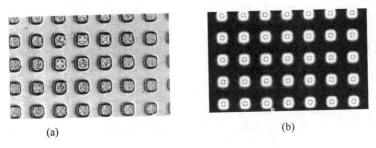

(a)   (b)

FIGURE 5. Diffraction image of the array. Observed pattern (a) and its calculated one (b)

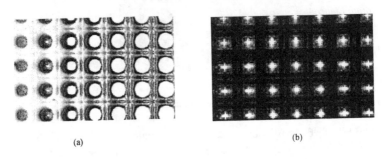

(a)   (b)

FIGURE 6. Diffraction image. Observed pattern (a) and its calculated one (b)

CONCLUSIONS

Soft x-ray diffraction has been demonstrated with SR at NSRL. The experimental result is in excellent agreement with the theoretical one. Although the array size is larger, it is very useful for us to understand the related experimental characteristic and techniques.

REFERENCES

1. Sayre D., "Diffraction-imaging possibilities with soft x-rays," In *X-ray Microscopy-Instrumentation and Biological Applications*, edited by Ping-chin Cheng and Gwo-jen Jan, Berlin: Spring-Verlag, 1987, pp.213-223.

2. Wen-bing Yun, Kirz J., and Sayre D., *Acta Cryst.*, **A43**,131-133(1987).

3. Sayre D. and Chapman H. N., *Acta Cryst.*, **A51**,237-252(1995).

4. Zhang Yunwu, Xu Chaoyin, Xu Xilin, et al., *Physica Scripta*, **41**,422-424(1990).

X-Ray Diffraction Imaging Provides Nanometer Spatial Resolution For Strain Determination

S. Di Fonzo[1], W. Jark[1], S. Lagomarsino[2], C. Giannini[3], L. De Caro[3], A. Cedola[2,4], M. Müller[4]

[1]SINCROTRONE TRIESTE, S.S. 14 km 163.5 in Area Science Park, 34012 Basovizza, Italy
[2]Istituto Elettronica Stato Solido (IESS) - CNR, V. Cineto Romano 42, 00156 Roma, Italy
[3]Centro Nazionale Ricerca e Sviluppo Materiali (PASTIS-CNRSM)
S. S. 7 Appia km 712, 72100 Brindisi, Italy
[4]European Synchrotron Radiation Facility, B. P. 220, 38043 Grenoble CEDEX, France

Abstract. This report presents x-ray waveguides as one-dimensional condensor lenses providing a source of submicron size for shadow projection and diffraction imaging techniques. A waveguide can compress an x-ray beam in one dimension to the level of 100 nm. This beam is spatially coherent. Consequently, the source size limited resolution in imaging experiments is expected to be at least the mentioned beam dimension.

PROJECTION IMAGING WITH X-RAY WAVEGUIDES

The simplest setup for obtaining a magnified image of an object is the shadow projection scheme, which requires, as the only elements, a small source of divergent radiation, the object and a two-dimensional detector. Even though very simple and without the need for complicated optical components, this projection setup is rarely used for high resolution microscopy in the x-ray range. An exception is the work of Newberry and colleagues (1). The reasons for the limited interest in such a scheme are manifold. State-of-the-art spatial resolution of film plates and sophisticated CCD camera set-ups (2) is approaching the submicron level, and thus submicron resolution can be achieved already by use of collimated x-rays. Imaging approaches with divergent light are thus only interesting, if they can provide even better submicron resolution of the order of 100 nm. However, as the resolution in projection schemes has its lower limit in the source size, submicron resolution requires a submicron source size. This cannot be provided directly neither by x-ray tubes nor by synchrotron radiation sources. Thus optical components, which can focus or compress the incident radiation to the submicron level, need to be employed. Many efforts have been made to develop optical components for this purpose, namely zone plates (3) and compact refractive lenses (4). However, until now beam sizes of 100 nm have been rarely achieved in the hard x-ray range. They were reported for capillary optics (5) and are routinely achieved with x-ray slab waveguides (6). Here the properties and

CP507, X-Ray Microscopy: Proceedings of the Sixth International Conference,
edited by W. Meyer-Ilse, T. Warwick, and D. Attwood
© 2000 American Institute of Physics 1-56396-926-2/00/$17.00

applications of projection schemes, employing very favorably the beam characteristics from the latter waveguides, will be discussed.

As presented elsewhere (6,7) and illustrated in figure 1, x-ray waveguides compress the incident radiation in only one direction, the vertical plane, to a dimension below 150 nm. This source provides spatially coherent and divergent radiation in the vertical plane (6,7). It will be shown here, that the lack of beam compression in the horizontal direction is not a drawback, but an advantage for several applications. As in this direction no magnification can be achieved, the spatial resolution in this direction will be only of the order of microns. Consequently for optimally utilizing the possible

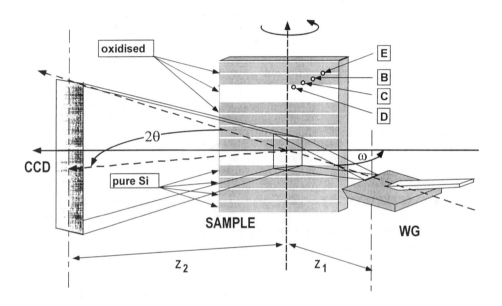

FIGURE 1. Experimental set-up for the diffraction imaging experiment: The beam is arriving from the right and is compresses by the waveguide (WG) in the vertical direction. The exiting vertically divergent beam is diffracted by the sample at an angle of grazing incidence ω towards the two-dimensional CCD detector positioned at a fixed scattering angle 2θ. The stripes at the sample are oriented parallel to the tangent plane of the waveguide. Gray areas are pure Si, while white areas are oxidised. The letters B, C, D and E indicate the positions for which the reciprocal space maps of figure 2 have been obtained.

submicron resolution in the vertical direction, the sample properties should be homogeneous in the horizontal direction at least on a length scale of several microns. This criterium is easily fulfilled in thin straight fibres and in narrow line structures on microchips. Before presenting the experimental data it is worthwhile to discuss the consequences of using a coherent beam. As long as small and thin objects are investigated, they will be partly transparent to the x-rays. Any discontinuity of the refractive index of the sample in the direction of beam compression, i.e. an edge or a crack, will lead to a locally varying phase retardation of the transmitted coherent light. Consequently, an in-line hologram and not just a magnified shadow projection of the sample will be observed at a detector, at some distance behind the sample. Even with opaque samples true shadow projection imaging is not possible with coherent light. Due to the diffraction of the coherent light at the edges, the projected edges show a superimposed diffraction pattern. We reported already experiments for the first configuration, the in-line holography set-up, using as source for the projection image the waveguide, and we could derive a spatial resolution of at least 140 nm (9). Under ideal conditions the resolution with incoherent illumination should be identical to the source size. Here, due to the coherence and the divergence of the beam, the source to consider is actually a virtual line source still inside the resonator, with a dimension of only a fraction of the resonator thickness.

X-ray Diffraction Imaging

With crystalline material an extension to micro-diffraction is quite straightforward and is presented in figure 1. The diffracting properties of the sample can be derived with high magnification in the vertical direction, if the horizontally collimated beam will be diffracted from the sample in the horizontal plane. We developed this idea in order to obtain 100 nm spatial resolution for strain in this micro-diffraction set-up. If a Si crystal is oxidised in narrow stripes of submicron width, several changes in its crystalline structure will occur. Expansion or compression of the crystal lattice under the stripe introduced by the strain in the interface may occur. Such a compression or expansion will extend over a limited depth. The strained volumes below the oxide stripes will then diffract at slightly different Bragg angles, according to their lattice spacing. As microelectronics devices are already produced routinely with about 180 nm feature width, strain analysis need to be made with better resolution for quality control. Such a resolution is not possible anymore with visible light and was not yet possible with focused x-ray beams. It is possible with electron beams, but their limited probing depth of the order of 10 nm does not allow to measure any buried interfaces beyond about 100 nm depth. If the sample is mounted as depicted in figure 1 with the stripes in the horizontal direction, a variation of the lattice parameter in the perpendicular direction, i.e. over the width of a stripe, can be measured. Scanning the rocking curve of the sample each pixel row at the detector receives the signal always from the same small area in the vertical direction at the sample. Local lattice plane expansions or contractions can then be detected by analysing the diffraction properties row by row. E.g. for an expanded lattice the Bragg condition will be found at more grazing angle compared to the unstrained crystal. With 10 micron pixel size of our detector already at 100 magnification each pixel row receives the information from a vertical area at the sample of only 100 nm width, corresponding to our source size.

Strain Analysis Under Oxidised Stripes

The most detailed analysis of the diffraction properties can be made by transferring the measured data into reciprocal space (S_X, S_Z) (10). In this space the areas with lattice planes expanded in the diffraction plane will have the Bragg peak at smaller S_Z. Peak enlargement in S_X can indicate either a bending of the sample in the direction of the stripes or the existence of mosaicity in the investigated area.

For the described sample we investigated the Si (400) Bragg reflection at an angle of grazing incidence around $\omega \approx 20.57°$ with $\Delta\omega = 0.0005°$ and with scattering angle resolution of $\Delta\theta = 0.001°$ (10 micron pixels at 573 mm distance between detector an the sample). Exposure times were 60 sec per image and about 20 images were taken per scan at a magnification of about 90, permitting a spatial resolution along the stripe width of 110 nm. For reference purposes we also measured a perfect Si crystal under

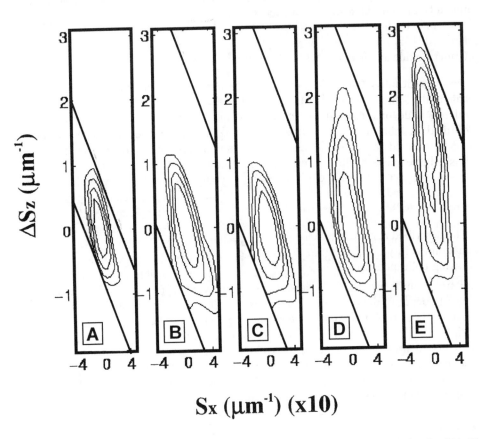

$$S_X \, (\mu m^{-1}) \, (x10)$$

FIGURE 2. Reciprocal space maps in a logarithmic intensity scale of the (004) reflection for 13 keV photon energy (λ @ 0.095 nm) characteristic for an untreated Si crystal (A), and for different positions at the patterned sample (see for reference figure 1): (B) is characteristic for the centre of the Si spacers, (C) for the Si spacer in the vicinity of the border with the wider oxidised stripe, (D) for the oxidised stripe very close to the same border and (E) for the centre of a sub-micron oxidised stripe. The 4 levels correspond to 1/16, 1/8, 1/4 and 1/2 of the maximum intensity in the centre. The shaded areas indicate the experimentally not accessed region in the S_X –S_Z space. The experimental resolution is 1/16 of the horizontal window size and 1/30 of the vertical window size, respectively.

identical conditions. The sample had several blocks with always 3 oxidised stripes of about 1 micron width, separated by 4 micron wide Si stripes. An oxidised stripe of at least 5.5 micron width separated the blocks. The reciprocal space maps for several points at the sample are presented in figure 2 and will now be discussed more in detail.

The maps for the treated sample do not show a significant broadening in S_x compared to the map for the Si crystal. Thus, the oxidation did not bend the crystal. However, depending on the position in the structure, we find peak broadening or even a displaced peak in the Sz direction. The development of only a small shoulder of the peak in figures 2B and 2C towards smaller Sz indicates that most of the lattice within the extinction depth remains unstrained. Instead, a compression in the direction perpendicular to the surface is clearly indicated in the maps for the D and E locations. The strain extends presumably over a certain depth leaving the bulk of the Si substrate unstrained. It is interesting to note that here the interface is buried under more than 300 nm of oxide and is thus not accessible with electron diffraction techniques.

The significantly different maps 2C and 2D are for areas, which are only 1 micron distant at the sample. However, differences are still seen between all 10 maps from the 1 micron wide oxide stripe, so that indeed the 110 nm spatial resolution can be confirmed. A more detailed data analysis especially in terms of strain values, which could be derived from these maps, is underway at the time of writing and will be published elsewhere (11).

ACKNOWLEDGEMENT

The described results were obtained at beamline ID13 at the European Synchrotron Radiation Facility ESRF in Grenoble (France). We would like to thank very much the following colleagues from ESRF for their help and for their input: Professor C. Kunz, C. Riekel, A. Freund, O. Hignette, W. Ludwig and E. Boller. We gratefully acknowledge as well the contributions of G. Sandrin from the Sincrotrone.

REFERENCES

1. Cheng, P. C. et al, *X-Ray Microscopy III*, eds. A. Michette, G. Morrison and C. Buckley, Berlin: Springer-Verlag, 1992, pp. 184-189

2. Koch, A., Raven, C., and Snigirev, A., *J. Opt. Soc. Am.* **A15**, 1940-1951 (1998)

3. see contribution of E. Di Fabrizio and of others in these proceedings

4. Snigirev, A. A., Kohn, V. G., Snigireva, I.I., and Lengeler, B., Nature 384, 49-51 (1996)

5. Bilderback, D. H., Hoffman, S. A., and Thiel, D. J., Science 263, 201-203 (1994)

6. Jark, W., Di Fonzo, S., Lagomarsino, S., Cedola, A., Di Fabrizio, E., Brahm, A., and Riekel, C., *J. Appl. Phys.* **80**, 4831-4836 (1996), see also our other contribution in these proceedings

7. Feng, Y. P., Sinha, S. K., Fullerton, E. E., Grübel, G., Abernathy, D., Siddons, D. P., and Hastings, J. B., *Appl. Phys. Lett.* **67,** 3647-3649 (1995)

8. Spiller, E., and Segmüller A., *Appl. Phys. Lett.* **24,** 60-62 (1974)

9. Lagomarsino, S., Cedola, A., Cloetens, P., Di Fonzo, S., Jark, W., Soullie, G., and Riekel, C., *Appl. Phys. Lett.* **71,** 2557-2559 (1997)

10. Fewster, P. F., *in X-ray and Neutron Dynamical Diffraction: Theory and Applications*, eds. A. Authier, S. Lagomarsino and B. K. Tanner, New York: Plenum, 1996, pp. 269-288

11. Di Fonzo, S., Jark, W., Lagomarsino, S., Giannini, C., De Caro, L., Cedola, A., and Müller, M., *to be published*

An Interferometric Microimaging System For Probing Laser Plasmas With An X-Ray Laser

D. Joyeux, R. Mercier, D. Phalippou, M. Mullot,[a]
S. Hubert, P. Zeitoun, A. Carillon, A. Klisnick, G. Jamelot,[b]
E. Béchir, G. de Lacheze-Murel [c]

[a] *Laboratoire Charles Fabry - CNRS, Institut d'Optique - BP 147 - 91403 Orsay, France*
[b] *Laboratoire de Spectroscopie Atomique et Ionique, Université Paris XI, 91405 Orsay, France*
[c] *CEN Bruyères-le-Chatel, CEA/DRIF/DCRE/SDE, BP 12, 91680 Bruyères-le-Chatel, France*

Abstract. The probing of dense (10^{21}-10^{22} electrons/cm^3) laser plasmas produced by high energy pulsed lasers usually makes use of X-ray emission imaging, or, more recently, of microradiography, i. e. of *absorption* imaging. However, it is well known that probing the *index of refraction* would bring much valuable informations. As far as index is concerned, interferometry is the choice method. However, the plasma context makes necessary to use for probing a very large brilliance, single pulse source, delivering energetic radiation, such as a soft X-ray laser. We report the design of a dedicated *interferometric microimaging system*, based on a Fresnel bimirror interferometer, associated with an off-axis spherical mirror, aspherized for diffraction limited (DL) imaging. In the first realization, the ideal mirror shape was approximated by a torus, making the system not diffraction limited. In a second step (currently in progress), a nearly perfect system will be realized, for diffraction limited resolution. The design and the realization techniques will be discussed and the first tests of the toroïdal system will be presented.

THE DESIGN PRINCIPLES

As the density of plasmas is linearly related to their index of refraction, interferometry is known to be an important tool for plasma studies, together with absorption and emission imaging. However, in the context of dense plasmas, such as those produced by high energy IR lasers, probing requires an energetic radiation, to avoid excessive absorption of the probing beam. Other constraints are the very short life time of these hot plasmas, and the fact that their self emission must not hide the signal from the probing beam. This makes necessary to use a X-ray laser as the probing source, for its very large brilliance, short pulse length, and good monochromaticity (which is also an important requirement for interferometry).

In this context, the starting idea is to associate an imaging system and an interferometer, both fed with the 21.2 nm X-ray laser beam operated by the LSAI group at Ecole Polytechnique (LULI, Palaiseau).[1] Although amplitude division interferometers were already demonstrated for plasma studies by Da Silva et al,[2] we choose to implement a wavefront division scheme, and work with a Fresnel bimirror interferometer, which does not require any beam splitter to work. The main requirement of such interferometers is spatial coherence, and the fact that only one

CP507, *X-Ray Microscopy: Proceedings of the Sixth International Conference,*
edited by W. Meyer-Ilse, T. Warwick, and D. Attwood
© 2000 American Institute of Physics 1-56396-926-2/00/$17.00

coherent étendue (1-dimensional) can be used. This requirement is in fact easily fulfilled by the X-ray laser, again due to its very large brilliance.[3]

As for the imaging system, several constraints must be taken into account. First, due to the debris which are projected by the tested plasma, a large enough working distance is required, thus forbidding the use of transmitting Fresnel zone plates as the imaging element. We choose to use a spherical mirror, working ~50 cm downstream from the plasma. Second, as the desired resolution is in the micrometer range, the system aperture is an issue: we choose a 8 mm (diameter) pupil, which provides a DL resolution of 1.6 µm at 21.2 nm. Finally, such a small object pixel makes necessary to introduce a large imaging magnification, as the intended detector is a CCD camera with 25 µm pixels. In summary, the imaging system is based on a spherical mirror with ~50 cm object distance, working with a large magnification (>15), for a diffraction limited resolution.

As is well known the only stigmatic conjugate points of a sphere is the center of curvature with itself. As this cannot be used for obvious reasons, it is necessary to design the system for off-axis imaging and for nearly focus-infinite conjugation. Even with a reduced pupil and aperture, this makes the image highly aberrant, due to the small wavelength, and it is necessary to aspherize the imaging system.

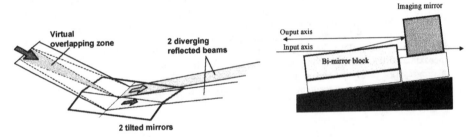

FIGURE 1. Left: the bimirror arrangement produces a virtual interferogram in the object plane. Right: association of the bimirror interferometer and the imaging mirror (side view).

Finally the interferometer is implemented as close as possible to the imaging mirror (system pupil). Its role is to separate the incident laser beam in two diverging halves, to get a *virtual* interferogram in the object (plasma) plane (figure 1). It consists in two grazing incidence (6 deg.) plane mirrors. The imaging mirror images this virtual interferogram in the magnified image plane. It is corrected for working at 6 deg. from normal incidence, in such a way that the optical axis towards image runs parallel to the optical axis from object. This gives the whole system a (vertical) plane of symmetry, for easier fabrication and assembly.

REALIZATION OF THE OPTICAL SYSTEM

The fabrication of the whole optical system consists of three parts, namely the interferometer, the imaging mirror and their assembly in a single solid optical block. All three tasks were performed by the Institut d'Optique's optical workshop.

The realization of the interferometer was made according to the principles already reported e.g. in ref. 4, except that the bimirror is illuminated along the common edge as shown in figure 1, instead of perpendicularly to it (the common way). Each mirror consists in a silica block, with a reflecting surface size of 80 mm along the beam by 20 mm transverse, and 20 mm thick. Flatness and low roughness surface is guaranteed up to the mirror edges. The bimirror tilt angle is 5.22 mrad (or 18 arcmin), and therefore (for 6 degree grazing incidence), the desired beam deviation is 1.09 mrad. Note the sign of the mirror tilt, which makes the beam diverge downstreams, thus producing the virtual overlap of the deviated half beams in the object plane, as already explained.

To be diffraction limited, the imaging mirror must meet a very severe specification in shape accuracy, i.e. $\lambda/32$ rms from the ideal shape. This is 0.66 nm rms, at $\lambda=21.2$ nm. The fabrication is made in two steps. First, a large sphere (100 mm aperture, 999.9 mm curvature) is polished for the highest possible accuracy and good roughness. This sphere is then tested in a high sensitivity phase shift interferometer, to select the best 30x30 mm² area, and this best zone is sawed to give a 30*30 mm² part. Note that sawing must be achieved very carefully in order not to distort the spherical shape. Finally, for aberration elimination, the central 8 mm disk of the spherical face is corrected by ion erosion, using an argon ion beam shaped by a suitable mask. More details on this process are available in reference 5.

Finally both elements are assembled to give a single optical block. This assembly process must fulfill the relative positional and angular tolerances of its parts, which are scaled by the corrected angular field size. The latter is given by the corrected image field size (3 mm), and the object distance, (close to 55 cm).

FIRST RESULTS

The first tests were performed with a torus approximation of the ideal imaging mirror. The expected resolution is therefore about 4 µm instead of 1.6 µm (DL). The general arrangement is depicted by the artist's view of figure 2.

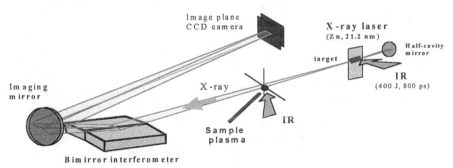

FIGURE 2. An artist's view of the general arrangement. Distances are: from X-ray source to sample plasma: 2.7 m; from sample plasma to imaging mirror: 55 cm; from imaging mirror to image: about 4.7 m.

Here the source is the Zn/Ne-like X-ray laser at 21.2 nm. The source is placed 2.7 m from the object plane, and the imaging magnification is 8.6. From the set-up geometrical parameters, it can be found that the fringe spacing computed in the object plane is 16.1 µm. The detector is a cooled, back illuminated, thinned CCD, with 25 µm pixels (thanks to prof. Hiroyuki Daido, University of Osaka, Japan).

The first results shown on figure 3 are an image of a grid, and an image-plane interferogram. Both the expected image resolution and the geometrical parameters correspond to the values expected at 21.2 nm.

A better, quasi-perfect correction is currently being realized, for DL imaging at 13.9 nm, in a future series of experiments.

 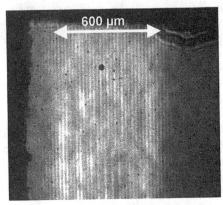

FIGURE 3. Left: image of a grid (no interference). Object period is 25 µm, and bar width is 7 µm. Right: interferogram in the image plane, with no object in the interference field. 600 µm is the corresponding object-plane size. (Printed right and left images have different magnification).

REFERENCES

1. Rus B., Carillon A., Dhez P., Jaeglé P., Jamelot G., Klisnick A., Nantel M., and. Zeitoun P., *Phys. Rev. A*, **55**, 3858-73 (1997).

2. Da. Silva L.B., Barbee T.W., Cauble R., Celliers P., Ciarlo D., Libby S., London R.A., Matthews D.L., Mrowka S., Moreno J.C., Ress D., Trebes J.E., Wan A.S.. and Weber F., *Phys. Rev. Letters*, **74**, 3991-94 (1995)

3. Albert F., Joyeux D., Jaeglé P., Carillon A., Chauvineau J. P., Jamelot G., Klisnick A., Lagron J. C. Ros D., and Phalippou. D., *Opt. Comm.*, **142**, 184-188 (1997)

4. Joyeux D., and Polack F., "Carbon index measurement near K edge, by interferometry, with optoelectronic detection" in *X-Ray Microscopy and Spectromicroscopy*, edited by J. Thieme, G. Schmahl, D. Rudolph, and E. Umbach, Springer, Berlin, 1998, pp II-103 - II-107

5. Mercier R., Mullot M., Lamare M., Tissot G.,"High precision cylindrical and quasi cylindrical aspherization of small surfaces by ion beam figuring", in *Optical Fabrication and testing*, SPIE Proceedings, **3739**, pp 155-160.

State of Art Micro-CT

A.Sassov

SkyScan, Aartselaar, Belgium

Abstract. By using high technology an X-ray microtomograph (micro-CT) was created as a simply usable, desktop instrument. The micro-CT is a laboratory system for non-destructive three-dimensional microscopy, giving true spatial resolution over a million times more detailed than the medical CT-scanners.

INTRODUCTION

In the last years one can find a strong reorientation of most microscopical methods to study objects in natural (or adjustable) conditions without preparation. Microscopical visualization without vacuum and coating allows maintaining the natural specimen structure as well as examining its behavior under external influences (loading, chemical reactions, interaction with other solids, liquids, gases etc.)

Another important issue for modern microscopy is the three-dimensional information. Most existing microscopes can visualize either the object surface or a transmission image through a thin section. That means the three-dimensional internal object structure can only be investigated destructively. Even with the most delicate preparation or cutting methods the specimen structure can change dramatically. For living or exceptional objects any cutting is not even possible.

One more significant aspect of modern microscopy is the quantitative interpretation of the images in terms of the microstructure of the object. Although most microscopes include or can be combined with powerful image processing systems, the interpretation of the contrast is still the main problem. On the other hand, reliable micromorphological information could be easily obtained from a set of thin flat cross sections which reveal only density information, from which case accurate two- and three-dimensional numerical parameters of the internal microstructure could be calculated.

Considering existing microscopical techniques, one can find that non-destructive information from the internal structure of an object in natural conditions can be obtained by transmission X-ray microscopy. Combination of X-ray transmission technique with tomographical reconstruction allows getting three-dimensional information about the internal microstructure [1-3]. In this case any internal area can be reconstructed as a set of flat cross sections which can be used to analyze the two- and three-dimensional morphological parameters [4]. For X-ray methods the contrast in the images is a mixed combination of density and compositional information. In some cases the compositional information can be separated from the density information [5]. Recently there has been a significant improvement in the

CP507, *X-Ray Microscopy: Proceedings of the Sixth International Conference,*
edited by W. Meyer-Ilse, T. Warwick, and D. Attwood
© 2000 American Institute of Physics 1-56396-926-2/00/$17.00

development of X-ray microscopes using synchrotron sources. However, these facilities are rather complicated and expensive and are not accessible for most researchers. On the other hand, the last few years have shown also a steady improvement in X-ray source technology so that now inexpensive compact sealed X-ray microfocus tubes can be produced with a very long lifetime. Because these sources emit polychromatic radiation one cannot use X-ray lenses for optical magnification. However, since the source spot size is small one can project the object over a large distance to the detector so as to obtain a geometrical magnification. In that case spatial resolution is limited by the X-ray spot size. At this moment, the attainable spot size in of the order of 5-10 micrometer but with the steady technological improvement one can expect submicron X-ray sources in the coming years.

PRINCIPLE OF OPERATION

Tomography is an established technique in the medical world. CT-scanners are used in many hospitals and a very valuable instrument in diagnosis. Microtomography or Micro-CT is a combination of tomographical algorithms and x-ray microscopy. Block-diagram of x-ray micro-CT is shown in the Fig.1.

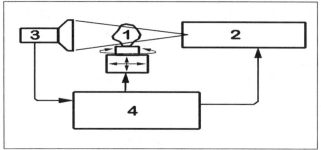

FIGURE 1. Block-diagram of micro-CT.

The object (1) is illuminated by an x-ray tube (2) with a spot size in a micrometer range. An x-ray camera (3) is then used to captures high-resolution shadow images. The object is rotated over at least 180 degrees in a number of discrete steps, often several hundreds, depending on the spatial resolution needed. At each position the shadow image is captured and stored by computer (4).

After the acquisition, with the information available in the shadow images, tomographical algorithms reconstruct the internal microstructure. This reconstruction can be done as a set of cross-sections or, with more powerful image processing, full 3D internal object can be visualized.

Obviously, for any tomographical reconstruction, the full object has to be inside the field of view and all shadow images have to contain all information on the reconstructed object parts. Naturally the object investigated has to transmit the x-rays to some extent.

The resolution is determined on one hand by the spot size of the x-ray source - the main limiting factor in the spatial resolution – and on the other by the division of the reconstructed image, for which the acquisition system is the most important factor. With the camera, there are two important elements that influence the reconstruction quality: the number of pixels in the image and the dynamic range. Higher dynamic ranges will give a better density resolution. Two materials with almost the same absorption for x-rays are easier to distinguish in an image captured with a wider dynamic range acquisition system. If a high absorbing part of the object and a low absorbing part are close together, it will be easier to resolve both in the same image.

DESCRIPTION OF THE SYSTEMS

A desktop X-ray micro-CT is available now with two instruments for the general-purpose non-destructive 3D-microscopy. First of them is a high-resolution microtomograph SkyScan-1072 with the best available specifications (Fig.2).

FIGURE 2. High-resolution desktop micro-CT.

It consist of a microfocus sealed X-ray source 20-80kV/100uA with 6-8um spot size and expected lifetime >10000hours, a precision object manipulator with two translations and one rotation, an X-ray CCD-detector consisting of special scintillator with fiber-optics coupling to 1024x1024 / 12bit CCD-camera. The computer processing and system control are done with internal Dual Pentium III 500MHz / 768Mb RAM / 16Gb HDD / CD-writer operated under Windows NT. The X-ray magnification range is between 10 – 120. For microtomographical reconstruction transmission X-ray images are acquired from up to 400 rotation views through 180 degrees (or 800 views through 360 degrees) of rotation. In order to study high-density materials the system can be supplied with a 130 KeV / 300 uA sealed tube with a focus spot size which can be selected at 10 um or 40 um. X-ray camera with 2048x1024 pixels / 14bit is an option for this system.

Another instruments for micro-CT is a compact, low-cost, medium resolution scanner SkyScan-1074 (Fig.3).

FIGURE 3. Portable micro-CT scanner.

This small scanner can be connected to any external IBM-compatible desktop or portable computer through supplied PCI or PCMCIA control card. It consists of a sealed X-ray source 20-40kV/1mA and can reach 3D spatial resolution <40um for the objects up to 30 mm in size.

A software package for both instruments has been developed for system control and microtomographical reconstruction. The microtomographical reconstruction algorithm is based on the filtered back-projection procedure for fan-beam geometry with specific noise-reduction corrections. The software optimized for fastest 32-bit execution and support multiprocessor architecture. The reconstruction time for one cross section of 1024x1024 float-point pixels from 200 projections takes near 20 sec in Pentium III 500MHz. The software package includes also image processing and analysis procedures, stereo-visualization, and realistic three-dimensional visualization with possibilities in software to rotate and to cut the object's image on the screen. The Windows-NT facilities allow for network connections.

APPLICATION EXAMPLES

The X-ray microtomography has been used for a variety of applications. Most spectacular applications can be found in those areas, where three-dimensional internal structures can only be visualized non-destructively and/or in normal environmental conditions.

A first example of application of microtomography is taken from life sciences. Here X-ray microscopy and microtomography allows to reconstruct the internal three-dimensional microstructure without any preparation and sometimes even of living objects. Fig.4 shows a 3D-reconstruction of the internal microstructure of an eye of a small fly (in natural conditions), created from the micro-CT results with high-resolution SkyScan-1072 system.

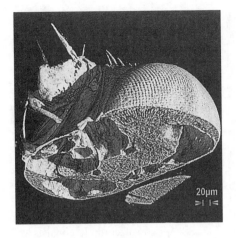

FIGURE 4. 3D reconstruction of the internal microstructure of a fly's eye.

Another important application area is the non-destructive defectoscopy of electronic components. Fig.5 shows the 3-dimensional results of reconstruction and cross section from the electrolytic capacitor. By the same way the internal microstructure of the most electronic components in plastic and thin metal cases can be visualized. Even small electronic assemblies like hybrid ICs, magnetic heads, microphones and ABS-sensors can be tested by microtomograpical methods.

FIGURE 5.Cross section and 3D-reconstruction of an electrolytic capacitor.

Another application areas of microtomography are biology, agriculture and food industry. Fig.6 shows a cross section in the top of full 3D-reconstruction from a roasted coffee bean obtained in normal conditions nondestructively. For this object the 3D information about the internal microstructure from micro-CT was used to correct roasting process.

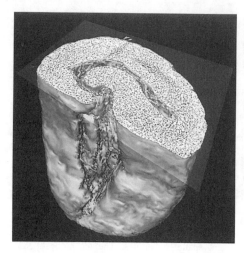

FIGURE 6. Cross-section and 3D-reconstruction of a coffee bean.

All shown examples obtained without any physical object cut of any other preparation.

CONCLUSIONS

The state of the art in X-ray technology and computer sciences allowed developing an inexpensive compact instrument for three-dimensional non-destructive microscopy. The technique does not require any specimen preparation. During the investigation the object can stay in environmental or special conditions. Even living objects can be investigated. At present the resolution is somewhat better than 8 micrometer and can be used for objects of centimeter size. Further progress in X-ray sources and cameras allows expecting improvements in spatial resolution and avoiding limitations in object size.

REFERENCES

1. A. Sasov. *J.of Microscopy*, v.147, 1987, 169-192.

2. P. Anderson, et al. *Micr.&Anal.*, March 1994, 35-37.

3. International symposium on computerized tomograpy for industrial applications, Berlin 1994.

4. T.Hildebrand, P.Ruegsegger. *CMBBE*, v.1, 1997, 15-23.

5. P. Trebbia. In: Proc 5th Eur.workshop on modern developments and applications in microbeam analysis, Torquay UK, 1997,149-173

Phase-Contrast Microtomography with Polychromatic Sealed Source

A.Sassov, D.Van Dyck

SkyScan, Aartselaar, Belgium

Abstract. Conventional X-ray microradiography and microtomography are based on X-ray attenuation inside an object. For light objects (in the terms of X-ray absorption) much better way would be to use phase contrast, rather than attenuation contrast. Recently it has been shown that one can obtain phase by using a polychromatic source provided the focal spot size and detector resolution are small enough to maintain sufficient spatial coherence. The technique opens perspectives for high-resolution micro-CT for the objects with low X-ray attenuation, such as diamonds, biomedical objects, etc.

INTRODUCTION

Non-destructive information from the internal structure of different materials in natural or adjustable conditions can be obtained by transmission X-ray microscopy. Combination of X-ray microscopy technique with tomographical reconstruction allows getting three-dimensional information about the internal microstructure nondestructively. The spatial resolution and detectability in the micron range of sizes can be dramatically improved by using of phase contrast.

THE EQUIPMENT

A desktop X-ray microscope-microtomograph [1] has been used for nondestructive 3-dimensional reconstruction of the internal microstructure of light materials with phase contrast enhancement. Block-diagram of equipment in shown on Fig.1.

It consist of a microfocus sealed X-ray source 20-80kV / 100uA with <8um spot size, a precision object manipulator for object rotation and movement inside the conical x-ray beam, an X-ray CCD-camera with field of view 25x25mm and an internal Dual Pentium III computer for system control and 3-dimensional reconstruction. System operated under Windows NT with software as a multiprocessor 32-bit application. Software package include possibilities for X-ray radiography, stereomicroscopy, three-dimensional microtomographical reconstruction, image correction, analysis and realistic 3D-visualization. The X-ray magnification is continuously variable in the range between 10 and 120. Reconstruction time is near 20 sec per cross section. A single cross section as well as full 3D internal structure can be reconstructed after one complete object scan. During investigation the specimen displaced in natural conditions with no needs for any preparation or coating.

CP507, *X-Ray Microscopy: Proceedings of the Sixth International Conference,*
edited by W. Meyer-Ilse, T. Warwick, and D. Attwood
© 2000 American Institute of Physics 1-56396-926-2/00/$17.00

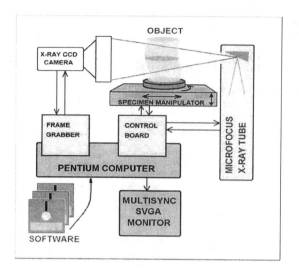

FIGURE 1. Block-diagram
of the X-ray
microtomograph

Powerful image processing and analysis possibilities integrated into the software package allows calculating the numerical characteristics of object's internal microstructure non-destructively. Visualization procedures contain of cross section by cross section displaying and realistic visualization of complete 3D-object onto the screen with possibilities for rotation, movements, zoom, recalculation of new cross section in any alternative direction and flight around or through the object's microstructure.

PHASE CONTRAST

Conventional X-ray radiography and tomography are based on X-ray attenuation inside an object. However, for light materials and hard X-rays, the attenuation can be small that it puts strong limits on the attainable resolution. Fortunately the attenuation is not only one physical phenomenon for X-ray transmission imaging. Another kind of information can be obtained by phase contrast. The phase contrast, rather than attenuation contrast, since the phase change, due to changes in index of refraction, can be up to 1000 times larger than the change in amplitude for visualization of light objects. In the same time the phase contrast have another geometrical behavior which allows using this kind of information for improvement in spatial resolution. However, phase contrast techniques require the disposal of monochromatic X-ray sources, such as synchrotrons, combined with special optics, such as double crystal monochromators and interferometers [2].

Recently it has been shown [3] that one can also obtain phase contrast by using a polychromatic X-ray source provided the source size and detector resolution are small enough to maintain sufficient spatial coherence. This opens perspectives for using phase contrast information in a microfocus tomographic system. Because the phase and absorption contrast presents in microfocus X-ray systems simultaneously, the optimization of geometry create the way for performance improvements by right combination of both of them.

Absorption contrast in any real system contains some unsharpeness as a result of finite spot size in the source. By another hand the phase contrast can be described as a small-angle reflection from the object surface which improve the sharpness of the image. Specific x-ray geometry has been found in which the blur produced by spot size in the source compensated by the phase contrast signal. It allows inside the system with 8 microns spot in the source visualize the objects with several microns in size. Improvement in visualization and reconstruction for the objects of 2-5 microns in size very important for the investigation of light materials, such as plastic foams, composites, diamonds, paper and others.

APPLICATION EXAMPLE

One of application examples for phase contrast improvements in micro-CT performances is shown in Fig.2. Object is a small piece of paper.

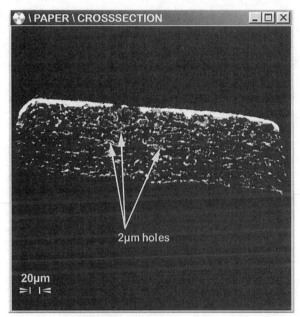

FIGURE 2. Reconstructed cross section through the paper sample.

In the micro-CT system with spot size of 8 microns in the source the object details up to 2 microns are clearly visible in the reconstructed cross section due to improvements by phase contrast.

CONCLUSIONS

Desk top microtomography system can be applied for nondestructive investigation of internal microstructure of light materials with small X-ray adsorption. Phase contrast improves resolution for details in the micron range.

REFERENCES

1. A.Sasov, D.Van Dyck. Microscopy and Analysis (European edition), March 1998, pp. 21-23.

2. V.N Ingal, E.A.Beliaevskaya, (1995), J. Phys. D: Applied Physics 28, 2314.

3. S.W.Wilkins, T.E.Gureyev, D.Gao, A.Pogamy and A.W.Stevenson. Nature, 1996, 384, 335.

Contrast Enhancement and Three-Dimensional Computed Tomography in Projection X-ray Microscopy

Hideyuki Yoshimura[1], Daisuke Shoutsu[1], Chiaki Kuzuryu[1], Chikara Miyata[1], Ryuji Sano[1], Takashi Obi[2], Nagaaki Ohyama[2]

[1]*Department of Physics, Meiji University*
Higashimita 1-1-1, Tama-ku, Kawasaki, 214-8571, Japan
[2]*Faculty of Engineering ,Tokyo Institute of Technology,*
4259 Nagatsuta, Midori-ku, Yokohama, 226-8503

Abstract. We have developed a projection X-ray microscope by modifying a scanning electron microscope (SEM), thus simplifying optical alignment and focusing process. When the object was much larger than the resolution of this microscope (0.1μm), the edge of the image was enhanced due to Fresnel diffraction of partially coherent light. Thus the projection images such as thin filaments in biological objects showed enough contrast even their absorption contrast was very low. Because there are no optical elements for X-ray (*e.g.* zone plate), this microscope has long focal depth. This means we can obtain the projection image of thick sample at one time without preparing thin sections of the sample. Taking this advantage we have applied this microscope for three-dimensional structure analysis by means of computed tomography.

INTRODUCTION

The projection X-ray microscope makes use of a very small X-ray light source excited by the focused electron beam of a scanning electron microscope (SEM) on a thin (0.1-1μm) metal target. If an object is placed just below the metal target, the expanding X-rays enlarge the shadow of the object with infinite focal depth. This type of X-ray microscope was developed as a shadow X-ray microscope in the 1950s (1) and was advanced further by Newberry (2), Yada *et al.* (3), and Thomas *et al.* (4). The long focal depth is advantageous for analysis of bulky samples (more than 5mm). An additional benefit is that images obtained by this technique show good contrast even when the contrast generated by absorption at a corresponding wavelength is very small. Nevertheless, this method was not developed into a major X-ray microscope instrument because of its limitations in the resolution and relatively low intensity of the X-ray light source. These problems have now, however, been overcome by recent progress in CCD camera technology for X-ray (5) and progress in the electron optics of SEMs, demanding the projection X-ray microscope to be re-considered as an analytical tool. The improvement in X-ray detection is beneficial when performing a computed tomographic study that requires many projection images to reconstruct a three-dimensional structure.

CP507, *X-Ray Microscopy: Proceedings of the Sixth International Conference,*
edited by W. Meyer-Ilse, T. Warwick, and D. Attwood
© 2000 American Institute of Physics 1-56396-926-2/00/$17.00

Here we report details of the edge enhancement effect in projection X-ray microscopy and of our experiment using the technique for performing computed tomography (CT).

X-RAY SPECTRA

The X-ray from the metal target was composed of continuous and characteristic X-rays. The spectrum depends on the acceleration voltage and the metal target material. For conventional purposes, we generally used a gold target with an acceleration voltage of 20 kV. Under these conditions, the characteristic X-rays of L_α, L_β, and M_α were observed with continuous X-rays (Fig.1). X-rays emitted from the metal target pass into the sample chamber, and are absorbed by the gas contained therein. The absorption effect was significant when the wavelength was longer than 0.2 nm in air (at 1 atm), and longer than 1nm in helium (at 1 atm).

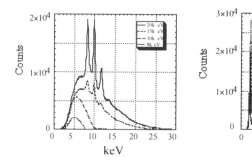

FIGURE 1. The spectra of the X-rays (left in air, right in helium). The gold target (thickness 0.3μm) was irradiated with electron beams of 8, 10, 15 and 20 kV acceleration voltage. The gold target emitted characteristic X-rays of L_α (9.71 keV, 0.128 nm), L_β (11.5 keV, 0.108 nm) and M_α (2.15 keV, 0.584 nm). The peak at 8.1 keV (0.15 nm) is probably the L_α of the copper that was used to frame the gold target. The spectrum was measured using an Si-PIN photodiode detector (XR-100T, AMPTEK Inc.) placed 100 mm from the target metal.

EDGE ENHANCEMENT

We generally used a gold target with an acceleration voltage of 20 kV as the resultant light intensity was relatively high. Under these conditions, the typical wavelength was around 0.13 nm, a range in which the absorption contrast of biological samples is low. However, the contrast was improved by edge enhancement effect when the distance between the light source and the object was much larger than the size of the light source. This effect can be explained by Fresnel diffraction of partially coherent light (6). Figure 2 shows the images of a section of aluminum foil (3 μm thickness) that was placed at 0.5 mm (A) and 5 mm (C) from the light source, and profiles of each image are shown in (B) and (D). Diffracted light at the edge is known to make interference fringes on the screen when the incident light is coherent. In this microscope, the light source was not monochromatic and it had finite size, so that neither coherency in time nor coherency in space was good. The interference fringes were weak and only the first fringe was observed in (D). When the distance between the light source and the object was close, the fringe almost vanished (B) because spatial coherency worsened.

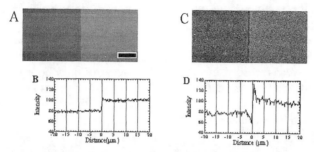

FIGURE 2. X-ray micrographs of the edge of aluminum foil against air (A, C) and profiles of them (B, D). A 3 μm aluminum foil was place at the left part of (A) and (C). Fringes at the edge were observed in (C) but only absorption contrast was observed in (A). The distance between the light source and the object was 0.5 mm (A) and 5 mm (C). The scale bar in (A) is 10 μm.

When the object was much larger than the resolution of this microscope (0.1μm), the fringe due to Fresnel diffraction helped to increase the contrast by edge enhancement. X-ray micrographs of the thin body hairs of an ant are shown in Figure 3. In (A) the sample was placed at 1.3 mm from the light source, whereas in (B) the sample was only 0.14 mm from the light source. In (B), the contrast of the resulting image was not good and only one hair can be barely seen in (B). In (A), however, two hairs are observed clearly. The smaller distance between the light source and the object in (B), meant that the spatial coherence was not sufficient to generate interference fringes. As the transmittance of a 2 μm section of carbon would be about 99% at 0.15 nm wavelength, it would be difficult to show clear contrast against background by absorption contrast alone.

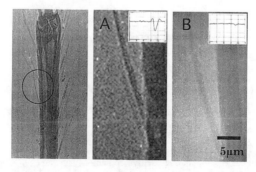

FIGURE 3. X-ray micrographs of a sample of thin body hairs from an ant. The area in the circle of the left image was magnified in (A). The distance between the light source and the object was 1.3mm (A). The experimental conditions in (B) were the same as (A) except the distance to the light source was 0.14 mm. The distance between the light source and negative film for image recording was fixed as 50 mm, thus original magnification of A and B were 38 and 360, respectively. Profiles of the hair are shown in the insets of (A) and (B).

527

COMPUTED TOMOGRAPHY

The projection X-ray microscope has a long focal depth, more than 5 mm, because there are no optical elements for X-rays (*e.g.* zone plate). This means we can obtain the projection image of a bulky sample with a single measurement without preparing thin sections of the sample. Here, we use this property to use this microscope for three-dimensional structure analysis using computed tomography.

Omiscus Porcellio was prepared by the critical point drying method (7) and fixed on thin adhesive tape. Subsequently, projection images of *Omiscus Porcellio* in all directions (360°) were recorded every 5° (72 images) using a cooled CCD camera (HAMAMATSU C4880) which detects X-rays directly. The three-dimensional image was reconstructed using the corn-beam algorithm (8). Figure 4 shows the cross sections (A~C) and the surface rendered image (D).

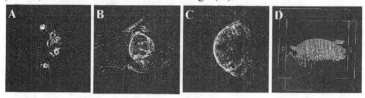

FIGURE 4. Reconstructed image of *Omiscus Porcellio* treated with critical point drying. (A-B-C) are the cross sections of the reconstructed image at head (A) to middle body (C). (D) represents the surface rendered image in three-dimensions.

ACKNOWLEDGEMENTS

We are grateful to Drs. K. Yada, K. Shinohara, T. Honda and A. Ito for their helpful discussion. This work was supported in part by a grant from the Japanese Ministry for Education and Meiji University Research Project Grant.

REFERENCES

1. Nixon, W. C., *Proc. Roy. Soc.*, **A232**, 475-485 (1955).

2. Newberry, S. P., *X-Ray Microscopy II*, Berlin; Springer-Verlag, 1988, pp.306-309.

3. Yada, K. and Takahashi, S., *X-Ray Microscopy II*, Berlin; Springer-Verlag, 1988, pp.323-326.

4. Thomas, X., Cazaux, J., Erre, D., Mouze. D., and Collard, P., *X-Ray Microscopy III*, Berlin; Springer-Verlag, 1992, pp.190-194.

5. Yoshimura, H. , Kumagai, S., Shoutsu, D., Sekiya, Y., and Mitsui, T., *Cell Vision*, **4**, 210-211 (1997).

6. Wilkins, S. W., Gureyev, T. E., Gao, D., Pogany, A., and Stevenson, A. W., *Nature*, **384**, 335-338 (1996).

7. Andersen, C. A. , *Trans. N. Y. Acad. Sci.*, **13**, 130-134 (1951).

8. Feldkamp, L. A., Davis, L. C., and Kress, J. W., *J. Opt. Soc. Am. A*, **1**, 612-619 (1984).

Hard X-ray Microscopy And Tomography At The ALS: Experiments And Plans

W. Yun[1], M. R. Howells[1], J. Feng[1], R. Celestree[1], C. –H. Chang[2], A. A. MacDowell[1], H. A. Padmore[1], J. Spence[3]

[1]Advanced Light Source, Lawrence Berkeley National Lab
[3]Pohang University of Science and Technology, Pohang 790-784, Republic of Korea
[3]Department of Physics, Arizona State University

Abstract. A hard x-ray imaging microscope with a spatial resolution of 0.12 μm was developed and tested using synchrotron radiation. The microscope can be operated in either dark-field or bright-field mode. Phase contrast is employed in the dark-field mode while absorption contrast is used in the bright-field mode. The objective of the x-ray microscope is a phase zone plate fabricated using a x-ray lithographic technique. We describe the hardware of the microscope and present the results obtained from the microscope. Its potential applications will also be discussed.

1. INTRODUCTION

An important advantage of an imaging microscope is the ability to image many resolution elements at the same time. This parallel 'processing' capability is not only important for many interesting applications, but makes the best use of the incoherent light source, such as a bending magnet source of synchrotron radiation. Imaging microscopes working in the soft x-ray spectral region (E<1 keV) have been developed by many groups using synchrotron as well as laboratory x-ray sources[1-3]. Spatial resolution of about 30 nm has been demonstrated and is still improving. These microscopes have been applied to a wide area of research, ranging from imaging biological cells to the study of dynamics of colloidal solutions. Phase contrast as well as absorption contrast has been employed to optimize image contrast and spatial resolution. Up to now, however, there are no similar dedicated imaging microscopes working in the multi-keV x-ray spectral region using a bending magnet source. We have initiated a program of developing an imaging microscope using 2-30 keV x-rays from a bending magnet source. Our goal is to develop a dedicated imaging microscope and tomography facility at the Advanced Light Source (ALS) with a spatial resolution better than 100 nm. Both phase contrast and absorption contrast will be utilized.

2. IMAGE CONTRAST

Both phase and absorption contrast can be used for x-ray imaging applications. However, until recently, only absorption contrast has been used. While absorption contrast is adequate for most imaging applications of medium to high Z materials at

CP507, *X-Ray Microscopy: Proceedings of the Sixth International Conference,*
edited by W. Meyer-Ilse, T. Warwick, and D. Attwood
© 2000 American Institute of Physics 1-56396-926-2/00/$17.00

multi-keV x-rays at relatively low resolution, it may become inadequate for imaging small features requiring high spatial resolution, especially for low Z materials. Phase contrast is generally larger than absorption contrast for nearly all materials in the hard x-ray spectral region, and thus it may provide adequate contrast for imaging smaller features, especially for low Z materials.

The relative figure of merit of x-ray phase contrast to amplitude contrast may be approximately given by the square of the ratio of the real (f_1) and imaginary (f_2) part of the atomic scattering factor. Because the real part of the atomic scattering factor is approximately equal to the number of free electrons in an atom while the imaginary part decreases with the cube of x-ray energy away from an absorption edge, the relative figure of merit can be significantly greater than unity. For example, the relative figure of merit is approximately equal to 600 for a sample consists of mainly carbon for 8 keV x-rays. The superiority of phase contrast compared to absorption contrast for low-Z objects was dramatically demonstrated in the work of Cloetens et al [4].

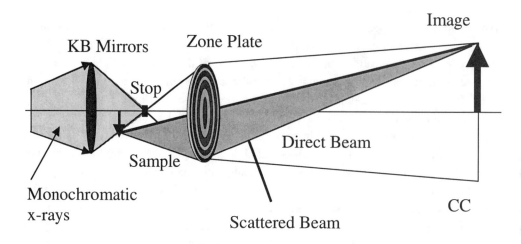

FIGURE 1. Schematic of the ALS microscope, showing operation in a dark-field mode. The bright-field mode is when the stop is removed from the beam. In phase contrast mode, the stop is replaced by a phase shifter.

3. THE ALS MICROSCOPE

The schematic of an imaging microscope is shown in Fig. 1. The microscope can be easily changed from a bright-field mode to a phase contrast mode or a dark-field mode. At this point, we have experimented with the dark-field and bright-field mode. The dark-field mode is essentially a special form of the phase contrast mode in which the non-perturbed part of the incident beam (zero order) is completely attenuated by the phase-shifter, which is only partially absorbing in the normal phase contrast mode. The image in the dark-field mode arises from the x-ray photons scattered out of the envelope of the incident beam which is then imaged by the zone plate. The image contrast is proportional to the scattering strength between two neighboring resolution elements, which is proportional to the square of the atomic scattering factor. Because of its low background, the dark field mode may be a preferred imaging technique for some special class of objects.

The experiment was conducted on the 7.3.3 beamline at the Advanced Light Source (ALS) of Lawrence Berkeley National Laboratory. The typical energy resolving power of the beamline is about 1500, which ensures that the chromatic aberration of the zone plate is smaller than its intrinsic resolution.

The key optical component in this setup is the zone plate and it was used as similar to the objective in a light microscope. The ultimate spatial resolution of this microscope is determined by the zone plate provided that the detector has adequate resolution. The zone plate was developed in a long-term collaboration of Argonne National Laboratory, University of Wisconsin at Madison, and Istituto di Electronica dello Stato Solid, Italy[5][6]. Its key parameters are given in table 1.

TABLE 1. Key Zone Plate Parameters

Parameter	Value
Diameter	145 μm
Outer most zone width	103 nm
Au thickness	900 nm
Focal length at 8 keV	10 cm
Zone profile	Square-wave (binary)

The theoretical spatial resolution of the phase zone plate is about 120 nm and an experimental resolution very close to this value was achieved. The magnified image of an object is recorded by a detector system consisting of a scintillation screen, a 20X microscope objective, and a CCD detector of 24-μm pixel size. The x-ray and optical magnification were 21 and 11 respectively and the combined total magnification is therefore 294. Each pixel of the CCD detector, therefore, represents a feature size of 103 nm at the object plane.

4. EXPERIMENTAL RESULTS

Imaging of many objects ranging from biological to materials objects has been made using either absorption or dark-field phase contrast, depending on the type of contrast of the particular object. High quality images of objects ranging from lichen, bone joints, metal grid, human hair, a shark tooth, and semiconductor integrated circuits have been obtained.

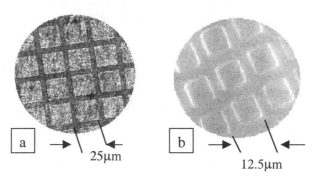

FIGURE 2. (a) Bright-field field image of a Cu 1000mesh/inch grid, and (b) dark-field image of a Cu 2000mesh/inch grid.

Fig. 2 shows images of Cu mesh grids obtained in bright-field (a) and dark-field mode (b). While the bright-field image (Fig. 2a) does represent a true image of the grid as image contrast is formed by the absorption of the 1000 mesh/inch Cu gird, the dark-field image only shows the boundary of the bars of the 2000 mesh/inch Cu grid. There is little difference in the image of the open area of the mesh grid and the centers of bars. Essentially, it may be considered as a derivative image of a regular phase or absorption image. This property is unique to dark-field imaging and may be particularly important to image features of sharp boundaries, such as cracks in materials. In fact, one of the key research areas of the microscope is the study of crack initiation and propagation in materials.

Fig. 3a shows the images of part of an integrated circuit chip supplied by Dr. Neogi of Intel Corp and in collaboration with Dr. Levin of NIST. The chip is made using the recent Cu technology for integrated devices. It consists of Cu interconnects at two layers connected by vias. All Si of the original integrated circuit (IC) chip was removed from the sample because it was originally prepared for low energy (~2 keV) imaging. A Si substrate of < 50 μm could have been left on the IC for the x-ray energy used in our experiment without affecting the exposure time significantly.

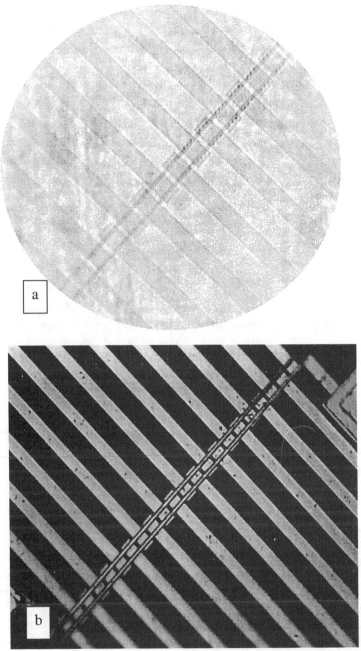

FIGURE 3. (a) a x-ray image of an integrated circuit showing images of vias connecting Cu wires in two different layers (obtained using the ALS microscope), and (b) an optical image of the same area. The vias in (a) are the small dots at the intersection between the thin and wide Cu wires.

Fig. 3 shows clearly the image of the vias whose dimension is about 0.4 μm. Note that the optical transmission image of the same IC (Fig. 3b) does not contain any information of the vias, indicating the unique capability of imaging optically opaque objects using x-rays. Preliminary analysis indicates that the spatial resolution of the x-ray image is about 120 nm, which is considerably better than obtainable in optical microscopes. It is also interesting to note that the edges around the Cu interconnects are markedly darker than the rest of the interconnect, indicating stronger absorption at the edges. This was resulting from 30-nm Ta liners used to prevent the diffusion of Cu into Si. The fact that the 30-nm Ta liners are clearly *detected* indicates that the feature can be detected by the microscope is much smaller than its spatial resolution.

5. The Future Plans

We have obtained sufficient funding for developing a dedicated high resolution tomography facility at the ALS based the imaging microscope described above. Phase contrast imaging will be implemented in the dedicated microscope. Potential scientific programs that will benefit from the capabilities include materials science, biology, and engineering such as IC chips.

ACKNOWLEDGEMENTS

This work was supported by the Director, Office of Energy Research, Office of Basic Energy Sciences, Materials Sciences Division of the U.S. Department of Energy, under Contract No. DE-AC03-76SF000098.

1.   Medenwaldt, R. and E. Uggerhoj, *Description of an x-ray microscope with 30 nm resolution.* Rev. Scien. Instrum., 1998. **69**: p. 2974.
2.   Meyer-Ilse, W., *et al. The High Resolution X-ray Microscope, XM-1.* in *VIth International Conference on X-ray Microscopy.* ca. Berkeley, California.
3.   Schneider, G., *et al. Visualization of 30 nm Structures in Frozen-Hydrated Biological Samples by Cryo Transmission X-ray Microscope.* in *VIth International Conference on X-ray Microscopy.* ca.
4.   Cloetens, P., *et al., Observation of microsctructure and damage in materials by phase sensitive radiography and tomography.* J. Appl. Phys., 1997. **81**: p. 5878.
5.   Yun, W., *et al., Development of Zone Plates with a Blazed Profile for Hard X-ray Applications.* Rev. Sci. Instrum., 1999. **70**: p. 3537.
6.   Yun, W., *et al., Nanometer Focusing of Hard X-rays by Phase Zone Plates.* Rev. Sci. Instrum., 1999. **70**: p. 2238.

X-ray microbeam with sputtered-sliced Fresnel zone plate at Spring-8 undulator beamline

Y. Suzuki,[1] M. Awaji,[1] Y. Kohmura,[1] A. Takeuchi,[1]

N. Kamijo,[1,2] S. Tamura,[1,3] and K. Handa,[4]

[1]SPring-8, Mikazuki, Hyogo 679-5198, Japan

[2]Kansai Medical University, Hirakata, Osaka, 573-1136, Japan

[3]Osaka National Research Institute, Ikeda, Osaka 563-8577, Japan

[4]Ritsumeikan University, Kusatsu, Siga 525-8577, Japan

Abstracts: Hard X-ray microbeam with sputtered-sliced Fresnel zone plate has been tested at SPring-8 undulator beam line. Focusing properties are evaluated in X-ray wavelength regions of 0.15-1.6 Å. The measured focal beam size is about 0.6 μm at an X-ray wavelength of 1.4 Å, and 0.7 μm at 0.45 Å.

INTRODUCTION

Many types of X-ray focusing devices are developed to generate micro-focus X-ray beam in hard X-ray regions, and sub-μm spot size has already been achieved. Among many types of optical devices, advantages of sputtered-sliced Fresnel zone plate (FZP) are narrow zone width and thickness of zone plate, ie., high aspect ratio of zone structure. Therefore, the sputtered-sliced FZP is considered to be useful for focusing high energy X-rays. In this paper, we describe the results of X-ray focusing test using sputtered-sliced Fresnel zone plates in X-ray wavelength regions of 0.15-1.6 Å.

EXPERIMENTAL SETUP

The experiment has been performed at the beamline 47XU in the SPring-8. The beamline 47XU is constructed for the purpose of research in X-ray optics and development of X-ray optical devices. The X-ray source is "in-vacuum" type permanent magnet planar undulator [1]. The periodic length is 32 mm, and the number of period is 139. The K-value of the undulator can be tuned within the K = 0.04-2.46 by changing the magnet gap from 50 mm to 8 mm. The first harmonics of the undulator radiation can be tuned within the energy range of 4.5-18.8 keV. Schematic diagram of the experimental setup is shown in Fig. 1. The undulator radiation is monochromatized through a "SPring-8 standard" Si 111 double-crystal monochromator. The energy range of the monochromator is 4-35 keV. The monochromator crystals are cooled by liquid nitrogen. Both the first crystal and the second crystal are kept at the same temperature to achieve constant exit beam position.

Sputtered-sliced FZPs [2-5] were used as the X-ray focusing device. The FZP consists of alternating multilayer zones constructed by magnetron sputtering. Fifty Al/Cu concentric multilayer structures are deposited onto a gold wire substrate of 47 μm

CP507, *X-Ray Microscopy: Proceedings of the Sixth International Conference,*
edited by W. Meyer-Ilse, T. Warwick, and D. Attwood
© 2000 American Institute of Physics 1-56396-926-2/00/$17.00

diameter. After the deposition, the wire sample is sliced normal to the wire axis onto a plate. The FZP is thinned up to about 20-40 μm by mechanical polishing. The gold core of the FZP works as a center beam stop for appropriate FZP thickness. We have tested two kind of FZPs. The width of the outermost zone is 0.25 μm and 0.15 μm, respectively. The diameter of the FZP is 80 μm for 0.25 μm outermost zone. Measured focal length is 158 mm at an X-ray wavelength of 1.4 Å for the FZP with outermost zone width of 0.25 μm. The measured focal length of the FZP with 0.15 μm outermost zone width is 234 mm at λ = 0.45 Å.

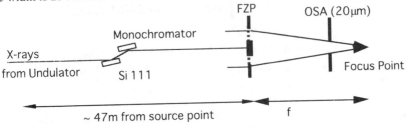

Figure 1. The optical system
FZP: Fresnel zone plate, OSA: order selecting aperture, f ~ 150 mm -700 mm,
Source size: ~ 40 μm (vertical) x 900 μm (horizontal).

Adjustment of tilting angle of the FZP to the optical axis is fairly important in the cases of thick zone plate because the aspect ratio of the finest zone is 100-300. The tilt angle are tuned while the profiles of diffracted beams and zero-th order beam are monitored by using high resolution X-ray image sensor (CCD camera coupled with phosphor screen and relay lens). The pixel size of the imgae sensor is 6 μm, and the spatial resolution is estimated to be about 10 μm. The focused beam profiles are measured by knife-edge scanning, and diffraction efficiency for the first order diffraction is also evaluated.

RESULTS AND DISCUSSION

The focused beam profiles measured by knife-edge scanning for the FZP with 0.25 μm outermost zone width are shown in Fig. 2. Thickness of the FZP is estimated to be about 20 μm. Edge-scanning was performed by using a gold wire (100 μm in diameter) as the "knife-edge" in transmission geometry. The full-width at half maximum of the focused beam is about 0.6 μm at an X-ray wavelength of 1.4 Å. This focused beam is generated by using pinhole X-ray source placed 9 m upstream from the FZP. Therefore, the focused beam profile is nearly symmetrical in vertical and horizontal direction. The magnification (M) is defined by the equation M = f/L, where f (= 158 mm) is the focal length of the FZP and L (=9 m) is the distance between the source point and the FZP. The diameter of the pinhole used as the pseudo-point source is estimated to be about 20 μm. When the pin-hole diameter is 20 μm, a focused spot size of 0.35 μm can be derived by geometrical optics. The diffraction limit of the FZP, $1.2 d_N$ (d_N: outermost zone width), is 0.3 μm for 0.25 μm outermost zone width. Therefore, the focused beam size of 0.6 μm is mainly determined by the geometrical optics and the diffraction limit of the FZP. Diffraction efficiency of the FZP was measured in the X-ray wavelength region of 0.9-1.6 Å. Measured efficiency and result of calculation are shown in Fig. 3. The Al/Cu FZP with thickness of 20 μm can operate as a phase-modulated FZP in these X-ray wavelength regions, and efficiency can be higher than 10% (theoretical limit of amplitude-modulated FZP). The deffraction efficiency measured at 1.4 Å is about 25% as shown in the figure. The discrepancy between measured efficiency and theoretical calculation is considered to be due to nonuniformity of thickness.

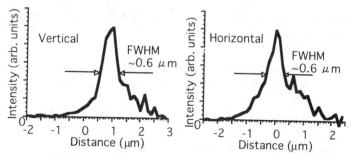

Figure2. Focused Beam Profile Measured by Edge-scan
X-ray wavelength: 1.4Å, f~158mm, Pinhole X-ray Source (20 μm in diameter).

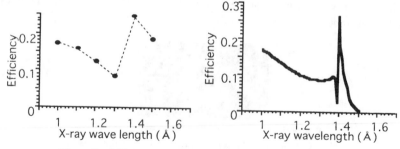

Figure3. Diffraction efficiency of Fresnel zone plate (first order)
Left: Measured efficiency. Cu/Al Sputtered-Sliced FZP, Thickness: ~20 μm
Right: Calculated diffraction efficiency of Fresnel zone plate with thickness of 20 μm

Focused beam profile for the FZP with 0.15 μm outermost zone and 40 μm thickness has been measured by edge scanning. The result is shown in Fig. 4. The gold wire (100 μm in diameter) was used as the "the knife edge", and Au L-fluorescent X-rays were monitored with a NaI scintillation counter. Measured beam size is 0.7 μm (vertically) x 4.7 μm (horizontally). In this experiment the focus spot is generated as a demagnified image of the undulator light source. Therefore the horizontal spot size is determined by the source size and magnification. Measured diffraction efficiency at 0.45 Å for 40 μm thick zone plate is about 15%. Therefore this FZP is also considered to be the phase-modulated FZP. The FZPs were not practical device in the second generation synchrotron light source. However, by using the SPring-8 undulator radiation, flux density of the focused beam has been much improved. The measured total intensity of the focused beam is 10^{10} photons/s for the focusing condition shown in Fig. 4.

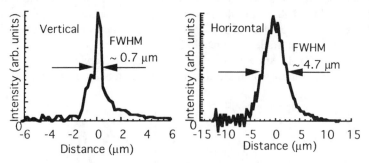

Figure4. Focused Beam Profile Measured by Edge-scan
X-ray wavelength: 0.45 Å (27.8 keV), f ~ 234 mm,

The efficiency for the 40 μm thick Al/Cu multilayer FZP is expected to take maximum at an X-ray wavelength of 0.15 Å. Therefore, we have tried to focus the 0.15 Å X-rays by using this FZP. Focused spot image taken by the image sensor is shown in Fig. 5. Apparently, the focal spot is observed at the center of image. Spot size is within 1 pixel (6 μm) width in vertical direction, and is within 2 pixel width (12 μm) in horizontal direction. Although the measured spot size is limited by the spatial resolution of the detector, it is confirmed that the sputtered-sliced FZP with appropriate thickness can work as the focusing element in the X-rays wavelength region of 0.15 Å.

Figure5. Focused Spot Image taken by Imgae Sensor
X-ray wavelength: 0.15 Å (82.7 keV), f ~ 690 mm.

REFERENCES

1) H. Kitamura, Present status of SPring-8 insertion devices, J. Synchrotron Radiation, **5**, 184-188 (1998).
2) Y. Suzuki, N. Kamijo, S. Tamura, K. Handa, A. Takeuchi, S. Yamamoto, and S. Sugiyama, X-ray Microscopy and Spectromicroscopy ed. by J. Thieme et al. Springer (1998) III-117-120.
3) Y. Suzuki, N. Kamijo, S. Tamura, K. Handa, A. Takeuchi, S. Yamamoto, H. Sugiyama, K. Ohsumi, and M. Ando, J. Synchrotron Radiation, **4**, 60-63 (1997).
4) N. Kamijo, S. Tamura, Y. Suzuki, and H. Kihara, Rev. Sci. Instrum., **68**, 14-16, (1997).
5) S. Tamura, K. Ohtani, M. Yasumoto, K. Murai, N. Kamijo, H. Kihara, K. Yoshida, and Y. Suzuki, Mat. Res. Soc. Proc., **524**, 31-35 (1998).

2D imaging by X-ray fluorescence microtomography

A. Simionovici[a], M. Chukalina[b], M. Drakopoulos[a], I. Snigireva[a],
A. Snigirev[a], Ch. Schroer[c], B. Lengeler[c], K. Janssens[d], F. Adams[d]

[a]ESRF, BP 220, 38043 Grenoble, FRANCE
[b]IMT-RAS, Chernogolovka, RUSSIA
[c]RWTH, Univ. of Aachen, GERMANY
[d]MITAC, Univ. of Antwerp, BELGIUM

Abstract. First experimental results of fluorescence microtomography in "pencil-beam" geometry with 6 µm resolution obtained at the ESRF/ID 22 are described. Image reconstructions are based on either a simplified algebraic reconstruction method (ART) or the filtered back-projection method (FBT). Simple cylindrical test objects are accurately reconstructed.

INTRODUCTION

During the past decade, X-ray imaging played an increasing role in micro-analysis. Recently, experiments using high-energy coherent phase contrast imaging were performed[1,2], which allow edge-enhanced contrast images of parameters such as refraction index or electron density. Reconstruction techniques are used for the retrieval of quantitative 2D/3D images of these parameters with sub-µm resolution in relatively short times[3,4]. Combining measurements of fluorescent signals with 2D/3D reconstruction procedures provides detailed qualitative and quantitative elemental information of the inside volume of an object with µm size resolution. Proposed in 1986 by Boisseau[5], SR fluorescence tomography for 2D/3D trace element distributions by the use of monochromatic X-ray microbeams was only recently turned into a precise and relatively simple method of microanalysis on 3rd generation synchrotrons. In this article we describe the first experimental measurements with this methodology on the ID22 beamline of ESRF using known test objects.

EXPERIMENT

Experimental set-up of the ID22 beamline of ESRF

The ID 22 beamline is dedicated to micro X-ray fluorescence spectrometry and X-ray absorption imaging (XRF, XAS), X-ray diffraction and phase contrast imaging of samples with micron resolution at high energy[6]. The basic design were the simplicity of the optics guaranteeing the high flux/brilliance necessary for a sensitive X-ray fluorescence microprobe set-up, while conserving the high degree of coherence used

CP507, *X-Ray Microscopy: Proceedings of the Sixth International Conference,*
edited by W. Meyer-Ilse, T. Warwick, and D. Attwood

for phase-sensitive imaging. The required optical quality of the beam was achieved by the use of highly polished and optically flat materials in the beam path.

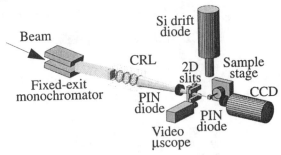

Figure 1. Experimental setup of the fluorescence microtomography apparatus, on beamline ID22 at the ESRF

The experimental arrangement of the optics on ID22 is presented in figure 1. The beam from the high ß undulator passes through polished Be and diamond windows to impinge onto a highly flat horizontally deflecting mirror with ≤ 1.5 µrad slope error and ≤ 1.5 Å micro-roughness, thus suppressing higher energy harmonics and lowering the heat load. A double crystal fixed-exit monochromator is used with Si crystals in either the 111 or 113 orientation. The microtomography set-up includes several normalization detectors (PIN diodes, ionization chambers...) and a lens assembly housing compound refractive lenses (CRL)[7]. A high precision sample stage with 7 degrees of freedom is used for positioning the sample and rotating the sample axis perpendicular to the beam. A SSD Si drift detector registers the fluorescence signal and an X-ray intensified CCD camera is used for alignment purposes.

Measurements

For the first experiment, a 20 keV monochromatic beam was used for the excitation of the elements in the sample, the energy cut-off of the mirror being set at 22 keV. The focusing element is a set of 31 parabolic CRL lenses in aluminium, which delivers a beam of approximately 2.10^9 ph/s in a beam spot of about 1.5 x 20 µm^2. A PIN diode detector operated in the current integration mode is positioned before the sample to monitor the flux of the focused beam. The sample is located at 2.3 m from the lens, slightly outside the optimal image plane in order to defocus the beam to about 5 µm vertical size to adapt the scanning interval to the sample dimensions. The sample slits are closed such that the horizontal beam size is limited to about 15 µm. The Si drift diode is set in the vertical plane at 90° with respect to the incident beam to minimize the scattering contribution. This particular geometry allows to make use of the horizontal rotation axis of the sample and to take advantage of the highly focused vertical beam size. An X-ray CCD is set behind the sample in transmission geometry and is used to align the sample with respect to the beam with a precision of a few µm. The sample is scanned in this "pencil-beam" geometry as follows: the sample is translated vertically with a 6 µm step in front of the Si drift diode detector. At each end of the travel, the sample is brought back to the incident position and rotated around the horizontal axis by an angle of approximately 3° and the translation scan is restarted. At each individual step a X-ray fluorescence spectrum is acquired during 5 seconds and the intensity associated with K and L lines of elements of interest are measured. The phantoms analyzed in these experiments consist of 3 adjoining quartz capillaries, of about 130 μm diameter and about 10 μm thick that were filled with 1% solutions of Fe, Ni, Cu, As and Cd whose K lines are measured. The capillaries themselves give rise to Sr, Zr K and Ba L lines. The count rates of all elements are normalized to the flux of the PIN diode and serve as the fluorescent signals used in the reconstruction algorithm described in the next section.

RECONSTRUCTION

The reconstruction algorithms used for the retrieval of the 2D fluorescent images follow the development of 3D absorption tomographic reconstruction. Filtered back-projection method[8] (FBT) and algebraic reconstruction techniques[9] (ART), the singular value decomposition[10] (SVD) algorithms were used with the attenuation effects correction to obtain medium resolution (20-500 μm) fluorescence tomographic images. In this work we used the FBT technique as well as a particular ART technique as described by Kak and Slaney[8]. This technique needs several simplifications and assumptions to reduce the complexity of the problem and to reduce the required calculation time[11]. The following approximations are made throughout the calculation:

• scattering from sample and scattering and absorption in the surrounding air is neglected,
• correction for the horizontal size of the beam is not applied i.e. the sample is considered to be two-dimensional,
• "enhancement effects " due to secondary/ternary fluorescence are not considered,
• the detector efficiency is considered 100%, regardless of the energy or angle of incidence. The detector-sample distance, the size of the detector and the angle between incident photons and the direction sample-detector are assumed to be constant.

The ART reconstructed image of the regions of elemental concentrations is calculated in a square grid of 480 x 480 μm in the x and y directions with 6 μm steps. Details of the reconstruction method as well as the "signal formation" algorithm are given in reference 12.

RESULTS

The first phantom measured contains Cu and As solutions in two of the three tubes. The fluorescent signals are obtained for the K_α lines of As, Cu, Zr, Sr and the $L_{\alpha,\beta}$ lines of Ba using regions of interest (ROI) directly obtained from integrating channels in the multichannel analyzer (MCA) spectra. The counting time is chosen such that several thousand counts are obtained per ROI, and varies between 2 to 10 sec/point. Prior to each tomographic scan, a high intensity spectrum of sufficient statistics is recorded to do a qualitative analysis and set up the ROI. Scanning the phantom in front of the beam using the translation/rotation scheme, sinograms of the fluorescence lines of different elements are obtained.

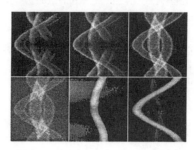

Figure 2. Sinograms of sample 1 (left to right, top to bottom) Ba-L_α and Ba-L_β, Sr-K_α, Zr-K_α, Cu-K_α, As-K_α

The sinograms are arrays of signal over the projection angle α and the position **t** in the beam. They are the primary images of the elemental data, and can be used to judge the uniformity and the smoothness of the signals. Figure 2 shows the sinograms obtained for the 5 elements in the phantom as arrays of 80 translations by 110 rotations. The presence of 3 stripes on the first 4 sinograms is indicative of the fluorescence of the 3 capillaries. The one signal belonging to the 1% As. solution is an indication that As is present at a trace level in the quartz glass substrate. Regions of uneven intensity and missing parts in the sinograms can be attributed to self-absorption of signals issuing from one part of the capillary on the opposite part and to measurement angles and positions that produce enhancement effects. Data treatment and reconstruction according to the method described yields the 2D tomographic cross-sections through the quartz capillaries presented in Figure 3. The relative differences between the capillary images obtained from Sr or Zr as opposed to those for Ba are apparent. Except for statistical differences and differences in concentrations this is due to the smaller absorption length of the low energy Ba L lines, which probes the more superficial layers of the quartz, than the K radiation measured for the other elements. Using the large translation steps (6 µm) the thin walls of the capillaries (10 µm) are partially imaged but the uniformity of the images proves the general validity of the self-absorption corrections. The two tubes containing As and Cu are very well rendered and their images appear uniform which validates the method for assessing the geometrical effects in this particular phantom.

To further test the validity of this method we performed a second experiment on a similar sample but at higher energy. The setup was modified to use a CRL composed of 79 lenses with a focusing distance of 1.8 m, at an energy of 28 keV. In this setup, the beam was focused to a spot of about 30 x 3 µm (H x V) and a Ø 5 µm pinhole was used to define the final beamspot, with a corresponding flux of about $3 \ 10^9$ ph/s. The larger flux is explained by the smaller distance CRL to source (undulator), which allows for a higher incident flux within the effective apperture of the lens. In this

542

setup, the horizontal scanning direction was used in front of the detector located horizontally in the plane of the storage ring and at 90° with respect to the beam.

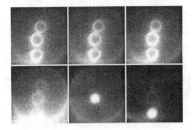

Figure 3. ART reconstructed images for sample 1

The second phantom was composed of the same three tubes but containing 1% solutions of Fe, Ni and Cd. The scanning step was 6 μm, however only 180° were covered with a 2° step. For the FBT reconstruction (see figure 4) which is a straightforward procedure, this is sufficient. However, for the fluorescence setup which records signal from the reflection geometry this appears to be incomplete: the full 360° projections should be recorded.

Figure 4. ART (a), FBT (b) reconstructed images for sample 2

The FBT reconstruction, shown in figure 4 appears of better quality for the contents of the tubes – these are homogenous cylinders, with little structure and self-absorption. In this case, the FBT yields fairly good images despite the incomplete collection of the projections. The overall resolution of the images seems lower than that of the ART reconstructed images and also some radial artifacts appear. The graininess of the FBT images is probably a consequence of the 180° collection region. The ART sinograms have been completed up to 360° by model simulated data and thus show better resolution. The tubes are well reconstructed by both techniques, however the ART image shows less artifacts and better resolution.

DISCUSSION AND CONCLUSION

The combination of a tomographic set-up in the "pencil-beam" parallel collection geometry and the two reconstruction methods: one, based on an modification of ART and the other on the FBT, provides high quality and precision of the reconstructed elemental images.

These results show a first attempt to use the ESRF ID 22 beamline for such 2D imaging. To further improve these results, smaller scanning steps could be used while also storing the full spectra for deconvolution in order to improve the statistics in relation to the presently used ROI integration. Furthermore, the addition of corrections for secondary fluorescence enhancement effects in the calculations of the reconstruction algorithm could improve the corrections for matrix effects. At present, using these homogenous phantoms which feature smoothly-varying structures and little self-absorption is is not obvious which reconstruction technique is more promising. It is necessary to go to extremes to test the accuracy of these methods: high resolution ($\approx$ 1 µm) and grainy, absorbing samples. Recent results of fluorescence tomography as well as absorption tomography we have obtained in high resolution ($\approx$ 1.2 µm) at 20 keV on inhomogenous samples[13] confirm the need of using sophisticated self-absorption corrections combined with adapted reconstruction techniques to fully use the potential of this type of measurements. Other mandatory items in order to fully exploit the capabilities of fluorescence-tomograhy are *limited-angle* as well as *partial volume* data collection, enabling the recording of a small volume inside a large object, with a limited number of projections and thus a relatively short acquisition time. Such improvements will be included in a future campaign of measurements that will be extended to samples of more complex composition and heterogeneity.

REFERENCES

1. A. Snigirev, I. Snigireva, V. Kohn, S. Kuznetsov, I. Shelokov, Rev. Sci. Instr. 66, 5486 (1995)

2. P. Cloetens, R. Barret, J. Baruchel, J. Guignay, M. Schlenker, J. Phys. D: App. Phys. 29, 133 (1996)

3. V. Kohn, Physica Scripta, vol.56, 14(1997)

4. T.E. Gureyev, C. Raven, A..Snigirev, I. Snigireva, S.W. Wilkins, J. of Physics D; Applied Physics, vol. 32, No. 5, March (1999)

5. P. Boisseau, "Determination of three-dimensional trace element distributions by the use of monochromatic X-ray microbeams", Ph.D Dissertation, M.I.T. Cambridge (1986)

6. http:// www.esrf.fr/exp_facilities/ID22/

7. B. Lengeler, C. G. Schroer, M. Richwin, J. Tummler, M. Drakopoulos, A. Snigirev, I. Snigireva, Applied Physics Letters 74, 3924, 1999

8. A.C. Kak, M. Slaney. Principles of computerized tomographic imaging, New York: IEEE Press, (1988)

9. R. Cesareo, S. Mascarenhas, Nucl. Instrum. Methods A 277, 669 (1989)

10. G.F. Rust, J. Weigelt, IEEE Trans. Nucl. Sci., vol. 45, Nr. 1, 75 (1998)

11. M.V. Chukalina, N.G. Ushakov, S.I. Zaitsev, Scan. Microscopy, vol. 11, Nr.2, 311 (1997)

12. A. Simionovici, M. Chukalina, M. Drakopoulos, I. Snigireva, A. Snigirev, Ch. Schroer, B. Lengeler, K. Janssens, F. Adams, Proceedings of the SPIE Conference, Denver, July 1999.

13. A. Simionovici, Ch. Schroer, T. Weitkamp, M. Chukalina, J. Tummler, B. Benner, F. Günzler, A. Snigirev, to be published.

Imaging Hard X-Ray Microscopy with a Multilayer Fresnel Zone Plate

Mitsuhiro Awaji[1], Yoshio Suzuki[1], Akihisa Takeuchi[1], Nagao Kamijo[1,2], Shigeharu Tamura[1,3], Masato Yasumoto[3] and Yoshiki Kohmura[4]

[1] Japan Synchrotron Radiation Research Institute (JASRI), SPring-8, Kouto 1-1-1, Mikazuki, Sayo-gun, Hyogo 679-5198, Japan
[2] Kansai Medical University, Uyamahigashi 18-89, Hirakata, Osaka 573-1136, Japan
[3] Osaka National Research Institute, Midorigaoka 1-8-31, Ikeda, Osaka 563-8577, Japan
[4] Harima Institute, The Institute of Physical and Chemical Research (RIKEN), SPring-8, Kouto 1-1-1, Mikazuki, Sayo-gun, Hyogo 679-5148, Japan

Abstract: Preliminary experiments by use of a multilayer Fresnel zone plate (FZP) have been performed for bright-field imaging at 25keV hard X-rays. The #1500 gold mesh was selected as a test specimen. The mesh structure has been successfully observed in the bright-field images.

INTRODUCTION

High-resolution observation of an inside of a living body or thick materials has been a challenging work for hard X-ray microscopists. In our group, X-ray focusing test using the multilayer FZP has been done in X-ray wavelength regions of 0.15-1.6 Å[1], and also the possibility of the FZP as an optical imaging element for hard X-rays has been found. In this paper, we introduce the results of a hard X-ray imaging (bright-field imaging) test using the multilayer FZP at 25keV.

INSTRUMENTS AND METHODS

Imaging hard X-ray microscopy with a multilayer FZP has been tested at an X-ray undulator beam line, 47XU of SPring-8. The X-ray source is in-vacuum type permanent magnet planar undulator. The first harmonics of the undulator radiation can be tuned within the energy range of 4.5-18.8keV. The undulator radiation is monochromatized by a SPring-8-Standard Si111 double crystal monochromator, and the crystals are cooled by liquid nitrogen. Both the first and the second crystals are kept at the same temperature to make exit beam position constant. The schematic diagram of the experimental setup for imaging hard X-ray microscopy is shown in Fig.1.

CP507, *X-Ray Microscopy: Proceedings of the Sixth International Conference,*
edited by W. Meyer-Ilse, T. Warwick, and D. Attwood
© 2000 American Institute of Physics 1-56396-926-2/00/$17.00

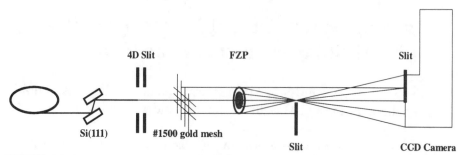

FIGURE 1. Schematic diagram of the experimental setup for bright-field imaging hard X-ray microscope.

The monochromatic incident beam was cut by a tandem two four-dimensional slits to 120μm x 120μm square to reduce the unnecessary scattered beams. The distance between the two slits of the tandem slit was about 50cm, which was the maximum distance in this setup condition. The #1500 gold mesh, whose wire diameter is 5.6μm, nominal aperture 11μm and the thickness about 3.8μm, was selected as a test specimen. The energy of the incident X-rays used in this imaging experiment was 25keV. In this energy for the gold mesh, the ratio of the incident and transmitted intensities $I/I_0 = \exp(-\mu t) = 0.72$, where linear absorption coefficient $\mu = 852.0 (cm^{-1})$ and t is the thickness of the mesh. The phase shift due to the gold mesh can be calculated as $\Delta\phi = 2.47$rad. with free electron approximation and that there is no anomalous dispersion. Therefore the contrast of the mesh can be observed by both in absorption and phase shift. Figure 2 shows the schematic diagram of the multilayer FZP. The sputtered-sliced multilayer FZP was fabricated by Kamijo and Tamura [2]. The distance between the light source and the FZP was about 43m. The specification of the FZP is: 50 layers of Cu/Al zones on Au wire substrate, 80μm in diameter, 0.25μm outermost zone width and about 36μm in thickness. The FZP is a phase modulation type, the focal length of the 1st-order diffracted beam is 508mm, the focused beam size is about 1.5μm, and the measured diffraction efficiency is about 15% at 25keV.

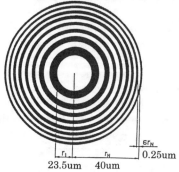

FIGURE 2. Schematic diagram of the multilayer FZP. Black and white rings are Al and Cu layers.

Object-Lens Distance

Bright Field Images

579(mm)
(a)

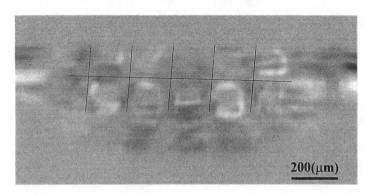

200(μm)

559(mm)
(b)

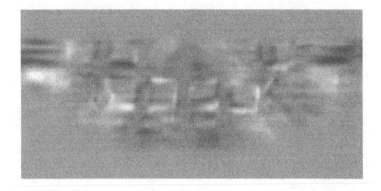

539(mm)
(c)

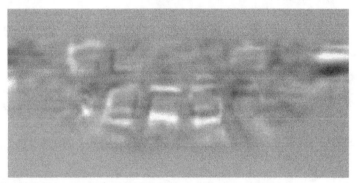

Figure 3. Bright-field images of the #1500 gold mesh wire. Magnification is 10 times. Object (Mesh) - Lens (FZP) distance: 579mm (a), 559mm (just focus) (b) and 539mm (c).

This FZP has been tested and used as an optical element of the imaging hard X-ray microscope. The tilting angle of the FZP to the optical axis was adjusted monitoring the focusing pattern, which is mainly formed by the 1st and 0th-order diffracted beams, using a cooled charge coupled device (CCD) camera coupled with relay lens and phosphor screen. The resolution of this CCD camera (HAMAMATSU Dual Mode Cooled CCD camera C-4880-10-14A) is about 10μm, while the pixel size of the image sensor is 6μm.

Bright-field images were taken using the 1st-order diffracted beam due to the FZP alone and the CCD camera. The 0th, 2nd, 3rd and higher-order diffracted beams due to the FZP were cut by a beam stop at the focal plane, and all the diffracted beams were cut by a beam stop at just in front of the CCD camera as shown in Fig.1, because the 0th, 2nd, 3rd and higher-order diffracted beams make disuse and strong background images. The distance between the FZP and the CCD camera was fixed at 5621mm, and the bright-field images were taken with three focus values changing the distance between the mesh and the FZP. In the imaging system, axial and parallel beam illumination, whose photon flux density is 4.2×10^{12}[photons/s/mm^2], was used.

RESULTS AND DISCUSSION

Figure 3 shows the bright-field images of the gold mesh with the magnification of 10 times. Figures (a) and (c) are the defocused images and Fig.(b) shows the focused image, whose mesh (object) - FZP (lens) distance is 579mm(a), 539mm(c) and 559mm(b), respectively. The exposure time was 30s in each image. The schematic position of the gold mesh wire is indicated by black narrow lines. It can be seen that the diameter of the magnified gold mesh wire images is about 56μm, and the contrast of the wire edge changes due to defoci. In the mesh wire moved, then the image also did. As the gold mesh wire itself has some wrinkles, the image of the gold mesh is slightly distorted. Thus it can be concluded that these bright-field images are successfully formed at 25keV by the 1st-order diffracted beam alone.

In this imaging experiment by use of the FZP and hard X-rays, the FZP with the larger zone area is desirable, and noise elimination technique for observing the high quality images is also important.

REFERENCES

1. Y. Suzuki, M. Awaji, Y.Kohmura, A. Takeuchi, N. Kamijo, S. Tamura, and K. Handa, " X-ray microbeam with sputtered-sliced Fresnel zone plate at Spring-8 undulator beamline," *Proceedings of the VIth International Conference on X-ray Microscopy*, Berkeley, California, August 1999. W. Meyer-Ilse, T. Warwick, and D. Attwood (eds.), American Institute of Physics, New York, NY.

2. S. Tamura, K. Mori, T. Yoshida, K. Ohtani, K. Kamijo, Y. Suzuki, and H. Kihara, Mat. Res. Symp. Proc. **441**, 779-783 (1997).

Fabrication and Hard X-ray Focusing of X-ray Refractive Lens from Liquid Materials

Yoshiki Kohmura, Mitsuhiro Awaji, Yoshio Suzuki, and Tetsuya Ishikawa,

SPring-8,

1-1-1, Kouto, Mikazuki, Sayo-gun, Hyogo, 679-5198, Japan

Yu.I.Dudchik, N.N.Kolchevsky and F.F.Komarov

Institute of Applied Physics Problems,

Kurchatova 7, Minsk 220064, Belarus

Abstract. X-ray refractive lens has been developed with liquid materials at SPring-8, with the expectation that smooth surfaces can be obtained by the natural surface tensions. The lenses developed consist of approximately spherical micro-lenses. Though the shapes are not well controlled and the parabola shapes are better in terms of spherical aberrations, some lenses developed had high tolerances to severe radiation damages (e.g. $\sim 5 \times 10^{12}$ photons/sec/0.03mm^2 of 18 keV X-rays) for an exposure time of one hour. The gain at the focal plane and the transmissivity for various types of the lens tested are also summarized.

Introduction

Recently various types of X-ray refractive lenses were fabricated, but most of them are made from metallic materials and precise processes are required.[1,2,3] We paid attention to the smooth surfaces at the interfaces between liquids and the gas due to the surface tension. Using bubbles or the spherical plastic ball with the inner diameter of 2mm and 1.7mm, two-dimensional focusing has been successfully done.[4]

The other lens was named, the "micro capillary lens"[5] and was developed at the Institute of Applied Physics Problems, Belarus and was tested for the first time at BL47XU, at SPring-8[6]. First type of the lens (No.1) was produced by forming bubbles in a glue in a thin glass capillary which was solidified to obtain a solid X-ray refractive

CP507, *X-Ray Microscopy: Proceedings of the Sixth International Conference,*
edited by W. Meyer-Ilse, T. Warwick, and D. Attwood
© 2000 American Institute of Physics 1-56396-926-2/00/$17.00

lens.[5] The bubbles were formed in glycerol in the second type of the lens (No.2, Fig.2). For both of these lenses, the surface tension of the liquid inside glass forces the surface to have an approximately spherical shape. The lens No.1 & No.2 have the length $L=59$ mm & 225 mm, inner radius $r=0.10$ mm & 0.4 mm with the number of micro-lenses $N=71$ & $N=185$.

Figure.1. (a, left panel) Container of the X-ray bubble lens and (b, right panel) the example of the spherical plastic ball for the X-ray plastic ball lens. The inner diameter of the bubbles and the plastic balls were around 2mm and 1.7mm, respectively.

Figure.2. Visible microscope image of the Micro-capillary lens (No.2, with glycerol). The inner diameter of the glass capillary is about 800 μmϕ.

Such lenses can be used also as the collimator of the beam when the focal distance is chosen to be the distance from the source to the lens. At SPring-8, one-dimensional lenses were evaluated as the vertical "collimator" of the undulator beam, and was found to be useful for high-angular and high-energy resolution experiments such as non-elastic scattering and nuclear scattering experiments.[7,8] These were fabricated on plastics or beryllium plates and their focal lengths were chosen to about 45 m.[7,8]

Focusing Capability of Liquid Refractive Lenses

The evaluation of the "micro capillary lens" was done at BL47 of SPring-8[9] using the standard in-vacuum undulator[10] and a cryogenically cooled silicon monochromator,[11] which could tolerate the full heat load from the undulator with small gap. The demagnified image of the undulator source was taken and the source image was evaluated in terms of the image size, the gain around the focal plane and the transmissivity of the lens was also measured. The following table.1 summarizes the results of such measurements.

TABLE.1. Comparison of the measured performances of X-ray refractive lenses made of liquids. The inequality for the performance of the X-ray plastic ball lens is due to the correction we have made for the water regions without the balls. The measured gain and the transmissivity are corrected so that separation of the plastic balls are negligible.

	X-ray Energy (keV)	Focus Gain	Transmissivity (%)	Demagnified Source Image Size at Focus (vertical FWHM, μm)
X-ray Bubble Lens	19.0	12	16%	48
X-ray Plastic Ball Lens	24.5	≦3.2	≦2.4%	53
Micro-Capillary Lens(No.1)	17.1	12	18%	8

The vartical undulator source size in FWHM was around 30 μm (assuming the coupling constant of 0.15 %). The focal length of the X-ray bubble lens, X-ray plastic ball lens, and the micro-capillary lens (No.1) were ~5 m, ~5 m, ~1 m, respectively, much shorter than the distance between the source and the lens, about 45 m. Therefore the expected geometrical demagnification factors were about ~1/9, ~1/9 and ~1/45, respectively. For the micro-capillary lens (No.1), ray trace simulations were made assuming spherical micro-lenses and parallel beam was assumed. The image size at focus observed (and calculated) in FWHM was expected to be 8μm (≦2 μm) and 16 μm (~18 μm) in the vertical and in the horizontal directions, respectively.

Here, the vertical and horizontal profile at L_2 of 0.8 m are shown which seemed to be reasonably near the focal plane (Fig.3(a,b)). A low frequency source movement in a few Hz appeared as oscillations at the tail around the focus spot. The observed focal length was around f=0.8 m, which suggests that the micro lenses has the "effective" radius of about 0.09 mm instead of the assumed radius of 0.10 mm.

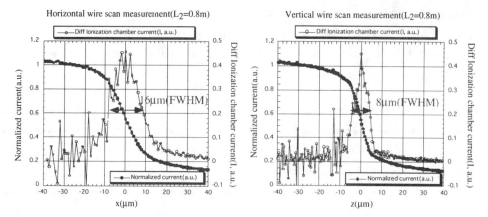

Horizontal wire scan measurement(L$_2$=0.8m)

Vertical wire scan measurement(L$_2$=0.8m)

Figure.3. Two-dimensional profiles near the focal plane using the Micro-capillary lens (No.1, containing glue). The left panel (a) and the right panel (b) show the horizontal and vertical profiles, respectively.

Test of Radiation Damage

The lens No.1 was formed with glues and the stability of the glue was not known. To check the stability in an extremely high X-ray flux and a high radiation damage, a lens with the same ingredient as the lens No.1 was placed into the optical axis. This experiment was done at BL47XU of the SPring-8 with the undulator gap of 10 mm (near the minimum of 8 mm). The monochromator was set so that the third order harmonics, 18 keV, was transmitted into the experimental station, The lens was adjusted to the X-ray optical axis and a flux density of 5×10^{12} photons/sec/0.03mm^2 was assumed to be illuminated on to the lens. The exposure time was about 1h.

After 1h exposure, the most upstream part of the lens, composed of glue, was discolored. The glue turned yellow from its initial color of dark green and small density fluctuations were seen as shown in Fig.4. Only two micro-lenses (the glue parts for a length of about 2 cm) were affected by this effect. However the focusing capability did not seem to be affected so much.

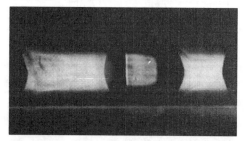

Figure. 4 . Visible microscope image of the Micro-capillary lens (similar type as No.1, with glue) after a radiation damage test. Inner diameter of the glass capillary is 350 μm.

Summary

Several different X-ray refractive lens from liquid materials had been developed. One of the Micro-Capillary lens (No.1 with glue) was especially promising due to the tolerance to a severe radiation damage. This lens was not severely damaged with a high X-ray flux $(5 \times 10^{12}$ photons/sec/0.03mm^2) of 18 keV X-rays for an exposure time of 1h.

References

[1] A.Snigirev, et al., *NATURE*, **384**, 49-51 (1996)

[2] P.Elleaume, *Journal of Synchrotron Radiation*, **5**, 1 (1997)

[3] B.Lengeler, et al., *Applied Physics Letters*, **74(26)**, 3924-3926 (1999)

[4] Y.Kohmura, et al., *Proc. of SPIE conference*, **3449**, 185-193 (1998)

[5] Yu.I.Dudchik, N.N.Kolchvsky, *Nucl. Instrum. and Methods*, **A, 421**, 361-364 (1999)

[6] Y.Kohmura, et al., *Rev.Sci.Instrum.*, in press

[7] A.Q.R.Baron, Y.Kohmura, et al., *Applied Physics Letters*, **74(10)**, 1492-1494 (1999)

[8] A.Q.R.Baron, V.V.Krishnamurthy, et al., *J. Synchrotron Radiation*, **6**, 953 (1999)

[9] Y.Kohmura, *SPring-8 User Infornation Newsletter (in Japanese)*, **3(4)**, 28 (1998)

[10] H.Kitamura, *J. Synchrotron Radiation*, **5**, 184-188 (1998)

[11] T.Ishikawa, et al., *Proc. of SPIE*, **3448**, 2-10 (1998)

X-ray optics for phase differential contrast: design, optimization, simulation, fabrication

V. Aristov[1], M. Chukalina[1], A. Firsov[1], T. Ishikawa[2], S. Kikuta[2], Y. Kohmura[2], A. Svintsov[1], S. Zaitsev[1].

[1] Institute of microelectronics technology and high purity materials RAS, Chernogolovka, Moscow district, 142432, Russia.
[2] SPring-8, 323-3, Mihara, Mikazuki-cho, Sayo-gun, Hyogo 679-5198, Japan.

Abstract. With increasing of X-ray energy an interesting situatuon appears when due to different dependence of refraction and absorption on X-ray energy a sample becomes transparent but still produces refraction. So such samples become unvisible for usual absorption methods but could be analyzed with variouse phase contrast methods. The situation becomes more actual for higher X-ray energy especially for range20-100keV. F. Polack and D. Joyeux described an extension of an interferential differential phase contrast to scanning X-ray microscopy. The principle is to illuminate two points of an object with coherent radiation and detect the fringe shift induced by small phase differences. Fresnel mirrors have been suggested to split a X-ray beam in two coherent ones. In this paper it is proposed instead to use bifocal (multi-focal) lenses. Design and fabrication process of the lenses are described.

Two different experimental schemes of phase differential contrast formation (contrast in scheme of Polack - Joyeux and Nomarsky) are considered. Evaluation of resolution and sensitivity for both schemes as well as experimental and model results are presented. The differential contrast is based on the phenomenon of interference which transforms phase shifts (from two or more sources) to intensity variations. F. Polack and D. Joyeux [1] have suggested an experimental scheme to measure the signal

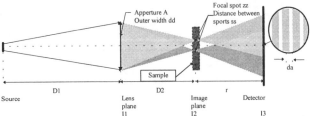

Fig.1. X-ray beam splitting.

corresponding to phase differences. Fresnel mirrors beam-splitter has been suggested in order to produce two coherent spots in the focal plane. But efficiency of that kind of splitting is not high, it is difficult to align two mirrows to provide small distance between spots. To avoid the difficulties of the two coherent spot creation it is suggested to use diffractive optical elements. The simplest example of such elements

CP507, *X-Ray Microscopy: Proceedings of the Sixth International Conference*,
edited by W. Meyer-Ilse, T. Warwick, and D. Attwood
© 2000 American Institute of Physics 1-56396-926-2/00/$17.00

is a lens with two focal spots. Operation of the lens is illustrated in Fig.1 in the scheme suggested in [1]. Two coherent spots produce wellknown interferencial picture on far screen in form of the strips, introducing a sample as shown in Fig.1 results to some phase difference for the two spots and as a consequence to shifting of the fringes . The strip shift da is related to variation of refraction index in the spot places and F. Polack and D. Joyeux have analyzed this dependence. The simplest scheme gives information on variation of refraction index along spot-spot direction. Usage of diffractive optics extends significantly possibilities in phase contrast formation. For example it is possible to produce two spots with phase shift equal to 180^0, at the first sight such source does not give any principal advantages in compare with bimirrow but used as imaging element this produces an an image where intensity is related refraction index gradient. Such scheme is very close to Nomarski contrast widely used in visible optics. Of course Nomarski contrast hardly could be realized with bimirrows. And more over consider element creating four (n) spots with phase shit increment $360^0/4$ $(360^0/n)$ such an element in Polack-Joyeux scheme will produce a signal related to gradient module of refraction index. Such a source is new even in visible optics and is considered below. And finaly one more but not the last example, let the diffractive elements create three spots belonging to one line with phase shift of the spots $(0,180^0,0)$ if intensity of the central one is two times higher then two outer ones then the signal measured in Polack-Joyeux scheme will be related to the second spatial derivative of the refraction index.

Detection of the differential phase signal.

Let us consider one of the focuses (Fig.2). Change of the phase $\Delta\varphi$ determined by a refraction index n inside a sample can be written:

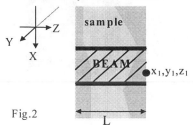

Fig.2

$$\Delta\varphi(x_1,y_1,z_1) = k \int_{z_1-L}^{z_1} n(x_1,y_1,z_1)dz$$

where k is wave vector, and L is a thickness of the sample. For simpicity we consider below situation when thickness of the sample is less then the focus depth. The complex amplitude of oscillations from small area dS_1 (centered at (x_1,y_1,z_1) and dS_2 (centered at corresponded point in the second beam (x_1+ss,y_1,z_1) (Fig.3) in the central point of the screen can be written:

$$A_3 = i\iiint_V \left(\exp\left(-\frac{2\pi}{\lambda}r_1(x,y,z)\right)n(x+ss,y,z) - \exp\left(-\frac{2\pi}{\lambda}r_2(x,y,z)\right)n(x,y,z)\right)dxdydz$$

Intensity detected is proportional to the amplitude squared

$$I\infty\left|\tilde{A}_3\right|^2 = I = \left|\tilde{A}_3\right|^2 = C\left(\underline{n}(r_1) - \underline{n}(r_2)\right)^2,$$

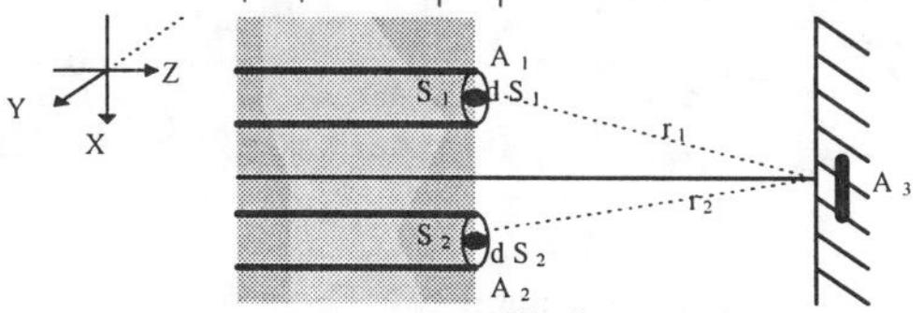

Fig.3.

where $\tilde{A}_3$ is a sum over all points of the detector area to take into account a size of the detector and $\underline{n}(r_l)$ is value of the refraction index averaged over volume V occupied by

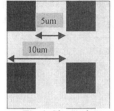

Fig.4. Si sample.   Fig.5. Model image of the sample Fig.6. Model image of the sample
generated by two-focal lens.   generated by four-focal lens.

beam. It is seen that signal proportional to gradient squared, in [1] have been showed that there is another detector position (shifted on a quater strip period) when signal is linear dependent on gradient, the conclusion is valid for our case as well. For future experiments consider a sample shown in Fig. 4. It is a Si grid of 5μm width with period 10μm and 5μm thickness (h). The model results are presented in Fig. 5-6. Wavelength used is 1A. Focal size is 0.4 micron. Distance between focal spots is 1 micron.

Phase differential contrast in Nomarsky scheme.

With diffractive optics quite different scheme (Fig.7a) could be realized with special bifocal lens when focal spots have phase shift equal to 180^0. Such a lens produces two coherent images of an object shifted on small distance, due to the phase shift electrical vector of the images are antiparallel the resulting to substracting of image amplitudes. So the resulting image will be related to gradient of object refraction index. (Fig.7 b)

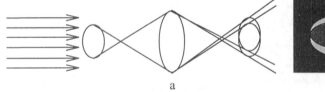

a

b

Fig.7.

The scheme is very close to Nomarski contrast widly used in visible optics. Some extention of the scheme is presented in Fig.8a where diffractive element for producing infinit number of focal spots arranged in a ring Fig.8b, Fig.8c shows that phase shift inscreases continuosely along the circle. Usage of the element produces an image related not to gradient along one direction but related to gradient module.

a.   b.   c.

Fig.8.

Results of image simulation in Nomarski sheme with element of Fig.8a (the sample submitted in Fig.4) are shown in Fig. 9. Usage of appodization known in visible optics as dark field mode improves image significantly due to suppressing of zero-th oder which could be too large even for phase diffractive optics. Fig.9a shows a model image of the sample formated by two-focal lense. Fig.9b shows a model image of the sample generated by lens shown in Fig 8a. Fig. 9c and 9d show the model images of the object by traditional lens (with one focal spot), Fig. 9c is one without appodization and Fig. 9d is one with appodization.

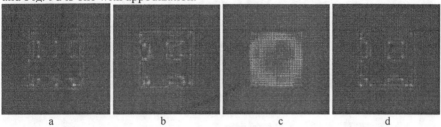

| a | b | c | d |

Fig.9

Experiments on Spring8 with the simplest bifocal lens were carried out. The lens was designed for Bragg reflection mode and optimized in focal distance, distance between spots, spots sizes etc. The lens was created by e-beam lithography with minimal feature about 0.3um. Plasma etching was used to produce relief of 1.25um depth. The lens is shown on Fig.10 where inverse part corresponds to 180^0 phase shift between two focal spots. Image of focal plane (Fig.11) shows two focal spots. Remarkable (but expected) feature of the image is sharp vertical light line (corresponding to low intensity) which is an evidence that the focal spots have electrical vectors of opposite directions so producing strong zero intensity in between spots.

Fig. 10.   Fig.11.

Acknowledgements.

The work is supported in part by RFBR grants 97-02-17318 and 98-02-17680.

References.

1. F. Polack and D. Joyeux Soft X-ray Interferential Scanning Microscopy: a feasibility assessment. Conf. X-Ray Microscopy, 1994 (433-438)

SILICON REFRACTIVE OPTICS FOR HARD X-RAYS

V. Aristov[1], A. Firsov[1], M. Grigoriev[1], T. Ishikawa[2], S. Kikuta[2], Y. Kohmura[2], S. Kuznetsov[1], L. Shabelnikov[1], V. Yunkin[1].

[1]Institute of microelectronics technology and high purity materials RAS, Chernogolovka, Moscow district, 142432, Russia.
[2]SPring-8, 323-3, Mihara, Mikazuki-cho, Sayo-gun, Hyogo 679-5198, Japan.

Abstract. X-rays refractive optics has made a real success in a hard radiation domain promising for many applications. An analysis made of refractive materials properties, which are most suitable for this new kind of X-ray focusing optics shows that in silicon refraction effects prevail on attenuation. A wide variety of technologies dedicated to microstructure formation on Si make it as a most suitable one. Using electron beam lithography and deep plasma etching planar parabolic profiles are produced with the focal distance 1 m on 17.4 keV radiation. They are tested on synchrotron radiation at beamline X47 SRPING-8. Experiments carried out open the way to construct kinoform refracting profiles, which exceed simple parabolas in gain and aperture. A number of kinoform profiles on silicon is created for energies up to 50 keV, which at apertures 2mm have calculated gain up to 5000. Some focusing properties of such planar refracting profiles are discussed.

Refractive lenses for "hard" X-ray radiation are an object of theoretical speculations and estimates [1]. The simplest refractive focusing system for "hard" X-ray radiation was suggested in [2] – compound refractive lenses (CRL). This is the mostly simple and improper focusing system for application in microfocus experiments because of a substantial absorption of radiation with energy, say, up to 20 keV, and low signal to noise ratio at higher energy. The characteristics of CRL can be considerably improved by constructing kinoform refractive profiles making use of the principles of X-ray diffraction optics [3]. This can be done by removing passive regions of the parabolic profile where the phase change of a passing wave is multiple

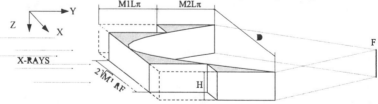

Fig.1. The scheme of X-ray focusing on element with a planar profile. D is the element aperture, H is the relief depth, M1 and M2 are the numbers of wave phase drops at the segment edge.

of 2π. A focusing element of this kind is a set of segments with even numbers of wave phase drops at the edges. The segments are proposed [4] to be formed as parabolic

CP507, X-Ray Microscopy: Proceedings of the Sixth International Conference,
edited by W. Meyer-Ilse, T. Warwick, and D. Attwood
© 2000 American Institute of Physics 1-56396-926-2/00/$17.00

ones (Fig.1). This configuration with the optical axis of the initial parabola in the plane of a focusing element allows virtually any arrangement of segments, without upsetting phase ratios. For a material with a refractive index of $n=1-\delta+i\beta$, the segment dimension in the direction of the element optical axis Y is multiple of the π-shift length $L_\pi=\lambda/2\delta$. For kinoform profiles the absorption of passing radiation with the linear attenuation coefficient $\mu=4\pi\beta/\lambda$ should be taken into account at a given length L_π by introducing the critical number $N_0=(\mu L\pi)^{-1}=\delta/4\pi\beta$ [1]. A data base on refractive properties of chemical elements with the atomic numbers Z from 1 to 92 and for more than 60 their compounds was earlier developed [4] where the energy range was extended to 1-100 keV for the calculation of kinoforms. Using the atomic scattering factors f_1 and f_2, decrements in the refractive indexes δ and β, and the values of μ, $L\pi$ and the critical number N_0 were calculated in it. The intensity of a passing wave focused by a refracting profile is determined by the Kirchhoff diffraction integral that can be explicitly calculated for parabolic profiles [1].

Kinoform profiles can be given as $\quad Y(x)= L_\pi x^2/ F\lambda-MnL_\pi$ $\qquad$ (1)

$R_F n^{1/2} < x < R_F (n+1)^{1/2}$, $R_F=(F\lambda)^{1/2}$ -Fresnel radius, n =0,1..N -current segment number. M - even number of phase drops. At $M\to\infty$, a kinoform profile transforms into parabolic. The computation of intensities focused by kinoform profiles involves numerical methods for the calculation of the diffraction integral, which can be done using the software program developed in this work. An important characteristic of refracting optics element is transmittance [5,6] that determines the fraction of an incident beam energy producing a focal spot

$$T =\frac{1}{A} \int_{-A/2}^{A/2} \exp(-\mu Y(x))dx \qquad (2)$$

where A - aperture of the said element. For kinoform profiles, the expression for T can be written as

$$T=\frac{2}{A}\sum_{n=0}^{N}\exp(\mu L_\pi Mn) \int_{x_n}^{x_{n+1}} \exp\left(-\frac{\mu L_\pi}{R_F^2}x^2\right)dx \qquad (3)$$

Taking: $\mu L_\pi Mn = \dfrac{Mn}{N_0}$ and introducing the function $erf(z) =\dfrac{2}{\sqrt{\pi}}\int_0^z dx\exp(-x^2)$ we

obtain

$$T =\frac{\sqrt{\pi N_0}R_F}{A}\sum_{n=0}^{N}\exp\left(\frac{Mn}{N_0}\right)\left[erf\left(\sqrt{\frac{M(n+1)}{N_0}}\right)-erf\left(\sqrt{\frac{Mn}{N_0}}\right)\right] \qquad (4)$$

Thus, the value of T is determined by the following parameters of a kinoform profile: the ratio of the Fresnel radius R_F to the aperture A and the number of phase drops M. The properties of a refracting material are given by N_0. The calculation of T also involves numerical methods. Another way to determine transmittance is to make use of the gain of a focusing element G when the source size s is variable [5] in the absence of diffraction affects

$$T =\frac{GsF}{A(L-F)} \qquad (5)$$

where F is The element focal distance and L is the distance between the source and the lens. The transmittance was found by calculating the dependence G(s) at L>F and then T(s) using (5). At s>100μm, when ratio (5) holds, the values of T do not depend on s and are the characteristics of a profile. Figure 2 shows the transmittance for Si-kinoform at a successive increase of the number of phase drops M. For Si-kinoform with the minimum number of phase drops M=2, the determined value T=0.9 is much higher than the value T=0.64 found in [5] for CRL from the Be with N_0=1028. The value G=5000 (Fig.3.) found in this work for

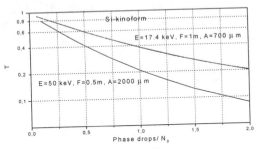

Fig.2. Transmittance silicon kinoform profile vs number of phase drops, computed for energies 17 keV (N_0=32) and 50 keV(N_0=291).

a point source also considerably exceeds those previously reported [5,6]. The curves in Fig.2 conform to the dependence of the exp(-M/N_0) kind. The calculations allow the determination of an optimal number of phase drops M in a single kinoform and the number of

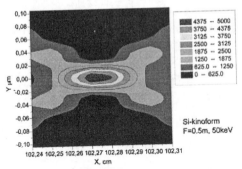

Fig.3. Calculated intensity distribution

kinoforms in a set in compound kinoform profiles. Planar refracting elements were fabricated by electron-beam lithography based on JSM-840 (JEOL) scanning electron microscope and deep plasmachemical etching of Si Fig.4. We used home constructed helicon ion source to realize plasma-chemical etching of silicon via the metal mask. The helicon source has been described in detail in [7]. By operating at lower pressure, we have a technique for etching that is inherently more anisotropic. At low pressures, your ability to control, and even eliminate, collisions in the plasma sheath is also increased. At low pressures, the gas species present may also dissociate more completely and this may simplify etch gas selection. Both parabolic profiles with minimized absorption and kinoform ones with the etching depth H=6μm and technological resolution Δ=0.4μm were obtained (Fig.4). A decrease in the contribution of distorted parabolic segments into the focusing intensity determines

Fig.4. SEM- images of plasma-etched planar refractive profiles.

their certain number $N_{eff}=0.5N_0$ and aperture $A_{eff}=2R_F(MN_{eff})^{1/2}$. Then these are exceeded, no increase in intensity occurs. The SI591 (Sentech Instruments GmbH) system of ion-plasma etching is going to be used for the purpose of the creation of profiles with a greater relief height H [8, 9]. The processes are optimized by matching the etching rates and trench profile shapes, characteristic for these gas system. The produced refracting elements were tested on focus distance 1.1m, whose fragment is shown in Fig.4, an image was obtained at a distance of 35 cm

depth 6 μm, it is a narrow band against the background of intense exposure from the

Fig.5. Image of Si-kinoform, taken on SR

rays along the element surface. Changes in intensity observed in the central part of the image (marked by arrows) are due to the parabolic segments focusing action. Blackening density measurements taken for this part of Fig.5 show good agreement with the calculated distribution for the given distance. Topologies of kinoform elements have been developed for energies to 50 keV with an aperture and transmittance much higher than those for refracting lenses.

Acknowledgements.

The work was supported by the Russian Foundation for Basic Research, grant no.98-02-16341.

References.

1. X.Yang, Nuclear Instr.& Meth. A 328 (1993), 578
2. A.Snigirev, V.Kohn, I.Snigireva, B.Lengeler, Nature, Vol. 384 (1996), 49
3. B.Vidal, A.I. Erko, V.V.Aristov, Diffraction X- ray optics, (Institute of Physics Publ., Bristol & Philadelphia, 1996), 385
4. V.V.Aristov, L.G.Shabelnikov, E.V.Shulakov et al, Surface investigation: X-ray, synchrotron and neutron techniques, No.1, (1999), 7.
5. B.Lengeler, J.Tummler, A.Snigirev, I.Snigireva, C.Raven, J.Appl.Phys., 84,(1998), 5855
6. P.Elleaume, J.Synchr.Rad., 5(1998), 1-5
7. R.W. Boswell and R.K. Porteos, Appl. Phys. Lett. 50, 1130, 1987), (R.W. Boswell, A.J. Perry, and M. Emami, J. Vac. Sci. Technol. A7, 3345, 1989
8. V.A Yunkin., D.Fischer., and E.Voges, Microelectron. Eng., 1994, **23**, 373.
9. V.A Yunkin., I.W. Rangelow, J.A. Schaefer, D. Fischer, E.Voges, and S. Sloboshanin , Microelectron. Eng. 1994, **23**, 361

New generation of diffraction optical elements to focus "hard" x-ray radiation.

V. Aristov[1], A. Firsov[1], T. Ishikawa[2], S. Kikuta[2], Y. Kohmura[2], A. Svintsov[1].

[1]Institute of microelectronics technology and high purity materials RAS, Chernogolovka, Moscow district, 142432, Russia.
[2]SPring-8, 323-3, Mihara, Mikazuki-cho, Sayo-gun, Hyogo 679-5198, Japan.

Abstract. Focusing of 20 keV x-ray radiation using elliptical diffraction zones fabricated inside a silicon crystal have been observed successfully for the first time. These are the lenses with the first and third diffraction orders. An alignment of focusing elements was made with a 10 micron resolution x-ray imaging system. Focusing quality of the lenses have been invesigated by using x-ray sensitive high resolution film and observing, under the light microscope, the shape of the registered focal spot first. And finally by scanning focal plane with pinhole and analysing three dimentional image. Experimens were made in a monohromatic beam. The signal/noise ratio of 300 have been obtained. These experiments open quite wide perspectives in focusing "hard" x-ray radiation up to 100 keV. Authors are hoping that this kind of focusing elements will find there real applications in microdiffraction experiments with energy about 20 keV and higher, in microfluorescence analysis experiments, near edge absorption experiments with micron spacial resolution. A new scheme to focus "hard" x-ray radiation have been suggested. Real geometry parameters for lens with elliptical diffraction zones was calculated for energy 60 keV.

INTRODUCTION.

Originally the idea of usage of elliptical diffraction zones have been directed to realyze the focusing "soft" x-ray radiation with wavelength of about some tens of Angstroms. Artificial multilayer structures have been used as mirrors for this wavelength range. Fabrication of elliptical diffraction zones inside the mirror had to allow to focus reflected radiation. In the first, the possibility to focus "soft" x-ray radiation with elliptical lenses based on multilayer structure was demonstrated at specially built for this purpose vacuum station and beamline at "Siberia I" synchrotron at Kurchatov Institute of Atomic Energy (Moscow) [1]. These lenses had a size of 300 microns by 200 microns and small number of diffraction zones. But bright success of the experiments had allow to put forward a new goal - to create giant elliptical lenses inside effectively reflecting multilayer structures with possibly small period [2]. The main aim was to create x-ray superprobe for microfluorescence analysis and microdiffraction. For these reasons, Sparks has shown in 1980 that the use of synchrotron source for X-ray fluorescence analysis leads to a very important reduction of the detection limits and a reduction of heat or damage to the sample [3]. Three lenses with the same focal length of 40 centimeters designed for 6, 10, and 14 keV were fabricated on one superpolished substrate coverd with a W/C multilayer structure

CP507, X-Ray Microscopy: Proceedings of the Sixth International Conference,
edited by W. Meyer-Ilse, T. Warwick, and D. Attwood

with a period of 28.8 Angstroms and 90 bilayers. The width of each lens was 400 microns. The 14 keV lens had the first, third, and fifth diffraction orders. The 6 keV and 10 keV lenses had the first and third orders. All the four lenses were 12 mm long. Minimum zone size is 0.3 micron. Up to now, this LURE-IMT microprobe is used for mineralogical studies such as micrometeorites, glass inclusions in volcanic rocks [4] and liquid inclusions in coesite-bearing rocks [5], plants investigations [6], microdiffraction experiments [7] and so on.

Experimental part.

This article represents the first experimental step in usage of elliptical diffraction zones

	Focus distance.	Last zone width.	Size of a lens.
Lens N1	18 cm.	0.24 micron	462 microns by 138 microns
Lens N2	29 cm	0.34 micron	534 microns by 160 microns
Lens N3	52 cm.	0.40 micron	800 microns by 237 microns

(Main parameters of a lenses: distance from source to the lens position – 48 meters, selected energy – 20 keV.)

Table 1.

made inside the crystal to focus 20 keV X-ray radiation. List of parameters for different lenses under investigation is shown in Table 1. The main goal was to investigate focusing properties of new lenses and to prove the ability of their utilizeing

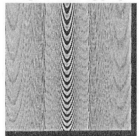
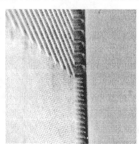

Figure 1.

for high energy resolution in microfluorescence analysis experiments, microdiffraction experiments and near edge absoption experiments. As a substrate a silicon crystal

Figure 2. (last zone size is 0.24 micron)

(111) was chosen because of plasmochemical etching process very good developed for it. At the standard substrates with thickness about 0.5 millimeter diffraction structures with elliptical profile of zones have been created. These are the lenses with first and third diffraction orders. Primary patterns of diffraction structures have been created by

the method of electron beam lithography with JEOL-840 SEM equiped with pattern generator. The diffraction element topology discription was created with ZON program, and later was used by PROXY-WIN writing program. Plasmochemical etching was done with home-constructed ion source based on helicon resonance in SF_6 + $CHClF_2$ plasma. Figure 1 represents the photos made in JEOL-840 SEM of diffraction focusing elemens. Figure 2 represents the photos which demonstrate the quality of plasmochemical etching. The smallest size of diffraction zones for different lenses is 0.24 micron, 0.34 micron, and 0.4 micron and the depth of etching is 1.25 microns. The experiments were made in a monochromatic X-ray beam at the beamline BL47XU at SPring-8. This is the X-ray undulator beamline which is to serve both scientific and technical R&D's for the novel utilization of X-ray undulator beam, especially in the hard X-ray region. applications, methodology for electron beam emittance measurement using emitted synchrotron radiation in the hard X-ray region, and the related instrumentation. An allignment of focusing elements were made with 10 a micron resolution X-ray imaging system wich includes luminofor screen and

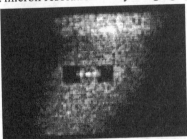

| Figure 3. | Figure 4. |

CCD camera. At figure 3 - the image obtained with this system. Focusing quality of the lenses have been investigated using X-ray sensitive high resolution film and observing, under the light microscope, the shape of the registered focal spot first Figure 4. For two dimensional scan of diffraction pattern at focus plane we used 10 microns pinhole made in a tantalum foil with 200 microns thickness. Signal after pinhole was regested with a detector. The signal to noise ratio of 300 was observed in these experiments.

Conclusion.

In the first focusing of 20 keV radiation with elliptical diffraction zones made inside silicon crystal have been received. The 300 signal/noise ratio have been achieved. The analys made with ZON program allow to conclude that convenient focusing elements can be created for the energy up to 100 keV. The main advantage of Bragg-Fresnel focusing elements is that they can work in a "white" beam, what give a possibility to increase a focused photon flux. Choosing appropriate reflecting medium (artificial multilayer structure; Si(111), Si(220), Ge(111) crystals) it is possible to change energy band width of a reflected beam. This kind of focusing elements will find there real applications in microdiffraction experiments with energy about 20 keV and higher, near edge absorption experiments with micron spacial resolution, in microfluorescence analysis experiments. At 4 keV a Si crystal reflects about 50% of incoming radiation. It means that usage of Bragg-Fresnel focusing elements with elliptical diffraction zones still can have sense especially if the fact that efficiency of a lens at this energy may be 33% is taken into account. The size of a source at BL47XU is 30 microns by 800 microns. This allow to receive high photon flux. To reach some demagnifing uniformity in both directions we propose to use double element optical scheme. As an example of a project lets take the energy 50 keV Fig.5. Bragg-Fresnel lens with elliptical diffraction zones built on Si (111) for this energy will have the following geometrical parameters: (for distance from source to lens 45 meters, and distance from lens to image plane 5 meters, Bragg angle is 2.3 degrees)

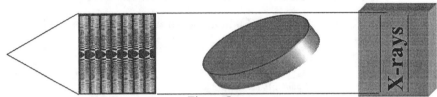

Figure 5.

Last zone size is 0.2 micron – geometry: 14100 microns by 562 microns, number of zones 704. Last zone size is 0.25 micron – geometry: 11300 microns by 450 microns, number of zones 450. From the technological point of view it is reasonable characteristics. Lenses with these parameters fit (overlope) quite well the incoming x-ray beam. Next lens have to be the two dimentional real refractive focusing element. Position with respect to the Bragg-Fresnel lens and its geometry can be calculated. If angle size of the Bragg-Fresnel lens will be larger than the full width at half maximum energy of X-ray beam it is possible to use the crystal with variable period [8].

REFERENCES

1. Aristov, V. V., Erko, A. I., Firsov, A. A., Gaponov, N. A. & Zabelin A. V. (1989). Focusing of soft X-ray radiation by an ellipsoidal Bragg-Fresnel lens. - 2nd European Conference on progress in X-ray synchrotron radiation research, XSR-89.
2. Firsov, A., Svintsov, A., Chevallier, P. & Dhez P. (1996). "Large aperture lenses" Abstracts P9, International Conference on X-Ray Microscopy and Spectromicroscopy, Wurzburg, August 19-23.
3. C. J. Sparks Junior in "Synchrotron Radiation Research", H. Winick & S. Doniach editors (N.Y.; Plenum Press, 1980) pp. 459-512
4. Mosbah, M., Clocchiatti, R., Michaud, V., Piccot, D., Chevallier, P., & Legrand, F. (1995). Als Nilsen G., and Grübel.G., N.I.M. , B104, 481.
5. Philippot, P., Chevallier, P., Chopin, C. & Dubessy, P. (1995). J.Contrib. Mineral. Petrol. , 121, 29.
6. Chevallier, P., Firsov, A., Populus, P. & Legrand, F. (1998). "Caracterisation et speciation chimique a l'echelle micrometrique utilisant le rayonnement synchrotron", Journal de physique IV, v. 8, Pr5-407 Pr5-412 (ISSN 1155 4339) Proceedings: Rayons X et Matiere 97, Ecole Nationale Superieure des Arts et Industries, Strasbourg, 14 - 16 Octobre 1997. J. Phys. IV France 8, Pr5-407 - Pr5-412.
7. Dillmann, P., Pupulus, P., Chevallier, P., Elkaim, E., Fluzin, P., Beranger, G. & Firsov, A. (1997) "Microdiffraction du rayonnement synchrotron. Identification de phases non metalliques dans les alliages ferreux", C. R. Acad. Sci. Paris, t. 324, Serie II b, p. 763-772, Physique appliquee/ Applied physics.
8. Erko A., Schafers F., Gudat W., Abrosimov N. V., Rossolenko S. N., Alex V., Groth S., Shcroder W., "On the feasibility of employing gradient crystals for high resolution synchrotron optics", NIM, A374, p. 408-412, 1996.

X-ray Imaging Microscopy using a Micro Capillary X-ray Refractive Lens

Yoshiki Kohmura, Kyoko Okada, Mitsuhiro Awaji, Yoshio Suzuki, and Tetsuya Ishikawa,

SPring-8,

1-1-1, Kouto, Mikazuki, Sayo-gun, Hyogo, 679-5198, Japan,

Yu.I.Dudchik, N.N.Kolchevsky and F.F.Komarov

Institute of Applied Physics Problems,

Kurchatova 7, Minsk 220064, Belarus

Abstract. A first X-ray imaging microscopy test using the two types of "micro-capillary lens" (kind of the X-ray refractive lenses) has been done at the undulator and bending magnet radiation of SPring-8. One (lens No.1) contained bubbles in glue whereas the other (lens No.2) contained bubbles in glycerol. The lenses had the inner diameters of 0.2 mm and 0.8 mm, respectively. The tests were mainly done using the lens No.2 and the 17-18 keV X-rays. By taking the images of a gold mesh, the spherical aberrations and the field distortions were carefully examined. This lens is superior with a large aperture, 0.8 mm, and a small field distortion, e.g., less than 10 % inside the diameter of 300 μm.

Introduction

Recently various types of X-ray refractive lenses were fabricated, but most of them are made from metallic materials and precise processes are required.[1,2,3] We paid attention to the smooth surfaces at the interfaces between liquids and the gas due to the surface tension. This will be especially powerful for the imaging microscopy experiments.

One of such lens named, the "micro capillary lens"[4] was developed at the Institute of Applied Physics Problems, Belarus and was tested for the first time at BL47XU, at SPring-8.[5] First type of the lens (No.1) was produced by forming bubbles in a glue in a thin glass capillary.[4] The bubbles were formed in glycerol in the second type of the lens (No.2). For both of these lenses, the surface tension of the liquid inside the glass forces the surface to have an approximately spherical shape. For the imaging microscopy experiment described here, the lens No.2 has been used which have the length L= 225 mm, inner radius r=0.4 mm with the number of micro-lenses N =185.

CP507, *X-Ray Microscopy: Proceedings of the Sixth International Conference,*
edited by W. Meyer-Ilse, T. Warwick, and D. Attwood

Setup for Imaging Microscopy experiments

The imaging microscopy test using the "micro capillary lens" was done at BL47XU[6] and BL20B2[7] of SPring-8, a undulator beamline and a bending magnet beamline, using the standard silicon monochromator[8,9].

Figure.1 Magnified image of the lens No.2 using the visible light microscope. The inner diameter of the capillary is 800 μm.

To visualize the spherical aberration and the field distortion, the images of a gold mesh were taken and the distortions from the lattice were evaluated. Figure 1 shows the magnified image of the lens No.2 using the visible light microscope and Fig.2 shows the experimental setup for taking images of the gold mesh using the lens at BL20B2. As for the mesh, a gold mesh manufactured by the Good Fellow Co., Britain was used which had the wire diameter of 5.6 μm with the thickness of ~ 5 μm with a pitch of 16.7 μm. The position-sensitive detector used is called 'beam monitor' and has a spatial resolution of ~10 μm.

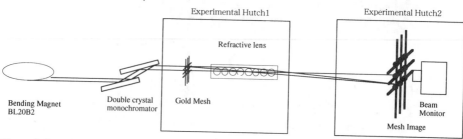

Figure.2. Experimental setup for taking images of a gold mesh at BL20B2.

Ray-Trace Simulations for Images by Micro-Capillary Lens

Experiment was carried out first at BL47XU, undulator beamline, with the magnification factor of around 3. The distance from the lens to the mesh and to the beam monitor were set to ~ 2 m and ~ 5 m, respectively. The focal length of the lens at 17.1 keV was about 1.2 m.

The results of the experiments were compared with the ray-trace simulations assuming perfectly spherical microlenses. For the calculations, parallel incident X-ray beams were assumed (the divergences of the undulator were $\sigma_x = 18$ μrad and $\sigma_y = 8$ μrad at the X-ray energy of ~18 keV). As for the experiment with the gold mesh, the diffraction and the scattering intensities are small at the gold wires and the mesh work, in a good approximation, as a mask along the parallel X-ray beam which does not change the direction of the incident X-rays.

Figure.3(a) shows the image obtained at 17.1 keV. The shadows of the wire became clear at 17.1 keV and the visibility for the near-axis wires approached the expected value of ~ 0.4 at this energy. The fine wires were barely seen without the lens but the magnification by ~ 3.2 enlarged the wire diameters and the pitches to about 16 μm and 53 μm, respectively, and made these wires to be easily resolved by the beam monitor. Figure.3(b) shows the observed image and the calculated one (assuming the focal length of the lens to be 1.2 m) overlapped. Figure.3(b) shows the resemblance of the observed mesh image and the simulated one. The focal lengths (calculated from the separation of the wires) was roughly proportional to the square of the X-ray energy as theoretically expected.

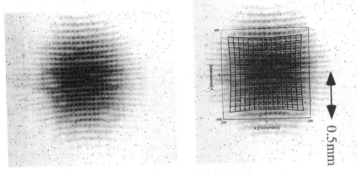

Figure.3. Image of a gold mesh (Good-Fellow Co.) using the lens No.2 with the magnification factor of ~3. The left panel (a) shows the observed image around the focus whereas the right panel (b) shows the shadows of the mesh calculated by ray-trace simulations overlapped on the observed image.

High-Magnification Imaging Microscopy

BL20B2 is a long beamline and experimental hutches 1 and 2 have the separation of about 160 m.[7] The aim of the experiment at BL20B2 was to achieve the high-magnification imaging microscopy. Since the focal length of the lens No.2 were around 2.1 m (at 16.5 keV), images with the magnification factor of around 80 could be obtained.

Mesh images were taken at the X-ray energies of 16.5 keV around the focus. Figures.4(a) and 4(b) show an image near the focus ($L_2 \sim 217$ cm) and an off-focus image

(L_2~187cm). The black part correspond to the area with high X-ray intensities. Fig.4(a) shows that the image near the focus is dominated by absorption contrast and the position of the wires appear white .Inside the black square corresponding to the aperture of the mesh, fine fringe structures were seen. This means that interference effect is slightly observed. In Fig.4(b), on the other hand, the center of the wires appear black and apparently a phase reversal is seen. The gold mesh is acting as a phase object and a kind of holographic image is obtained.

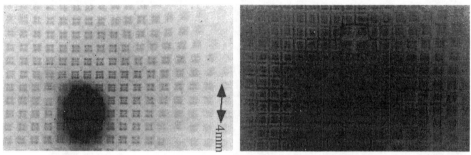

Figure.4. Image of a gold mesh (Good-Fellow Co.) using the lens No.2 with the magnification factor of ~80. The left panel (a) shows the image obtained around the focus (L_2~217cm) whereas the right panel (b) shows the image at off-focus position (L_2~187cm).

The bright spots appearing on the left in both images are due to the 51 keV X-rays by the 3rd order reflection plane of the monochromator crystals. In Fig.4(b), there are two square areas where the contrasts of the wires are nearly reversed (from black to white). These are probably due to the inhomogenuity inside the glue for sealing the liquid or inside the glycerol. Similar contrast reversal is also seen on the right hand side of Fig.4(b). At this position, wires show slight field distortions from the lattice and the images were formed by the off-center flux of the lens.

Summary

First X-ray imaging microscopy test using the micro-capillary lens (X-ray refractive lens) has been successfully done at SPring-8. The one used (lens No.2) contained bubbles in glycerol inside a glass capillary of the inner diameter of 0.8 mm. The spherical aberrations and the field distortions were carefully examined by taking the images of the gold mesh.

The mesh image showed close resemblance to the image by ray-trace simulation which suggested that "micro-capillary-lens" reported here is suitable for fabricating a series of spherical microlenses.

The images at focus was absorption-dominated while the off-focus image showed strong interference effect due to the phase shift caused by the gold. However the inhomogenuity in the lens material need to be improved for better images.

References

(1) A.Snigirev, et al., *NATURE*, **384**, 49-51 (1996)

(2) P. Elleaume, *Journal of Synchrotron Radiation*, **5**, 1 (1997)

(5) B.Lengeler, et al., *Applied Physics Letters*, **74(26)**, 3924-3926 (1999)

(4) Yu.I.Dudchik, N.N.Kolchvsky, *Nucl. Instrum. and Methods.*, **A, 421**, 361-364 (1999)

(5) Y.Kohmura, et al., *Rev.Sci.Instrum.,* in press

(6) Y.Kohmura, *SPring-8 User Infornation Newsletter (in Japanese)*, **3(4)**, 28 (1998)

(7) K.Umetani, et al., *SPring-8 User Information Newsletter (in Japanese)*, **4(3)**, 50 (1999)

(8) T.Uruga, et al., *Rev. Sci. Instrum.*, **66(2)**, 2254-2256 (1995)

(9) T.Ishikawa, et al., *Proc. of SPIE*, **3448**, 2-10 (1998)

NOVEL APPROACHES TO COHERENT IMAGING

Demonstration Of Phase Contrast In Scanning Transmission X-ray Microscopy: Comparison Of Images Obtained At NSLS X1-A With Numerical Simulations

François Polack[1], Denis Joyeux[2], Michael Feser[3], Daniel Phalippou[2], Mary Carlucci-Dayton[3], Konstantyn Kaznacheyev[3], and Chris Jacobsen[3]

[1] *LURE, bât. 209 D , Centre Universitaire Paris XI, BP 34, F-91898 ORSAY Cedex - FRANCE*
[2] *Laboratoire Charles Fabry, Institut d'Optique - BP 147 - 91403 Orsay, FRANCE*
[3] *Physics Department, State University of New York at Stony Brook, Stony Brook, NY , USA*

Abstract. Unlike transmission X-ray microscopy, scanning transmission X-ray microscopy (STXM), till now, does not allow phase contrast. Several method have been suggested but no proof of practical feasibility has been yet given. Here we analyze the methods based on the detection of the small beam deflection induced by the object phase gradient, by a segmented detector. It is shown that structuring the zone plate illumination potentially improves the detection. A diffractive beam profiler has been constructed to condition the beam of the NSLS X1A STXM. Recent images are shown, which, compared to numerical simulations, indicate the presence of phase contrast

INTRODUCTION

Compared to transmission X-ray microscopy (TXM), scanning transmission microscopy (STXM) has till now the obvious disadvantage that it does not allow phase contrast. Phase contrast has however been successfully applied to TXM, and the published images clearly reveal the potential advantages of this type of imaging[1]. Biological materials have usually very similar chemical compositions. Small objects embedded in a closely related matrix may not appear by absorption contrast only, while adding a contrast related to the phase difference may reveal their presence.

Moreover, specimen damage due to the absorbed radiation dose is one of the big issues in X-ray microscopy. The only way to reduce this dose is to base the image formation on a contrast mechanism different from absorption. Phase contrast should allow to increase the energy above the usual soft X-ray range, because absorption decreases more rapidly than phase shift.

For all these reasons any method to extend the phase contrast mode to STXM is worth consideration. In this paper we analyze the principles of phase detection in scanning microscopy, and describe the progress made on an approach based on structured illumination of the pupil and segmented detection in the object far field.

CP507, *X-Ray Microscopy: Proceedings of the Sixth International Conference,*
edited by W. Meyer-Ilse, T. Warwick, and D. Attwood

PRINCIPLE OF DIFFERENTIAL PHASE CONTRAST

The detection of the phase shift induced by an object can be done by two main ways, Zernike phase contrast and differential phase contrast (also often called Nomarski phase contrast). In the first case, one detects the phase difference between the object-diffracted and undiffracted light. This is possible with an imaging microscope because in the backfocal plane of the objective, i.e. the Fourier plane, the contributions of different spatial frequencies are separated and can be selectively phase shifted. It cannot of course be applied to a scanning microscope where the signal is detected directly behind the object. In the second case, local phase gradients are detected through the small phase shift between two close points of the sample. This principle can still be applied to a scanning microscope without an imaging objective, though the implementation has to be rather different from the original Nomarski scheme.

Detection Of Phase Contrast In A Scanning Microscope

In a scanning microscope only a small region of the object is illuminated. If the object has a local phase gradient, it acts as a small prism and the emerging beam is slightly deflected from the position where it stands when no object is present (figure 1.). This deflection can be detected by segmented detector as proposed in references 2 to 4. However, the deflection angle is usually small with respect to the aperture angle that is required to achieve a high resolution. Considering an image, critically sampled with a pixel size roughly equal to the resolution element, the phase gradient required to shift the beam of half its diameter is $\lambda/4$ per pixel. To give an idea of the object modulation, this value corresponds, for a biological object at $\lambda \approx 3nm$, to a thickness variation around 0.5 μm per pixel. It is therefore reasonable to look for maximum sensitivity to small deflections.

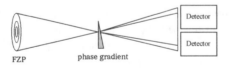

FIGURE 1. Detection of phase gradient in the STXM. The beam may have a particular intensity profile to help the detection of small beam shift by the segmented detector.

Another approach is to go back the Nomarski scheme and consider two close points of the object. We suppose that we illuminate them with two wavefronts having a constant phase relationship. This implies that in the far field, as well upstream as downstream from the focal plane, the complex amplitude inside the pupil aperture is modulated by a cosine fringe. Now, if we introduce a sample which slightly changes the phase relationship by $\Delta\varphi$, then in the far field the fringe is shifted by a fraction of the period equal to $\Delta\varphi/2\pi$. Small shifts are better detected on the sides of the fringe where the intensity gradient is the larger[5,6]. In référence 6, we previously suggested to modulate the amplitude incoming on the zone plate by a cosine fringe having a maximum in the center of the zone plate and a period of about 1.6 times the zone plate diameter (period of the intensity pattern, p =0.8 D). The detection is made differentially with two detectors of width p/3, separated by a blind region of same width.

It is obvious that the two-point description is a view of the mind and that there is a limited but continuous zone of illumination in the focal plane. If the object phase gradient is uniform in this region, the global beam deflection and the fringe shift are the same. If the gradient is not locally uniform the pupil and the fringe pattern are distorted and become the convolution of the pupil illumination by the diffraction pattern of the local modulation. The two detection scheme actually differ by the way the useful signal is distributed in the pupil plane as shown in the next paragraph.

First Order Signal Analysis

We consider here that by use of a convenient beam profiler, the zone plate is illuminated by a definite complex amplitude. We denote $P(\theta)$ the complex amplitude distribution in the zone plate pupil plane. The object illumination is the Fourier transform of $P(\theta)$ that we denote $\tilde{P}(\mathbf{x})$. We assume that the $P(\theta)$ spatial modulation is slow enough that the spot size is not significantly increased with respect to the uniform illumination. We also assume that the object is weakly modulated and this modulation is also slowly varying, so that the object transmittance can be developed to the first order in the illuminated region as

$$T(\mathbf{x}) = 1 - O(\mathbf{x}_0 + \mathbf{x}) = 1 - O(\mathbf{x}_0) + \mathbf{x} \cdot \nabla O(\mathbf{x}_0) + 2^{nd}\, order \qquad 1$$

where $\mathbf{x}$ is the space variable in the zone plate focal plane, $O(\mathbf{x})$ is the object modulation in amplitude, $O(\mathbf{x}) \ll 1$, and $\mathbf{x}_0$ is the point of the object which is located at the center of the X-ray spot during the considered step of the scanning process.

Just behind the object, the complex amplitude is the product $\tilde{P}(\mathbf{x}) T(\mathbf{x})$, and we obtain the amplitude $D(\theta)$ in the detector plane by a Fourier transform with respect to the variable $\mathbf{x}$:

$$D(\theta) = TF\left\{\tilde{P}(\mathbf{x})\left[1 - O(\mathbf{x}_0)\right] - \tilde{P}(\mathbf{x})\,\mathbf{x} \cdot \nabla O(\mathbf{x}_0)\right\} \qquad 2$$

Which, using the properties of derivation of the Fourier transform can be rewritten as:

$$D(\theta) = P(\theta)\left[1 - O(\mathbf{x}_0)\right] - \frac{\lambda}{2 i \pi} \nabla P(\theta) \cdot \nabla O(\mathbf{x}_0) \qquad 3$$

Finally the intensity is obtained by squaring the amplitude and, with a weak object, we can neglect second order terms in $O(\mathbf{x}_0)$ and $\nabla O(\mathbf{x}_0)$, and get :

$$|D(\theta)|^2 = |P(\theta)|^2 \left\{1 - 2\,Re\left[O(\mathbf{x}_0)\right]\right\} - \frac{\lambda}{\pi} Im\left[P^*(\theta)\,\nabla P(\theta) \cdot \nabla O(\mathbf{x}_0)\right] + 2^{nd}\, order \quad 4$$

In order to better distinguish the behavior of the real and imaginary parts of the object modulation, we let appear the amplitude and phase of the pupil function and their gradients with :

$$P(\theta) = |P(\theta)| \, e^{i\phi(\theta)}; \qquad \nabla P(\theta) = \nabla |P(\theta)| \, e^{i\phi(\theta)} + i \, P(\theta) \nabla \phi(\theta) \qquad\qquad 5$$

Hence :

$$|D(\theta)|^2 = |P(\theta)|^2 \left\{ 1 - 2 \operatorname{Re}[O(\mathbf{x}_0)] - \frac{\lambda}{\pi} \operatorname{Re}[\nabla O(\mathbf{x}_0)] \cdot \nabla \phi(\theta) \right\}$$
$$- \frac{\lambda}{2\pi} \nabla |P(\theta)|^2 \cdot \operatorname{Im}[\nabla O(\mathbf{x}_0)] \qquad\qquad 6$$

Thus, the signal received in the detection plane has three components. The first one is proportional to the local intensity transmission factor of the object and the pupil intensity, it could be called the normal signal. The second signal is proportional to the real part of the object gradient, the pupil phase gradient and the pupil intensity. This signal appears for instance when the zone plate is out of focus or if individual zones are not perfectly registered. The third signal is the phase signal we are looking for; it is proportional to the object phase gradient and the gradient of the pupil intensity.

Optimum Detection Scheme

The signal distribution is quite different in the two detection schemes we have considered. In the first case, the pupil is uniform and the phase signal is only located on the pupil edge. Zone plates are usually not as perfect on the outside as in the center, because placement errors are more likely to happen at high frequency. These phase gradients may bring in some parasitic amplitude signal.

On the contrary if the pupil intensity is modulated by a sine fringe with a maximum on the zone plate axis, the intensity gradient is more evenly distributed. The signal can be integrated on a larger area, therefore a larger fraction of the energy contributes to the signal. The regions of maximum signal are also located in the mid-frequency range of the zone plate where the pattern is expected to be more accurate. The functions $|P(\theta)|^2$ and $\nabla |P(\theta)|^2$ which modulate the local contribution of the intensity and phase signals are respectively even and odd. Two identical detectors are symmetrically disposed in the detection plane and therefore the intensity signal is obtained as the sum of the detector signals and the phase signal as their difference.

The geometry of the detection can be optimized to get the maximum phase signal. When the circular shapes of the zone plate and of its central obstruction are taken into account, one obtains the parameters we have already given. These parameters however are not critical.

BEAM PROFILING TECHNIQUES

Young's fringes

In a STXM, the zone plate is illuminated by a coherent wavefront. It is thus possible to produce the desired sine fringe by the diffraction from a pair of slits, inserted at some distance upstream the zone plate. This set up was used for the first experiments we did on the X1A microscope at NSLS. In order to obtain the proper fringe period, the two slits were set at 570 mm from the zone plate. The slit width were 18 μm and they were separated by a 18 μm wide opaque region. Though it was possible to see the fringe shift with this set-up, it was not stable enough to get differential images. The X1A STXM is supported by an air suspended table, but due to the distance it was not possible to attach the slits to the same table. With the set up parameters a relative displacement of 1 μm between the slit and the zone plate produces a spurious signal equivalent to a phase shift of λ/40.

Diffractive Beam Profiler

For the most recent experiments, a new beam profiler has been constructed. It is based on the variation of the 0th order efficiency of a binary grating, versus the groove duty cycle.[7] When the phase shift due to the groove depth is π, this efficiency is related to the duty cycle, c (defined as the groove width to period ratio) by :

$$E_0 = \frac{1-2c}{2\pi}, \quad c \in [0,1] \qquad 7$$

The wanted continuous and signed amplitude modulation, can be obtained by modulating the line width (LW) of a high carrier frequency binary grating. The construction parameters are the following: Carrier period 46 μm, period of LW modulation 7.4 mm, ruled length 24 mm. The grating is used in grazing reflection at 1 deg, in the 2 to 4.5 nm wavelength range. The minimum distance at which the orders are separated is 100 mm; it is actually set at 400 mm from the zone plate, but is rigidly linked to the microscope table. The grating was designed in order that under parallel illumination the period of projected modulation is 66 μm, adapted to a 80 μm diameter ZP. Unfortunately the beam from the monochromator is diverging, so that the projected period is actually 100 μm.

The grating beam-profiler is incorporated in an assembly of four reflectors (figure 2). The grating is RIE etched in a layer of amorphous silicon. It does not occupy the whole width of the blank, but leaves a 5 mm reflecting track which can be used to reject the higher harmonics from the monochromator. The thickness of the Si layer varies from 28 nm on one side of the grating to 74 nm on the other. For each wavelength, the best modulation (π phase shift) is achieved by the translation of the whole device. The 15 mm tracks of the grating and mirrors which are used with the beam profiler are nickel coated. On the harmonic rejection path, the external. mirrors (1deg grazing) are chromium coated, the internal ones (3deg grazing) are bare silica.

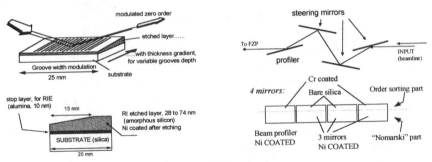

FIGURE 2. Principle of the beam profiler and schematics of the harmonics rejection device

EXPERIMENTS

Demonstrating a phase contrast in soft X-ray is not an easy task because materials are both absorber and phase shifter. The influence of parasitic amplitude signal can be reduced if the sample is immersed in a medium of comparable absorption coefficient but different refractive index. Such a close match occurs for instance with silica and toluene in the 2.5 to 2.6 nm wavelength range.

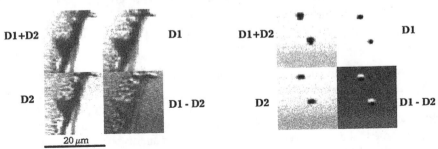

FIGURE 3. STXM images of 2μm diameter silica spheres in toluene. Left group: cluster of several spheres; right group: two isolated spheres. Individual images show the D1 and D2 detector signals and their combinations, D1+D2 is the amplitude signal, D1-D2 is the phase signal.

We used silica beads (1- 2 μm diameter), and maintained them in toluene between two sealed Si_3N_4 membranes. 2μm thick silica alone has a transmission factor of ~ 10% and a 1.1 λ optical thickness. In toluene, the transmission difference is only a few percent, but the difference in optical thickness is still 0.5 λ. The sample however is difficult to manipulate. Capillary forces tends to exclude toluene from the center of the window. It is useful to reduce the thickness of the liquid layer, but the membranes cannot withstand the stress for a long time. We have been able to make only a few images, with a limited spatial resolution. Figure 4 shows two regions of the sample, one with a cluster of silica spheres, the other with two isolated spheres. Both images are

100x100 pixels, 0.2 μm each. The four views of each region correspond respectively to the signal from each detector, their sum and their difference. On the difference one can see the contrast reversal which is characteristic of differential phase contrast.

COMPARISON WITH NUMERICAL SIMULATIONS

To convince ourselves that the contrast we see is actually due to phase shifts, we made computer simulation of the image which are expected from these simple objects in various situations. The computation procedure is rather straightforward.
1) The pupil illumination defined by the zone plate geometry and the fringe period is sampled on a 128 x128 complex array, and Fourier transformed to obtain the sample illumination
2) For one particular position of the object with respect to the probe, the sample illumination is multiplied by the complex transmission factor of the object, and the result is Fourier transformed to obtain the complex amplitude in the detection plane.
3) The intensity is computed and integrated on each detector window.
The sample is then moved to the next pixel and step 2 and 3 are repeated until the whole image is computed.

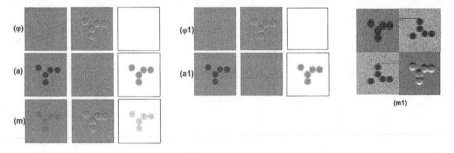

FIGURE 4. Simulated images of a pattern of 5 spheres of 2μm diameter in a 12 μm field. In the two left groups (φ to a1) the 1st column is the detector sum D1+D2 (amplitude), the 2nd column is the difference D1-D2 (phase), and the 3rd column is the image in normal transmission mode: φ) pure phase object, λ/2 maximum phase shift; a) pure amplitude, object minimum transmission 50%; m) mixed contrast as silica spheres in toluene, max. phase shift λ/2, minimum transmission 85%; φ1) and a1) same as φ) and a), but the probe is out of focus by 3μm; m1) shows the simulated 4 signals from silica spheres not immersed in toluene, out of focus by 6 μm; same image arrangement as in figure 3.

Figure 4 shows the results in some particular conditions. First we look at pure phase and amplitude objects with a probe in perfect focus, phase and amplitude images are perfectly discriminated as well with pure phase or amplitude objects (φ/ and a/ images) as with a sample presenting a mixed contrast like silica spheres in toluene (m). The contrast reversed features which appears in the phase images have a higher resolution in the simulation than on the STXM image. This is an indication the STXM was out of focus.

The second, set of images simulates what happen when the probe is defocused by

3μm. While the 1st order analysis says the signal remains perfectly discriminated, the simulations show a weak cross-talk signal. This signal, however, cannot explain by itself the observed contrast in the phase image even for a larger defocusing.

The last image simulate the case of a dried-up sample of silica spheres without toluene and a strong 6 μm defocusing. The phase image looks quite similar to the STXM image but the contrast of the amplitude image is stronger on the simulation. We therefore conclude that our images are actual phase images, taken with a rather out of focus probe and that our sample suffered from some loss of toluene.

CONCLUSION

It has been shown that in the first order approximation, three object related terms are present in the detection plane of a STXM, the intensity transmission factor, and the real and imaginary part of the gradient of the complex transmission. According to the illumination of the zone plate, these terms have different symmetry of distribution and can be separately recorded by segmented detectors. An optimized detection scheme was described which includes, for stability, a 0th order diffractive beam profiler to modulate the zone plate illumination. Preliminary images have been compared to computer simulations. The contrast observed can only be explained by the presence of a phase contribution.

REFERENCES

1. Schmahl, G., Guttmann, P., Schneider, G., Thieme, J., Niemann, B., and Wilhein, T., *Synchrotron radiation News*, **7**, 19-22,. (1994)

2. Morrison, G. R., « Phase Contrast and Darkfiel Imaging in X-ray Microscopy »,in *Soft X-ray Microscopy*, edited by C. J. Jacobsen and J. E. Trebes, Proc.SPIE 1741, 1992, pp. 186-193

3. Morrison, G. R. and Niemann, B., « Differential Phase Contrast X-Ray Microscopy», in: *X-ray Microscopy and Spectromicroscopy,* edited by J. Thieme et al., Springer Verlag, 1998, pp. I-855 - I-94

4. Kirz, J., Jacobsen, C., and Howells, M., *Quarterly Reviews of Biophysics* **28**, 33-130 (1995)

5. Polack, F., and Joyeux, D., « Soft X-ray interferential microscopy: a feasibilityassessment » in: *X-ray Microscopy IV*, edited by A. I. Erko and V. V. Aristov, Bogorodski Pechatnik, Chernogolovka, 1994

6. Polack, F., Joyeux D., and Phalippou,D.,« Phase Contrast Experiments on the NSLS-X1A Scanning Microscope », in: *X-ray Microscopy and Spectromicroscopy,* edited by J. Thieme et al., Springer Verlag, 1998, pp. I-105 - I-110

7 Joyeux, D., and Polack, F., « A Wavefront Profiler as an Insertion Device for Scanning Phase Contrast Microscopy », in: *X-ray Microscopy and Spectromicroscopy,* edited by J. Thieme et al., Springer Verlag, 1998, pp. II-201 - II-206

Extending The Methodology Of X-ray Crystallography To Allow X-ray Microscopy Without X-ray Optics

Jianwei Miao[*], Pambos Charalambous[†], Janos Kirz[*] & David Sayre[*]

[*]Department of Physics & Astronomy, State University of New York,
Stony Brook, New York 11794-3800, USA
[†]Kings College, Strand, London WC2R 2LS, UK

Abstract. We demonstrate that the soft X-ray diffraction pattern from a micron-size non-crystalline specimen can be recorded and inverted to form a high-resolution image. The phase problem is overcome by oversampling the diffraction pattern. The image is obtained using an iterative algorithm. The technique provides a method for X-ray microscopy requiring no high-resolution X-ray optical elements or detectors. In the present work, a resolution of approximately 60 nm was obtained, but we believe that considerably higher resolution can be achieved.

1. INTRODUCTION

X-ray microscopy is a well-established technique to image micron-size specimens at submicron resolution[1,2]. The resolution of the technique is limited by the resolution of the X-ray optics -- commonly a zone plate -- and, for biological specimens, the radiation damage. While the radiation damage problem can be mitigated somewhat by using cryogenic techniques, the resolution of the zone plate is limited by fabrication difficulties. At present, the highest resolution achievable is around 30 - 50 nm[3,4]. To get considerably higher resolution, much shorter wavelength particles such as electrons are needed for imaging. While an electron microscope can achieve less than 1 nm resolution, it can only study thin (< 0.5 μm) specimens due to the penetration length and contrast mechanism of the electrons. To obtain even higher resolution, X-ray crystallography is employed which needs not only short wavelength X-rays, but also crystalline specimens. Constructive interference among the large number of identical unit cells generates strong Bragg peaks by which X-ray crystallography can achieve atomic resolution without serious radiation damage to the specimens. Although X-ray crystallography becomes such a powerful technique to image specimens both in material science and structural biology, it is only applicable to crystalline specimens, while most biological specimens, for example, can not be or are too big to be crystallized. That the methodology of X-ray crystallography may be extended to non-crystals was first proposed by Sayre in 1980[5]. This extension, by combining X-ray Crystallography with X-ray Microscopy, eliminates the necessity of crystallization and high resolution X-ray optics. Although in principle this technique can obtain three-dimensional image of whole biological cells and complex sub-cellular biological specimens at high resolution, it faces two challenges. Firstly, when the specimen is non-crystalline, the diffraction pattern is faint and continuous, which poses the challenge to record such a diffraction pattern[6,7]. The second challenge is the

CP507, *X-Ray Microscopy: Proceedings of the Sixth International Conference,*
edited by W. Meyer-Ilse, T. Warwick, and D. Attwood
© 2000 American Institute of Physics 1-56396-926-2/00/$17.00

well-known 'phase problem', the usually unavoidable loss of the phase information in the diffraction intensity. It was not until recently that we reported the first successful recording and reconstruction of the diffraction pattern from a non-crystalline specimen[8]. We report here detailed information of our experiment and a second successful reconstruction at higher resolution with less reconstruction time.

2. RECORDING THE DIFFRACTION PATTERN

Our experiment was performed at the X1A beamline at the National Synchrotron Light Source (NSLS). We used an entrance slit, grating and exit slit to select monochromatic X-rays. The widths of the entrance and exit slits were set at 40 and 23 μm, respectively, which corresponds to a resolving power of about 850 for soft X-rays with λ = 1.7 nm. Our experimental chamber, as Fig. 1 shows, was mounted about 80 cm downstream of the exit slit. The first element inside the chamber was a 10 μm pinhole which was used to generate a small, collimated beam and also assured spatial coherence. The pinhole could be manually adjusted by an X-Y stage mounted outside of the chamber. To limit the effect of the scatter from the edge of this pinhole, the specimen was placed only about 30 μm from the corner of the silicon nitride membrane, allowing the silicon support to protect three quadrants of the detector from the scatter. The silicon nitride membrane was mounted on a commercial cryogenic sample holder. The holder could be manually adjusted in both the X and Y direction and could also be rotated by the stages outside the chamber. The detector, a back-thinned, liquid nitrogen cooled CCD with 512 x 512 pixels and a 24 μm x 24 μm pixel size, was placed downstream of the specimen at a distance of 25 cm. To position the specimen to the small beam spot, an optical microscope mounted above the chamber was used for coarse alignment and the CCD detector for fine alignment. In front of the CCD detector was a 220 μm in diameter wire which was used as beam stop and could be adjusted both in the X and Y direction. A photodiode could be inserted between the beam stop and the CCD to monitor the beam intensity in the presence or absence of pinhole, beam stop and specimen, and to help to align these components in the beam. The entire chamber was in vacuum with pressure around 10^{-5} torr and an air-lock for rapid sample change.

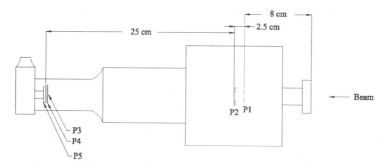

FIGURE 1. Schematic layout of the diffraction chamber. P1: Pinhole, P2: Specimen, P3: Central Stop, P4: Photodiode, and P5: CCD.

By using this setup, we studied a non-crystalline test specimen. The specimen was a collection of gold dots, each ~ 100 nm in diameter and 80 nm thick, deposited on a 100 nm thick silicon nitride membrane, to form a set of six letters. Fig. 2a shows a Scanning Transmission X-ray Microscopy (STXM) image of the specimen. The image was recorded on the X1A beamline at the NSLS. Fig. 2b is a diffraction pattern of the specimen, in which the fourth-quadrant data were obtained by using central symmetry. Since the central region was obscured by the beam stop, we replaced the central area of the X-ray diffraction pattern by a patch consisting of the low-resolution part of the squared magnitude of the Fourier transform of the STXM image of Fig. 2a. The patch, a circular area with a 19-pixel radius area, occupies less than 0.5% of the whole diffraction pattern. The pattern with exposure time of about 5 min extends to the edge of the CCD, suggesting that a larger detector would directly lead to higher spatial resolution.

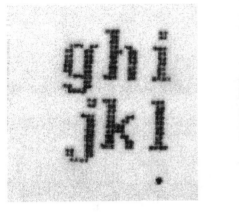

a

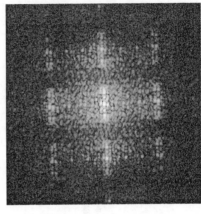

b

FIGURE 2 a. A STXM image of the specimen, b. A diffraction pattern of the specimen (using a logarithmic scale).

3. RECONSTRUCTING THE DIFFRACTION PATTERN

The intensity of the diffraction pattern provides a record of the magnitude, but not the phase, of the structure information. To invert an image from the diffraction pattern, one faces therefore the phase problem. The situation for the non-crystalline specimens is different, however, in that the diffraction pattern is faint and continuous instead of a collection of discrete Bragg peaks. This continuous pattern can therefore be sampled on a finer scale. That the oversampling technique may lead to the phase information was first suggested by Bates[9]. Most recently, we made progress in understanding the applicability of the oversampling technique to the phase problem. We proposed a theory to explain the oversampling technique and showed that Bates's criteria can be relaxed somewhat for the higher-dimensional cases[10-12]. To apply the oversampling technique for the phase retrieval, we developed an iterative algorithm by modifying

Fienup's[13]. Each iteration of the algorithm consists of the following steps. From the magnitude of the Fourier transform and a guessed set of phases (a random set phase for the initial iteration), a Fourier transform pattern can be obtained. By applying inverse Fourier transform to the pattern, we get an image in the real space. We then enforce two kinds of constraints on the image; (i). If the diffraction pattern is oversampled, we define a finite support based on the oversampling degree, and drive the pixel value outside of the support close to zero. (ii). Inside the finite support, positivity constraints are enforced either on the real part or the imaginary part. After these processes, a new image is generated. By applying Fourier transform on the new image, we get a new pattern in the Fourier space. We adopt the phases from the pattern (also restoring the phase of the central pixel to zero) and thereby obtain a new guessed phase set. Usually, after a few hundred iterations, the correct phase set is retrieved. For detailed information, one may refer to previous publications[6, 10-12].

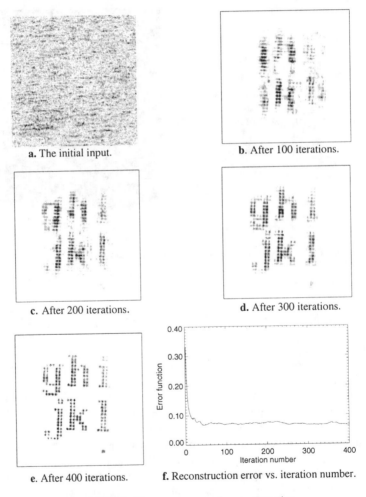

a. The initial input.

b. After 100 iterations.

c. After 200 iterations.

d. After 300 iterations.

e. After 400 iterations.

f. Reconstruction error vs. iteration number.

FIGURE 3. The convergence of a reconstruction.

By employing the algorithm mentioned above, we reconstructed the diffraction pattern of Fig. 2b. Fig. 3 shows the convergence of a reconstruction with a finite support of a 7.5 μm x 7.5 μm square. Fig. 3a is an image processed from the experimental diffraction pattern and a random phase. Figs. 3b, 3c, 3d and 3e show the reconstructions after 100, 200, 300 and 400 iterations, respectively. Interestingly, the image was rotated 180° between Figs. 3b and 3c. This is due to the fact that one can not distinguish a phase and its conjugate from a diffraction pattern alone. Fig. 3f shows the convergence of the reconstruction error vs. the iteration number. The error function was defined as the ratio of the sum of the pixel values outside the finite support to the sum of the total pixel value[10]. Since the image reconstructed from an oversampled diffraction pattern was confined inside the finite support, the error function of a correct phase should be zero for a noise free diffraction pattern, but was only close to zero for an experimental pattern. Fig. 3f implies that the reconstruction

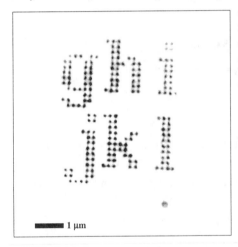

converged very fast during the first 50 iterations, and more slowly after that. After 400 iterations, a high quality image (Fig. 4) was obtained. The computing time of 400 iteration is ~ 15 min on a 450 MHz Pentium II workstation. Fig. 4, the same image as Fig. 3e but interpolated for display purposes, is consistent with the resolution limit, ~ 60 nm, set by the angular extent of the CCD detector. We also performed a few more reconstructions from the diffraction pattern with different initial phases, and found that the convergence speed is somewhat different in each case.

FIGURE 4. A high quality reconstruction from the diffraction pattern (after 400 iterations).

4. CONCLUSION

We believe that the successful recording and reconstruction of the test object opens a door for high-resolution three dimensional imaging of biological cells and complex sub-cellular biological structure without high resolution X-ray optics. To get 3D images, we have to record a series of diffraction patterns by rotating the specimen perpendicular to the beam, which will increase radiation damage to the specimen. We hope to circumvent this problem by using cryogenic technique of cooling the specimen down to liquid nitrogen temperature. Previous experiments have shown that biological specimens at this temperature can stand up to 10^{10} Gy dosage without observable morphological damage[14,15].

ACKNOWLEDGEMENTS

The decision to try oversampling as a phasing technique was arrived at in a conversation in the late 1980s with G. Bricogne. W. Yun and H. N. Chapman also participated in early parts of this experiment. These contributions are gratefully acknowledged. We are grateful to C. Jacobsen for his help and advice especially with the numerical reconstruction. We thank him and M. Howells for use of the apparatus in which the exposures were made, and S. Wirick for her assistance in the data acquisition. P. Charalambous thanks the Leverhulme Trust Great Britain for supporting the nanofabrication program at King's College, London. This work was carried out at the National Synchrotron Light Source, which is supported by the Department of Energy. Our work is supported in part by grant # DE-FG02-89ER60858 from the Department of Energy.

REFERENCES

1. Kirz, J., Jacobsen, C., and Howells, M., *Q. Rev. Biophys.* **28**, 33-130 (1995).

2. Sayre, D., and Chapman, H. N., *Acta Cryst.* A **51**, 237-252 (1995).

3. Jacobsen, C., Kirz, J., and Williams, S., *Ultramicroscopy* **47**, 55-79 (1992).

4. Thieme, J., Schmahl, G., Umbach, E., and Rudolph, D. (eds), *X-ray Microscopy and Spectromicroscopy,* Berlin: Springer-Verlag, 1998.

5. Sayre, D., "Prospects for long-wavelength x-ray microscopy and diffraction" in *Imaging Processes and Coherence in Physics*, edited by Schlenker, M. et al, Springer, Berlin, 1980, pp. 229-235.

6. Sayre, D., Chapman, H. N., and Miao, J., *Acta Cryst.* A **54**, 233-239 (1998).

7. Yun, W. B., Kirz, J., and Sayre, D., *Acta Cryst.* A **43**, 131-133 (1987).

8. Miao, J., Charalambous, C., Kirz, J., and Sayre, D., *Nature* **400**, 342-344 (1999).

9. Bates, R. H. T., *Optik* **61**, 247-262 (1982).

10. Miao, J., Sayre, D., and Chapman, H. N., *J. Opt. Soc. Am.* A **15**, 1662-1669 (1998).

11. Miao, J, Chapman, H.N., and Sayre, D., *Microscopy and Microanalysis* **3** (supplement 2), 1155-1156 (1997).

12. Miao, J., and Sayre, D., "Solving the Phase Problem by Using the Oversampling Technique," in preparation.

13. Fienup, J. R., *Appl. Opt.* **21**, 2758-2769 (1982).

14. Schneider, G., and Niemann, B., "Cryo X-ray microscopy experiment with the X-ray microscope at BESSY." in *X-ray Microscopy and Spectromicroscopy,* edited by Thieme, J., Schmahl, G., Umbach, E., and Rudolph, D., Berlin: Springer-Verlag, 1998, pp. 25-34.

15. Maser, J. et al., "Development of a cryo scanning x-ray microscope at the NSLS." in *X-ray Microscopy and Spectromicroscopy,* edited by Thieme, J. et al., Berlin: Springer, 1998, pp. 35-44.

A modern approach to x-ray holography

M. R. Howells[1], B. Calef[2], C. J. Jacobsen[3], J. H. Spence[4], and W. Yun[1]

[1]Advanced Light Source, Lawrence Berkeley National Laboratory, Berkeley, CA 94720, USA
[2]Department of Mathematics, University of California, Berkeley, CA 94720, USA
[3]Department of Physics, State University of New York, Stony Brook, NY 11794, USA
[4]Department of Physics and Astronomy, Arizona State University, Tempe, AZ 85287, USA

Abstract. We consider technical approaches to the problem of making 3-D images of large (>10 µm) life-science samples with about 10 nm resolution . We find no existing methods and discuss the possibility of using soft x-ray diffractive techniques especially holography. We propose a new form of Fourier transform x-ray holography using a reference object consisting of an array of small pinholes formed by etching nuclear tracks. This scheme promises high resolution and a simple way to determine both the phase and amplitude of the diffracted wavefield.

INTRODUCTION

Soft x-rays already offer various useful capabilities for imaging life-science samples[21]. In this paper we discuss their possible use in the difficult problem of imaging the so-called molecular machines inside the cell. These objects are large assemblies of proteins which function as a group. They cannot normally be crystallized so their structure is hard to determine. The detailed atomic-resolution structure of the individual proteins may or may not be known from crystallographic studies but the way the macromolecules fit together, which is important to their function, is not known. This poses a microtomography problem at a resolution of 3-12 nm which we discuss in this paper.

The usual approaches to determining structure in life-science investigations all face fundamental resolution limitations in addressing this type of problem. The light microscope faces a wavelength limit. The electron microscope is limited by multiple scattering because the cells in which the interesting objects are embedded are too large (> 10 µm) even for high voltage microscopes. The lenses of zone plate x-ray microscopes do not reach this level of resolution. The solution we propose here, which we have already developed to some degree[17, 22], is soft x-ray diffraction tomography[6, 34]. This method requires a recording of both the amplitude and phase of the diffracted wavefield at each view direction. In what follows we propose ways to accomplish this using soft x-ray diffractive methods, especially holography.

FOURIER-TRANSFORM X-RAY HOLOGRAPHY

Although our previous x-ray holographic experiments have been in Gabor (in-line) geometry[18, 23], we believe that Fourier-transform holography is needed here. In this scheme, first described by Stroke and Falconer[32], the sample and reference object (usually point-like) are located in the same plane as shown in Fig. 1. The experiment can be lensless and has a CCD-based (i. e. automated) detection system suitable for

CP507, *X-Ray Microscopy: Proceedings of the Sixth International Conference,*
edited by W. Meyer-Ilse, T. Warwick, and D. Attwood
© 2000 American Institute of Physics 1-56396-926-2/00/$17.00

tomography. This type of x-ray holography was first proposed in 1966[33] and was demonstrated with carbon K radiation by Kikuta et al[20] and (with a zone plate) by Reuter and Mahr[28]. It was the x-ray first holographic geometry to be used with synchrotron radiation, in the 1972 experiment of Aoki et al[1], and later it was adopted by McNulty et al[24] who used a zone-plate optical system, CCD detector and undulator beam to reach a resolution of about 50 nm. The principle limitation until now has been the resolution of the zone plate used to create the nominal point source.

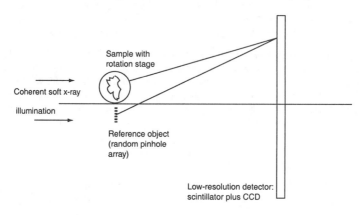

FIGURE 1. Fourier transform holography experiment with generalized reference object

One solution is a reference object which is not a point but whose autocorrelation function is point-like[4, 31]. However, even this limitation is unnecessary and any object will work as a reference provided it is has sharp spatial structure on the desired resolution scale (i. e. a roughly flat power spectrum out to the desired frequencies) and a high x-ray transparency. We propose to realize the reference object as a random array of pinholes constructed by etching of charged-particle tracks[15]. In mica, for example, the exit apertures of the tracks are oriented identical diamond shapes[19] and after light etching, track widths below 10 nm in size have been observed[15]. The resolution of this scheme will ultimately be limited by the size of the exit apertures of the etched tracks and the sharpness of their edges. If these are indeed the principal limitations, then a resolution in the target range of 3-12 nm can be projected.

SOFT X-RAY DIFFRACTION

Holography is the not the only possible way to determine the amplitude and phase of a wavefield. An important alternative method known as soft x-ray diffraction was proposed in 1980 by Sayre[29]. Sayre's scheme is essentially the experiment shown in Fig. 1 without the reference object. The diffraction pattern of the object is simply recorded using an intensity detector and the determination of the phases is addressed using the iterative Fourier transform algorithm[8] with a support constraint[10]. A large step forward in soft x-ray diffraction was reported recently by Miao et al[25] who successfully reconstructed a fairly complicated object from its diffraction pattern. In view of this success using an experimental setup which is simpler than the corresponding holographic setup, it is very important to understand the relationship between these two methods. As we shall see in the next sections, the essence is that holography is more powerful but more difficult .

DIFFRACTIVE SCHEMES: GROUND RULES

Let us represent the unknown object by a transparency function $f(\mathbf{x})$ with Fourier transform $F(\mathbf{s})$. The autocorrelation function $R(\mathbf{x})$ of the object will then be $\int_{-\infty}^{+\infty} f(\mathbf{x}' + \mathbf{x}) f^*(\mathbf{x}') \, d\mathbf{x}'$. Since $R(\mathbf{x})$ is the transform of the measured function $|F(\mathbf{s})|^2$, the problem in both holography and diffraction is equivalent to reconstructing a complex object from its autocorrelation. Now if the object has support of width a, $R(\mathbf{s})$ will have support of width $2a$ and the recorded pattern $|F(\mathbf{s})|^2$ will have a Nyquist sampling interval of $1/(2a)$. If a were the lattice constant of a crystal, the sampling interval imposed by the reciprocal lattice would be $1/a$ which implies that in crystallography $|F(\mathbf{s})|^2$ is undersampled by a factor of two[27]. On the other hand, in optics there is no restriction on the sampling interval and one would naturally sample the diffraction pattern at its Nyquist rate. If $f(\mathbf{x})$ is a real function then it can be shown that both $|F(\mathbf{s})|^2$ and $R(\mathbf{x})$[3] are real and centrosymmetric (Friedel's law). However, although a real $f(\mathbf{x})$ is a sufficient condition for the diffraction pattern to be symmetrical, it is not a necessary one and we return to this point later.

ITERATIVE FOURIER-TRANSFORM ALGORITHMS ETC

This method of phase retrieval was developed extensively during the 1980's by Fienup and others[30]. It uses the measured intensity in the $\mathbf{s}$ plane and knowledge of the object support (i. e. the region over which it is non-zero) in the $\mathbf{x}$ plane. In many situations the algorithm converges reliably to a unique result in reasonable time even in the presence of some noise[5]. However, complex objects are much more difficult than real ones and can only be reconstructed in favorable cases[2, 10, 11] (see Table 1).

The holographic approach to the inverse-diffraction problem is to add a known reference object in the plane of the unknown object. To see the potential benefits of this, consider the idealized object $f(\mathbf{x}) = f_1(\mathbf{x}) + A\delta(\mathbf{x} - \mathbf{b})$ where $\mathbf{b}$ is such that the spacing between f_1 and δ is greater than a where a is the support width; a condition known in the phase-retrieval literature as the "holography condition". Substituting for f in the expression for the autocorrelation function we immediately obtain[4, 9]

$$R(\mathbf{x}) = |A|^2 \delta(\mathbf{x}) + R_1(\mathbf{x}) + A f_1^*(\mathbf{b} - \mathbf{x}) + A^* f_1(\mathbf{b} + \mathbf{x}).$$

The primary image and its conjugate are separated from each other and from the autocorrelation term provided the holography condition is satisfied. Thus, in this idealized case, there is no phase problem. The unknown object is obtained in one step by taking the Fourier transform of the diffraction pattern. The phenomenon of the separation of the autocorrelation into parts because the object support is in parts is also important in enabling a good estimate of the object support when only the diffraction pattern (and thence the autocorrelation) is known[12]. More generally, the delta function

Table 1: Effectiveness of various supports for complex phase retrieval

Support of entire object	Unique?	Complex object retrievable?
Any shape plus a point reference obeying HC* (Fourier holography)	Yes	Yes: in one step by inspection of the autocorrelation[9]
Certain known shapes e. g. triangles with reference point(s) not satisfying HC*	Yes	Yes[9]
Rectangular object with one reference point on a diagonal spaced by one increment	Yes***	Yes[7, 9]
Known shape in two parts** satisfying HC*		Yes: works well[10]
The same but not satisfying HC*		Yes[10]
Unknown support in two well-separated parts		Yes: using a support inferred from the autocorrelation[11]
Support initially unknown and determined from a low-resolution image		Yes if tight enough: using an expanding weighting function on $\|F(\mathbf{s})\|$[13]
Object of known support with a convex hull having no parallel sides	Yes	Yes if tight enough[10]
Known centrosymmetrical supports such as ellipses and rectangles		Usually not: encourages stagnation at a mixture of the image and conjugate[10, 14]
Known supports which are loose or have tapered edges		Usually not[10]
Unknown support of any simple shape		Usually not because the support inferred from the autocorrelation is too loose[11]

*HC=holography condition. **It seems that a simple shape with a hole (preferably unsymmetrically placed) behaves similarly to a support in two parts[10, 26]. ***among all functions

is replaced by a known function $g(\mathbf{x})$, as we propose here. The standard hologram reconstruction procedure then yields a signal $\psi(\mathbf{x})$ in the object plane consisting of the autocorrelations of f_1 and g in the center of the field with the primary image of f_1 above them and the conjugate image of f_1 below them (see for example Fig 8.24 in Collier et al[4]). The primary and conjugate images are convolved with the autocorrelation and self convolution respectively of g. If the latter functions are reasonably sharp, as they would be in our experiment, then this worsens the resolution somewhat but does not change the capability for a one-step retrieval of the phases provided the holography condition continues to be respected. To recover the diffraction-limited resolution (provided the power spectrum of g is free of zeros) one can make the deconvolution $f(\mathbf{x} + \mathbf{b}) = \mathcal{F}^{-1}\{\mathcal{F}(\psi)/|G|^2\}$.

Thus pure holography with a known reference object can give the phases immediately. Examination of Table 1 also shows that the most favorable support geometries for complex phase retrieval are quite similar to holography. The best case is a support in two separated parts satisfying the holography condition. Even when the holography condition is not satisfied and neither object is known, the separated support is still quite powerful. On the other hand, simple symmetrical supports like squares and circles are the least effective. This is especially true if the support is unknown a priori.

We now turn to the question of the implications of the success of the experiment of Miao et al[25]. We note that the gold layer used in the experiment had an intensity transmission of 25% and a phase change of 0.9 radians indicating that it was certainly a complex object. Nevertheless, the diffraction pattern was observed to be symmetric over half of its area and symmetry was enforced over the other half. This raises two questions. First how could the object be complex and still have a symmetric diffraction pattern? Second why did the reconstruction work so well when it was an unfavorable case for complex phase retrieval? Now we noted above that the diffraction pattern can be symmetrical, even if $f(\mathbf{x})$ is complex. In fact the object used by Miao et al[25] was a pattern made of pure gold dots of constant thickness. Such an object can be written $1-(1-c)f(\mathbf{x})$ where the complex transparency of the dots is represented by c and $f(\mathbf{x})$ is real. Now in this type of experiment the illumination is essentially restricted to the sample area and the diffraction pattern is only observed outside that area. Therefore the effective sample function is $-(1-c)f(\mathbf{x})$ and the autocorrelation is that of the real function $f(\mathbf{x})$ multiplied by the real constant $|1-c|^2$. Thus the phase problem that was solved by Miao et al was special[16] and was essentially that of a real object, which answers both of the questions.

CONCLUSION

These considerations show that, although the experiment of Miao et al represents a big step forward in x-ray imaging, it still does not demonstrate a pathway to phasing the diffraction pattern of a general complex object. Moreover, there is strong evidence from the phase retrieval literature that the way to facilitate such a reconstruction is to adopt one of the favorable support geometries. This means either using a holographic reference object or at least an intermediate holography-like geometry where the overall object support is in two separated parts. Therefore, our general conclusion is that although soft x-ray diffraction has made important strides it will not put holography out of business. Rather, the next generation of diffractive imaging measurements, which will feed amplitude and phase data to the 3-D algorithms of diffraction-tomography experiments, are likely to have a decidedly holographic look.

ACKNOWLEDGEMENTS

This work was supported by the Director, Office of Energy Research, Office of Science, Materials Sciences Division of the U. S. Department of Energy, under contract No. DE-AC03-76SF000098.

REFERENCES

1. Aoki, S., Ichahara, Y., and Kikuta, S., *Jpn. J. Appl. Phys.*, **11**, 1857 (1972).
2. Bates, R. H. T., and Tan, D. G. H., "Fourier phase retrieval when the image is complex", in *Inverse Optics II*, Devaney, A. J., (Ed), Proc. SPIE, Vol. **558**, Bellingham, SPIE, 1985.
3. Bracewell, R. N., *The Fourier Transform and its Applications*, New York, McGraw-Hill, 1978.
4. Collier, R. J., Burckhardt, C. B., and Lin, H. L., *Optical Holography*, Academic, New York, 1971.
5. Dainty, J. C., and Fienup, J. R., "Phase retrieval and image reconstruction for astronomy", in *Image Recovery: Theory and Application*, Stark, H., (Ed), Orlando, Academic Press, 1987.
6. Devaney, A. J., *Inverse Problems*, **2**, 161-183 (1986).
7. Fiddy, M. A., Brames, B. J., and Dainty, J. C., *Opt. Lett.*, **8**, 96-98 (1983).
8. Fienup, J. R., *Appl. Opt.*, **21**, 2758-2769 (1982).
9. Fienup, J. R., *J. Opt. Soc. Am.*, **73**, 1421-1426 (1983).
10. Fienup, J. R., *J. Opt. Soc. Am. A*, **4**, 118-123 (1987).
11. Fienup, J. R., "Phase-retrieval imaging problems", in *International Trends in Optics*, Goodman, J. W., (Ed), Boston, Academic Press, 1991.
12. Fienup, J. R., Crimmins, T. R., and Holsztynski, W., *J. Opt. Soc. Am.*, **4**, 610-624 (1982).
13. Fienup, J. R., and Kowalczyk, A. M., *J. Opt. Soc. Am. A*, **7**, 450-458 (1990).
14. Fienup, J. R., and Wackerman, C. C., *!. Opt. Soc. Am. A*, **3**, 1897-1907 (1986).
15. Fleischer, R. L., Price, P. B., and Walker, R. M., *Nuclear Tracks in Solids*, Berkeley, University of California Press, 1975.
16. If each dot is an even function $e(\mathbf{x})$, so that any pattern of repetitions of e can be written $e(\mathbf{x}) * \sum_m \delta(\mathbf{x} - \mathbf{x_m})$, then the diffraction pattern is $|E(\mathbf{s})|^2 \left| \sum_m e^{2\pi i \mathbf{x}_m \cdot \mathbf{s}} \right|^2$ and is seen to be even.
17. Howells, M., Jacobsen, C., Kirz, J., McQuaid, K., and Rothman, S., "Progress and Prospects in Soft X-ray Holographic Microscopy", in *Modern Microscopies: Techniques and Applications*, Duke, P., A. G. Michette, (Ed), New York, Plenum, 1990.
18. Jacobsen, C., Howells, M. R., Kirz, J., and Rothman, S. S., *J. Opt. Soc. Am.*, **7**, 1849-1861 (1990).
19. Khan, H. A., Khan, N. A., and Spohr, R., *Nucl. Instrum. Meth.*, **189**, 577-581 (1981).
20. Kikuta, S., Aoki, S., Kosaki, S., and Kohra, K., *Opt. Comm*, **5**, 86-9 (1972).
21. Kirz, J., Jacobsen, C., and Howells, M., *Quarterly Reviews of Biophysics*, **28**, 33-130 (1995).
22. Lindaas, S., Calef, B., Downing, K., Howells, M., Magowan, C., Pinkas, D., and Jacobsen, C., "X-ray holography of fast-frozen hydrated biological samples", in *X-ray Microscopy and Spectroscopy*, Thieme, J., G. Schmahl, E. Umbach, D. Rudolph, (Ed), Heidelberg, Springer-Verlag, 1997.
23. Lindaas, S., Howells, M., Kalinovsky, A., and Jacobsen, C., *J. Opt. Soc. Am. A*, **13**, 1788-1800 (1996).
24. McNulty, I., Kirz, J., Jacobsen, C., Anderson, E., and Howells, M.., *Science*, **256**, 1009 (1992).
25. Miao, J., Charalambous, P., Kirz, J., and Sayre, D., *Nature*, **400**, 342-344 (1999).
26. Miao, J., Sayre, D., and Chapman, H. N., *J. Opt. Soc. Am. A*, **15**, 1662-1669 (1998).
27. Millane, R. P., *J. Opt. Soc. Am. A*, **7**, 394-411 (1990).
28. Reuter, B., Mahr, H.., *J. Phys. E*, **9**, 746-751 (1976).
29. Sayre, D., "Prospects for long-wavelength x-ray microscopy and diffraction", in *Imaging Processes and Coherence in Physics*, Schlenker, M., M. Fink, J. P. Goedgebuer, C. Malgrange, J. C. Viénot, R. H. Wade, (Ed), Lecture Notes in Physics, Vol. **112**, Berlin, Springer-Verlag, 1980.
30. Stark, H., ed., *Image Recovery: Theory and Application*, Academic Press, Orlando, 1987.
31. Stroke, G. W., "Attainment of high resolutions in image-forming x-ray microscopy with `lensless' Fourier-transform holograms and correlative source-effect compensation", in *Optique des Rayons X et Microanalyse*, Castaing, R., P. Deschamps, J. Philibert, (Ed), Paris, Hermann, 1966.
32. Stroke, G. W., and D. G. Falconer, *Phys. Lett.*, **13**, 306-309 (1964).
33. Winthrop, J. T., and C. R. Worthington, *Phys. Lett.*, **15**, 124-126 (1965).
34. Wolf, E., *Opt. Comm.*, **1**, 153-6 (1969).

NANOMETER X-RAY OPTICS

The PS/PDI: a high accuracy development tool for diffraction limited short-wavelength optics

Patrick Naulleau[a], Kenneth A. Goldberg[a], Sang H. Lee[a,b], Chang Chang[a,b], Phillip Batson[a], David Attwood[a,b], and Jeffrey Bokor[a,b]

[a] Center for X-Ray Optics, Lawrence Berkeley National Laboratory, Berkeley, CA 94720
[b] EECS Department, University of California, Berkeley, CA 94720

Abstract. The extreme ultraviolet (EUV) phase-shifting point diffraction interferometer (PS/PDI) was developed and implemented at Lawrence Berkeley National Laboratory to meet the significant measurement challenge of characterizing EUV projection lithography optics. The PS/PDI has been in continuous use and under ongoing development since 1996. This unique and flexible tool is applicable to any imaging system with real conjugate points, including Schwarschild objectives, Fresnel zone plates, and Kirkpatrick-Baez systems. Here we describe recent improvements made to the interferometer, and we summarize metrology results from state-of-the-art 10×-reduction EUV Schwarschild objective.

INTRODUCTION

The quest to develop extreme ultraviolet (EUV) optics for use in next-generation projection lithography systems providing sub-100-nm resolution has led to various innovations in EUV wavefront metrology,[1,2] including the development of the EUV phase-shifting point diffraction interferometer (PS/PDI).[3-5] Not being limited to EUV lithographic systems, the metrology capabilities of the PS/PDI are directly applicable to the development of high-resolution X-ray microscopy tools.

The PS/PDI is a diffraction-class interferometer,[6-8] in which the illumination and reference waves are created by diffraction from small apertures. Furthermore, a diffraction grating is used as the beam-splitting and phase-shifting element. This diffraction configuration allows the PS/PDI to attain high reference-wavefront accuracy which has recently been measured to be better than $\lambda_{EUV}/350$ (0.4 Å) within a numerical aperture (NA) of 0.082.[9]

The PS/PDI is a flexible, system-level tool that may be applied to characterizing any imaging system with real conjugate points. Such systems, commonplace in X-ray microscopy, include Schwarschild objectives, Fresnel zone plates, and Kirkpatrick-Baez optics. Another advantage of the PS/PDI is that its diffraction characteristics, in principle, allow it to be implemented at virtually any wavelength. Here we describe the PS/PDI as implemented to test EUV 10×-reduction Schwarschild systems.

When characterizing imaging systems, it is important to consider flare in addition to wavefront error. Flare is the *halo* of light surrounding the optical system point-

CP507, *X-Ray Microscopy: Proceedings of the Sixth International Conference*,
edited by W. Meyer-Ilse, T. Warwick, and D. Attwood
© 2000 American Institute of Physics 1-56396-926-2/00/$17.00

spread function (PSF) and is caused by roughness of the optical components. While conventional wavefront errors directly lead to a loss of resolution, flare negatively affects the image contrast. In order to enable direct measurement of flare using the PS/PDI, the PS/PDI has recently been modified to support a greatly extended spatial-frequency measurement range.[10] This new capability also enables the PS/PDI to be used to qualify profilometry- and scatterometry-based flare measurement techniques.[11]

DESCRIPTION OF THE PS/PDI

The PS/PDI is briefly described here; more complete descriptions have been previously published.[3-5] The PS/PDI is a variation of the conventional point diffraction interferometer[6,7] in which a transmission grating is added to improve the optical throughput of the system and to add phase-shifting capability. In the PS/PDI (Fig. 1), the optical system under test is coherently illuminated by a spherical wave generated by diffraction from a pinhole placed in the object plane. To guarantee the quality of the illumination, the pinhole diameter is chosen to be smaller than the resolution limit of the optical system. A grating placed either before or after the optic is used to split the illuminating beam, creating the requisite test and reference beams. A mask (the *PS/PDI mask* in Fig. 1) is placed in the image plane of the optic to block unwanted diffracted orders generated by the grating. The mask also serves to spatially filter the reference beam using a second pinhole (the *reference pinhole*), thereby, removing the aberrations imparted by the optical system. The test beam, which also contains the

aberrations imparted by the optical system, is largely undisturbed by the image-plane mask: it passes through a window that is large relative to the diameter of the optical system PSF. The test and reference beams propagate to the detector where they overlap to create an interference pattern. The recorded interferogram yields information on the deviation of the test beam from the nominally spherical reference beam.

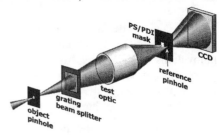

FIGURE 1. Schematic of the phase-shifting point diffraction interferometer (PS/PDI).

CHARACTERIZING ACCURACY

Significant effort has been directed toward characterizing the accuracy of the PS/PDI.[9] The two primary sources of measurement error limiting its accuracy are reference-wave imperfections and systematic effects that arise from the geometry of the system. Noting that the systematic geometric effects can be removed, provided they can be measured, the accuracy of the PS/PDI is typically limited by the reference-pinhole-induced errors.

In order to characterize the errors described above, and hence calibrate the PS/PDI, null tests have been performed. Analogous to Young's two-slit experiment, the null test is implemented by replacing the image-plane window with a pinhole. In this test, two reference waves are generated by diffraction from the image-plane mask, creating a null-test interferogram (Fig. 2). Aberrations calculated from the null-test interferogram quantify the systematic and random errors in the interferometer.

Implementation of this test shows that the primary error results from the hyperbolic fringe pattern produced by the two, laterally displaced, nominally spherical waves. Because this error is easily measured and subtracted during analysis,[5] we consider the reference-wavefront-limited accuracy to be the residual error after its removal. Table 1 enumerates the measured accuracy as a function of pinhole size over a NA of 0.082. The image-side NA of the optic used for this test was 0.08. As expected, the reference-wavefront accuracy improves with a reduction in pinhole size, and a resultant improvement in spatial filtering.

FIGURE 2. Null-test interferogram using 100-nm pinholes ($\lambda = 13.5$ nm).

TABLE 1. Reference wave rms accuracy as a function of null-mask pinhole size.

Pinhole Size (nm)	Systematic-error-limited rms Accuracy (waves)
140	0.012 ± 0.001 ($\lambda/83$)
120	0.010 ± 0.001 ($\lambda/100$)
100	0.0041 ± 0.0003 ($\lambda/244$)
80	0.0028 ± 0.0001 ($\lambda/357$)

The measurements described here were performed using an undulator beamline[12] at the Advanced Light Source synchrotron radiation facility at Lawrence Berkeley National Laboratory. The beamline provides a tunable source of EUV radiation with a coherence area that is significantly larger than the 0.75-μm diameter object pinhole.[12]

CHARACTERIZING AND ALIGNING EUV OBJECTIVES

During the past year, three newly fabricated 10×-reduction EUV Schwarzschild objectives have been characterized with the EUV PS/PDI. Furthermore, two of the objectives underwent at-wavelength alignment, significantly improving the system wavefronts.[13] All three cameras were fabricated to the same optical design specifications and were built within the past two years. The optics have an image-side NA of 0.088 and utilize molybdenum/silicon multilayer coatings designed for peak reflectivity at 13.4 nm wavelength.

Figure 3 shows the PS/PDI measured wavefronts along with the rms (σ) and peak-to-valley (PV) wavefront error magnitudes within the full 0.088 NA. EUV alignment was performed on objectives **B2** and **A**. These results demonstrate that the PS/PDI is well suited to characterizing and, more importantly, optimizing short-wavelength high-resolution optical systems.

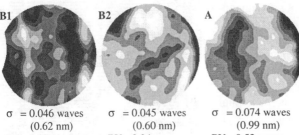

B1	B2	A
σ = 0.046 waves (0.62 nm)	σ = 0.045 waves (0.60 nm)	σ = 0.074 waves (0.99 nm)
PV = 0.27 waves (3.6 nm)	PV = 0.34 waves (4.6 nm)	PV = 0.52 waves (7.0 nm)

FIGURE 3. PS/PDI-measured wavefronts of three recently fabricated 10×-reduction EUV (λ=13.4 nm) Schwarzschild objectives with an image side NA of 0.088. The wavefront statistics are based on 37-term Zernike polynomial fitting. The displayed wavefronts, however, include higher spatial-frequency features.

MEASURING FLARE

The original design of the PS/PDI was directed towards high-accuracy wavefront characterization. For lithographic printing, however, it is equally important to consider flare. The capabilities of the PS/PDI have recently been extended, allowing it to measure both wavefront and flare simultaneously.[10] Flare may be characterized by recording an extended-field image of the PSF. Noting that the PS/PDI can be viewed as producing an off-axis Fourier-transform hologram of the PSF, it is well suited to measuring flare. From this point of view it is evident that the area over which the flare can be measured in the image plane is simply the area of the test window through which the PSF is effectively viewed. Using elongated windows it is possible to characterize the flare over significant distances in the image plane.[10]

The PS/PDI-based flare measurement technique has been demonstrated[10] using the **B1** EUV objective described above. This optical system developed to meet a flare specification of less than 5% in a 4-μm line. The measurements were performed using 30×3-μm test windows. The narrow window size in the direction of the beam separation is necessary to meet the beam isolation requirements imposed by off-axis holography.[14]

Figure 4 shows a logarithmically scaled image of the holographically reconstructed image of the PSF with flare. The image contains the customary twin images and intermodulation image.[14] Because the reconstructed image is as viewed through the image-plane window, we simultaneously get an image of the window itself. The bars seen in Fig. 4 are support features added to the window to prevent the thin, open-stenciled membrane from rupturing. The small (0.3-μm wide) protrusions in the center window portion are alignment aids. We note that the resolution in the reconstruction is determined by the reference pinhole size, which in this case is approximately 100 nm.

From the measured PSF, the normalized scatter-energy density as a function of radial distance from the PSF peak can be found (Fig. 5). The imperfect Airy lobes are caused by aberrations in the optic (figure error). To predict the flare in a typical imaging situation, the scatter-energy density must be known over the full radial extent of the field. For the optics considered here the field size is 250-μm radius in the image plane. The extended-range data can be obtained by extrapolation of the PS/PDI data or by use of data derived from profilometry performed on the individual substrates.[11] In order to avoid possible extrapolation errors, we choose the latter. The plot in Fig. 5 shows an overlay of the scatter-energy density predicted from profilometry and the PS/PDI measurement. The two measurement methods overlap in the radial range from 1 to 16 μm where good agreement is evident. Considering an isolated, dark 4-μm line in a 250-μm-radius bright field, the flare is calculated to be (3.9$\pm$0.1)%. The flare value predicted by profilometry alone is (4.0$\pm$0.1)%.[11]

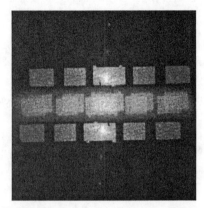

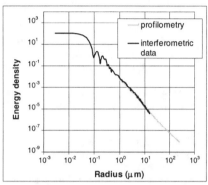

FIGURE 4. Holographically reconstructed point-spread function with flare. Image has been logarithmically scaled for display.

FIGURE 5. Comparison of the scatter-energy density as a function of radial separation from the PSF peak determined by the PS/PDI- and profilometry-based methods respectively.

CONCLUSION

A high accuracy PS/PDI, capable of characterizing both wavefront and flare, has been developed. This versatile diffractive interferometer may be implemented at virtually any wavelength to test a wide range of imaging systems characterized by real conjugate points. As implemented to test 10× Schwarschild objectives, the PS/PDI has been demonstrated to have a reference wavefront accuracy of better than λ/350 (0.4 nm) and has been used to align several lithographic-quality systems. A second implementation of the PS/PDI is currently being developed to test high-resolution Fresnel zone plates to be used in soft X-ray microscopy applications.

ACKNOWLEDGEMENTS

The authors are greatly indebted to Hector Medecki and Edita Tejnil for their pivotal roles in the early development of the PS/PDI. We are also grateful to Erik Anderson for nanofabrication of pinholes and windows, and to members of the CXRO staff, including Bill Bates, Rene Delano, Keith Jackson, Gideon Jones, Drew Kemp, David Richardson, and Senajith Rekewa for facilitating this research. Special thanks are due to Paul Denham for expert assistance with experimental control systems. This research was supported by the Extreme Ultraviolet Limited Liability Company, the Semiconductor Research Corporation, DARPA Advanced Lithography Program, and the Department of Energy Office of Basic Energy Science.

REFERENCES

1. J. E. Bjorkholm, A. A. MacDowell, O. R. Wood II, Z. Tan, B. LaFontaine, and D. M. Tennant, "Phase-measuring interferometry using extreme ultraviolet radiation," *J. Vac. Sci. & Technol. B* **13**, 2919-2922 (1995).

2. A. K. Ray-Chaudhuri, W. Ng, F. Cerrina, Z. Tan, J. Bjorkholm, D. Tennant, and S. J. Spector, "Alignment of a multilayer-coated imaging system using extreme ultraviolet Foucault and Ronchi interferometric testing," *J. Vac. Sci. Technol. B* **13**, 3089-3093 (1995).

3. H. Medecki, E. Tejnil, K. A. Goldberg, and J. Bokor, "Phase-shifting point diffraction interferometer," *Opt. Lett.* **21**, 1526-1528 (1996).

4. E. Tejnil, K. A. Goldberg, S. H. Lee, H. Medecki, P. J. Batson, P. E. Denham, A. A. MacDowell, J. Bokor, and D. Attwood, "At-wavelength interferometry for EUV lithography," *J. Vac. Sci. & Technol. B* **15**, 2455-2461 (1997).

5. K. A. Goldberg, "Extreme Ultraviolet Interferometry," Ph.D. dissertation (University of California, Berkeley, 1997).

6. W. Linnik, "A simple interferometer to test optical systems," *Proceedings of the Academy of Science of the USSR* **1**, 210-212 (1933).

7. R. N. Smartt and W. H. Steel, "Theory and application of point-diffraction interferometers," *Jap. J. Appl. Phys.* **14**, Suppl.14-1, 351-356 (1975).

8. G. E. Sommargren, "Phase shifting diffraction interferometry for measuring extreme ultraviolet optics," OSA Trends in Optics and Photonics Vol. 4, *Extreme Ultraviolet Lithography*, G. D. Kubiak and D. R. Kania, eds. (Optical Society of America, Washington, DC 1996), pp. 108-112.

9. P. Naulleau, K. A. Goldberg, S. Lee, C. Chang, D. Attwood, and J. Bokor, "The EUV phase-shifting point diffraction interferometer: a sub-angstrom reference-wave accuracy wavefront metrology tool," *Appl. Opt.,* to be published (1999).

10. P. Naulleau, K. A. Goldberg, E. Gullikson, and J. Bokor, "Interferometric at-wavelength flare characterization of EUV optical systems," *J. Vac. Sci. & Technol. B*, to be published (1999).

11. E. Gullikson, S. Baker, J. Bjorkholm, J. Bokor, K. Goldberg, J. Goldsmith, C. Montcalm, P. Naulleau, E. Spiller, D. Stearns, J. Taylor, and J. Underwood, "EUV scattering and flare from 10× projection cameras," in *Emerging Lithographic Technologies III*, Y. Vladimirski, ed., Proc. SPIE, **3676**, 717-723 (1999).

12. D. Attwood, P. Naulleau, K. Goldberg, E. Tejnil, C. Chang, R. Beguiristain, P. Batson, J. Bokor, E. Gullikson, M. Koike, H. Medecki, and J. Underwood, "Tunable coherent radiation in the soft X-ray and extreme ultraviolet spectral regions," *IEEE J. Quantum Electron,* **35**, 709-720 (1999).

13. K. A. Goldberg, P. Naulleau, and J. Bokor, "EUV interferometric measurements of diffraction-limited optics," *submitted to J. Vac. Sci. & Technol. B* (6/99).

14. E. N. Leith and J. Upatnieks, "Reconstructed wavefronts and communication theory," *J. Opt. Soc. Am.*, **52**, 1123-1130 (1962).

Electron Beam Lithography of Fresnel Zone Plates Using A Rectilinear Machine And Trilayer Resists

D. Tennant[1], S. Spector[2,3], A. Stein[2], C. Jacobsen[2]

[1] Lucent Technologies Bell Laboratories
[2] Department of Physics and Astronomy, SUNY Stony Brook
[3] Present address: MIT Lincoln Laboratory

Abstract. We describe the use of a commercial e-beam lithography system (JEOL JBX-6000FS) to fabricate Fresnel zone plates for x-ray microscopy. The machine is capable of controlling the pitch of optical gratings with sub-nanometer precision, so its beam placement properties are more than adequate for zone plate fabrication. The zone plate pattern is written into a thin top layer (PMMA or Calixarene) of a trilayer resist, and transferred into thick nickel zones using reactive ion etching (RIE) followed by electroplating. Zone plates with outermost zone widths of 30 nm have exhibited efficiencies up to 10.0% at a 390 eV photon energy and with diameters in the range 80 to 120 μm. Zone plates with outer zones of 18 to 20 nm were also fabricated in thinner Ni with correspondingly lower efficiencies of 2.6%. Zone plates with outermost zone widths of 45 nm have been fabricated with larger diameters up to 160 μm. All results reported were obtained with a 50 kV system with 80 μm field deflection size; future efforts will make use of a 100 kV, 500 μm field size system.

INTRODUCTION

Fresnel zone plates are the key optic in most x-ray microscopes, allowing high-resolution investigation of a wide range of samples. The X1A beamline at the National Synchrotron Light Source (NSLS) at Brookhaven National Laboratory uses zone plates as the focussing elements in scanning transmission x-ray microscopes (STXM) (see *e.g.*, [1, 2]). The zone plate is illuminated by x-rays from an undulator source and focussed to a diffraction-limited spot through which the sample is scanned. The transmitted x-rays are used to form an image of the object. By tuning the monochromator, the microscope can take absorption contrast images, elemental and chemical maps and spectroscopic data. The STXM plays a role in biology, polymer science, colloidal science, environmental science, and geochemistry research.

To achieve the high resolution patterning required for fabrication, the zone plates are exposed with a JEOL JBX-6000FS electron beam lithography (EBL) system at Lucent Technologies, Bell Laboratories. The system uses a ZrO/W field emission gun with a current density of about 2,000 A/cm². This machine delivers 1 nA of beam current at 50 keV into a 7 nm spot with 16 bit control of the beam position within a 80 μm writing field. A λ/1024 (~0.6 nm) interferometer is incorporated for deflector calibration and stage positioning to enable field stitching over larger areas. The precision of such an

CP507, *X-Ray Microscopy: Proceedings of the Sixth International Conference,*
edited by W. Meyer-Ilse, T. Warwick, and D. Attwood
© 2000 American Institute of Physics 1-56396-926-2/00/$17.00

instrument for production of diffractive optical elements has been demonstrated in other applications such as optical gratings for wavelength selectable sources [3] needed for future lightwave communication systems. In that work, arrays of gratings with periods centered near 240 nm were produced in which adjacent grating were designed and measured to have consecutive period changes of 0.13 nm.

Fresnel zone plates are diffractive optics that give the highest resolution of any optic presently available in the soft x-ray region. The transverse spatial resolution, δ_t, is equal to $1.22\delta_{rN}$, where δ_{rN} is the outermost zone width. As with all imaging techniques, the highest resolution is the ultimate goal. Thus, the first fabrication challenge is to produce the smallest δ_{rN} while maintaining proper zone placement.

The focal length f of a zone plate with diameter d is given by $f=d\delta_{rN}/\lambda$. Thus, a decrease in δ_{rN} leads directly to a decrease in the focal length if the diameter d is not increased proportionally. Since the zone plates are diffractive, they have multiple orders, which are eliminated from 1st order imaging by an order-sorting aperture (OSA) - a pinhole placed between the zone plate and the sample (see Figure 1). Certain applications of STXM require large focal lengths for working with thick sample holders. The focal length is also inversely proportional to the wavelength of the x-rays. This means that certain spectral regions (e.g. the carbon edge) imply a smaller, i.e. more difficult, focal length. The second fabrication challenge, therefore, is to produce large diameter zone plates while maintaining proper zone placement and small δ_{rN}.

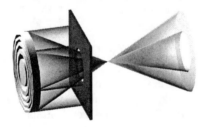

Figure 1: Schematic view of a Fresnel zone plate used in a scanning microscope. The first order focus is isolated from the non-diverging 0th order (and the diverging -1st order, not shown here) by the use of a central stop on the zone plate, and a pinhole called the Order Sorting Aperture (OSA) located at a distance of about $(2/3)f$ from the zone plate of focal length f.

Lastly, the diffraction efficiency must be as high as possible, so that the sample is well illuminated in as short a time necessary for high-resolution imaging. This is achieved by fabricating zones which are thick enough to attenuate (for absorption zone plates) or phase-shift (for phase shifting zone plates) the incoming x-rays. Also, the zones should be made from a material that gives the desirable phase-shifting/absorption properties; a good choice for soft x rays is nickel [4].

To meet these challenges with a rectilinear scanned commercial electron beam lithography tool exploratory materials and software modifications were employed. Zone plates with minimum zone widths of $\delta_{rN}=18$ nm, $d=80$ μm were fabricated in nickel [5].

Zone plates with δ_{rN}=45 nm, d=160 μm were also fabricated in nickel to give larger focal lengths while maintaining high resolution.

FABRICATION

Trilayer Resist and Electroplating

The zone plates are fabricated using a trilayer (Figure 2) scheme [6] on a Si_3N_4 membrane. The trilayer allows high-resolution e-beam lithography in a thin resist while still allowing higher aspect ratios to be achieved. Presently, the trilayer consists of a plating form of AZ 4110 resist deposited on a plating base of chrome and gold. A thin layer of germanium is evaporated for use as a hard etching mask upon which the resist is deposited. After exposure and development, the pattern is transferred by RIE and then electroplated with nickel.

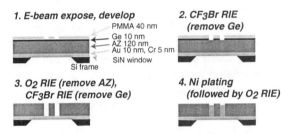

Figure 2: The use of trilevel resist schemes allows one to fabricate high aspect ratio structures at the resolution limits of electron beam lithography. The electron beam exposure is done in a thin resist to minimize linewidth degradation by electron forward scattering. This pattern is then transferred into a thicker resist by reactive ion etching, and that resist is then used as a plating mask. The particular scheme shown here is that used by Spector *et al.*, *J. Vac. Sci. Tech.* **B 15**, 2872 (1997).

Resists

Initially, poly(methyl methacrylate) (PMMA) was used for the zone plate patterning. This has proven to be adequate for δ_{rN} > 30 nm [7]. As the need for higher resolution (and thus smaller δ_{rN}) has grown, the performance of PMMA was not acceptable. The resist Calixarene, with its increased resolution [8], was used to fabricate test zone plates with δ_{rN} down to 18 nm in 60 nm thick nickel [5]. A δ_{rN}=20 nm test zone plate is shown in Figure 3 along with its diffraction efficiency measurements. The figure shows that the efficiency falls off only rather slowly as the zone width is decreased to 20 nm.

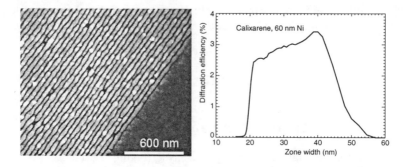

Figure 3: At left is shown 20 nm outermost zones in a test zone plate fabricated in 60 nm nickel using the photoresist Calixarene. The diffraction efficiency of this zone plates at 400 eV is shown at right; it is encouraging to note that the diffraction efficiency falls off only slowly as the zone width is decreased from 50 to 20 nm with this process.

Larger Diameter Zone Plates

Certain applications of the STXM and the Cryo-STXM [9], such as tomographic imaging [10] and wet specimen holder work [11, 12], require a working distance between order sorting aperture and specimen to be in the 300-1000 μm range. At the same time, the diameter of the central stop used for order sorting should be no more than about half the zone plate diameter to maintain good optical performance. As a result, it is important for many applications to have the focal length of the zone plate be significantly larger than 600-2000 μm.

This requirement for long focal lengths means that the zone plate diameter must be increased. For example, a δ_{rN}=30 nm, d=80 μm diameter zone plate has a focal length of only 560 μm at the carbon absorption edge (290 eV). Figure 4 shows how the diameter must increase as zone width is decreased to maintain a 1 mm focal length.

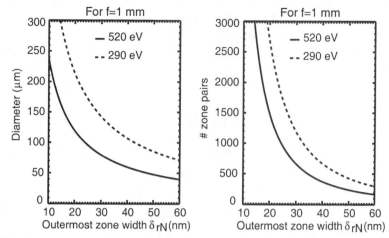

Figure 4 : Requirements on zone plate diameter and number of zone pairs to maintain a 1 mm focal length as outermost zone width is decreased, for 520 eV (oxygen edge) and 290 eV (carbon edge). At present, the largest zone plates we have fabricated have d=160 μm and δ_{rN} = 45nm. Future efforts will be directed towards increasing d for finer outermost zone widths δ_{rN}.

Because the JEOL's deflectors have an 80 μm range, the larger zone plates require movement of the stage and the stitching of multiple fields. Figure 5 shows examples of good and bad stitching as well as a 160 μm zone plate with 45 nm outer zone width. Good field stitching requires accurate alignment, thermal stability and good electrical conductivity on the sample during pattern writing.

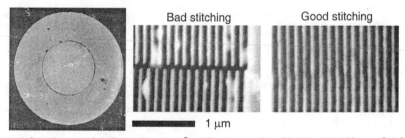

Figure 5: At left is shown a d=160 μm diameter, δ_{rN}=45 nm zone plate fabricated in 180 nm of nickel (no zones can be seen at this magnification scale, but the central stop can be distinguished). With the JEOL JBX-6000FS, zone plates of this type require the stitching of four 80 μm fields. Initial efforts in field stitching were less than fully successful (middle), but in subsequent work (right) we were able to obtain excellent field stitching by improving thermal equilibration and electrical conduction of the zone plate substrate so as to prevent electrostatic charging.

Future developments

A new EBL tool, the JEOL JBX-9300FS, is due to be delivered and installed at Bell Labs during 1Q 2000. This machine will provide several improvements over the present system that will facilitate an improvement in zone plate fabrication. The operating acceleration voltage is increased from 50 to 100 keV, thereby further reducing the forward scattering of electrons in thin resists [13] and increasing exposure resolution due to improved absorbed energy contours. The beam scanning resolution is increased to 19 bits (or 1.0 nm within a 500 μm field) and the spot size is expected to drop from 7 to 4 nm. The deflection range increases from 80 to 500 μm at the highest resolution setting so fabrication of large diameter zone plates will be possible without the movement of the stage, and therefore without field stitching.

REFERENCES

1. Jacobsen, C., *et al. New developments in scanning microscopy at Stony Brook.* These proceedings.

2. Feser, M., *et al. Instrumentation advances and detector development with the Stony Brook scanning transmission x-ray microscope.* These proceedings.

3. Tennant, D., *et al., Multiwavelength distributed Bragg reflector laser array fabricated using near field holographic printing with an electron-beam generated phase grating mask.* Journal of Vacuum Science and Technology, 1993. **B 11**(6): p. 2509-2513.

4. Anderson, E. and D. Kern. *Nanofabrication of zone plate lenses for high resolution x-ray microscopy.* In A.G. Michette, G.R. Morrison, and C.J. Buckley, ed.,*X-ray Microscopy III.* (Springer-Verlag, 1992), pp. 75--78.

5. Spector, S., C. Jacobsen, and D. Tennant, *Process optimization for production of sub-20 nm soft x-ray zone plates.* Journal of Vacuum Science and Technology, 1997. **B 15**(6): p. 2872--2876.

6. Tennant, D., *et al., 25 nm features patterned with trilevel e-beam resist.* Journal of Vacuum Science and Technology, 1981. **19**: p. 1304-1307.

7. Spector, S.J., *Diffractive optics for soft x rays.* PhD dissertation, Department of Physics, State University of New York at Stony Brook: Stony Brook, NY, 1997.

8. Fujita, J., *et al., Ultrahigh resolution of calixarene negative resist in electron beam lithography.* Applied Physics Letters, 1996. **68**: p. 1297--1299.

9. Maser, J., *et al., Soft x-ray microscopy with a cryo STXM: I. Instrumentation, imaging, and spectroscopy.* Journal of Microscopy (in press).

10. Wang, Y., *et al., Soft x-ray microscopy with a cryo STXM: II. Tomography.* Journal of Microscopy (in press).

11. Neuhäusler, U., *et al., Soft x-ray spectromicroscopy on solid stabilized emulsions.* Colloid & Polymer Science, 1999. **277**: p. 719-726.

12. Neuhäusler, U., *et al., A specimen chamber for soft x-ray spectromicroscopy on aqueous and liquid samples.* Journal of Synchrotron Radiation (in press).

13. Kyser, D.F., *Spatial-resolution limits in electron beam nanolithography.* Journal of Vacuum Science and Technology B, 1983. **1**: p. 1391--1397.

Fabrication Of Thick Zone Plates For Multi-Kilovolt X-Rays

A. Düvel, D. Rudolph and G. Schmahl

Georg-August-Universität Göttingen, Institut für Röntgenphysik (IRP),
Geiststraße 11, 37073 Göttingen, Germany

Abstract. Phase zone plates for multi-kilovolt X-rays with an outermost zone width of 16.9-30 nm and aspect ratios up to 177 were generated by a sputtered-sliced technology. The generated lenses of alternating Ni80Cr20 and SiO_2 consist of 188-365 layers on borosilicate glass wires and were polished by mechanical and ion processes to a thickness of 3-4 µm without carrier foil. The groove efficiency was measured at the ESRF beamline ID 21 up to 3.8% in first order. But small parts of a zone plate show an efficiency of about 10-15%.

1. INTRODUCTION

A major requirement for the application of X-ray microscopy are zone plates with high diffraction efficiency and a resolution in the 20 nm range or better. Therefore, large aspect ratios (height/width of zones) have to be generated. To overcome the limitation of the aspect ratio inherent to usual lithographic techniques, particularly for X-ray optics in the multi-kilovolt range, the sputtered sliced technology was introduced at the IRP [1] and is also adopted by other labs [2,3]. In this process, layers of NiCr and SiO_2 are alternately deposited onto a microwire of borosilicate glass (Duran) by magnetron sputtering. The deposited wires are reinforced by nickel-electroplating to 3-4 mm diameter, fixed with a resin into a metallic tube and sliced perpendicular to its axis (Fig. 1). The discs with a thickness of about 0.5 mm are polished by mechanical processes and ion-milling to the required thickness of a few µm.

FIGURE 1. Principle of sputtered-sliced technology.

CP507, *X-Ray Microscopy: Proceedings of the Sixth International Conference,*
edited by W. Meyer-Ilse, T. Warwick, and D. Attwood
© 2000 American Institute of Physics 1-56396-926-2/00/$17.00

2. DEPOSITION PROCESS

The layer systems were deposited by magnetron sputtering onto very smooth glass wires, drawn from softened Duran glass tubes [4,5]. During the deposition process the revolving wires were hanging in the center of a shutter cylinder between two magnetrons, which were both continuously powered during the whole process. The flux of the sputtered atoms reaches the wire through a cylinder window (Fig. 2). The target diameter of the magnetrons is 90 mm and the target-substrate spacing 100 mm. To change the layer material, the cylinder can be turned by a stepper motor, so that the window can face both targets alternately. The thickness of each layer is controlled by the sputtering time. Multilayers with an outermost zone width from 16.9-30 nm and 188-365 zones (total width of the layer systems 4.7-6.6 µm) were generated (see Fig. 3). We found that the smoothness of the zones mainly depends on the diameter of the wires, it is independent from the argon pressure in the used range of 0.18-0.4 Pa or a negative voltage of 5-20 V applied to the wire. It was not possible to prepare smooth layer systems on wires thinner than 50 µm. The cause is possibly a shadowing effect due to the small radius of too thin wires that expose only small portions of the surface perpendicular to the sputtering source.

3. POLISHING

The polishing process is the most time consuming of all preparation steps. It comprises mechanical polishing followed by ion milling. First the thickness of the slices is reduced to some 10 µm with sandpaper and one side is polished by alumina and argon-ions and coated with a metal layer. The actual thinning process can be done by polishing the sample from the second side over its whole area with sandpaper and alumina in a handholder to a thin foil of a few µm thickness. Another possibility is to polish only the region close to the zone plate to a thickness of some µm by using a ball-grinder or a dimpler and alumina or diamond suspension as abrasives. The result is a relatively stable sample. First tests with a ball-grinder showed unsatisfactory results, so that a special dimpler build at the IRP was used for the preparation of the most samples (Fig. 4). In a next step the samples were milled with 10 keV argon ions under an angle of 10 degrees -to the sample surface- to a final thickness of 3-4 µm (Fig. 5). The thickness of the zone plates was measured by focusing in a light microscope onto the front side and the metal coated back side.

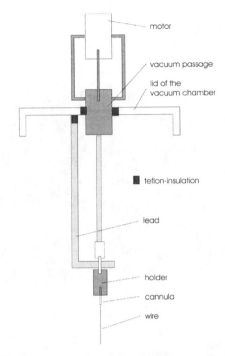

motor

vacuum passage

lid of the
vacuum chamber

teflon-insulation

lead

holder

cannula

wire

FIGURE 2. Sputter configuration: During the deposition process, the revolving wire is placed behind the cylinder window between the two magnetrons (bottom). The material of the layers can be selected by turning the cylinder, so that the cylinder window is faced to the used target. The complete wire drive (top) is mounted into the lid of the vacuum chamber.

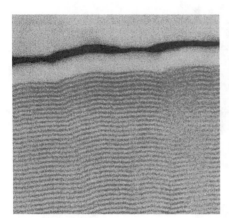

a) DRn: 22 nm
   wire: 35 μm ⌀
   270 zones

c) DRn: 30 nm
   wire: 64.1 μm ⌀
   188 zones

b) DRn: 16.9 nm
   wire: 57.4 μm ⌀
   365 zones

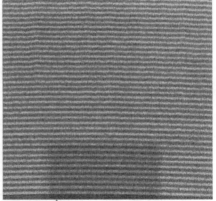

d) inner zones
   of c):
   DR1: 35.7 nm

FIGURE 3. Scanning electron micrographs of sputtered Ni80Cr20/SiO$_2$ systems.

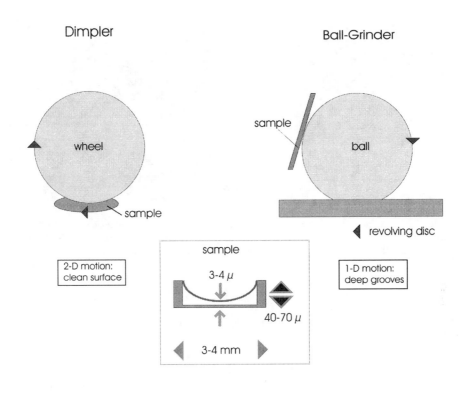

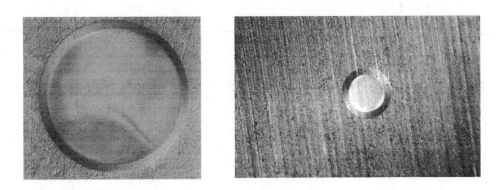

FIGURE 4. Schemes and results of dimpler (left) and ball-grinder (right) thinning. The dimpled zone plate has a diameter of 88.8 μm, the zone plate thinned with the ball-grinder 48.2 μm.. Both samples are 3 μm thick.

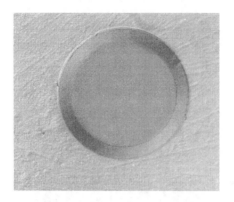

FIGURE 5. Light microscope images of mechanically thinned and ion-milled zone plates. **Left:** Layer systems with smallest zones of 16.9 nm on a wire of 57.4 μm diameter (thickness 3 μm). **Right:** Layer systems with smallest zones of 30 nm on a wire of 64.1 μm diameter (thickness 4 μm).

4. EFFICIENCY MEASUREMENT

The efficiency measurement was done in the scanning transmission X-ray microscope (STXM) at ESRF beamline ID 21 [6]. The zone plates were scanned with a 5 μm-pinhole mounted on the sample holder of the STXM. The sample holder which is placed between the zone plate and a detector for the X-ray measurement was moved by a piezo stage. Groove efficiencies up to 3.8% in first order (theoretical value in the range of 15-20%, calculated according to the coupled wave theory [7]) and 0.7% in third order of a 3 μm thick zone plate consisting of 365 layers and an outermost zone width of 16.9 nm (aspect ratio: 177) were measured at 4.1 keV. Figure 6 shows a vertical scan in first order of this lens (FWHM 5 μm, 4.1 keV). Wire, zone system and nickel reinforcement are clearly separated by their different transmission. A 2-D scan of the defocused first order (same lens and energy used for the vertical scan) showed instead of an uniform bright first order ring two spots on a dark ring (Fig. 7). The dark regions of the zone system which diffract photons to the spots have a groove efficiency of 10-15%. One possible explanation for this behavior is a damage caused by the mechanical polishing process. If the zone plate is bent in one direction with the bending line along the bright spots, most of the zones are tilt in direction to the beam. The efficiency of zones with a thickness of some μm will rapidly decrease when they are bent in this way. However, parts of the zone system parallel to the beam will still have a high efficiency. So it would be better to reduce the thickness of the zone plates by mechanical processes only to 15-20 μm and then polish by ion milling under an angle of 1-2 degrees to the calculated final thickness. Even with an angle of 5 degrees –the smallest angle that can be obtained with the ion mill used for this experiments– we found different etching rates of nickel reinforcement, zone system and the glass wire which result in uneven surfaces and edge effects by milling more than 1-2 μm from the sample surface.

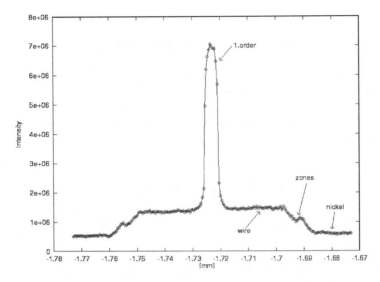

FIGURE 6. First order peak of a sputtered-sliced zone plate with an outermost zone width of 16.9 nm and a thickness of 3 μm at 4.1 keV. The zone plate was scanned with a 5 μm pinhole which was moved in vertical direction. The measured zone efficiency at 4.1 keV was 3.8% in first order and 0.7% in third order. The FWHM of 5 μm correspondence with the pinhole diameter. Derivatives of the flanks give a resolution limit below 0.8-1.0 μm (Scan: B. Kaulich, STXM-ID 21/ESRF).

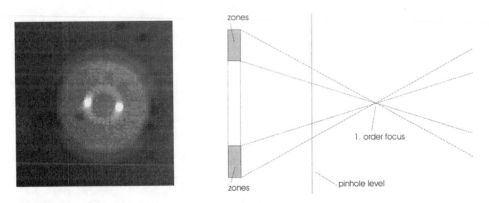

FIGURE 7. Left: Defocused first order of a sputtered-sliced zone plate at 4.1 keV (same optic as in Fig. 6). Instead of an uniform bright first order ring, two spots on the ring are visible. The dark regions of the layer system which diffract photons to the spots have an efficiency about 10-15% (Scan: B. Kaulich, STXM-ID 21/ESRF). **Right:** Schematic first order.

5. ACKNOWLEDGEMENTS

This work was supported by the Deutsche Forschungsgemeinschaft under contract number Schm 1118/1-1. The authors are indebted to the persons whose skill and competence made these results possible. We are grateful to the staff of the ESRF beamline ID 21, especially we want to thank B. Kaulich for the measurement at the ESRF beamline ID 21.

6. REFERENCES

1. D. Rudolph, G. Schmahl: Annals of the New York Academy of Sciences, Vol. 342, 1980,

   pp. 94-104.

2. R.M. Bionta, K.M. Skulina, J. Weinberg: Appl. Phys. Lett. 64 (8), 1994, pp. 945-947.

3. N. Kamijo, S. Tamura, Y. Suzuki, K. Handa, A. Takeuchi, S. Yamamoto, M. Ando, K. Ohsumi, H. Kihara: Rev. Sci.Instr., Vol 68 (1), 1997, pp. 14-16.

4. D. Rudolph, G. Schmahl: SPIE Proc, Vol. 316, 1981, pp. 103-105.

5. P. Witt, in: X-RAY MICROSCOPY IV, V.V. Aristov, A.I. Erko (Eds), Bogorodskii Pechatnik Publishing Company, Chernogolovka, Moscow Region, 1994, pp. 500-503.

6. R. Barrett, B. Kaulich, S. Oestreich, J. Susini: SPIE Proc, 1998, pp. 3449-3459.

7. G. Schneider: Appl. Phys. Lett., Vol. 73 (5), 1998, pp. 599-601.

Phase Zone Plates for Hard X-ray Focusing

Yuli Vladimirsky

Singapore Synchrotron Light Source, National University of Singapore, 5 Research Link, Singapore
Center for X-ray Lithography, UW-Madison, 3731 Schneider Dr., Stoughton, WI 53589

Abstract. The major obstacle for constructing a deep UV or X-ray lens is the low refraction and high absorption of materials in this region. Only a few Fresnel zones are utilized, so the diameter of an X-ray lens is of the order few microns, and the resolution is hardly better than that of a pinhole. The thickness of a phase zone plates (ZP) is determined by the refraction of the material. To provide π phase shift and form a hard x-ray phase ZP a thickness of a few microns is required. Approaches and techniques developed at the Center for X-ray Lithography (CXrL) for producing thick phase ZPs are presented.

The success of Fresnel ZPs as excellent focusing elements in soft X-ray region has stimulated strong interest and efforts to develop hard X-ray ZP fabrication techniques[1]. These efforts have been pursued at CXrL in order to support the Argonne National Laboratory (ANL) in development of advanced hard X-ray (8-30 keV) high spatial resolution (< 0.1 μm) microprobe techniques, such as microscopy, microanalysis, in-situ micro-spectroscopy and micro-diffraction.

I. REFRACTIVE X-RAY LENSES

A microscope in a projection or a probe mode requires a focusing element. A lens made of a refractive material is used for this purpose in visible and near UV light. Generally, the refractive index is a complex quantity $n = 1 - \delta - i\beta$, with a real or 'refractive' part $1-\delta$ and an imaginary part – extinction coefficient β (see Table 1.). The focusing condition of a lens (see Fig.1a)

$$\sqrt{y^2 + (f - x)^2} + x \operatorname{Re}(n) = f \tag{1}$$

can be expressed in terms of a single surface lens profile:

$$y^2 = 2xR_0 - 2x^2\delta + x^2\delta^2, \tag{2}$$

where $R_0 = f\delta$ is the radius of curvature [2] at the pole, y is the radial coordinate, x is the corresponding lens thickness, and f is the focal length. When the lens thickness is small ($x_{max} \ll f$) Eq. (2) is reduced to a simple parabola:

$$y^2 = 2xR_0. \tag{3}$$

CP507, *X-Ray Microscopy: Proceedings of the Sixth International Conference,*
edited by W. Meyer-Ilse, T. Warwick, and D. Attwood
© 2000 American Institute of Physics 1-56396-926-2/00/$17.00

Table 1. Comparison Of Optical And X-Ray Lenses								
Material	λ	δ	β	N_0	f	R_0	R_A	R_C
quartz	*500nm*	*0.45*	*$<10^{-6}$*	*$>5\cdot10^{+4}$*	*0.1 m*	*45 mm*	*>50 mm*	*~3.7 mm*
Au	1.5 Å	$4.6\cdot10^{-5}$	$4.7\cdot10^{-6}$	1.5	1 m	46.1 µm	15.5µm	~29 µm
Al	1.5 Å	$8.5\cdot10^{-6}$	$1.5\cdot10^{-7}$	8.7	1 m	8.5 µm	36.6µm	~8.4 µm
Be	1.5 Å	$5.3\cdot10^{-6}$	$2.4\cdot10^{-9}$	354	1 m	5.3 µm	233µm	~4.1 µm

The absorption limited lens thickness $x_{max}=1/\mu=\frac{1}{4}\pi\beta$ defines the effective aperture:

$$R_A^2 = \frac{2\delta f}{\mu} = \frac{f\lambda}{2\pi\beta} = N_0 f\lambda .\qquad(4)$$

The Eq. 4 can be interpreted as the lens "ZP" formula with $N_0=\frac{1}{2}\delta/\pi\beta$ being number of zones. Usually lens surfaces are approximated by a spherical shape with $R_0=f\delta$

$$y^2 = 2xR_0 - x^2 ,\qquad(5)$$

what will limit the lens aperture compared with the parabolic shape given by Eq.2. or Eq.3. (optical path difference should not exceed $\lambda/4$):

$$R_C^2 = \sqrt{2\lambda f}R_0 - \frac{\lambda f}{2}.\qquad(6)$$

The useful lens radius is the smaller of the values from (4) and (6). Parameters for single surface lenses are presented in Table 1. For visible light the refraction is $1-\delta>1$

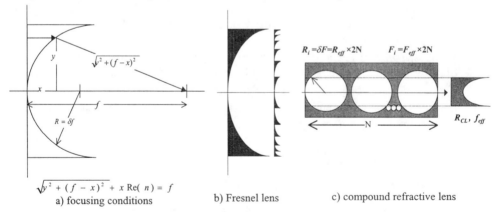

$$\sqrt{y^2 + (f - x)^2} + x\,Re(n) = f$$
a) focusing conditions   b) Fresnel lens   c) compound refractive lens

FIGURE 1. Refractive X-ray Lenses

and extinction is $\beta \to 0$ - a very favorable situation, unlike that in the X-ray region. The value of $N_0=\delta/2\pi\beta$ also represents optical quality of a material [3], and for an acceptable lens it is $\delta\,|\,2\pi\beta>10$. As it can be seen from Table 1, Be could be a suitable material for a lens with a large number of zones and aperture. However, even for would-be-efficient

X-ray lenses the surface curvature and the aperture are of the order of several microns and only a few Fresnel zones are utilized (see Table 1.). With the resolution of the order of the diameter, this lens is hardly more useful than a pinhole[4]. By reducing the thickness and improving transmission a Fresnel Lens [2] (Fig.1 b.) can be formed. When thickness is controlled in each zone it is called a *Coherent* or *Phase Fresnel Lens*, or *Blazed ZP*[4,5].

To utilize the potential of low Z materials like Be, a compound lens[6] has been proposed. It consists of a linear array of circular lenses (holes) with large individual radii of curvature $R_i=R\times 2N$ and focal lengths $F_i=R/\delta$ (Fig.1 c). The focal length of the compound lens is $f_{eff}=f=F_i/2N$, and the radius is larger compared to a simple lens:

$$R_{CL}^2 = R_i \sqrt{2\lambda f_{eff}} - \frac{\lambda f_{eff}}{2}. \tag{7}$$

A compound Be lens with $f_{eff}=1$ m and $R_{CL}=233\mu m$ would require ~300 circular well aligned holes with $R_i \approx 3.1$mm . The quality of such lenses has yet to be demonstrated.

II. Phase Fresnel Zone Plates

In general, the zone boundary radii of a ZP can be presented [1,7] as

$$R_n^2 = nf\lambda + \frac{n^2\lambda^2}{4}\frac{M^3+1}{(M+1)^3} \cdots \tag{8}$$

A zone pate acts like a thin lens, following the lens formula $a^{-1}+b^{-1}=f^{-1}$, with magnification $M=b/a$ and a and b being distances from the ZP to the object and image planes respectively. Two special cases are usually identified[1]:

1. $R_n^2 = nf\lambda + \frac{n^2\lambda^2}{16}$ - corresponding to $M=1$, and

2. $R_n^2 = nf\lambda + \frac{n^2\lambda^2}{4}$ - a plane ($M=\infty$) or a converging ($M=0$) waves are formed.

Clearly, the radius of ZP and the outermost zone width are $R_N \approx Nf\lambda$ and $d_N \approx \sqrt{f\lambda}/2\sqrt{N}$ respectively. The focusing ability of a ZP as a diffractive element is characterized by multiple foci and chromaticity.

Multiple order foci. In the general case, the phase relations in the diffracted wave provide constructive interference not only for optical path difference of $\pm\lambda$, but also for $\pm 2\lambda, \pm 3\lambda, \ldots \pm m\lambda, \ldots$ The corresponding focal lengths are

$$f_{\pm 1} = f, f_{\pm 2} = \pm\frac{f}{2}, f_{\pm 3} = \frac{f}{3}, \ldots f_{\pm m} = \pm\frac{f}{m}, \ldots, \tag{9}$$

where m is the diffraction order and the negative sign indicates virtual foci. It is necessary to block the undesirable spatial components, especially the 0th order. This is achieved by forming a central stop [1,4] and letting the desirable order pass through an aperture, or to use only the outer ZP portions[4,8]. Phase ZPs, optimized for the 1st order diffraction, exhibit partial suppression of the 0th and higher orders, and for an optimized blazed ZP the undesirable orders could be absent altogether[9].

Chromaticity. The ZP is inherently chromatic and the strong dispersion of a ZP is reflected in the wavelength dependence of the focal length. For a ZP with a small numerical aperture this relation is reasonably simple [1]:

$$f \approx R_N^2 \Big/ N\lambda \,. \tag{10}$$

The restriction on spectral bandwidth $\Delta\lambda/\lambda < 1/2N$ requires an appropriate degree of monochromatization, usually provided by a condenser ZP[10] or a double-crystal monochromator [11] positioned upstream.

Resolution. The radial resolution of a ZP is very close to that of a lens[1], and can be expressed explicitly through the outermost zone width:

$$\omega \approx \frac{f\lambda}{2R_N} \approx 1.22 \frac{\sqrt{f\lambda}}{2\sqrt{N}} = 1.22 \, d_N \,. \tag{11}$$

The resolution of a ZP in higher orders is: $\omega_m \approx \omega_1/m \approx 1.22 \, d_N/m$. To determine the resolution a knife-edge scanning technique[11] was used. A 200Å thick Cr film has been formed on a Si, and by scanning this "knife-edge" across the focal spot the Cr K_α fluorescence intensity distribution profile was obtained(see Fig.2a). A derivative of this profile produces bell shaped curves with half-width values of 150 nm and < 90 nm at the 1st and at the 3rd order respectively. The values of $\omega_1 = 80$ nm, and $\omega_3 = 52$ nm were calculated by subtracting instrumental broadening. The 1st order value agrees with the

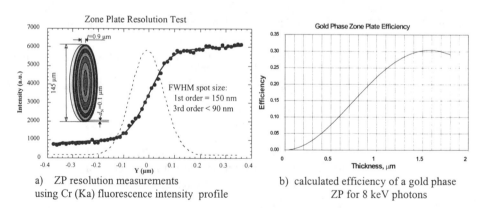

a) ZP resolution measurements
using Cr (Ka) fluorescence intensity profile

b) calculated efficiency of a gold phase
ZP for 8 keV photons

FIGURE 2. Resolution and Efficiency of a Phase ZP

expected one, but the 3rd order value is bigger than theoretical: $\omega_{3theor} = 27$nm. The possible reasons are: mechanical vibration, inaccuracy in the focal plane determination, and small errors in zone placements[11].

Efficiency. The efficiency of the ZP is a normalized focused intensity. For equal open and obstructed zone widths the efficiency can be presented[1,7] as:

$$\eta_m = \frac{1}{m^2 \pi^2}(1 + e^{-2\kappa\phi} - 2e^{-\kappa\phi}\cos\phi), \, m = \pm 1, \pm 3, \pm 5, \ldots, \tag{12}$$

where $\kappa = \beta/\delta$, $\phi = 2\pi t\delta/\lambda$, and t is the thickness of the ZP material (see Fig. 2b). For opaque zones ($\kappa = \infty$) the 1st and 3rd orders efficiencies are $\eta_1 = 1/\pi^2 \approx 10.1$ %, and $\eta_3 = 1/9\pi^2 \approx 1.1$ % respectively. With weak absorption ($\kappa \to 0$) and $t \to \lambda/2\pi\delta$ the efficiency increases (Fig. 2 b), and can be as high as $\eta_1 = 4/\pi^2 \approx 40.5$ %. A 33% efficiency at 8 keV has been reported for a phase ZP [12].

The ZP performance optimization requires a blazed profile. Arbitrary blaze shape fabrication could be very difficult. To approximate the blaze shape by a "staircase" [1,13] each pair of zones is sectioned into p ($v = 1,2...p$) bands with radii expressed by

$$r_n(v) = \sqrt{(n + 2v - 2)f\lambda}, \tag{13}$$

The efficiency of a non-absorbing phase ZP with a staircase profile is

$$\eta_m(p) = \frac{1}{m^2}\left(\frac{\sin(\pi/p)}{\pi/p}\right)^2. \tag{14}$$

In case of $p = 2$ we have a familiar binary phase ZP, and the efficiency is $\eta_1(2) = 40.5\%$. When $p = 3$, $p = 4$ and $p = \infty$ the efficiency will be $\eta_1(3) = 68.5.5\%$ and $\eta_1(4) = 81.2\%$ and $\eta_1(\infty) = 100\%$ respectively.

III. ZONE PLATE ABERRATIONS AND INACCURACIES

The resolution and efficiency of a ZP depends on the utilized optical scheme[14], on the accuracy of the shape and relative position of the zones[1,15]. In this respect, the ZP aberrations are caused by off-axis imaging and by the pattern inaccuracies.

Off-axis aberrations. In addition to chromaticity, the ZP imaging, analogously to a lens, is affected by off-axis aberrations [1,14]. The wave front deformation caused by a marginal ray propagating at an angle α from the axial ray can be expressed by:

$$\partial W = \frac{\alpha R_N^3}{2f^2}\cos\phi - \frac{\alpha^2 R_N^2}{2f}\cos^2\phi - \frac{\alpha^2 R_N^2}{4f}. \tag{15}$$

The first term in Eq. (15) represents coma, the second astigmatism, and the third field curvature. But, unlike a lens, the ZP is completely free of linear distortion [1,14]. Spherical aberration is also absent when ZP is used for a design wavelength. When the number of zones N is large, the diffraction-limited field is small, coma dominates, and the acceptable field is $2\alpha = 1/2(f/\lambda)^{1/2} N^{-3/2}$. For a ZP with few zones the astigmatism and field curvature dominate, but the diffraction-limited field is large $\alpha = (3N)^{-1/2}$

ZP pattern errors. The performance of a ZP depends on the accuracy and positions of zones [1,15]. The likely types of imperfections are [15] ellipticity, radial error, and non-concentricity of the zones (see Fig. 3a), causing astigmatism, spherical aberration, and coma respectively. Defining ε as the eccentricity, γ as the non-concentricity in the Φ direction, and σ as the radial error, Eq. (7) can be generalized [1,15] as:

$$R_n^2 (1 - \varepsilon^2 \cos^2 \Theta) - \frac{1}{\sqrt{f\lambda}} \gamma R_n^3 \cos(\Theta + \Phi) - \frac{1}{f\lambda}(\sigma + \frac{\lambda}{4f})R_n^4 = nf\lambda \cdot \quad (16)$$

The primary aberrations can be balanced to a certain degree by appropriate defocusing[1] and tilting. The aberration balancing occurs naturally when the "best" focus is obtained during alignment. The astigmatism of an elliptical ZP can be compensated also by tilting, but this will introduce coma. A detailed analysis of ZP pattern inaccuracies and their detection is given elsewhere[1,15].

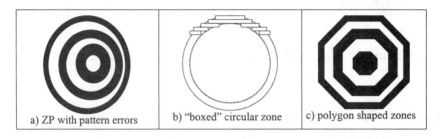

a) ZP with pattern errors   b) "boxed" circular zone   c) polygon shaped zones

FIGURE 3. ZP Pattern Errors and Common Representations of Circular Zones

Rectangle and polygon zone approximation. When a circular pattern is mapped onto the orthogonal coordinates accuracy of circle "fractioning" and the resulting data file size must be considered. The reduction of the data volume can be achieved by using rectangular primitives[16], or a fast polar coordinate pattern generator[17]. A simple approximation (Fig. 3 b), defines a zone as a union of N_{box} rectangles[16]

$$N_{box} = 2kN^2, \quad (17)$$

where k is the ratio of zone width to "box" width. The "staircase" shaped edge formed by rectangles does not degrade ZP resolution, but the efficiency decreases. The effect can be made small by increasing the k value:

$$\eta_{box} = \left(1 - \frac{1}{4k^2}\right)^2, \quad (18)$$

and for $\eta_{box} = 0.81$ $k > 1.6$. However, this approximation tends to generates extremely big files[16]: for k=3 and N=1860 the data file size is 500 MB. A substantial reduction in a data file size can be achieved by an approximation of a zone with an m-sided polygon (Fig.3 c). The total number of all sides in a ZP[16] is:

$$N_{poly} = \frac{mN}{2} \quad (19)$$

When m=180 and N=1860 the data file size is only 2 MB. The resolution of a ZP is not affected, but the efficiency is reduced, depending on a number m of polygon sides:

$$\eta_{poly} = 1 - \frac{N^2 \pi^4}{3m^4}. \tag{20}$$

For η_{poly} >0.81, an acceptable polygon side number is $m_{min} > (5\beta)^{1/4} \pi N^{1/2} \approx 1.15 N^{1/2}$. For 1860 zones m_{min} >154, and m =180 was used in our design[16].

IV. ZONE PLATE FABRICATION

Remarkable progress in ZP fabrication has been made in the last 15-20 years by the development of advanced lithographic techniques[1,18]. Interferometric [10,19] lithography has been used to produce ZP patterns with many hundreds of zones with the outermost zone width as small as 60 nm. But nearly all of the ZPs in use are produced by direct write e-beam lithography, allowing fabrication of ZPs with outer zone widths as small as 40-30 nm [1,11]. The ZP is formed by electroplating of gold [1], Ni [19] or another suitable metal. High-resolution ZPs require the use of relatively thin resist (~250 nm) which limits the absorber thickness. Thin silicon nitride films (50-500 nm thick) are used to support the ZP. The most recent progress in ZP fabrication was demonstrated by producing a high performance phase and blazed ZPs [1,12,13] for hard X-rays. To increase absorber thickness the gold ZP patterns are used as masks in a variety of x-ray lithographic replication processes [8,11-13].

Multi-step replication process. The most common approach is one-step resist printing, followed by gold electro-plating[1,11,20]. The thickness of phase shifter produced this way depends on thickness of the absorber on the mask, which is limited to 100-200 nm of gold, as a consequence of e-beam writing. Due to low contrast of these masks, only ~0.5 µm thick gold pattern can be formed in a first step replication. This

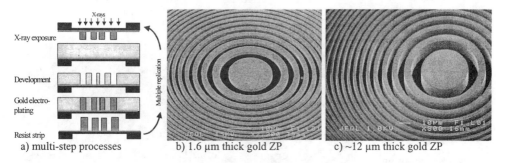

a) multi-step processes   b) 1.6 µm thick gold ZP   c) ~12 µm thick gold ZP

FIGURE 4. Multi-Step Phase ZP Fabrication Process

thickness is sufficient to provide ~⅓ π phase shift for 8 keV photons in gold. To produce thicker phase shifter a multi-step process is used (see Fig 4a), using replicas from the previous step as masks. A three-step process produced a 1.6µm thick phase ZP (see Fig. 4b) with 0.25-µm outermost zone width[16,20]. In the 4th replication step, using the

1.6 μm thick phase ZP as the mask, a ~12 μm thick ZP pattern (see Fig. 4c) with features as small as 0.8 μm has been formed.

Multi-level fabrication process. In this process the same or complementary masks can be used to print one level on the top of the previous, in the same way as it is done in microcircuit fabrication. A highly precise alignment between the levels is essential. This

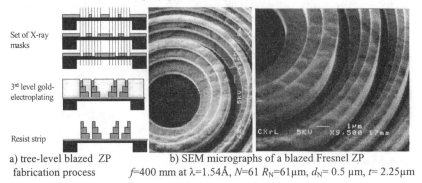

a) tree-level blazed ZP
fabrication process

b) SEM micrographs of a blazed Fresnel ZP
f=400 mm at λ=1.54Å, N=61 R_N=61μm, d_N= 0.5 μm, t= 2.25μm

FIGURE 5. Four –Band Blazed Fresnel ZP for Hard X-rays

method was utilized to fabricate a four-band zone blazed ZP[1,13,21] with a gold phase shifter. A three-level lithographic process, requiring a set of three masks (see Fig. 5a), and high precision alignment was utilized. Each level includes x-ray exposure, development, gold micro-electroplating, and resist removal steps. SEM micrographs of a blazed ZP, fabricated using this process[13,21], are shown in Fig. 5b.

Multi-level self-aligned process. For application of phase ZPs in hard X-ray region (20 keV photons) a 3.5-μm thick gold pattern is required. The limitations for high-aspect ratio patterning are imposed by mechanical properties of the resist and surface tension effects during development[20]. To compensate for these limitations a multilevel self-aligned process [22] was developed (see Fig.6). Two variations of this method are possible: topside and backside resist application and phase shifter forming.

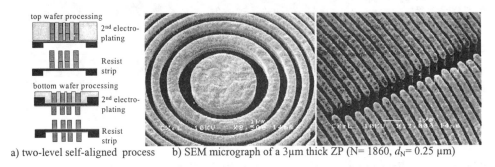

a) two-level self-aligned process   b) SEM micrograph of a 3μm thick ZP (N= 1860, d_N= 0.25 μm)

FIGURE 6. Gold Phase ZP Fabricated by Two-level Self-Aligned Process.

Negative resist is applied on the top or on the back of a previously produced ZP (see Fig 6a). A blanket X-ray exposure is performed from the backside or from the topside of the membrane respectively. The first layer gold pattern serves here as a contact X-ray mask. After development the negative resist remains in the exposed areas (between the gold lines) providing alignment for the following electroplating. Two-layer 3µm thick gold phase zones plate with $d_N=0.25$µm have been formed[22], as shown in Fig.6 b. As the first level a 1.6 µm thick ZP (see Fig. 4 b) was used.

V. SUMMARY

The phase Fresnel ZPs are excellent focusing elements in soft and hard X-ray regions. Several approaches to produce thick ZPs and to form proper phase shifting have been developed: multi-step X-ray lithographic pattern replication, multi-level processes, suitable also for blazed profiles, and self-aligned process.

High performance phase ZPs have been produced and demonstrated the focusing capabilities < 90 nm for 8 keV X-rays. They have been used for the 8-30 keV X-ray region in a state-of-the-art x-ray microprobe at the ANL (Table 2.).

Table 2. CXrL produced Phase Zone Plates						
f, cm	Photons, keV	Diameter, µm	d_N, µm	t, µm	N	Quantity
5	8	51-77	0.15-0.1	0.5-1.6	85-190	15
10	8	100-150	0.15-0.1	0.8,1.7	160-365	5
40	8	308	0.2-0.25	1.2-1.5,2.4	380	9
100	8	616	0.25	0.4-1.5	620	5
300	8	1860	0.25	1.6	1860	5
25	20	60	0.25	3-3.5	60	4
40	20	82	0.3	3.3	70	2
75	20	1860	0.25	3	1860	1

VI. ACKNOWLEDGEMENTS

This work is based on the combined efforts of many workers (affiliations are at the time when work was performed): Michelle Brown, Franco Cerrina, Zheng Chen, Azalia Krasnoperova, Quinn Leonard, Fred More, Collin Tan, and Olga Vladimirsky (CXrL, University of Wisconsin-Madison) Enzo Di Fabrizio, Massimo Gentili (Istituto di Elettronica dello Stato Solido, Consiglio Nazionale delle Ricerche Rome, Italy.), Wenbing Yun, Barry Lai, Efim Gluskin (Advanced Photon Source, AN L, U.S.A.). The support for this work was provided in part by grants from ANL/DOE under Grant No 062242-02. CXrL is supported in part by DARPA/ONR Grant No N00014-97-1-0460.

VII. REFERENCES

1. Vladimirsky, Y., "Zone Plates" in *Vacuum Ultraviolet Spectroscopy I*, edited by. J.A.Samson and D.L.Ederer, (Experimental Methods in Physical Sciences, **32**) Academic Press, 1998, pp.289-303

2. Jenkins, F.A. and White, H.E., *Fundamentals of Optics*, McGraw-Hill, New York, 1957, p.51,

3. Yang, B.X., *Nucl. Instrum. Methods A*, **328**, 1993, pp. 578-587

4. Spiller, E., *Soft X-ray Optics,* SPIE Engineering Press, Bellingham, Washington, USA, 1994

5. Miyamoto, K., *J. Opt. Soc. Am.* **51**(1), 1961, pp.17-20

6. Snigirev, A., Kohn, V., Snigireva, I., and Lengeler, B., *Nature*, **384**, 1996, pp.49-51

7. Kirz, J., *J. Opt. Soc. Am.,* **64**, 1974, pp.301-309

8. Vladimirsky, Y., Källne E. and Spiller, E., *SPIE Proc.,* **448**, 1984, pp.25-37

9. Tatchyn, R. O., Springer Series in Optical Sciences, **43**, *X-Ray Microscopy*, edited by G. Schmahl and D. Rudolph, (Springer Verlag), Berlin, 1984 , pp.40-50

10. Schmahl, G., Rudolph, D., Niemann, B., *Scanned Image Microscopy*, edited by E. A. Ash, Academic Press, London, 1980, pp.393-411

11. Yun, W., Lai, B., Cai, Z., Maser, J., Gluskin, E., Chen, Z., Krasnoperova, A.A., Vladimirsky, Y., Cerrina, F., DiFabrizio, E., Gentili, M., *Rev. Sci. Instrum.,* **70**, No 5, 1999, pp.2238-2241

12. Krasnoperova, A.A., Xiao, J., Cerrina, F., Di Fabrizio, E., Luciani, L., Figliomeni, M., Gentili, M., Yun, W., Lai, B., Gluskin, E., *J. Vac. Sci. Technolog.*, **B 11** (6), 1993, pp. 2588-2591

13. Di Fabrizio, E., Gentili, M., Grella, L., Baciocchi, M., Krasnoperova, A. A., Cerrina, F., Yun, W., Lai, B., Gluskin, E., *J. Vac. Sci. Technolog.,* **B 12** (6, 1994), pp. 3979-3985

14. Young, M., *J. Opt. Soc. Am.* **62**(8), 1972, pp.972-976

15. Y. Vladimirsky, H.W.P.Koops, *J. Vac. Sci. Technolog.,* **B 6** (6), 1988 , pp. 2142-2146

16. Chen, Z., Vladimirsky, Y., Brown, M., Leonard, Q., Vladimirsky, O., Moore, F., Cerrina, F., Lai, B., Yun, W., Gluskin, E., *J.Vac.Sci.Technol.*, **B 15**(6), 1997, pp. 2522-2527

17. Vladimirsky, Y., Kern, D. P., Chang, T. H. P., Attwood, D. T., Ade, H., Kirz, J. , McNulty, I. , Rarback, H. , and Shu, D., *J.Vac.Sci.Technol.* **B 6** (1), 1988, pp. 311-315

18. Vladimirsky, Y., "Lithography", in *Vacuum Ultraviolet Spectroscopy II*, eds. J.A.Samson and D.L.Ederer, (Experimental Methods in Physical Sciences, **31**) Academic Press, 1998, pp.205-223

19. Schneider, G., Wilhein, T., Nieman, B., Guttmann, P., Schliebe, T., Lehr, J., Aschoff, H., Thieme, J., Rudolph, D. and Schmahl, G., *SPIE Proc.,* **2516**, 1995, pp.90-101

20. Chen, Z., Vladimirsky, Y., Cerrina, F., Lai, B., Yun, W., Gluskin, E., *SPIE Proc.,* **3331**, 1998, p. 591

21. Krasnoperova, A.A., Chen, Z., Cerrina, F., DiFabrizio, E., Gentili, M., Yun, W., Lai, B., and Gluskin, E., *SPIE Proc.,* **2516**, 1995, pp.15-26

22. Krasnoperova, A.A., Chen, Z., DiFabrizio, E., Gentili, M., Cerrina, F., *J. Vac. Sci. Technol.* **B 13**, 1995, p. 3061

Fabrication and Characterization of Tungsten Zone Plates for Multi KeV X-rays

P. Charalambous

Department of Physics, King's College,
The Strand, London WC2R 2LS, U.K.

Abstract. Tungsten is a material well suited for the fabrication of diffractive optics intended for use at high (multi KeV) X-ray energies. It is however a particularly difficult material to deposit, especially when relatively thick (>500 nm) films on thin membranes are needed. In such a case, deposition stress and granularity lead to membrane deformation or destruction. We have devoted a lot of time and effort in identifying and controlling the critical deposition parameters, and we are now in a position to successfully deposit up to 1 μm thick Tungsten films, which can be processed and structured with Fluorine based RIE. This has enabled us to fabricate a number of ZPs suitable for operation in the 0.3 – 8 KeV regime. In addition, through the use of custom metrology algorithms, we have endeavored to assess the general accuracy of our lithographic patterns, and more specifically the circularity of our ZPs.

INTRODUCTION

The proliferation of X-ray sources and more specifically Synchrotron Radiation Facilities, has led to a very dramatic increase in the available X-ray microscopy experiments world wide. In the early years of X-ray microscopy, there was a definite interest in the so called "water window" region of the energy spectrum, as it was felt that it would allow the recording of images from "living" and untreated biological samples. More recently however, there has been a shift in the energy region of interest, as it became apparent that using "harder" X-rays (2 – 8 KeV), would make possible the imaging of much thicker samples, and more importantly, would allow the recording of phase contrast images. [1]

Zone Plates are still the preferred focussing element for X-rays, especially for high spatial resolution imaging, but the fabrication of these optical elements for operation at the higher energies is a very challenging technological task.

We have been involved in the fabrication of Zone Plates, diffraction gratings, as well as test samples for X-ray microscopy for many years, and more recently have concentrated our efforts in all aspects of the design and fabrication of Zone Plates for harder X-rays. In this paper, we will present some aspects of our fabrication program, as well as some examples of our latest results, and possible future directions of our research.

CP507, *X-Ray Microscopy: Proceedings of the Sixth International Conference,*
edited by W. Meyer-Ilse, T. Warwick, and D. Attwood
© 2000 American Institute of Physics 1-56396-926-2/00/$17.00

ZONE PLATE FABRICATION FACILITY

The Zone Plate fabrication facility at King's College has been described in detail on a number of occasions in the past [2],[3], but for the sake of completeness a very brief description will be given again here.

Our lithography is carried out on a customized SEM (JEOL JSM820), which is controlled by a simple PENTIUM 233 PC. The original SEM stage has been replaced by a BURLEIGH INCHWORM stage, with HEIDENHEIN position encoders, allowing a range of movement of 30 x 30 mm at 0.1 µm resolution. The electron source is a simple Tungsten "hairpin" filament, and the electron optics of the system can shape the electron probe to a few nanometers at typically 10-1000 pA. We have also installed a beam blanking system, with a blanking resolution of 50 nS. This is particularly important for accurate exposure control for Chemically Amplified Resists (CARs). The electron energy can be set from 1 to 30 KeV.

The only aspect of the SEM which is somewhat limiting for accurate pattern generation, is the level of scan field distortion. Commercial E-beam systems offer guaranteed linearity of scan fields, typically of the order of 20 nm in a 200 µm field. This coupled with electron energies of >100 KeV, and interferometric stage control with < 0.6 nm resolution, makes these machines ideal for accurate pattern generation (like ZPs). However, the price tag for these systems is quite prohibitive for educational institutions. The scan field distortion levels on a typical SEM are of the order of 1%, which is perfectly adequate for imaging, but quite unacceptable for accurate lithography. This leaves us with the only option of constantly striving to improve the performance of our SEM in all aspects that affect the specific task at hand, which is mainly the creation of accurate circles, or zones. In this respect, we have developed a suite of algorithms, which specifically assess the circularity of our Zone Plates, and which will be described in more detail in the next section.

Both the pattern generation hardware, and software, have been designed and implemented by the author, with particular emphasis in critical optimization of both in order to achieve the highest possible speed of pattern generation. This is imperative in view of the fact that there is no stage registration at present, and any possible stage drift could severely degrade our patterns. The Pattern Generator is based on a pair of 16 bit DACs, which are optically coupled to the control PC. All software is written in C, with some critical sections re-written in Assembly. One of our priorities in the near future, is to introduce registration during pattern generation. The motivation for this, is to plot multiple level patterns on the same ZPs for the creation of blaze type zone profiles. Registration will also compensate and correct any effects due to stage drift.

ASSESSING ZONE PLATE CIRCULARITY

The first step in the fabrication of ZPs and other optical elements, is that of lithography. The electron beam is driven by the Pattern Generator (PG) under software control, and the pattern is plotted onto the resist. The accuracy of the pattern however, can be severely degraded by a number of distortions that are present to some degree in all E-beam systems. There are broadly two categories of distortions; Static, which are mainly the various types of scan field distortion, and dynamic, like mechanical vibrations, stage drift, and EM interference, especially mains [3].

There are mainly two types of static scan field distortions that can affect the circularity of a ZP. These are the cross calibration of, and orthogonality between the X and Y scans. We have devised a method of assessing the circularity of any object, which allows the interactive calibration of both these relationships between the X and Y scans, and the subsequent measurement of the circularity of any object. We can routinely plot objects with circularity better than 100 nm in diameters of 200 μm.

Circular Raster Scan (CRS)

The diagram of fig. 1 outlines the principle of assessment of circularity. What is displayed on the computer screen is a "live" raster image of a square format. On the sample however, each line represents a *radial* scan at some distance from the scan field center (arrow lines), and centered about the mean position were the zone under observation is physically placed. Both the number and the length of these radial scans is under operator control, and define the frame size of the image on the computer screen. As an example, the choice of 1 degree angular resolution between scans will produce an image of 360 lines/frame on the screen, each line representing an arbitrary physical distance perpendicular to the zone under observation, depending on the chosen pixel step. The scanning rate is > 10 Hz with the present PENTIUM 233 PC.

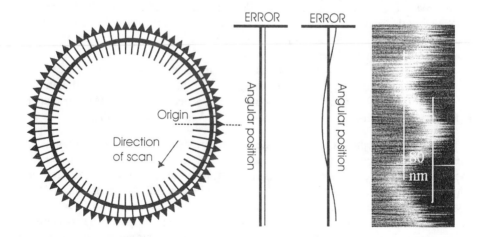

FIGURE 1. The principle of CRS. The two graphs to the right show what could be displayed on the computer monitor as an image. The first graph (straight line) will be produced if the physical zone is perfectly circular but larger than the CRS. The sine wave will be the result of the zone being equal to the CRS but off center to the right (like the diagram). The micrograph is a real CRS scan of 300 lines (1.2 degree steps), showing an orthogonality induced ellipticity of 80 nm in a 180 μm diameter zone.

Before using the CRS, it must be calibrated against a known (*"perfect"*) circular object. Both the amplitudes of the X and Y scans, as well as the angle between them is adjusted from the keyboard until the plot on the screen is as near as possible to a straight line. When this is achieved, the sample can be rotated by $90°$, and the scan repeated, in order to eliminate the possibility of ellipticity in the object. Once calibrated, the scans can be trusted to plot objects which should in principle be as circular as the calibration circle. We use ZPs (160 μm diameter), plotted on a LEICA system as our calibration circles. The use of a "perfect" circle for calibration is not strictly necessary but simply speeds up the calibration process.

The next stage is to plot a series of concentric circles extending in the whole scan field using the calibrated scans. This usually takes a few seconds, eliminating any possible effects of stage drift on the shape of the circles. After development and possibly metallization, the test circles can be *examined* using the CRS. Looking at circular objects with the scan that created them may seem rather pointles, but in fact there is a good reason for doing so. If there is pincussion or barrel distortions in the can field, then circles will be less circular depending on where they are in the scan field. Again, if the circles are rotated by $90°$ from the orientation that they were created, it will reveal any ellipticity.

WAFER DESIGN FOR ZONE PLATE FABRICATION

We have experimented with a number of different techniques for the fabrication of high energy ZPs. The first choice was based on Electroplating, using single level resist as the mold. Following exposure and development of the pattern, we could deposit any of a number of different materials, like Gold, Nickel, or Platinum. This technique works well with aspect ratios in the resist of less than 3:1. Deeper molds lead to Zone collapse either during resist development, or even during Electroplating. Because of this, the resolution or thickness of ZPs made in this way, is rather limited. The next obvious development for Electroplating is to create the mold using two or three level resist, followed by RIE in order to transfer the pattern into the required mold material. This technique could create mold features with aspect ratios of 10:1 or better! Preliminary experiments using this approach, yielded encouraging results.

The alternative method for ZP fabrication, is to pre-deposit the Zone material at the required thickness on a suitable substrate (in our case Si_3N_4), followed by the deposition of multi level resist materials finishing with the resist itself at the very top, with a thickness rarely exceeding 100 nm. The obvious advantage of this, is that forward scattering of 30 KeV electrons is negligible, and the exposure of sub 20 nm lines is readily achieved. After exposure and development, the pattern is transferred into the underlying resist materials, until the eventual mask material is reached, which is usually Nickel. The final RIE stage is always fluorine based, and Nickel is an exceptionally resistant material to this chemistry. This approach, with Tungsten as the final zone material, has yielded the best high energy ZPs so far both in terms of thickness of zones, and resolution (outermost zone width).

The Deposition of Tungsten Films

Tungsten has a very high melting point, and cannot be evaporated, so sputtering is the only real option for its deposition. We have a home build single head Magnetron deposition system, as well as thermal evaporation equipment for the preparation of our wafers. A search of the literature shows that it is exceptionally difficult to deposit *stress free* Tungsten films. The situation is even more critical if the films need to be deposited on relatively thin, X-ray transparent substrates, in our case 100-200 nm thick Si_3N_4. The details of deposition control are outside the scope of this paper, but after many hours of experimentation with deposition conditions (deposition rate, chamber pressure, substrate temperature, etc) the main conclusion is that there seems to be a strong relationship between grain size and film stress. There is also a very sharp transition from compressive to tensile stress, depending on the chamber pressure. The critical window for stress free depositions is only < 2 mbar. Within this window, there is still a detectable grain size in the films, of around 10 – 20 nm, which is perfectly acceptable for the fabrication of high energy ZPs with zones down to 50 nm. Lower grain size can be achieved at slightly lower deposition pressures, but the films are deposited at considerably more compressive stress, which limits the thickness that can be usefully achieved, to < 300 nm. These films could be used for the fabrication of higher resolution ZPs, (<20 nm) with good diffraction efficiency from the water window, up to 1.5 KeV and beyond. Finally, we have found that the use of pulsed deposition, can also help in the control of film stress.

Within the critical "stress free" pressure regime, we are able to successfully deposit Tungsten films up to 1 μm thick. The micrographs of figure 2 show two examples of zones transferred in both grainy, and optimally deposited Tungsten films.

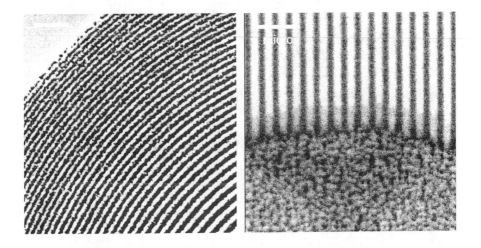

FIGURE 2. The micrograph to the left shows 150 nm outer zones of a ZP transferred into 500 nm thick Tungsten, under severe tensile stress, with mean grain size >100 nm. In contrast, the micrograph to the right shows 150 nm zones, in 750 nm thick Tungsten, deposited under optimum conditions. In this case, the mean grain size is less than 20 nm, and hardly visible.

CONCLUSIONS

We have made considerable progress towards the fabrication of Tungsten ZPs suitable for operation at high X-ray energies. Table 1 below shows a sample of Tungsten ZPs that we have tested to date. In addition we have a number of ZPs transferred into 750 nm thick Tungsten, which we will test with X-rays in December 99. The parameters of some of the ZPs are optimized for use with our Scanning Near Field X-ray Microscope (SNXM)[4], which is based at ID21 at the ESRF. The optical arrangement of the SNXM would be greatly simplified if we could eliminate the use of an OSA. This could be possible if the zones of the condenser ZPs had a blazed profile, minimizing the presence of zero[th] order. Next year, we are hoping to fabricate blazed ZPs in Tungsten, Tantalum, Gold, or combinations of these.

Table 1. Sample of High Energy Tungsten Zone Plates tested to date

Diameter (μm)	Outermost dr_n (nm)	Thickness (nm)	Energy (KeV)	Groove eff. (%)	COMMENTS
50	125-160	450	3.5	7.5	Array of four
70	148 , 194	500	3.5	11	Array of two
70	185	500	6.0	8.8	Single
92, 200	90,200	360	3.0	10	Array of two
300	116	360	3.0	8.4	Single

ACKNOWLEDGEMENTS

The author would like to thank the Leverhulme Trust for supporting this work

REFERENCES

1. Eaton, W. J., Morrison, G.R., Waltham, N. R., "Configured Detector System for STXM Microscopy", This Volume

2 Charalambous, P., Anastasi, P., Burge, R. E., Popova, K., "The Fabrication of High Resolution X-ray diffractive Optics at King's College, London", SPIE95 Vol 2516/3, San Diego (1995).

2. Charalambous, P., Burge, R. E.,"Zone Plate Fabrication at King's College London", Proceedings of XRM96, Wurzburg 1997.

3. Burge, R. E., Browne, M. T., Charalambous, P., Yuan, X. C., "Combined Near-Field STXM with Cylindrical Collimator and AFM", This Volume

X-Ray Microbeams From Waveguide Optics

S. Di Fonzo[1], W. Jark[1], S. Lagomarsino[2], A. Cedola[2,3],
C. Riekel[3]

[1]SINCROTRONE TRIESTE, S.S. 14 km 163.5 in Area Science Park, 34012 Basovizza, Italy
[2]Istituto Elettronica Stato Solido (IESS) - CNR, V. Cineto Romano 42, 00156 Roma, Italy
[3]European Synchrotron Radiation Facility, B. P. 220, 38043 Grenoble CEDEX, France

Abstract. This report will review the important properties as efficiency, energy tunability and the spatial coherence and divergence of x-ray microbeams exiting at the termination of thin film or slab x-ray waveguides of about 0.15 μm dimension.

THE PRINCIPLE OF X-RAY WAVEGUIDES

The properties of thin film or slab waveguides are well understood in the optical range[1]. As a condition for mode guiding in them the refractive index of the slab n_1 has to be larger than the refractive indeces n_2 and n_3 of the materials which enclose it[1] (see figure 1), i.e. $n_1 > n_2 \geq n_3$. Such a relation between the refractive indeces can also be realized in the hard x-ray regime and should then give rise to traveling x-ray modes. In the optical range with $n > 1$ a larger refractive index n_1 corresponds usually a to "heavier" material. Now in the x-ray range all materials have refractive indeces $n_j < 1$, which are close to unity and thus the index is usually written as $n_j = 1 - \delta_j + i \beta_j$, with δ_j and β_j being small positive numbers. In the x-ray range away from absorption edges δ is proportional to the atomic number Z and thus the "lighter" elements are now those with larger refractive index, i.e. with refractive index closer to unity. As the lighter elements have also a smaller extinction coefficient β_j, x-ray waveguides are feasible with e.g. a carbon film enclosed between "heavier" metal layers. The coupling of the x-rays to the traveling modes can be made in a very efficient way: if the slab is coated with only a thin metal overcoat as indicated in figure 1, then the evanescant x-ray wave

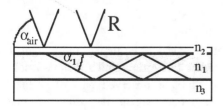

FIGURE 1. Scheme of an x-ray slab waveguide.

CP507, *X-Ray Microscopy: Proceedings of the Sixth International Conference,*
edited by W. Meyer-Ilse, T. Warwick, and D. Attwood
© 2000 American Institute of Physics 1-56396-926-2/00/$17.00

in the metal layer can couple intensity into the internal modes. The permitted modes have nodes at the slab interfaces and thus their propagation angle can be calculated readily from $2 \sin(\alpha_{1,m}) = (m+1) \lambda/D$, with m being the mode index and D being the slab thickness. The corresponding incidence angle in air onto the waveguide is calculated as $\alpha_{air,m} = SQRT(\alpha_{1,m}^2 + 2 \delta_1)$. The refraction at the top interface of the slab is thus refracting the intensity towards a more grazing internal angle. The result is a significant increase in the internal intensity in case an internal mode can be exited. For this reason this coupling scheme is nominated as resonant beam coupling (RBC). The amplitude of the internal intensity depends on the slab absorption and can exceed 100 times the incident intensity[2]. The following properties of the guided x-ray modes are of interest in this study: a) the spatial coherence of the traveling mode in the direction of its confinement and b) its small extent in the same direction perpendicular to its propagation direction. Indeed the optimum slab thickness for x-rays is of the order of 150 nm. This is a beam size, which is rarely achieved with other devices and very interesting for microbeam experiments with x-rays.

While the traveling of x-ray modes in thin films was already observed by Spiller and Segmüller in 1974[3], only recently in 1995 our collaboration[4] (Lagomarsino) and Feng et al[5], detected at different beamlines at the ESRF almost simultaneously for the first time the beam exiting from an x-ray waveguide terminal. The intensity of the traveling mode was found to be limited to the vertical slab extent[6], and the modulation in the Fraunhofer diffraction pattern measured far from the waveguide exit[5,6], pointed onto a high degree of spatial coherence of the exiting beam.

Intensity Gain At The Exit Of An X-ray Waveguide

The obvious question at this point is whether the introduction of an x-ray waveguide into an x-ray beam provides advantages over the use of a simple slit. As the boundary condition the operation in monochromatic unfocused synchrotron radiation at a low emittance storage ring like the ESRF is considered (relative energy bandpass $\Delta E/E \cong 2 \cdot 10^{-4}$ i.e. standard channelcut Si(111) crystal monochromator). Prepared on a highly polished flat substrate an x-ray waveguide acts as a high quality slit with opening in the submicron range. Slits constructed with knife-edges neither provide similar straightness nor similar parallelity. Due to the beam compression into the waveguide in the resonant coupling process an x-ray waveguide should provide an intensity gain (intensity = flux per unit area). Figure 2 presents the gain observed behind a waveguide with 162 nm thick slab, which was illuminated with 17 keV photons up to the waveguide terminal. For simplicity the exiting flux was considered to be distributed homogeneously over the slab thickness, even though the fundamental mode has the intensity maximum in the slab center with a fullwidth at halfmaximum of only half of the thickness. In the light of this the observed gain of 4 for the fundamental mode and of 6 for the second mode exceeds the slit performance significantly. Similar performance could be achieved consistently with different waveguides of similar parameters. Moreover, at the time of writing this, first tests at further improved waveguides point onto an almost 10fold improvement in the gain.

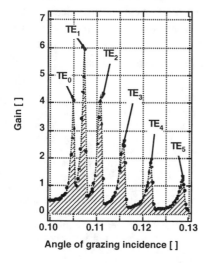 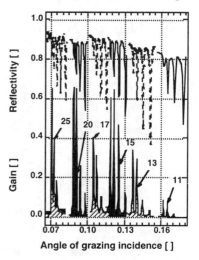

FIGURE 2. Gain (i.e. the flux per unit area at the waveguide exit divided by flux per unit area of the incident beam) for the first 6 guided modes (transverse electric TE_0 to TE_5) in a waveguide with 162 nm carbon slab on Cr and covered with 4 nm of Cr measured for 17 keV photon energy in the unfocused X-ray beam at the undulator beamline ID13 at ESRF.

FIGURE 3. Reflectivity (top curves) and corresponding gain (filled curves at bottom) depending on angle of grazing incidence for a second waveguide with the parameters of figure 2 for the different indicated photon energies (from left to right: 25, 20, 17, 15, 13 and 11 keV). As in this case the last 3 mm in front of the waveguide termination were covered with a thick opaque Cr guiding layer (20 nm) the gain is reduced compared to the sample in figure 2.

The energy tunability of such a waveguide structure, here with a 3 mm wide section in front of the waveguide terminal, in which no further beam coupling will occur, is presented in figure 3. Photon energies between 11 keV and more than 25 keV can be compressed to the same submicron dimension with a change in angle of incidence of only about 0.1°. In any case only one guided mode is excited at a time, as the incident beam divergence ($\cong 5$ µrad) is significantly smaller than the mode separation (> 35 µrad). On the other hand the beam divergence is larger than the angular interval in which the fundamental mode can be excited (< 4 µrad).

Spatial Coherence Of The Exiting Beam

In the direction of confinement a guided mode in the slab is producing a standing wave. In this direction the intensity is thus spatially coherent at the waveguide termination and as this it will produce a Fraunhofer diffraction pattern far from this exit. Figure 4 presents these patterns for three guided modes registered with a CCD camera at 890 mm distance. The measured profiles are in very good agreement with the simulations taking as the source the corresponding standing waves at the waveguide termination[5,6]. The profile of the diffraction pattern of the fundamental mode TE_0 is approximated well with a Gaussian profile. Thus the waveguide provides a coherent beam, which is "cleaner" as far as the principle beam trajectories are concerned than the beam passing a conventional slit, in which almost 10% of the radiation is diffracted into secondary peaks. In addition it is slightly divergent (appr. 1 mrad).

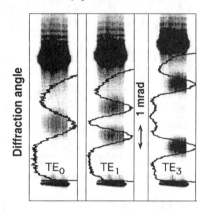

Figure 4. CCD images of the diffraction patterns produced al large distance (890 mm) from the waveguide termination by the fundamental and the first two odd modes TE_0 (left), TE_1 (center) and TE_3 (right). The photon energy was 13 keV and the slab thickness 137 nm. The upper intense spot is the unblocked reflected beam and the lower spot is caused by the not intercepted part of the incident beam. The waveguided beams are found between these spots. Their intensity integrated over 5 pixel columns is overlayed in the image as a black line with the zero at the left border of each image (exposure time 2 s).

ACKNOWLEDGEMENT

The described results were obtained at the beamlines ID13 and BM05 at the European Synchrotron Radiation Facility ESRF in Grenoble (France). We would like to thank very much the following colleagues from ESRF for their help and for their input: Professor C. Kunz, M. Müller, P. Wattecamps, A. Freund, O. Hignette, R. Hustache, A. Souvorov, P. Cloetens and W. Ludwig. We gratefully acknowledge as well the contributions of G. Sandrin, G. Sostero and G. Soullie' from the Sincrotrone.

REFERENCES

1. Marcuse, D., *Light Transmission Optics*, *2nd edition*, Malabar, Florida: Krieger Publ. Co., 1989

2. Wang, J., Bedzyk, M. J., and Caffrey, M., *Science* **258**, 775-778 (1992)

3. Spiller, E., and Segmüller A., *Appl. Phys. Lett.* **24**, 60-62 (1974)

4. Lagomarsino, S., Jark, W., Di Fonzo, S., Cedola, A., Müller, B. R., Riekel, C., and Engstrom, P., *J. Appl. Phys.* **79**, 4471-4473 (1996)

5. Feng, Y. P., Sinha, S. K., Fullerton, E. E., Grübel, G., Abernathy, D., Siddons, D. P., and Hastings, J. B., *Appl. Phys. Lett.* **67**, 3647-3649 (1995)

6. Jark, W., Di Fonzo, S., Lagomarsino, S., Cedola, A., Di Fabrizio, E., Brahm, A., and Riekel, C., *J. Appl. Phys.* **80**, 4831-4836 (1996)

Toward High Resolution and High Efficiency Zone Plate for X-ray Applications

Enzo Di Fabrizio*, Filippo Romanato*, Massimo Gentili**

*INFM-TASC Elettra Synchrotron Light Source S.S. 14 Km 163.5, Area Science Park, Padriciano 99 34012 Basovizza-Trieste Italy

** Istituto di Elettronica dello Stato Solido Via Cineto Romano 42 00156 Rome-Italy

Abstract. High resolution and high efficiency Zone Plate for X-rays in the energy range of 300 eV and 12 KeV fabricated by means of electron beam and X-ray lithography are presented. Regarding the high resolution issue gold zone plate with 70 nm resolution and thickness of 0.4 μm are shown. When the efficiency is more important, multilevel zone plate can provide an increase of the first diffraction order while suppressing higher ones. The optical test of four level zone plate made by Gold or Nickel was performed at the European Synchrotron Radiation Facility, beamline ID21. At 7 KeV an efficiency of 55% was measured.

INTRODUCTION

X-ray microscopy is living a period of fast development, opening new avenues in many domains of science and technology. This progress was triggered by the performances of the focusing optical elements for X-rays, which have recently pushed their resolution limit to the nanometer scale, thus fully exploiting the brightness of the third generation synchrotron radiation sources. To drive the development it is the fabrication by electron beam and X-ray lithography of high resolution Fresnel Zone Plates[1,2] (ZP). The attained performances already have increased ZP to the role of main optical devices for high resolution beam focusing. When the efficiency is more important than the resolution, it can be possible to fabricate a new type of multilevel zone plate. This geometry offers the double advantage of increasing the efficiency and of introducing selection rules that redistributes useless high diffraction order. However, the fabrication of a multilevel FZP includes difficulties beyond that for a binary ZP, requiring the alignment of the subsequent levels deposition with an accuracy of few nanometers.

EXPERIMENTAL

The FZPs were patterned by using electron beam lithography at an accelerating voltages of 50 KeV and a beam current of 0.5 nA. The exposure field was set at 327.68 μm. The resist used was PMMA 950 K. The standard substrates used are

CP507, X-Ray Microscopy: Proceedings of the Sixth International Conference,
edited by W. Meyer-Ilse, T. Warwick, and D. Attwood
© 2000 American Institute of Physics 1-56396-926-2/00/$17.00

silicon nitride membranes about 100 nm thick, made by means of standard wet etching on silicon wafers with chemically vapor deposited layers of silicon nitride. The sample was developed in 1:3 methyl-isobutyl-ketone:isopropyl-alcohol (MIBK:IPA) at 20 °C. The multilevel FZP was realized by 3-step process consisting in exposure and electroplating growth of each level. In order to align the subsequent levels a non standard alignment procedure was used reaching an accuracy better than 50 nm. The material was deposited by electroplating Gold or Nickel.

HIGH RESOLUTION ZONE PLATES

Lateral resolution is one of the most important optical characteristics of Zone Plates. The optical resolution is related to the outermost zone width of any ZP by the relationship: $R=1.22\ dr_n$, where R is the focusing resolution and dr_n is the outermost zone width. Any micro fabrication procedure must comply with the requirements of having adequate resolution whilst maintaining flexibility for the generation and the placement of the many circular rings composing the ZP. This is the case of the electron beam lithography machine used for ZP exposure. In our case the machine is a Leica Microsystems Lithography EBMF 10, working at 40-50 kV accelerating voltage and equipped with an exaboride lanthanum electron emitter.

Since most of the commercial e-beam systems have not a specific polar pattern generator, we have developed a software that is able to approximate the circular ZP patterning features with the available set of feature primitives which include rectangles and polygons. In order to achieve the highest lithographic resolution the software must also compensates for both forward and backward scattering effects in resist and substrate. This correction, which is mandatory in most of the high resolution application, is called *proximity effect compensation*[3]. The home-developed algorithm assigns the single zone exposure dose as a result of all neighbored zone-zone interactions and thus averaging locally the resist absorbed dose. Otherwise, when passing from the center of the ZP to its outermost, densely patterned part, the electron scattering effects would result very different, leading to a non-uniform absorbed dose.

Forward and backward scattering influence the lateral resolution of resist impression not only through dose proximity effects. In the case of very thin substrate (100 nm silicon nitride), a weak effect of electron back scattering from the substrate is expected. Monte Carlo electron scattering simulation was carried out to evaluate quantitatively this amount; only 2% of electrons are reflected back and their contribute is negligible in terms of a proximity effect. More important are forward scattering effects (FS) related to the spreading of the electron beam passing through the resist. At 40 kV, Monte Carlo simulation indicates that beam spreading caused by FS is about 50 nm for a 250 nm thick resist. Figure 1 shows the Monte Carlo normalized exposure dose for an infinitely small beam spot size incoming on substrate plus resist system. The calculated point spread function takes in account both the forward and back-scattering effects. The most intense and collimated peak has to be related to the electron forward scattering, whereas the broader distribution, much less in intensity, to the weak electron back-scattering effect from the substrate. The ultimate resolution in this experimental system is related to the e-beam spot size to the forward scattering effect and, in part, also on the degree of approximation of the circular features with rectangular primitives. A mere 30 nm beam spot size increase results in a change of about 40 % of the absorbed dose modulation (peak-to-valley distance).

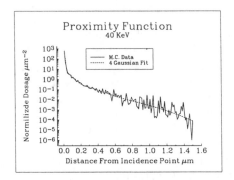

FIGURE 1. Monte Carlo simulation of the normalized exposure dose for a 40 kV electron beam with an infinitely small spot size incoming on substrate plus resist system.

Figure 2.a shows an high resolution 70 micron diameter ZP used for spectromicroscopy at ELETTRA in Trieste fabricated taking into account the previously described lithographic processes. The interference patterns visible on the picture are the so-called Moire' figures. Figure 2.b an high-resolution image on the 70 nm outermost zone. Microscopy test have confirmed a lateral resolution of 85nm with an efficiency very close to 10% (theoretical limit) at 700 eV photon energy.

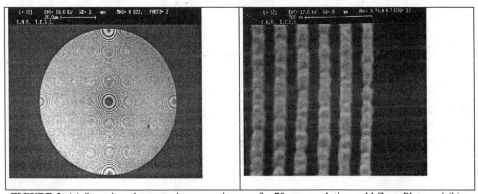

FIGURE 2. (a) Scanning electron microscope image of a 70 nm resolution gold Zone Plate and (b) a zoom of the outermost zones.

HIGH EFFICIENCY ZONE PLATES

Together with lateral resolution, high efficiency and wide energy operational range represent the other most important optical characteristics for a Zone Plates. Physical and fabricated limits, however, force often to find a trade-off between these competitive requirements The thickness dependence of phase shift Zone Plates is strictly related to the working energy range. In order to achieve a π shift, a thickness equal to $t = \lambda/2\delta$ is necessary, where δ is the difference by the unity of the real part of the diffraction index. A good approximation of δ is given by $\delta = \dfrac{N_e r_r}{2\pi} \lambda^2$, where N_e and r_e are the effective number of electrons and the classical electron radius respectively. These functional behaviors show that the thickness of phase shifter is proportional to the inverse of the X-ray wavelength λ, e.g. proportional to the beam energy. In the case of ZP fabrication process for soft X-ray microscopy, in order to have half tone contrast and hence good efficiency too, gold-plated structures should be 120-130 nm thick. Under these conditions the outermost zone with desired width below 40 nm should have aspect ratio larger than 3. It is well known that high resolution resist features made of high (larger than 3) aspect ratio can collapse due to the lack of mechanical strength. This therefore represents a fabrication challenge to the energy operational range of the high resolution ZP[4].

Binary phase shift ZP are moreover limited by quite low efficiency. The theoretical limit depends on the beam energy but in the range of soft X-ray is of the order of 10%. When lateral resolution requirements are not stringent, then multilevel ZPs can at the time strongly increase both the efficiency and the working energy range[5,6].

In the case of multilevel ZP the position of the zone radii determined by:

$$r_{n,l} = \sqrt{\lambda f\left(2\frac{l}{L} + n - 2\right)}, \qquad (1)$$

where λ is the wavelength, f is the focus length and n is the zone index number (in this formula it can have only even values) and where each zone is divided in L levels whose index is $1 \le l \le L$. Note that in the case of a binary FZP ($L=2$), the zone radius law reduces to the well known binary FZP relation, $r_{n,2} = \sqrt{\lambda f n}$. The total radiation amplitude delivered by the FZP to the focus, can be calculated by using the phasor method by integrating the contribution of the phase shift and absorption provided by each level. The integrated amplitude then results:

$$A^m = C\frac{e^{i2\pi\frac{m}{L}} - 1}{i2\pi m} \sum_{l=1}^{L} e^{-2\pi\frac{\kappa}{\delta}\frac{(l-1)}{L}} e^{-i2\pi\frac{(l-1)}{L}(m-1)} \qquad (2)$$

where $e^{-2\pi\kappa\Delta t/\lambda}$ is the field attenuation factor and κ the imaginary part of the refractive index, $n_r = 1 - \delta + i\kappa$. From equation 2 it can be derived the FZP efficiency, i.e. the fraction of the incident intensity delivered to the focus, of the mth diffraction order that is defined as: $\eta^m = |A_p^n|^2/C^2$. The groove asymmetry of FZP multilevel geometry introduces many phase contributions that, according to the diffraction order, can interfere constructively or destructively. This allows to condense the photon flux on the first diffraction order, almost completely suppressing the other orders. The imaginary term of the sum in Eq. 2 accounts for the selection rules of the diffractive order selection. In the limit of zero attenuation ($\kappa=0$), the efficiency is $\eta^m = 2(1 - \cos(2\pi m/L))(L/m)^2$. It results that mostly the *first order is active*(82%) and that the other active orders have periodicity of L. For a four level FZP, the first active order beyond the first is when $n=5$ whose efficiency is decreased for a factor 0.04 compared to the first order. Respect to a binary lens the 3-th order is suppressed. For $m<0$ (these are not focusing orders and contribute to the background) the first active order is $m=-3$ whose efficiency is decreased by a factor of 0.11 compared to the first order. Higher orders are completely negligible.

When absorption is considered, the efficiency of the first order slightly decreases. For a photon energy of 7 KeV, it changes from 81.5 % to 72% for a FZP made by Nickel and to 55% for a Gold FZP. On the other hand, absorption allows forbidden orders to appear, however leaving their intensity negligible. The higher is the zero order whose intensity is <4%. On the base of these considerations we have fabricated a Gold (figure. 1 a) and a Nickel (figure 1.b) FZP lenses tested for the energy range between 5.5 and 8 KeV. The geometrical characteristics of our FZP at energy of 8 KeV are the following: focus length=1 meter, diameter=150 µm, number of levels L=4, outermost zone-width for the fourth level=500 nm.

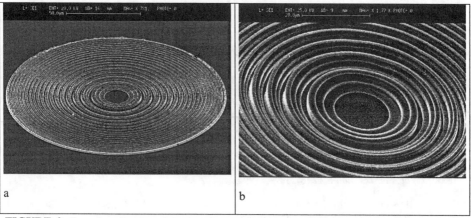

FIGURE 3. Scanning electron microscope image of quaternary Gold (a) and Nickel (b) FZPs. The geometrical characteristics designed for an energy of 8 keV are: focal length=1 meter, diameter=150 µm, number of levels=4 and outermost zone-width of the fourth level=500 nm.

The zone plate were tested optically at the X-ray microscopy beam line ID21 at ESRF (European Synchrotron Radiation Facility) which can operate over a large energy range (2-8keV). The zone-plate efficiency measurements were performed at 5.5, 6.0, 7.0 and 8 keV. At the energy of 7 KeV, Gold FZP provided an efficiency of 38%. The highest efficiency of 55% was obtained by the Nickel FSZ at 7 KeV. At 5.5 and 8 KeV the efficiency is, however, still higher than 40%. The difference between the theoretical and experimental efficiency measurements has to be attributed to the fabrication errors: electroplating growth, line width errors, alignment errors. From fabrication point of view, the most severe error is that deriving from alignment of the levels. As an example, in the case of the zone plate presented in this paper, the alignment accuracy δ necessary to limit the efficiency reduction to a 5% δ must be lower than 0.05 μm. One of the most interesting figure of merit for microscopy is the contrast of a focusing optical device. In the case of the four level Nickel ZP, we have measured a contrast equal to 680 downstream to a 10μm of a order sorting aperture. This result is again due to the multilevel geometry of the FZP that, suppressing the spurious diffraction orders and reinforcing the first order peak, makes realistic the use of a quaternary FZP quite similar to standard optical lens.

CONCLUSIONS

A fabrication process by electron beam nano lithography for sub-100 nm resolution large diameter amplitude soft x-ray ZP has been developed. In particular the issues of resolution, control of line-to-mark ratio, have been addressed in detail by means of Monte Carlo electron scattering modeling, proximity effect correction and experiments. Under optimized conditions, 70 nm resolution ZP have been fabricated. Measured efficiency for amplitude ZP was 9.5 % very close to the theoretical limit. Moreover, quaternary FZP were fabricated by using electron beam lithography, their optical performances showed efficiency and signal to noise ratio that resulted, in our knowledge, the highest ever obtained at keV regime. With the consolidation of the advent of synchrotrons of third generation, multilevel FZP will gain a relevant position as focusing element in research fields where high efficiency and high signal to noise ratio are needed in the X-ray wavelength.

REFERENCES

[1] E. Di Fabrizio, M. Gentili, L. Grella, M. Baciocchi, A. Krasnoperova, F. Cerrina, W. Yun, B. Lai, E. Gluskin, *J. Vac. Sci. Technol. B* **12** (6), 3979, (1994)

[2] C. David, B. Kaulich, R. Medenwaldt, M. Hettwer, N. Fay, M. Diehl, J. Thieme and G. Smahl: , *J. Vac. Sci. Techonol. B* **13**(6), 2762-2766 (1995)

[3] E. Di Fabrizio, M. Gentili, M. Kiskinova, M. Marsi, Brown, *Synchrotron Radiation News* **12**, 37-46 (1999).

[4] W. Yun, B. Lai, Z. Cai, J. Maser, D. Legnini, E. Gluskin, Z. Chen, A. A. Kraspenorova, Y. Vladimirsky, F. Cerrina, E. Di Fabrizio and M.Gentili, *Rev. Sci. Inst.* **70**, 2238-2241 (1999)

[5] W. Yun, B. Lai, A. A. Kraspenorova, E. Di Fabrizio Z. Cai, F. Cerrina, Z. Chen, M. Gentili and E. Gluskin, *Rev. Sci. Inst.* **70**, 3537-3541 (1999)

[6] E.Di Fabrizio, F. Romanato, M. Gentili , S. Cabrini, B. Kaulich, J. Susini, R. Barrett, Nature, 625, Oct. 28th 1999.

Interferometric Soft X-Ray Index Measurements: Progress of System Design and Recent Results

D. Joyeux, F. Polack, and D. Phalippou

Laboratoire Charles Fabry, Institut d'Optique - BP 147 - 91403 Orsay, FRANCE

Abstract. We present the most recent evolution of our interferometric system for index measurements. It was tested by recording the index dispersion of carbon and silver on both sides of absorption edges, in the water window region. Its current performances and limitations are discussed in light of these results.

THE EXPERIMENTAL ARRANGEMENT

The current experimental arrangement is based on the same principles as the system previously reported.[1] A Fresnel bimirror interferometer produces a linear fringe pattern. Upon insertion of a sample in front of one of the two mirrors, a fringe shift is generated, which is measured directly by means of a dedicated optoelectronic detection system. However, to improve the performances, the numerical values of some parameters have been modified, and a piezoelectric device has been introduced into the interferometer, to time-modulate the fringe pattern position. We therefore summarize the present design and parameters of the system hereafter (fig. 1).

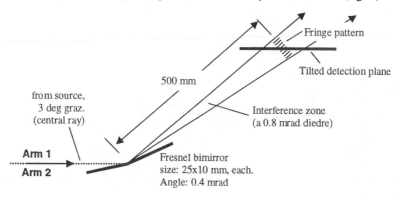

FIGURE 1. The Fresnel bimirror arrangement. A sample can be inserted into one arm, up to the central ray, i.e. the ray incident onto the common edge of the reflecting planes. The distance from sample to mirror ridge is 50 mm, small enough to keep the spread out of the sample beam onto the reference mirror (due to Fresnel diffraction) negligible.

The Fresnel bimirror (uncoated silica) is illuminated at 3 deg grazing incidence, by a quasi parallel beam, issued by the exit slit of a monochromator placed 10 m

CP507, *X-Ray Microscopy: Proceedings of the Sixth International Conference,*
edited by W. Meyer-Ilse, T. Warwick, and D. Attwood
© 2000 American Institute of Physics 1-56396-926-2/00/$17.00

upstream. (The primary source is the SU7 undulator of the SuperACO synchrotron radiation ring). With respect to the previous setup, the bimirror tilt has been changed to 0.4 mrad and the distance to fringe detection has been increased to 500 mm. Therefore, more and wider fringes are produced, namely 70 with 5.5 μm spacing at λ=4.4 nm. This modification was realized to increase the detection accuracy and efficiency of the moiré optoelectronic system described in ref. 1. As a matter of fact, more fringes improves the rejection of the parasitic modulation of fringes by Fresnel diffraction, and a larger spacing decreases the modulation losses in the fluorescent conversion layer, and allows a better relative accuracy of the detection mask.

These changes also increase the total width of the input beam that participates to interference to 2x400 μm. Therefore, the required spatial coherence width is also 400 μm, and the active reflecting zone on each mirror is 400/sin(3) = 8 mm from the common edge, which makes the accuracy requirements for the mirrors more difficult to fulfill. On the contrary, this does note change the average detected flux, as long as the source slit is adjusted to match the spatial coherence width.

As for the detection system itself, it still consists of a moiré based, spatial-phase demodulator applied to the aerial fringe pattern (fig. 2). In short, it consists in *optically tracking the fringe position* by screening the fringe pattern with a binary grating mask having the same spatial frequency, and *measuring the mask position* by a high sensitivity (0.01 μm) displacement transducer. To get flexibility, the X-ray pattern is converted into a visible fringe pattern through a thin fluorescent layer deposited on the mask. For a better accuracy, the mask-layer plate is tilted with respect to fringes, which increases the apparent fringe spacing by a factor of 5 to 10, depending on the precise grazing angle onto the conversion layer. This also allows to accomodate for wavelength changes, making spectral dispersion studies easy. In order to allow the use of a lock-in technique for fringe tracking, the relative position of fringe and mask is time-modulated.

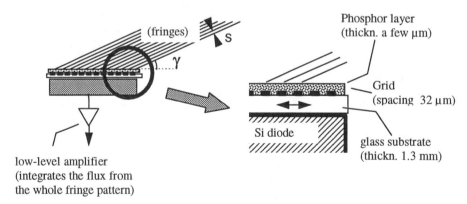

FIGURE 2. The aerial x-ray fringe pattern is first converted into a visible pattern, with stretched spacing (due to the grazing incidence). The total flux passing from the fringe pattern through the grating mask is integrated by a single detector. Once the grid and fringe pattern are matched (spacing and orientation), the total moiré flux monitors the relative position of these patterns (modulo one spacing).

This time modulation was previously realized by vibrating the mask itself. As it was responsible for a spurious detected signal, arising from the finite size of the mask ("window effect"), this was replaced by vibrating the fringe pattern itself, by means of piezoelectric plates introduced in the interferometer assembly (fig. 3). This essentially introduces a phase shift in each wavefront, before they interfere. Note that the absolute calibration of the piezos is not important, because it acts as a tracking signal. However, as the system also introduces a small *rotation* of each mirrors, it is in principle required that both piezos produce the same displacement (with opposite signs), in order not to change the bimirror angle. In fact, this symmetry is not critical, because the rotation applied to each mirror is very small, about 0.5 μrad for a 10 nm vibrating amplitude, to be compared to the 400 μrad tilt. Therefore any reasonable (even as large as 10 %) lack of symmetry produces a negligible change of the fringe spacing and of the associated fringe shift error, taking into account the very small *maximum* interference order (less than a few 10's in any case, because the interferometer is adjusted to work on both sides of the zero path difference).

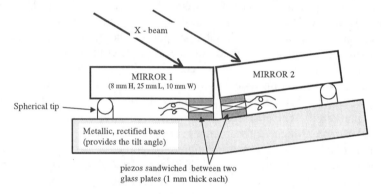

FIGURE 3. The mirror system assembly showing the piezoelectric parts inserted for phase modulation. The different parts are rigidly bound by using drops of optical UV-bond, at selected positions (see ref. 1 for details on the bonding technique and the required mounting accuracy). As only one piezo is placed under each mirror's end, each mirror also rotates around the fixed points defined by the spherical tips. To keep the bimirror angle constant, the mirrors are driven with equal amplitudes and opposite phase. A mirror translation of 10 nm produces a half fringe shift of the interference pattern (3 deg. grazing incidence, λ ≈ 4 nm)

The detection mask (spacing 32 μm) was realized by e-beam lithography, for good regularity and *symmetry*. Although the exact value of the spacing is unimportant, the symmetry of the duty cycle (black=white) provides a rejection of the fringes that might be produced from the second harmonic spectral pollution of the monochromator. Taking into account the other relative attenuation factors (smaller brightness of the second harmonic, loss of coherence, loss of modulation transfer inside the fluorescent conversion layer), we consider that no significant spurious signal can be produced by the second harmonic of the illumination wavelength. This is to be compared to the strong perturbation that harmonic pollution brings to absorption measurements.

SYSTEM PERFORMANCES

The main issue of system operation is the alignment of the grating mask and of the fringe pattern, with equal spacing, and the stability of this adjustment during measurements, and even over days. The first system alignment (after assembly) is performed by exploration of the adjustment parameter range (fringe orientation and fringe grazing incidence) from a "good" starting position, searching for a modulated signal from the moiré detector. This is generally not difficult, because it is possible to realize a good preadjustment during system assembly, thus reducing the range of parameters to explore. Once aligned, the system remains close enough to the ideal adjustment, so that only small correction are required, even after days. Also note that changing the wavelength only requires to adjust the grazing incidence onto the detector plane, in a predictable way.

The intrinsic sensitivity of the detection system is given by the smallest shift that can be detected, taking into account the noise level. This was found to be 0.05 μm, for fair absorption conditions, i.e. $0.05/32 \approx 1/600$ of the fringe spacing, or about $\lambda/600$ in terms of optical path. However the accuracy is probably not as good, due to a residual mechanical instability of the whole table supporting the system, which disturbs the two position measurements required for the shift determination. To illustrate the required degree of stability, let us consider that the whole optical system (interferometer and detection system as a block) is submitted to an unwanted rotation ε with respect to the input beam. At a distance D of the bimirror, this produces a spurious translation of the fringe pattern of $\varepsilon.D$, *with respect to the detector coordinates system* (which is also rotated). With D=500 mm and a fringe spacing of 5.5 μm, it is found that an unwanted rotation of 0.02 μrad produces a signal equivalent to a variation of optical path equal to $\lambda/550$, i. e. detectable. Accuracy is presently estimated to be $\lambda/150$-$\lambda/200$ (optical path). We expect raising this figure to $\lambda/400$ after correction of the stability problem.

The spectral resolution $\lambda/\Delta\lambda$ is not better than about 120-200, depending on the absorption conditions. This poor resolution is related to the fact that, due to the relatively small spectral brightness of our source, it was necessary to open as much as possible the input slit of the monochromator, to increase the signal level. This is directly related to the "coherent flux" available from the source, which is completely determined by its spectral luminance multiplied by the wavelength. Therefore, to improve the ultimate measurement accuracy (arising from the detection signal to noise ratio) and/or the spectral resolution, the most radical approach is to use a brighter, third generation source, which should by itself improve the signal level by at least 2 orders of magnitude. However, even with the superACO source, other efficient solutions could be implemented. The first one is to take advantage (by means of a specific astigmatic optics) of the full beam emittance *along* the fringes, where coherence is of no concern. The second solution, already mentioned, is to suppress the x-ray to visible conversion layer, and have the demodulation grating directly etched onto the detector surface. This should improve the signal level by at least one order of magnitude.

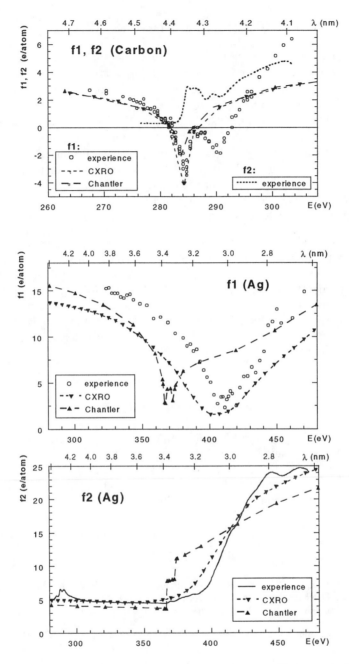

FIGURE 4. Measured and tabulated atomic scattering factors for carbon (K edge) and silver (M edge).

RECENT RESULTS AND DISCUSSION

This experimental system was tested in 1998 with two different samples. The first sample is a free standing evaporated carbon layer, with mass surface $52\,\mu g/cm^2$, supplied by Goodfellow (Great Britain). Its optical thickness was measured against vacuum from $\lambda=4.1$ to 4.65 nm, i.e. *across* the K edge. The second sample is a layer of silver, 150 nm thick, deposited on half the surface of a Si_3N_4 membrane. This sample was measured from $\lambda=2.6$ nm to 3.8 nm, i.e. also *across* an absorption edge (M). Results are shown on figure 4, in terms of the atomic diffusion factors f_1 (for the refractive index) and f_2 (for absorption, measured on same samples, with same detector). Note that the fringe shifts corresponding to f_1 are almost always smaller than 1/4 fringe. We also show the corresponding data from two different data sets, namely the Henke tables[2] (from CXRO web site) and the tables published by Chantler.[3]

As for the physical significance of the f_1 data presented, it must be pointed out that the 3 sets of data are not exactly comparable (and comparison was not the main goal of this series of experiments). This emphasizes strongly the need of being able to measure independently (and accurately) f_1 for "real" materials, as well for X-ray optics applications as for material sciences studies. Note that our f_1/f_2 data were not cross checked with respect to Kramers-Kronig relations. This was not done here essentially because the spectral range of the new data was not wide enough (an advantage of interferometric method under other respects...).

ACKNOWLEDGEMENTS

We are pleased to acknowledge here the participation of Jan Svatos (during his PhD thesis), Jean-Pierre Chauvineau and Françoise Bridou, for the realization and calibration of the silver sample. Finally, the polishing and assembly of the soft X-ray interferometer is a masterwork, that only the skill and experience of the Institut d'Optique's optical workshop have made possible.

REFERENCES

1. Joyeux D., Polack F., "Carbon Index Measurement Near K-Edge, by Interferometry with Optoelectronic Detection" in *X-Ray Microscopy and Spectromicroscopy*, edited by J. Thieme, G. Schmahl, D. Rudolph, and E. Umbach, Springer, Berlin, 1998, pp II.103-II.112

2. B.L. Henke, E.M. Gullikson and J.C. Davis, Atomic Data and Nuclear Tables 54, 181 (1993)

3. C. T. Chantler, J. Phys. Chem. Ref. Data, **24** 1 (1995)

Zone-Plate-Array Lithography (ZPAL): Simulations for System Design

Rajesh Menon[*], D. J. D. Carter[+], Dario Gil[*], and Henry I. Smith[*]

[*]*Department of Electrical Engineering and Computer Science, Massachusetts Institute of technology, Cambridge, MA 02139.*
[+]*Research Laboratory of Electronics, Massachusetts Institute of Technology, Cambridge, MA 02139.*

Abstract. We present a simulation study which examines the use of zone plates for lithography. Zone-Plate-Array Lithography (ZPAL) is a maskless lithography scheme that uses an array of shuttered zone plates to print arbitrary patterns in resist on a substrate. We have demonstrated a working ZPAL system in the UV regime, and are pursuing further experiments with the 4.5 nm X-ray to obtain smaller feature sizes. A general numerical simulation tool, based on the Fresnel-Kirchhoff diffraction theory, has been developed. A pattern will consist of many pixels exposed independently in the resist. Various zone plate and system parameters will affect the intensity distribution at the focal plane. We present simulation results which show the effect of these parameters on both the individual spots and exposed patterns.

INTRODUCTION

Zone-plate-array lithography (ZPAL) is a maskless lithography scheme that employs an array of shuttered zone plates to expose patterns of arbitrary geometry on a resist-coated substrate [1-4]. It is illustrated schematically in Fig. 1. By using an array of zone plates, and independently controlling their illumination while moving the substrate, one can achieve parallel writing in a dot-matrix fashion.

ZPAL borrows heavily from the field of X-ray microscopy, which over the last two decades has greatly advanced the technology of fabricating zone plates. Zone plates with minimum outer zone widths of sub-25 nm have been fabricated [5], and it is not unreasonable to expect that this will be further reduced in the future. Because the focal spot or point-spread-function of a zone plate is approximately equal to the width of the outermost zone, we believe that ZPAL can approach the limits of the lithographic process. For the lithography application we believe the optimal wavelength is 4.5 nm, i.e. just beyond the carbon-K edge. The 4.5 nm photon can be used to expose thick films of carbonaceous resist, while minimizing the proximity effects due to photoelectrons [6,7]. In fact, back in the late 1970's D. C. Flanders demonstrated that lines and spaces of 18 nm can be exposed in PMMA using C_K X-ray lithography in a contact mode [8]. Similar results have been obtained in the intervening years [9-12].

The main disadvantage of ZPAL at 4.5 nm wavelength is the necessity of using an undulator or similar collimated source of narrow-band radiation. However, such sources are clearly feasible [13]. The main challenges to developing ZPAL are the

CP507, *X-Ray Microscopy: Proceedings of the Sixth International Conference,*
edited by W. Meyer-Ilse, T. Warwick, and D. Attwood
© 2000 American Institute of Physics 1-56396-926-2/00/$17.00

multiplexed shuttering of the illumination to individual zone plates, and the problem of matching the efficiencies of all the zone plates of a large array. To address this problem of multiplexing, and our inaccessibility to an undulator, we have instead pursued ZPAL at UV wavelengths. In addition, we have developed ZPAL simulation tools which enable us to evaluate a variety of tradeoffs among such system and zone plate parameters as: source bandwidth, number of zones, fabrication errors, effects of order-sorting apertures, and number of zones per array vs. multiplexing rate.

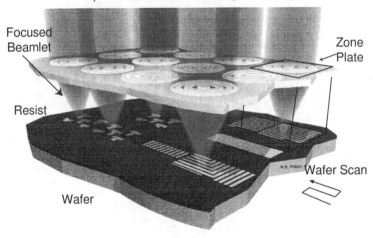

FIGURE 1. Schematic of zone-plate-array lithography (ZPAL). An array of zone plates focuses radiation beamlets onto a substrate. The individual beamlets are turned on and off by upstream micromechanics as the substrate is scanned under the array. In this way, patterns of arbitrary geometry can be created with a minimum linewidth equal to the minimum width of the outermost zone of the zone plates. Using 4.5 nm radiation, we estimate that lines and spaces of 20 nm should be achievable, provided that the zone plates can be fabricated.

THEORY

Fresnel-Kirchhoff diffraction theory [14-16] is used to calculate the point-spread-function of a zone plate. Scalar theory is suitable at the 4.5 nm wavelength since zone widths are much larger than the wavelength. However, at UV wavelengths, where we have done most of our experiments, the use of scalar theory is subject to question. Because the exposure of any arbitrary pattern is made up of separately exposed focal spots (pixels) we calculate the intensity distribution for any given pattern by adding the intensities of point-spread-functions. For simplicity, a binary clipping level was used to model the resist development.

The solution of the exposure/development simulation was implemented on an IBM-SP2 with 10 parallel processors. A Single Instruction Multiple Data (SIMD) model of parallelization was utilized in these simulations [17]. We can break up the problem spatially into numerous points, which are divided up among the available processors.

Each processor then computes the phase and amplitude of the diffracted wave at its assigned group of points. In this scalar model, the field at each point is independent of its neighbours and hence, inter-processor message passing is not necessary. Thus our problem is an obvious candidate for the SIMD parallel-processing model.

SIMULATONS

Fig. 2 compares a simulated spot compared to spots exposed in resist. Phase zone plates, with an outer zone width of 331 nm were used at an exposure wavelength of λ = 442 nm. The simulation tools took into account a 10% measured phase-shift error due to over-etching of the quartz during fabricaton. Despite the well-known inadequacy of scalar diffraction theory when wavelengths are comparable to or smaller than zone widths, our simulations come quite close to experimental results. This may reflect that fact that line-to-space ratios are close to unity near the outer zones [18].

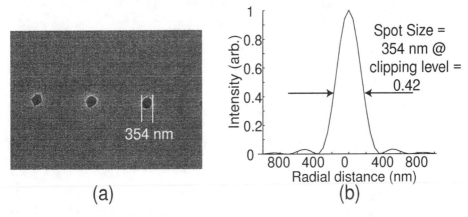

(a)   (b)

FIGURE 2. (a) Focal spots exposed and developed in photoresist, using the 442 nm wavelength HeCd laser, and zone plates fabricated in fused silica using direct-write e-beam lithography and reactive-ion etching. Zone plates of the array have 76 zones. They were etched within 10% of the π-phase depth. (b) Simulated point-spread-function for these zone plates, illustrating that a clipping level of 0.42 would produce the 354 nm diameter spot. The zone plates have a numerical aperture of 0.66.

In Fig.3, we show the comparison of our simulation to the micrograph of a pattern written in resist using the above-mentioned set of zone plates. The agreement is surprisingly good. Other patterns show similar agreement.

Fig. 4 shows a simulation assuming a collimated source of 4.5 nm x rays having a bandwidth of 1/35 (e.g., an undulator on a synchrotron) [13]. The zone plates have 35 zones, the outer ones with a 50 nm pitch (i.e., 25 nm outer zone). Note that the proximity effect (i.e., the widening of linewidths in densely patterned areas) due to exposure contributions from nearby pixels is relatively small. In this case, no order-sorting apertures were assumed; they would significantly reduce the proximity effect.

Fig. 5 illustrates a possible order-sorting aperture configuration, with the higher-order blocker located about half way between the zone plate and the substrate.

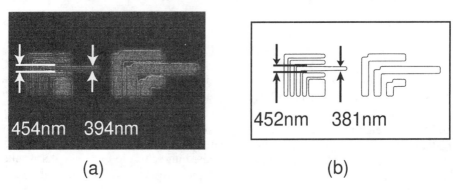

(a)   (b)

FIGURE 3.: (a) Scanning electron micrograph of a "nested L" pattern exposed and developed in Shipley 1613 photoresist, using UV ZPAL. (b) Simulation of the "nested L" exposure.

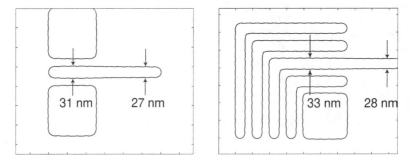

FIGURE 4.: Simulations of X-ray ZPAL exposures using a zone plate of 35 zones and a collimated source of 4.5 nm radiation, with a bandwidth of 1/35. The minimum zone width was 25 nm, corresponding to a focal distance of 19 μm. Line widening is due to background radiation from orders other than +1. (Order-sorting apertures would signifi-cantly reduce this proximity effect.)

Spot size, depth of focus, contrast and fidelity of patterns are some of the most important lithographic figures-of-merit. We can use our simulation tools to study how the system parameters such as source size, source bandwidth, demagnification and zone plate geometry affect these figures-of-merit.

Fig. 6(a) shows the intensity distribution for a grating, exposed using the same source and zone plate as in Fig. 4. Linewidths of 25 nm would be achieved by "clipping" at the intensity level of ~0.65. Perhaps more important is the exposure latitude. This is commonly expressed by means of the so-called Modulation Transfer Function (MTF), defined for a grating as

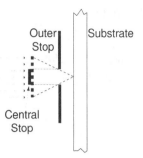

Outer Stop

Substrate

Central Stop

Figure 5.: Schematic of a zone plate, modified by the addition of a zero-order central stop, and an outer stop about mid way between the zone plate and the substrate. This stop effectively eliminates background effects due to the diffracted orders other than +1. The latter is focused on the substrate.

$$MTF = \frac{Max - Min}{Max + Min} \tag{1}$$

Fig. 6(b) plots the MTF as a function of the distance from optimal focus. In this case the focal length is 19 μm.

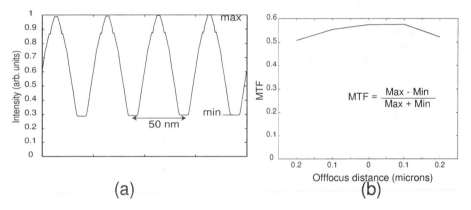

(a)

(b)

FIGURE 6.: (a) Intensity distribution for a grating of 50 nm pitch, exposed with the zone plate of Fig. 4 at the focal plane. (b) Modulation-transfer function (MTF) of the grating in (a) as a function of deviation from the focal distance of 19 μm.

CONCLUSION

We have developed a suite of general simulation tools for ZPAL and compared the results with patterns, exposed with UV ZPAL. The feasibility of transferring 25 nm features by X-ray ZPAL was also indicated. We plan to use these tools to facilitate the improvement our existing UV-ZPAL system and to design the optimal X-ray ZPAL system.

ACKNOWLEDGEMENTS

The authors would like to thank D. Foss, J. Hastings, J. Ferrera, E. Moon, K. Murooka, M. Schattenberg, T. Murphy and A. Bernshteyn for helpful discussions on various aspects of this work. This work was jointly sponsored by DARPA grant MDA972-97-1-0010 and SRC contract 96-LC-460.

REFERENCES

1. Smith, H. I., *J. Vac. Sci. Technol. B* **14**, 4318-4322, (1996).

2. Djomehri, I. J., Savas, T. A., and Smith, H. I., *J. Vac. Sci. Technol. B* **16**, 3426-3429, (1998).

3. Carter, D. J. D., Gil, D., Menon, R., Djomehri, I. J., and Smith, H. I., *SPIE* **3676**, 324-332 (1999).

4. Carter, D. J. D., Gil, D., Menon, R., Mondol, M. K., Smith, H. I., and Anderson, E. H., *"Maskless, Parallel Patterning with Zone-Plate-Array Lithography (ZPAL)."* To be published in *J. Vac. Sci. Technol. B*, (1999).

5. Anderson, E. H., Harteneck, B., Olynick, D., "Nanofabrication of X-ray Zone Plates with the Nanowriter electron-Beam Lithography System" in Proceedings, Sixth International Conference on X-ray Microscopy (XRM99), edited by D. Atwood., Berkeley, CA. (1999).

6. Ocola, L. E., and Cerrina, F., *J. Vac. Sci. Technol. B* **11**, 2839-2844, (1993).

7. Carter, D. J. D., Pepin, A., Schweizer, M. R., Smith, H. I., and Ocola, L. E., *J. Vac. Sci. Technol. B* **15**, 2509-2513, (1997).

8. Smith, H. I, and Flanders D. C., *J. Vac. Sci. Technol.* **17**, 533-535, (1980).

9. Early, K., Schattenburg, M. L., and Smith, H. I., *Microelectronic Engineering* **11**, 317-321 (1990).

10. Carter, D. J., Ph.D. Thesis, MIT, 1998.

11. Smith, H. I., and Craighead, H. G., "Nanofabrication", *Physics Today*, pp.24-30, February (1990).

12. Smith, H. I., and Cerrina, F., *Microlithography World* **6**, 10-15, (1997).

13. Private communication, E. Toyota, Sunitomo Heavy Industries, Lt., Tokyo, Japan.

14. Born, M., and Wolf, E., *Principles of Optics*, Elmsford, NY: Pergamon Press, 1975.

15. Michette, A. G., *Optical Systems for Soft X-rays*, NY and London: Plenum Press, 1986.

16. Thieme, J., Dissertation, Universität zu Göttingen, 1988.

17. Chandy, M. K., *Parallel program design: a foundation*, Reading MA: Addison-Wesley Pub. Co., 1988.

18. Pommet, D. A., Moharam, M. G., and Grann, E. B., *J. Opt. Soc. Am. A* **11**, 1827-1834, June (1994).

"Two-color" reflection multilayers for He-I and He-II resonance lines for micro-UPS using Schwarzschild Objective

Takeo EJIMA, Yuzi KONDO, and Makoto WATANABE

Research Institute for Scientific Measurements, Tohoku University
2-1-1 Katahira, Aoba-ku, Sendai, 980-8577 JAPAN

Abstract. "Two-color" multilayers reflecting both He-I (58.4 nm) and He-II (30.4 nm) resonance lines have been designed and fabricated for reflection coatings of Schwarzschild objectives of micro-UPS instruments. They are designed so that their reflectances for both He-I and He-II resonance lines are more than 20%. The "two-color" multilayers are piled double layers coated with top single layers. Fabricated are multilayers of SiC(top layer)-Mg/SiC(double layers) and SiC(top layer)-Mg/Y_2O_3 (double layers), and their reflectances for the He-I and the He-II are 23% and 17%, and 20% and 23%, respectively.

INTRODUCTION

Photoelectron micro-spectroscopy is powerful method to investigate electronic structures of condensed matters of small size and those consisting of grains [1]. In laboratories, He-I (58.4 nm) and He-II (30.4 nm) resonance lines are used for ultraviolet photoelectron spectroscopy (UPS). Changing the exciting photon energy for photoelectron spectroscopy is important to obtain partial density of states (p-DOS) of valence bands [2]. Even if the energies of the initial states are same, symmetries of final states excited with different photon energies are different. The difference is due to the dependence of the transition matrix element on the exciting photon energies [3,4]. Therefore, the p-DOS with different symmetries can be seen even if the photoelectrons are emitted from valence states with the same energy. For example, the Ce $4f$ p-DOS in $CeSi_2$ was obtained using He-I and He-II lines [5]. Therefore, the micro-UPS instruments of the practical use are very promising tool if the suitable demagnifying optical elements for both resonance lines are available.

The Schwarzschild objective (SO) is one of micro-focusing optical elements in ultraviolet and soft x-ray regions [6]. In the SO unit, the light from the source is reflected almost normally by a convex mirror, succeedingly by a concave mirror, and finally focussed. Therefore, the mirror surfaces should be coated with high reflectance materials for normal incidence. One of such materials is the multilayer of the piled double layers consisting of two materials with period of more than 10 (usual multilayer). However, the usual multilayer has a high reflectivity at only a certain wavelength for a fixed angle of incidence.

CP507, *X-Ray Microscopy: Proceedings of the Sixth International Conference*,
edited by W. Meyer-Ilse, T. Warwick, and D. Attwood
© 2000 American Institute of Physics 1-56396-926-2/00/$17.00

Therefore, in this paper, special "two-color" multilayers reflecting both He-I and He-II resonance lines have been designed and fabricated for reflection coatings of the SO units in micro-UPS instruments.

DESIGN AND FABRICATION OF "TWO-COLOR" MULTILAYERS

There are several materials which reflect He-I line efficiently (>20%) by their single layers. However, there are no such materials for He-II line, so that usual multilayers are required. Extinction coefficients of materials are generally larger for He-I line than for He-II line. Therefore, the "two-color" multilayer was designed to be composed of the top single layer reflecting He-I line with high transmittance for He-II line and the usual multilayer reflecting He-II line under the top layer.

To obtain high reflectance multilayers, combinations of two materials were selected before the fabrication according to the following principle. The principle is that, the absolute value of the difference of Fresnel reflection coefficients (abs.) between two materials has to be as large as possible for both resonance lines [7]. Adding to this, the imaginary parts of these Fresnel coefficients (im.) which are proportional to the extinction coefficients have to be as small as possible to minimize the absorption of the light by the layers.

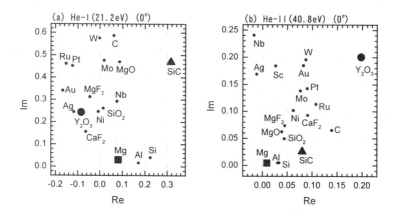

FIGURE 1. Fresnel reflection coefficients of various materials for He-I (58.4nm) and He-II (30.4nm) resonance lines.

In Figs. 1(a) and (b), Fresnel reflection coefficients of various materials for He-I and He-II resonance lines at the angle of incidence of 0° are plotted in the complex-plane using the Henke's and the Palik's data [8,9]. The single layer of SiC was chosen for the He-I reflection, because in SiC the abs. is large for He-I line as shown in Fig.

1(a) (The Fresnel coefficient of the vacuum is 0.), and the im. is small for He-II line as shown in Fig. 1(b). The piled double layers of Mg/Y_2O_3 and Mg/SiC were chosen for the He-II reflection. In the case of Mg/Y_2O_3 multilayer, the abs. between Mg and Y_2O_3 is large as shown in Fig. 1(b), so that the high reflectance can be achieved in small number of layers. In the case of Mg/SiC multilayer, the abs. between Mg and SiC is small, but the im. of each material is small, so that the high reflectance can be achieved by piling both materials in large number of layers.

Calculations of reflectance of the multilayers are performed using the recurrent method with extension as for the number of materials [7]. The "two-color" multilayers were designed to have reflectances of more than 20% for both He-I and He-II lines. Through the calculations, the thickness of the layers in the SiC (top layer)-Mg/SiC (double layers) and the SiC (top layer)-Mg/Y_2O_3 (double layers) were determined. The former multilayer has an advantage of being composed of only two materials. Calculated reflectances are presented as dashed curves in figure 2. The angle of incidence to all multilayers is 10° in the figure.

All samples were fabricated by magnetron sputtering. The sputtering system (ANELVA SPL-500) is of a vertical type with the targets and substrates placed vertically. Mg was dc-sputtered with an input power of 100 W, and SiC and Y_2O_3 were rf-sputtered with input power of 200 W. The Ar pressure was 2.0 mTorr for the all depositions, while the base pressure was 2.0×10^{-6} Torr. The multilayers were deposited onto commercially available 4 inch (10.16 cm) Si wafers. Effective deposition rate of Mg, SiC, and Y_2O_3 are 0.12 nm/s, 0.02 nm/s, and 0.04 nm/s, respectively. The film thickness was measured by an x-ray diffractometer within 5% error for the top layers and 2 % error for the double layers.

RESULTS AND DISCUSSION

Reflectances of "two-color" multilayers were measured with a reflectometer at the beamline BL5B, the Ultraviolet Synchrotron Orbital Radiation Facility (UVSOR), the Institute for Molecular Science (IMS), Japan. This beamline has a plane-grating monochromator having combinations of interchangeable three gratings and seven mirrors to cover a wide wavelength range of 1.8-238 nm [10]. In this experiments, we used the combinations of the G3 grating and the M25 mirror for the 27-54 nm range and the G3 grating and the M26 mirror for the 50-110 nm range. An average resolving power $\lambda/d\lambda$ was about 500. The resolution of the angle of incidence was about 1° in reflection measurements.

In Fig. 2, reflectances of SiC(14.4 nm)-Mg(10.4 nm)/SiC(6.1 nm) and SiC(10.4 nm)-Mg(12.4 nm)/Y_2O_3 (4.1 nm) multilayers at the angle of incidence of 10° are presented. In all figures, solid curves represent experimental results and dashed curves, calculated ones. Furthermore, left and right solid curves are the results of G3-M25 and G3-M26 combinations, respectively. In both figures, reflectances show the rise around 28 nm, become maximum at 30 nm and decrease, and increase gradually toward the longer wavelength. The measured reflectance obtained by the G3-M25 combination did not connect smoothly to that obtained by the G3-M26 combination. We supposed these discrepancies may be due to the higher order light and scattered light in the output light in the G3-M25 combination. Therefore, we took the results of the G3-M26 combination as the obtained values of the reflectances for He-I line. Comparing experimental results with calculated ones, spectral shapes and absolute values of reflectances resemble to each other. This suggests that the "two-color" multilayers were fabricated well. The reflectance of the SiC-Mg/SiC for the He-I is 23% and for the He-II, 17%. The reflectance of the SiC-Mg/Y_2O_3 for the He-I is 20% and for the He-II, 23%.

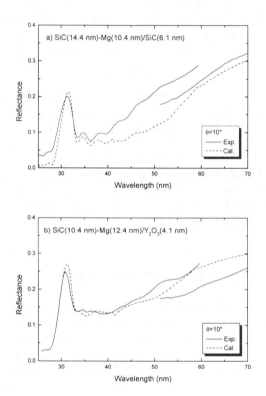

FIGURE 2. Measured reflectance of SiC-Mg/SiC multilayers, (a), and SiC-Mg/Y_2O_3 multilayers, (b), for 10° angle of incidence.

SUMMARY

The SiC-Mg/SiC and SiC-Mg/Y_2O_3 multilayers were designed and fabricated as "two-color" multilayers which can reflect both He-I and He-II resonance lines. Reflectances of the SiC(14.4 nm)-Mg(10.4 nm)/SiC(6.1 nm) and the SiC(10.4 nm)-

Mg(12.4 nm)/Y_2O_3 (4.1 nm) were 23% for the He-I and 17% for the He-II, and 20% for the He-I and 23% for the He-II, respectively. These "two-color" multilayers will be used for a Schwarzschild objective of a micro-UPS with a He discharge lamp, which we are constructing.

ACKNOWLEDGMENTS

The authors appreciate very much Profs. M. Yanagihara and M. Yamamoto for helpful discussions. The present work was financially supported by the Grant-in-Aid of *Reimei* research from JAERI.

REFERENCES

1. *see for instance*, H. W. Ade, Nucl. Instrum. Methods Phys. Res. **A319**, 311-319 (1992).

2. *Photoemission in Solids I*, edited by M.Cardona and L. Ley, New York: Springer-Verlag, 1978, pp. 84-93.

3. J. J. Yeh and I. Lindau, At. Data Nucl. Data Tables **32**, 1 (1985).

4. *Photoemission in Solids I*, edited by M.Cardona and L. Ley, New York: Springer-Verlag, 1978, pp. 135-163.

5. M. Grioni, D. Malterre, P. Weibel, B. Dordel and Y. Baer, Physica **B 186-188**, 38-43 (1993).

6. *see for instance*, F. Cerrina, J. Imag. Sci. **30**, 80-87 (1986).

7. M. Yamamoto, S. Nakayama, and T. Namioka, SPIE **Vol. 984**, 160-165 (1988).

8. *Hand book of Optical Constants of Solids*, edited by E. D. Palik, San Diego: Acad. Press. Inc., 1985.

9. B. L. Henke, P. Lee, T. J. Tanaka, R. L. Shimabukuro, and B. K. Fujikawa, At. Data Nucl. Data Tables **27**, 1 (1985), and B. L. Henke, J. C. Davis, E. M. Gullikson and R. C. C. Perera, Lawrence Berkley Laboratory LBL-26259 and a diskette containing the data (1988).

10. M. Sakurai, S. Morita, J. Fujita, H. Yonezu, K. Fukui, K. Sakai, E. Nakamura, M. Watanabe, E. Ishiguro, and K. Yamashita, Rev. Sci. Instrum. **60**, 2089-2092 (1989).

The Design of Broadband Multilayer Optics using Simulated Annealing

J.M.Tait, A.K. Powell, P. Edmundson and A.G. Michette

King's College London, Physics Department, Strand, London, WC1R 2LS

Abstract. A technique is presented for the broadband design optimisation of multilayered reflective optic coatings. The optimisation algorithm converges onto a stable configuration giving a high reflectivity for the wavelength range 10–12 nm and 13–16 nm. The silicon absorption edge at 12.4 nm limits the performance over the whole 10–16 nm range.

INTRODUCTION

A particular wavelength range of interest for laser generated plasma x-ray lithography is 10–16 nm, which is suited to the production of feature sizes required for the next generation of integrated circuits. The achievement of good contrast masks for patterning circuit designs is an additional possibility. With this in mind, a technique is presented for the broadband design optimisation of multilayered reflective optical coatings; these have enhanced performance over traditional multilayers, which are wavelength selective.

A 50 layer pair dielectric stack was created by combining sub-layers of molybdenum and silicon. Typically, 10 or 20 of these sub-layers make up a quarter wavelength period or layer. Multi-layers were designed using 30 iterations of simulated annealing, where one iteration is an attempt to change the material of every sub-layer in the initially random dielectric stack.

SIMULATED ANNEALING

Simulated annealing technique models the thermodynamic annealing of a metallic lattice. Such a system can be considered constantly to strive for the lowest energy by determining the states of its constituent particles using a Boltzmann probability distribution. As the temperature reduces, the system becomes locked in to its lowest energy lattice. In an analogous way, the reflectivity R of a multilayer mirror may be maximised (or $1-R$ minimised) when the criterion for accepting sub-layer flips is Boltzmann probabilistic. This probability is governed by a control temperature which is gradually reduced until a fixed configuration is achieved. Initially, at high temperatures, all sub-layer flips are accepted. Gradually, as the temperature is reduced,

CP507, *X-Ray Microscopy: Proceedings of the Sixth International Conference,*
edited by W. Meyer-Ilse, T. Warwick, and D. Attwood
© 2000 American Institute of Physics 1-56396-926-2/00/$17.00

only those flips which increase the reflectivity are allowed, leading to a global optimum solution. This is illustrated schematically in figure 1.

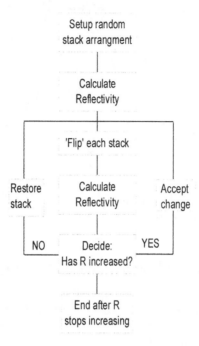

FIGURE 1. Flow diagram for the simulated annealing process

OPTIMISATION PROCEDURES

The reflectivity is calculated at particular wavelengths, using the standard recurrence relation in the optimisation procedure. How the reflectivity is used to control the optimisation depends upon the desired result.

Optimisation for Spot Wavelengths in the Range 17–20 nm

Where high reflectivity is required at one or more spot wavelengths, the algorithm can be designed to achieve maximum reflectivity at these wavelengths rather than giving a flat response (figure 2). In some cases, especially at wavelengths removed from the silicon absorption edge at 12.4 nm, this spot wavelength optimisation may even achieve a flat response. However, in other cases, one wavelength may be 'rejected' whilst others are optimised.

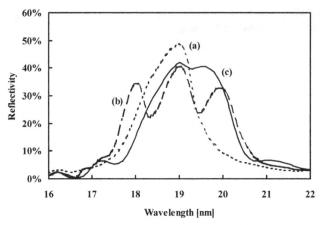

FIGURE 2. A comparison of different spot optimisations for Mo/Si multilayers for the range 17–20nm, optimised for (a) one wavelength (19 nm), (b) three wavelengths (18, 19 and 20 nm) and (c) five wavelengths (18, 18.5, 19, 19.5 and 20 nm).

Total Integrated Reflectivity in the Range 10 –16 nm

In cases where a flat response is more desirable than peak reflectivity performance the algorithm can be expanded to optimise for an approximated integrated reflectivity. Nevertheless this does not always create flat reflectivity curves as spot wavelengths can be favoured at the expense of others.

Absorption Edges Of Stack Materials

Note that due to the choice of Silicon as spacer material with an absorption edge at 12.4 nm there is negligible performance below the edge. This shows the importance of the choice of stack materials.

Target Reflectivity in the Range 13 –16 nm

Perhaps one of the most effective methods for the design of broadband multilayers is the use of a target reflectivity. A flatter response can be achieved by restricting the optimisation of wavelengths which have reached this target (figure 3). As a result the algorithm has greater freedom to pursue the optimisation at other wavelengths.

CONCLUSIONS AND FURTHER WORK

The results have shown that computer optimisation of multi-level mirrors for x-ray wavelengths can give relatively high reflectivities over a broad range of wavelengths. This has uses in areas such as astronomy, x-ray microscopy and x-ray lithography. Further work includes a comprehensive investigation into the choice of

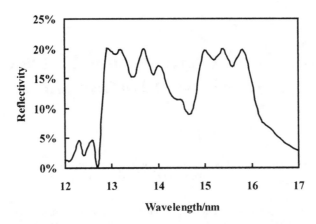

FIGURE 3. Optimisation over 10–16nm for a target of 20% reflectivity.

materials and the range of wavelengths for which they are particularly suited. In addition, combinations including more that two materials are being investigated.

ACKNOWLEDGEMENTS

The majority of this work was carried out during a King's College London summer studentship held by one of us (JMT), and the authors wish to acknowledge Prof Gordon Davies for organising these studentships and the King's College London Physics Department for funding them.

REFERENCES

1. Howells, M. R., and Jacobsen, C., *J. Appl. Phys.* **71**(6), 2993–3001(1992) .

2. Fienup, J.R., *Opt. Eng.* **19**(3), 297–305 (1980).

3. Bennink,R.S., Powell, A. K., and Fish, D. A., *Opt. Commun.* **141,** 194–202 (1997).

4. Underwood, J. H., and Barbee, T. W., *App. Optics* **20**, 17 (1981)

High Numerical Aperture Zone Plates Using High Orders Of Diffraction

Dirk Hambach[*]

Institut für Röntgenphysik, Georg-August-Universität Göttingen
Geiststraße 11, D-37073 Göttingen, Germany

Abstract. The implementation of laboratory X-ray microscopes with isotropic sources like laser generated plasmas requires condenser elements which collect photons from large solid angles, match the aperture of the micro zone plate objective to insure high spatial frequency transfer and allow to work at different wavelengths. All these requirements can be met by a zone plate condenser used in a high order of diffraction. Recent progress in nanostructuring allows the production of galvanoforms with aspect ratios >14, where the structure size is below 90 nm. A first linear nickel test zone plate with an outermost zone width of 75 nm and 500 nm high zones showed a diffraction efficiency of 1% at 3.4 nm wavelength in the sixth order of diffraction, which is significantly above the value predicted by the theory of thin gratings. These results open the way towards zone plate condensers with numerical apertures well beyond those of first order diffractive elements.

INTRODUCTION

Laser generated plasmas are promising candidates as sources for tabletop X-ray microscopes. As these sources emit isotropically, the photons have to be collected efficiently from a large solid angle, to ensure short exposure times. Furthermore, the

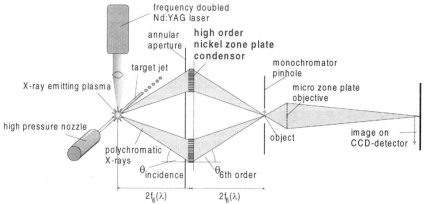

FIGURE 1. Planned setup for a laboratory transmission X-ray microscope with a zone plate condenser working in the sixth order of diffraction and with 1x magnification of the source.

[*] Electronic mail: dhambac@gwdg.de

CP507, *X-Ray Microscopy: Proceedings of the Sixth International Conference,*
edited by W. Meyer-Ilse, T. Warwick, and D. Attwood
© 2000 American Institute of Physics 1-56396-926-2/00/$17.00

apertures of the object illumination and the micro zone plate should be matched for high spatial frequency transfer. For good flexibility of the microscope, the condenser should also allow to work at different wavelengths. Multilayer mirrors, which could be used as condenser optics, are difficult to manufacture for the water window wavelength region. In addition they are restricted to a single design wavelength. Alternatively, zone plate condensors can be used.

The numerical aperture NA of zone plates is given by:

$$NA = \frac{\lambda \times m}{2dr_n} \tag{1}$$

where λ denotes the wavelength, m the order of diffraction and $2dr_n$ the outermost, smallest zone period. Up to now, zone plates for the X-ray region have been used in the first order of diffraction and their aperture has been improved by decreasing dr_n. This requires an outermost zone width of $dr_n^{cond} = dr_n^{MZP} m/(M+1)$ for the condenser, if the source is imaged on the object with an magnification M. However, for point sources like laser generated plasmas, the magnification of the condenser is in the order of $M \approx 1$. Thus, a condenser working in the first order of diffraction would be required to have $dr_n^{cond} = dr_n^{MZP}/2$ in order to match its aperture to that of the micro zone plate. However, on the comparetively large areas required for condenser zone plates, structure sizes below 40 nm are very difficult to fabricate. The utilization of high orders of diffraction circumvents the need to fabricate very small structures on large areas, if large NA is required. Calculations with coupled-wave-theory (CWT) predict, that a fraction of up to 50% of the incident light can be diffracted into a single high order by high aspect zone structures. The combination of high NA, a sufficiently high efficiency and the possibility to perform experiments at different wavelengths makes such high order zone plate condensers interesting for experiments with isotropic X-ray sources[4]. Figure 1 shows a possible setup of a laboratory X-ray microscope with a laser generated plasma source and a high order zone plate condenser.

EFFICIENCY OF HIGH ASPECT RATIO ZONE PLATES

If volume effects of the zone structures are neglected, as it is done in the theory of thin gratings, the diffraction efficiency decreases with $1/m^2$. However, for zone structures with high aspect ratio, the zone plates have to regarded as volume gratings [3]. Such objects can be analyzed with more general electrodynamic theories, e.g. CWT, which showed new effects for high aspect

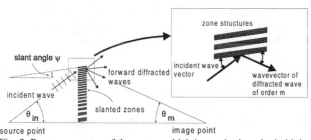

Fig. 2. Bragg-geometry of the zones, which is required to obtain high diffraction efficiency in high orders.

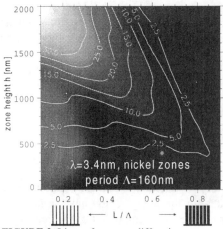

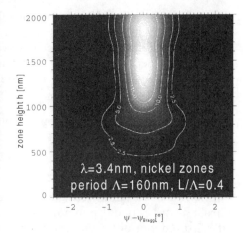

FIGURE 2. Lines of constant diffraction efficiency [%] in the (L/Λ, h)-plane for the sixth order of diffraction (ψ=ψ_Bragg).

FIGURE 3. Lines of constant diffraction efficiency [%] in the (ψ-ψ_Bragg, h)-plane for the sixth order of diffraction.

ratio grating structures. A result of particular interest is the high efficiency for high orders of diffraction, which was predicted for zone plates used in Bragg-Geometry (Fig. 2) having zones which exceed an aspect ratio of 10 [4]. The fraction of the incident photons which is diffracted into a single order m depends on the height of the structures h, the slant angle ψ of the zones against the optical axis and the ratio L/Λ, where L denotes the width of the zone bars and Λ is the zone period. Figure 2 shows lines of constant sixth order efficiency for a nickel grating with $\Lambda = 160$ nm zone period in the $(L/\Lambda, h)$-plane as calculated with CWT. The calculations have been performed for radiation with 3.4 nm wavelength. Furthermore, the influence of slanting the zones against the optical axis by an angle ψ has been studied. Figure 3. shows lines of constant efficiency in the (ψ, h) - plane for a nickel grating with $L/\Lambda = 0.4$. It can be seen that the efficiency is maximized if the zones are slanted by an angle $\psi_{Bragg} = (\theta_{in} - \theta_m) / 2$. In this geometry the Bragg-condition is satisfied, i.e. the wavevector of the incident wave forms the same angle with the local lattice vector of the zone structures as the wavevector of the diffracted wave m. The angular width of high efficiency region is sufficient to image sources which have an angular extension of less than 1°. Currently used planar nanostructuring technologies do not allow to etch circular zones with $\psi \neq 0$. Thus $\theta_{in} = \theta_m$ is required in to fulfil the Bragg-condition. This can be realized, if the condenser images the source at 1x magnification on the object. For laser generated plasma sources with a typical diameter of about 20 µm and a 20 µm wide field of view in transmission X-ray microscopes a 4f - setup for the condenser is suited to meet all requirements. For example, an annular condenser zone plate with 150 nm < Λ < 180 nm and a focal length $f_6 = 20$ mm has a NA = 0.136 in the sixth order of diffraction at $\lambda = 3.4$ nm and collects in a 4f - setup photons from a solid angle $\Omega = 4.4 \times 10^{-3}$ sr. The angular extension of sources with less than 100 µm diameter is then < 0.15°. Thus all source points can be imaged with nearly equal efficiency.

NANOSTRUCTURE FABRICATION PROCESS

To obtain the high aspect ratio nickel structures required for a reasonable efficiency in high orders of diffraction, a three level process is used to structure a radiation enhanced copolymer which serves in the following process step as galvanoform for nickel plating [5,6] (Fig. 4). A plating base of 10 nm Cr and 15 nm Ge is electron beam evaporated on the substrate. The cross - linked copolymer of 80% phenylethylen and 20% divinylbenzene monomers is deposited, which will later form the galvanoform. To enhance the degree of cross - linking, this layer is irradiated with soft X-rays, where a dose of $5 \times 10^9 Gy$ is applied. A 10 nm thick Ti layer is used as an intermediate mask in the three-level process. Finally 75 nm of PMMA (Dupont Elvacite 2041, 450kg/mol) are spun on the samples.

The zone pattern is generated by electron beam lithography with a LEICA LION-LV1 system. Its continuous path control mode allows to expose the large curved condenser patterns without the need of field stitching. The width of the lines is adjusted by controlled defocusing of the electron beam [7]. After development the pattern is transferred into the Ti layer by a RIE step with BCl_3. The Ti intermediate mask provides an etch selectivity >120 between the Ti and the copolymer in the following O_2 etch step. An optimized O_2 RIE etch process allows to etch copolymer galvanoforms with aspect ratios >14 at 90 nm feature size [8]. After etching the galvanoform is filled with Ni using a commercial nickel sulfamate bath (Fig.5). Further BCl_3 and O_2 etch steps are used to remove the galvanoform.

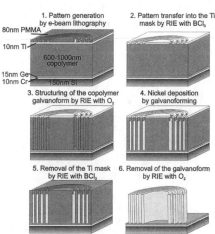

FIGURE 4. The nanofabrication process used to generate high aspect ratio nickel zones

FIGURE 5. SEM micrograph of electro-plated nickel structures before the removal of the galvanoform. zone period:180nm; height of the nickel zones: 1000nm; height of copolymer form :1380nm

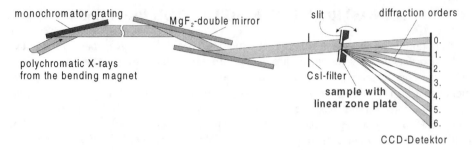

FIGURE 6: Setup for the measurement of the high order efficiency for a linear zone plate in the X-ray test chamber at BESSY. The zone plate can be tilted around an axis parallel to the zones. The diffraction orders are recorded simultaneously on the CCD-detector.

RESULTS AND DISCUSSION

Zone plates optimized for the sixth order of diffraction have been produced. It was possible to obtain galvanoforms which are 1380 nm high and have 90 nm wide zones (Fig. 5). A linear zone plate with zone periods from 150 nm-180 nm, a focal length f_6 of 20 mm in the sixth order of diffraction and a zone height of 500 nm was tested in the X-ray test chamber of the TXM-beamline at BESSY. The polychromatic radiation from a bending magnet was monochromatized by a nickel grating with 600 lines/mm. A MgF_2 was used to supress shorter wavelength in the beam, especially the third order of the grating. The radiation of the diffraction orders m = 0...6 has been recorded simultaneously on a CCD-detector. The linear zone plate, which was illuminated with the quasiparallel synchrotron beam, was tilted around an axis parallel to the zones, to determine the dependence of the diffraction efficiency on the zone slant angle ψ. A maximum diffraction efficiency of 1% in the sixth order of diffraction was obtained. Furthermore, the efficiency decreased at angles $\psi \neq \psi_{Bragg}$ as was expected (Fig. 7).

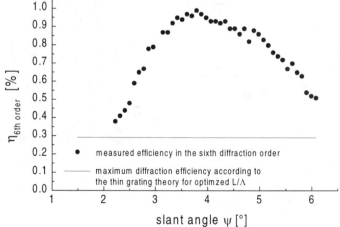

FIGURE 7: Sixth order diffraction efficiency of a nickel linear zone plate with 500 nm high zones and $dr_n = 75$ nm at $\lambda = 3.4$ nm.

By comparison, according to the theory of thin gratings the efficiency scales as:

$$\eta_m \propto \frac{1}{\pi^2 m^2} \sin^2(m\pi L / \Lambda) \tag{2}$$

This means that even in the cases of an optimal L/Λ - ratio, i.e. $L/\Lambda = (2n+1)/2m$, n=0, 1, ..., 6 , only η_6=0.29% is predicted for zones of 500 nm height. Therefore, the measured efficiency is significantly higher than the value expected from the thin grating theory. The CWT predicts an efficiency of 2.7% for this zone plate, but deviations of the zone profile from the ideal rectangular shape have to be taken into account. Further investigations are necessary in order to characterize the high order diffraction with respect to zone profile imperfections.

ACKNOWLEDGEMENTS

This work was funded by the German Federal Minister for Education and Research under contract No. 05 SL8MG11. The author would like to thank G. Schmahl, G. Schneider, D. Rudolph, B. Niemann, M. Peuker, P. Guttmann, J. Herbst, and T. Gronemann. Furthermore, the excellent working conditions at BESSY are gratefully acknowledged.

REFERENCES

1 . M. Berglund, L. Rymell, and H.M. Hertz, *Appl. Phys. Lett.* **69** (12), 1683 (1996)

2 . T.K. Gaylord and M.G. Moharam, *Proc. of the IEEE* **75** (5), 894 (1985)

3 . J. Maser and G. Schmahl, *Opt. Commun.* **89**, 355 (1992)

4. G. Schneider, *Appl. Phys. Lett.*, **71**, (16) , 2242 (1997)

5 . G. Schneider, T. Schliebe, H. Aschoff, *J. Vac. Sci. Technol. B* **13** (6), 2809 (1995)

6. D. Weiss, M. Peuker, G. Schneider, *Appl. Phys. Lett.* **72** (15), 1805 (1998)

7. C. David and D. Hambach, *Microelectronic Engineering*, **46** (1-4), 219 (1999)

8. D. Hambach and G. Schneider, *J. Vac. Sci. Technol. B*, Nov/Dec (1999), in press

Focusing Properties of Tantalum Phase Zone Plate and Its Application to Hard X-Ray Microscope

Yasushi Kagoshima, Kengo Takai, Takashi Ibuki, Kazushi Yokoyama, Shingo Takeda, Masafumi Urakawa, Yoshiyuki Tsusaka and Junji Matsui

Faculty of Science, Himeji Institute of Technology
3-2-1 Kouto, Kamigori, Ako, Hyogo 678-1297, Japan

Abstract. A phase zone plate made of tantalum has been designed and fabricated for focusing hard x-rays. It is designed to optically match to the undulator radiation of the BL24XU at the SPring-8. Its focusing properties were measured. The focused beam size smaller than 1 μm was obtained at the photon energy of 10 keV. The diffraction efficiency was measured around 10 keV and the maximum efficiency of 20.7 % was obtained at 9.8 keV. Photon flux was estimated to be $\sim 4 \times 10^8$ photons/s. A system for a scanning x-ray microscope has been constructed and the transmission and/or fluorescence microscopy has also been performed.

INTRODUCTION

Fresnel zone plates (ZP's) are known to be the most powerful focusing devices in a soft x-ray region. They have been recently used in a hard x-ray region and there have been some papers reporting sub-micron resolution.[1, 2] Our ZP's are made of tantalum (Ta). They were originally used in the soft x-ray region.[3] Since Ta is a relatively heavy metal and gives an appropriate phase shift in the hard x-ray region, Ta ZP's with thicker zones can work as phase zone plates (PZP's) even in the hard x-ray region. The fabrication process is described elsewhere.[4] In this paper, the focusing properties of a Ta PZP, and its application to a scanning x-ray microscope is presented.

PARAMETERS OF TANTALUM PZP

Parameters of the Ta PZP were designed optically match to the undulator radiation of BL24XU [5] at the SPring-8. Main design criteria are (1) the operation photon energy is 10 keV, which determines the optimized thickness of Ta 'phase' zones to be 2.4 μm, (2) the outermost zone width is 0.25 μm, which is limited by the present fabrication technology (the maximum aspect ratio is 10 at present), (3) the PZP is coherently illuminated, which determines the diameter of the PZP to be 100 μm at the distance of ~70 m from the source. Considering the above conditions, the parameters were designed and are shown in Table 1. A center stop of a deposited gold mask was also

CP507, *X-Ray Microscopy: Proceedings of the Sixth International Conference,*
edited by W. Meyer-Ilse, T. Warwick, and D. Attwood

made for the order sorting. The diffraction efficiency was calculated and that for 10, 15, 20 and 25 keV is shown in Table 2. The Ta PZP has the efficiency higher than 10 % even in 25 keV.

TABLE 1. Parameters of Ta PZP.

Radius of Innermost Zone (μm)	5.0
Number of Zones	100
Outer Diameter (μm)	100
Outermost Zone Width (μm)	0.25
Focal Length @ 10 keV (mm)	201.6
Radius of Center Stop (μm)	22.5
Material of Center Stop	Au
Thickness of Center Stop (μm)	5.6 (measured)

TABLE 2. Calculated Diffraction Efficiency of 2.4 μm-thick Ta PZP with Center Stop.

Photon Energy (keV)	10	15	20	25
Diffraction Efficiency (%)	20.8	17.4	15.9	11.7

BEAM SIZE AND DIFFRACTION EFFICIENCY

An optical system of a scanning x-ray microscope is shown in Fig.1. The undulator radiation was monochromatized by a silicon double-crystal monochromator. The beam was next collimated by a 100 μm-diameter pinhole placed in front of the PZP. A 20 μm-diameter pinhole was used as an order-sorting aperture (OSA). The focal spot size was determined using a slit scanning technique. The transmission intensity was measured by a Si PIN photodiode. The focal spot size was evaluated to be 0.97 μm in horizontal and 0.89 μm in vertical directions as shown in Fig. 2(a). The beam size was larger than the diffraction limited size of 0.3 μm. The reason is under investigation.

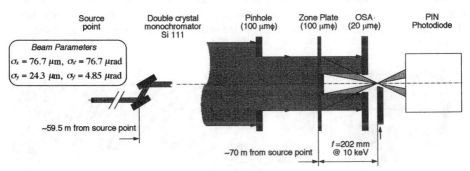

FIGURE 1. Optical System of Scanning X-Ray Microscope (Top View). An ion chamber as I_0 monitor was placed between PZP and OSA. The fine scanning stage was a piezo-driven stage (PI: P-731.20) capable of 10-nm resolution. The fine scanning stage was mounted on the pulse-motor-driven x-y-z coarse stages (KOHZU Mont Blanc series: capable of 0.25-μm resolution). Fluorescent x-rays were also measured with a small SSD (Amptek: XR-100CR).

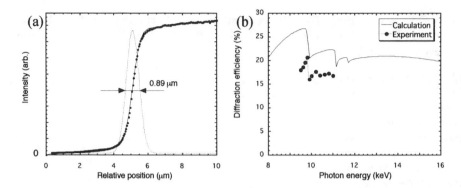

FIGURE 2. (a) Result of Vertical Focal Spot Size Measurement. Dots (raw data), solid line (fitted curve) and dashed line (derivative of fitted curve) are shown. The raw data were fitted with an error function assuming a uniform background. (b) Calculated and Measured Diffraction Efficiency. The efficiency was defined as the ratio of the output photocurrent to the incident one. The discrete structure corresponds to the L-absorption edges of Ta.

The diffraction efficiency was also measured around the photon energy of 10 keV, and the result is shown in Fig. 2(b). The maximum efficiency was 20.7 % at 9.8 keV. Photon flux was estimated to be ~4×10^8 photons/s, which was almost comparable to the calculated flux. As the PZP was placed after a 4-m long air path in the hutch, the photon flux about 10 times higher can be expected by replacing the air path with a vacuum or a helium path.

APPLICATION OF THE MICROSCOPE TO REAL SAMPLES

The feasibility test of the microscope system was made. Images of a copper #2000 mesh and a large Ta zone plate were taken as shown in Fig. 3. The system was capable of taking signals both of transmission and fluorescent x-rays simultaneously. The transmission intensity was measured as a photocurrent of the Si photodiode (HAMAMATSU: S3590-06). The energy of fluorescent x-rays was analyzed with an MCA.

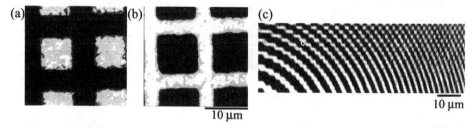

FIGURE 3. (a) Transmission and (b) Fluorescent X-Ray Images of Cu #2000 Mesh. Pixel size and number are 0.5 μm × 0.5 μm and 50 × 50, respectively. Counting gate time was 1.0 s/pixel. (b) Fluorescent X-ray Image of Large Ta ZP (Ta Lα). Pixel size and number are 0.5 μm × 2.0 μm and 200 × 20, respectively. Counting gate time was 1.0 s/pixel.

Figure 4 shows fluorescent intensity distributions of a coating layer cross section. The images clearly demonstrate the applicability of the microscope system to the metal samples. Biological samples will also be tried.

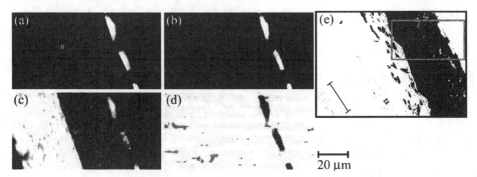

FIGURE 4. Fluorescent X-Ray Intensity Distributions of the Cross Section of a Coating Layer. (a) Fe, (b) Zn and (c) Ti. Transmission intensity distribution is shown in (d). Pixel size and number are 1 μm × 2 μm and 100 × 25, respectively. Counting gate time was 1.0 s/pixel. Optical micrograph is also shown in (e). A gray rectangle in (e) corresponds to the x-ray image area of (a)-(d).

ACKNOWLEDGEMENTS

The authors would like to express their gratitude to Hyogo prefecture for supporting the projects. They also thank to all SPring-8 staffs for the operation of the storage ring. They also thank to Dr. Ninomiya, Forensic Science Laboratory of Hyogo Police, for providing a coating layer sample. This work has been carried out according to the proposal number of C99A24XU-034N.

REFERENCES

1. Suzuki, Y., Kamijo, N., Tamura, S., Handa, K., Takeuchi, A., Yamamoto, S., Sugiyama, H., Ohsumi, K. and Ando, M., *J. Synchrotron Rad.* **4**, 60-63 (1997).

2. Yun, W., Lai, B., Cai, Z., Maser, J., Legnini, D., Gluskin, E., Chen, Z., Krasnoperova, A. A., Vladimirsky, Y., Cerrina., F., Di Fabrizo, E. and Gentilli, M., *Rev. Sci. Insturm.* **70**, 2238-2241(1999).

3. Kagoshima, Y., Aoki, S., Kakuchi, M., Sekimoto, M., Maezawa, H., Hyodo, K. and Ando, M., *Rev. Sci. Instrum.* **60**, 2448-2451 (1989).

4. Ozawa, A., Tamamura, T., Ishii, T., Yoshihara, H. and Kagoshima, Y., *Microelectronic Engineering* **35**, 525-529 (1997).

5. Matsui, J., Kagoshima, Y., Tsusaka, Y., Katsuya, Y., Motoyama, M., Watanabe, Y., Yokoyama, K., Takai, K., Takeda, S. and Chikawa, J., "Hyogo BL(BL24XU)" in *SPring-8 Annual Report 1997*: Japan Synchrotron Radiation Research Institute, 1998, pp. 125-130.

Fabrication of high energy X-ray Fresnel phase zone plate

N. Kamijo[3,1], Y. Suzuki[1], S. Tamura[2,1], M. Awaji[1],
M. Yasumoto[2], Y. Kohmura[1], K. Handa[4], A.Takeuchi[1]

1. SPring-8, 1-1-1 Kohto, Mikazuki, Sayo-gun, Hyo-go 679-5198 Japan
2. Osaka National Research Institute, 1-8-31, Midorigaoka, Ikeda Osaka, 563-8577 Japan
3. Kansai Medical University 18-89, Uyama-higashi, Hirakata, Osaka, 573-1136 Japan
4. Ritsmeikan University 1-1-1, Noji-higashi, Kusatsu, Shiga, 525-8577 Japan

Abstract. Zone plates for high energy X-rays with an outermost zone width as small as ~ 0.1 μm and thickness with 10 ~ 50 μm have been fabricated using sputtered-sliced technology. With the brilliant x-ray beam from the undulator at the SPring-8 the minimum focal spot size attained was 0.7 μm at 27.5 keV. The light collecting efficiency was determined to be ~15 %. We have also succeeded to focus much higher energy x-rays as 80 keV with these zone plates.

INTRODUCTION

Third generation synchrotron light source such as the SPring-8 can produce the x-ray beam of unprecedented quality: small emitting source size, coupled with high brilliance and high spatial coherence, open new oppotunities for microfocusing studies in the hard x-ray regions. Some expected applications are hard x-ray microscopy, micro-fluorescence analysis for trace element mapping, microdiffraction and microspectroscopy. X-ray microscopy which has been extensively used in soft x-ray region can be extended to higher x-ray photon energies. This extension will enable new investigations: study of thicker biological samples in their natural environment and that of material sciences.

Fresnel phase zone plate (FPZP) is the promising microfocusing optical devices and widely used in the soft x-ray region, and fabricated using electron-beam lithographic patternning techniques. This patternning technique can produce high-resolution structures, but is limited in its aspect ratios (hight/width) to around unity though the thicker zone plates usable for 4 keV ~ 10 keV x-rays are recently fabricated by electron and x-ray beam lithography technique [1~6]

To overcome the aspect ratio limitation the sputtered-sliced zone plate method has been applied.[7~13] In this technology, a thin microwire is deposited alternatively with two different materials. The zone plate is generated from the mutilayer rod by slicing it perpendicular to its axis and thinning the slices down to the required thickness.

In this article we describe the fabrication of the sputtered-sliced FPZP for higher x-ray energies (20 keV~80 keV) with a brief focusing test using the high brilliant synchrotron radiation from the SPring-8 undulator beam line (R&D1, BL47XU).

CP507, X-Ray Microscopy: Proceedings of the Sixth International Conference,
edited by W. Meyer-Ilse, T. Warwick, and D. Attwood
© 2000 American Institute of Physics 1-56396-926-2/00/$17.00

ZONE PLATE FABRICATION

Our sputtered-sliced Fresnel zone plate (FZP) were made by physical vapor deposition (dc or rf planar magnetron sputtering) by alternating transparent (Al) and half transparent (Cu) layers (total 50 ~ 100 layers) on a fine gold wire core of radius r_0 (23.5 μm and 49 μm) at a rotation speed of 15 rpm.[10] The radius r_n of the nth zone is given by

$$r_n^2 = r_0^2 + n\lambda f \quad (1)$$

where λ and f are the wave length and the focal length respectively. The parameters of the FZPs obtained are given in Table 1. The outermost zone width is between 0.25 μm and 0.09 μm at present. Details of fabrication process of multi-layers wire samples are given in the previous papers.[10-11]

After deposition of the final layer, an additional Cu layer of 3 μm in thickness was overcoated on the surface of the outermost Cu layer for protection against damage that might be caused by the slicing, thinning and polishing process of the wire sample. Interface roughness of the outer zone layers was observed in all the FZPs fabricated in this method. In general a lower Ar gas pressure (0.2 Pa) was found to result in smoother zone boundaries.

The wire sample was then fixed into melt solder (Sn: 60%, Pb:40%), thightned and sliced into a plate (10 mm x 10 mm x 1 mm) using a micro-cutting machine perpendicular to the wire axis. Fixing the sample onto a pyrolytic graphite plate (grain free and ~1 mm in thickness (Union Carbide Co. Ltd.)), the sample was then polished by a micro-grinding machine using the sand papers. After polishing one side, the sample was turned over, and finally fixed on the graphite plate by the resin glue "LOCTITE 420" and the other side polished.

TABLE 1. zone plate parameters fabricated (Cu/Al system)

zone plates	zp(1)	zp(2)	zp(3)	zp(4)
outer most zone width	0.25 μm	0.19 μm	0.15 μm	0.09 μm
center stop diameter (gold wire core diameter)	47 μm	47 μm	47 μm	97 μm
primary focal length (observed at 12.4 keV)	220 mm ($\lambda = 1$ Å)	235 mm ($\lambda = 1$ Å)	105 mm ($\lambda = 1$ Å)	95 mm ($\lambda = 1$ Å)
number of zones	50	100	50	70
zone plate diameter	80 μm	100 μm	70 μm	113 μm
thickness of zone plate	10 ~50 μm	10 ~50 μm	10 ~50 μm	10 ~50 μm

A scanning electron micrograph of zp(2) is shown in Fig. 1 as an example.

The merit of FPZP as a focusing device such as high efficiency and low background at the focus point has long been recognized.[14] The diffraction efficiency of a FPZP with square zone profile and equal adjacent zones areas can be calculated.[14,15]

The calculations of the diffraction efficiencies in the first order light were performed for our Cu/Al zone plate at the different photon energies. Fig. 2 shows the diffraction efficiencies without the influence of roughness and interdifffusion of zones in the photon energy range 8 ~ 100 keV. Here, the central gold core is not taken into acount. High diffraction efficiencies up to 40 % can be theoretically achieved at the photon energy range of 50 ~ 100 keV. Then, to fabricate the high efficiency FPZP it is necessary to polish the FZP to required thickness.

The zone plate, zp(3) in Table 1, was polished, and the thickness obtained was ~43 μm.[16] This thickness is expected to give high focusing efficiency at the energy of 25 keV and 80~90 keV x-rays respectively.

FIGURE 1.
The scanning micrograph of the zone plate, zp(2).
(100 zones)

FIGURE 2. The diffraction efficiency of the zone plate (Cu/Al) calculated.

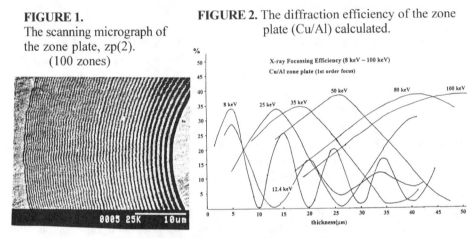

BRIEF OF THE FOCUSING TEST

Characterization, heat load and radiation damage tests at various x-ray energies were made for the FPZPs obtained. For this experiment, SR beam from the undulator at the BL47XU of the SPring-8 was used with the beam monochromatized by Si(111) double crystal monochromator. An ion chamber or NaI scitillation counter was used in the intensity measurements. The beam profile was measured by knife-edge scan (for this a thin gold wire was used). With the FPZP (zp(3)) the minimum focal beam size [full width at half maximum (FWHM)] of 0.7 μm in vertical scan was obtained for first order focal beam at 27.5 keV as shown in Fig. 3(a). Here, the focus spot was directly generated by the zone plate demagnifying the undulator source image. The focal length and the light collecting efficiency were determined to be 233 mm and 15 % respectively. Details of the microbeam experiment at the SPring-8 are given in this volume.[16]

Using this FPZP, we have succeeded in focusing much higher x-ray energy like 83 keV. Preliminary focusing image on CCD camera is shown in Fig. 3(b).

During full power x-ray irradiation (14.4 keV, ring current: 70 mA) the surface temperature upon the FPZP was measured which was between 27 ~34℃. It was found that the resin glue "LOCTITE 420" fixing the FPZP on the graphite plate was chemically damaged during full power illumination. To avoid the radiation damage it might be necessary to remove the resin glue just on the FPZP or to use other type glue.

FIGURE 3(a).
Hard x-ray microbeam at 27.5 keV.

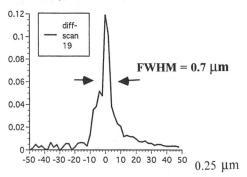

FWHM = 0.7 μm

0.25 μm

FIGURE 3(b).
Focusing image of 83 keV x-ray obtained on CCD camera.

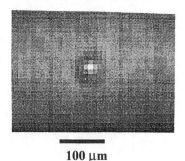

100 μm

ACKNOWLEDGEMENTS

The synchrotron radiation experiments were performed at the SPring-8 with approval of the Japan Synchrotron Radiation Institute (JASRI) (proposal No. 1999A0091,NM-np)

REFERENCES

1. E.D. Fabrizio, M. Gentili, F. Romanto, this volume.
2. J. Maser, B. Lai, Z. Cai, W. Rodgriges, D. Legnini, P.Ilinski, W.B. Yun, Z. Chen, A.A.Kranoperova, Y Vladimirsky, F. Cerrina, E.D. Fabrizio, M. Gentili, this volume.
3. M. Panitz, G. Schneider, M. Peuker, Hambuch, B. Kaulich, S. Ostreich, J. Susini, G. Schmahl, this volume.
4. Y. Kagoshima, this volume.
5. Y. Vladimirsky, this volume.
6. P. Charalambus, this volume.
7. D. Rudolph and G. Schmahl, SPIE **342**, 94(1980); 316, 103(1981).
8. K. Saitoh, K. Inagawa, K. Kohra, C. Hayashi, A. Iida, and N. Kato, Rev. Sci. Instrum. **60**, 1519(1989).
9. R.M. Bionta, E. Ables, O. Clamp, O.D. Edwards, P.C. Gabriele, K. Miller, L.L. Ott, K.M.Sukulina, R. Tilley, and T. Viada, Opt. Eng. **29**, 576(1990).
10. S. Tamura, K. Ohtani, and N. Kamijo, Appl. Surf. Sci. **79/80**, 514(1994).
11. N. Kamijo, S. Tamura, Y. Suzuki, K. Handa, A. Takeuchi, S. Yamamoto, M. Ando, K. Ohsumi, and H. Kihara, Rev. Sci. Instrum. **68**, 14(1997).
12. B. Kaulich, SPIE's 43rd Annual Meeting, San Diego, July, 19-24(1998).
13. A. Duevel, M. Panitz, D. Rudolph, G. Schmahl, this volume.
14. J. Kirz, J. Opt. Soc. Am, **64**, 301(1974).
15. W.B. Yun, J. Chzas, and P.J. Viccaro, SPIE Proc. **1345**, 146(1990).
16. Y. Suzuki, N. Kamijo, S. Tamura, M. Yasumoto, M. Awaji, Y. Kohmura, and K. Handa, this volume.

Electroplated Gold Zone Plates As X-Ray Objectives For Photon Energies Of 2-8 keV

M. Panitz, G. Schneider, M. Peuker, D. Hambach, B. Kaulich*,
S. Oestereich*, J. Susini*, G. Schmahl

Institut für Röntgenphysik, Georg-August-Universität Göttingen, Geiststr. 11, 37073 Göttingen, Germany
* European Synchrotron Radiation Facility, Beamline ID21, BP220, 38043 Grenoble Cedex, France

Abstract. Zone plates with high spatial resolution and high diffraction efficiency are required as objectives for X-ray microscopes. In the energy region of 2-8 keV gold is a well suited material for highly efficient zone plates. In this work a process for Ni zone plates is extended for the fabrication of sub-100 nm gold zone structures. This tri-level process is based on pattern generation with electron beam lithography followed by dry etching to transfer the zone pattern into a highly cross-linked polymer which serves as galvanoform. After the electrodeposition of a thin intermediate layer of Ni, the galvanoform is filled with gold by electroplating. With this process zone plates with outermost zone widths of 70 nm and zone heights of about 500 nm, yielding a first-order diffraction efficiency of 10 % at a photon energy of 4.1 keV, were manufactured.

INTRODUCTION

X-ray microscopy in the energy range from 2-8 keV allows to image hydrated biological samples with thicknesses up to 100 μm. Furthermore, the K-absorption edges of elements like Ca, P, S and the M- or L- absorption edges of heavy metals are accessible. In addition, X-ray fluorescence can be used for trace element mapping. Zone plates (ZPs) are the most suitable type of X-ray optics to image the samples with high spatial resolution. According to the Rayleigh criterion for incoherent object illumination the obtainable resolution in the first focussing order of a ZP is given by $\delta \approx 1.22 \, dr_n$, where dr_n denotes the outermost zone width. Using gold as ZP material, zone heights of more than 400 nm have to be structured to achieve efficiencies ≥ 10 % at 4.1 keV photon energy(see Fig. 1).

In the past several processes have been developed to build ZPs. One method is the sputtered-sliced technique [1]. A circular wire is coated alternately with two materials with suited indices of refraction. Afterwards, the wire is cut in thin slices

CP507, *X-Ray Microscopy: Proceedings of the Sixth International Conference,*
edited by W. Meyer-Ilse, T. Warwick, and D. Attwood
© 2000 American Institute of Physics 1-56396-926-2/00/$17.00

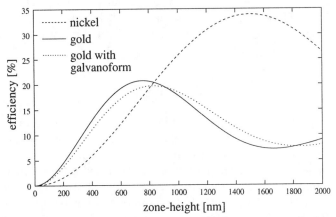

FIGURE 1. Theoretical first-order diffraction efficiency of gold and nickel zone plates as a function of the zone height for 4.1 keV photon energy. The coupled wave theory was applied to perform the calculations for a ZP with smallest outermost zone width of 70 nm, line-to-space ratio of 1:1, imaging magnification of 1000 and zones parallel to the optical axis [3,4].

according to the required ZP height for optimum first-order diffraction efficiency [2]. Several other methods of manufacturing ZPs make use of planar nanofabrication technologies:

1. The ZP pattern generated with electron beam lithography (EBL) is transferred into the ZP-material by several steps of reactive ion etching (RIE). With such a direct etch process ZPs from tungsten were fabricated for the keV-range [5].

2. The ZP pattern is exposed by EBL in a layer of the resist PMMA. After development the PMMA structures serve directly as galvanoform for the electrodeposition of the ZP material [6]. ZPs with outermost zone widths of about 30 nm were built for the water window wavelength region. However, the achievable linewidth in thick resist layers is limited by the scattering of the electrons in the PMMA.

3. In other types of processes the required high aspect ratio zone structures are fabricated with X-ray lithography, using a ZP with low aspect ratio structures as X-ray mask. Zone widths down to 100 nm were achieved with this technique. First-order diffraction efficiencies of about 13 % were obtained for 8 keV photon energy [7].

4. Furthermore a tri-level process can be used to achieve the necessary aspect ratio of the structures [8]. A thin layer of PMMA is exposed by EBL. Then the ZP pattern is transferred into a titanium intermediate layer by RIE. Afterwards, this layer serves as an etch mask for the subsequent RIE step with

O_2 to transfer the pattern into a polymer. Finally the polymer structures are filled with the ZP material by electroplating. Nickel ZPs with outermost zone widths of 25 nm yielding a groove efficiency of about 15 % at 2.4 nm wavelength were built with this process [9].

PROCESS

In this work the process described in item 4 has been extended to manufacture ZPs with outermost zone widths below 100 nm for photon energies of about 4 keV. The layer system and the processing steps are shown in figure 2.

Silicon membranes with a thickness of about 150 nm and a transmission of about 98% at 4.1 keV photon energy were used as substrates. On the membrane a plating base from Cr and Ge is deposited. These materials are chosen due to their low sputter rate. Thereby the formation of a conducting coating on the galvanoform sidewalls during the RIE step with oxygen can be avoided. The highly cross-linked copolymer PEDVB consisting of the monomeres styrene and divinylbenzene is deposited by spin coating as material for the galvanoform. To increase its mechanical stability it was X-ray hardened with a radiation dose of $4.5 \cdot 10^9$ Gy of polychromatic soft X-rays generated in a bending magnet at the electron storage ring BESSY [10]. Finally 7 nm Ti are deposited by electron beam evaporation followed by the spin coating of 55 nm PMMA.

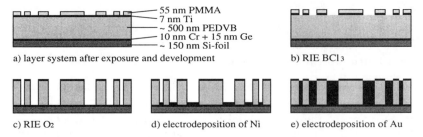

a) layer system after exposure and development
55 nm PMMA
7 nm Ti
~ 500 nm PEDVB
10 nm Cr + 15 nm Ge
~ 150 nm Si-foil

b) RIE BCl₃

c) RIE O₂

d) electrodeposition of Ni

e) electrodeposition of Au

FIGURE 2. The processing steps for manufacturing gold zone plates. The pattern is generated by electron beam lithography and transferred by RIE into the copolymer PEDVB. A 100 nm thick intermediate layer of nickel is deposited, which serves as plating base for the gold electrodeposition.

The ZP pattern is generated by EBL with an electron energy of 40 keV. After exposure and development, the pattern is transferred into a 7 nm thick titanium layer by RIE with BCl_3 and then into the about 500 nm thick layer of PEDVB by RIE with O_2. Galvanoforms with aspect ratios of 8:1 for a structure size of 70 nm have been achieved. After the RIE steps the galvanoform is immersed in a nickel sulfamate electrolyte to deposit a 100 nm thick intermediate layer of nickel which serves as plating base for electrodeposition of the gold in a sulphitic electrolyte. It is difficult to control the height of the deposited gold in the galvanoform zone structures during electroplating. In the optimal case one has to stop the plating process at the height of the galvanoform. Otherwise the gold zone height is too low or gold is growing outside the galvanoform which leads to a gold layer covering the zone structures. Such a covering of the ZP can be removed by sputtering with Ar$^+$ ions in normal incidence. The sputter process leads to a rather rough surface (see Fig. 3), but this roughness has only a negligible effect on the efficiency. Note that it is unnecessary to remove the PEDVB galvanoform, because it does not significantly affect the diffraction efficiency of the ZPs in the multi keV range.

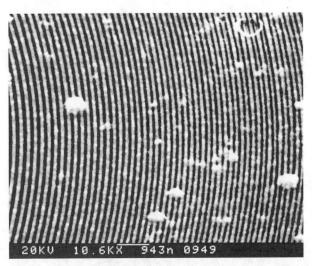

FIGURE 3. SEM micrograph of a gold zone plate after sputtering with Ar. Smallest zone widths of 80 nm are visible. The PEDVB galvanoform (black) was not removed between the gold structures (white).

EFFICIENCY MEASUREMENTS

The efficiency measurements were done at the ESRF beamline ID21 [11]. To measure the intensity diffracted into an order, the ZPs were illuminated with monochromatic radiation with a photon energy of 4.1 keV. A 5 μm pinhole was used to select a diffraction order in the ZPs focal plane. This pinhole was scanned perpendicular to the optical axis. The intensity of the transmitted radiation was detected with a Si-photodiode (see Fig. 4). The peak visible in this intensity distribution corresponds to the first diffraction order. First-order diffraction efficiencies of up to 10% were measured for ZPs with a smallest zone width of 70 nm and up to 5 % for ZPs with a smallest zone width of 60 nm.

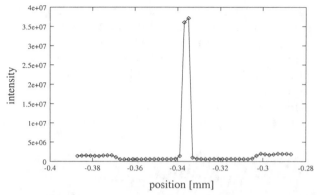

FIGURE 4. Intensity distribution in the first-order focal plane. The peak corresponds to a first diffraction order of a ZP with an absolute diffraction efficiency of 8.7 %.

IMAGING

Gratings with an increasing line density up to 11,000 lines/mm corresponding to a smallest line width of 90 nm were used as objects to determine the imaging properties of the ZPs which have been generated in this work. These gratings consist of gold and were manufactured with the same process as described above for the gold

FIGURE 5. Micrograph of a gold grating taken at 4.1 keV photon energy with the TXM at the ESRF beamline ID21. The smallest visible line width is 90 nm (arrow). The ZP used as objective has an outermost zone width of 70 nm and an absolute first-order diffraction efficiency of 8.7 % at 4.1 keV photon energy.

ZPs. Micrographs of these gratings were taken with the TXM (Transmission X-ray Microscope) of the ESRF beamline ID21 at 4.1 keV photon energy. The finest structure sizes of the used grating of 90 nm are visible in the TXM micrograph (see Fig. 5).

SUMMARY

A tri-level process for manufacturing gold ZPs with outermost zone widths of 70 nm and 60 nm has been developed. The ZP pattern was generated by electron beam lithography and transferred into the highly cross-linked copolymer PEDVB by RIE with O_2. The resulting galvanoform has been used for the electrodeposition of nickel and gold. Galvanoforms with aspect ratios of about 8:1 were achieved. Absolute first-order diffraction efficiencies of up to 10 % for 70 nm smallest zones and 5 % for 60 nm smallest zones were measured at 4.1 keV photon energy. Gold gratings used as objects were imaged with the TXM at the ESRF beamline ID21 using a ZP with an outermost zone width of 70 nm. Line widths of 90 nm are visible in TXM micrographs taken at 4.1 keV photon energy.

REFERENCES

1. Rudolph D., Schmahl, G., and Niemann, B.: Status of the sputtered sliced zone plates for X-ray microscopy, *SPIE Proc.* **316**, 103-105 (1981).
2. Düvel, A., Rudolph, D., and Schmahl, G.: Fabrication Of Thick Zone Plates For Multi-Kilovolt X-Rays, this volume.
3. Maser, J.: Theoretical description of the diffraction properties of zone plates with small outermost zone widh, *X-Ray Microscopy IV*, Aristov, V.V., and Erko, A.I. (Eds.), Begorodski Pechatnik Publishing Company, 1994, pp. 523-530.
4. Schneider, G., *Appl. Phys. Lett.* **71**, 2242-2244 (1997).
5. Charalambous, P., and Burge, R.E.: Zone Plate Fabrication at King's College, London, *X-Ray Microscopy and Spectromicroscopy*, Thieme, J., Schmahl, G., Rudolph, D., and Umbach, E. (Eds.), Springer-Verlag Berlin Heidelberg, 1998, 57-64.
6. Anderson, E.H., and Kern, D.: Nanofabrication of Zone Plates for X-Ray Microscopy, *X-Ray Microscopy III*, Michette, A., Morrison, G., and Buckley, C. (Eds.), Springer Series in Optical Sciences **67**, 1992, pp. 75-78.
7. Yun, W., Lai, B., Cai, Z., Maser, J., Legnini, D., Gluskin, E., Chen, Z., Krasnoperova, A.A., Vladimirsky, Y., Cerrina, F., Di Fabrizio, E., and Gentili, M., *Rev. Sci. Instrum.* **70**, 2238-2241 (1999).
8. Schneider, G., Schliebe, T., and Aschoff, H., *J. Vac Sci. Technol. B* **13**, 2809-2812 (1995).
9. Peuker, M.: Nickel Zone Plates for Soft X-Ray Microscopy, this volume.
10. Weiss, D., Peuker,M., and Schneider, G., *Appl. Phys. Lett.* **72**, 1805-1807 (1998).
11. Barrett, R., Kaulich, B., Oestreich, S., and Susini, J.: The scannning microscopy end-station at the ESRF X-ray microscopy beamline, SPIE Proc. **3449**, 80-91 (1998).

Nickel Zone Plates for Soft X-Ray Microscopy

M. Peuker

*Georg August Universität Göttingen, Institut für Röntgenphysik, Geiststraße 11,
37073 Göttingen, Germany*

Abstract. Soft X-ray microscopes require high-spatial-resolution Fresnel zone plates with high
and uniform diffraction efficiency as imaging optics. The combination of the zone height for
optimal diffraction efficiency with the fact that the first-order spatial resolution of a zone plate
scales linearly with its outermost zone width leads to high aspect ratios of the nanostructures.
In the water window wavelengths region nickel is a well suited zone plate material, yielding high
diffraction efficiency at moderate aspect ratios of the zones. Since it cannot be structured by
reactive ion etching a tri-layer process making use of electrodeposition techniques was
developed.
All nickel zone plates presented in this work have demonstrated excellent first-order diffraction
efficiencies at 2.4 nm wavelength. Zone plates with outermost zone widths of 40 nm, 30 nm and
25 nm were fabricated with the tri-level process yielding 21.6%, 18.0% and 14.7% first-order
groove efficiency, respectively. These measured efficiencies correspond to maximum obtainable
efficiencies determined by the galvanoform height of 100%, 95% and 85%.

INTRODUCTION

Modern high-spatial-resolution X-ray microscopes use Fresnel micro zone plates
(MZP) as objectives. Essentially, these optics have to combine high spatial resolution,
i.e. minimum outermost zone width, with optimal diffraction efficiency to minimize
exposure time and the X-ray dose applied to the specimen.

Germanium and especially nickel are well suited materials for phase zone plates in the
water window wavelengths region. Figure 1 shows the first-order diffraction
efficiency η_1 at 2.4 nm wavelength for MZPs made of germanium and nickel as a
function of zone height. The calculations are based on coupled wave theory which has
been applied to MZPs[1,2]. This dynamical theory allows diffraction efficiency
calculations for a wide range of MZP parameters and is valid even for high aspect
ratios (AR) . The maximum theoretical η_1 for Ge MZPs with 25 nm outermost zone
width is 19.0% at a zone height of 400 nm. The crucial parameter in the
nanostructuring process is the AR to be manufactured. The corresponding AR for the
maximum theoretical η_1 is 16:1. Using Ni the same η_1 can be achieved by realizing an
AR of just 7:1. Additionally, a higher maximum theoretical η_1 of 23.5% can be
obtained with Ni at an AR of 10.4:1. Further calculations yield that for Ni MZPs η_1 of
~15% can be achieved for all water window wavelengths with zone heights of 150 -
200 nm, corresponding to an AR of 6 - 8 for MZPs with 25 nm outermost zone width
– see figure 2.

CP507, *X-Ray Microscopy: Proceedings of the Sixth International Conference,*
edited by W. Meyer-Ilse, T. Warwick, and D. Attwood
© 2000 American Institute of Physics 1-56396-926-2/00/$17.00

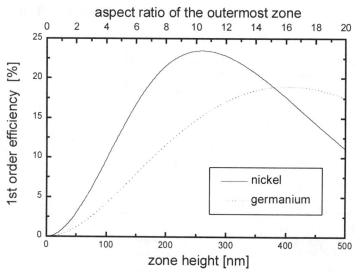

FIGURE 1. Theoretical first-order diffraction efficiency at λ = 2.4 nm of nickel and germanium zone plates as a function of the zone height in the 25 nm zone width region. Calculations were performed for a rectangular zone profile, unslanted zone structures, a line-to-space ratio of 1:1 and a typical MZP imaging magnification of 2500x. On the upper horizontal scale the corresponding AR of 25 nm zone width is indicated.

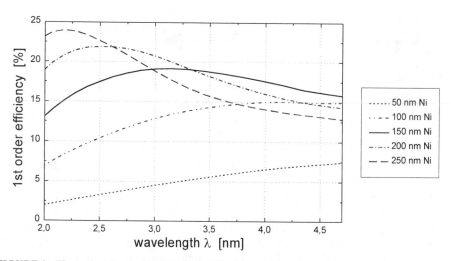

FIGURE 2. Theoretical first-order diffraction efficiency of nickel zone plates with 25 nm outermost zone width as a function of the wavelength for different nickel zone heights. Calculations were performed for a rectangular zone profile, unslanted zone structures, a line-to-space ratio of 1:1 and a typical MZP imaging magnification of 2500x.

683

TRI-LEVEL NICKEL NANOSTRUCTURING PROCESS

Nickel nanostructures cannot be manufactured directly by reactive ion etching (RIE), since Ni does not form volatile compounds with any etching gas. Therefore, electrodeposition techniques have to be used for pattern transfer. Resist structures can serve as galvanoform[3] but the resist thickness necessary to build MZPs with adequate Ni height limits the achievable resolution in e-beam lithography. To overcome these limitations a tri-level process has been developed allowing an independent choice of resist and galvanoform material and height[4,5,6,7].

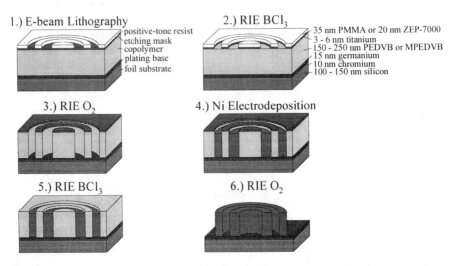

1.) E-beam Lithography
- positive-tone resist
- etching mask
- copolymer
- plating base
- foil substrate

2.) RIE BCl$_3$
- 35 nm PMMA or 20 nm ZEP-7000
- 3 - 6 nm titanium
- 150 - 250 nm PEDVB or MPEDVB
- 15 nm germanium
- 10 nm chromium
- 100 - 150 nm silicon

3.) RIE O$_2$

4.) Ni Electrodeposition

5.) RIE BCl$_3$

6.) RIE O$_2$

FIGURE 3. Layer Sequence and processing steps for the manufacture of nickel zone plates by electrodeposition.

The layer sequence and the processing steps for Ni MZP manufacture by electrodeposition are shown in figure 3. The investigations are focused on the galvanoform material since its quality determines the obtainable structure width and height of the electroplated material. This material has to fulfill the following criteria: high electrical resistance, high mechanical stability, structurable by RIE, and non-swelling in nickel electroplating bath. These criteria are met by the copolymer PEDVB consisting of the monomers phenylethylene (PE) and divinylbenzene (DVB). PEDVB is made by radical copolymerisation and the insertion of DVB into the PE chains leads to a highly cross-linked copolymer[5,6]. To ensure the process reproducibility technical grade DVB has been replaced by pure 1,4-DVB which is synthesized by a Wittig substitution[6,7]. Furthermore, 4-methyl phenylethylene (MPE) as a more reactive second copolymer constituent is introduced – MPEDVB. Finally, the degree of cross-linking which is a measure for the mechanical stability of the copolymer structures is limited by the diffusion of the monomers in the cross-linked network during baking. To overcome this limitation the baked copolymer is irradiated with high doses of X-rays inducing further cross-linking[6,7].

E-beam lithography is performed with an AKASHI DS 130C SEM which is equipped with an ELPHY III vector scan lithography device from RAITH Co. at 40 kV primary electron energy[8]. PMMA is used as resist for line widths down to 30 nm. For smaller zone width ZEP-7000 from Nippon Zeon Co. is employed in the process. The pattern is transferred by RIE with BCl_3 into a Ti layer and with O_2 into the copolymer. Ni is electroplated from a sulphamate type bath on a plating base made of Ge and Cr. The Ti mask and the copolymer are removed by RIE. Finally, the ZP rests on the plating base layers and the silicon foil substrate which have an overall soft X-ray transmission of ~70% at λ = 2.4 nm. The absolute diffraction efficiency of a ZP is less than its groove efficiency by this factor.

RESULTS

Using PEDVB and MPEDVB exact pattern replication in the galvanoform is possible for ARs up to 4:1 and 5:1, respectively. Additional cross-linking induced by irradiation with high doses of X-rays results in higher ARs. Samples irradiated with 6.7×10^8 Gy – exposure was performed at BESSY – exhibit no pattern distortions for ARs up to 8.2:1 at 30 nm line width[6,7]. With these galvanoforms exact pattern replication in Ni was achieved at zone widths down to 25 nm – see figure 4.

The MZP diffraction efficiency is measured at BESSY storage ring in the X-ray test chamber at 2.4 nm wavelength. The radiation diffracted by the MZP into its first order is focused into a 5 µm pinhole. A CCD camera is used to detect the cone of first-order radiation emerging from the pinhole. The total incident radiation is measured by replacing the MZP with a 50 µm pinhole.

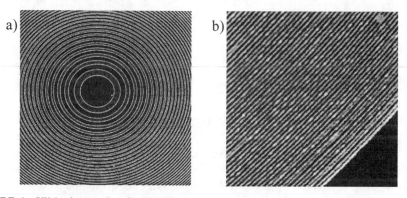

FIGURE 4. SEM micrographs of a Ni MZP showing (a) the inner zones and (b) the outermost zone with minimum linewidth of 25 nm.

Nickel MZPs with 40 nm and 30 nm outermost width with 21.6% and 18.0% uniform first-order groove diffraction efficiency were manufactured. Furthermore, MZPs with 25 nm outermost zone width yielding up to 14.7% integral groove efficiency were built. Detailed analysis of the intensity distribution of the first diffraction order reveals that even the outer zones yield more than 8% groove efficiency – see figure 5.

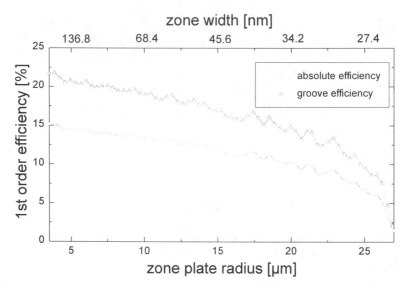

FIGURE 5. Absolute and groove first-order diffraction efficiency of a ZP with 25 nm outermost zone width and 53.8 μm diameter calculated from the intensity distribution of the first diffraction order.

The maximum obtainable efficiency is determined by the galvanoform height. The copolymer height is adapted to the outermost zone width to guarantee perfect pattern replication in nickel. Consequently, the measured first-order groove diffraction efficiencies correspond to 100% achievable efficiency for the MZP with 40 nm outermost zone width, and to 95% and 85% for the MZPs with 30 nm and 25 nm outermost zone width, respectively.

All these nickel MZPs yield more than 14% first-order groove diffraction efficiency for all waterwindow wavelengths. They have been successfully employed in the transmission X-ray microscope at BESSY for imaging cells, colloid samples and magnetic structures.

ACKNOWLEDGEMENTS

The author would like to thank G. Schmahl, G. Schneider, D. Rudolph, D. Weiß, T. Schliebe, and the staff at BESSY. This work was funded by the German Federal Minister for Education and Research (BMBF) under contract no. 05 SL8MG1.

REFERENCES

1. Maser, J., "Theoretical description of the diffraction properties of zone plates with small outermost zone width" in *X-Ray Microscopy IV*, edited by V. V. Aristov and A. I. Erko, Bogorodskii Pechatnik Publishing Company, Chernogolovka, Moscow region, 142432, Russia, 1992, pp. 523-530.

2. Schneider, G., *Appl. Phys. Letters* **71**, 2242-2244 (1997).

3. Anderson, J., and Kern, D., "Nanofabrication of Zone Plates for X-Ray Microscopy" in *X-Ray Microscopy III*, edited by A. G. Michette et al., Berlin: Springer Verlag, 1990, pp. 75-78.

4. Schneider, G., Schliebe, T., and Aschoff, H., *J. Vac. Sci. Technol. B* **13**, 2809-2812 (1995).

5. Schliebe, T., Schneider, G., and Aschoff, H., *Microelectronic Engineering* **30**, 513-516 (1996).

6. Weiss, D., Peuker, M., and Schneider, G., *Appl. Phys. Letters* **72**, 1805-1807 (1998).

7. Peuker, M., Schneider, G., and Weiss, D., "High resolution phase zone plates for water window wavelengths" in *X-Ray Microfocusing: Applications and Techniques*, edited by I. Mc Nulty., Proceedings of SPIE 3449, Bellingham, WA, 1998, pp. 118-128.

8. David, C., Kaulich, B., Medenwaldt, R., Hettwer, M., Fay, N., Diehl, M., Thieme, J., and Schmahl, G., *J. Vac. Sci. Technol. B* **13**, 2762-2766 (1995).

Nanostructuring of Free-Standing Zone Plates

S. Rehbein[1], R.B. Doak[2,3,4], R.E. Grisenti[3], G. Schmahl[1],
J.P. Toennies[3], and Ch. Wöll[4]

[1] *Institut für Röntgenphysik, Georg-August-Universität Göttingen, D-37073 Göttingen*
[2] *Department of Physics and Astronomy, Arizona State University, Tempe, AZ 85287-1504 USA*
[3] *Max-Planck-Institut für Strömungsforschung, Bunsenstrasse 10, D-37073 Göttingen*
[4] *Physikalische Chemie I, Ruhr-Universität Bochum, D-44780 Bochum*

Abstract. A method for the fabrication of free-standing silicon zone plates is presented. Free-standing zone plates with 100 nm and 50 nm outermost zone width and a diameter of 270 μm and 540 μm, respectively, have been fabricated. The zone plates are designed for focusing a neutral, ground state helium atom beam at thermal energy. The design focal length of the first diffraction order is 150 mm for a de Broglie wavelength of 0.18 nm. Helium beam focusing experiments have been carried out with the zone plates with 100 nm outermost zone width.

INTRODUCTION

Zone plates are well suited as optical elements for X-ray beams and have been used with great success in X-ray microscopy. Analogous zone plate applications are now emerging with neutral, ground state, thermal energy atomic beams. These include the focusing of helium beams, for the eventual actualization of an atom beam microscope, and velocity dispersion, for the purpose of monochromatizing an atom beam by a zone plate linear monochromator [1], [2]. Zone plates with small outermost zones are required both for high spatial resolution in the case of an atom beam microscope and for a zone plate monochromator of high energy resolution. Since thermal energy atoms cannot penetrate even the thinnest support foil, the zone plate structures for helium atom diffraction must be completely free-standing. Therefore a support structure is necessary to keep the opaque annuli of the zone plate in place. Circular free-standing zone plates have already been fabricated by other groups, but to our knowledge only with feature sizes larger than 100 nm [3], [4]. We present a method for the fabrication of free-standing silicon zone plates with an outermost zone width down to 50 nm.

CP507, *X-Ray Microscopy: Proceedings of the Sixth International Conference*,
edited by W. Meyer-Ilse, T. Warwick, and D. Attwood
© 2000 American Institute of Physics 1-56396-926-2/00/$17.00

FABRICATION PROCESS

A two part fabrication process is employed, creating separately the zone plate pattern and the strut structure needed to support the free-standing zone plate annuli. By exposing these two patterns individually, the amount of data in the exposure file is reduced significantly. This reduces the writing time markedly compared to a single-exposure fabrication, from ca. 50 hours to ca. 3 hours for the zone plate with 50 nm outermost zone width [7].

The fabrication process is depicted in Fig. 1. A silicon wafer with a ca. 200 nm thick boron doped layer is spin coated with a 115 nm thick PMMA layer. The PMMA layer is exposed with the support structure pattern using the electron beam lithography system LION LV1 from Leica Lithographie Systeme Jena. Alignment marks (not illustrated in Fig.1) are also placed on the wafer in this exposure. These cross marks lie outside the central area of the wafer and are later used to align the second lithography step. After development, the structured PMMA layer (Fig.1a)) is coated with a 10 nm thick chromium layer by electron beam evaporation (Fig.1b)). The chromium support structure mask and the cross marks are then obtained by

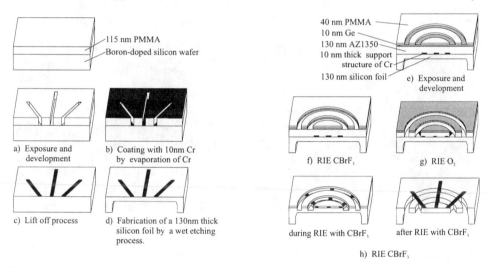

FIGURE 1. Layer sequence and processing steps for the fabrication of free-standing zone plates.

a lift-off process in acetone (Fig.1c)). Afterwards the wafer is mounted in a special holder to protect the chromium pattern during a subsequent boron-selective wet chemical etch [5]. This etch thins the wafer from the backside to finally yield a silicon foil of ca. 130 nm thickness in the central area of the wafer (Fig.1d)). The silicon foil with the chromium support structure pattern on top is then coated with a layer sequence of 130 nm AZ1350, 10 nm Ge and 40 nm PMMA [6]. The layer sequence is removed in the area of the chromium marks by RIE, while covering the rest of the wafer with a metal sheet to protect the deposited layers during the RIE.

The chromium serves as an etch mask for structuring silicon with $CBrF_3$. A final RIE step with $CBrF_3$ transfers the pattern of cross marks into the silicon. The marks are subsequently about 150 nm high with sharp edges, providing sufficient contrast for a reliable mark detection in the LION LV1. The next step is to expose the zone plate pattern, correctly positioned with respect to the chromium support structure mask. To this end, the positions of the alignment marks are detected by the LION LV1 and the correct exposure position for the zone plate is calculated by the lithography system. After exposure and development the zone plate pattern (Fig.1e)) is transferred into the Ge layer by reactive ion etching (RIE) with $CBrF_3$ (Fig.1f)). The AZ layer is then structured by RIE with O_2 (Fig.1g)). Subsequently, the structured AZ-mask serves, in combination with the chromium structures, as an etch mask in the final RIE step for structuring the silicon foil with $CBrF_3$ (Fig.1h)).

FABRICATED FREE-STANDING ZONE PLATES

Two types of zone plates have been fabricated. The first type is designed for focusing a helium atom beam for an eventual actualization of an atom beam microscope [8] with sub-micron spatial resolution. This free-standing zone plate is

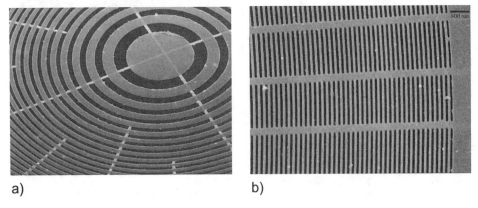

a)   b)

FIGURE 2. SEM micrographs of a free-standing silicon zone plate of 270 μm diameter. a) center of the zone plate. b) 100 nm wide outermost zones.

shown in Figure 2. The zone plate has a diameter of 270 μm and 100 nm outermost zone width. The support strut structure can be seen in the micrographs. The support bars are 450 nm wide.

The second variety of zone plate is designed to be the dispersive optical element of a helium zone plate monochromator. This zone plate is 540 μm in diameter and has an outermost zone width of 50 nm. Its design focal length (first diffraction order) is 150 mm for a de Broglie wavelength of 0.18 nm. Fig. 3a) shows a differential interference contrast (DIC) image of the zone plate pattern transferred into the AZ layer. As evident in the image, the innermost zones have not been exposed in

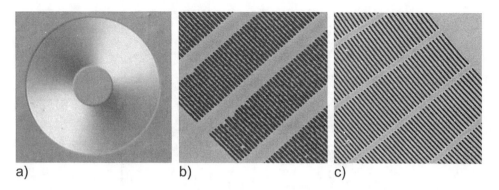

FIGURE 3. a) DIC Image of the 540 μm diameter zone plate pattern transferred in the AZ layer. b) SEM micrograph. 50 nm wide outermost free-standing zones of a silicon zone plate of 540 μm diameter. c) SEM micrograph of the inner free-standing zones about 180 nm wide.

order to yield an opaque central stop for the helium beam in the monochromator application. Fig. 3b) shows the 50 nm wide free-standing outermost zones and Fig. 3c) the about 180 nm wide inner free standing zones of the final silicon zone plate.

MEASUREMENTS

Initial measurements of the focusing properties have been carried out with the zone plates of 270 μm diameter [8]. A scheme of the set-up is shown in Fig. 4. The helium atom beam was generated by a standard supersonic free-jet expansion. The

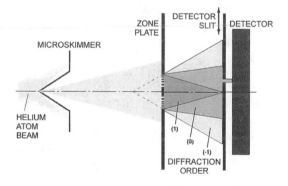

FIGURE 4. Scheme of the set-up.

de Broglie wavelength of the helium atoms was 0.088 nm. At this wavelength the zone plate has a focal length of 307 mm. Mikroskimmer [9] of diameters ranging

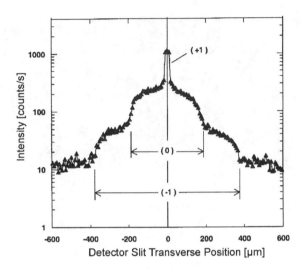

FIGURE 5. Transverse intensity distribution in the detector plane. A microskimmer of 4 μm diameter and a 25 μm wide detector slit were used for this measurement [8].

from 1 to 14 μm were used to obtain a small source diameter. The respective skimmer aperture was imaged by the zone plate with a magnification of 0.4 to generate an image in the plane of a 5 mm high detector slit. Two different slit widths, 15 μm and 25 μm, were used for the measurements. The slit was scanned through the focused beam to yield a transverse intensity distribution as illustrated in Fig. 5 [8]. The sharp central peak in the logarithmic plot is the signal of the $(+1)^{st}$ diffraction order of the zone plate. The background signal arises essentially from the $(0)^{th}$ and defocused $(-1)^{st}$ diffraction order. The background level is of course much larger with the slit detector than would be the case with a circular aperture. Scans with higher resolution through the focused $(+1)^{st}$ diffraction order of the zone plate have also been made. Under the conditions of the measurement, the focused beam spot size is much smaller than the detector slit width. Accordingly, a high resolution transverse scan through the beam spot yields a trapezoidal intensity profile. The information on the spot size is contained in the intensity rise at the edge of the $(+1)$ peak and can be extracted by fitting the peak profile. In this manner a minimum spot diameter of the focused beam of 2 μm was measured, as described elsewhere [8]. Since the skimmer aperture is nominally 1 μm diameter and the zone plate magnification is 0.4, this is a somewhat larger spot size (2 μm) than expected (0.4 μm). Further experiments are necessary to fully understand the cause of the discrepancy between the measured and expected spot size. Nevertheless, this already represents an improvement in comparison to previous work [4] by a factor of 10 in resolution and 10^3 in signal intensity.

CONCLUSION

Free-standing zone plates of 100 nm and 50 nm outermost zone width have been fabricated for helium atom beam applications by means of a two-step exposure process. Initial measurements of the focusing properties have been carried out with the zone plates of 270 μm diameter and 100 nm outermost zone width. Significant improvements in both spatial resolution and signal intensity of a focused atom beam have been demonstrated.

ACKNOWLEDGEMENTS

This work was funded in part by the Deutsche Forschungsgemeinschaft (DFG) under contract SCHM 1118/3-1 and SCHM 1118/3-3.
We wish to thank D. Hambach and T. Gronemann of the Institut für Röntgenphysik, as well as C. David of the Paul Scherrer Institut in Switzerland. R.B. Doak is grateful to the German-American Fulbright Program for support.

REFERENCES

1. G. Schmahl, D. Rudolph, *Optik*, Vol. **29**/6, 577-585 (1969)
2. B. Niemann, D. Rudolph, and G. Schmahl, *Optics Communications*, Vol. **12**, No. 2, 160-163 (1974)
3. D.M. Tennant, J.E. Bjorkholm, M.L. O'Malley, M.M. Becker, J.A. Gregus, and R.W. Epworth, *J. Vac. Sci. Technol.* B **8** 1975-1979 (1990)
4. O. Carnal, M. Sigel, T. Sleator, H. Takuma, and J. Mlynek, *Physical Review Letters*, Vol. **67**, No. 23, 3231-3234 (1991)
5. R. Medenwaldt and M. Hettwer, *Journal of X-Ray Science and Technology* **5** (2), S.202-206 (1995)
6. C. David, D.R. Kayser, H.U. Müller, B. Völkel, M. Grunze, in *X-Ray Microscopy and Spectromicroscopy*, Eds.: J. Thieme, G. Schmahl, D. Rudolph, E. Umbach, Springer-Verlag Berlin Heidelberg New York (1998), Part IV-83
7. S. Rehbein, R.B. Doak, R.E. Grisenti, G. Schmahl, J.P. Toennies, and Ch. Wöll, *Proceedings of the Micro- and Nano-Engineering'99 Conference*, 21-23 Sept 1999, Rome, Italy, eds. M. Gentili, E.Di Fabrizio, and M. Meneghini, Elsevier, Amsterdam, 2000
8. R.B. Doak, R.E. Grisenti, S. Rehbein, G. Schmahl, J.P. Toennies, and Ch. Wöll, *Physical Review Letters*, in print
9. J. Braun, P.K. Day, J.P. Toennies, and G. Witte, *Review of Scientific Instruments* **68**, 3001-3009 (1997)

Compound Refractive Lenses for X-Ray Microanalysis

C. G. Schroer*, B. Lengeler*, B. Benner*, J. Tümmler*, F. Günzler*,
M. Drakopoulos†, A. S. Simionovici†, A. Snigirev†, and I. Snigireva†

*2. Physikalisches Institut, RWTH Aachen, D-52056 Aachen, Germany
†European Synchrotron Radiation Facility ESRF, BP220, F-38043 Grenoble, France

Abstract. We describe parabolic compound refractive X-ray lenses made of aluminum that are genuine imaging devices, similar to glass lenses for visible light. When used to image a synchrotron radiation source onto a sample in a strongly demagnifying setup, these lenses allow to produce a small pencil beam of high intensity that can be used for (scanning) microprobe experiments, such as microdiffraction and microfluorescence. The aluminum lenses are most effective above 10keV and are particularly suited for hard x-rays up to at least 60keV. The pencil beam has a typical lateral size in the micrometer range with a gain in intensity of two to three orders of magnitude. The small beam convergence in the spot (typically $\Delta k/k < 10^{-4}$) is ideal for microdiffraction and micro-SAXS experiments. We give examples of microprobe experiments including microdiffraction, microfluorescence, and fluorescence microtomography.

INTRODUCTION

There is a growing demand for hard X-ray microbeam techniques in fundamental science and technology, in particular in bio-medicine, environmental and earth sciences, microelectronics, material science, chemistry and physics. However, producing hard x-ray microbeams of high intensity is a challenging task and many complementary optics have been designed over time, such as curved mirrors [1], bent crystals [2], multilayers [3], single and multiple capillaries [4], Fresnel zone plates [5], and Bragg Fresnel optics [6]. Recently, cylindrical and crossed cylindrical *compound refractive lenses* (CRLs) have been introduced that have focal distances in the meter range and produce focal line and point spots in the micrometer range [7–10].

Here, we describe CRLs with parabolic profile and rotational symmetry that are *genuine imaging devices* like glass lenses in visible light [11,12]. They are free of spherical aberration, and have excellent imaging properties. The straight optical path allows for easy alignment of the setup and for efficient switching between scanning and full field imaging applications. Recently, a hard x-ray full field mi-

CP507, X-Ray Microscopy: Proceedings of the Sixth International Conference,
edited by W. Meyer-Ilse, T. Warwick, and D. Attwood
© 2000 American Institute of Physics 1-56396-926-2/00/$17.00

(a)

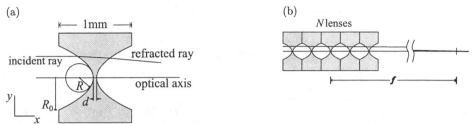

(b)

FIGURE 1. Schematic sketch of a parabolic compound refractive lens. The individual lenses (a) are stacked behind each other to form a CRL (b).

croscope based on these CRLs has been built [11] (see contribution 87 to XRM99). Due to the modularity of the CRLs (assembly and adjustment of a CRL take no more than 15 minutes), a lens can be assembled to meet the specific requirements of a given experiment, and microbeams from several $10\mu m$ down to $0.5\mu m$ can be routinely produced. At $E = 19.5$keV, $1.1 \cdot 10^{10}$ ph/s where recorded in a spot of $0.55 \times 5.5\mu m^2$ FWHM (V × H) with a intensity gain of 1120. Micro beams with energies up to 55keV ($N = 250$, $f = 2.18$m) have been used. CRLs are robust and withstand the white beam of an ESRF undulator source. Therefore, they can be used in "pink" beam (radiation from one undulator harmonic ($\Delta E/E = 10^{-2}$)) increasing the flux up to two orders of magnitude with respect to the monochromatic beam. This might be useful where particularly high fluxes are required such as for microfluorescence measurements of very dilute systems. The development of the CRLs is at a stage, where scanning microprobe experiments can be carried out routinely.

COMPOUND REFRACTIVE LENSES

The main physical constraints to build refractive optics for hard x-rays are the weak refraction and the strong absorption of x-rays in matter. The weak refraction can be accounted for by using both a strong curvature ($R = 0.2$mm, see Fig. 1(a)) of the lens surfaces and a stacking of many lenses (N is typically between 20 and 150) behind each other forming a compound refractive lens (Fig. 1(b)). To avoid strong absorption, the lenses must be as thin as possible ($d \approx 10\mu m$) and made from low Z materials, such as Be, B, C, or Al. Since the refractive index of hard x-rays in matter is smaller than one, focussing lenses have concave shape. A parabolic profile avoids spherical aberration.

The lenses described here are made of polycrystalline aluminium and have been designed and manufactured at the University of Technology in Aachen, Germany. The single lenses (Fig. 1(a)) whose surface shape is a paraboloid of rotation are made by a pressing technique and are aligned in a stack (Fig. 1 (b)) of variable length on two high-precision shafts. For a detailed description and discussion see [9,11,12].

FIGURE 2. Pencil beam recorded on a high resolution ccd-camera at beamline ID22 of the ESRF. ($E = 15$keV, $N = 33$, $f = 1.26$m, $L_1 = 63$m, $L_2 = 1.28$m, magn.$= 1/49$, source size: $35 \times 700\mu m^2$). Beam size: $0.9 \times 14.3\mu m^2$ FWHM (V $\times$ H).

The focal distance is given by $f = R/(2\delta N)$ (index of refraction $n = 1-\delta-i\beta$) and ranges from 0.4m to 2.5m in typical applications. Due to the stronger absorption in the outer parts of the lens, a CRL has an effective aperture D_{eff} smaller than the geometric aperture R_0 (Fig. 1(a)). See [12] for details. It is D_{eff} that determines the diffraction at the lens and typically ranges from 100 to $250\mu m$ for aluminium lenses.

An ESRF undulator source (high-β section) illuminates an area of about 1mm^2 in a distance of 40m with a monochromatic ($\Delta E/E = 10^{-4}$) flux in the range from 10^{11} to 10^{13} ph/s depending on the photon energy. An aluminium CRL typically transmits and focuses about 0.1% to 5% of this incident radiation.

The use of more transparent lens materials will considerably increase the transmission of a CRL, allowing for faster scans. For Be, efficiencies of up to 40% for a lens with 1mm aperture are expected. Since D_{eff} increases, diffraction effects decrease allowing for smaller beam sizes.

SCANNING X-RAY MICROSCOPE USING CRLS

To produce a microbeam, the synchrotron source is imaged through the lens onto the sample in a strongly demagnifying setup [12]. Usually, monochromatic radiation is used, but the "pink" beam can be used without significant deterioration of the beam size. For a typical source to lens distance L_1 of 40m to 60m, magnifications between $m = 1/20$ and $1/120$ are achieved. The sample is placed a distance $L_2 = L_1 f/(L_1 - f)$ behind the lens. The setup requires no vacuum chamber and samples can be kept in their natural environment. No order sorting pinhole close to the sample is needed.

The lateral size of the pencil beam depends on the size of the source, the demagnification of the setup and the diffraction broadening due to the lens [12]. An example of a pencil beam is shown in Fig. 2. The size of the Airy disc is typically between $0.2\mu m$ and $0.5\mu m$ [12], limiting the beam size by diffraction at that level.

With $f \sim 1$m and $D_{eff} \sim 0.1$mm the k-vector of the incident photons is well defined ($\Delta k/k = D_{eff}/f \sim 10^{-4}$), which allows for microdiffraction experiments as well as micro-SAXS.

The following experiments were carried out at beamline ID22 of the ESRF.

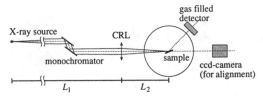

FIGURE 3. Experimental setup for microdiffraction.

(a) Bragg reflection of a single grain

(b) dislocation density: preliminary results

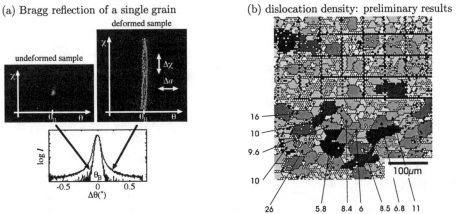

FIGURE 4. Microdiffraction on single grains. (a) Bragg reflection and line profile of a single grain. (b) EBSD-Image of the sample. Dislocation densities are indicated in arbitrary units. The dislocation density of an undeformed grain is 0.16 for comparison.

At an ESRF user experiment (beamline ID22) by O. Castelnau, et al., [13], the dislocation density of single grains in a cold rolled IF-Ti steel sample have been determined by *single grain line profile analysis*. The goal of this experiment was the determination of the local dislocation density as a function of orientation of single grains with respect to the direction of deformation.

A schematic sketch of the experimental setup is shown in Fig. 3. The monochromatic beam ($E = 16\text{keV}$) was focused into the center of rotation of a six circle diffractometer. A two dimensional gas filled detector was used to record the full shape of a single grain Bragg reflection at a time (Fig. 4(a)). The typical grain size in the sample was about $20\mu\text{m}$ to $50\mu\text{m}$, and a microbeam (gain= 150m $4.5 \cdot 10^9$ ph/s) with a size of $3 \times 12\mu\text{m}^2$ FWHM (V $\times$ H) was used. ($N = 40$, $f = 1.18\text{m}$, $L_1 = 45\text{m}$, $L_2 = 1.21\text{m}$). The angular resolution $\Delta k/k = 1.3 \cdot 10^{-4}$ of the incident beam is over ten times higher than required by the breadths of the Bragg reflections. The line profile of up to 7 different Bragg reflections ({110}, {200}, {211}, {220}, {310}, {222}, {400}) was measured for each individual grain (Fig. 4(a)). Preliminary results are shown in Fig. 4(b). To navigate different grains of the sample to the center of rotation the sample was placed on a xyz-stage inside the diffractometer. Note that the straight optical path was crucial to the alignment of

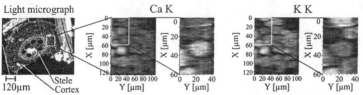

FIGURE 5. Maps of K and Ca in the root of a corn plant (resolution $1.2 \times 5\mu m^2$ (V × H)).

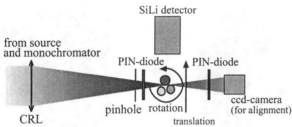

FIGURE 6. Fluorescence microtomography setup

the setup [13].

In this example of micro fluorescence mapping the distribution of K and Ca in a microtome (thickness $2\mu m$) of the root of a corn plant has been investigated (sample: W. Schroeder, FZ-Jülich). This distribution is of interest to biologists to understand the transport mechanisms in the plant as a function of changes in its genom. At $E = 15$keV a CRL ($N = 63$) was used to produce a microbeam with a lateral size of $1.1 \times 11.6\mu m^2$ FWHM (V × H) and a gain of 20 ($f = 0.67$m, $L_1 = 41.3$m, $L_2 = 0.68$m). The sample was scanned through the microbeam and the fluorescence signal was recorded with a SiLi-detector. Fig. 5 shows the measured distributions of K and Ca together with a light micrograph of the plant. Single cells are clearly resolved. The typical cell size is $25\mu m$.

Fluorescence microtomography allows to determine the elemental distribution inside thick samples with a minimum of sample preparation. A detailed description of the method is presented in contribution 26 to XRM99 [14]. The typical experimental setup is shown in Fig. 6. For different rotation angles of the sample tomographic projections are acquired by scanning the sample across the pencil beam and recording the fluorescence spectrum with the energy dispersive detector. Two beam monitors before and behind the sample record the incident and transmitted intensity, respectively.

Fig. 7(a) shows the sinogram of a test sample consisting of three glass capillaries filled with 1% solutions of different elements. At $E = 28$keV a CRL ($N = 79$) was used to illuminate a $5\mu m$ pinhole to produce a pencil beam of $5 \times 3\mu m^2$ FWHM (H × V) with a gain of 167 and $6.2 \cdot 10^9$ ph/s behind the pinhole ($f = 1.75$m, $L_1 = 40.8$m, $L_2 = 1.82$m). The horizontal scanning resolution was $6\mu m$ with a scanning time of 2s. The sample was rotated through 180° with an angular resolution of 2°. Fig. 7(b) shows the reconstruction by filtered back projection of

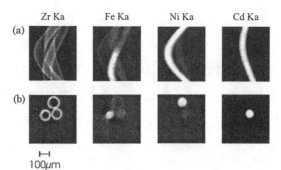

Zr Ka   Fe Ka   Ni Ka   Cd Ka

(a)

(b)

⊢⊣
100μm

FIGURE 7. (a) Fluorescence sinograms for Zr Kα, Fe Kα, Ni Kα, and Cd Kα. (b) Reconstruction of the elemental distributions in the sample using filtered back projection .

the elemental distribution in the sample [14].

We would like to thank H. Schlösser for his excellent work in manufacturing the pressing tools for the lenses, and J.-M. Rigal for his excellent technical support during the measurements made at beamline ID22 of the ESRF.

REFERENCES

1. Kirkpatrick, P., Baez, A., *J. Opt. Soc. Am.* **38**, 766 (1948).
2. Lienert, U., Schulze, C., Honkimaki, V., Tschentscher, Th., Garbe, S., Hignette, O., Horsewell, A., Lingham, M., Poulsen, H. F., Thomsen, N. B., Ziegler, E., *J. Synchrotron Rad.*, **5**, 226 (1998).
3. Underwood , J., Barbee, T., Jr., Frieber, C., *Appl. Opt.* **25**, 1730 (1986).
4. Bilderback, D., Hoffman, S., Thiel, D., *Science* **263**, 201 (1994).
5. Lai, B., et al., *Appl. Phys. Lett.* **61**, 1877 (1992).
6. Chevallier, P., Dhez, P., Legrand, F., Erko, A., Agafonov, Yu., Panchenko, L. A., Yakshin, A., *J. of Trace and Microprobe Techniques*, **14**, 517 (1996).
7. Snigirev, A., Kohn, V., Snigireva, I., Lengeler, B., *Nature* **384**, 49 (1996).
8. Elleaume, P., *Nucl. Instrum. Methods Phys. Res.* **A 412**, 483 (1998).
9. Lengeler, B., Tümmler, J., Snigirev, A., Snigireva, I., Raven, C., *J. Appl. Phys.* **84**, 5855 (1998).
10. Snigirev, A., Kohn, V., Snigireva, I., Souvorov, A., Lengeler, B., *Appl. Opt.* **37**, 653 (1998).
11. Lengeler, B., Schroer, C., Richwin, M., Tümmler, J., Drakopoulos, M., Snigirev, A., Snigireva, I., *Appl. Phys. Lett.* **74**(26), 3924 (1999).
12. Lengeler, B., Schroer, C., Tümmler, J., Benner, B., Richwin, M., Snigirev, A., Snigireva, I., Drakopoulos, M., *J. Synchro. Rad.* **6**(6), 1153 (1999).
13. Castelnau, O., Chauveau, T., Drakopoulos, M., Snigirev, A., Snigireva, I., Schroer, C., Ungar, T., *Proc. ECRS5 conf.*, Delft (Holland), 28-30. Sep. 1999, to be published
14. Simionovici, A., Chukalina, M., Drakopoulos, M., Snigireva, I., Snigirev, A., Schroer, Ch., Lengeler, B., Janssens, K., Adams, F., *Rev. Sci. Instrum.*, unpublished.

High Aspect Ratio Zone Plates: Numerical Simulation of Imaging Performance

A.V. Popov[1], Yu.V. Kopylov[1], A.N. Kurokhtin[1], A.V. Vinogradov[2]

[1]- *Institute of Terrestrial Magnetism, Ionosphere and Radiowave Propagation, 142092 Troitsk, Russia.*
[2]- *Lebedev Physical Institute, Leninsky Pt., 53, Moscow, Russia.*

Abstract. Semi-analytical approach based on the parabolic wave equation (PWE) has been developed for simulation of X-ray optical elements. Using it one can easily obtain spatial resolution, diffraction efficiency and field of view of realistic X-ray zone plates. Calculations reveal new unexpected features, absent for idealized thin zone plates.

Parabolic Wave Equation

Our goal was to develop mathematical methods and computer codes for numerical simulation of the high aspect ratio zone plates (ZP) used in X-ray microscopy. The approach is based on the direct numerical solution of the parabolic wave equation

$$2ik\frac{\partial u}{\partial z} + \Delta_\perp u + k^2(\varepsilon - 1)u = 0 \tag{1}$$

for slowly varying complex wave amplitude $u(x, y, z) = E(x, y, z)\exp(-ikz)$ [1,2]. This gives a consistent full-wave description of ZP considered as bulk dielectric structure with given optical constants (refractivity and absorption $\delta, \beta \ll 1$), realistic shape and dimensions. The applicability of PWE is granted by the smallness of numerical apertures and diffraction angles in typical X-ray transmission optical schemes - Fig.1a. To solve PWE, we use Crank-Nicolson finite difference scheme in a cylinder $r < A$ containing the optical element and surrounded with a fully transparent artificial boundary: $\frac{\partial u}{\partial r}(A, z) = -\sqrt{\frac{2k}{\pi i}} \int_0^z \frac{\partial u}{\partial \zeta}(A, \zeta) \frac{d\zeta}{\sqrt{z - \zeta}}$ [1] Amazing computational efficiency of the PWE enables one to calculate in detail very fine and complicated diffraction patterns produced by realistic X-ray Fresnel zone plates [2].

Simulation of Realistic X-ray Zone Plates

First of all, experimentalists are interested in focusing efficiency, spatial resolution, aberrations and field of view of their optical elements. Here, we consider a realistic model of soft X-ray zone plate actually used at the Center for X-Ray Optics, LBNL. It contains 300 zones formed of 300 nm thick nickel plate, with external zone radius 18 μm and is designed for focal length 0.45 mm at $\lambda = 2.4$ nm. As an example, we consider 1:1 imaging of an on-axis point source. Our calculations confirm a priori estimates for spatial resolution and the main focus diffraction efficiency. However, they reveal some new unexpected features, absent for idealized thin ZP, depending on geometry and material properties.

CP507, X-Ray Microscopy: Proceedings of the Sixth International Conference,
edited by W. Meyer-Ilse, T. Warwick, and D. Attwood
© 2000 American Institute of Physics 1-56396-926-2/00/$17.00

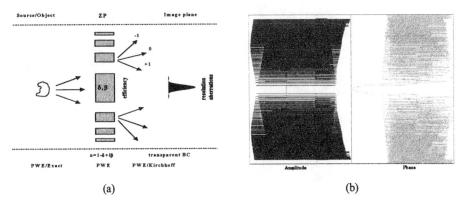

(a)   (b)

FIGURE 1. (a) Scheme of numerical simulation. (b) Output wave field of thick nickel ZP.

Diffraction efficiency

Fourier analysis of the calculated output wave field on the ZP back side (Fig. 1b) gives a reliable estimate of the energy flux directed towards different diffraction orders: about 22% to the main focus, 10% - undiffracted background, 1,5% and 3% accordingly to the second and third order foci, see Fig. 2a. Unlike an idealized thin Fresnel ZP, the second diffraction order is well pronounced.

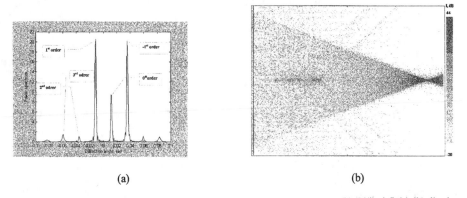

(a)   (b)

FIGURE 2. (a) Output field in segment 10μm<r<11μm power spectrum. (b) Global field distribution.

Spatial Resolution

Continuing finite-difference calculation up to the image plane we obtain the global wave field distribution (Fig. 2b), including fine structure of the focal spot. We see that the field in the image plane (Fig. 3a) is Airy pattern with the first dark ring radius about 75 nm, which agrees with the Rayleigh criterion $Y = 0.61\lambda/NA$. For accurate calculation of spatial resolution, the Kirchhoff integral representation starting from the back side of the ZP is preferable. As an example we present the amplitude distribution along the optical axis (Fig. 3b), demonstrating high intensity and longitudinal resolution in second and third order foci.

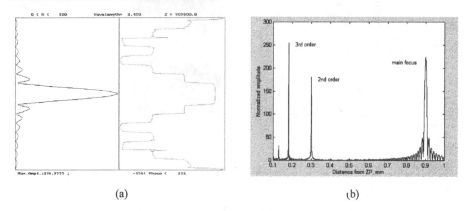

(a)   (b)

FIGURE 3. (a) Amplitude and phase in image plane. (b) Axial amplitude distribution.

Chromatic Aberration

As real synchrotron radiation is not chromatic, dependence of the focal distance on the wavelength $f = R_N^2/N\lambda$ affects the image quality. For incoherent power spectrum $A(\lambda)$, intensity in the focal plane can be estimated by the approximate formula:

$$I(r,z) = \int I_\lambda(r,z)A(\lambda)d\lambda \approx \int I_{\lambda_0}(r,z+\frac{\lambda}{\lambda_0}f_0)A(\lambda)d\lambda \qquad (2)$$

Off-axis Aberrations of Thick ZP

In order to estimate off-axis aberrations of a zone plate, the lens formula for wave front deviation $\Delta L = \frac{1}{8}\frac{\rho^4}{f^3}+\frac{1}{2}\frac{\rho^3}{f^2}\eta\cos\varphi+\frac{1}{2}\frac{\rho^2}{f}\eta^2\cos^2\varphi+...$ [3] is usually applied (here, ρ is the current radius in the lens plane, f - focal distance and η is off-axis angular displacement of the point source). It results into the following amplitude distribution in the image plane (spherical aberration is neglected):

$$U(v,\varphi) \approx C_0\frac{J_1(v)}{v}+C_1\frac{J_2(v)}{v}\cos\varphi+C_2\frac{J_3(v)}{v}\cos2\varphi, \; v = \frac{kR_N}{z}r \qquad (3)$$

being normalized radial coordinate centered at the paraxial image point. This is true for realistic X-ray zone plates. However, the coefficients C_1 and C_2 corresponding to coma and astigmatism depend on ZP thickness and material, and therefore must be calculated in frames of wave theory.

We start from 3D PWE (1), and look for a solution in the form of perturbation series in small parameter η:

$$u(x, y, z) = [w_0(r, z) + \eta w_1(r, \varphi, z) + \eta^2 w_2(r, \varphi, z) + ...] \exp(ik[\eta^2(l - z) - \eta x]), \qquad (4)$$

Substitution into (1) yields a chain of recurrent parabolic wave equations,

$$2ik\frac{\partial w_n}{\partial z} + \Delta_\perp w_n + k^2(\varepsilon(r) - 1)w_n = 2ik\left(\frac{\partial w_{n-1}}{\partial r}\cos\varphi - \frac{1}{r}\frac{\partial w_{n-1}}{\partial \varphi}\sin\varphi\right) \qquad (5)$$

to be solved successively. This is not difficult as the angular dependence of these terms could be easily obtained and interpreted: $w_1(r, \varphi, z) = u_1(r, z)\cos\varphi$ - coma, $w_2(r, \varphi, z) = u_0(r, z) + u_2(r, z)\cos 2\varphi$ - field curvature and astigmatism. 2D equations for radial factors $u_m(r, z)$ are similar to PWE (1) and can be solved numerically in the same way. In the image plane we get the field structure (3) with coefficients C_n being functionals of ZP geometry and material properties. If ZP thickness b is small, $C_n \sim b^n$. For the considered above ZP, Rayleigh criterion [3] for calculated C_n gives such limitation on the field of view due to coma (see Fig 4a): $\eta \le 0.054 \approx 1.3NA$.

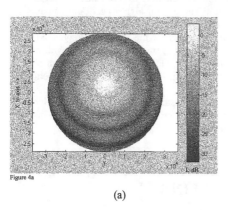

Figure 4a

(a)

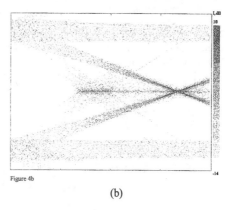

Figure 4b

(b)

FIGURE 4. (a) Off-axis PSF of thick soft X-ray ZP. Magnification 240, η=0.1 rad. Axes scaled in nm $\times 10^4$. (b) Plane wave focusing by Cu-Cr hard X-ray ZP, global field intensity.

Simulation of Hard X-ray Zone Plates

The developed methods are also helpful in the design of perspective ZP in hard X-ray spectral range. As an example we demonstrate global field distribution in the case of plane wave focusing by 6 μm thick Cu-Cr ZP at λ=0.23 nm (Fig. 4b). It contains 60 zone pairs, with external zone radius 72 μm, central stop 50μm and is designed for focal length 10 cm. Sharp main focus unaffected by background is clearly seen.

References

1. Kopylov, Yu.V., Popov, A.V., and Vinogradov, A.V., "Application of the Parabolic Wave Equation to X-ray Diffraction Optics", *Opt. Comm.* **118**, pp. 619-636 (1995).

2. Vinogradov, A.V., Kopylov, Yu.V., Popov, A.V., and Kurokhtin, A.N., "Numerical Simulation of X-ray Diffraction Optics", Moscow: A&B Publishing House, 1999.

3. Born, M., and Wolf, E., "Principles of Optics", Oxford: Pergamon, 1980.

Nanofabrication of Custom X-ray Optical Components

C. David[a], C. Musil[a], A. Souvorov[b], B. Kaulich[b]

a) Laboratory for Micro- and Nanotechnology, Paul Scherrer Institute, CH 5232 Villigen-PSI, Switzerland

b) European Synchrotron Radiation Facility, B. P. 220, F-38043 Grenoble Cedex, France

Abstract. A variety of nanostructuring processes have been developed at the Paul Scherrer Institute (PSI) in order to provide optical instrumentation for beamline experiments. We present methods for the fabrication of custom elements such as sub-micron high aspect ratio tantalum slits and pinholes, a calcium fluoride test object for element mapping experiments at the Ca-K-edge, and transmission zone plates with germanium structures for the multi-keV region. Furthermore, we report on the generation of linear silicon Bragg-Fresnel Lenses (BFLs) with unprecedented dimensions. Lenses with 100nm outermost zone width, lengths up to 10mm and widths up to 1mm were tested at the optics beam line of the ESRF. We measured a diffraction efficiency of 26% at a beam energy of 13.25keV.

The purpose of this contribution to the conference is to demonstrate the flexibility of the micro- and nanostructuring facilities at PSI and to encourage scientific groups at synchrotron beam lines to join future collaborations.

FABICATION OF PINHOLES AND SLITS USING FOCUSED ION BEAM (FIB) MILLING

A number of experiments at synchrotron beam lines require apertures and slits with dimensions that are not commercially available. We used focused 30keV Ga^+-ions to mill the desired structures through 5-10µm thick tantalum foils. The transmitted ion current was monitored during the milling to in order determine when the FIB has completely removed the foil material. The pattern generator of our FIB allows for the generation of complex patterns within a 200µm x 200µm deflection field. The lateral resolution is limited by two factors. Firstly, the minimum spot size of our machine was

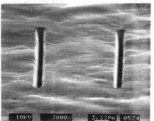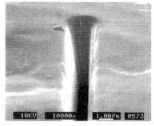

FIGURE 1. SEM images of a 0.5µm diameter pinhole (left picture) and of a 1µm wide double slit (middle and right pictures) with 50µm length and 20µm separation milled into a 5µm thick Ta foil.

CP507, *X-Ray Microscopy: Proceedings of the Sixth International Conference,*
edited by W. Meyer-Ilse, T. Warwick, and D. Attwood
© 2000 American Institute of Physics 1-56396-926-2/00/$17.00

on the order of 200nm. Secondly, we found that the redeposition of the sputtered material on the side walls of the structure limited maximum obtainable aspect ratio to about 10. In figure 1, two devices milled into 5µm thick tantalum foil are shown. The close up of a double slit which was designed for coherence measurements gives an impression of the achievable side wall angle. Layers thicker than 10µm or other materials are also possible to structure by FIB milling.

A TEST OBJECT FOR CALCIUM ELEMENT MAPPING

By taking x-ray transmission images just below and just above the Ca absorption edge at approx. 4keV photon energy, a quantitative determination of the calcium concentration in the object can be calculated (element mapping) [1]. For testing purposes of this analytic method at the ID 21 beamline at the ESRF, we generated test objects with known dimensions and Ca content. Figure 2 depicts a CaF_2 Siemens star on a capton foil generated by electron beam lithography and subsequent lift off. The outermost structures are 650nm wide, and the structure height is 200nm.

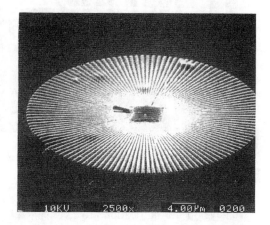

FIGURE 2. SEM image of a CaF_2. test object for Ca element mapping. The outermost structures are 650nm wide and 200nm high. Number of spokes:128.

SILICON BRAGG FRESNEL LENSES WITH 100nm OUTERMOST ZONE WIDTH

Bragg-Fresnel Lenses (BFLs) have been applied successfully in a number of experiments in order to collimate hard x-rays. They combine the Bragg reflection of a crystal or multilayer with the focusing properties of a Fresnel zone plate. The most common devices are linear zone plates etched into silicon <111> crystals [2]. For maximum diffraction efficiency, the zone structures should be etched 1.24µm deep independent of the photon energy [3]. Furthermore, it is desirable to have large lenses in order to collect a large fraction of the incident radiation and short focal lengths to demagnify the source as much as possible. The small desired outermost zone widths makes the fabrication of BFLs with sufficient efficiency very challenging due to the extreme aspect ratio of the zone structures.

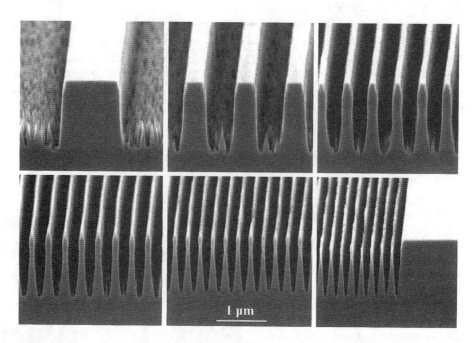

FIGURE 3. Scanning electron micrographs of the cross section through the Si structures of a BFL. The outermost zone width is 100nm (200nm period). The depth of the zone structures is 1.2μm for the central zones; it gradually decreases to 1.05μm for the outermost zones.

We developed a fabrication process using a LION-LV1 low voltage electron beam lithography system (Leica Microsystems Jena). Its unique continuous path control mode allows for pattern definition over large distances without the need for field stitching. The line width of each zone was controlled by a controlled defocusing of the electron beam [4]. The resist pattern was transferred into a 20nm thick chromium layer, which then served as a highly resistive mask for the patterning of the silicon substrate by reactive ion etching (RIE). Figure 3 illustrates the resulting structures of a linear BFL. The extreme roughness between the zone structures is caused by micro-masking of sputtered Cr particles. The optical performance of the lenses, however, should not affected by this phenomenon. We fabricated a variety of lenses with widths of up to 1mm and lengths of up to 10mm, which is larger than any other lens of this kind which has been previously reported.

The diffraction efficiency of the lenses for 13.25keV photon energy was measured at the optics beam line at ESRF. The fraction of the radiation within the rocking curve of the Si <111> Bragg reflex that is diffracted into the first diffraction order was measured to be 26±0.5% for the lens shown in figure 3 [5]. Considering the fact, that the outermost zone widths of our lenses is a factor of three smaller than any other silicon BFL reported so far, this is an excellent value. The combination of wide aperture angle, high efficiency and large size of our devices results in a significant gain in achievable x-ray flux. Furthermore, the higher numerical aperture of our lenses together with the excellent placement accuracy of the zone structures improves the diffraction limit of spatial resolution by a factor of three.

GERMANIUM ZONE PLATES FOR HARD X-RAYS

In order to monitor the source size at the ESRF diagnostics beam line, it is planned to image the source at 25keV photon energy using Bragg-Fresnel lenses, compound refractive lenses and transmission phase zone plates. We fabricated the latter by electron beam lithography and reactive ion etching. In order to provide sufficient diffraction efficiency, we chose germanium as the phase shifting material. For 5μm high Ge structures, we calculated a maximum achievable efficiency of 9%. Figure 4 shows images of a zone plate with a diameter of 500μm and an outermost zone width of 480nm. The pattern was defined by electron beam lithography and reactive ion etching. Significant deposition of polymers between the outermost structures by the reactive etching process can be observed; however, these should not affect the optical properties of the device at high photon energy. The substrate is a 380μm thick silicon wafer which has a transmission of approximately 80% at 25keV photon energy. First experiments with these optics are scheduled for the end of 1999.

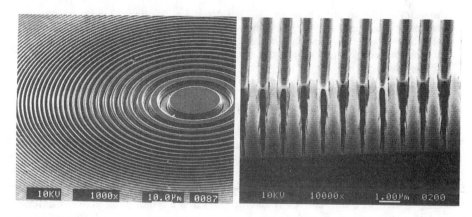

FIGURE 4. Scanning electron micrographs of a Ge phase zone plate for beam monitoring at 25keV photon energy. The diameter is 500μm, and the outermost zones (right) are 480nm wide and 5μm high. The focal length at 25keV photon energy was designed to be 4.8m.

REFERENCES

1. Kaulich, B., Niemann, B., Rostaing, S., Oestereich, S., Salome, M., Barrett, R., and Susini, J., this volume.

2. Aristov, V.V., Basov, Y.A., Snigirev, A.: Rev. Sci. Instr. .**60**, no.7, 1989, p.1517-18.

3. Babin, S.V.,. Erko, A.I.: Nuc. Instrum. Methods Phys. Res. A, **282**, no.2-3, 1989, p.529-31

4. David, C., Hambach, D.: Microelectronic Engineering **46** no.1-4, 1999, pp. 219-222

5. David, C., Souvorov, A.: Rev. Sci. Instr (accepted for publication)

Performance of Hard X-ray Zone Plates at the Advanced Photon Source

J. Maser[a], B. Lai[a], Z. Cai[a], W. Rodrigues[a], D. Legnini[a], P. Ilinski[a], W. Yun[b], Z. Chen[c], A.A.Krasnoperova[c], Y. Vladimirsky[c], F. Cerrina[c], E. Di Fabrizio[d], M. Gentili[d]

[a]Advanced Photon Source, Argonne National Laboratory, Argonne, IL 60439
[b]Advanced Light Source, Lawrence Berkeley National Laboratory, Berkeley, CA 94720
[c]Center for X-ray Lithography, University of Wisconsin, Madison, WI 53706
[d]Istituto di Elettronica dello Stato Solido, Consiglio Nazionale delle Ricerche (CNR),
Italy-00156

Abstract. Fresnel zone plates have been highly successful as focusing and imaging optics for soft x-ray microscopes and microprobes. More recently, with the advent of third-generation high-energy storage rings, zone plates for the hard x-ray regime have been put to use as well. The performance of zone plates manufactured using a combination of electron-beam lithography and x-ray lithography is described.

INTRODUCTION

Fresnel zone plates have proven to be superior optical systems for x-ray microscopy in the soft x-ray region [1,2]. With the advent of third-generation hard x-ray synchrotrons, microfocusing experiments at high spatial resolution have become feasible in the intermediate to hard x-ray region as well. We use Fresnel zone plates as focusing elements for our hard x-ray microprobes at the sector 2 undulator beamline at the Advanced Photon Source for x-ray energies between 6 keV and 30 keV [3] and will report on performance measurements.

With increasing photon energy, the interaction of x-rays with matter decreases, and the material thickness required to obtain the phase shift and/or absorption necessary to obtain significant diffraction efficiency increases even if heavier elements are chosen. The typical thickness t ranges from hundreds of nanometers in the soft x-ray range to micrometers in the multi-keV range to tens of micrometers for harder energies (Fig. 1). On the other hand, the numerical aperture, and thereby the resolution limit of the zone plate, are determined by the outermost zone width dr_n. The aspect ratio t/dr_n therefore increases significantly with increasing energy, making the manufacture of zone plates with a small outermost zone width more difficult. With most manufacturing techniques, high resolution can therefore be achieved only at the cost of reducing the thickness of the zone plate, thereby limiting the usable energy range and/or reducing the diffraction efficiency.

CP507, X-Ray Microscopy: Proceedings of the Sixth International Conference,
edited by W. Meyer-Ilse, T. Warwick, and D. Attwood
© 2000 American Institute of Physics 1-56396-926-2/00/$17.00

We show data obtained for zone plates with an outermost zone width of 0.1 μm at energies around 8 keV, from stacked zone plates with an outermost zone width of 0.25 μm at energies of 12 and 40 keV, and of blazed zone plates with an outermost period of 0.5 μm.

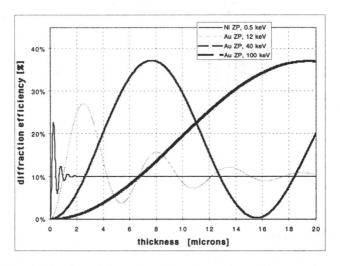

FIGURE 1 Calculated diffraction efficiency (from [4]) for zone plates with rectangular grating profile at x-ray energies between 0.5 keV and 100 keV.

RESOLUTION OF MULTI-KEV ZONE PLATES

Fig. 2 shows resolution tests performed at a photon energy of 8 keV. We used a gold zone plate with an outermost zone width of 0.1 μm and a diameter of 145 μm, corresponding to a focal length of 10 cm at 8 keV. The resolution was determined by scanning a knife edge with a 300Å thick layer of Cr through the focal spot and collecting Cr K_α fluorescence radiation with an energy dispersive detector (Fig. 2). We performed vertical scans, thus taking advantage of the small vertical particle beam source size (approx. 52 μm at 1% coupling). The zone plate was located 72 m from the source, yielding a geometrical demagnification of 1:720 in the first diffraction order. Convolution of the demagnified image of the source of 72 nm FWHM with the diffraction-limited spot size of 122 nm yields an expected spot size of 142 nm in the vertical direction. From the knife edge scan, we obtain a obtain a FWHM spot size of 150 nm, well in agreement with our expectation.

To obtain a smaller spot size, we also investigated focusing the third diffraction order of the zone plate. As previously, we placed a Cr knife edge into the third-order focal plane and scanned it through the focal spot. In the third diffraction order, the numerical aperture is three times larger than in the first order, and a spot size of 47 nm would be expected. We measured a spot size of just below 90 nm. We attribute the difference mostly to mechanical vibrations, coupled with long dwell times at reduced third order flux, and a smaller signal/noise ratio due to the absence of a central stop.

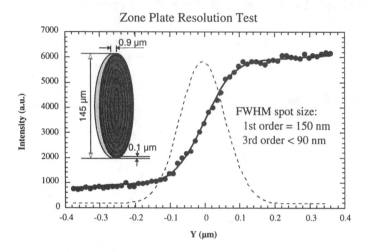

FIGURE 2 Measured resolution of a zone plate with an outermost zone width of 0.1 μm. Measurements were performed for the first as well as for the third diffraction order of the zone plate. A Cr knife edge was scanned vertically through the focal spot.

BLAZED ZONE PLATES

One of the factors that determines the diffraction efficiency is the zone profile. A pure phase zone plate with rectangular profile can achieve a diffraction efficiency of 41%. By changing the zone profile, the distribution of energy into the different diffraction orders is changed. By approaching an idealized, parabolic profile, diffraction efficiencies approaching 100% can be achieved.

A good approach to the ideal zone profile is a step profile. A zone plate with three electroplated gold levels, representing a 4-step profile, a thickness of 2.5 μm and an outermost period of 0.5 μm was manufactured using x-ray lithography (Fig. 3) [5,6]. We measured a diffraction efficiency of 44% at a photon energy of 8 keV in the first diffraction order. The flux density in the zero order is reduced, yielding a flux density ratio between focus and background of > 1000.

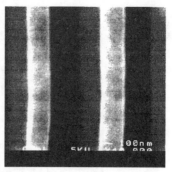

FIGURE 3 Gold zone plate with stepped profile consisting of 4 steps. The left image shows the center, the right image the outermost zones. The outermost period is 0.5 μm.

NEAR-FIELD STACKING OF ZONE PLATES

A relatively simple way of multiplying the zone plate thickness to increase the efficiency and use zone plates at higher energies is by stacking several identical zone plates on one axis. To achieve constructive interference, all zone plates have to be aligned laterally to better than a zone width (typically 1/3 dr_n), and longitudinally such that they stay within near-field diffraction, within a proximity p of each other:

$$p < dr_n^2/\lambda \qquad \text{(i.e. } p \ll \text{depth of focus).}$$

For example, for zone plates with an outermost zone width of 200 nm, the required minimum proximity p for the stack of zone plates is 1.3 mm at 40 keV and 3.3 mm at 100 keV. Accordingly, a stack of three zone plates, aligned to within 0.9 mm, and with an individual thickness of 1.5 μm would achieve a diffraction efficiency of 25% at 40 keV. A stack of six such zone plates would achieve a focusing efficiency of 15% at a photon energy of 100 keV and could have a focal length between 50 cm and 100 cm.

We have used stacked zone plates at different energies. Fig. 4 shows alignment of two stacked zone plates with an thickness of 3 μm and an outermost zone width of 250 μm at a photon energy of 40 keV [3]. We measured a diffraction efficiency of 25%. The flux density of the aligned stack is approximately 60% larger than the flux density achieved with a the single zone plate, demonstrating the effect of multiplying the thickness. Alignment of the two zone plates is achieved by maximizing the fringe spacing of the Moire patterns resulting from interference between the two zone plates. The Moire patterns are observed in transmission on a scintillator screen. Alignment is usually accomplished in few minutes.

Fig. 5 shows stacking of two zone plates with a thickness of 1.5 μm and an outermost zone width of 0.25 μm at an energy of 12 keV. An improved stage that allows stacking of up to three zone plates was used, and an increase of 47% in the focused flux is observed. The aligned stack was stable over a 12 hour testing period and has allowed us to use stacked zone plates as standard optical elements in microfocusing experiments. We are currently commissioning a stage that allows alignment of two zone plates with small outermost zone width in close proximity.

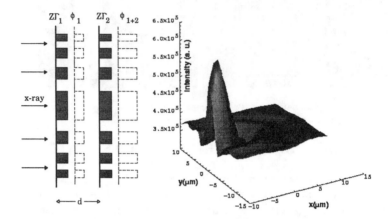

Figure 4. Near-field stacking of two gold zone plates with thickness 3 μm at a photon energy of 40 keV. Measurement of the interference pattern is obtained by scanning one zone plate with respect to the other one. An increase of focused flux by 60% is achieved when both zone plates are aligned (cf. fig. 1).

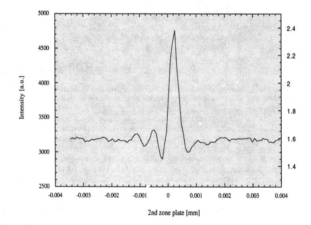

Figure 5. Near-field stacking of two gold zone plates with a thickness of 1.5 μm at a photon energy of 12 keV. A 47% increase of efficiency is obtained when the two zone plates are aligned (cf. fig. 1).

CONCLUSIONS

We have discussed enhancements of our capabilities in using zone plates in the multi-keV and hard x-ray range.

1) We use zone plates with focal lengths between 5 cm and 3 m (at 8 keV) and outermost zone widths of 100 nm – 250 nm for a wide range of applications. We obtain a focused flux of $5*10^7 - 10^{10}$ hv/s, and achieve a flux density gain of $10^4 - 10^5$.

2) For standard operation between 4 keV and 12 keV, zone plates with an outermost zone width of 100 nm and a focal length between 5 cm (4 keV) and 15 cm (12 keV) are used. A resolution of 150 nm in first order is routinely achieved. A resolution of < 90 nm in third order at 8 keV has been measured

3) We measured a diffraction efficiency of 44% for blazed zone plates with an outermost period of 0.5 µm and demonstrated a reduction of background from the zero diffraction order.

4) Near-field stacking of two zone plates has been demonstrated to increase the diffraction efficiency by 50% - 60% for energies between 12 keV - 40 keV.

ACKNOWLEDGEMENTS

This work is supported by the U. S. Department of Energy, Office of Basic Energy Sciences, under contract W-31-109-ENG-38.

REFERENCES

1. See papers in *"X-ray Microscopy and Spectromicroscopy"*, edited by Thieme, J., Schmahl, G., Rudolph, D. E. Umbach, E., Springer, Berlin, 1998.

2. Kirz, J., Jacbsen, C., Howells, M., *Quart. Rev. Biophys.* **28**, 33-130 (1995).

3. Lai, B., Yun, W.,. Maser, J.,. Cai, Z., Rodrigues, W., Legnini, D., Chen, Z.,. Krasnoperova, A., Vladimirsky, Y., Cerrina, F., Di Fabrizio, E., Gentili, M, *SPIE Proc.* **3449**, 133 (1998).

4. Kirz, J., *J. Opt. Soc. Am.* **64**, 301 - 309 (1974).

5. DiFabrizio, E., Gentili, M., Grella, L., Baciocchi, M., Krasnoperova, A., Cerrina, F., Yun, W., Lai, B., Gluskin, E., *J. Vac. Sci. Technol.* **B 12(6)**, 3979-95 (1994).

6. DiFabrizio, E., this volume.

Soft X-ray Reflectivity and Structure Evaluation of Ni/Ti and Ni-N/Ti-N Multilayers

Hisataka Takenaka, Yasuji Muramatsu*, Shigeki Hayashi**, Hisashi Ito,
Yuko Ueno*, Naoji Moriya**, Eric M. Gullikson***, and Rupert C. C.
Perera**

NTT Advanced Technology Co., Musashino, Tokyo 180-8585, Japan
*NTT Lifestyle and Environmental Technology Laboratories, Musashino, Tokyo 180-8585, Japan
**Shimadzu Co., Hadano, Kanagawa 239-1304, Japan
*** Lawrence Berkeley National Laboratory (LBNL), Berkeley, CA 94720, USA

Abstract. We fabricated Ni/Ti and Ni-N/Ti-N multilayers structures containing N mirrors use as a grazing-incident angle reflector to focus x-ray micro-beams. The multilayer structures were fabricated by magnetron sputtering. The layer structures were evaluated by using an x-ray diffractometer and the soft x-ray reflectivities were measured at Beamline 6.3.2 at the Advanced Light Source (ALS). Although the Ni/Ti multilayer has a high interface roughness of about 1.36 nm, the soft-x-ray peak reflectivity of this mirror showed fairly high reflectivity of 39%, at just above the Ti-absorption edge with a 9-degree incident angle. The peak reflectivity of Ni-N/Ti-N multilayer mirrors was 36% in almost the same conditions as the reflectivity measurement of the Ni/Ti multilayer.

INTRODUCTION

The development of highly-reflective multilayer mirrors for use in the water-window region has been desired for x-ray microscopy and x-ray photoemission spectroscopy, for example. For these applications, reflectivity is one of the most critical parameters determining the performances of multilayer mirrors. Ni/Ti-based multilayers are appropriate candidates for such mirrors because the combination of Ni and Ti theoretically has high reflectivity at just above the Ti absorption edge (around 2.8 nm) because of the optical constants of Ni and Ti. The reflectivity of multilayer mirrors is also related to their structures. The interface roughness, intermixed widths, and the thickness ratio between each layer are important facts affecting the performances of multilayer mirrors.

We have designed Ni/Ti and Ni-N/Ti-N multilayer mirrors to make an x-ray micro-beam with high reflectivity, We used grazing-incident type optics, such as ellipsoidal mirrors, and have fabricated the mirrors using a sputtering technique. We evaluated the structures and reflectivity of these multilayers using a soft x-ray reflectometer.

CP507, *X-Ray Microscopy: Proceedings of the Sixth International Conference,*
edited by W. Meyer-Ilse, T. Warwick, and D. Attwood
© 2000 American Institute of Physics 1-56396-926-2/00/$17.00

DESIGN AND FABRICATION OF NI/TI-BASED MULTILAYERS

We calculated the soft x-ray reflectivities of many kinds of multilayers (Ni/Ti, Ni/Sc, W/Ti etc), assuming that the multilayers had an ideal structure. The calculations were performed using the Fresnel equation and Henke's optical data. Each of these multilayers has 20 layer pairs, at a wavelength of around 2.76 nm, with the peak reflectivity at a 9 degree grazing incidence. The periodic length was around 9.2 nm and the thickness ratio between the high-Z layer and the low-Z layer was almost 0.35 : 0.65. The calculated peak reflectivities of the high-Z /Ti multilayers have high reflectivities at 2.76 nm. For example, the peak reflectivity of a Ni/Ti multilayer is a 78%, and that of a Co/Ti multilayer is also almost the same value. Figure 1 shows the reflectivity profiles for five multilayer material combinations as examples.

A d.c. magnetron sputtering deposition system was used to fabricate the Ni/Ti-based multilayers. It basically consists of three fixed-source targets, a rotating substrate table, a substrate rotation-speed control system, and a shutter-opening and -closing controller synchronized with the substrate rotation. The Ni/Ti multilayer was deposited on Si wafers under an Ar gas atmosphere. In the Ni/Ti multilayer, the periodic length of this multilayer was about 9.2 nm, and the thickness ratio between the Ni layer and the Ti layer was about 0.35 : 0.65. The layer pairs were 20. In addition to this multilayer, we had also fabricated an Ni-N/Ti-N multilayer under in an Ar +N2 (80% / 20%) gas atomosphere.

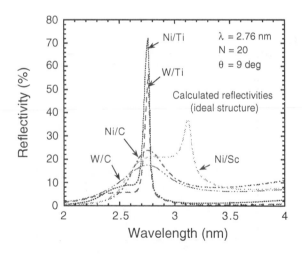

FIGURE 1. Calculated reflectivies of multilayers at wavelengths around 2.76 nm at grazing incidence.

SOFT X-RAY REFLECTIVITIES

Reflectivity for soft x-rays was measured at Beamline 6.3.2 [1] at the Advanced Light Source at LBNL. In this beamline the reflected beam from the multilayer sample is measured as the current from a Si photodiode. The absolute reflectivities are obtained by dividing the reflected beam intensity by the full beam intensity.

Figure 2 shows the measured and ideal reflectivities of the fabricated Ni/Ti multilayer. The measured peak reflectivity is 39% at a 2.74 nm wavelength at the incident angle of around 9°. This peak reflectivity is sufficient for our grazing incidence x-ray optics.

We estimated the periodic length, the layer thickness ratio, the interface roughness, the intermixed layer thickness, and the density of the fabricated multilayer by using this measured reflectivity. The fitting curve of the measured reflectivity is also shown in Fig. 2. The curve was calculated using the following parameters: a periodic length of 9.21 nm, a layer thickness ratio of 0.43 : 0.57, an interface roughness of 1.36 nm, an intermixed width of 1 nm, and a density of 0.9 times the bulk of Ni and Ti. In this calculation, we used NiTi as the intermixed layer, because x-ray diffractometer measurements of the annealed (at 500°C, for 1 hour, in an Ar atmosphere) Ni/Ti multilayer showed the existence of a NiTi phase. There is large deviation in the reflectivity between the measured and fitting reflectivity under 2.74 nm. We assume that this is because of the Ti- and Ni-oxide layers on the surface of this multilayer.

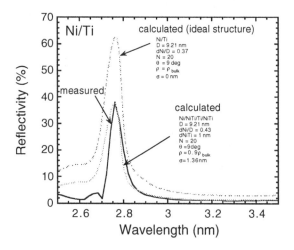

FIGURE 2. The measured and ideal reflectivities of the fabricated Ni/Ti multilayer, and the fitting reflectivity.

Figure 3 shows the measured and ideal reflectivities of the fabricated Ni-N/Ti-N multilayer. The measured peak reflectivity is 36% at a 2.74-nm wavelength and an incident angle of around 9°. This peak reflectivity is slightly smaller than that of the Ni/Ti multilayer. This slightly smaller peak reflectivity may explain the difference in the optical constants between the Ni and Ni-N layers, and of that between the Ti and Ti-N layers.

We estimated the periodic length, the layer thickness ratio, the interface roughness, and the density of the fabricated multilayer by using this measured reflectivity. The fitting curve of the measured reflectivity is also shown in Fig. 3. The curve was calculated using the following parameters: a periodic length of 8.84 nm, and a layer thickness ratio of 0.36 : 0.66, an interface roughness of 0.76 nm, and a density of 0.9 times the bulk of Ni and Ti. We did not consider intermixed layer in this calculation, because x-ray diffraction of the annealed (at 500°C, for 1 hour, in an Ar atmosphere) Ni-N/Ti-N multilayer only revealed Ni and Ti phases. The interface of this multilayer is smoother than that of the Ni/Ti multilayer.

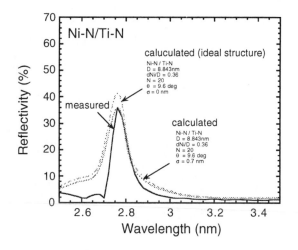

Figure 3. The measured and ideal reflectivities of the fabricated Ni-N/Ti-N multilayer, and fitting reflectivity.

This work was supported by the "Advanced Photon Processing and Measurement Technologies" program, which was assighed to the R&D Institute for Photonics Engineering by the New Energy and Industrial Technology Development Organization (NEDO) of Japan.

REFERENCES

1. J. H. Underwood, E. M. Gullikson, M. Koike et al.,Review of Scientific Instruments 67 (9) 1996

COMPACT SOURCES

Compact water-window x-ray microscopy with a droplet laser-plasma source

H. M. Hertz, M. Berglund, G. A. Johansson, M. Peuker*, T. Wilhein*, and H. Brismar[+]

Biomedical and X-Ray Physics, Royal Institute of Technology, SE-10044 Stockholm, Sweden
*Forschungseinrichtung Röntgenphysik, Georg-August Universität Göttingen, Göttingen, Germany
[+]Dept. of Woman and Child Health, Karolinska Institutet, SE-171 76 Stockholm, Sweden

Abstract. We summarize the development of the Stockholm compact water-window x-ray microscope (CXM-1) and show the first image of a biological object. The microscope is based on a liquid-droplet laser-plasma source, which is combined with normal-incidence water-window condenser optics for the object illumination. High-resolution imaging is performed with zone-plate optics and CCD detection. We demonstrate sub-100-nm resolution imaging with good signal-to-noise ratio and exposure times of minutes for dry diatoms and fixed cells.

INTRODUCTION

X-ray microscopy utilize the natural contrast between water and, e.g., proteins, in the water-window region (λ = 2.4 - 4.4 nm). Such microscopes allow high-resolution imaging of thick unstained biological objects in an aqueous environment.[1,2] Current operational transmission water-window x-ray microscopes rely on high-brightness synchrotron radiation sources in order to compensate for the low-efficiency of the diffractive x-ray optical elements (zone plates). New table-top x-ray sources (primarily laser plasmas) have paved the way for more compact microscopes. By combining such sources with new x-ray optical elements, sufficient object-plane photon densities can be achieved to allow compact x-ray microscopy with reasonable exposure times. The wider spread of such compact microscopes would significantly increase the accessibility and applicability of this imaging technology.

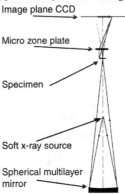

Image plane CCD

Micro zone plate

Specimen

Soft x-ray source

Spherical multilayer mirror

FIGURE 1. Experimental arrangement for the Stockholm compact x-ray microscope.

CP507, X-Ray Microscopy: Proceedings of the Sixth International Conference,
edited by W. Meyer-Ilse, T. Warwick, and D. Attwood

In the present paper we summarize the building blocks of the Stockholm compact x-ray microscope. This is the first compact x-ray microscope demonstrating sub-optical resolution with good signal-to-noise ratio.[3] An outline of the microscope is depicted in Fig. 1. The microscope consists of a negligible-debris droplet-target laser-plasma source, a normal-incidence multilayer condenser, a high-resolution zone plate and a back-illuminated CCD detector. These elements and their integration into a working instrument are briefly described in the following paragraphs. Finally, the first images of biological samples with the compact x-ray microscope are presented.

LASER-PLASMA SOURCE

The laser-produced plasma offers a compact and relatively inexpensive high-brightness x-ray source. The major drawback has been emission of debris, i.e., small fragments and particles, which may destroy sensitive components positioned close to the plasma. We have developed a new type of laser-plasma source based on a microscopic liquid-jet or liquid-droplet target. Compared to conventional solid-target laser plasmas this arrangement has several advantages. By practically eliminating the debris problem, sensitive components such as condenser optics and filters can be positioned close to the source without risking contamination or damages. Another important advantage is that the liquid target is regenerative. Thereby, uninterrupted full-day operation at high laser repetition rates, i.e., for high average x-ray flux, may be performed.

The liquid-target laser plasma is thoroughly described elsewhere.[4,5] Below the arrangement in the microscopy experiment is briefly described. The beam from a 3 ns, 100 mJ SHG Nd:YAG laser is focused to a small spot (~12 μm diameter) on the target liquid. The target is generated by forcing ethanol through a 10 μm glass capillary nozzle. Drops are formed by minimisation of surface energy in the continuous microscopic liquid jet. By piezoelectrically vibrating the nozzle and synchronising the laser with this piezoelectrical vibration, it possible to hit a single droplet with each laser pulse. The system emits proper line-emission in the water window with sufficient narrow bandwidth (typically $\lambda/\Delta\lambda > 500$)[6] to allow high-resolution zone plate imaging. The flux from the plasma is approx. $1 \cdot 10^{12}$ photons/(sr×line×pulse).

In the present work we employ the carbon line at $\lambda = 3.37$ nm (C VI, 1s-2p). This line was chosen due to the present difficulty to fabricate a multilayer condenser mirror for shorter wavelengths. However, imaging of thick samples would benefit from a shorter wavelength due to lower absorption. For this purpose we have developed sources based on nitrogen emission from ammonium hydroxide[7] and liquid nitrogen,[8] which operates at lower wavelength of 2.48 and 2.88 nm. With progress in fabrication of small-d-spacing multilayers, these sources will become useful.

THE OPTICS

There are two major optical elements in the compact x-ray microscope: the condenser mirror and the high-resolution zone plate. These will be discussed below.

The purpose of the condenser arrangement is to illuminate the sample with as high a photon density as possible. Furthermore, the condenser should monochromatize the radiation to allow for aberration-free high-resolution zone plate imaging. We employ a normal-incident spherical multilayer mirror.[9]

The technology for fabricating large-diameter multilayer mirrors with the short d-spacing and low roughness necessary for substantial normal-incidence reflectivity in the water-window is now emerging.[10,11,12] Compared to, e.g., grazing-incidence condensers, spherical normal-incidence condensers produce much smaller aberrations, resulting in a significantly relaxed alignment accuracy. This is especially significant when the small high-brightness droplet-target source is employed. For microscopy, an additional advantage of multilayer-coated condensers is that such mirrors provide wavelength selectivity, which is important since chromatic aberration in the microscope zone plate objective is avoided. The droplet-target source provides several emission lines with the necessary narrow line width for zone plate imaging. With the multilayer condenser mirror one such line is selected.

The spherical condenser mirror was coated with a multilayer coating of W/B_4C, where the layer period was selected for maximum normal-incidence reflectivity at $\lambda = 3.37$ nm (C VI, 1s-2p). The combined source/condenser arrangement is depicted in the lower part of Fig. 1. The mirror parameters are chosen so that the source is magnified 1.8× in the object plane and the numerical aperture matches a zone plate objective having an outer zone width of ~30 nm.

Figure 2 shows the spectrum emitted from the plasma as well as the reflected spectrum from the mirror. It is clear that the mirror works as an efficient monochromator for the λ=3.37 nm line. The reflectivity varies across the mirror surface but is on the average ~0.5%. Unfortunately, the mirror exhibit some flare, resulting in that only ~15% of the reflected radiation is focussed into the central 35 μm spot in the object plane that is used for the imaging.

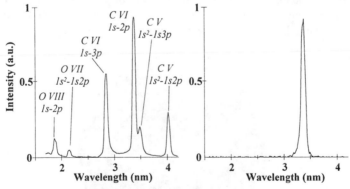

FIGURE 2. Water-window spectra from direct plasma emission (left) and the reflected spectrum from the condenser mirror showing a single line at λ=3.37 nm (right).

High-resolution imaging is performed with a 56.1 μm diameter nickel phase zone with an outermost zone width of 30 nm. The focal length is 498 μm at λ=3.37 nm.[13,14]

The first-order efficiency at $\lambda=3.37$ nm is estimated from BESSY measurements to 7.3 %.

COMPACT X-RAY MICROSCOPY

The source and optics described above are combined with a 1024×1024 pixel back-illuminated CCD camera to a compact microscope as outlined in Fig. 1. The sample is inserted on a 50 nm thick Si_3N_4 foil. A wet cell has not yet been implemented. The system is thermally stable to allow 5 minutes exposures. Below a few typical images are shown.

Figure 3 shows test imaging of a dry diatom. The magnification is 650× and the exposure time 2 min. Structures below 100 nm are visible. Similar images of zone-plate test objects show structures below 60 nm.[3]

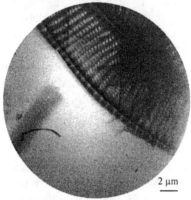

2 µm

FIGURE 3. X-ray image of diatom.

Figure 4 shows the first image of a cell with the compact x-ray microscope. A COS-7 cell was incubated on a Si_3N_4 membrane for 2 hours before it was fixed in gradually increasing ethanol concentration. The image shows considerably more detail than is visible in a light microscope. The fibrous structure is interpreted as actin filaments.

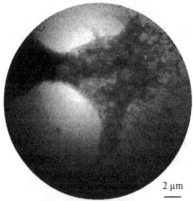

2 µm

FIGURE 4. X-ray image of fixed COS-7 cell.

CONCLUSIONS

We have developed the first water-window compact x-ray microscope, which allows sub-optical imaging with good signal-to-noise ratio. We have also demonstrated its applicability to the imaging of dry fixed cells. Our work will now focus on improving the exposure time, mechanical stability and wet-imaging possibilities of the microscope.

ACKNOWLEDGEMENTS

We thank G. Schmahl, D. Rudolph, J. Thieme and B. Niemann for their interest and engagement in the project.

REFERENCES

1. J. Kirz, C. Jacobsen, and M. Howells, Q. Rev. Biophys. **28**, 33 (1995); G. Schmahl, D. Rudolph, B. Niemann, P. Guttmann, J. Thieme, G. Schneider, C. David, M. Diehl, and T. Wilhein, Optik **93**, 95 (1993).

2. See several papers in "X-ray Microscopy and Spectromicroscopy", edited by J. Thieme, G. Schmahl, D. Rudolph, and E. Umbach (Springer, Heidelberg, 1998).

3. M. Berglund, L. Rymell, M. Peuker, T. Wilhein, and H. M. Hertz, subm. to J. Microscopy.

4. L. Rymell and H. M. Hertz, Opt. Commun. **103**, 105 (1993).

5. H. M. Hertz, L. Rymell, M. Berglund, and L. Malmqvist, in "X-ray Microscopy and Spectromicroscopy", Eds. J. Thieme, G. Schmahl, E. Umbach, and D. Rudolph (Springer, Heidelberg 1997), V-3.

6. T.Wilhein, D. Hambach, B. Niemann, M. Berglund, L. Rymell, and H. M. Hertz, Appl. Phys. Lett. **71**, 190 (1997).

7. L. Rymell, M. Berglund, and H. M. Hertz, Appl. Phys. Lett. 66, 2625 (1995).

8. M. Berglund, L. Rymell, T. Wilhein, and H.M. Hertz, Rev. Sci. Instrum. **69**, 2361 (1998).

9. H. M. Hertz, L. Rymell, M. Berglund, G. A. Johansson, T. Wilhein, Y. Platonov, and D. Broadway, Proc. SPIE **3766**, 247 (1999).

10. T. W. Barbee, Opt. Eng. **25**, 898 (1986).

11. G. Gutman, J. X-Ray Sci. Technol. **4**, 142 (1994).

12. N. N. Salashchenko, Yu. Ya. Platonov, and S. Yu. Zuev, Nucl. Instrum. Methods **A 359**, 114 (1995).

13. M. Peuker, G. Schneider, and D. Weiss, Proc. SPIE **3449**, 118 (1998).

14. G. Schneider, T. Schliebe, and H. Aschoff, J. Vac. Sci. Technol. **B13**, 2809 (1995).

Development And Characterization Of A Soft X-Ray Source Using Room-Temperature And Cryogenic Liquid Jets As Low Debris Target

M. Wieland, U. Vogt, M. Faubel[a], T. Wilhein, D. Rudolph and G. Schmahl

University Georgia Augusta at Göttingen, Institute for X-Ray Physics, Geiststraße 11, 37073 Göttingen, Germany
[a]*Max-Planck-Institute for Fluid Dynamics, Bunsenstraße 10, 37073 Göttingen, Germany*

Abstract. We describe the development and the characterization of a laser plasma soft X-ray source, especially for the "water-window" region. Liquid jets are used as low debris targets. To produce the jets two different nozzle-systems exist, one for room-temperature liquids and one for cryogenic liquids, respectively. The source is characterized using a slit-grating spectrograph to measure absolute photon numbers and a zone plate to determine its size. The highest measured photon numbers in single spectral lines surpass 10^{12} Photons sr^{-1} pulse^{-1} for both target systems. The source sizes were found to be about 20 µm in diameter.

INTRODUCTION

Table-top laser-plasma soft X-ray sources become more and more important for applications using this spectral range due to the relatively simple experimental setup. The recent development in this field was successful and had come out with brilliant, nearly debris-free soft X-ray sources that can be operated over long periods to serve an experiment. Liquid jets as targets have proven to be appropriate due to debris-reduction and source-stability [1]. The experiments described below show the setup of a variable target chamber that can be equipped with two nozzle systems to offer a wide range of accessible target liquids that determine the spectral line emission from the plasma. Conventional liquids at room-temperature as, e.g. ethanol, and cryogenic liquids, i.e. liquefied gases, especially liquid nitrogen, were investigated as target.

EXPERIMENTAL SETUP

The target chamber out of stainless steel is evacuated by a turbo-drag pump (880 l/s) and a rotary-vane pump (16m^3/h). Besides the nozzle systems it contains a stage for a plano-convex lens (focal length 60 mm) to focus the laser onto the liquid jets, a photodiode covered with a thin metal filter to observe the X-ray emission from the plasma and a newly developed jet-monitoring system. In particular, this monitoring systems consists of two perpendicular mounted CCD-cameras on which a magnified image of the jet or the plasma is generated with achromatic lenses. The images are then displayed on monitors. The resolution of the systems is about 5μm.

A frequency doubled Nd:YAG laser with a pulse length of t=3 ns, is used to produce the plasma. The maximum average power is 23 W, i.e. 230 mJ pulse energy at 100 Hz repetition rate. A plano-convex lens produces a focal spot of 10 μm in diameter, thus the intensity in the focus reaches 10^{14} W cm^{-2}. The lens is mounted on a stage which can be moved from outside.

The nozzle system for cryogenic jets consists of a cryostat filled with liquid nitrogen and a stainless steel capillary system led through the cryostat to liquefy gas from a high pressure gas bottle. The capillary ends up in the nozzle where the thin liquid jet is formed using electron-microscopy apertures with diameters between 10...30 μm. To obtain stable, laminar jets, it is necessary to use dry and clean gas, since at liquid nitrogen temperatures very small amounts of impurities result in a clogged nozzle. Moreover the vapor pressure of the cooling nitrogen has to be controlled in order to minimize thermal instabilities and to reduce evaporation from the jet. Besides, changing the vapor pressure of the nitrogen allows us to liquefy different gases than nitrogen simply by adjusting the pressure in the cryostat. This method works well within the temperature range accessible with liquid nitrogen as cooling material, i. e. from 65-100 K. The nozzle can be aligned with a motorized stage from outside to obtain the desired jet-position with an accuracy of several μm. To remove the liquid from the vacuum chamber after it is used for plasma production a differential pumping stage is applied, since freezing-out is not a suitable method to remove liquid or frozen nitrogen [3].

The nozzle system for conventional liquids is similar to the one described above. A stainless steel capillary from a high pressure store tank for the liquids ends up in the nozzle made of PEEK plastic where an electron-microscopy aperture is mounted to form the jets with diameters between 30...50 μm. The high pressure is provided by a nitrogen gas bottle [4]. The nozzle can be aligned from outside.

Room-temperature jets can easily be removed from the vacuum chamber using a cooling trap filled with liquid nitrogen. Such a cooling trap is mounted a few cm away from the nozzle orifice. For the work presented in this paper this system was used with ethanol but it is also suitable for other liquids like, e.g. water.

To measure absolute photon numbers a spectrograph has been attached to the target chamber. It consists of a 10000 lines/mm free-standing transmission grating and an entrance slit that is covered with a thin Al-filter to absorb visible light. A thinned, back-illuminated CCD-camera is used as detector. Both, the grating and the camera are calibrated [2,7]. A complete description of the spectrograph is given elsewhere [5]. The achieved spectral resolution is limited by the geometrical setup of the spectrograph and was found to be $\lambda/\Delta\lambda = 234$ at $\lambda = 2.88$ nm (time integrated).

To determine the size of the plasma source a transmission zone plate of type KZP7 was used. This type of zone plate functions as condensor in the X-ray microscope at BESSY I [6]. Imaging the plasma source with a zone plate follows the lens formula

$1/f=1/g+1/b$ and according to the wavelength dependence of the focal length of the zone plate $f_{KZP7}=201$mm x 2.4 nm/ λ [nm], a monochromaticity of $\lambda/\Delta\lambda > 200$ can be achieved (20 µm source size). The achieved spatial resolution is mainly limited by the CCD-pixel size and found to be $\delta \approx 2$ µm (single shot) at magnifications of up to 24 .

RESULTS

Both target materials were investigated by taking spectra for different laser parameters as well as taking images of the plasma source at different wavelengths, in particular for the He-α– and Ly- α-line of carbon and nitrogen. The highest spectral emission was achieved at the maximum pulse energy of 230 mJ for both targets. For ethanol, the He-series of C V and O VII and the Ly-series of C VI and O VIII, for nitrogen the He- series for N VI and the Ly-series of N VII, can clearly be seen in the spectra. Figure 1 and 2 show the calibrated spectra. The highest measured photon numbers determined by integrating over the line width of a single spectral line are given in table 1, in addition the corresponding conversion efficiency and the measured linewidth (FWHM) are given.

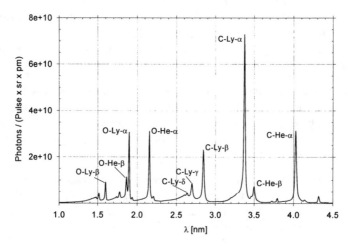

FIGURE 1. Calibrated ethanol spectrum, pulse energy 230 mJ

TABLE 1. Measured photon numbers, line width and conversion efficiencies

Spectral Lines	Photon Numbers [1 Puls⁻¹ sr⁻¹]	$\lambda/\Delta\lambda$	η [%]
C V: He-α (4.03 nm)	$9.4 \ 10^{11}$	156	0.3
C VI: Ly-α (3.37 nm)	$1.9 \ 10^{12}$	156	0.6
N VI: He-α (2.88 nm)	$6.9 \ 10^{11}$	234	0.3
N VII:Ly-α (2.48 rm)	$1.8 \ 10^{12}$	202	0.7

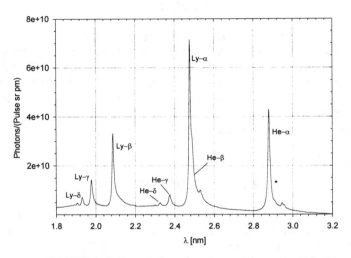

FIGURE 2. Calibrated nitrogen spectrum, pulse energy 230 mJ

The results of the source size measurements are shown in Figure 3. For both targets the source diameter was found to be the same in the He- and the Ly-α lines taking the uncertainty of the measurement of δ ≈ 2 µm into account. However, the ethanol plasma source appears a bit smaller (19 µm) than the nitrogen source (24 µm). The sources seem to be symmetric, the size parallel and perpendicular to the liquid jet is almost the same. Spatial stability of the source over longer periods can be achieved for laser repetition rates of up to 50 Hz. At higher frequencies the liquid jets become instable and affected by the impact of the laser pulses. This may lead to a change in flow direction and the jet has to be realigned. This effect was mainly seen at high pulse energies and high repetition rates of the laser.

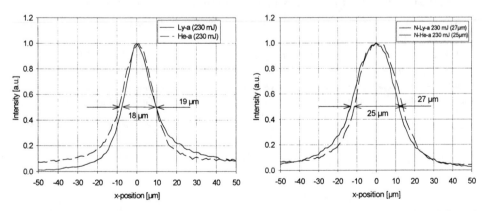

FIGURE 3. Normalized line scans of the sources for the Ly-α and He-α-Line (carbon (left) and nitrogen (right))

DISCUSSION

With the new developed target chamber using room-temperature and cryogenic liquid jets as target for laser generated plasmas it is possible to create radiation in the spectral range of the water-window and serve applications that work with soft X-rays, e.g. X-ray microscopes. With a suitable condensor it should be possible to build up a laboratory X-ray microscope and achieve exposure times of several seconds. Future work will be on the stabilization of the plasma source and on other liquids like, e.g. liquid argon and water. The use of a liquid droplet source is planned.

ACKNOWLEDGEMENTS

We would like to thank the staff of the Institute for X-ray Physics, Goettingen, especially S. Rehbein and D. Hambach for manufacturing the grating and J. Thieme for his support, and Hans Hertz and his group in Stockholm for continuos and helpful discussions. This work has been funded by the Deutsche Forschungsgemeinschaft under contract number Schm 1118/4-1.

REFERENCES

1. Hertz et al., "Debris-free soft X-ray generation using a liquid droplet laser-plasma target", in: *X-ray microscopy and spectromicroscopy*, edited by J. Thieme et al., Berlin, Springer, 1998, V 3

2. Vogt, U., "Laserinduziertes Plasma als Quelle weicher Roentgenstrahlung", Diploma Thesis, Institute for X-Ray Physics, University Goergia Augusta at Göttingen, 1999

3. Wieland, M., "Kryogene Flüssigkeitsstrahlen als Target einer laserinduzierter Plasmaquelle", Diploma Thesis, Institute for X-Ray Physics, University Goergia Augusta at Göttingen, 1999

4. Wagner, U., "Ultraduenne Flüssigkeitsstrahlen als Targets laserinduzierter Plasmaquellen", DiplomaThesis, Institute for X-Ray Physics, University Goergia Augusta at Göttingen, 1997

5. Wilhein et al., "A slit grating spectrograph for quantitative soft X-ray spectroscopy", *Rev. Sci. Instr.*, 1999, **69**, 1694-1699

6. Hettwer, M., Rudolph, D., "Fabrication of the X-Ray Condensor Zone Plate KZP7", in: *X-ray microscopy and spectromicroscopy*, edited by J. Thieme, Berlin, Springer, 1998, IV-21

7. Wilhein, T., "Geduennte CCDs: Charakterisierung und Anwendungen im Bereich weicher Roentgenstrahlen", PHD Thesis, Institute for X-Ray Physics, University Goergia Augusta at Göttingen, 1994

Soft X-Ray Contact Microscopy using
the Asterix Laser Source

A. D. Stead[a], T. W. Ford[a], A. Marranca[a], D. Batani[b], C. Botto[b], A. Masini[b],
F. Bortolotto[b] and K. Eidmann[c]

[a]Biological Sciences, Royal Holloway, Univ. London, Egham, Surrey, UK. TW20 0EX.
[b]Università degli Studi di Milano Bicocca and INFM, Italy
[c]Max Planck Institute für Quantenoptik, Garching, Germany

Abstract. The use of a high-powered laser (Asterix) has permitted comparisons of the soft x-ray emission from various target materials and the quality of images of hydrated biological specimens produced by soft x-ray contact microscopy to be performed. Although targets such as Zr produced lower conversion efficiencies the soft x-ray flux was still sufficient to produce good quality images. However, with gold foil targets the proportion of hard x-rays was high and detracted from image quality. Images of Chlamydomonas showed the x-ray absorbing spheres reported previously, and in yeast the changes in carbon density within the cytoplasm have been related to the cell cycle. Using spematozoa the paired microtubules of the flagella were imaged thus confirming a resolution for the technique of about 40-50nm.

INTRODUCTION

Soft x-ray contact microscopy (SXCM) was performed at the Max Planck Institute für Quantenoptik (Garching, Germany) using the high-energy laser Asterix. The aim was to test the limit and the reproducibility of SXCM as a technique allowing hydrated biological specimens to be imaged with high resolution and, unlike electron microscopy, without the need for any sample preparation. The biological samples were imaged on PMMA photoresists and analysed by atomic force microscopy.

The use of a very high-energy laser system (up to 300J) produced very high x-ray fluxes, thus giving many advantages over smaller, commercial laser plasma sources. Indeed it was possible to expose several (up to 10) biological samples simultaneously so that image quality could be compared when specimens were placed at various distances from the source (up to 30cm) thus varying the x-ray exposure and minimizing penumbral blurring. Moreover, the combined use of different plasma diagnostics (X-ray streak camera, pinhole camera, PIN diodes) allowed the source characteristics to be monitored on each laser shot. Comparisons were also made using different target materials that, according to the available literature, should produce spectra with a high x-ray flux in the water window region as the laser-plasma x-ray source. Given the high power output of the laser it was even possible to use those with low laser to x-ray conversion efficiencies. All these features are essential in order to characterise the plasma, assess the reproducibility of the system, and test the performance of SXCM.

CP507, X-Ray Microscopy: Proceedings of the Sixth International Conference,
edited by W. Meyer-Ilse, T. Warwick, and D. Attwood
© 2000 American Institute of Physics 1-56396-926-2/00/$17.00

MATERIAL AND METHODS

The X-Ray Source

X-rays were generated from a laser plasma source by focussing a laser pulse onto foil targets. Emitted spectra were strongly dependent on the target material used, ranging from simple K-shell spectra for low Z materials, to more complex ones for higher Z materials [1-2]. The laser (Asterix), used during the experiment at the Max Planck Institute, is an iodine laser operating at 1.314µm, this was converted to the third harmonic (0.438µm). The temporal shape of a pulse is gaussian with a FWHM of 0.45ns [3]. Each shot provided about 250J thus the laser intensity varied between $2x10^{15}$ W/cm^2 and $2x10^{14}$ W/cm^2 when the focal spot diameter was changed from 200µm up to 650µm by defocusing the beam (i.e. by moving the target out of the lens focus).

The Interaction Chamber

The chamber had eight apertures and a flat ground plate for mounting the targets and the biological sample holders. Up to ten sample holders were placed in the chamber per shot, thus the same biological samples were imaged with differing x-ray fluxes or different biological specimens were imaged under identical experimental. The source was monitored on each laser shot and a full range of diagnostic data is available for each exposure. The chamber was evacuated to a pressure of 10^{-4} Torr within a 30min pump time, thus specimens remained in their holders for this period of time. This time could have been significantly reduced if the diagnostics had been omitted since such pressure is mainly required for the diagnostics.

Diagnostics

The x-ray source was monitored by two PIN diodes and two grating spectrometers, one coupled with a streak camera, the other with a pinhole camera. X-ray spectra were obtained by a streak camera [4] coupled with a grating spectrometer. Conversion efficiencies for different targets were calculated from PIN diode data. The plasma source size was imaged with a pinhole camera coupled with a grating spectrometer [5]. The diameter of the source was found to be the same as the size of the diameter of the laser beam focused on the target.

The Biological Samples

Samples were chosen for their biological interest and their capability to test the source performance with respect to SXCM. The green alga (*Chlamydomonas reinhardtii*), yeast cells (*Saccharomyces cerevisiae*) and pig spermatozoa were all imaged under various conditions (different target materials and distance of target to specimen). Both *Chlamydomonas* and yeast cells are good test samples for SXCM performance because organelles within each cell are of a variety of sizes and the carbon density of their varies markedly [6]. For example, the vacuole(s) contain relatively dilute concentrations of carbon when compared to the surrounding cytoplasm. Furthermore, *Chlamydomonas* cells have a cell membrane and very thin cellulose wall whilst yeast cells have thick, carbon dense, cell walls, which could make it difficult to image the inner structures. The flagella of pig spermatozoa are comprised of a series of doublets of microtubules and in previous studies these have proved an ideal test of the resolution limit. [7-9].

Hydrated biological specimens were placed in suitable holders; a droplet (2-5µl) of the culture was placed on a silicon nitride window (Fastec, Silverstone, UK) and a PMMA photoresist placed on top. The holder was assembled and tightened to reduce the distance between the photoresist and the window to a minimum (about the diameter of the cell samples), thereby reducing the water thickness. The silicon nitride window, which has more than 60% transmittance to so-called water-window x-rays, ensured that the holder could be placed inside the chamber without damaging the specimen. The time to prepare each specimen was kept to a minimum so that the specimens did not have the chance to dry out or suffer unduly from oxygen stress prior to being imaged.

Exposed photoresists were cleaned (sonication in water, or in extreme cases in IPA), and developed in 50:50 MIBK:IPA and the development monitored by interference light microscopy. Finally chemically developed photoresists were examined by atomic force microscopy using either a Burleigh Personal SPM or a Park Scientific Instrument SA.

Expected Resolution Limit in SXCM

The main factors affecting resolution in SXCM are penumbral blurring and Fresnel diffraction[10, 11]. The resolution limit p due to penumbral blurring is given by

$$p = d\left(\frac{S}{D}\right) \qquad (1)$$

where d = distance between the sample and the photoresist, D = distance between the source and the sample and S = source diameter. When the samples are in contact with the photoresist d equals the cell diameter. For example with a spot size of 650µm, holders 10cm from the source and the cell diameter 6µm, penumbral blurring prevents imaging details <40nm.

The resolution limit f due to Fresnel diffraction is given by:

$$f = (\lambda d)^{\frac{1}{2}} \qquad (2)$$

where λ = radiation wavelength and d = sample to photoresist distance or cell diameter. Thus with yeast cells (diameter about 6µm) the blurring due to Fresnel diffraction is c.100nm. Lower values for the sperm or algae are expected because they are smaller and also the flagella may be closer to the photoresist surface than the rest of the cell body. Moreover, for all biological samples, the use of formulas based on Fresnel diffraction are an overestimation of blurring, since specimens are not opaque to x-ray radiation.

RESULTS AND DISCUSSION

X-Ray Emission Spectra

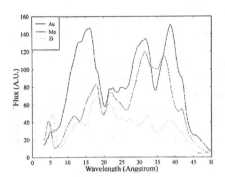

FIGURE 1. Soft x-ray emission spectra obtained from Mo, Zr and Au.

The spectra (Fig. 1) showed that, under the conditions used, Zirconium produced a lower flux within the water window however, the flux was sufficient to get high quality images. Gold and molybdenum had nearly the same flux within the water window. Gold, however, had a stronger emission in the 1 KeV region and, for this reason, it did not seem optimal for SXCM, indeed images produced using gold targets produced images that lacked detail.

It is assumed that hard x-rays produce uniform damage on the photoresist and cause a significant loss of contrast between zones with different carbon contents. Hard x-ray emission is not only related to the choice of target, but also to the laser intensities. A larger focal spot, which gives a lower intensity and a lower plasma temperature, could be of some advantage, as it should result in a lower x-ray emission at high photon energies. During the present experiment the laser focal spot on the target, was adjusted between 400μm and 650μm, to produce a stronger emission in the water window.

Biological observations

Images of *Chlamydomonas* (Fig. 2a,b) revealed similar features to those previously seen using both SXCM and soft x-ray transmission microscopy [12]. In particular the x-ray absorbing spherical inclusions that have no obvious parallel when similar cells are studied by transmission electron microscopy. The ability to image these suggests that the technique has given good discrimination between adjacent areas of the cells that differ only in carbon density. In addition the two flagella are clearly distinguishable.

Yeast cells posses a much thicker, denser, carbon-containing cell wall, as compared to *Chlamydomonas*, nevertheless high quality images of these cells were also obtained albeit that development times were greater. The cell contents of undividing were fairly uniform, although some denser organelles were distinguishable (Fig. 2c). The carbon-dense cell wall was also distinct, and in some cells small protrusions were evident; these correspond to the beginning of the vegetative budding process by which these cells multiply. In the later stages of the budding process the constriction between the cell is very obvious but the cell contents appear different (Fig. 2d). In these cells there were numerous areas (1-2μm in diameter) in which the carbon density was significantly lower than the surrounding cytoplasm and these are probably vacuoles. This difference is that by taking up liquid from the media, the expanding vacuoles provide a driving force for cell expansion.

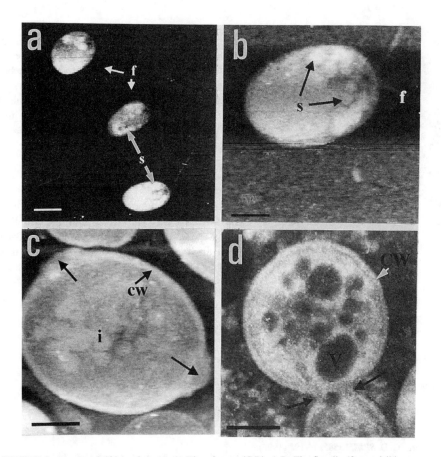

FIGURE 2. Images of *Chlamydomonas* (A,B) and yeast (C,D). A,B. The flagella (f) are visible as are the x-ray absorbing spherical inclusions (s) reported previously. In yeast the beginnings of budding can be seen in C (arrowed) and at this time the cells contain x-ray absorbing organelles (i). Later the constriction between mother and daughter cell is clear (D –arrowed) and the cells contain x-ray lucent areas (v). Scale bars a=5μm; b=2μm; c=2.5μm; d=2μm.

The images of pig sperm provided images with the highest apparent resolution since in the area behind the mid-piece linear structures believed to correspond to paired microtubules were evident. The diameter of individual microtubules, as determined by electron microscopy, is about 25nm, pairs of microtubules are of the order of 40-50nm and these have clearly been imaged (Fig. 3).

CONCLUSIONS

Exposure of many samples per shot, combined with the use of diagnostics to monitor the plasma source, allowed comparisons of the images to understand how optimal, reproducible, conditions can be achieved. Due to the high-energy laser flux, samples were exposed up to a distance of 30cm from the x-ray source, thus reducing the effects of penumbral blurring.

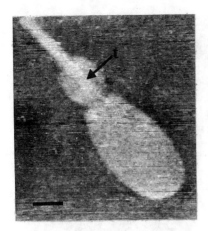

FIGURE 3. AFM of an image of pig spermatazoa. Pairs of microtubules (t) are visible in the mid piece region. Scale = 2μm.

Images obtained with molybdenum, zirconium and yttrium targets, with the samples at distances of about 10-15cm from the source, produced the best images of biological specimens. In particular, pig spermatozoa images showed a resolution of about 50nm, which we calculate as the resolution limit for SXCM.

ACKNOWLEDGEMENTS

The work was supported by the European Union with the Program "Human Capital and Mobility: Access to Large Scale Facility", contract ERB-CI-PDCT0083. We would like to thank all the staff at MPQ in Garching and from Fastec, Peter Anastasi.

REFERENCES

1. Eidmann, K., and Kishimoto, T. *Applied Phys. Lett.* **49**, 377-378 (1986).

2. Vinogradov, A. V., and Shlyaptsev, V. N. *Sov. J. Quantum Electron.* **17**, 1-13 (1987).

3. Baumhacker, H., Brederlow, G., Eidmann, K., Fill, E., Volk, R., Witkowski, S., and Witte, K. *J. Appl. Phys. B*. 61, 325- 332 (1995).

4. Tsakiris, G. D. *SPIE*. **1358**, 174-192 (1990).

5. Eidmann, K., and Tsakiris, G. D. *J. of X-Ray Science and Tech.* **2**, 259-273 (1990).

6. Stead, A. D., Page, A. M., Cotton, R. A., Neely, D., Bagby, R., Miura, E., Tomie, T., Shimizu, S., Majima, T.,Anastasi, P. A. F., and Ford, T. W. *SPIE* **2523**, 202-211 (1995).

7. Castellaniceresa, L., Colombo, R., Cotelli, F., and Loralamia, C. *Expl. Cell. Biol.* **54**, 112-118 (1986).

8. Loralamia, C., Castellaniceresa, L., Andreetta, F., Cotelli, F. and Brivio, M. *J. Ultras. & Mol. Struct. Res.* **96**, 12-21 (1986).

9. Tomie, T., Shimizu, H., Majima, T., Yamada, M., Kanayama, T., Kondo, H, Yano, M. and Ono, M. *Science* **252**, 691-693 (1991).

10. Anastasi, P. A. F., and Burge, R. E. *X-Ray Microscopy III*, Berlin: Springer, 1992, pp. 341-343.

11. Kondo, H., and Tomie, T. *J. Appl. Phys.* **75**, 3798-3805 (1994).

12. Ford, T. W., Page, A. M., Meyer-Ilse, W., Brown, J. T., Heck, J., and Stead, A. D. *X-Ray Microscopy and Spectromicroscopy*, Berlin: Springer, 1998, pp.185-190.

Study of Compact X-ray Laser with a Cavity Pumped by Pulse-Train YAG Laser

Naohiro Yamaguchi[*], Chiemi Fujikawa, Tadayuki Ohchi[**], Kozo Ando
and Tamio Hara

Toyota Technological Institute, 2-12-1 Hisakata, Tempaku, Nagoya 468-8511, Japan

Abstract. We have been developing a tabletop x-ray laser based on the recombination plasma scheme. X-ray lasing of the 15.47 nm transition line of Li-like Al in an x-ray laser cavity has been demonstrated. The cavity output has been characterized to have a beam divergence of about 3 mrad with an absolute intensity of approximately 1×10^8 photons/shot. The double-target configuration has been introduced to improve cavity output intensity substantially. Preliminary results will be presented.

1. INTRODUCTION

Tabletop x-ray lasers which operate at shorter wavelengths less than 20 nm, are promising tools for many important applications such as x-ray photoelectron spectroscopy, x-ray microscopy, x-ray holography and x-ray lithography. An essential requirement for the realization of tabletop x-ray lasers is to reduce the input energy for pumping. Introducing a resonant cavity for the x-ray laser could improve the effective gain-length product of the system, which would contribute to reducing the required pumping energy. Furthermore, the development of a resonant cavity is important in producing x-rays with small divergence and high coherence. The first experiment on an x-ray laser cavity was carried out by Ceglio et al. (1). In this experiment, multipass amplification was limited to only three passes by the short duration of the gain medium.

As a method of achieving compact and efficient soft x-ray lasers, we proposed plasma production by the irradiation of pulse-train lasers (2,3). The advantages of pulse-train laser irradiation are that highly charged ions such as Al^{11+} are produced efficiently through successive plasma heating, and that the electron temperature drops rapidly as soon as the laser irradiation ceases. Therefore, the use of the pulse-train laser enables us to achieve high gain through recombination processes. Recently, we have observed the amplification of Li-like Al soft x-ray transitions in recombining Al plasmas produced by a pulse-train YAG laser with an input energy of only 1.5-2 J/cm (4).

---

[*] E-mail Address: yamagch@toyota-ti.ac.jp.
[**] Present Address: National Institute of Materials and Chemical Research, 1-1 Higashi, Tsukuba, Ibaraki 305-8565, Japan. E-mail Address: ohchi@nimc.go.jp.

CP507, *X-Ray Microscopy: Proceedings of the Sixth International Conference,*
edited by W. Meyer-Ilse, T. Warwick, and D. Attwood

We have performed cavity experiments using multilayer mirrors for the Al XI 3d-4f (15.47 nm) transition line. Clear enhancement of x-ray from the x-ray laser cavity has been confirmed for the first time (5). In this paper we will report on the x-ray laser cavity experiment and characteristics of the cavity output x-ray and some preliminary results of an advanced experiment in a double-target scheme.

2. CAVITY EXPERIMENT

2.1 Experimental Setup

A 100 ps laser pulse from a mode-locked YAG oscillator, 1.064 μm, was transformed into a linearly polarized 16-pulse-train through an optical stacker and a delay line component. The interval of each pulse was 200 ps. The envelope of the pulse-train was shaped so that the former eight pulses were more than four times as intense as the latter ones. The shaped pulse-train YAG laser of only 1.5-2 J was irradiated on an Al slab target via a segmented lens assembly for line focusing. The lens assembly consists of a segmented prism, a beam expander and a cylindrical lens, and forms an 11-mm-long focused-line on an Al slab target. In detail, the irradiation pattern had small dots aligned in a line, which were about 50 μm diameter with a spacing of 140 μm (6). So, the power density was $1.0-1.3 \times 10^{12}$ W/cm^2 for the first half pulses.

A schematic of the cavity is shown in Fig. 1. The cavity consists of a concave mirror with a radius of curvature of 60 mm and a flat mirror with an output coupler of 0.1×0.1 mm^2. Two mirrors were set 50 mm apart, in which the center of the plasma was placed on the cavity midplane. The cavity mirrors were coated with Mo/Si multilayers with a layer-pair thickness of 8.1 nm and a calculated reflectivity of 56%

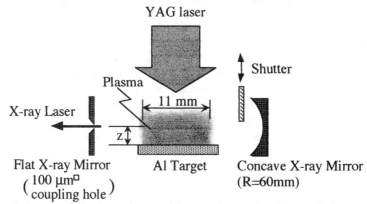

FIGURE 1. Schematic of experimental setup of the x-ray laser cavity. The z-axis denotes the spatial coordinate along the target normal.

for the 15.47 nm x-ray at the normal incidence, taking into account the measured layer roughness of 0.04 nm rms. A movable shutter was settled between the plasma and the rear mirror, that was used to confirm the effect of the cavity.

The x-ray emission from the cavity was analyzed using a space-resolving flat-field grazing incidence soft x-ray spectrograph with an aberration-corrected concave grating (1200 grooves/mm, Hitachi 001-0437). The viewing angle subtended by this spectrograph was about 10 mrad. The detector used was either an x-ray streak camera for time-resolved measurements or a back-illuminated x-ray CCD camera for time-integrated and space-resolved spectral measurements.

2.2 Characteristics of Cavity Output

When the cavity was removed, amplified spontaneous emissions (ASE's) were observed in Li-like Al ion transitions, the 3d-4f (15.47 nm) and 3d-5f (10.57 nm) lines. The spatial distribution of the gain for these soft x-rays was measured to extend z=0.2-0.4 mm from the target surface (7). Therefore, the opening of the front mirror was set near or slightly beyond this region during the cavity experiment.

Spectroscopic observations showed that the 15.47 nm line appeared very bright for the cavity-on case, where the shutter was opened, while the line intensity was almost the same as that of the other lines for the cavity-off case, the shutter was closed. Its enhancement was up to 10 times as much as the intensity of the cavity-off case. When the front mirror was reversed, a half-cavity configuration, the enhancement was not so large.

Mirrors in the cavity suffered no serious damage during the experimental shot, because the input laser energy was small. Though, only the deposition of target materials on the mirror surface would lead to deteriorate the reflectivity. In our experimental setup, mirrors can be used for more than thousand shots.

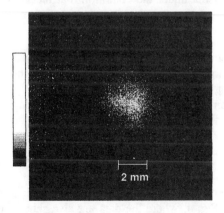

FIGURE 2. Beam profile of the x-ray laser output at 1 m distant from the cavity detected by an x-ray CCD camera.

Figure 2 indicates the beam profile of the cavity output, that was measured using x-ray CCD camera, through a Si filter of 1 μm thickness, 1 m distant from the cavity, when the spectrograph was removed. The divergence of the x-ray laser beam was determined to be 3 mrad, that was smaller than that of the normal x-ray emission from the laser-produced plasma. The absolute intensity of the cavity output x-ray can be estimated from the CCD count integrated over the spot image in Fig. 2. The estimated photon number was 0.5-1.0×10^8 photons/shot.

3. DOUBLE-TARGET EXPERIMENT

We have started an advanced experiment to improve x-ray laser output substantially, promising an x-ray lasing in the saturated-gain regime and to produce a highly coherent x-ray radiation. For this purpose, we have prolonged the x-ray laser medium in a cavity by using a double-target configuration. Because our experiment is now under half way toward a full cavity experiment, the preliminary experimental results on a half cavity configuration will be given.

3.1 Experimental Setup

The schematic drawing of the double-target experiment is shown in Fig. 3. An Nd:glass amplifier having a 25 mm diameter rod was added to the YAG laser system. The output laser energy was increased up to 6 J. The laser beam was divided into two beams through a half mirror and each beam impinged onto a slab target via the segmented lens system. Two 11-mm long plasmas were produced aligning in a line with a gap of 4 mm distance. A Mo/Si multilayer mirror was set 20 mm distant from one end of the target 1. A movable shutter was also inserted between the mirror and target 1. Observation of x-rays was done by using the soft x-ray spectrogaph along the axis of the double-plasma under the condition that the view point was set at the position with the same distance from each target surface along the target surface normal.

In the single-pass geometry where the shutter was closed, the 15.47 nm x-ray intensity in the double-target configuration was higher than the summation of that in the individual target shot, the single target configuration, only when the observation point was set at z=0.25 mm. The data shown further in this paper were taken with the fixed observation point at z=0.25 mm.

3.2 Experimental Results

The 15.47 nm line intensity in the half-cavity experiment is plotted as a function of the total input energy in Fig. 4, where the data are also shown for the shutter-closed case, the single-pass case. The enhancement of x-ray due to the double-target configuration appears when the input energy is larger than 4.5 J, and its factor is 1.5-3 while the value is 1.3-2 for the single-target configuration, the target 1. Providing that the beam divergence of x-rays from each plasma end and the reflectivity of the multilayer mirror used are 5 mrad and 50 %, respectively, we can estimate the gain-length product for each plasma from the data shown in Fig. 4. Furthermore we can estimate the intensity in the double-target and half-cavity configuration by using the above results and assumptions for the input energy of 4.9 J. The estimation was consistent with the measured results shown in Fig. 4.

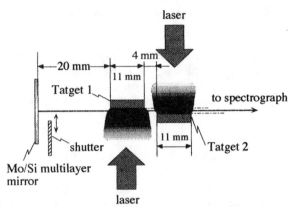

FIGURE 3. Schematic of the double-target configuration. Each target is denoted as target 1 or target 2 in the figure.

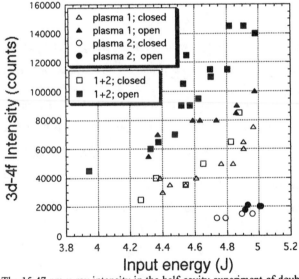

FIGURE 4. The 15.47 nm x-ray intensity in the half-cavity experiment of double-target configuration plotted as a function of total input energy (rectangular mark). The data of single-target configuration are also shown by triangle and circle marks.

4. CONCLUSIONS

We have been developing a tabletop x-ray laser based on the recombining plasma scheme by using the pulse-train laser irradiation method. Furthermore, by applying a micro-dots pattern irradiation, we have performed a series of compact x-ray laser experiments using a small YAG laser that brings out a high-repetition rate operation. X-ray lasing of the 3d-4f transition of Li-like Al ion, 15.47 nm, has been demonstrated using a plane-concave cavity for the first time, however, the overall efficiency is still small, 10^{-9}. It is necessary to improve the performance of x-ray laser cavity.

We have started the double-target experiment expecting substantially intense x-ray laser output. The effect of the double-target scheme has been observed. The results in the half-cavity experiment were consistent with a simple model estimation. We will proceed our research to realize a compact x-ray laser system for practical uses.

ACKNOWLEDGEMENTS

This work is supported by the X-ray Laser Research Consortium, Toyota Technological Institute, and also supported in part by a Grant-in-Aid for Scientific Research from the Ministry of Education, Science, Sports and Culture (Nos. 08458112 and 09680469).

REFERENCES

1. Ceglio, N. M., Stearns, D. G., Gaines, D. P., Hawryluk, A. M., and Trebes, J. E., *Opt. Lett.* **13** (1988) 108-110.

2. Hara, T., Ando, K., Negishi, F., Yashiro, H., and Aoyagi, Y., *Proc. 2nd X-ray Laser Conference* (IOP Publishing, Bristol, Philadelphia, 1990) pp. 263-266.

3. Hirose, H., Hara, T., Ando, K., Negishi, F., and Aoyagi, Y., *Jpn. J. Appl. Phys.* **32** (1993) L1538-L1541.

4. Yamaguchi, N., Hara, T., Fujikawa, C., and Hisada, Y., *Jpn. J. Appl. Phys.* **36** (1997) L1297-L1300.

5. Yamaguchi, N., Hara, T., Ohchi, T., Fujikawa, C., and Sata, T., *Jpn. J. Appl. Phys.* **38** (1999) 5114-5116.

6. Yamaguchi, N., Ohchi, T., Fujikawa, C., Ogata, A., Hisada, Y., Okasaka, K., Hara, T., Tsunashima, T., and Iizuka, Y., *Rev. Sci. Instrum.* **70** (1999) 1285-1287.

7. Yamaguchi, N., Ogata, A., Fujikawa, C., Ohchi, T., Okasaka, K., Hara, T., *J. Electron Spectrosc. Rel. Phenomena* **101-103** (1999) 907-912.

743

747